电子信息前沿技术丛书

Pearson

FIELD AND WAVE ELECTROMAGNETICS

2nd Edition

电磁场与电磁波

第 2 版

[美] 郑钧（David K. Cheng）著

清華大學出版社

北 京

北京市版权局著作权合同登记号　图字：01-2007-2036

图书在版编目(CIP)数据

电磁场与电磁波：第 2 版＝Field and Wave Electromagnetics，2nd Edition：英文/(美)郑钧著.—2 版.—北京：清华大学出版社，2020.1(2024.12重印)
　(电子信息前沿技术丛书)
　ISBN 978-7-302-54664-1

Ⅰ.①电…　Ⅱ.①郑…　Ⅲ.①电磁场—英文②电磁波—英文　Ⅳ.①O441.4

中国版本图书馆 CIP 数据核字(2019)第 284478 号

责任编辑：文　怡
封面设计：王昭红
责任校对：白　蕾
责任印制：杨　艳

出版发行：清华大学出版社
　　　　网　　址：https://www.tup.com.cn, https://www.wqxuetang.com
　　　　地　　址：北京清华大学学研大厦 A 座　　　　　邮　　编：100084
　　　　社 总 机：010-83470000　　　　　　　　　　邮　　购：010-62786544
　　　　投稿与读者服务：010-62776969，c-service@tup.tsinghua.edu.cn
　　　　质量反馈：010-62772015，zhiliang@tup.tsinghua.edu.cn
　　　　课件下载：https://www.tup.com.cn，010-83470236
印 装 者：三河市铭诚印务有限公司
经　　销：全国新华书店
开　　本：185mm×230mm　　印　张：45.25　　　　字　　数：833 千字
版　　次：2007 年 7 月第 1 版　2020 年 1 月第 2 版　印　　次：2024 年 12 月第 4 次印刷
定　　价：99.00 元

产品编号：085771-01

Field and Wave Electromagnetics 2nd Edition

影 印 版 序

　　电磁场与电磁波是十分重要又十分难学的课程之一。电磁场是宇宙中普遍存在的一种物质形态,但是人们用肉眼只能看见以光波存在的电磁场,却看不到其他状态的电磁场,这就使得广大学习者感到电磁场理论的抽象。为了方便思维和构造电磁模型,科学家发明了电力线和磁力线,采用形象的几何图形结合数学公式的方式理解电磁场与电磁波。在大学本科和研究生教学中,电磁场与电磁波常常成为教与学都令人感到很困难的课程。其实,掌握这门学问的关键是建立和培养关于电磁场与电磁波的形象思维能力以及将物理结构和数学相结合的建模能力,这也成为有关电磁场与电磁波的课程中教师和学生应该共同完成的基本任务。

　　本书是关于电磁场与电磁波的一本很有特色的教材,取材新颖,笔法灵活,逻辑性强。教材从矢量分析和场论入手,以简洁清晰的方式建立了电磁模型。紧接着,全面地阐述了电磁场和电磁波的基础理论,包括静电场、静磁场、稳恒电流的场、边值问题的经典解法、时变电磁场与麦克斯韦方程组、平面电磁波及其传播、传输线、阻抗圆图、微带线、波导与谐振腔、天线与电磁辐射、电磁屏蔽等内容。本书除了内容全面之外,还具有很大的灵活性,这是因为全书的内容安排具有较完整的模块性,从而可以灵活地取舍和组合成适合不同行业和不同对象的电磁场与电磁波的教材。

　　本书配备了大量插图,版面整洁漂亮,非常有利于读者理解电磁场与电磁波过程,有利于读者潜移默化地建立起电磁场与电磁波的形象化的概念。书中采用的符号都是非常标准的,特别是采用了各种微分算符,使得公式的表达简洁、形象,便于理解。哲学家告诉我们,人类智力的成长要归功于语言能力和符号的艺术,面对电磁场理论的抽象性,在电磁场教科书中更应该突出相应的符号艺术。此外,将电磁场理论与电磁工程实际紧密结合也是本书的一大特点,书中配合电磁学理论的阐述列举了大量例题和习题,这些例题和习题都是取之于实际的电磁设备或电磁工程,具有相当强的实用性。通过这些例题和习题也介绍和剖析了许多电磁工程中的具体结构和原理,加之书中配备的各种参数的表格,使得该书对从事相关研究的读者来讲也是一本很有价值的参考书。

　　在全球一体化的今天,英语教材和英文授课对培养高水平人才有着不可估量的意义,本书特别适宜作为理工科院校高年级本科生和研究生基础课的英文授课教材或英文参考书。

　　Addison-Wesley 图书公司出版了许多非常优秀的科技图书,该公司的出版物一直是广大科技工作者关注的热点。本书是 1983 年出版的,1989 年再版,目前影印的版本是1992 年印刷的订正版本,这些都说明该书得到了极广泛的认可,已经成为世界性的经典教材。笔者读了本书竟爱不释手,因而推荐给同行专家和广大读者共享为快。现在,清华大学出版社出版本书影印版以飨读者,希望广大读者能从中受益。

吕英华　教授

北京邮电大学

Preface

The many books on introductory electromagnetics can be roughly divided into two main groups. The first group takes the traditional development: starting with the experimental laws, generalizing them in steps, and finally synthesizing them in the form of Maxwell's equations. This is an inductive approach. The second group takes the axiomatic development: starting with Maxwell's equations, identifying each with the appropriate experimental law, and specializing the general equations to static and time-varying situations for analysis. This is a deductive approach. A few books begin with a treatment of the special theory of relativity and develop all of electromagnetic theory from Coulomb's law of force; but this approach requires the discussion and understanding of the special theory of relativity first and is perhaps best suited for a course at an advanced level.

Proponents of the traditional development argue that it is the way electromagnetic theory was unraveled historically (from special experimental laws to Maxwell's equations), and that it is easier for the students to follow than the other methods. I feel, however, that the way a body of knowledge was unraveled is not necessarily the best way to teach the subject to students. The topics tend to be fragmented and cannot take full advantage of the conciseness of vector calculus. Students are puzzled at, and often form a mental block to, the subsequent introduction of gradient, divergence, and curl operations. As a process for formulating an electromagnetic model, this approach lacks cohesiveness and elegance.

The axiomatic development usually begins with the set of four Maxwell's equations, either in differential or in integral form, as fundamental postulates. These are equations of considerable complexity and are difficult to master. They are likely to cause consternation and resistance in students who are hit with all of them at the beginning of a book. Alert students will wonder about the meaning of the field vectors and about the necessity and sufficiency of these general equations. At the initial stage students tend to be confused about the concepts of the electromagnetic model, and they are not yet comfortable with the associated mathematical manipulations. In any case, the general Maxwell's equations are soon simplified to apply to static fields, which allow the consideration of electrostatic fields and magnetostatic fields separately. Why then should the entire set of four Maxwell's equations be introduced at the outset?

It may be argued that Coulomb's law, though based on experimental evidence, is in fact also a postulate. Consider the two stipulations of Coulomb's law: that the charged bodies are very small compared with their distance of separation, and that the force between the charged bodies is inversely proportional to the square of their distance. The question arises regarding the first stipulation: How small must the charged bodies be in order to be considered "very small" compared with their distance? In practice the charged bodies cannot be of vanishing sizes (ideal point charges), and there is difficulty in determining the "true" distance between two bodies of finite dimensions. For given body sizes the relative accuracy in distance measurements is better when the separation is larger. However, practical considerations (weakness of force, existence of extraneous charged bodies, etc.) restrict the usable distance of separation in the laboratory, and experimental inaccuracies cannot be entirely avoided. This leads to a more important question concerning the inverse-square relation of the second stipulation. Even if the charged bodies were of vanishing sizes, experimental measurements could not be of an infinite accuracy no matter how skillful and careful an experimentor was. How then was it possible for Coulomb to know that the force was *exactly* inversely proportional to the *square* (not the 2.000001th or the 1.999999th power) of the distance of separation? This question cannot be answered from an experimental viewpoint because it is not likely that during Coulomb's time experiments could have been accurate to the seventh place. We must therefore conclude that Coulomb's law is itself a postulate and that it is a law of nature discovered and assumed on the basis of his experiments of a limited accuracy (see Section 3–2).

This book builds the electromagnetic model using an *axiomatic approach in steps*: first for static electric fields (Chapter 3), then for static magnetic fields (Chapter 6), and finally for time-varying fields leading to Maxwell's equations (Chapter 7). The mathematical basis for each step is Helmholtz's theorem, which states that a vector field is determined to within an additive constant if both its divergence and its curl are specified everywhere. Thus, for the development of the electrostatic model in free space, it is only necessary to define a single vector (namely, the electric field intensity **E**) by specifying its divergence and its curl as postulates. All other relations in electrostatics for free space, including Coulomb's law and Gauss's law, can be derived from the two rather simple postulates. Relations in material media can be developed through the concept of equivalent charge distributions of polarized dielectrics.

Similarly, for the magnetostatic model in free space it is necessary to define only a single magnetic flux density vector **B** by specifying its divergence and its curl as postulates; all other formulas can be derived from these two postulates. Relations in material media can be developed through the concept of equivalent current densities. Of course, the validity of the postulates lies in their ability to yield results that conform with experimental evidence.

For time-varying fields, the electric and magnetic field intensities are coupled. The curl **E** postulate for the electrostatic model must be modified to conform with

Faraday's law. In addition, the curl **B** postulate for the magnetostatic model must also be modified in order to be consistent with the equation of continuity. We have, then, the four Maxwell's equations that constitute the electromagnetic model. I believe that this gradual development of the electromagnetic model based on Helmholtz's theorem is novel, systematic, pedagogically sound, and more easily accepted by students.

In the presentation of the material, I strive for lucidity and unity, and for smooth and logical flow of ideas. Many worked-out examples are included to emphasize fundamental concepts and to illustrate methods for solving typical problems. Applications of derived relations to useful technologies (such as ink-jet printers, lightning arresters, electret microphones, cable design, multiconductor systems, electrostatic shielding, Doppler radar, radome design, Polaroid filters, satellite communication systems, optical fibers, and microstrip lines) are discussed. Review questions appear at the end of each chapter to test the students' retention and understanding of the essential material in the chapter. The problems in each chapter are designed to reinforce students' comprehension of the interrelationships between the different quantities in the formulas, and to extend their ability of applying the formulas to solve practical problems. In teaching, I have found the review questions a particularly useful device to stimulate students' interest and to keep them alert in class.

Besides the fundamentals of electromagnetic fields, this book also covers the theory and applications of transmission lines, waveguides and cavity resonators, and antennas and radiating systems. The fundamental concepts and the governing theory of electromagnetism do not change with the introduction of new electromagnetic devices. Ample reasons and incentives for learning the fundamental principles of electromagnetics are given in Section 1–1. I hope that the contents of this book, strengthened by the novel approach, will provide students with a secure and sufficient background for understanding and analyzing basic electromagnetic phenomena as well as prepare them for more advanced subjects in electromagnetic theory.

There is enough material in this book for a two-semester sequence of courses. Chapters 1 through 7 contain the material on fields, and Chapters 8 through 11 on waves and applications. In schools where there is only a one-semester course on electromagnetics, Chapters 1 through 7, plus the first four sections of Chapter 8 would provide a good foundation on fields and an introduction of waves in unbounded media. The remaining material could serve as a useful reference book on applications or as a textbook for a follow-up elective course. Schools on a quarter system could adjust the material to be covered in accordance with the total number of hours assigned to the subject of electromagnetics. Of course, individual instructors have the prerogative to emphasize and expand certain topics, and to deemphasize or delete certain others.

I have given considerable thought to the advisability of including computer programs for the solution of some problems, but have finally decided against it. Diverting students' attention and effort to numerical methods and computer software would distract them from concentrating on learning the fundamentals of electromagnetism. Where appropriate, the dependence of important results on the value of a parameter is stressed by curves; field distributions and antenna patterns are illustrated by graphs; and typical mode patterns in waveguides are plotted. The computer programs for obtaining these curves, graphs, and mode patterns are not always simple. Students in science and engineering are required to acquire a facility for using computers; but the inclusion of some cookbook-style computer programs in a book on the fundamental principles of electromagnetic fields and waves would appear to contribute little to the understanding of the subject matter.

This book was first published in 1983. Favorable reactions and friendly encouragements from professors and students have provided me with the impetus to come out with a new edition. In this second edition I have added many new topics. These include Hall effect, d-c motors, transformers, eddy current, energy-transport velocity for wide-band signals in waveguides, radar equation and scattering cross section, transients in transmission lines, Bessel functions, circular waveguides and circular cavity resonators, waveguide discontinuities, wave propagation in ionosphere and near earth's surface, helical antennas, log-periodic dipole arrays, and antenna effective length and effective area. The total number of problems has been expanded by about 35 percent.

The Addison-Wesley Publishing Company has decided to make this second edition a two-color book. I think the readers will agree that the book is handsomely produced. I would like to take this opportunity to express my appreciation to all the people on the editorial, production, and marketing staff who provided help in bringing out this new edition. In particular, I wish to thank Thomas Robbins, Barbara Rifkind, Karen Myer, Joseph K. Vetere, and Katherine Harutunian.

Chevy Chase, Maryland   D. K. C.

Contents

3   Static Electric Fields   72

4   Solution of Electrostatic Problems   152

5 Steady Electric Currents   198

6 Static Magnetic Fields   225

7   Time-Varying Fields and Maxwell's Equations   307

8   Plane Electromagnetic Waves   354

9 Theory and Applications of Transmission Lines 427

10   Waveguides and Cavity Resonators   520

11 Antennas and Radiating Systems 600

Appendixes

A Symbols and Units 671

B Some Useful Material Constants 674

C   Index of Tables   677

General Bibliography   679

Answers to Selected Problems   681

Index   693

Some Useful Items

Some Useful Vector Identities
Gradient, Divergence, Curl, and Laplacian Operations in Cartesian
Coordinates
Gradient, Divergence, Curl, and Laplacian Operations in Cylindrical
and Spherical Coordinates

1

The Electromagnetic
Model

1–1 Introduction

Stated in a simple fashion, *electromagnetics* is the study of the effects of electric charges at rest and in motion. From elementary physics we know that there are two kinds of charges: positive and negative. Both positive and negative charges are sources of an electric field. Moving charges produce a current, which gives rise to a magnetic field. Here we tentatively speak of electric field and magnetic field in a general way; more definitive meanings will be attached to these terms later. A *field* is a spatial distribution of a quantity, which may or may not be a function of time. A time-varying electric field is accompanied by a magnetic field, and vice versa. In other words, time-varying electric and magnetic fields are coupled, resulting in an electromagnetic field. Under certain conditions, time-dependent electromagnetic fields produce waves that radiate from the source.

The concept of fields and waves is essential in the explanation of action at a distance. For instance, we learned from elementary mechanics that masses attract each other. This is why objects fall toward the earth's surface. But since there are no elastic strings connecting a free-falling object and the earth, how do we explain this phenomenon? We explain this action-at-a-distance phenomenon by postulating the existence of a gravitational field. The possibilities of satellite communication and of receiving signals from space probes millions of miles away can be explained only by postulating the existence of electric and magnetic fields and electromagnetic waves. In this book, *Field and Wave Electromagnetics*, we study the principles and applications of the laws of electromagnetism that govern electromagnetic phenomena.

Electromagnetics is of fundamental importance to physicists and to electrical and computer engineers. Electromagnetic theory is indispensable in understanding the principle of atom smashers, cathode-ray oscilloscopes, radar, satellite communication, television reception, remote sensing, radio astronomy, microwave devices, optical fiber communication, transients in transmission lines, electromagnetic compatibility

FIGURE 1-1
A monopole antenna.

problems, instrument-landing systems, electromechanical energy conversion, and so on. Circuit concepts represent a restricted version, a special case, of electromagnetic concepts. As we shall see in Chapter 7, when the source frequency is very low so that the dimensions of a conducting network are much smaller than the wavelength, we have a quasi-static situation, which simplifies an electromagnetic problem to a circuit problem. However, we hasten to add that circuit theory is itself a highly developed, sophisticated discipline. It applies to a different class of electrical engineering problems, and it is important in its own right.

Two situations illustrate the inadequacy of circuit-theory concepts and the need for electromagnetic-field concepts. Figure 1-1 depicts a monopole antenna of the type we see on a walkie-talkie. On transmit, the source at the base feeds the antenna with a message-carrying current at an appropriate carrier frequency. From a circuit-theory point of view, the source feeds into an open circuit because the upper tip of the antenna is not connected to anything physically; hence no current would flow, and nothing would happen. This viewpoint, of course, cannot explain why communication can be established between walkie-talkies at a distance. Electromagnetic concepts must be used. We shall see in Chapter 11 that when the length of the antenna is an appreciable part of the carrier wavelength,[†] a nonuniform current will flow along the open-ended antenna. This current radiates a time-varying electromagnetic field in space, which propagates as an electromagnetic wave and induces currents in other antennas at a distance.

In Fig. 1-2 we show a situation in which an electromagnetic wave is incident from the left on a large conducting wall containing a small hole (aperture). Electromagnetic fields will exist on the right side of the wall at points, such as P in the figure, that are not necessarily directly behind the aperture. Circuit theory is obviously inadequate here for the determination (or even the explanation of the existence) of the field at P. The situation in Fig. 1-2, however, represents a problem of practical importance as its solution is relevant in evaluating the shielding effectiveness of the conducting wall.

[†] The product of the wavelength and the frequency of an a-c source is the velocity of wave propagation.

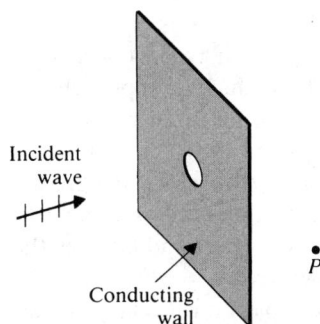

FIGURE 1–2
An electromagnetic problem.

Generally speaking, circuit theory deals with lumped-parameter systems—circuits consisting of components characterized by lumped parameters such as resistances, inductances, and capacitances. Voltages and currents are the main system variables. For d-c circuits the system variables are constants, and the governing equations are algebraic equations. The system variables in a-c circuits are time-dependent; they are scalar quantities and are independent of space coordinates. The governing equations are ordinary differential equations. On the other hand, most electromagnetic variables are functions of time as well as of space coordinates. Many are vectors with both a magnitude and a direction, and their representation and manipulation require a knowledge of vector algebra and vector calculus. Even in static cases the governing equations are, in general, partial differential equations. It is essential that we be equipped to handle vector quantities and variables that are both time- and space-dependent. The fundamentals of vector algebra and vector calculus will be developed in Chapter 2. Techniques for solving partial differential equations are needed in dealing with certain types of electromagnetic problems. These techniques will be discussed in Chapter 4. The importance of acquiring a facility in the use of these mathematical tools in the study of electromagnetics cannot be overemphasized.

Students who have mastered circuit theory may initially have the impression that electromagnetic theory is abstract. In fact, electromagnetic theory is no more abstract than circuit theory in the sense that the validity of both can be verified by experimentally measured results. In electromagnetics there is a need to define more quantities and to use more mathematical manipulations in order to develop a logical and complete theory that can explain a much wider variety of phenomena. The challenge of field and wave electromagnetics is not in the abstractness of the subject matter but rather in the process of mastering the electromagnetic model and the associated rules of operation. Dedication to acquiring this mastery will help us to meet the challenge and reap immeasurable satisfaction.

1–2 The Electromagnetic Model

There are two approaches in the development of a scientific subject: the inductive approach and the deductive approach. Using the inductive approach, one follows the historical development of the subject, starting with the observations of some simple experiments and inferring from them laws and theorems. It is a process of reasoning from particular phenomena to general principles. The deductive approach, on the other hand, postulates a few fundamental relations for an idealized model. The postulated relations are axioms, from which particular laws and theorems can be derived. The validity of the model and the axioms is verified by their ability to predict consequences that check with experimental observations. In this book we prefer to use the deductive or axiomatic approach because it is more elegant and enables the development of the subject of electromagnetics in an orderly way.

The idealized model we adopt for studying a scientific subject must relate to real-world situations and be able to explain physical phenomena; otherwise, we would be engaged in mental exercises for no purpose. For example, a theoretical model could be built, from which one might obtain many mathematical relations; but, if these relations disagreed with observed results, the model would be of no use. The mathematics might be correct, but the underlying assumptions of the model could be wrong, or the implied approximations might not be justified.

Three essential steps are involved in building a theory on an idealized model. *First*, some basic quantities germane to the subject of study are defined. *Second*, the rules of operation (the mathematics) of these quantities are specified. *Third*, some fundamental relations are postulated. These postulates or laws are invariably based on numerous experimental observations acquired under controlled conditions and synthesized by brilliant minds. A familiar example is the circuit theory built on a circuit model of ideal sources and pure resistances, inductances, and capacitances. In this case the basic quantities are voltages (V), currents (I), resistances (R), inductances (L), and capacitances (C); the rules of operations are those of algebra, ordinary differential equations, and Laplace transformation; and the fundamental postulates are Kirchhoff's voltage and current laws. Many relations and formulas can be derived from this basically rather simple model, and the responses of very elaborate networks can be determined. The validity and value of the model have been amply demonstrated.

In a like manner, an electromagnetic theory can be built on a suitably chosen electromagnetic model. In this section we shall take the first step of defining the basic quantities of electromagnetics. The second step, the rules of operation, encompasses vector algebra, vector calculus, and partial differential equations. The fundamentals of vector algebra and vector calculus will be discussed in Chapter 2 (Vector Analysis), and the techniques for solving partial differential equations will be introduced when these equations arise later in the book. The third step, the fundamental postulates, will be presented in three substeps in Chapters 3, 6, and 7 as we deal with static electric fields, steady magnetic fields, and electromagnetic fields, respectively.

The quantities in our electromagnetic model can be divided roughly into two categories: source quantities and field quantities. The source of an electromagnetic field is invariably electric charges at rest or in motion. However, an electromagnetic field may cause a redistribution of charges, which will, in turn, change the field; hence the separation between the cause and the effect is not always so distinct.

We use the symbol q (sometimes Q) to denote *electric charge*. Electric charge is a fundamental property of matter and exists only in positive or negative integral multiples of the charge on an electron, $-e$.[†]

$$e = 1.60 \times 10^{-19} \qquad (C), \qquad\qquad\qquad (1-1)$$

where C is the abbreviation of the unit of charge, coulomb.[‡] It is named after the French physicist Charles A. de Coulomb, who formulated Coulomb's law in 1785. (Coulomb's law will be discussed in Chapter 3.) A coulomb is a very large unit for electric charge; it takes $1/(1.60 \times 10^{-19})$ or 6.25 million trillion electrons to make up -1 C. In fact, two 1 C charges 1 m apart will exert a force of approximately 1 million tons on each other. Some other physical constants for the electron are listed in Appendix B–2.

The principle of *conservation of electric charge*, like the principle of conservation of momentum, is a fundamental postulate or law of physics. It states that electric charge is conserved; that is, it can neither be created nor be destroyed. This is a law of nature and cannot be derived from other principles or relations. Its truth has never been questioned or doubted in practice.

Electric charges can move from one place to another and can be redistributed under the influence of an electromagnetic field; but the algebraic sum of the positive and negative charges in a closed (isolated) system remains unchanged. *The principle of conservation of electric charge must be satisfied at all times and under any circumstances*. It is represented mathematically by the *equation of continuity*, which we will discuss in Section 5–4. Any formulation or solution of an electromagnetic problem that violates the principle of conservation of electric charge *must be* incorrect. We recall that the Kirchhoff's current law in circuit theory, which maintains that the sum of all the currents leaving a junction must equal the sum of all the currents entering the junction, is an assertion of the conservation property of electric charge. (Implicit in the current law is the assumption that there is no cumulation of charge at the junction.)

Although, in a microscopic sense, electric charge either does or does not exist at a point in a discrete manner, these abrupt variations on an atomic scale are unimportant when we consider the electromagnetic effects of large aggregates of charges. In constructing a macroscopic or large-scale theory of electromagnetism we find that the use of smoothed-out average density functions yields very good results. (The same approach is used in mechanics where a smoothed-out mass density function is defined, in spite of the fact that mass is associated only with elementary particles in a discrete

---

[†] In 1962, Murray Gell-Mann hypothesized *quarks* as the basic building blocks of matter. Quarks were predicted to carry a fraction of the charge of an electron, and their existence has since been verified experimentally.

[‡] The system of units will be discussed in Section 1–3.

manner on an atomic scale.) We define a **volume charge density**, ρ, as a source quantity as follows:

$$\rho = \lim_{\Delta v \to 0} \frac{\Delta q}{\Delta v} \qquad (\text{C/m}^3), \tag{1-2}$$

where Δq is the amount of charge in a very small volume Δv. How small should Δv be? It should be small enough to represent an accurate variation of ρ but large enough to contain a very large number of discrete charges. For example, an elemental cube with sides as small as 1 micron (10^{-6} m or 1 μm) has a volume of 10^{-18} m^3, which will still contain about 10^{11} (100 billion) atoms. A smoothed-out function of space coordinates, ρ, defined with such a small Δv is expected to yield accurate macroscopic results for nearly all practical purposes.

In some physical situations an amount of charge Δq may be identified with an element of surface Δs or an element of line $\Delta \ell$. In such cases it will be more appropriate to define a **surface charge density**, ρ_s, or a **line charge density**, ρ_ℓ:

$$\rho_s = \lim_{\Delta s \to 0} \frac{\Delta q}{\Delta s} \qquad (\text{C/m}^2), \tag{1-3}$$

$$\rho_\ell = \lim_{\Delta \ell \to 0} \frac{\Delta q}{\Delta \ell} \qquad (\text{C/m}). \tag{1-4}$$

Except for certain special situations, charge densities vary from point to point; hence ρ, ρ_s, and ρ_ℓ are, in general, point functions of space coordinates.

Current is the rate of change of charge with respect to time; that is,

$$I = \frac{dq}{dt} \qquad (\text{C/s or A}), \tag{1-5}$$

where I itself may be time-dependent. The unit of current is coulomb per second (C/s), which is the same as ampere (A). A current must flow through a finite area (a conducting wire of a finite cross section, for instance); hence it is not a point function. In electromagnetics we define a vector point function **volume current density** (or simply **current density**) **J**, which measures the amount of current flowing through a unit area normal to the direction of current flow. The boldfaced **J** is a vector whose magnitude is the current per unit area (A/m^2) and whose direction is the direction of current flow. We shall elaborate on the relation between I and **J** in Chapter 5. For very good conductors, high-frequency alternating currents are confined in the surface layer as a current sheet, instead of flowing throughout the interior of the conductor. In such cases there is a need to define a **surface current density** **J**$_s$, which is the current per unit width on the conductor surface normal to the direction of current flow and has the unit of ampere per meter (A/m).

There are four fundamental *vector* field quantities in electromagnetics: *electric field intensity* **E**, *electric flux density* (or *electric displacement*) **D**, *magnetic flux*

TABLE 1-1
Fundamental Electromagnetic Field Quantities

Symbols and Units for Field Quantities	Field Quantity	Symbol	Unit
Electric	Electric field intensity	**E**	V/m
	Electric flux density (Electric displacement)	**D**	C/m^2
Magnetic	Magnetic flux density	**B**	T
	Magnetic field intensity	**H**	A/m

density **B,** and *magnetic field intensity* **H.** The definition and physical significance of these quantities will be explained fully when they are introduced later in the book. At this time we want only to establish the following. Electric field intensity **E** is the only vector needed in discussing electrostatics (effects of stationary electric charges) in free space; it is defined as the electric force on a unit test charge. Electric displacement vector **D** is useful in the study of electric field in material media, as we shall see in Chapter 3. Similarly, magnetic flux density **B** is the only vector needed in discussing magnetostatics (effects of steady electric currents) in free space and is related to the magnetic force acting on a charge moving with a given velocity. The magnetic field intensity vector **H** is useful in the study of magnetic field in material media. The definition and significance of **B** and **H** will be discussed in Chapter 6.

The four fundamental electromagnetic field quantities, together with their units, are tabulated in Table 1-1. In Table 1-1, V/m is volt per meter, and T stands for tesla or volt-second per square meter. When there is no time variation (as in static, steady, or stationary cases), the electric field quantities **E** and **D** and the magnetic field quantities **B** and **H** form two separate vector pairs. In time-dependent cases, however, electric and magnetic field quantities are coupled; that is, time-varying **E** and **D** will give rise to **B** and **H**, and vice versa. All four quantities are point functions; they are defined at every point in space and, in general, are functions of space coordinates. Material (or medium) properties determine the relations between **E** and **D** and between **B** and **H**. These relations are called the *constitutive relations* of a medium and will be examined later.

The principal objective of studying electromagnetism is to understand the interaction between charges and currents at a distance based on the electromagnetic model. Fields and waves (time-dependent and space-dependent fields) are basic conceptual quantities of this model. Fundamental postulates will relate **E, D, B, H,** and the source quantities; and derived relations will lead to the explanation and prediction of electromagnetic phenomena.

TABLE 1–2
Fundamental SI Units

Quantity	Unit	Abbreviation
Length	meter	m
Mass	kilogram	kg
Time	second	s
Current	ampere	A

1–3   SI Units and Universal Constants

A measurement of any physical quantity must be expressed as a number followed by a unit. Thus we may talk about a length of three meters, a mass of two kilograms, and a time period of ten seconds. To be useful, a unit system should be based on some fundamental units of convenient (practical) sizes. In mechanics, all quantities can be expressed in terms of three basic units (for length, mass, and time). In electromagnetics a fourth basic unit (for current) is needed. The SI (*International System of Units* or *Le Système International d'Unités*) is an *MKSA system* built from the four fundamental units listed in Table 1–2. All other units used in electromagnetics, including those appearing in Table 1–1, are derived units expressible in terms of meters, kilograms, seconds, and amperes. For example, the unit for charge, coulomb (C), is ampere-second (A·s); the unit for electric field intensity (V/m) is kg·m/A·s³; and the unit for magnetic flux density, tesla (T), is kg/A·s². More complete tables of the units for various quantities are given in Appendix A.

The official SI definitions, as adopted by the International Committee on Weights and Measures, are as follows:[†]

> *Meter.* Once the length between two scratches on a platinum-iridium bar (and originally calculated as one ten-millionth of the distance between the North Pole and the equator through Paris, France), is now defined by reference to the *second* (see below) and the speed of light, which in a vacuum is 299,792,458 meters per second.

> *Kilogram.* Mass of a standard bar made of a platinum-iridium alloy and kept inside a set of nested enclosures that protect it from contamination and mishandling. It rests at the International Bureau of Weights and Measures in Sèvres, outside Paris.

> *Second.* 9,192,631,770 periods of the electromagnetic radiation emitted by a particular transition of a cesium atom.

[†] P. Wallich, "Volts and amps are not what they used to be," *IEEE Spectrum*, vol. 24, pp. 44–49, March 1987.

Ampere. The constant current that, if maintained in two straight parallel conductors of infinite length and negligible circular cross section, and placed one meter apart in vacuum, would produce between these conductors a force equal to 2×10^{-7} newton per meter of length. (A newton is the force that gives a mass of one kilogram an acceleration of one meter per second squared.)

In our electromagnetic model there are three universal constants, in addition to the field quantities listed in Table 1–1. They relate to the properties of the free space (vacuum). They are as follows: *velocity of electromagnetic wave* (including light) in free space, c; *permittivity* of free space, ϵ_0; and *permeability* of free space, μ_0. Many experiments have been performed for precise measurement of the velocity of light, to many decimal places. For our purpose it is sufficient to remember that

$$\boxed{c \cong 3 \times 10^8 \quad \text{(m/s).}}\qquad \text{(in free space)} \qquad (1\text{–}6)$$

The other two constants, ϵ_0 and μ_0, pertain to electric and magnetic phenomena, respectively: ϵ_0 is the proportionality constant between the electric flux density **D** and the electric field intensity **E** in free space, such that

$$\boxed{\mathbf{D} = \epsilon_0 \mathbf{E};}\qquad \text{(in free space)} \qquad (1\text{–}7)$$

μ_0 is the proportionality constant between the magnetic flux density **B** and the magnetic field intensity **H** in free space, such that

$$\boxed{\mathbf{H} = \frac{1}{\mu_0}\mathbf{B}.}\qquad \text{(in free space)} \qquad (1\text{–}8)$$

The values of ϵ_0 and μ_0 are determined by the choice of the unit system, and they are not independent. In the *SI system* (rationalized[†] MKSA system), which is almost universally adopted for electromagnetics work, the permeability of free space is chosen to be

$$\boxed{\mu_0 = 4\pi \times 10^{-7} \quad \text{(H/m),}}\qquad \text{(in free space)} \qquad (1\text{–}9)$$

where H/m stands for henry per meter. With the values of c and μ_0 fixed in Eqs. (1–6) and (1–9) the value of the permittivity of free space is then derived from the following

---

[†] This system of units is said to be *rationalized* because the factor 4π does not appear in the Maxwell's equations (the fundamental postulates of electromagnetism). This factor, however, will appear in many derived relations. In the unrationalized MKSA system, μ_0 would be 10^{-7} (H/m), and the factor 4π would appear in the Maxwell's equations.

TABLE 1–3
Universal Constants in SI Units

Universal Constants	Symbol	Value	Unit
Velocity of light in free space	c	3×10^8	m/s
Permeability of free space	μ_0	$4\pi \times 10^{-7}$	H/m
Permittivity of free space	ϵ_0	$\dfrac{1}{36\pi} \times 10^{-9}$	F/m

relationships:

$$c = \frac{1}{\sqrt{\epsilon_0\mu_0}} \quad \text{(m/s)} \tag{1–10}$$

or

$$\epsilon_0 = \frac{1}{c^2\mu_0} \cong \frac{1}{36\pi} \times 10^{-9}$$
$$\cong 8.854 \times 10^{-12} \quad \text{(F/m)}, \tag{1–11}$$

where F/m is the abbreviation for farad per meter. The three universal constants and their values are summarized in Table 1–3.

Now that we have defined the basic quantities and the universal constants of the electromagnetic model, we can develop the various subjects in electromagnetics. But, before we do that, we must be equipped with the appropriate mathematical tools. In the following chapter we discuss the basic rules of operation for vector algebra and vector calculus.

Review Questions

R.1–1 What is electromagnetics?

R.1–2 Describe two phenomena or situations, other than those depicted in Figs. 1–1 and 1–2, that cannot be adequately explained by circuit theory.

R.1–3 What are the three essential steps in building an idealized model for the study of a scientific subject?

R.1–4 What are the four fundamental SI units in electromagnetics?

R.1–5 What are the four fundamental field quantities in the electromagnetic model? What are their units?

R.1–6 What are the three universal constants in the electromagnetic model, and what are their relations?

R.1–7 What are the source quantities in the electromagnetic model?

2

Vector
Analysis

2-1 Introduction

As we noted in Chapter 1, some of the quantities in electromagnetics (such as charge, current, and energy) are scalars; and some others (such as electric and magnetic field intensities) are vectors. Both scalars and vectors can be functions of time and position. At a given time and position, a *scalar* is completely specified by its magnitude (positive or negative, together with its unit). Thus we can specify, for instance, a charge of -1 μC at a certain location at $t = 0$. The specification of a *vector* at a given location and time, on the other hand, requires both a magnitude and a direction. How do we specify the direction of a vector? In a three-dimensional space, three numbers are needed, and these numbers depend on the choice of a coordinate system. Conversion of a given vector from one coordinate system to another will change these numbers. However, physical laws and theorems relating various scalar and vector quantities certainly must hold irrespective of the coordinate system. The general expressions of the laws of electromagnetism, therefore, do not require the specification of a coordinate system. A particular coordinate system is chosen only when a problem of a given geometry is to be analyzed. For example, if we are to determine the magnetic field at the center of a current-carrying wire loop, it is more convenient to use rectangular coordinates if the loop is rectangular, whereas polar coordinates (two-dimensional) will be more appropriate if the loop is circular in shape. The basic electromagnetic relation governing the solution of such a problem is the same for both geometries.

Three main topics will be dealt with in this chapter on vector analysis:

1. Vector algebra—addition, subtraction, and multiplication of vectors.
2. Orthogonal coordinate systems—Cartesian, cylindrical, and spherical coordinates.
3. Vector calculus—differentiation and integration of vectors; line, surface, and volume integrals; "del" operator; gradient, divergence, and curl operations.

11

Throughout the rest of this book we will decompose, combine, differentiate, integrate, and otherwise manipulate vectors. It is *imperative* to acquire a facility in vector algebra and vector calculus. In a three-dimensional space a vector relation is, in fact, three scalar relations. The use of vector-analysis techniques in electromagnetics leads to concise and elegant formulations. A deficiency in vector analysis in the study of electromagnetics is similar to a deficiency in algebra and calculus in the study of physics; and it is obvious that these deficiencies cannot yield fruitful results.

In solving practical problems we always deal with regions or objects of a given shape, and it is necessary to express general formulas in a coordinate system appropriate for the given geometry. For example, the familiar rectangular (x, y, z) coordinates are, obviously, awkward to use for problems involving a circular cylinder or a sphere because the boundaries of a circular cylinder and a sphere cannot be described by constant values of x, y, and z. In this chapter we discuss the three most commonly used orthogonal (perpendicular) coordinate systems and the representation and operation of vectors in these systems. Familarity with these coordinate systems is essential in the solution of electromagnetic problems.

Vector calculus pertains to the differentiation and integration of vectors. By defining certain differential operators we can express the basic laws of electromagnetism in a concise way that is invariant with the choice of a coordinate system. In this chapter we introduce the techniques for evaluating different types of integrals involving vectors, and we define and discuss the various kinds of differential operators.

2–2 Vector Addition and Subtraction

We know that a vector has a magnitude and a direction. A vector $\mathbf{A}$ can be written as

$$\mathbf{A} = \mathbf{a}_A A, \tag{2–1}$$

where A is the magnitude (and has the unit and dimension) of $\mathbf{A}$,

$$A = |\mathbf{A}|, \tag{2–2}$$

and $\mathbf{a}_A$ is a dimensionless unit vector[†] with a unity magnitude having the direction of $\mathbf{A}$. Thus,

$$\mathbf{a}_A = \frac{\mathbf{A}}{|\mathbf{A}|} = \frac{\mathbf{A}}{A}. \tag{2–3}$$

The vector $\mathbf{A}$ can be represented graphically by a directed straight-line segment of a length $|\mathbf{A}| = A$ with its arrowhead pointing in the direction of $\mathbf{a}_A$, as shown in Fig. 2–1. Two vectors are equal if they have the same magnitude and the same direction, even

---

[†] In some books the unit vector in the direction of $\mathbf{A}$ is variously denoted by $\hat{\mathbf{A}}$, $\mathbf{u}_A$, or $\mathbf{i}_A$. We prefer to write $\mathbf{A}$ as in Eq. (2–1) instead of as $\mathbf{A} = \hat{\mathbf{A}}A$. A vector going from point P_1 to point P_2 will then be written as $\mathbf{a}_{P_1P_2}(\overline{P_1P_2})$ instead of as $\widehat{P_1P_2}(P_1P_2)$, which is somewhat cumbersome. The symbols $\mathbf{u}$ and $\mathbf{i}$ are used for velocity and current, respectively.

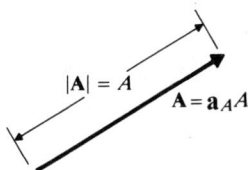

FIGURE 2–1
Graphical representation of vector **A**.

though they may be displaced in space. Since it is difficult to write boldfaced letters by hand, it is a common practice to use an arrow or a bar over a letter (Å or Ā) or a wiggly line under a letter (A̰) to distinguish a vector from a scalar. This distinguishing mark, once chosen, *should never be omitted* whenever and wherever vectors are written.

Two vectors **A** and **B**, which are not in the same direction nor in opposite directions, such as given in Fig. 2–2(a), determine a plane. Their sum is another vector **C** in the same plane. **C** = **A** + **B** can be obtained graphically in two ways.

1. By the parallelogram rule: The resultant **C** is the diagonal vector of the parallelogram formed by **A** and **B** drawn from the same point, as shown in Fig. 2–2(b).

2. By the head-to-tail rule: The head of **A** connects to the tail of **B**. Their sum **C** is the vector drawn from the tail of **A** to the head of **B**; and vectors **A**, **B**, and **C** form a triangle, as shown in Fig. 2–2(c).

It is obvious that vector addition obeys the commutative and associative laws.

$$\text{Commutative law:} \quad \mathbf{A} + \mathbf{B} = \mathbf{B} + \mathbf{A}. \tag{2–4}$$

$$\text{Associative law:} \quad \mathbf{A} + (\mathbf{B} + \mathbf{C}) = (\mathbf{A} + \mathbf{B}) + \mathbf{C}. \tag{2–5}$$

Vector subtraction can be defined in terms of vector addition in the following way:

$$\mathbf{A} - \mathbf{B} = \mathbf{A} + (-\mathbf{B}), \tag{2–6}$$

where $-\mathbf{B}$ is the negative of vector **B**; that is, $-\mathbf{B}$ has the same magnitude as **B**, but its direction is opposite to that of **B**. Thus

$$-\mathbf{B} = (-\mathbf{a}_B)B. \tag{2–7}$$

The operation represented by Eq. (2–6) is illustrated in Fig. 2–3.

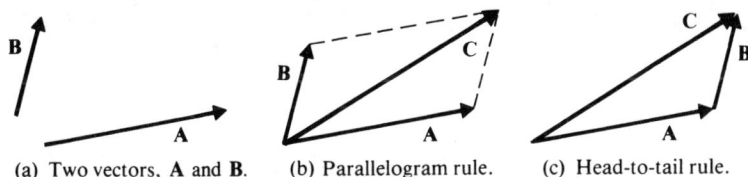

(a) Two vectors, **A** and **B**.   (b) Parallelogram rule.   (c) Head-to-tail rule.

FIGURE 2–2
Vector addition, **C** = **A** + **B**.

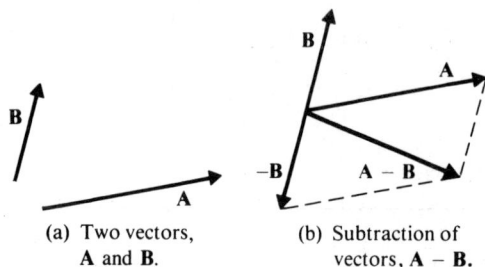

(a) Two vectors,   (b) Subtraction of   **FIGURE 2-3**
    A and **B**.   vectors, **A** − **B**.   Vector subtraction.

2–3 Products of Vectors

Multiplication of a vector **A** by a positive scalar k changes the magnitude of **A** by k times without changing its direction (k can be either greater or less than 1).

$$k\mathbf{A} = \mathbf{a}_A(kA). \qquad (2\text{-}8)$$

It is not sufficient to say "the multiplication of one vector by another" or "the product of two vectors" because there are two distinct and very different types of products of two vectors. They are (1) scalar or dot products, and (2) vector or cross products. These will be defined in the following subsections.

2–3.1 SCALAR OR DOT PRODUCT

The scalar or dot product of two vectors **A** and **B**, denoted by **A** · **B**, is a scalar, which equals the product of the magnitudes of **A** and **B** and the cosine of the angle between them. Thus,

$$\boxed{\mathbf{A} \cdot \mathbf{B} \triangleq AB \cos \theta_{AB}.} \qquad (2\text{-}9)$$

In Eq. (2–9) the symbol $\triangleq$ signifies "equal by definition," and θ_{AB} is the *smaller* angle between **A** and **B** and is less than π radians (180°), as indicated in Fig. 2–4. The dot product of two vectors (1) is less than or equal to the product of their magnitudes; (2) can be either a positive or a negative quantity, depending on whether the angle between them is smaller or larger than $\pi/2$ radians (90°); (3) is equal to the product

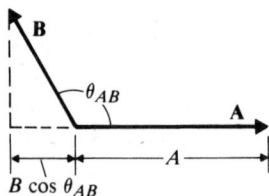

FIGURE 2-4
Illustrating the dot product of **A** and **B**.

of the magnitude of one vector and the projection of the other vector upon the first one; and (4) is zero when the vectors are perpendicular to each other. It is evident that

$$\mathbf{A} \cdot \mathbf{A} = A^2 \tag{2-10}$$

or

$$A = \sqrt[+]{\mathbf{A} \cdot \mathbf{A}}. \tag{2-11}$$

Equation (2-11) enables us to find the magnitude of a vector when the expression of the vector is given in any coordinate system.

The dot product is commutative and distributive.

Commutative law: $\mathbf{A} \cdot \mathbf{B} = \mathbf{B} \cdot \mathbf{A}.$ (2-12)

Distributive law: $\mathbf{A} \cdot (\mathbf{B} + \mathbf{C}) = \mathbf{A} \cdot \mathbf{B} + \mathbf{A} \cdot \mathbf{C}.$ (2-13)

The commutative law is obvious from the definition of the dot product in Eq. (2-9), and the proof of Eq. (2-13) is left as an exercise. The associative law does not apply to the dot product, since no more than two vectors can be so multiplied and an expression such as $\mathbf{A} \cdot \mathbf{B} \cdot \mathbf{C}$ is meaningless.

EXAMPLE 2-1 Prove the law of cosines for a triangle.

Solution The law of cosines is a scalar relationship that expresses the length of a side of a triangle in terms of the lengths of the two other sides and the angle between them. Referring to Fig. 2-5, we find the law of cosines states that

$$C = \sqrt{A^2 + B^2 - 2AB \cos \alpha}.$$

We prove this by considering the sides as vectors; that is,

$$\mathbf{C} = \mathbf{A} + \mathbf{B}.$$

Taking the dot product of $\mathbf{C}$ with itself, we have, from Eqs. (2-10) and (2-13),

$$C^2 = \mathbf{C} \cdot \mathbf{C} = (\mathbf{A} + \mathbf{B}) \cdot (\mathbf{A} + \mathbf{B})$$
$$= \mathbf{A} \cdot \mathbf{A} + \mathbf{B} \cdot \mathbf{B} + 2\mathbf{A} \cdot \mathbf{B}$$
$$= A^2 + B^2 + 2AB \cos \theta_{AB}.$$

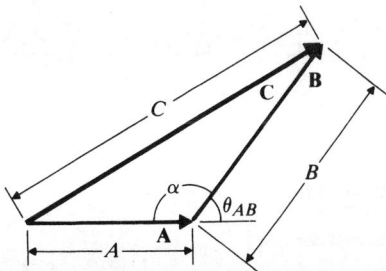

FIGURE 2-5
Illustrating Example 2-1.

Note that θ_{AB} is, by definition, the *smaller* angle between **A** and **B** and is equal to $(180° - \alpha)$; hence $\cos \theta_{AB} = \cos (180° - \alpha) = -\cos \alpha$. Therefore,

$$C^2 = A^2 + B^2 - 2AB \cos \alpha,$$

and the law of cosines follows directly.   ▬

2–3.2   VECTOR OR CROSS PRODUCT

The vector or cross product of two vectors **A** and **B**, denoted by **A** × **B**, is a vector perpendicular to the plane containing **A** and **B**; its magnitude is $AB \sin \theta_{AB}$, where θ_{AB} is the *smaller* angle between **A** and **B**, and its direction follows that of the thumb of the right hand when the fingers rotate from **A** to **B** through the angle θ_{AB} (the right-hand rule).

$$\boxed{\mathbf{A} \times \mathbf{B} \triangleq \mathbf{a}_n |AB \sin \theta_{AB}|.}$$ (2–14)

This is illustrated in Fig. 2–6. Since $B \sin \theta_{AB}$ is the height of the parallelogram formed by the vectors **A** and **B**, we recognize that the magnitude of **A** × **B**, $|AB \sin \theta_{AB}|$, which is always positive, is numerically equal to the area of the parallelogram.

   Using the definition in Eq. (2–14) and following the right-hand rule, we find that

$$\mathbf{B} \times \mathbf{A} = -\mathbf{A} \times \mathbf{B}.$$ (2–15)

Hence the cross product is *not* commutative. We can see that the cross product obeys the distributive law,

$$\mathbf{A} \times (\mathbf{B} + \mathbf{C}) = \mathbf{A} \times \mathbf{B} + \mathbf{A} \times \mathbf{C}.$$ (2–16)

Can you show this in general without resolving the vectors into rectangular components?

   The vector product is obviously *not* associative; that is,

$$\mathbf{A} \times (\mathbf{B} \times \mathbf{C}) \neq (\mathbf{A} \times \mathbf{B}) \times \mathbf{C}.$$ (2–17)

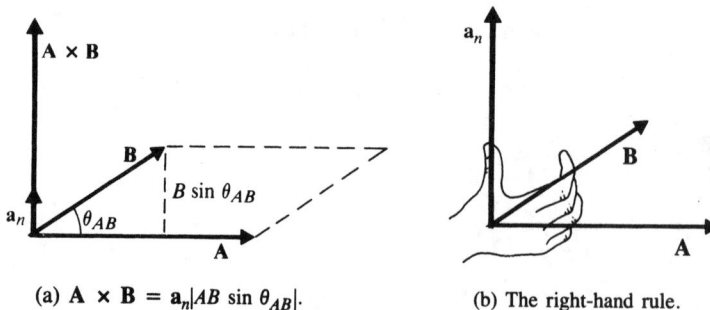

(a) **A** × **B** = $\mathbf{a}_n |AB \sin \theta_{AB}|$.   (b) The right-hand rule.

FIGURE 2–6
Cross product of **A** and **B**, **A** × **B**.

The vector representing the triple product on the left side of the expression above is perpendicular to **A** and lies in the plane formed by **B** and **C**, whereas that on the right side is perpendicular to **C** and lies in the plane formed by **A** and **B**. The order in which the two vector products are performed is therefore vital, and *in no case should the parentheses be omitted.*

EXAMPLE 2-2 The motion of a rigid disk rotating about its axis shown in Fig. 2-7(a) can be described by an angular velocity vector ω. The direction of ω is along the axis and follows the right-hand rule; that is, if the fingers of the right hand bend in the direction of rotation, the thumb points to the direction of ω. Find the vector expression for the lineal velocity of a point on the disk, which is at a distance d from the axis of rotation.

Solution From mechanics we know that the magnitude of the lineal velocity, **v**, of a point P at a distance d from the rotating axis is ωd and the direction is always tangential to the circle of rotation. However, since the point P is moving, the direction of **v** changes with the position of P. How do we write its vector representation?

Let O be the origin of the chosen coordinate system. The position vector of the point P can be written as **R**, as shown in Fig. 2-7(b). We have

$$|\mathbf{v}| = \omega d = \omega R \sin \theta.$$

No matter where the point P is, the direction of **v** is always perpendicular to the plane containing the vectors ω and **R**. Hence we can write, very simply,

$$\mathbf{v} = \omega \times \mathbf{R},$$

which represents correctly both the magnitude and the direction of the lineal velocity of P.

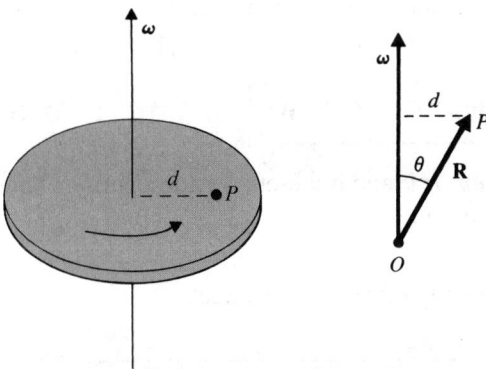

(a) A rotating disk.   (b) Vector representation.

FIGURE 2-7
Illustrating Example 2-2.

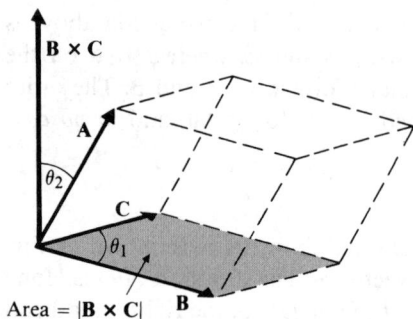

FIGURE 2–8
Illustrating scalar triple product $\mathbf{A} \cdot (\mathbf{B} \times \mathbf{C})$.

2–3.3 PRODUCT OF THREE VECTORS

There are two kinds of products of three vectors; namely, the *scalar triple product* and the *vector triple product*. The scalar triple product is much the simpler of the two and has the following property:

$$\mathbf{A} \cdot (\mathbf{B} \times \mathbf{C}) = \mathbf{B} \cdot (\mathbf{C} \times \mathbf{A}) = \mathbf{C} \cdot (\mathbf{A} \times \mathbf{B}). \qquad (2\text{–}18)$$

Note the cyclic permutation of the order of the three vectors $\mathbf{A}$, $\mathbf{B}$, and $\mathbf{C}$. Of course,

$$\begin{aligned}
\mathbf{A} \cdot (\mathbf{B} \times \mathbf{C}) &= -\mathbf{A} \cdot (\mathbf{C} \times \mathbf{B}) \\
&= -\mathbf{B} \cdot (\mathbf{A} \times \mathbf{C}) \qquad (2\text{–}19) \\
&= -\mathbf{C} \cdot (\mathbf{B} \times \mathbf{A}).
\end{aligned}$$

As can be seen from Fig. 2–8, each of the three expressions in Eq. (2–18) has a magnitude equal to the volume of the parallelepiped formed by the three vectors $\mathbf{A}$, $\mathbf{B}$, and $\mathbf{C}$. The parallelepiped has a base with an area equal to $|\mathbf{B} \times \mathbf{C}| = |BC \sin \theta_1|$ and a height equal to $|A \cos \theta_2|$; hence the volume is $|ABC \sin \theta_1 \cos \theta_2|$.

The vector triple product $\mathbf{A} \times (\mathbf{B} \times \mathbf{C})$ can be expanded as the difference of two simple vectors as follows:

$$\mathbf{A} \times (\mathbf{B} \times \mathbf{C}) = \mathbf{B}(\mathbf{A} \cdot \mathbf{C}) - \mathbf{C}(\mathbf{A} \cdot \mathbf{B}). \qquad (2\text{–}20)$$

Equation (2–20) is known as the *"back-cab" rule* and is a useful vector identity. (Note "BAC-CAB" on the right side of the equation!)

■■■■ **EXAMPLE 2–3[†]** Prove the back-cab rule of vector triple product.

[†] The back-cab rule can be verified in a straightforward manner by expanding the vectors in the Cartesian coordinate system (Problem P.2–12). Only those interested in a general proof need to study this example.

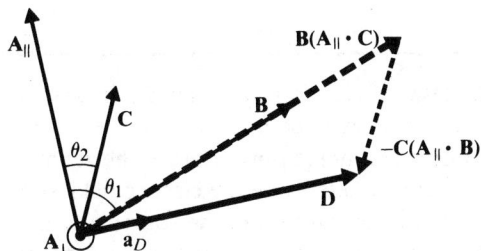

FIGURE 2–9
Illustrating the back-cab rule of vector triple product.

Solution In order to prove Eq. (2–20) it is convenient to expand $\mathbf{A}$ into two components:

$$\mathbf{A} = \mathbf{A}_{\parallel} + \mathbf{A}_{\perp},$$

where $\mathbf{A}_{\parallel}$ and $\mathbf{A}_{\perp}$ are parallel and perpendicular, respectively, to the plane containing $\mathbf{B}$ and $\mathbf{C}$. Because the vector representing $(\mathbf{B} \times \mathbf{C})$ is also perpendicular to the plane, the cross product of $\mathbf{A}_{\perp}$ and $(\mathbf{B} \times \mathbf{C})$ vanishes. Let $\mathbf{D} = \mathbf{A} \times (\mathbf{B} \times \mathbf{C})$. Since only $\mathbf{A}_{\parallel}$ is effective here, we have

$$\mathbf{D} = \mathbf{A}_{\parallel} \times (\mathbf{B} \times \mathbf{C}).$$

Referring to Fig. 2–9, which shows the plane containing $\mathbf{B}$, $\mathbf{C}$, and $\mathbf{A}_{\parallel}$, we note that $\mathbf{D}$ lies in the same plane and is normal to $\mathbf{A}_{\parallel}$. The magnitude of $(\mathbf{B} \times \mathbf{C})$ is $BC \sin(\theta_1 - \theta_2)$, and that of $\mathbf{A}_{\parallel} \times (\mathbf{B} \times \mathbf{C})$ is $A_{\parallel}BC \sin(\theta_1 - \theta_2)$. Hence,

$$\begin{aligned}
D = \mathbf{D} \cdot \mathbf{a}_D &= A_{\parallel}BC \sin(\theta_1 - \theta_2) \\
&= (B \sin \theta_1)(A_{\parallel}C \cos \theta_2) - (C \sin \theta_2)(A_{\parallel}B \cos \theta_1) \\
&= [\mathbf{B}(\mathbf{A}_{\parallel} \cdot \mathbf{C}) - \mathbf{C}(\mathbf{A}_{\parallel} \cdot \mathbf{B})] \cdot \mathbf{a}_D.
\end{aligned}$$

The expression above does not alone guarantee the quantity inside the brackets to be $\mathbf{D}$, since the former may contain a vector that is normal to $\mathbf{D}$ (parallel to $\mathbf{A}_{\parallel}$); that is, $\mathbf{D} \cdot \mathbf{a}_D = \mathbf{E} \cdot \mathbf{a}_D$ does not guarantee $\mathbf{E} = \mathbf{D}$. In general, we can write

$$\mathbf{B}(\mathbf{A}_{\parallel} \cdot \mathbf{C}) - \mathbf{C}(\mathbf{A}_{\parallel} \cdot \mathbf{B}) = \mathbf{D} + k\mathbf{A}_{\parallel},$$

where k is a scalar quantity. To determine k, we scalar-multiply both sides of the above equation by $\mathbf{A}_{\parallel}$ and obtain

$$(\mathbf{A}_{\parallel} \cdot \mathbf{B})(\mathbf{A}_{\parallel} \cdot \mathbf{C}) - (\mathbf{A}_{\parallel} \cdot \mathbf{C})(\mathbf{A}_{\parallel} \cdot \mathbf{B}) = 0 = \mathbf{A}_{\parallel} \cdot \mathbf{D} + kA_{\parallel}^2.$$

Since $\mathbf{A}_{\parallel} \cdot \mathbf{D} = 0$, then $k = 0$ and

$$\mathbf{D} = \mathbf{B}(\mathbf{A}_{\parallel} \cdot \mathbf{C}) - \mathbf{C}(\mathbf{A}_{\parallel} \cdot \mathbf{B}),$$

which proves the back-cab rule, inasmuch as $\mathbf{A}_{\parallel} \cdot \mathbf{C} = \mathbf{A} \cdot \mathbf{C}$ and $\mathbf{A}_{\parallel} \cdot \mathbf{B} = \mathbf{A} \cdot \mathbf{B}$. ◼

Division by a vector is not defined, and expressions such as $k/\mathbf{A}$ and $\mathbf{B}/\mathbf{A}$ are meaningless.

2–4 Orthogonal Coordinate Systems

We have indicated before that although the laws of electromagnetism are invariant with coordinate system, solution of practical problems requires that the relations derived from these laws be expressed in a coordinate system appropriate to the geometry of the given problems. For example, if we are to determine the electric field at a certain point in space, we at least need to describe the position of the source and the location of this point in a coordinate system. In a three-dimensional space a point can be located as the intersection of three surfaces. Assume that the three families of surfaces are described by u_1 = constant, u_2 = constant, and u_3 = constant, where the u's need not all be lengths. (In the familiar Cartesian or rectangular coordinate system, u_1, u_2, and u_3 correspond to x, y, and z, respectively.) When these three surfaces are mutually perpendicular to one another, we have an ***orthogonal coordinate system***. Nonorthogonal coordinate systems are not used because they complicate problems.

Some surfaces represented by u_i = constant (i = 1, 2, or 3) in a coordinate system may not be planes; they may be curved surfaces. Let $\mathbf{a}_{u_1}$, $\mathbf{a}_{u_2}$, and $\mathbf{a}_{u_3}$ be the unit vectors in the three coordinate directions. They are called the ***base vectors***. In a general right-handed, orthogonal, curvilinear coordinate system the base vectors are arranged in such a way that the following relations are satisfied:

$$\mathbf{a}_{u_1} \times \mathbf{a}_{u_2} = \mathbf{a}_{u_3}, \tag{2–21a}$$

$$\mathbf{a}_{u_2} \times \mathbf{a}_{u_3} = \mathbf{a}_{u_1}, \tag{2–21b}$$

$$\mathbf{a}_{u_3} \times \mathbf{a}_{u_1} = \mathbf{a}_{u_2}. \tag{2–21c}$$

These three equations are not all independent, as the specification of one automatically implies the other two. We have, of course,

$$\mathbf{a}_{u_1} \cdot \mathbf{a}_{u_2} = \mathbf{a}_{u_2} \cdot \mathbf{a}_{u_3} = \mathbf{a}_{u_3} \cdot \mathbf{a}_{u_1} = 0 \tag{2–22}$$

and

$$\mathbf{a}_{u_1} \cdot \mathbf{a}_{u_1} = \mathbf{a}_{u_2} \cdot \mathbf{a}_{u_2} = \mathbf{a}_{u_3} \cdot \mathbf{a}_{u_3} = 1. \tag{2–23}$$

Any vector $\mathbf{A}$ can be written as the sum of its components in the three orthogonal directions, as follows:

$$\boxed{\mathbf{A} = \mathbf{a}_{u_1} A_{u_1} + \mathbf{a}_{u_2} A_{u_2} + \mathbf{a}_{u_3} A_{u_3}.} \tag{2–24}$$

From Eq. (2–24) the magnitude of $\mathbf{A}$ is

$$A = |\mathbf{A}| = (A_{u_1}^2 + A_{u_2}^2 + A_{u_3}^2)^{1/2}. \tag{2–25}$$

■■■■■ **EXAMPLE 2–4**   Given three vectors $\mathbf{A}$, $\mathbf{B}$, and $\mathbf{C}$, obtain the expressions of (a) $\mathbf{A} \cdot \mathbf{B}$, (b) $\mathbf{A} \times \mathbf{B}$, and (c) $\mathbf{C} \cdot (\mathbf{A} \times \mathbf{B})$ in the orthogonal curvilinear coordinate system (u_1, u_2, u_3).

Solution   First we write **A**, **B**, and **C** in the orthogonal coordinates (u_1, u_2, u_3):

$$\mathbf{A} = \mathbf{a}_{u_1}A_{u_1} + \mathbf{a}_{u_2}A_{u_2} + \mathbf{a}_{u_3}A_{u_3},$$
$$\mathbf{B} = \mathbf{a}_{u_1}B_{u_1} + \mathbf{a}_{u_2}B_{u_2} + \mathbf{a}_{u_3}B_{u_3},$$
$$\mathbf{C} = \mathbf{a}_{u_1}C_{u_1} + \mathbf{a}_{u_2}C_{u_2} + \mathbf{a}_{u_3}C_{u_3}.$$

a) $\mathbf{A} \cdot \mathbf{B} = (\mathbf{a}_{u_1}A_{u_1} + \mathbf{a}_{u_2}A_{u_2} + \mathbf{a}_{u_3}A_{u_3}) \cdot (\mathbf{a}_{u_1}B_{u_1} + \mathbf{a}_{u_2}B_{u_2} + \mathbf{a}_{u_3}B_{u_3})$

$$= A_{u_1}B_{u_1} + A_{u_2}B_{u_2} + A_{u_3}B_{u_3}, \qquad (2\text{–}26)$$

in view of Eqs. (2–22) and (2–23).

b) $\mathbf{A} \times \mathbf{B} = (\mathbf{a}_{u_1}A_{u_1} + \mathbf{a}_{u_2}A_{u_2} + \mathbf{a}_{u_3}A_{u_3}) \times (\mathbf{a}_{u_1}B_{u_1} + \mathbf{a}_{u_2}B_{u_2} + \mathbf{a}_{u_3}B_{u_3})$

$$= \mathbf{a}_{u_1}(A_{u_2}B_{u_3} - A_{u_3}B_{u_2}) + \mathbf{a}_{u_2}(A_{u_3}B_{u_1} - A_{u_1}B_{u_3}) + \mathbf{a}_{u_3}(A_{u_1}B_{u_2} - A_{u_2}B_{u_1})$$

$$= \begin{vmatrix} \mathbf{a}_{u_1} & \mathbf{a}_{u_2} & \mathbf{a}_{u_3} \\ A_{u_1} & A_{u_2} & A_{u_3} \\ B_{u_1} & B_{u_2} & B_{u_3} \end{vmatrix}. \qquad (2\text{–}27)$$

Equations (2–26) and (2–27) express the dot and cross products, respectively, of two vectors in orthogonal curvilinear coordinates. They are important and should be remembered.

c) The expression for $\mathbf{C} \cdot (\mathbf{A} \times \mathbf{B})$ can be written down immediately by combining the results in Eqs. (2–26) and (2–27):

$$\mathbf{C} \cdot (\mathbf{A} \times \mathbf{B}) = C_{u_1}(A_{u_2}B_{u_3} - A_{u_3}B_{u_2}) + C_{u_2}(A_{u_3}B_{u_1} - A_{u_1}B_{u_3}) + C_{u_3}(A_{u_1}B_{u_2} - A_{u_2}B_{u_1})$$

$$= \begin{vmatrix} C_{u_1} & C_{u_2} & C_{u_3} \\ A_{u_1} & A_{u_2} & A_{u_3} \\ B_{u_1} & B_{u_2} & B_{u_3} \end{vmatrix}. \qquad (2\text{–}28)$$

Eq. (2–28) can be used to prove Eqs. (2–18) and (2–19) by observing that a permutation of the order of the vectors on the left side leads simply to a rearrangement of the rows in the determinant on the right side.   ▬

In vector calculus (and in electromagnetics work) we are often required to perform line, surface, and volume integrals. In each case we need to express the differential length-change corresponding to a differential change in one of the coordinates. However, some of the coordinates, say u_i ($i = 1, 2,$ or 3), may not be a length; and a conversion factor is needed to convert a differential change du_i into a change in length $d\ell_i$:

$$d\ell_i = h_i\, du_i, \qquad (2\text{–}29)$$

where h_i is called a **_metric coefficient_** and may itself be a function of u_1, u_2, and u_3. For example, in the two-dimensional polar coordinates $(u_1, u_2) = (r, \phi)$, a differential change $d\phi\ (= du_2)$ in $\phi\ (= u_2)$ corresponds to a differential length-change $d\ell_2 = r\, d\phi$ $(h_2 = r = u_1)$ in the $\mathbf{a}_\phi\ (= \mathbf{a}_{u_2})$-direction. A directed differential length-change in an arbitrary direction can be written as the vector sum of the component length-changes:

$$d\boldsymbol{\ell} = \mathbf{a}_{u_1}\, d\ell_1 + \mathbf{a}_{u_2}\, d\ell_2 + \mathbf{a}_{u_3}\, d\ell_3 \qquad (2\text{--}30)^\dagger$$

or

$$d\boldsymbol{\ell} = \mathbf{a}_{u_1}(h_1\, du_1) + \mathbf{a}_{u_2}(h_2\, du_2) + \mathbf{a}_{u_3}(h_3\, du_3). \qquad (2\text{--}31)$$

In view of Eq. (2–25) the magnitude of $d\boldsymbol{\ell}$ is

$$\begin{aligned} d\ell &= [(d\ell_1)^2 + (d\ell_2)^2 + (d\ell_3)^2]^{1/2} \\ &= [(h_1\, du_1)^2 + (h_2\, du_2)^2 + (h_3\, du_3)^2]^{1/2}. \end{aligned} \qquad (2\text{--}32)$$

The differential volume dv formed by differential coordinate changes du_1, du_2, and du_3 in directions $\mathbf{a}_{u1}$, $\mathbf{a}_{u2}$, and $\mathbf{a}_{u3}$, respectively, is $(d\ell_1\, d\ell_2\, d\ell_3)$, or

$$dv = h_1 h_2 h_3\, du_1\, du_2\, du_3. \qquad (2\text{--}33)$$

Later we will have occasion to express the current or flux flowing through a differential area. In such cases the cross-sectional area perpendicular to the current or flux flow must be used, and it is convenient to consider the differential area a vector with a direction normal to the surface; that is,

$$d\mathbf{s} = \mathbf{a}_n\, ds. \qquad (2\text{--}34)$$

For instance, if current density $\mathbf{J}$ is not perpendicular to a differential area of a magnitude ds, the current, dI, flowing through ds must be the component of $\mathbf{J}$ normal to the area multiplied by the area. Using the notation in Eq. (2–34), we can write simply

$$\begin{aligned} dI &= \mathbf{J} \cdot d\mathbf{s} \\ &= \mathbf{J} \cdot \mathbf{a}_n\, ds. \end{aligned} \qquad (2\text{--}35)$$

In general orthogonal curvilinear coordinates the differential area ds_1 normal to the unit vector $\mathbf{a}_{u_1}$ is

$$ds_1 = d\ell_2\, d\ell_3$$

or

$$ds_1 = h_2 h_3\, du_2\, du_3. \qquad (2\text{--}36)$$

Similarly, the differential areas normal to unit vectors $\mathbf{a}_{u_2}$ and $\mathbf{a}_{u_3}$ are, respectively,

$$ds_2 = h_1 h_3\, du_1\, du_3 \qquad (2\text{--}37)$$

---

† The $\boldsymbol{\ell}$ here is the symbol of a vector of length ℓ.

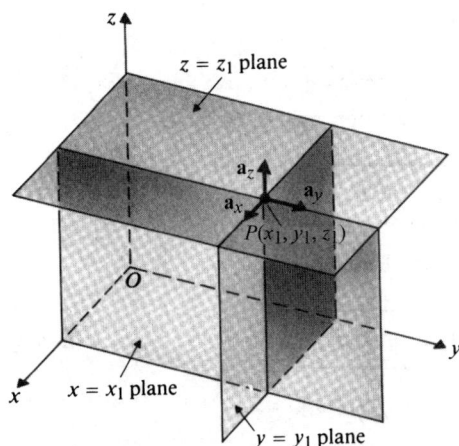

FIGURE 2–10
Cartesian coordinates.

and

$$ds_3 = h_1 h_2 \, du_1 \, du_2.$$

$$(2\text{--}38)$$

Many orthogonal coordinate systems exist; but we shall be concerned only with the three that are most common and most useful:

1. Cartesian (or rectangular) coordinates.[†]

2. Cylindrical coordinates.

3. Spherical coordinates.

These will be discussed separately in the following subsections.

2–4.1 CARTESIAN COORDINATES

$$(u_1, u_2, u_3) = (x, y, z)$$

A point $P(x_1, y_1, z_1)$ in Cartesian coordinates is the intersection of *three planes* specified by $x = x_1$, $y = y_1$, and $z = z_1$, as shown in Fig. 2–10. It is a right-handed system with base vectors $\mathbf{a}_x$, $\mathbf{a}_y$, and $\mathbf{a}_z$ satisfying the following relations:

$$\mathbf{a}_x \times \mathbf{a}_y = \mathbf{a}_z,$$

$$(2\text{--}39a)$$

$$\mathbf{a}_y \times \mathbf{a}_z = \mathbf{a}_x,$$

$$(2\text{--}39b)$$

$$\mathbf{a}_z \times \mathbf{a}_x = \mathbf{a}_y.$$

$$(2\text{--}39c)$$

[†] The term "Cartesian coordinates" is preferred because the term "rectangular coordinates" is customarily associated with two-dimensional geometry.

The position vector to the point $P(x_1, y_1, z_1)$ is

$$\vec{OP} = \mathbf{a}_x x_1 + \mathbf{a}_y y_1 + \mathbf{a}_z z_1. \tag{2-40}$$

A vector **A** in Cartesian coordinates can be written as

$$\boxed{\mathbf{A} = \mathbf{a}_x A_x + \mathbf{a}_y A_y + \mathbf{a}_z A_z.} \tag{2-41}$$

The dot product of two vectors **A** and **B** is, from Eq. (2-26),

$$\boxed{\mathbf{A} \cdot \mathbf{B} = A_x B_x + A_y B_y + A_z B_z,} \tag{2-42}$$

and the cross product of **A** and **B** is, from Eq. (2-27),

$$\mathbf{A} \times \mathbf{B} = \mathbf{a}_x (A_y B_z - A_z B_y) + \mathbf{a}_y (A_z B_x - A_x B_z) + \mathbf{a}_z (A_x B_y - A_y B_x)$$

$$= \begin{vmatrix} \mathbf{a}_x & \mathbf{a}_y & \mathbf{a}_z \\ A_x & A_y & A_z \\ B_x & B_y & B_z \end{vmatrix}. \tag{2-43}$$

Since x, y, and z are lengths themselves, all three metric coefficients are unity; that is, $h_1 = h_2 = h_3 = 1$. The expressions for the differential length, differential area, and differential volume are—from Eqs. (2-31), (2-36), (2-37), (2-38), and (2-33)—respectively,

$$\boxed{d\boldsymbol{\ell} = \mathbf{a}_x \, dx + \mathbf{a}_y \, dy + \mathbf{a}_z \, dz;} \tag{2-44}$$

$$\boxed{\begin{aligned} ds_x &= dy \, dz, \\ ds_y &= dx \, dz, \\ ds_z &= dx \, dy; \end{aligned}} \qquad \begin{aligned} &\text{(2-45a)} \\ &\text{(2-45b)} \\ &\text{(2-45c)} \end{aligned}$$

and

$$\boxed{dv = dx \, dy \, dz.} \tag{2-46}$$

A typical differential volume element at a point (x, y, z) resulting from differential changes dx, dy, and dz is shown in Fig. 2-11. The differential surface areas ds_x, ds_y, and ds_z normal to the directions $\mathbf{a}_x$, $\mathbf{a}_y$, and $\mathbf{a}_z$ are also indicated.

EXAMPLE 2-5   Given $\mathbf{A} = \mathbf{a}_x 5 - \mathbf{a}_y 2 + \mathbf{a}_z$, find the expression of a unit vector **B** such that

a) $\mathbf{B} \| \mathbf{A}$.

b) $\mathbf{B} \perp \mathbf{A}$, if **B** lies in the xy-plane.

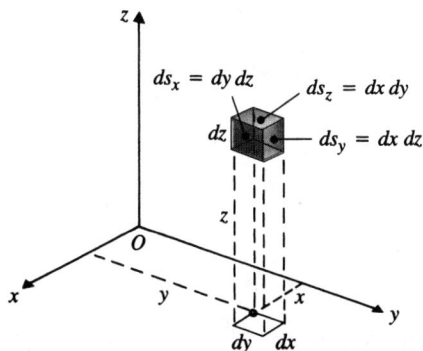

FIGURE 2–11
A differential volume in Cartesian coordinates.

Solution   Let $\mathbf{B} = \mathbf{a}_x B_x + \mathbf{a}_y B_y + \mathbf{a}_z B_z$. We know that

$$|\mathbf{B}| = (B_x^2 + B_y^2 + B_z^2)^{1/2} = 1. \qquad (2\text{–}47)$$

a) $\mathbf{B} \,\|\, \mathbf{A}$ requires $\mathbf{B} \times \mathbf{A} = 0$. From Eq. (2–43) we have

$$-2B_z - B_y = 0, \qquad (2\text{–}48a)$$

$$B_x - 5B_z = 0, \qquad (2\text{–}48b)$$

$$5B_y + 2B_x = 0. \qquad (2\text{–}48c)$$

The above three equations are not all independent. For instance, subtracting Eq. (2–48c) from twice Eq. (2–48b) yields Eq. (2–48a). Solving Eqs. (2–47), (2–48a), and (2–48b) simultaneously, we obtain

$$B_x = \frac{5}{\sqrt{30}}, \qquad B_y = -\frac{2}{\sqrt{30}} \qquad \text{and} \qquad B_z = \frac{1}{\sqrt{30}}.$$

Therefore,

$$\mathbf{B} = \frac{1}{\sqrt{30}} (\mathbf{a}_x 5 - \mathbf{a}_y 2 + \mathbf{a}_z).$$

b) $\mathbf{B} \perp \mathbf{A}$ requires $\mathbf{B} \cdot \mathbf{A} = 0$. From Eq. (2–42) we have

$$5B_x - 2B_y = 0, \qquad (2\text{–}49)$$

where we have set $B_z = 0$, since $\mathbf{B}$ lies in the xy-plane. Solution of Eqs. (2–47) and (2–49) yields

$$B_x = \frac{2}{\sqrt{29}} \qquad \text{and} \qquad B_y = \frac{5}{\sqrt{29}}.$$

Hence,

$$\mathbf{B} = \frac{1}{\sqrt{29}} (\mathbf{a}_x 2 + \mathbf{a}_y 5).$$

EXAMPLE 2–6   (a) Write the expression of the vector going from point $P_1(1, 3, 2)$ to point $P_2(3, -2, 4)$ in Cartesian coordinates. (b) What is the length of this line?

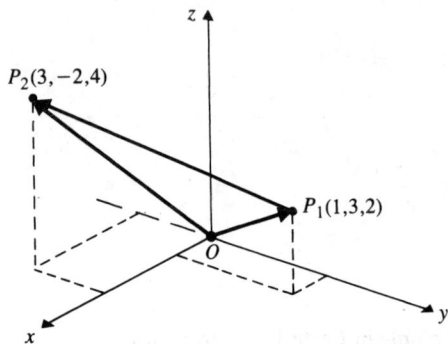

FIGURE 2–12
Illustrating Example 2–6.

Solution

a) From Fig. 2–12 we see that

$$\overrightarrow{P_1P_2} = \overrightarrow{OP_2} - \overrightarrow{OP_1}$$
$$= (\mathbf{a}_x 3 - \mathbf{a}_y 2 + \mathbf{a}_z 4) - (\mathbf{a}_x + \mathbf{a}_y 3 + \mathbf{a}_z 2)$$
$$= \mathbf{a}_x 2 - \mathbf{a}_y 5 + \mathbf{a}_z 2.$$

b) The length of the line is

$$P_1P_2 = |\overrightarrow{P_1P_2}|$$
$$= \sqrt{2^2 + (-5)^2 + 2^2}$$
$$= \sqrt{33}.$$

■

■■■■■ **EXAMPLE 2–7**   The equation of a straight line in the xy-plane is given by $2x + y = 4$.

a) Find the vector equation of a unit normal from the origin to the line.

b) Find the equation of a line passing through the point $P(0, 2)$ and perpendicular to the given line.

Solution   It is clear that the given equation $y = -2x + 4$ represents a straight line having a slope -2 and a vertical intercept $+4$, shown as L_1 (solid line) in Fig. 2–13.

a) If the line is shifted down four units, we have the dashed parallel line L_1' passing through the origin whose equation is $2x + y = 0$. Let the position vector of a point on L_1' be

$$\mathbf{r} = \mathbf{a}_x x + \mathbf{a}_y y.$$

The vector $\mathbf{N} = \mathbf{a}_x 2 + \mathbf{a}_y$ is perpendicular to L_1' because

$$\mathbf{N} \cdot \mathbf{r} = 2x + y = 0.$$

Obviously, $\mathbf{N}$ is also perpendicular to L_1. Thus, the vector equation of the unit normal at the origin is

$$\mathbf{a}_N = \frac{\mathbf{N}}{|\mathbf{N}|} = \frac{1}{\sqrt{5}}(\mathbf{a}_x 2 + \mathbf{a}_y).$$

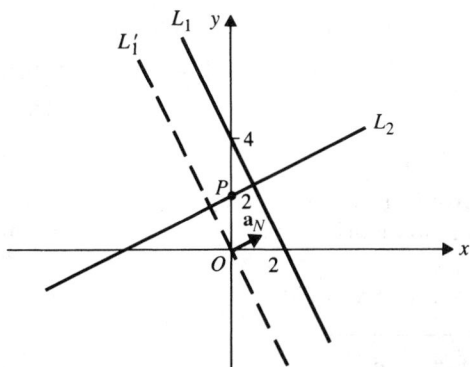

FIGURE 2–13
Illustrating Example 2–7.

Note that the slope of $\mathbf{a}_N$ ($=\frac{1}{2}$) is the negative reciprocal of that of lines L_1 and L_1' ($=-2$).

b) Let the line passing through the point $P(0, 2)$ and perpendicular to L_1 be L_2. L_2 is parallel to and has the same slope as $\mathbf{a}_N$. The equation of L_2 is then

$$y = \frac{x}{2} + 2 \qquad \text{or} \qquad x - 2y = -4,$$

since L_2 is required to pass through the point $P(0, 2)$. ▬

2–4.2 CYLINDRICAL COORDINATES

$$(u_1, u_2, u_3) = (r, \phi, z)$$

In cylindrical coordinates a point $P(r_1, \phi_1, z_1)$ is the intersection of a circular cylindrical surface $r = r_1$, a half-plane containing the z-axis and making an angle $\phi = \phi_1$ with the xz-plane, and a plane parallel to the xy-plane at $z = z_1$. As indicated in Fig. 2–14, angle ϕ is measured from the positive x-axis, and the base vector $\mathbf{a}_\phi$ is

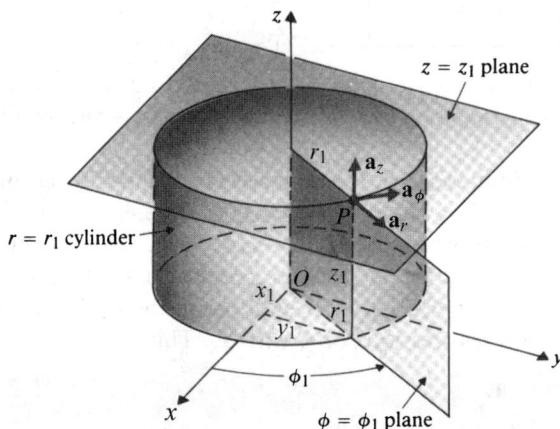

FIGURE 2–14
Cylindrical coordinates.

tangential to the cylindrical surface. The following right-hand relations apply:

$$\mathbf{a}_r \times \mathbf{a}_\phi = \mathbf{a}_z, \tag{2-50a}$$

$$\mathbf{a}_\phi \times \mathbf{a}_z = \mathbf{a}_r, \tag{2-50b}$$

$$\mathbf{a}_z \times \mathbf{a}_r = \mathbf{a}_\phi. \tag{2-50c}$$

Cylindrical coordinates are important for problems with long line charges or currents, and in places where cylindrical or circular boundaries exist. The two-dimensional polar coordinates are a special case at $z = 0$.

A vector in cylindrical coordinates is written as

$$\mathbf{A} = \mathbf{a}_r A_r + \mathbf{a}_\phi A_\phi + \mathbf{a}_z A_z. \tag{2-51}$$

The expressions for the dot and cross products of two vectors in cylindrical coordinates follow from Eqs. (2-26) and (2-27) directly.

Two of the three coordinates, r and z (u_1 and u_3), are themselves lengths; hence $h_1 = h_3 = 1$. However, ϕ is an angle requiring a metric coefficient $h_2 = r$ to convert $d\phi$ to $d\ell_2$. The general expression for a differential length in cylindrical coordinates is then, from Eq. (2-31),

$$d\ell = \mathbf{a}_r \, dr + \mathbf{a}_\phi r \, d\phi + \mathbf{a}_z \, dz. \tag{2-52}$$

The expressions for differential areas and differential volume are

$$ds_r = r \, d\phi \, dz, \tag{2-53a}$$

$$ds_\phi = dr \, dz, \tag{2-53b}$$

$$ds_z = r \, dr \, d\phi, \tag{2-53c}$$

and

$$dv = r \, dr \, d\phi \, dz. \tag{2-54}$$

A typical differential volume element at a point (r, ϕ, z) resulting from differential changes dr, $d\phi$, and dz in the three orthogonal coordinate directions is shown in Fig. 2-15.

A vector given in cylindrical coordinates can be transformed into one in Cartesian coordinates, and vice versa. Suppose we want to express $\mathbf{A} = \mathbf{a}_r A_r + \mathbf{a}_\phi A_\phi + \mathbf{a}_z A_z$ in Cartesian coordinates; that is, we want to write $\mathbf{A}$ as $\mathbf{a}_x A_x + \mathbf{a}_y A_y + \mathbf{a}_z A_z$ and determine A_x, A_y, and A_z. First of all, we note that A_z, the z-component of $\mathbf{A}$, is not changed by the transformation from cylindrical to Cartesian coordinates. To find A_x, we equate the dot products of both expressions of $\mathbf{A}$ with $\mathbf{a}_x$. Thus

$$A_x = \mathbf{A} \cdot \mathbf{a}_x$$
$$= A_r \mathbf{a}_r \cdot \mathbf{a}_x + A_\phi \mathbf{a}_\phi \cdot \mathbf{a}_x.$$

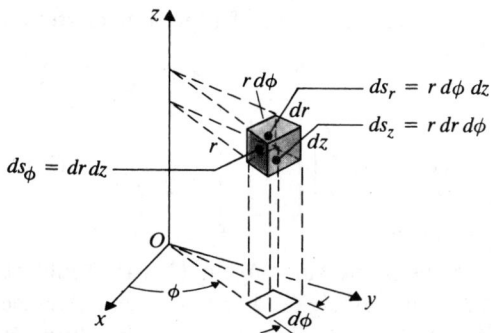

FIGURE 2–15
A differential volume element in cylindrical coordinates.

The term containing A_z disappears here because $\mathbf{a}_z \cdot \mathbf{a}_x = 0$. Referring to Fig. 2–16, which shows the relative positions of the base vectors $\mathbf{a}_x$, $\mathbf{a}_y$, $\mathbf{a}_r$, and $\mathbf{a}_\phi$, we see that

$$\mathbf{a}_r \cdot \mathbf{a}_x = \cos \phi \tag{2-55}$$

and

$$\mathbf{a}_\phi \cdot \mathbf{a}_x = \cos\left(\frac{\pi}{2} + \phi\right) = -\sin \phi. \tag{2-56}$$

Hence,

$$A_x = A_r \cos \phi - A_\phi \sin \phi. \tag{2-57}$$

Similarly, to find A_y, we take the dot products of both expressions of $\mathbf{A}$ with $\mathbf{a}_y$:

$$A_y = \mathbf{A} \cdot \mathbf{a}_y$$
$$= A_r \mathbf{a}_r \cdot \mathbf{a}_y + A_\phi \mathbf{a}_\phi \cdot \mathbf{a}_y.$$

From Fig. 2–16 we find that

$$\mathbf{a}_r \cdot \mathbf{a}_y = \cos\left(\frac{\pi}{2} - \phi\right) = \sin \phi \tag{2-58}$$

and

$$\mathbf{a}_\phi \cdot \mathbf{a}_y = \cos \phi. \tag{2-59}$$

It follows that

$$A_y = A_r \sin \phi + A_\phi \cos \phi. \tag{2-60}$$

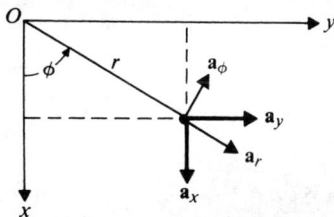

FIGURE 2–16
Relations between $\mathbf{a}_x$, $\mathbf{a}_y$, $\mathbf{a}_r$, and $\mathbf{a}_\phi$.

It is convenient to write the relations between the components of a vector in Cartesian and cylindrical coordinates in a matrix form:

$$
\begin{bmatrix} A_x \\ A_y \\ A_z \end{bmatrix} = \begin{bmatrix} \cos\phi & -\sin\phi & 0 \\ \sin\phi & \cos\phi & 0 \\ 0 & 0 & 1 \end{bmatrix} \begin{bmatrix} A_r \\ A_\phi \\ A_z \end{bmatrix}.
\qquad (2\text{--}61)
$$

Our problem is now solved except that the $\cos\phi$ and $\sin\phi$ in Eq. (2–61) should be converted into Cartesian coordinates. Moreover, A_r, A_ϕ, and A_z may themselves be functions of r, ϕ, and z. In that case, they too should be converted into functions of x, y, and z in the final answer. The following conversion formulas are obvious from Fig. 2–16. From cylindrical to Cartesian coordinates:

$$
x = r\cos\phi, \qquad (2\text{--}62a)
$$
$$
y = r\sin\phi, \qquad (2\text{--}62b)
$$
$$
z = z. \qquad (2\text{--}62c)
$$

The inverse relations (from Cartesian to cylindrical coordinates) are

$$
r = \sqrt{x^2 + y^2}, \qquad (2\text{--}63a)
$$
$$
\phi = \tan^{-1}\frac{y}{x}, \qquad (2\text{--}63b)
$$
$$
z = z. \qquad (2\text{--}63c)
$$

EXAMPLE 2–8   The cylindrical coordinates of an arbitrary point P in the $z = 0$ plane are $(r, \phi, 0)$. Find the unit vector that goes from a point $z = h$ on z-axis toward P.

Solution   Referring to Fig. 2–17, we have

$$
\overrightarrow{QP} = \overrightarrow{OP} - \overrightarrow{OQ}
$$
$$
= (\mathbf{a}_r r) - (\mathbf{a}_z h).
$$

Hence,

$$
\mathbf{a}_{QP} = \frac{\overrightarrow{QP}}{|\overrightarrow{QP}|} = \frac{1}{\sqrt{r^2 + h^2}}\,(\mathbf{a}_r r - \mathbf{a}_z h).
$$

EXAMPLE 2–9   Express the vector

$$
\mathbf{A} = \mathbf{a}_r(3\cos\phi) - \mathbf{a}_\phi 2r + \mathbf{a}_z 5
$$

in Cartesian coordinates.

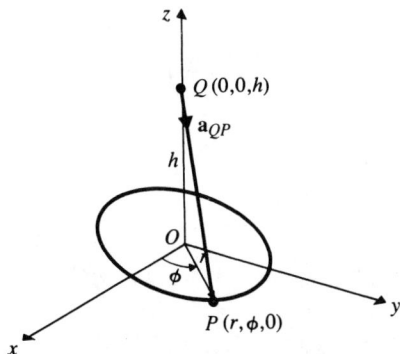

FIGURE 2–17
Illustrating Example 2–8.

Solution Using Eq. (2–61) directly, we have

$$\begin{bmatrix} A_x \\ A_y \\ A_z \end{bmatrix} = \begin{bmatrix} \cos\phi & -\sin\phi & 0 \\ \sin\phi & \cos\phi & 0 \\ 0 & 0 & 1 \end{bmatrix} \begin{bmatrix} 3\cos\phi \\ -2r \\ 5 \end{bmatrix}$$

or

$$\mathbf{A} = \mathbf{a}_x(3\cos^2\phi + 2r\sin\phi) + \mathbf{a}_y(3\sin\phi\cos\phi - 2r\cos\phi) + \mathbf{a}_z5.$$

But, from Eqs. (2–62) and (2–63),

$$\cos\phi = \frac{x}{\sqrt{x^2 + y^2}}$$

and

$$\sin\phi = \frac{y}{\sqrt{x^2 + y^2}}.$$

Therefore,

$$\mathbf{A} = \mathbf{a}_x\left(\frac{3x^2}{x^2 + y^2} + 2y\right) + \mathbf{a}_y\left(\frac{3xy}{x^2 + y^2} - 2x\right) + \mathbf{a}_z5,$$

which is the desired answer. ■

2–4.3 SPHERICAL COORDINATES

$$(u_1, u_2, u_3) = (R, \theta, \phi)$$

A point $P(R_1, \theta_1, \phi_1)$ in spherical coordinates is specified as the intersection of the following three surfaces: a spherical surface centered at the origin with a radius $R = R_1$; a right circular cone with its apex at the origin, its axis coinciding with the $+z$-axis and having a half-angle $\theta = \theta_1$; and a half-plane containing the z-axis and making an angle $\phi = \phi_1$ with the xz-plane. *The base vector $\mathbf{a}_R$ at P is radial from the origin and is quite different from $\mathbf{a}_r$ in cylindrical coordinates, the latter being perpendicular to the z-axis. The base vector $\mathbf{a}_\theta$ lies in the $\phi = \phi_1$ plane and is tangential to the*

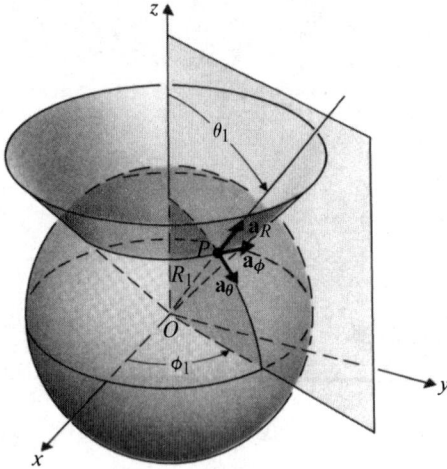

FIGURE 2–18
Spherical coordinates.

spherical surface, whereas the base vector $\mathbf{a}_\phi$ is the same as that in cylindrical coordinates. These are illustrated in Fig. 2–18. For a right-handed system we have

$$\mathbf{a}_R \times \mathbf{a}_\theta = \mathbf{a}_\phi, \tag{2–64a}$$

$$\mathbf{a}_\theta \times \mathbf{a}_\phi = \mathbf{a}_R, \tag{2–64b}$$

$$\mathbf{a}_\phi \times \mathbf{a}_R = \mathbf{a}_\theta. \tag{2–64c}$$

Spherical coordinates are important for problems involving point sources and regions with spherical boundaries. When an observer is very far from the source region of a finite extent, the latter could be considered as the origin of a spherical coordinate system; and, as a result, suitable simplifying approximations could be made. This is the reason that spherical coordinates are used in solving antenna problems in the far field.

A vector in spherical coordinates is written as

$$\boxed{\mathbf{A} = \mathbf{a}_R A_R + \mathbf{a}_\theta A_\theta + \mathbf{a}_\phi A_\phi.} \tag{2–65}$$

The expressions for the dot and cross products of two vectors in spherical coordinates can be obtained from Eqs. (2–26) and (2–27).

In spherical coordinates, only $R(u_1)$ is a length. The other two coordinates, θ and ϕ (u_2 and u_3), are angles. Referring to Fig. 2–19, in which a typical differential volume element is shown, we see that metric coefficients $h_2 = R$ and $h_3 = R \sin \theta$ are required to convert $d\theta$ and $d\phi$ into $d\ell_2$ and $d\ell_3$, respectively. The general expression for a differential length is, from Eq. (2–31),

$$\boxed{d\boldsymbol{\ell} = \mathbf{a}_R\, dR + \mathbf{a}_\theta R\, d\theta + \mathbf{a}_\phi R \sin \theta\, d\phi.} \tag{2–66}$$

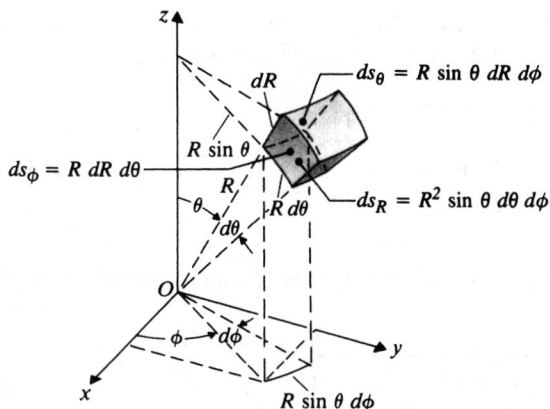

FIGURE 2–19
A differential volume element in spherical coordinates.

The expressions for differential areas and differential volume resulting from differential changes dR, $d\theta$, and $d\phi$ in the three coordinate directions are

$$ds_R = R^2 \sin \theta \, d\theta \, d\phi, \qquad (2\text{–}67\text{a})$$
$$ds_\theta = R \sin \theta \, dR \, d\phi, \qquad (2\text{–}67\text{b})$$
$$ds_\phi = R \, dR \, d\theta, \qquad (2\text{–}67\text{c})$$

and

$$dv = R^2 \sin \theta \, dR \, d\theta \, d\phi. \qquad (2\text{–}68)$$

For convenience the base vectors, metric coefficients, and expressions for the differential volume are tabulated in Table 2–1.

TABLE 2–1
Three Basic Orthogonal Coordinate Systems

Coordinate System Relations		Cartesian Coordinates (x, y, z)	Cylindrical Coordinates (r, ϕ, z)	Spherical Coordinates (R, θ, ϕ)
Base vectors	$\mathbf{a}_{u_1}$	$\mathbf{a}_x$	$\mathbf{a}_r$	$\mathbf{a}_R$
	$\mathbf{a}_{u_2}$	$\mathbf{a}_y$	$\mathbf{a}_\phi$	$\mathbf{a}_\theta$
	$\mathbf{a}_{u_3}$	$\mathbf{a}_z$	$\mathbf{a}_z$	$\mathbf{a}_\phi$
Metric coefficients	h_1	1	1	1
	h_2	1	r	R
	h_3	1	1	$R \sin \theta$
Differential volume	dv	$dx \, dy \, dz$	$r \, dr \, d\phi \, dz$	$R^2 \sin \theta \, dR \, d\theta \, d\phi$

A vector given in spherical coordinates can be transformed into one in Cartesian or cylindrical coordinates, and vice versa. From Fig. 2–19 it is easily seen that

$$x = R \sin \theta \cos \phi, \qquad (2\text{–}69\text{a})$$
$$y = R \sin \theta \sin \phi, \qquad (2\text{–}69\text{b})$$
$$z = R \cos \theta. \qquad (2\text{–}69\text{c})$$

Conversely, measurements in Cartesian coordinates can be transformed into those in spherical coordinates:

$$R = \sqrt{x^2 + y^2 + z^2}, \qquad (2\text{–}70\text{a})$$
$$\theta = \arctan^{-1} \frac{\sqrt{x^2 + y^2}}{z}, \qquad (2\text{–}70\text{b})$$
$$\phi = \arctan^{-1} \frac{y}{x}. \qquad (2\text{–}70\text{c})$$

EXAMPLE 2–10   The position of a point P in spherical coordinates is $(8, 120°, 330°)$. Specify its location (a) in Cartesian coordinates, and (b) in cylindrical coordinates.

Solution   The spherical coordinates of the given point are $R = 8$, $\theta = 120°$, and $\phi = 330°$.

a) *In Cartesian coordinates.* We use Eqs. (2–69a, b, c):
$$x = 8 \sin 120° \cos 330° = 6,$$
$$y = 8 \sin 120° \sin 330° = -2\sqrt{3},$$
$$z = 8 \cos 120° = -4.$$

Hence the location of the point is $P(6, -2\sqrt{3}, -4)$, and the **position vector** (the vector going from the origin to the point) is
$$\overrightarrow{OP} = \mathbf{a}_x 6 - \mathbf{a}_y 2\sqrt{3} - \mathbf{a}_z 4.$$

b) *In cylindrical coordinates.* The cylindrical coordinates of point P can be obtained by applying Eqs. (2–63a, b, c) to the results in part (a), but they can be calculated directly from the given spherical coordinates by the following relations, which can be verified by comparing Figs. 2–14 and 2–18:
$$r = R \sin \theta, \qquad (2\text{–}71\text{a})$$
$$\phi = \phi, \qquad (2\text{–}71\text{b})$$
$$z = R \cos \theta. \qquad (2\text{–}71\text{c})$$

We have $P(4\sqrt{3}, 330°, -4)$; and its position vector in cylindrical coordinates is
$$\overrightarrow{OP} = \mathbf{a}_r 4\sqrt{3} - \mathbf{a}_z 4.$$

We note here that the position vector of a point in cylindrical coordinates does not contain the angle $\phi = 330°$ explicitly. However, the exact direction of $\mathbf{a}_r$ depends on ϕ. In terms of spherical coordinates the position vector (the vector from the origin to the point P) consists of only a single term:

$$\overrightarrow{OP} = \mathbf{a}_R 8.$$

Here the direction of $\mathbf{a}_R$ changes with the θ and ϕ coordinates of the point P.

EXAMPLE 2–11 Convert the vector $\mathbf{A} = \mathbf{a}_R A_R + \mathbf{a}_\theta A_\theta + \mathbf{a}_\phi A_\phi$ into Cartesian coordinates.

Solution In this problem we want to write $\mathbf{A}$ in the form of $\mathbf{A} = \mathbf{a}_x A_x + \mathbf{a}_y A_y + \mathbf{a}_z A_z$. This is very different from the preceding problem of converting the coordinates of a point. First of all, we assume that the expression of the given vector $\mathbf{A}$ holds *for all points* of interest and that all three given components A_R, A_θ, and A_ϕ may be functions of coordinate variables. Second, at a given point, A_R, A_θ, and A_ϕ will have definite numerical values, but these values that determine the direction of $\mathbf{A}$ will, in general, be entirely different from the coordinate values of the point. Taking dot product of $\mathbf{A}$ with $\mathbf{a}_x$, we have

$$A_x = \mathbf{A} \cdot \mathbf{a}_x$$
$$= A_R \mathbf{a}_R \cdot \mathbf{a}_x + A_\theta \mathbf{a}_\theta \cdot \mathbf{a}_x + A_\phi \mathbf{a}_\phi \cdot \mathbf{a}_x.$$

Recalling that $\mathbf{a}_R \cdot \mathbf{a}_x$, $\mathbf{a}_\theta \cdot \mathbf{a}_x$, and $\mathbf{a}_\phi \cdot \mathbf{a}_x$ yield, respectively, the component of unit vectors $\mathbf{a}_R$, $\mathbf{a}_\theta$, and $\mathbf{a}_\phi$ in the direction of $\mathbf{a}_x$, we find, from Fig. 2–19 and Eqs. (2–69a, b, c):

$$\mathbf{a}_R \cdot \mathbf{a}_x = \sin \theta \cos \phi = \frac{x}{\sqrt{x^2 + y^2 + z^2}}, \tag{2–72}$$

$$\mathbf{a}_\theta \cdot \mathbf{a}_x = \cos \theta \cos \phi = \frac{xz}{\sqrt{(x^2 + y^2)(x^2 + y^2 + z^2)}}, \tag{2–73}$$

$$\mathbf{a}_\phi \cdot \mathbf{a}_x = -\sin \phi = -\frac{y}{\sqrt{x^2 + y^2}}. \tag{2–74}$$

Thus,

$$A_x = A_R \sin \theta \cos \phi + A_\theta \cos \theta \cos \phi - A_\phi \sin \phi$$
$$= \frac{A_R x}{\sqrt{x^2 + y^2 + z^2}} + \frac{A_\theta xz}{\sqrt{(x^2 + y^2)(x^2 + y^2 + z^2)}} - \frac{A_\phi y}{\sqrt{x^2 + y^2}}. \tag{2–75}$$

Similarly,

$$A_y = A_R \sin \theta \sin \phi + A_\theta \cos \theta \sin \phi + A_\phi \cos \phi$$
$$= \frac{A_R y}{\sqrt{x^2 + y^2 + z^2}} + \frac{A_\theta yz}{\sqrt{(x^2 + y^2)(x^2 + y^2 + z^2)}} + \frac{A_\phi x}{\sqrt{x^2 + y^2}}. \tag{2–76}$$

and

$$A_z = A_R \cos \theta - A_\theta \sin \theta = \frac{A_R z}{\sqrt{x^2 + y^2 + z^2}} - \frac{A_\theta \sqrt{x^2 + y^2}}{\sqrt{x^2 + y^2 + z^2}}. \quad (2\text{--}77)$$

If A_R, A_θ, and A_ϕ are themselves functions of R, θ, and ϕ, they too need to be converted into functions of x, y, and z by the use of Eqs. (2–70a, b, c). Equations (2–75), (2–76), and (2–77) disclose the fact that when a vector has a simple form in one coordinate system, its conversion into another coordinate system usually results in a more complicated expression.   ▄

EXAMPLE 2–12   Assuming that a cloud of electrons confined in a region between two spheres of radii 2 and 5 (cm) has a charge density of

$$\frac{-3 \times 10^{-8}}{R^4} \cos^2 \phi \quad (C/m^3),$$

find the total charge contained in the region.

Solution   We have

$$\rho = -\frac{3 \times 10^{-8}}{R^4} \cos^2 \phi,$$

$$Q = \int \rho \, dv.$$

The given conditions of the problem obviously point to the use of spherical coordinates. Using the expression for dv in Eq. (2–68), we perform a triple integration:

$$Q = \int_0^{2\pi} \int_0^\pi \int_{0.02}^{0.05} \rho R^2 \sin \theta \, dR \, d\theta \, d\phi.$$

Two things are of importance here. First, since ρ is given in units of coulombs per cubic meter, the limits of integration for R must be converted to meters. Second, the full range of integration for θ is from 0 to π radians, *not* from 0 to 2π radians. A little reflection will convince us that a half-circle (not a full-circle) rotated about the z-axis through 2π radians (ϕ from 0 to 2π) generates a sphere. We have

$$Q = -3 \times 10^{-8} \int_0^{2\pi} \int_0^\pi \int_{0.02}^{0.05} \frac{1}{R^2} \cos^2 \phi \sin \theta \, dR \, d\theta \, d\phi$$

$$= -3 \times 10^{-8} \int_0^{2\pi} \int_0^\pi \left(-\frac{1}{0.05} + \frac{1}{0.02} \right) \sin \theta \, d\theta \cos^2 \phi \, d\phi$$

$$= -0.9 \times 10^{-6} \int_0^{2\pi} (-\cos \theta) \Big|_0^\pi \cos^2 \phi \, d\phi$$

$$= -1.8 \times 10^{-6} \left(\frac{\phi}{2} + \frac{\sin 2\phi}{4} \right) \Big|_0^{2\pi} = -1.8\pi \quad (\mu C).$$
                                                                                    ▄

2–5 Integrals Containing Vector Functions

In electromagnetics work we have occasion to encounter integrals that contain vector functions such as

$$\int_V \mathbf{F}\, dv, \tag{2-78}$$

$$\int_C V\, d\boldsymbol{\ell}, \tag{2-79}$$

$$\int_C \mathbf{F} \cdot d\boldsymbol{\ell}, \tag{2-80}$$

$$\int_S \mathbf{A} \cdot d\mathbf{s}. \tag{2-81}$$

The volume integral in (2–78) can be evaluated as the sum of three scalar integrals by first resolving the vector $\mathbf{F}$ into its three components in the appropriate coordinate system. If dv denotes a differential volume, then (2–78) is actually a shorthand way of representing a triple integral over three dimensions.

In the second integral, in (2–79), V is a scalar function of space, $d\boldsymbol{\ell}$ represents a differential increment of length, and C is the path of integration. If the integration is to be carried out from a point P_1 to another point P_2, we write $\int_{P_1}^{P_2} V\, d\boldsymbol{\ell}$. If the integration is to be evaluated around a closed path C, we denote it by $\oint_C V\, d\boldsymbol{\ell}$. In Cartesian coordinates, (2–79) can be written as

$$\int_C V\, d\boldsymbol{\ell} = \int_C V(x, y, z)[\mathbf{a}_x\, dx + \mathbf{a}_y\, dy + \mathbf{a}_z\, dz], \tag{2-82}$$

in view of Eq. (2–44). Since the Cartesian unit vectors are constant in both magnitude and direction, they can be taken out of the integral sign, and Eq. (2–82) becomes

$$\int_C V\, d\boldsymbol{\ell} = \mathbf{a}_x \int_C V(x, y, z)\, dx + \mathbf{a}_y \int_C V(x, y, z)\, dy + \mathbf{a}_z \int_C V(x, y, z)\, dz. \tag{2-83}$$

The three integrals on the right-hand side of Eq. (2–83) are ordinary scalar integrals; they can be evaluated for a given $V(x, y, z)$ around a path C.

EXAMPLE 2–13 Evaluate the integral $\int_O^P r^2\, d\mathbf{r}$, where $r^2 = x^2 + y^2$, from the origin to the point $P(1, 1)$: (a) along the direct path OP, (b) along the path OP_1P, and (c) along the path OP_2P in Fig. 2–20.

Solution

a) Along the direct path OP:

$$\int_O^P r^2\, d\mathbf{r} = \mathbf{a}_r \int_0^{\sqrt{2}} r^2\, dr = \mathbf{a}_r \frac{2\sqrt{2}}{3}$$

$$= \frac{2\sqrt{2}}{3}\,(\mathbf{a}_x \cos 45° + \mathbf{a}_y \sin 45°)$$

$$= \mathbf{a}_x \tfrac{2}{3} + \mathbf{a}_y \tfrac{2}{3}.$$

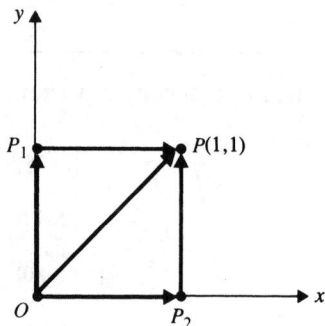

FIGURE 2–20
Illustrating Example 2–13.

b) Along the path OP_1P:

$$\int_O^P (x^2 + y^2)\, d\mathbf{r} = \mathbf{a}_y \int_O^{P_1} y^2\, dy + \mathbf{a}_x \int_{P_1}^P (x^2 + 1)\, dx$$

$$= \mathbf{a}_y \tfrac{1}{3}y^3 \Big|_0^1 + \mathbf{a}_x(\tfrac{1}{3}x^3 + x)\Big|_0^1$$

$$= \mathbf{a}_x \tfrac{4}{3} + \mathbf{a}_y \tfrac{1}{3}.$$

c) Along the path OP_2P:

$$\int_O^P (x^2 + y^2)\, d\mathbf{r} = \mathbf{a}_x \int_O^{P_2} x^2\, dx + \mathbf{a}_y \int_{P_2}^P (1 + y^2)\, dy$$

$$= \mathbf{a}_x \tfrac{1}{3}x^3 \Big|_0^1 + \mathbf{a}_y(y + \tfrac{1}{3}y^3)\Big|_0^1$$

$$= \mathbf{a}_x \tfrac{1}{3} + \mathbf{a}_y \tfrac{4}{3}.$$

Obviously, the value of the integral depends on the path of integration, since the results in parts (a), (b), and (c) are all different.   ▬

The integrals in (2–80) and (2–81) are mathematically of the same form; they both lead to a scalar result. The expression in (2–80) is a line integral, in which the integrand represents the component of the vector **F** along the path of integration. This type of scalar line integral is of considerable importance in both physics and electromagnetics. (If **F** is a force, the integral is the work done by the force in moving an object from an initial point P_1 to a final point P_2 along a specified path C; if **F** is replaced by **E**, the electric field intensity, then the integral represents the work done by the electric field in moving a unit charge from P_1 to P_2.) We will encounter it again later in this chapter and in many other parts of this book.

▬   **EXAMPLE 2–14**   Given $\mathbf{F} = \mathbf{a}_x xy - \mathbf{a}_y 2x$, evaluate the scalar line integral

$$\int_A^B \mathbf{F} \cdot d\boldsymbol{\ell}$$

along the quarter-circle shown in Fig. 2–21.

Solution We shall solve this problem in two ways: first in Cartesian coordinates, then in cylindrical coordinates.

a) *In Cartesian coordinates.* From the given **F** and the expression for $d\boldsymbol{\ell}$ in Eq. (2–44) we have

$$\mathbf{F} \cdot d\boldsymbol{\ell} = xy\, dx - 2x\, dy.$$

The equation of the quarter-circle is $x^2 + y^2 = 9$ $(0 \le x, y \le 3)$. Therefore,

$$\int_A^B \mathbf{F} \cdot d\boldsymbol{\ell} = \int_3^0 x\sqrt{9 - x^2}\, dx - 2\int_0^3 \sqrt{9 - y^2}\, dy$$

$$= -\frac{1}{3}(9 - x^2)^{3/2}\Big|_3^0 - \left[y\sqrt{9 - y^2} + 9\sin^{-1}\frac{y}{3} \right]_0^3$$

$$= -9\left(1 + \frac{\pi}{2}\right).$$

b) *In cylindrical coordinates.* Here we first transform **F** into cylindrical coordinates. Inverting Eq. (2–61), we have

$$\begin{bmatrix} A_r \\ A_\phi \\ A_z \end{bmatrix} = \begin{bmatrix} \cos\phi & -\sin\phi & 0 \\ \sin\phi & \cos\phi & 0 \\ 0 & 0 & 1 \end{bmatrix}^{-1} \begin{bmatrix} A_x \\ A_y \\ A_z \end{bmatrix}$$

$$= \begin{bmatrix} \cos\phi & \sin\phi & 0 \\ -\sin\phi & \cos\phi & 0 \\ 0 & 0 & 1 \end{bmatrix} \begin{bmatrix} A_x \\ A_y \\ A_z \end{bmatrix}. \qquad (2\text{--}84)$$

With the given **F**, Eq. (2–84) gives

$$\begin{bmatrix} F_r \\ F_\phi \\ F_z \end{bmatrix} = \begin{bmatrix} \cos\phi & \sin\phi & 0 \\ -\sin\phi & \cos\phi & 0 \\ 0 & 0 & 1 \end{bmatrix} \begin{bmatrix} xy \\ -2x \\ 0 \end{bmatrix},$$

which leads to

$$\mathbf{F} = \mathbf{a}_r(xy\cos\phi - 2x\sin\phi) - \mathbf{a}_\phi(xy\sin\phi + 2x\cos\phi).$$

For the present problem the path of integration is along a quarter-circle of a radius 3. There is no change in r or z along the path $(dr = 0$ and $dz = 0)$; hence

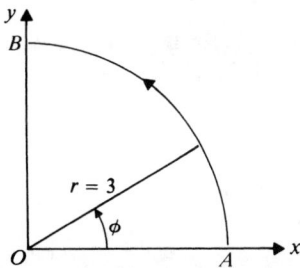

FIGURE 2–21
Path for line integral (Example 2–14).

Eq. (2–52) simplifies to

$$d\boldsymbol{\ell} = \mathbf{a}_\phi 3\, d\phi$$

and

$$\mathbf{F} \cdot d\boldsymbol{\ell} = -3(xy \sin \phi + 2x \cos \phi)\, d\phi.$$

Because of the circular path, F_r is immaterial to the present integration. Along the path, $x = 3 \cos \phi$ and $y = 3 \sin \phi$. Therefore

$$\int_A^B \mathbf{F} \cdot d\boldsymbol{\ell} = \int_0^{\pi/2} - 3(9 \sin^2 \phi \cos \phi + 6 \cos^2 \phi)\, d\phi$$

$$= -9(\sin^3 \phi + \phi + \sin \phi \cos \phi)\Big|_0^{\pi/2}$$

$$= -9\left(1 + \frac{\pi}{2}\right),$$

which is the same as before. ∎

In this particular example, $\mathbf{F}$ is given in Cartesian coordinates, and the path is circular. There is no compelling reason to solve the problem in one or the other co-ordinates. We have shown the conversion of vectors and the procedure of solution in both coordinates.

The expression in (2–81), $\int_S \mathbf{A} \cdot d\mathbf{s}$, is a surface integral. It is actually a double integral over two dimensions; but it is written with a single integral sign for simplicity. The integral measures the flux of the vector field $\mathbf{A}$ flowing through the area S. In the integral the vector differential surface element $d\mathbf{s} = \mathbf{a}_n\, ds$ has a magnitude ds and a direction denoted by the unit vector $\mathbf{a}_n$. The conventions for the positive direction of $d\mathbf{s}$ or $\mathbf{a}_n$ are as follows:

1. If the surface of integration, S, is a *closed surface* enclosing a volume, then the positive direction for $\mathbf{a}_n$ is always in the *outward* direction from the volume. This is illustrated in Fig. 2–22(a). We see that the positive direction of $\mathbf{a}_n$ depends on the location of $d\mathbf{s}$. A small circle is added over the integral sign if the integration is to be performed over an enclosed surface:

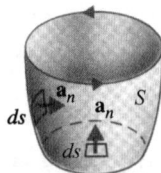

$$\oint_S \mathbf{A} \cdot d\mathbf{s} = \oint_S \mathbf{A} \cdot \mathbf{a}_n\, ds.$$

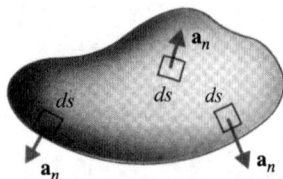

(a) A closed surface.   (b) An open surface.   (c) A disk.

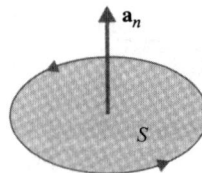

FIGURE 2–22
Illustrating the positive direction of $\mathbf{a}_n$ in scalar surface integral.

2. If S is an open surface, the positive direction for $\mathbf{a}_n$ depends on the direction in which the perimeter of the open surface is traversed. This is illustrated in Fig. 2–22(b), in which a cup-shaped surface (with no lid) is shown. We apply the right-hand rule: If the fingers of the right hand follows the direction of travel around the perimeter, then the thumb points in the direction of positive $\mathbf{a}_n$. Here again, the positive direction of $\mathbf{a}_n$ depends on the location of ds. A plane, such as the disk in Fig. 2–22(c), is a special case of an open surface where $\mathbf{a}_n$ is a constant.

EXAMPLE 2–15   Given $\mathbf{F} = \mathbf{a}_r k_1/r + \mathbf{a}_z k_2 z$, evaluate the scalar surface integral

$$\oint_S \mathbf{F} \cdot d\mathbf{s}$$

over the surface of a closed cylinder about the z-axis specified by $z = \pm 3$ and $r = 2$.

Solution   The specified surface of integration S is that of a closed cylinder shown in Fig. 2–23. The cylinder has three surfaces: the top face, the bottom face, and the side wall. We write

$$\oint_S \mathbf{F} \cdot d\mathbf{s} = \oint_S \mathbf{F} \cdot \mathbf{a}_n \, ds$$
$$= \int_{\substack{\text{top} \\ \text{face}}} \mathbf{F} \cdot \mathbf{a}_n \, ds + \int_{\substack{\text{bottom} \\ \text{face}}} \mathbf{F} \cdot \mathbf{a}_n \, ds + \int_{\substack{\text{side} \\ \text{wall}}} \mathbf{F} \cdot \mathbf{a}_n \, ds,$$

where $\mathbf{a}_n$ is the unit normal *outward* from the respective surfaces. The three integrals on the right side can be evaluated separately.

a) *Top face.* $z = 3$, $\mathbf{a}_n = \mathbf{a}_z$,
$$\mathbf{F} \cdot \mathbf{a}_n = k_2 z = 3k_2,$$
$$ds = r \, dr \, d\phi \quad \text{(from Eq. 2–53c)};$$
$$\int_{\substack{\text{top} \\ \text{face}}} \mathbf{F} \cdot \mathbf{a}_n \, ds = \int_0^{2\pi} \int_0^2 3k_2 r \, dr \, d\phi = 12\pi k_2.$$

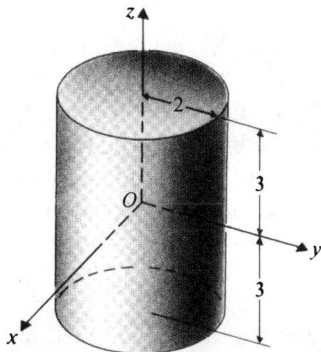

FIGURE 2–23
A cylindrical surface (Example 2–15).

b) *Bottom face.* $z = -3$, $\mathbf{a}_n = -\mathbf{a}_z$,

$$\mathbf{F} \cdot \mathbf{a}_n = -k_2 z = 3k_2,$$
$$ds = r\, dr\, d\phi;$$

$$\int_{\substack{\text{bottom} \\ \text{face}}} \mathbf{F} \cdot \mathbf{a}_n\, ds = 12\pi k_2,$$

which is exactly the same as the integral over the top face.

c) *Side wall.* $r = 2$, $\mathbf{a}_n = \mathbf{a}_r$,

$$\mathbf{F} \cdot \mathbf{a}_n = \frac{k_1}{r} = \frac{k_1}{2},$$

$$ds = r\, d\phi\, dz = 2\, d\phi\, dz \text{ (from Eq. 2–53a)};$$

$$\int_{\substack{\text{side} \\ \text{wall}}} \mathbf{F} \cdot \mathbf{a}_n\, ds = \int_{-3}^{3} \int_{0}^{2\pi} k_1\, d\phi\, dz = 12\pi k_1.$$

Therefore,

$$\oint_S \mathbf{F} \cdot d\mathbf{s} = 12\pi k_2 + 12\pi k_2 + 12\pi k_1$$
$$= 12\pi(k_1 + 2k_2).$$

This surface integral gives the net *outward flux* of the vector $\mathbf{F}$ through the closed cylindrical surface. ▬

2–6 Gradient of a Scalar Field

In electromagnetics we have to deal with quantities that depend on both time and position. Since three coordinate variables are involved in a three-dimensional space, we expect to encounter scalar and vector fields that are functions of four variables: (t, u_1, u_2, u_3). In general, the fields may change as any one of the four variables changes. We now address the method for describing the space rate of change of a scalar field at a given time. Partial derivatives with respect to the three space-coordinate variables are involved, and, inasmuch as the rate of change may be different in different directions, a vector is needed to define the space rate of change of a scalar field at a given point and at a given time.

Let us consider a scalar function of space coordinates $V(u_1, u_2, u_3)$, which may represent, say, the temperature distribution in a building, the altitude of a mountainous terrain, or the electric potential in a region. The magnitude of V, in general, depends on the position of the point in space, but it may be constant along certain lines or surfaces. Figure 2–24 shows two surfaces on which the magnitude of V is constant and has the values V_1 and $V_1 + dV$, respectively, where dV indicates a small change in V. We should note that constant-V surfaces need not coincide with any of the surfaces that define a particular coordinate system. Point P_1 is on surface V_1; P_2 is the corresponding point on surface $V_1 + dV$ along the normal vector $d\mathbf{n}$; and P_3 is a point close to P_2 along another vector $d\boldsymbol{\ell} \neq d\mathbf{n}$. For the same change dV in V, the space rate of change, $dV/d\ell$, is obviously greatest along $d\mathbf{n}$ because dn is the

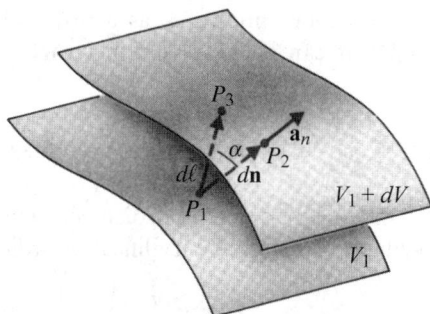

FIGURE 2–24
Concerning gradient of a scalar.

shortest distance between the two surfaces.[†] Since the magnitude of $dV/d\ell$ depends on the direction of $d\ell$, $dV/d\ell$ is a directional derivative. *We define the vector that represents both the magnitude and the direction of the maximum space rate of increase of a scalar as the gradient of that scalar.* We write

$$\textbf{grad } V \triangleq \mathbf{a}_n \frac{dV}{dn}. \tag{2–85}$$

For brevity it is customary to employ the operator *del*, represented by the symbol ∇ and write ∇V in place of **grad** V. Thus,

$$\nabla V \triangleq \mathbf{a}_n \frac{dV}{dn}. \tag{2–86}$$

We have assumed that dV is positive (an increase in V); if dV is negative (a decrease in V from P_1 to P_2), ∇V will be negative in the $\mathbf{a}_n$ direction.

The directional derivative along $d\ell$ is

$$\frac{dV}{d\ell} = \frac{dV}{dn}\frac{dn}{d\ell} = \frac{dV}{dn}\cos\alpha$$

$$= \frac{dV}{dn}\mathbf{a}_n \cdot \mathbf{a}_\ell = (\nabla V) \cdot \mathbf{a}_\ell. \tag{2–87}$$

Equation (2–87) states that the space rate of increase of V in the $\mathbf{a}_\ell$ direction is equal to the projection (the component) of the gradient of V in that direction. We can also write Eq. (2–87) as

$$dV = (\nabla V) \cdot d\ell, \tag{2–88}$$

[†] In a more formal treatment, changes ΔV and $\Delta\ell$ would be used, and the ratio $\Delta V/\Delta\ell$ would become the derivative $dV/d\ell$ as $\Delta\ell$ approaches zero. We avoid this formality in favor of simplicity.

where $d\ell = \mathbf{a}_\ell d\ell$. Now, dV in Eq. (2–88) is the total differential of V as a result of a change in position (from P_1 to P_3 in Fig. 2–24); it can be expressed in terms of the differential changes in coordinates:

$$dV = \frac{\partial V}{\partial \ell_1} d\ell_1 + \frac{\partial V}{\partial \ell_2} d\ell_2 + \frac{\partial V}{\partial \ell_3} d\ell_3, \tag{2–89}$$

where $d\ell_1$, $d\ell_2$, and $d\ell_3$ are the components of the vector differential displacement $d\ell$ in a chosen coordinate system. In terms of general orthogonal curvilinear coordinates (u_1, u_2, u_3), $d\ell$ is (from Eq. 2–31),

$$\begin{aligned} d\ell &= \mathbf{a}_{u_1} d\ell_1 + \mathbf{a}_{u_2} d\ell_2 + \mathbf{a}_{u_3} d\ell_3 \\ &= \mathbf{a}_{u_1}(h_1\, du_1) + \mathbf{a}_{u_2}(h_2\, du_2) + \mathbf{a}_{u_3}(h_3\, du_3). \end{aligned} \tag{2–90}$$

We can write dV in Eq. (2–89) as the dot product of two vectors, as follows:

$$\begin{aligned} dV &= \left(\mathbf{a}_{u_1} \frac{\partial V}{\partial \ell_1} + \mathbf{a}_{u_2} \frac{\partial V}{d\ell_2} + \mathbf{a}_{u_3} \frac{\partial V}{\partial \ell_3} \right) \cdot (\mathbf{a}_{u_1} d\ell_1 + \mathbf{a}_{u_2} d\ell_2 + \mathbf{a}_{u_3} d\ell_3) \\ &= \left(\mathbf{a}_{u_1} \frac{\partial V}{\partial \ell_1} + \mathbf{a}_{u_2} \frac{\partial V}{\partial \ell_2} + \mathbf{a}_{u_3} \frac{\partial V}{\partial \ell_3} \right) \cdot d\ell. \end{aligned} \tag{2–91}$$

Comparing Eq. (2–91) with Eq. (2–88), we obtain

$$\nabla V = \mathbf{a}_{u_1} \frac{\partial V}{\partial \ell_1} + \mathbf{a}_{u_2} \frac{\partial V}{\partial \ell_2} + \mathbf{a}_{u_3} \frac{\partial V}{d\ell_3} \tag{2–92}$$

or

$$\boxed{\nabla V = \mathbf{a}_{u_1} \frac{\partial V}{h_1\, \partial u_1} + \mathbf{a}_{u_2} \frac{\partial V}{h_2\, \partial u_2} + \mathbf{a}_{u_3} \frac{\partial V}{h_3\, \partial u_3}.} \tag{2–93}$$

Equation (2–93) is a useful formula for computing the gradient of a scalar, when the scalar is given as a function of space coordinates.

In Cartesian coordinates, $(u_1, u_2, u_3) = (x, y, z)$ and $h_1 = h_2 = h_3 = 1$, we have

$$\boxed{\nabla V = \mathbf{a}_x \frac{\partial V}{\partial x} + \mathbf{a}_y \frac{\partial V}{\partial y} + \mathbf{a}_z \frac{\partial V}{\partial z}} \tag{2–94}$$

or

$$\nabla V = \left(\mathbf{a}_x \frac{\partial}{\partial x} + \mathbf{a}_y \frac{\partial}{\partial y} + \mathbf{a}_z \frac{\partial}{\partial z} \right) V. \tag{2–95}$$

In view of Eq. (2–95), it is convenient to consider ∇ *in Cartesian coordinates* as a vector differential operator.

$$\boxed{\nabla \equiv \mathbf{a}_x \frac{\partial}{\partial x} + \mathbf{a}_y \frac{\partial}{\partial y} + \mathbf{a}_z \frac{\partial}{\partial z}.} \tag{2–96}$$

From Eq. (2–93), we see that we can define ∇ as

$$\nabla \equiv \left(\mathbf{a}_{u_1} \frac{\partial}{h_1 \, \partial u_1} + \mathbf{a}_{u_2} \frac{\partial}{h_2 \, \partial u_2} + \mathbf{a}_{u_3} \frac{\partial}{h_3 \, \partial u_3} \right) \tag{2–97}$$

in general orthogonal coordinates. As we shall see later in this chapter, the same vector differential operator is also used to signify *divergence* ($\nabla \cdot$) and *curl* ($\nabla \times$) operations on a vector. In these cases it is important to remember that the differentiation of a base vector in a curvilinear coordinate system may lead to a new vector in a different direction. (For instance, $\partial \mathbf{a}_r / \partial \phi = \mathbf{a}_\phi$ and $\partial \mathbf{a}_\phi / \partial \phi = -\mathbf{a}_r$.) Proper care must be exercised when the ∇ defined in Eq. (2–97) is used to operate on vectors in curvilinear coordinate systems.

EXAMPLE 2–16   The electrostatic field intensity $\mathbf{E}$ is derivable as the negative gradient of a scalar electric potential V; that is, $\mathbf{E} = -\nabla V$. Determine $\mathbf{E}$ at the point $(1, 1, 0)$ if

a) $V = V_0 e^{-x} \sin \dfrac{\pi y}{4}$,

b) $V = E_0 R \cos \theta$.

Solution   We use Eq. (2–93) to evaluate $\mathbf{E} = -\nabla V$ in Cartesian coordinates for part (a) and in spherical coordinates for part (b).

a) $\mathbf{E} = -\left[\mathbf{a}_x \dfrac{\partial}{\partial x} + \mathbf{a}_y \dfrac{\partial}{\partial y} + \mathbf{a}_z \dfrac{\partial}{\partial z} \right] E_0 e^{-x} \sin \dfrac{\pi y}{4}$

$\qquad = \left(\mathbf{a}_x \sin \dfrac{\pi y}{4} - \mathbf{a}_y \dfrac{\pi}{4} \cos \dfrac{\pi y}{4} \right) E_0 e^{-x}.$

Thus, $\mathbf{E}(1, 1, 0) = \left(\mathbf{a}_x - \mathbf{a}_y \dfrac{\pi}{4} \right) \dfrac{E_0}{\sqrt{2}} = \mathbf{a}_E E,$

where

$$E = E_0 \sqrt{\frac{1}{2} \left(1 + \frac{\pi^2}{16} \right)},$$

$$\mathbf{a}_E = \frac{1}{\sqrt{1 + (\pi^2/16)}} \left(\mathbf{a}_x - \mathbf{a}_y \frac{\pi}{4} \right).$$

b) $\mathbf{E} = -\left[\mathbf{a}_R \dfrac{\partial}{\partial R} + \mathbf{a}_\theta \dfrac{\partial}{R \partial \theta} + \mathbf{a}_\phi \dfrac{\partial}{R \sin \theta \, \partial \phi} \right] E_0 R \cos \theta$

$\qquad = -(\mathbf{a}_R \cos \theta - \mathbf{a}_\theta \sin \theta) E_0.$

In view of Eq. (2–77), the result above converts very simply to $\mathbf{E} = -\mathbf{a}_z E_0$ in Cartesian coordinates. This is not surprising, since a careful examination of the given V reveals that $E_0 R \cos \theta$ is, in fact, equal to $E_0 z$. In Cartesian coordinates,

$$\mathbf{E} = -\nabla V = -\mathbf{a}_z \frac{\partial}{\partial z} (E_0 z) = -\mathbf{a}_z E_0.$$

2–7   Divergence of a Vector Field

In the preceding section we considered the spatial derivatives of a scalar field, which led to the definition of the gradient. We now turn our attention to the spatial derivatives of a vector field. This will lead to the definitions of the **divergence** and the **curl** of a vector. We discuss the meaning of divergence in this section and that of curl in Section 2–9. Both are very important in the study of electromagnetism.

In the study of vector fields it is convenient to represent field variations graphically by directed field lines, which are called **flux lines** or **streamlines**. They are directed lines or curves that indicate at each point the direction of the vector field, as illustrated in Fig. 2–25. The magnitude of the field at a point is depicted either by the density or by the length of the directed lines in the vicinity of the point. Figure 2–25(a) shows that the field in region A is stronger than that in region B because there is a higher density of equal-length directed lines in region A. In Fig. 2–25(b), the decreasing arrow lengths away from the point q indicate a radial field that is strongest in the region closest to q. Figure 2–25(c) depicts a uniform field.

The vector field strength in Fig. 2–25(a) is measured by the number of flux lines passing through a unit surface normal to the vector. The flux of a vector field is analogous to the flow of an incompressible fluid such as water. For a volume with an enclosed surface there will be an excess of outward or inward flow through the surface only when the volume contains a source or a sink, respectively; that is, a net positive divergence indicates the presence of a source of fluid inside the volume, and a net negative divergence indicates the presence of a sink. The net outward flow of the fluid per unit volume is therefore a measure of the strength of the enclosed source. In the uniform field shown in Fig. 2–25(c) there is an equal amount of inward and outward flux going through any closed volume containing no sources or sinks, resulting in a zero divergence.

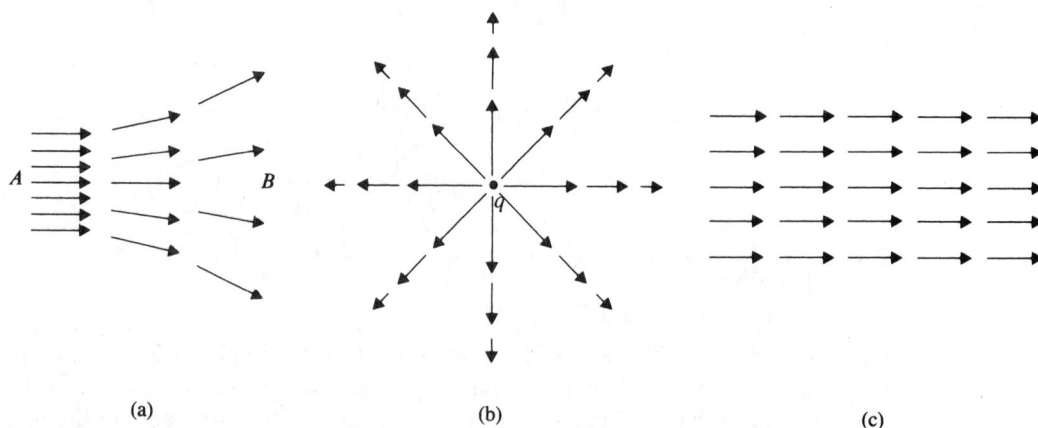

(a)   (b)   (c)

FIGURE 2–25
Flux lines of vector fields.

We define the divergence of a vector field **A** *at a point, abbreviated div* **A**, *as the net outward flux of* **A** *per unit volume as the volume about the point tends to zero:*

$$\text{div } \mathbf{A} \triangleq \lim_{\Delta v \to 0} \frac{\oint_S \mathbf{A} \cdot d\mathbf{s}}{\Delta v}. \tag{2-98}$$

The numerator in Eq. (2–98), representing the net outward flux, is an integral over the *entire* surface S that bounds the volume. We were exposed to this type of surface integral in Example 2–15. Equation (2–98) is the general definition of div **A** which is a *scalar quantity* whose magnitude may vary from point to point as **A** itself varies. This definition holds for any coordinate system; the expression for div **A**, like that for **A**, will, of course, depend on the choice of the coordinate system.

At the beginning of this section we intimated that the divergence of a vector is a type of spatial derivative. The reader might perhaps wonder about the presence of an integral in the expression given by Eq. (2–98); but a two-dimensional surface integral divided by a three-dimensional volume will lead to spatial derivatives as the volume approaches zero. We shall now derive the expression for div **A** in Cartesian coordinates.

Consider a differential volume of sides Δx, Δy, and Δz centered about a point $P(x_0, y_0, z_0)$ in the field of a vector **A**, as shown in Fig. 2–26. In Cartesian coordinates, $\mathbf{A} = \mathbf{a}_x A_x + \mathbf{a}_y A_y + \mathbf{a}_z A_z$. We wish to find div **A** at the point (x_0, y_0, z_0). Since the differential volume has six faces, the surface integral in the numerator of Eq. (2–98) can be decomposed into six parts:

$$\oint_S \mathbf{A} \cdot d\mathbf{s} = \left[\int_{\substack{\text{front} \\ \text{face}}} + \int_{\substack{\text{back} \\ \text{face}}} + \int_{\substack{\text{right} \\ \text{face}}} + \int_{\substack{\text{left} \\ \text{face}}} + \int_{\substack{\text{top} \\ \text{face}}} + \int_{\substack{\text{bottom} \\ \text{face}}} \right] \mathbf{A} \cdot d\mathbf{s}. \tag{2-99}$$

On the front face,

$$\int_{\substack{\text{front} \\ \text{face}}} \mathbf{A} \cdot d\mathbf{s} = \mathbf{A}_{\substack{\text{front} \\ \text{face}}} \cdot \Delta \mathbf{s}_{\substack{\text{front} \\ \text{face}}} = \mathbf{A}_{\substack{\text{front} \\ \text{face}}} \cdot \mathbf{a}_x (\Delta y \, \Delta z)$$

$$= A_x \left(x_0 + \frac{\Delta x}{2}, y_0, z_0 \right) \Delta y \, \Delta z. \tag{2-100}$$

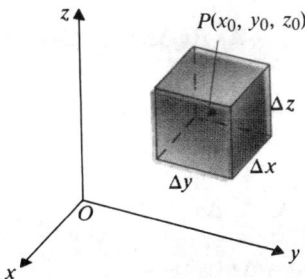

FIGURE 2–26
A differential volume in Cartesian coordinates.

The quantity $A_x([x_0 + (\Delta x/2), y_0, z_0])$ can be expanded as a Taylor series about its value at (x_0, y_0, z_0), as follows:

$$A_x\left(x_0 + \frac{\Delta x}{2}, y_0, z_0\right) = A_x(x_0, y_0, z_0) + \frac{\Delta x}{2}\frac{\partial A_x}{\partial x}\bigg|_{(x_0, y_0, z_0)} + \text{higher-order terms,}$$

$$(2\text{-}101)$$

where the higher-order terms (H.O.T.) contain the factors $(\Delta x/2)^2$, $(\Delta x/2)^3$, etc. Similarly, on the back face,

$$\int_{\substack{\text{back}\\\text{face}}} \mathbf{A} \cdot d\mathbf{s} = \mathbf{A}_{\substack{\text{back}\\\text{face}}} \cdot \Delta\mathbf{s}_{\substack{\text{back}\\\text{face}}} = \mathbf{A}_{\substack{\text{back}\\\text{face}}} \cdot (-\mathbf{a}_x \Delta y \Delta z)$$

$$(2\text{-}102)$$

$$= -A_x\left(x_0 - \frac{\Delta x}{2}, y_0, z_0\right)\Delta y \Delta z.$$

The Taylor-series expansion of $A_x\left(x_0 - \dfrac{\Delta x}{2}, y_0, z_0\right)$ is

$$A_x\left(x_0 - \frac{\Delta x}{2}, y_0, z_0\right) = A_x(x_0, y_0, z_0) - \frac{\Delta x}{2}\frac{\partial A_x}{\partial x}\bigg|_{(x_0, y_0, z_0)} + \text{H.O.T.} \qquad (2\text{-}103)$$

Substituting Eq. (2–101) in Eq. (2–100) and Eq. (2–103) in Eq. (2–102) and adding the contributions, we have

$$\left[\int_{\substack{\text{front}\\\text{face}}} + \int_{\substack{\text{back}\\\text{face}}}\right]\mathbf{A} \cdot d\mathbf{s} = \left(\frac{\partial A_x}{\partial x} + \text{H.O.T.}\right)\bigg|_{(x_0, y_0, z_0)}\Delta x \Delta y \Delta z. \qquad (2\text{-}104)$$

Here a Δx has been factored out from the H.O.T. in Eqs. (2–101) and (2–103), but all terms of the H.O.T. in Eq. (2–104) still contain powers of Δx.

Following the same procedure for the right and left faces, where the coordinate changes are $+\Delta y/2$ and $-\Delta y/2$, respectively, and $\Delta s = \Delta x \Delta z$, we find

$$\left[\int_{\substack{\text{right}\\\text{face}}} + \int_{\substack{\text{left}\\\text{face}}}\right]\mathbf{A} \cdot d\mathbf{s} = \left(\frac{\partial A_y}{\partial y} + \text{H.O.T.}\right)\bigg|_{(x_0, y_0, z_0)}\Delta x \Delta y \Delta z. \qquad (2\text{-}105)$$

Here the higher-order terms contain the factors $\Delta y, (\Delta y)^2$, etc. For the top and bottom faces we have

$$\left[\int_{\substack{\text{top}\\\text{face}}} + \int_{\substack{\text{bottom}\\\text{face}}}\right]\mathbf{A} \cdot d\mathbf{s} = \left(\frac{\partial A_z}{\partial z} + \text{H.O.T.}\right)\bigg|_{(x_0, y_0, z_0)}\Delta x \Delta y \Delta z, \qquad (2\text{-}106)$$

where the higher-order terms contain the factors $\Delta z, (\Delta z)^2$, etc. Now the results from Eqs. (2–104), (2–105), and (2–106) are combined in Eq. (2–99) to obtain

$$\oint_S \mathbf{A} \cdot d\mathbf{s} = \left(\frac{\partial A_x}{\partial x} + \frac{\partial A_y}{\partial y} + \frac{\partial A_z}{\partial z}\right)\bigg|_{(x_0, y_0, z_0)}\Delta x \Delta y \Delta z$$

$$(2\text{-}107)$$

$$+ \text{ higher-order terms in } \Delta x, \Delta y, \Delta z.$$

Since $\Delta v = \Delta x \Delta y \Delta z$, substitution of Eq. (2–107) in Eq. (2–98) yields the expression of div **A** in Cartesian coordinates:

$$\text{div } \mathbf{A} = \frac{\partial A_x}{\partial x} + \frac{\partial A_y}{\partial y} + \frac{\partial A_z}{\partial z}. \qquad (2\text{-}108)$$

The higher-order terms vanish as the differential volume $\Delta x \, \Delta y \, \Delta z$ approaches zero. The value of div **A**, in general, depends on the position of the point at which it is evaluated. We have dropped the notation (x_0, y_0, z_0) in Eq. (2–108) because it applies to any point at which **A** and its partial derivatives are defined.

With the vector differential operator del, ∇, defined in Eq. (2–96) we can write Eq. (2–108) alternatively as $\nabla \cdot \mathbf{A}$; that is,

$$\nabla \cdot \mathbf{A} \equiv \text{div } \mathbf{A}. \qquad (2\text{-}109)$$

In general orthogonal curvilinear coordinates (u_1, u_2, u_3), Eq. (2–98) will lead to

$$\nabla \cdot \mathbf{A} = \frac{1}{h_1 h_2 h_3} \left[\frac{\partial}{\partial u_1} (h_2 h_3 A_1) + \frac{\partial}{\partial u_2} (h_1 h_3 A_2) + \frac{\partial}{\partial u_3} (h_1 h_2 A_3) \right]. \qquad (2\text{-}110)$$

EXAMPLE 2-17 Find the divergence of the position vector to an arbitrary point.

Solution We will find the solution in Cartesian as well as in spherical coordinates.

a) *Cartesian coordinates.* The expression for the position vector to an arbitrary point (x, y, z) is

$$\overrightarrow{OP} = \mathbf{a}_x x + \mathbf{a}_y y + \mathbf{a}_z z. \qquad (2\text{-}111)$$

Using Eq. (2–108), we have

$$\nabla \cdot (\overrightarrow{OP}) = \frac{\partial x}{\partial x} + \frac{\partial y}{\partial y} + \frac{\partial z}{\partial z} = 3.$$

b) *Spherical coordinates.* Here the position vector is simply

$$\overrightarrow{OP} = \mathbf{a}_R R. \qquad (2\text{-}112)$$

Its divergence in spherical coordinates (R, θ, ϕ) can be obtained from Eq. (2–110) by using Table 2–1 as follows:

$$\nabla \cdot \mathbf{A} = \frac{1}{R^2} \frac{\partial}{\partial R} (R^2 A_R) + \frac{1}{R \sin \theta} \frac{\partial}{\partial \theta} (A_\theta \sin \theta) + \frac{1}{R \sin \theta} \frac{\partial A_\phi}{\partial \phi}. \qquad (2\text{-}113)$$

Substituting Eq. (2–112) in Eq. (2–113), we also obtain $\nabla \cdot (\overrightarrow{OP}) = 3$, as expected.

EXAMPLE 2–18 The magnetic flux density **B** outside a very long current-carrying wire is circumferential and is inversely proportional to the distance to the axis of the wire. Find $\mathbf{V} \cdot \mathbf{B}$.

Solution Let the long wire be coincident with the z-axis in a cylindrical coordinate system. The problem states that

$$\mathbf{B} = \mathbf{a}_\phi \frac{k}{r}.$$

The divergence of a vector field in cylindrical coordinates (r, ϕ, z) can be found from Eq. (2–110):

$$\boxed{\mathbf{V} \cdot \mathbf{B} = \frac{1}{r} \frac{\partial}{\partial r} (rB_r) + \frac{1}{r} \frac{\partial B_\phi}{\partial \phi} + \frac{\partial B_z}{\partial z}.} \qquad (2\text{–}114)$$

Now $B_\phi = k/r$, and $B_r = B_z = 0$. Equation (2–114) gives

$$\mathbf{V} \cdot \mathbf{B} = 0.$$

We have here a vector that is not a constant, but whose divergence is zero. This property indicates that the magnetic flux lines close upon themselves and that there are no magnetic sources or sinks. A divergenceless field is called a *solenoidal field*. More will be said about this type of field later in the book.

2–8 Divergence Theorem

In the preceding section we defined the divergence of a vector field as the net outward flux per unit volume. We may expect intuitively that *the volume integral of the divergence of a vector field equals the total outward flux of the vector through the surface that bounds the volume*; that is,

$$\boxed{\int_V \mathbf{V} \cdot \mathbf{A} \, dv = \oint_S \mathbf{A} \cdot d\mathbf{s}.} \qquad (2\text{–}115)$$

This identity, which will be proved in the following paragraph, is called the *divergence theorem*.[†] It applies to any volume V that is bounded by surface S. The direction of $d\mathbf{s}$ is always that of the *outward normal*, perpendicular to the surface ds and directed away from the volume.

For a very small differential volume element Δv_j bounded by a surface s_j, the definition of $\mathbf{V} \cdot \mathbf{A}$ in Eq. (2–98) gives directly

$$(\mathbf{V} \cdot \mathbf{A})_j \, \Delta v_j = \oint_{s_j} \mathbf{A} \cdot d\mathbf{s}. \qquad (2\text{–}116)$$

[†] It is also known as *Gauss's theorem*.

In case of an arbitrary volume V, we can subdivide it into many, say N, small differential volumes, of which Δv_j is typical. This is depicted in Fig. 2–27. Let us now combine the contributions of all these differential volumes to both sides of Eq. (2–116). We have

$$\lim_{\Delta v_j \to 0} \left[\sum_{j=1}^{N} (\mathbf{\nabla} \cdot \mathbf{A})_j \Delta v_j \right] = \lim_{\Delta v_j \to 0} \left[\sum_{j=1}^{N} \oint_{s_j} \mathbf{A} \cdot d\mathbf{s} \right]. \tag{2–117}$$

The left side of Eq. (2–117) is, by definition, the volume integral of $\mathbf{\nabla} \cdot \mathbf{A}$:

$$\lim_{\Delta v_j \to 0} \left[\sum_{j=1}^{N} (\mathbf{\nabla} \cdot \mathbf{A})_j \Delta v_j \right] = \int_V (\mathbf{\nabla} \cdot \mathbf{A}) \, dv. \tag{2–118}$$

The surface integrals on the right side of Eq. (2–117) are summed over all the faces of all the differential volume elements. The contributions from the internal surfaces of adjacent elements will, however, cancel each other, because at a common internal surface the outward normals of the adjacent elements point in opposite directions. Hence the net contribution of the right side of Eq. (2–117) is due only to that of the external surface S bounding the volume V; that is,

$$\lim_{\Delta v_j \to 0} \left[\sum_{j=1}^{N} \int_{s_j} \mathbf{A} \cdot d\mathbf{s} \right] = \oint_S \mathbf{A} \cdot d\mathbf{s}. \tag{2–119}$$

The substitution of Eqs. (2–118) and (2–119) in Eq. (2–117) yields the divergence theorem in Eq. (2–115).

The validity of the limiting processes leading to the proof of the divergence theorem requires that the vector field $\mathbf{A}$, as well as its first derivatives, exist and be continuous both in V and on S. The divergence theorem is an important identity in vector analysis. *It converts a volume integral of the divergence of a vector to a closed surface integral of the vector, and vice versa.* We use it frequently in establishing other theorems and relations in electromagnetics. We emphasize that, although a single integral sign is used on both sides of Eq. (2–115) for simplicity, the volume and surface integrals represent triple and double integrations, respectively.

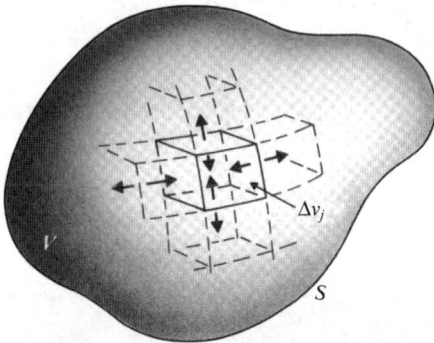

FIGURE 2–27
Subdivided volume for proof of divergence theorem.

EXAMPLE 2–19 Given $\mathbf{A} = \mathbf{a}_x x^2 + \mathbf{a}_y xy + \mathbf{a}_z yz$, verify the divergence theorem over a cube one unit on each side. The cube is situated in the first octant of the Cartesian coordinate system with one corner at the origin.

Solution Refer to Fig. 2–28. We first evaluate the surface integral over the six faces.

1. Front face: $x = 1$, $d\mathbf{s} = \mathbf{a}_x \, dy \, dz$;

$$\int_{\substack{\text{front} \\ \text{face}}} \mathbf{A} \cdot d\mathbf{s} = \int_0^1 \int_0^1 dy \, dz = 1.$$

2. Back face: $x = 0$, $d\mathbf{s} = -\mathbf{a}_x \, dy \, dz$;

$$\int_{\substack{\text{back} \\ \text{face}}} \mathbf{A} \cdot d\mathbf{s} = 0.$$

3. Left face: $y = 0$, $d\mathbf{s} = -\mathbf{a}_y \, dx \, dz$;

$$\int_{\substack{\text{left} \\ \text{face}}} \mathbf{A} \cdot d\mathbf{s} = 0.$$

4. Right face: $y = 1$, $d\mathbf{s} = \mathbf{a}_y \, dx \, dz$;

$$\int_{\substack{\text{right} \\ \text{face}}} \mathbf{A} \cdot d\mathbf{s} = \int_0^1 \int_0^1 x \, dx \, dz = \tfrac{1}{2}.$$

5. Top face: $z = 1$, $d\mathbf{s} = \mathbf{a}_z \, dx \, dy$;

$$\int_{\substack{\text{top} \\ \text{face}}} \mathbf{A} \cdot d\mathbf{s} = \int_0^1 \int_0^1 y \, dx \, dy = \tfrac{1}{2}.$$

6. Bottom face: $z = 0$, $d\mathbf{s} = -\mathbf{a}_z \, dx \, dy$;

$$\int_{\substack{\text{bottom} \\ \text{face}}} \mathbf{A} \cdot d\mathbf{s} = 0.$$

Adding the above six values, we have

$$\oint_S \mathbf{A} \cdot d\mathbf{s} = 1 + 0 + 0 + \tfrac{1}{2} + \tfrac{1}{2} + 0 = 2. \qquad (2\text{–}120)$$

Now the divergence of **A** is

$$\nabla \cdot \mathbf{A} = \frac{\partial}{\partial x}(x^2) + \frac{\partial}{\partial y}(xy) + \frac{\partial}{\partial z}(yz) = 3x + y.$$

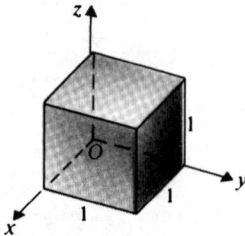

FIGURE 2–28
A unit cube (Example 2–19).

Hence,

$$\int_V \mathbf{V} \cdot \mathbf{A} \, dv = \int_0^1 \int_0^1 \int_0^1 (3x + y) \, dx \, dy \, dz = 2, \tag{2-121}$$

which is the same as the result of the closed surface integral in (2–120). The divergence theorem is therefore verified. ∎

EXAMPLE 2–20   Given $\mathbf{F} = \mathbf{a}_R kR$, determine whether the divergence theorem holds for the shell region enclosed by spherical surfaces at $R = R_1$ and $R = R_2 (R_2 > R_1)$ centered at the origin, as shown in Fig. 2–29.

Solution   Here the specified region has two surfaces, at $R = R_1$ and $R = R_2$. At the outer surface: $R = R_2$, $d\mathbf{s} = \mathbf{a}_R R_2^2 \sin \theta \, d\theta \, d\phi$;

$$\int_{\substack{\text{outer} \\ \text{surface}}} \mathbf{F} \cdot d\mathbf{s} = \int_0^{2\pi} \int_0^{\pi} (kR_2) R_2^2 \sin \theta \, d\theta \, d\phi = 4\pi k R_2^3.$$

At the inner surface: $R = R_1$, $d\mathbf{s} = -\mathbf{a}_R R_1^2 \sin \theta \, d\theta \, d\phi$;

$$\int_{\substack{\text{inner} \\ \text{surface}}} \mathbf{F} \cdot d\mathbf{s} = -\int_0^{2\pi} \int_0^{\pi} (kR_1) R_1^2 \sin \theta \, d\theta \, d\phi = -4\pi k R_1^3.$$

Actually, since the integrand is independent of θ or ϕ in both cases, the integral of a constant over a spherical surface is simply the constant multiplied by the area of the surface ($4\pi R_2^2$ for the outer surface and $4\pi R_1^2$ for the inner surface), and no integration is necessary. Adding the two results, we have

$$\oint_S \mathbf{F} \cdot d\mathbf{s} = 4\pi k (R_2^3 - R_1^3). \tag{2-122}$$

To find the volume integral, we first determine $\mathbf{V} \cdot \mathbf{F}$ for an $\mathbf{F}$ that has only an F_R component. From Eq. (2–113), we have

$$\mathbf{V} \cdot \mathbf{F} = \frac{1}{R^2} \frac{\partial}{\partial R} (R^2 F_R) = \frac{1}{R^2} \frac{\partial}{\partial R} (kR^3) = 3k.$$

Since $\mathbf{V} \cdot \mathbf{F}$ is a constant, its volume integral equals the product of the constant and the volume. The volume of the shell region between the two spherical surfaces with

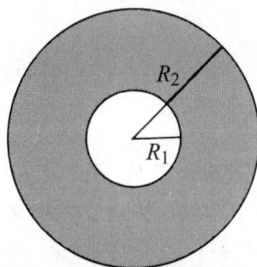

FIGURE 2–29
A spherical shell region (Example 2–20).

radii R_1 and R_2 is $4\pi(R_2^3 - R_1^3)/3$. Therefore,

$$\int_V \mathbf{V} \cdot \mathbf{F} \, dv = (\mathbf{V} \cdot \mathbf{F})V = 4\pi k(R_2^3 - R_1^3), \qquad (2\text{--}123)$$

which is the same as the result in Eq. (2–122).

This example shows that the divergence theorem holds even when the volume has holes inside—that is, even when the volume is enclosed by a multiply connected surface. ━

2–9   Curl of a Vector Field

In Section 2–7 we stated that a net outward flux of a vector $\mathbf{A}$ through a surface bounding a volume indicates the presence of a source. This source may be called a *flow source*, and div $\mathbf{A}$ is a measure of the strength of the flow source. There is another kind of source, called *vortex source*, which causes a circulation of a vector field around it. The *net circulation* (or simply *circulation*) of a vector field around a *closed path* is defined as the scalar line integral of the vector over the path. We have

$$\text{Circulation of } \mathbf{A} \text{ around contour } C \triangleq \oint_C \mathbf{A} \cdot d\boldsymbol{\ell}. \qquad (2\text{--}124)$$

Equation (2–124) is a mathematical definition. The physical meaning of circulation depends on what kind of field the vector $\mathbf{A}$ represents. If $\mathbf{A}$ is a force acting on an object, its circulation will be the work done by the force in moving the object once around the contour; if $\mathbf{A}$ represents an electric field intensity, then the circulation will be an electromotive force around the closed path, as we shall see later in the book. The familiar phenomenon of water whirling down a sink drain is an example of a *vortex sink* causing a circulation of fluid velocity. A circulation of $\mathbf{A}$ may exist even when div $\mathbf{A} = 0$ (when there is no flow source).

Since circulation as defined in Eq. (2–124) is a line integral of a dot product, its value obviously depends on the orientation of the contour C relative to the vector $\mathbf{A}$. In order to define a point function, which is a measure of the strength of a vortex source, we must make C very small and orient it in such a way that the circulation is a maximum. We define[†]

$$\boxed{\begin{aligned} \text{curl } \mathbf{A} &\equiv \mathbf{V} \times \mathbf{A} \\ &\triangleq \lim_{\Delta s \to 0} \frac{1}{\Delta s} \left[\mathbf{a}_n \oint_C \mathbf{A} \cdot d\boldsymbol{\ell} \right]_{\max}. \end{aligned}} \qquad (2\text{--}125)$$

In words, Eq. (2–125) states that *the curl of a vector field* $\mathbf{A}$, *denoted by curl* $\mathbf{A}$ *or* $\mathbf{V} \times \mathbf{A}$, *is a vector whose magnitude is the maximum net circulation of* $\mathbf{A}$ *per unit*

[†] In books published in Europe, the curl of $\mathbf{A}$ is often called the rotation of $\mathbf{A}$ and written as rot $\mathbf{A}$.

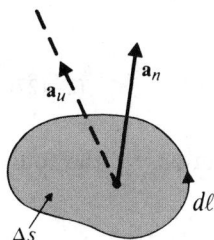

FIGURE 2–30
Relation between $\mathbf{a}_n$ and $d\ell$ in defining curl.

area as the area tends to zero and whose direction is the normal direction of the area when the area is oriented to make the net circulation maximum. Because the normal to an area can point in two opposite directions, we adhere to the right-hand rule that when the fingers of the right hand follow the direction of $d\ell$, the thumb points to the $\mathbf{a}_n$ direction. This is illustrated in Fig. 2–30. Curl $\mathbf{A}$ is a vector point function and is conventionally written as $\mathbf{\nabla} \times \mathbf{A}$ (del cross $\mathbf{A}$). The component of $\mathbf{\nabla} \times \mathbf{A}$ in any other direction $\mathbf{a}_u$ is $\mathbf{a}_u \cdot (\mathbf{\nabla} \times \mathbf{A})$, which can be determined from the circulation per unit area normal to $\mathbf{a}_u$ as the area approaches zero.

$$(\mathbf{\nabla} \times \mathbf{A})_u = \mathbf{a}_u \cdot (\mathbf{\nabla} \times \mathbf{A}) = \lim_{\Delta s_u \to 0} \frac{1}{\Delta s_u} \left(\oint_{C_u} \mathbf{A} \cdot d\ell \right), \qquad (2\text{–}126)$$

where the direction of the line integration around the contour C_u bounding area Δs_u and the direction $\mathbf{a}_u$ follow the right-hand rule.

We now use Eq. (2–126) to find the three components of $\mathbf{\nabla} \times \mathbf{A}$ in Cartesian coordinates. Refer to Fig. 2–31, in which a differential rectangular area parallel to the yz-plane and having sides Δy and Δz is drawn about a typical point $P(x_0, y_0, z_0)$. We have $\mathbf{a}_u = \mathbf{a}_x$ and $\Delta s_u = \Delta y \, \Delta z$, and the contour C_u consists of the four sides 1, 2, 3,

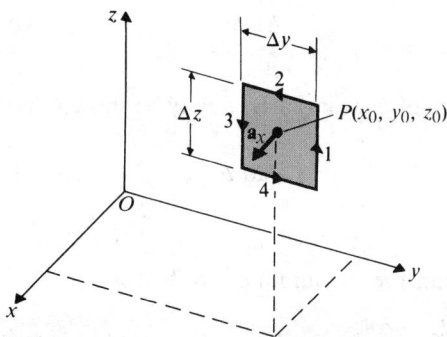

FIGURE 2–31
Determining $(\mathbf{\nabla} \times \mathbf{A})_x$.

and 4. Thus,

$$(\nabla \times \mathbf{A})_x = \lim_{\Delta y \, \Delta z \to 0} \frac{1}{\Delta y \, \Delta z} \left(\oint_{\substack{\text{sides} \\ 1, 2, 3, 4}} \mathbf{A} \cdot d\boldsymbol{\ell} \right). \qquad (2\text{–}127)$$

In Cartesian coordinates, $\mathbf{A} = \mathbf{a}_x A_x + \mathbf{a}_y A_y + \mathbf{a}_z A_z$. The contributions of the four sides to the line integral are as follows.

Side 1: $d\boldsymbol{\ell} = \mathbf{a}_z \, \Delta z, \, \mathbf{A} \cdot d\boldsymbol{\ell} = A_z \left(x_0, y_0 + \dfrac{\Delta y}{2}, z_0 \right) \Delta z,$

where $A_z \left(x_0, y_0 + \dfrac{\Delta y}{2}, z_0 \right)$ can be expanded as a Taylor series:

$$A_z \left(x_0, y_0 + \frac{\Delta y}{2}, z_0 \right) = A_z(x_0, y_0, z_0) + \frac{\Delta y}{2} \frac{\partial A_z}{\partial y} \bigg|_{(x_0, y_0, z_0)} + \text{H.O.T.}, \qquad (2\text{–}128)$$

where H.O.T. (higher-order terms) contain the factors $(\Delta y)^2$, $(\Delta y)^3$, etc. Thus,

$$\int_{\text{side 1}} \mathbf{A} \cdot d\boldsymbol{\ell} = \left\{ A_z(x_0, y_0, z_0) + \frac{\Delta y}{2} \frac{\partial A_z}{\partial y} \bigg|_{(x_0, y_0, z_0)} + \text{H.O.T.} \right\} \Delta z. \qquad (2\text{–}129)$$

Side 3: $d\boldsymbol{\ell} = -\mathbf{a}_z \, \Delta z, \, \mathbf{A} \cdot d\boldsymbol{\ell} = A_z \left(x_0, y_0 - \dfrac{\Delta y}{2}, z_0 \right) \Delta z,$

where

$$A_z \left(x_0, y_0 - \frac{\Delta y}{2}, z_0 \right) = A_z(x_0, y_0, z_0) - \frac{\Delta y}{2} \frac{\partial A_z}{\partial y} \bigg|_{(x_0, y_0, z_0)} + \text{H.O.T.}; \qquad (2\text{–}130)$$

$$\int_{\text{side 3}} \mathbf{A} \cdot d\boldsymbol{\ell} = \left\{ A_z(x_0, y_0, z_0) - \frac{\Delta y}{2} \frac{\partial A_z}{\partial y} \bigg|_{(x_0, y_0, z_0)} + \text{H.O.T.} \right\} (-\Delta z). \qquad (2\text{–}131)$$

Combining Eqs. (2–129) and (2–131), we have

$$\int_{\substack{\text{sides} \\ 1 \,\&\, 3}} \mathbf{A} \cdot d\boldsymbol{\ell} = \left(\frac{\partial A_z}{\partial y} + \text{H.O.T.} \right) \bigg|_{(x_0, y_0, z_0)} \Delta y \, \Delta z. \qquad (2\text{–}132)$$

The H.O.T. in Eq. (2–132) still contain powers of Δy. Similarly, it may be shown that

$$\int_{\substack{\text{sides} \\ 2 \,\&\, 4}} \mathbf{A} \cdot d\boldsymbol{\ell} = \left(-\frac{\partial A_y}{\partial z} + \text{H.O.T.} \right) \bigg|_{(x_0, y_0, z_0)} \Delta y \, \Delta z. \qquad (2\text{–}133)$$

Substituting Eqs. (2–132) and (2–133) in Eq. (2–127) and noting that the higher-order terms tend to zero as $\Delta y \to 0$, we obtain the x-component of $\nabla \times \mathbf{A}$:

$$(\nabla \times \mathbf{A})_x = \frac{\partial A_z}{\partial y} - \frac{\partial A_y}{\partial z}. \qquad (2\text{–}134)$$

A close examination of Eq. (2–134) will reveal a cyclic order in x, y, and z and enable us to write down the y-component and z-component of $\mathbf{V} \times \mathbf{A}$. The entire expression for the curl of $\mathbf{A}$ in Cartesian coordinates is

$$\mathbf{V} \times \mathbf{A} = \mathbf{a}_x \left(\frac{\partial A_z}{\partial y} - \frac{\partial A_y}{\partial z} \right) + \mathbf{a}_y \left(\frac{\partial A_x}{\partial z} - \frac{\partial A_z}{\partial x} \right) + \mathbf{a}_z \left(\frac{\partial A_y}{\partial x} - \frac{\partial A_x}{\partial y} \right). \qquad (2\text{–}135)$$

Compared to the expression for $\mathbf{V} \cdot \mathbf{A}$ in Eq. (2–108), that for $\mathbf{V} \times \mathbf{A}$ in Eq. (2–135) is more complicated, as it is expected to be, because it is a vector with three components, whereas $\mathbf{V} \cdot \mathbf{A}$ is a scalar. Fortunately, Eq. (2–135) can be remembered rather easily by arranging it in a determinantal form in the manner of the cross product exhibited in Eq. (2–43).

$$\mathbf{V} \times \mathbf{A} = \begin{vmatrix} \mathbf{a}_x & \mathbf{a}_y & \mathbf{a}_z \\ \dfrac{\partial}{\partial x} & \dfrac{\partial}{\partial y} & \dfrac{\partial}{\partial z} \\ A_x & A_y & A_z \end{vmatrix}. \qquad (2\text{–}136)$$

The derivation of $\mathbf{V} \times \mathbf{A}$ in other coordinate systems follows the same procedure. However, it is more involved because in curvilinear coordinates not only $\mathbf{A}$ but also $d\ell$ changes in magnitude as the integration of $\mathbf{A} \cdot d\ell$ is carried out on opposite sides of a curvilinear rectangle. The expression for $\mathbf{V} \times \mathbf{A}$ in general orthogonal curvilinear coordinates (u_1, u_2, u_3) is given below:

$$\mathbf{V} \times \mathbf{A} = \frac{1}{h_1 h_2 h_3} \begin{vmatrix} \mathbf{a}_{u_1} h_1 & \mathbf{a}_{u_2} h_2 & \mathbf{a}_{u_3} h_3 \\ \dfrac{\partial}{\partial u_1} & \dfrac{\partial}{\partial u_2} & \dfrac{\partial}{\partial u_3} \\ h_1 A_1 & h_2 A_2 & h_3 A_3 \end{vmatrix}. \qquad (2\text{–}137)$$

The expressions of $\mathbf{V} \times \mathbf{A}$ in cylindrical and spherical coordinates can be easily obtained from Eq. (2–137) by using the appropriate u_1, u_2, and u_3 and their metric coefficients h_1, h_2, and h_3 listed in Table 2–1.

▬▬ **EXAMPLE 2–21**   Show that $\mathbf{V} \times \mathbf{A} = 0$ if

a) $\mathbf{A} = \mathbf{a}_\phi (k/r)$ in cylindrical coordinates, where k is a constant, *or*

b) $\mathbf{A} = \mathbf{a}_R f(R)$ in spherical coordinates, where $f(R)$ is any function of the radial distance R.

Solution

a) In cylindrical coordinates the following apply: $(u_1, u_2, u_3) = (r, \phi, z)$; $h_1 = 1$, $h_2 = r$, and $h_3 = 1$. We have, from Eq. (2–137),

$$\mathbf{\nabla} \times \mathbf{A} = \frac{1}{r} \begin{vmatrix} \mathbf{a}_r & \mathbf{a}_\phi r & \mathbf{a}_z \\ \dfrac{\partial}{\partial r} & \dfrac{\partial}{\partial \phi} & \dfrac{\partial}{\partial z} \\ A_r & r A_\phi & A_z \end{vmatrix}, \tag{2–138}$$

which yields, for the given $\mathbf{A}$,

$$\mathbf{\nabla} \times \mathbf{A} = \frac{1}{r} \begin{vmatrix} \mathbf{a}_r & \mathbf{a}_\phi r & \mathbf{a}_z \\ \dfrac{\partial}{\partial r} & \dfrac{\partial}{\partial \phi} & \dfrac{\partial}{\partial z} \\ 0 & k & 0 \end{vmatrix} = 0.$$

b) In spherical coordinates the following apply: $(u_1, u_2, u_3) = (R, \theta, \phi)$; $h_1 = 1, h_2 = R$, and $h_3 = R \sin \theta$. Hence,

$$\mathbf{\nabla} \times \mathbf{A} = \frac{1}{R^2 \sin \theta} \begin{vmatrix} \mathbf{a}_R & \mathbf{a}_\theta R & \mathbf{a}_\phi R \sin \theta \\ \dfrac{\partial}{\partial R} & \dfrac{\partial}{\partial \theta} & \dfrac{\partial}{\partial \phi} \\ A_R & R A_\theta & R \sin \theta A_\phi \end{vmatrix}, \tag{2–139}$$

and, for the given $\mathbf{A}$,

$$\mathbf{\nabla} \times \mathbf{A} = \frac{1}{R^2 \sin \theta} \begin{vmatrix} \mathbf{a}_R & \mathbf{a}_\theta R & \mathbf{a}_\phi R \sin \theta \\ \dfrac{\partial}{\partial R} & \dfrac{\partial}{\partial \theta} & \dfrac{\partial}{\partial \phi} \\ f(R) & 0 & 0 \end{vmatrix} = 0. \qquad \blacksquare$$

A curl-free vector field is called an **irrotational** or a **conservative field**. We will see in the next chapter that an electrostatic field is irrotational (or conservative). The expressions for $\mathbf{\nabla} \times \mathbf{A}$ given in Eqs. (2–138) and (2–139) for cylindrical and spherical coordinates, respectively, will be useful for later reference.

2–10 Stokes's Theorem

For a very small differential area Δs_j bounded by a contour C_j, the definition of $\mathbf{\nabla} \times \mathbf{A}$ in Eq. (2–125) leads to

$$(\mathbf{\nabla} \times \mathbf{A})_j \cdot (\Delta \mathbf{s}_j) = \oint_{C_j} \mathbf{A} \cdot d\boldsymbol{\ell}. \tag{2–140}$$

In obtaining Eq. (2–140), we have taken the dot product of both sides of Eq. (2–125) with $\mathbf{a}_n \Delta s_j$ or $\Delta \mathbf{s}_j$. For an arbitrary surface S, we can subdivide it into many, say N, small differential areas. Figure 2–32 shows such a scheme with Δs_j as a typical differential element. The left side of Eq. (2–140) is the flux of the vector $\mathbf{V} \times \mathbf{A}$ through the area $\Delta \mathbf{s}_j$. Adding the contributions of all the differential areas to the flux, we have

$$\lim_{\Delta s_j \to 0} \sum_{j=1}^{N} (\mathbf{V} \times \mathbf{A})_j \cdot (\Delta \mathbf{s}_j) = \int_S (\mathbf{V} \times \mathbf{A}) \cdot d\mathbf{s}. \tag{2-141}$$

Now we sum up the line integrals around the contours of all the differential elements represented by the right side of Eq. (2–140). Since the common part of the contours of two adjacent elements is traversed in opposite directions by two contours, the net contribution of all the common parts in the interior to the total line integral is zero, and only the contribution from the external contour C bounding the entire area S remains after the summation:

$$\lim_{\Delta s_j \to 0} \sum_{j=1}^{N} \left(\oint_{c_j} \mathbf{A} \cdot d\boldsymbol{\ell} \right) = \oint_C \mathbf{A} \cdot d\boldsymbol{\ell}. \tag{2-142}$$

Combining Eqs. (2–141) and (2–142), we obtain ***Stokes's theorem***:

$$\boxed{\int_S (\mathbf{V} \times \mathbf{A}) \cdot d\mathbf{s} = \oint_C \mathbf{A} \cdot d\boldsymbol{\ell},} \tag{2-143}$$

which states that *the surface integral of the curl of a vector field over an open surface is equal to the closed line integral of the vector along the contour bounding the surface.*

As with the divergence theorem, the validity of the limiting processes leading to Stokes's theorem requires that the vector field $\mathbf{A}$, as well as its first derivatives, exist and be continuous both on S and along C. Stokes's theorem converts a surface integral of the curl of a vector to a line integral of the vector, and vice versa. Like the divergence theorem, Stokes's theorem is an important identity in vector analysis, and we will use it frequently in establishing other theorems and relations in electromagnetics.

If the surface integral of $\mathbf{V} \times \mathbf{A}$ is carried over a closed surface, there will be no surface-bounding external contour, and Eq. (2–143) tells us that

$$\oint_S (\mathbf{V} \times \mathbf{A}) \cdot d\mathbf{s} = 0 \tag{2-144}$$

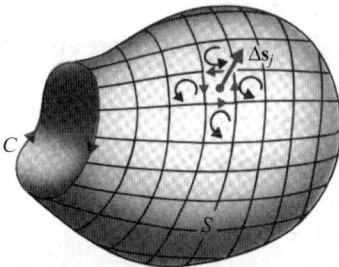

FIGURE 2–32
Subdivided area for proof of Stokes's theorem.

for any closed surface S. The geometry in Fig. 2–32 is chosen deliberately to emphasize the fact that a nontrivial application of Stokes's theorem always implies *an open surface with a rim*. The simplest open surface would be a two-dimensional plane or disk with its circumference as the contour. We remind ourselves here that the directions of $d\boldsymbol{\ell}$ and $d\mathbf{s}(\mathbf{a}_n)$ follow the right-hand rule.

EXAMPLE 2–22   Given $\mathbf{F} = \mathbf{a}_x xy - \mathbf{a}_y 2x$, verify Stokes's theorem over a quarter-circular disk with a radius 3 in the first quadrant, as was shown in Fig. 2–21 (Example 2–14, page 38).

Solution   Let us first find the surface integral of $\nabla \times \mathbf{F}$. From Eq. (2–136),

$$\nabla \times \mathbf{F} = \begin{vmatrix} \mathbf{a}_x & \mathbf{a}_y & \mathbf{a}_z \\ \dfrac{\partial}{\partial x} & \dfrac{\partial}{\partial y} & \dfrac{\partial}{\partial z} \\ xy & -2x & 0 \end{vmatrix} = -\mathbf{a}_z(2 + x).$$

Therefore,

$$\begin{aligned} \int_S (\nabla \times \mathbf{F}) \cdot d\mathbf{s} &= \int_0^3 \int_0^{\sqrt{9-y^2}} (\nabla \times \mathbf{F}) \cdot (\mathbf{a}_z \, dx \, dy) \\ &= \int_0^3 \left[\int_0^{\sqrt{9-y^2}} -(2+x) \, dx \right] dy \\ &= -\int_0^3 [2\sqrt{9-y^2} + \tfrac{1}{2}(9-y^2)] \, dy \\ &= -\left[y\sqrt{9-y^2} + 9 \sin^{-1}\frac{y}{3} + \frac{9}{2}y - \frac{y^3}{6} \right]\Big|_0^3 \\ &= -9\left(1 + \frac{\pi}{2}\right). \end{aligned}$$

It is *important* to use the proper limits for the two variables of integration. We can interchange the order of integration as

$$\int_S (\nabla \times \mathbf{F}) \cdot d\mathbf{s} = \int_0^3 \left[\int_0^{\sqrt{9-x^2}} -(2+x) \, dy \right] dx$$

and get the same result. But it would be quite wrong if the 0 to 3 range were used as the range of integration for both x and y. (Do you know why?)

For the line integral around $ABOA$ we have already evaluated the part around the arc from A to B in Example 2–14.

From B to O: $x = 0$, and $\mathbf{F} \cdot d\boldsymbol{\ell} = \mathbf{F} \cdot (-\mathbf{a}_y \, dy) = 2x \, dy = 0$.

From O to A: $y = 0$, and $\mathbf{F} \cdot d\boldsymbol{\ell} = \mathbf{F} \cdot (\mathbf{a}_x \, dx) = xy \, dx = 0$. Hence

$$\oint_{ABOA} \mathbf{F} \cdot d\boldsymbol{\ell} = \int_A^B \mathbf{F} \cdot d\boldsymbol{\ell} = -9\left(1 + \frac{\pi}{2}\right),$$

from Example 2–14, and Stokes's theorem is verified.

Of course, Stokes's theorem has been established in Eq. (2–143) as a general identity; there is no need to use a particular example to prove it. We worked out the example above for practice on surface and line integrals. (We note here that both the vector field and its first spatial derivatives are finite and continuous on the surface as well as on the contour of interest.)

2–11 Two Null Identities

Two identities involving repeated del operations are of considerable importance in the study of electromagnetism, especially when we introduce potential functions. We shall discuss them separately below.

2–11.1 IDENTITY I

$$\boxed{\mathbf{\nabla} \times (\mathbf{\nabla} V) \equiv 0} \tag{2–145}$$

In words, *the curl of the gradient of any scalar field is identically zero.* (The existence of V and its first derivatives everywhere is implied here.)

Equation (2–145) can be proved readily in Cartesian coordinates by using Eq. (2–96) for $\mathbf{\nabla}$ and performing the indicated operations. In general, if we take the surface integral of $\mathbf{\nabla} \times (\mathbf{\nabla} V)$ over any surface, the result is equal to the line integral of $\mathbf{\nabla} V$ around the closed path bounding the surface, as asserted by Stokes's theorem:

$$\int_S [\mathbf{\nabla} \times (\mathbf{\nabla} V)] \cdot d\mathbf{s} = \oint_C (\mathbf{\nabla} V) \cdot d\boldsymbol{\ell}. \tag{2–146}$$

However, from Eq. (2–88),

$$\oint_C (\mathbf{\nabla} V) \cdot d\boldsymbol{\ell} = \oint_C dV = 0. \tag{2–147}$$

The combination of Eqs. (2–146) and (2–147) states that the surface integral of $\mathbf{\nabla} \times (\mathbf{\nabla} V)$ over any surface is zero. The integrand itself must therefore vanish, which leads to the identity in Eq. (2–145). Since a coordinate system is not specified in the derivation, the identity is a general one and is invariant with the choices of coordinate systems.

A converse statement of Identity I can be made as follows: *If a vector field is curl-free, then it can be expressed as the gradient of a scalar field.* Let a vector field be $\mathbf{E}$. Then, if $\mathbf{\nabla} \times \mathbf{E} = 0$, we can define a scalar field V such that

$$\mathbf{E} = -\mathbf{\nabla} V. \tag{2–148}$$

The negative sign here is unimportant as far as Identity I is concerned. (It is included in Eq. (2–148) because this relation conforms with a basic relation between *electric field intensity* $\mathbf{E}$ and *electric scalar potential* V in electrostatics, which we will take up in the next chapter. At this stage it is immaterial what $\mathbf{E}$ and V represent.) We know from Section 2–9 that a curl-free vector field is a conservative field; hence *an irrotational (a conservative) vector field can always be expressed as the gradient of a scalar field.*

2–11.2   IDENTITY II

$$\boxed{\mathbf{V} \cdot (\mathbf{V} \times \mathbf{A}) \equiv 0} \tag{2–149}$$

In words, *the divergence of the curl of any vector field is identically zero.*

Equation (2–149), too, can be proved easily in Cartesian coordinates by using Eq. (2–96) for $\mathbf{V}$ and performing the indicated operations. We can prove it in general without regard to a coordinate system by taking the volume integral of $\mathbf{V} \cdot (\mathbf{V} \times \mathbf{A})$ on the left side. Applying the divergence theorem, we have

$$\int_V \mathbf{V} \cdot (\mathbf{V} \times \mathbf{A})\, dv = \oint_S (\mathbf{V} \times \mathbf{A}) \cdot d\mathbf{s}. \tag{2–150}$$

Let us choose, for example, the arbitrary volume V enclosed by a surface S in Fig. 2–33. The closed surface S can be split into two open surfaces, S_1 and S_2, connected by a common boundary that has been drawn twice as C_1 and C_2. We then apply Stokes's theorem to surface S_1 bounded by C_1, and surface S_2 bounded by C_2, and we write the right side of Eq. (2–150) as

$$\oint_S (\mathbf{V} \times \mathbf{A}) \cdot d\mathbf{s} = \int_{S_1} (\mathbf{V} \times \mathbf{A}) \cdot \mathbf{a}_{n1}\, ds + \int_{S_2} (\mathbf{V} \times \mathbf{A}) \cdot \mathbf{a}_{n2}\, ds$$
$$= \oint_{C_1} \mathbf{A} \cdot d\boldsymbol{\ell} + \oint_{C_2} \mathbf{A} \cdot d\boldsymbol{\ell}. \tag{2–151}$$

The normals $\mathbf{a}_{n1}$ and $\mathbf{a}_{n2}$ to surfaces S_1 and S_2 are *outward* normals, and their relations with the path directions of C_1 and C_2 follow the right-hand rule. Since the contours C_1 and C_2 are, in fact, one and the same common boundary between S_1 and S_2, the two line integrals on the right side of Eq. (2–151) traverse the same path in opposite directions. Their sum is therefore zero, and the volume integral of $\mathbf{V} \cdot (\mathbf{V} \times \mathbf{A})$ on the left side of Eq. (2–150) vanishes. Because this is true for any arbitrary volume, the integrand itself must be zero, as indicated by the identity in Eq. (2–149).

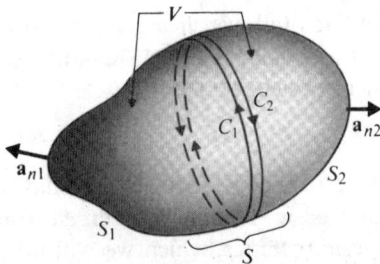

FIGURE 2–33
An arbitrary volume V enclosed by surface S.

A converse statement of Identity II is as follows: *If a vector field is divergenceless, then it can be expressed as the curl of another vector field.* Let a vector field be **B**. This converse statement asserts that if $\mathbf{V} \cdot \mathbf{B} = 0$, we can define a vector field **A** such that

$$\mathbf{B} = \mathbf{V} \times \mathbf{A}. \tag{2–152}$$

In Section 2–7 we mentioned that a divergenceless field is also called a solenoidal field. Solenoidal fields are not associated with flow sources or sinks. The net outward flux of a solenoidal field through any closed surface is zero, and the flux lines close upon themselves. We are reminded of the circling magnetic flux lines of a solenoid or an inductor. As we will see in Chapter 6, *magnetic flux density* **B** is solenoidal and can be expressed as the curl of another vector field called *magnetic vector potential* **A**.

2–12 Helmholtz's Theorem

In previous sections we mentioned that *a divergenceless field is solenoidal* and *a curl-free field is irrotational.* We may classify vector fields in accordance with their being solenoidal and/or irrotational. A vector field **F** is

1. Solenoidal and irrotational if
$$\mathbf{V} \cdot \mathbf{F} = 0 \quad \text{and} \quad \mathbf{V} \times \mathbf{F} = 0.$$
   EXAMPLE: A static electric field in a charge-free region.

2. Solenoidal but not irrotational if
$$\mathbf{V} \cdot \mathbf{F} = 0 \quad \text{and} \quad \mathbf{V} \times \mathbf{F} \neq 0.$$
   EXAMPLE: A steady magnetic field in a current-carrying conductor.

3. Irrotational but not solenoidal if
$$\mathbf{V} \times \mathbf{F} = 0 \quad \text{and} \quad \mathbf{V} \cdot \mathbf{F} \neq 0.$$
   EXAMPLE: A static electric field in a charged region.

4. Neither solenoidal nor irrotational if
$$\mathbf{V} \cdot \mathbf{F} \neq 0 \quad \text{and} \quad \mathbf{V} \times \mathbf{F} \neq 0.$$
   EXAMPLE: An electric field in a charged medium with a time-varying magnetic field.

The most general vector field then has both a nonzero divergence and a nonzero curl, and can be considered as the sum of a solenoidal field and an irrotational field.

Helmholtz's Theorem: A vector field (vector point function) is determined to within an additive constant if both its divergence and its curl are specified everywhere. In an unbounded region we assume that both the divergence and the curl of the vector field vanish at infinity. If the vector field is confined within a region bounded by a surface, then it is determined if its divergence and curl throughout the region, as well as the normal component of the vector over the bounding surface, are given.

Here we assume that the vector function is single-valued and that its derivatives are finite and continuous.

Helmholtz's theorem can be proved as a mathematical theorem in a general way.[†] For our purposes, we remind ourselves (see Section 2–9) that the divergence of a vector is a measure of the strength of the flow source and that the curl of a vector is a measure of the strength of the vortex source. When the strengths of both the flow source and the vortex source are specified, we expect that the vector field will be determined. Thus, we can decompose a general vector field $\mathbf{F}$ into an irrotational part $\mathbf{F}_i$ and a solenoidal part $\mathbf{F}_s$:

$$\mathbf{F} = \mathbf{F}_i + \mathbf{F}_s, \tag{2–153}$$

with

$$\begin{cases} \mathbf{\nabla} \times \mathbf{F}_i = 0 & \tag{2–154a} \\ \mathbf{\nabla} \cdot \mathbf{F}_i = g & \tag{2–154b} \end{cases}$$

and

$$\begin{cases} \mathbf{\nabla} \cdot \mathbf{F}_s = 0 & \tag{2–155a} \\ \mathbf{\nabla} \times \mathbf{F}_s = \mathbf{G}, & \tag{2–155b} \end{cases}$$

where g and $\mathbf{G}$ are assumed to be known. We have

$$\mathbf{\nabla} \cdot \mathbf{F} = \mathbf{\nabla} \cdot \mathbf{F}_i = g \tag{2–156}$$

and

$$\mathbf{\nabla} \times \mathbf{F} = \mathbf{\nabla} \times \mathbf{F}_s = \mathbf{G}. \tag{2–157}$$

Helmholtz's theorem asserts that when g and $\mathbf{G}$ are specified, the vector function $\mathbf{F}$ is determined. Since $\mathbf{\nabla}\cdot$ and $\mathbf{\nabla}\times$ are differential operators, $\mathbf{F}$ must be obtained by integrating g and $\mathbf{G}$ in some manner, which will lead to constants of integration. The determination of these additive constants requires the knowledge of some boundary conditions. The procedure for obtaining $\mathbf{F}$ from given g and $\mathbf{G}$ is not obvious at this time; it will be developed in stages in later chapters.

The fact that $\mathbf{F}_i$ is irrotational enables us to define a scalar (potential) function V, in view of identity (2–145), such that

$$\mathbf{F}_i = -\mathbf{\nabla}V. \tag{2–158}$$

Similarly, identity (2–149) and Eq. (2–155a) allow the definition of a vector (potential) function $\mathbf{A}$ such that

$$\mathbf{F}_s = \mathbf{\nabla} \times \mathbf{A}. \tag{2–159}$$

Helmholtz's theorem states that a general vector function $\mathbf{F}$ can be written as the sum of the gradient of a scalar function and the curl of a vector function. Thus

$$\mathbf{F} = -\mathbf{\nabla}V + \mathbf{\nabla} \times \mathbf{A}. \tag{2–160}$$

[†] See, for instance, G. Arfken, *Mathematical Methods for Physicists*, Section 1.15, Academic Press, New York, 1966.

In following chapters we will rely on Helmholtz's theorem as a basic element in the axiomatic development of electromagnetism.

EXAMPLE 2–23   Given a vector function

$$\mathbf{F} = \mathbf{a}_x(3y - c_1 z) + \mathbf{a}_y(c_2 x - 2z) - \mathbf{a}_z(c_3 y + z).$$

a) Determine the constants c_1, c_2, and c_3 if $\mathbf{F}$ is irrotational.

b) Determine the scalar potential function V whose negative gradient equals $\mathbf{F}$.

Solution

a) For $\mathbf{F}$ to be irrotational, $\nabla \times \mathbf{F} = 0$; that is,

$$\nabla \times \mathbf{F} = \begin{vmatrix} \mathbf{a}_x & \mathbf{a}_y & \mathbf{a}_z \\ \dfrac{\partial}{\partial x} & \dfrac{\partial}{\partial y} & \dfrac{\partial}{\partial z} \\ 3y - c_1 z & c_2 x - 2z & -(c_3 y + z) \end{vmatrix}$$

$$= \mathbf{a}_x(-c_3 + 2) - \mathbf{a}_y c_1 + \mathbf{a}_z(c_2 - 3) = 0.$$

Each component of $\nabla \times \mathbf{F}$ must vanish. Hence $c_1 = 0$, $c_2 = 3$, and $c_3 = 2$.

b) Since $\mathbf{F}$ is irrotational, it can be expressed as the negative gradient of a scalar function V; that is,

$$\mathbf{F} = -\nabla V = -\mathbf{a}_x \frac{\partial V}{\partial x} - \mathbf{a}_y \frac{\partial V}{\partial y} - \mathbf{a}_z \frac{\partial V}{\partial z}$$

$$= \mathbf{a}_x 3y + \mathbf{a}_y(3x - 2z) - \mathbf{a}_z(2y + z).$$

Three equations are obtained:

$$\frac{\partial V}{\partial x} = -3y, \tag{2–161}$$

$$\frac{\partial V}{\partial y} = -3x + 2z, \tag{2–162}$$

$$\frac{\partial V}{\partial z} = 2y + z. \tag{2–163}$$

Integrating Eq. (2–161) with respect to x, we have

$$V = -3xy + f_1(y, z), \tag{2–164}$$

where $f_1(y, z)$ is a function of y and z yet to be determined. Similarly, integrating Eq. (2–162) with respect to y and Eq. (2–163) with respect to z leads to

$$V = -3xy + 2yz + f_2(x, z) \tag{2–165}$$

and

$$V = 2yz + \frac{z^2}{2} + f_3(x, y). \tag{2–166}$$

Examination of Eqs. (2–164), (2–165), and (2–166) enables us to write the scalar potential function as

$$V = -3xy + 2yz + \frac{z^2}{2}. \qquad (2\text{--}167)$$

Any constant added to Eq. (2–167) would still make V an answer. The constant is to be determined by a boundary condition or the condition at infinity.   ▬

Review Questions

R.2–1 Three vectors **A**, **B**, and **C**, drawn in a head-to-tail fashion, form three sides of a triangle. What is **A** + **B** + **C**? What is **A** + **B** − **C**?

R.2–2 Under what conditions can the dot product of two vectors be negative?

R.2–3 Write down the results of **A** · **B** and **A** × **B** if (a) **A**∥**B**, and (b) **A** ⊥ **B**.

R.2–4 Which of the following products of vectors do not make sense? Explain.

    a) (**A** · **B**) × **C**   b) **A**(**B** · **C**)   c) **A** × **B** × **C**
    d) **A**/**B**   e) **A**/a_A f) (**A** × **B**) · **C**

R.2–5 Is (**A** · **B**)**C** equal to **A**(**B** · **C**)?

R.2–6 Does **A** · **B** = **A** · **C** imply **B** = **C**? Explain.

R.2–7 Does **A** × **B** = **A** × **C** imply **B** = **C**? Explain.

R.2–8 Given two vectors **A** and **B**, how do you find (a) the component of **A** in the direction of **B**, and (b) the component of **B** in the direction of **A**?

R.2–9 What makes a coordinate system (a) orthogonal? (b) curvilinear? and (c) right-handed?

R.2–10 Given a vector **F** in orthogonal curvilinear coordinates (u_1, u_2, u_3), explain how to determine (a) F, and (b) a_F.

R.2–11 What are metric coefficients?

R.2–12 Given two points $P_1(1, 2, 3)$ and $P_2(-1, 0, 2)$ in Cartesian coordinates, write the expressions of the vectors $\overrightarrow{P_1P_2}$ and $\overrightarrow{P_2P_1}$.

R.2–13 What are the expressions for **A** · **B** and **A** × **B** in Cartesian coordinates?

R.2–14 What is the difference between a scalar quantity and a scalar field? Between a vector quantity and a vector field?

R.2–15 What is the physical definition of the gradient of a scalar field?

R.2–16 Express the space rate of change of a scalar in a given direction in terms of its gradient.

R.2–17 What does the del operator **∇** stand for in Cartesian coordinates?

R.2–18 What is the physical definition of the divergence of a vector field?

R.2–19 A vector field with only radial flux lines cannot be solenoidal. True or false? Explain.

R.2–20 A vector field with only curved flux lines can have a nonzero divergence. True or false? Explain.

R.2–21 State the divergence theorem in words.

R.2–22 What is the physical definition of the curl of a vector field?

R.2–23 A vector field with only curved flux lines cannot be irrotational. True or false? Explain.

R.2–24 A vector field with only straight flux lines can be solenoidal. True or false? Explain.

R.2–25 State Stokes's theorem in words.

R.2–26 What is the difference between an irrotational field and a solenoidal field?

R.2–27 State Helmholtz's theorem in words.

R.2–28 Explain how a general vector function can be expressed in terms of a scalar potential function and a vector potential function.

Problems

P.2–1 Given three vectors **A**, **B**, and **C** as follows,

$$\mathbf{A} = \mathbf{a}_x + \mathbf{a}_y 2 - \mathbf{a}_z 3,$$
$$\mathbf{B} = -\mathbf{a}_y 4 + \mathbf{a}_z,$$
$$\mathbf{C} = \mathbf{a}_x 5 - \mathbf{a}_z 2,$$

find

a) $\mathbf{a}_A$
b) $|\mathbf{A} - \mathbf{B}|$
c) $\mathbf{A} \cdot \mathbf{B}$
d) θ_{AB}
e) the component of **A** in the direction of **C**
f) $\mathbf{A} \times \mathbf{C}$
g) $\mathbf{A} \cdot (\mathbf{B} \times \mathbf{C})$ and $(\mathbf{A} \times \mathbf{B}) \cdot \mathbf{C}$
h) $(\mathbf{A} \times \mathbf{B}) \times \mathbf{C}$ and $\mathbf{A} \times (\mathbf{B} \times \mathbf{C})$

P.2–2 Given

$$\mathbf{A} = \mathbf{a}_x - \mathbf{a}_y 2 + \mathbf{a}_z 3,$$
$$\mathbf{B} = \mathbf{a}_x + \mathbf{a}_y - \mathbf{a}_z 2,$$

find the expression for a unit vector **C** that is perpendicular to both **A** and **B**.

P.2–3 Two vector fields represented by $\mathbf{A} = \mathbf{a}_x A_x + \mathbf{a}_y A_y + \mathbf{a}_z A_z$ and $\mathbf{B} = \mathbf{a}_x B_x + \mathbf{a}_y B_y + \mathbf{a}_z B_z$, where all components may be functions of space coordinates. If these two fields are parallel to each other everywhere, what must be the relations between their components?

P.2–4 Show that, if $\mathbf{A} \cdot \mathbf{B} = \mathbf{A} \cdot \mathbf{C}$ *and* $\mathbf{A} \times \mathbf{B} = \mathbf{A} \times \mathbf{C}$, where **A** is not a null vector, then $\mathbf{B} = \mathbf{C}$.

P.2–5 An unknown vector can be determined if both its scalar product and its vector product with a known vector are given. Assuming that **A** is a known vector, determine the unknown vector **X** if both p and **B** are given, where $p = \mathbf{A} \cdot \mathbf{X}$ and $\mathbf{B} = \mathbf{A} \times \mathbf{X}$.

P.2–6 The three corners of a triangle are at $P_1(0, 1, -2)$, $P_2(4, 1, -3)$, and $P_3(6, 2, 5)$.
  a) Determine whether $\triangle P_1 P_2 P_3$ is a right triangle.
  b) Find the area of the triangle.

P.2–7 Show that the two diagonals of a rhombus are perpendicular to each other. (A rhombus is an equilateral parallelogram.)

P.2–8 Prove that the line joining the midpoints of two sides of a triangle is parallel to and half as long as the third side.

P.2–9 Unit vectors $\mathbf{a}_A$ and $\mathbf{a}_B$ denote the directions of two-dimensional vectors **A** and **B** that make angles α and β, respectively, with a reference x-axis, as shown in Fig. 2–34. a) Obtain a formula for the expansion of the cosine of the difference of two angles, $\cos(\alpha - \beta)$, by taking the scalar product $\mathbf{a}_A \cdot \mathbf{a}_B$. b) Obtain a formula for $\sin(\alpha - \beta)$.

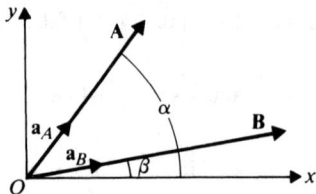

FIGURE 2–34
Graph for Problem P.2–9.

P.2–10 Prove the law of sines for a triangle.

P.2–11 Prove that an angle inscribed in a semicircle is a right angle.

P.2–12 Verify the back-cab rule of the vector triple product of three vectors, as expressed in Eq. (2–20) in Cartesian coordinates.

P.2–13 Prove by vector relations that two lines in the xy-plane (L_1: $b_1 x + b_2 y = c$; L_2: $b_1' x + b_2' y = c'$) are perpendicular if their slopes are the negative reciprocals of each other.

P.2–14
   a) Prove that the equation of any plane in space can be written in the form $b_1 x + b_2 y + b_3 z = c$. (*Hint:* Prove that the dot product of the position vector to any point in the plane and a normal vector is a constant.)
   b) Find the expression for the unit normal passing through the origin.
   c) For the plane $3x - 2y + 6z = 5$, find the perpendicular distance from the origin to the plane.

P.2–15 Find the component of the vector $\mathbf{A} = -\mathbf{a}_y z + \mathbf{a}_z y$ at the point $P_1(0, -2, 3)$, which is directed toward the point $P_2(\sqrt{3}, -60°, 1)$.

P.2–16 The position of a point in cylindrical coordinates is specified by $(4, 2\pi/3, 3)$. What is the location of the point
   a) in Cartesian coordinates?
   b) in spherical coordinates?

P.2–17 A field is expressed in spherical coordinates by $\mathbf{E} = \mathbf{a}_R(25/R^2)$.
   a) Find $|\mathbf{E}|$ and E_x at the point $P(-3, 4, -5)$.
   b) Find the angle that **E** makes with the vector $\mathbf{B} = \mathbf{a}_x 2 - \mathbf{a}_y 2 + \mathbf{a}_z$ at point P.

P.2–18 Express the base vectors $\mathbf{a}_R$, $\mathbf{a}_\theta$, and $\mathbf{a}_\phi$ of a spherical coordinate system in Cartesian coordinates.

P.2–19 Determine the values of the following products of base vectors:
   a) $\mathbf{a}_x \cdot \mathbf{a}_\phi$ **b)** $\mathbf{a}_\theta \cdot \mathbf{a}_y$ **c)** $\mathbf{a}_r \times \mathbf{a}_x$
   d) $\mathbf{a}_R \cdot \mathbf{a}_r$ **e)** $\mathbf{a}_y \cdot \mathbf{a}_R$ **f)** $\mathbf{a}_R \cdot \mathbf{a}_z$
   g) $\mathbf{a}_R \times \mathbf{a}_z$ **h)** $\mathbf{a}_\theta \cdot \mathbf{a}_z$ **i)** $\mathbf{a}_z \times \mathbf{a}_\theta$.

P.2–20 Given a vector function $\mathbf{F} = \mathbf{a}_x xy + \mathbf{a}_y (3x - y^2)$, evaluate the integral $\int \mathbf{F} \cdot d\boldsymbol{\ell}$ from $P_1(5, 6)$ to $P_2(3, 3)$ in Fig. 2–35
   a) along the direct path $P_1 P_2$,
   b) along path $P_1 A P_2$.

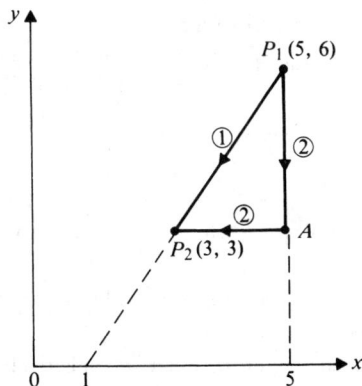

FIGURE 2-35
Paths of integration for Problem P.2-20.

P.2-21 Given a vector function $\mathbf{E} = \mathbf{a}_x y + \mathbf{a}_y x$, evaluate the scalar line integral $\int \mathbf{E} \cdot d\ell$ from $P_1(2, 1, -1)$ to $P_2(8, 2, -1)$
   a) along the parabola $x = 2y^2$,
   b) along the straight line joining the two points.
Is this $\mathbf{E}$ a conservative field?

P.2-22 For the $\mathbf{E}$ of Problem P.2-21, evaluate $\int \mathbf{E} \cdot d\ell$ from $P_3(3, 4, -1)$ to $P_4(4, -3, -1)$ by converting both $\mathbf{E}$ and the positions of P_3 and P_4 into cylindrical coordinates.

P.2-23 Given a scalar function

$$V = \left(\sin \frac{\pi}{2} x\right)\left(\sin \frac{\pi}{3} y\right)e^{-z},$$

determine
   a) the magnitude and the direction of the maximum rate of increase of V at the point $P(1, 2, 3)$,
   b) the rate of increase of V at P in the direction of the origin.

P.2-24 Evaluate

$$\oint_S (\mathbf{a}_R 3 \sin \theta) \cdot d\mathbf{s}$$

over the surface of a sphere of a radius 5 centered at the origin.

P.2-25 The equation in space of a plane containing the point (x_1, y_1, z_1) can be written as

$$\ell(x - x_1) + m(y - y_1) + p(z - z_1) = 0,$$

where ℓ, m, and p are direction cosines of a unit normal to the plane:

$$\mathbf{a}_n = \mathbf{a}_x \ell + \mathbf{a}_y m + \mathbf{a}_z p.$$

Given a vector field $\mathbf{F} = \mathbf{a}_x + \mathbf{a}_y 2 + \mathbf{a}_z 3$, evaluate the integral $\int_S \mathbf{F} \cdot d\mathbf{s}$ over the square plane surface whose corners are at $(0, 0, 2)$, $(2, 0, 2)$, $(2, 2, 0)$, and $(0, 2, 0)$.

P.2-26 Find the divergence of the following radial vector fields:
   a) $f_1(\mathbf{R}) = \mathbf{a}_R R^n$,
   b) $f_2(\mathbf{R}) = \mathbf{a}_R \dfrac{k}{R^2}$.

P.2-27 Show that $\frac{1}{3} \oint_S \mathbf{R} \cdot d\mathbf{s} = V$, where $\mathbf{R}$ is the radial vector and V is the volume of the region enclosed by surface S.

P.2–28 For a scalar function f and a vector function $\mathbf{A}$, prove that

$$\nabla \cdot (f\mathbf{A}) = f\nabla \cdot \mathbf{A} + \mathbf{A} \cdot \nabla f$$

in Cartesian coordinates.

P.2–29 For vector function $\mathbf{A} = \mathbf{a}_r r^2 + \mathbf{a}_z 2z$, verify the divergence theorem for the circular cylindrical region enclosed by $r = 5$, $z = 0$, and $z = 4$.

P.2–30 For the vector function $\mathbf{F} = \mathbf{a}_r k_1/r + \mathbf{a}_z k_2 z$ given in Example 2–15 (page 41) evaluate $\int \nabla \cdot \mathbf{F}\, dv$ over the volume specified in that example. Explain why the divergence theorem fails here.

P.2–31 Use the definition in Eq. (2–98) to derive the expression of $\nabla \cdot \mathbf{A}$ for a vector field $\mathbf{A} = \mathbf{a}_r A_r + \mathbf{a}_\phi A_\phi + \mathbf{a}_z A_z$ in cylindrical coordinates.

P.2–32 A vector field $\mathbf{D} = \mathbf{a}_R (\cos^2 \phi)/R^3$ exists in the region between two spherical shells defined by $R = 1$ and $R = 2$. Evaluate

a) $\oint \mathbf{D} \cdot d\mathbf{s}$,
b) $\int \nabla \cdot \mathbf{D}\, dv$.

P.2–33 For two differentiable vector functions $\mathbf{E}$ and $\mathbf{H}$, prove that

$$\nabla \cdot (\mathbf{E} \times \mathbf{H}) = \mathbf{H} \cdot (\nabla \times \mathbf{E}) - \mathbf{E} \cdot (\nabla \times \mathbf{H}).$$

P.2–34 Assume the vector function $\mathbf{A} = \mathbf{a}_x 3x^2 y^3 - \mathbf{a}_y x^3 y^2$.

a) Find $\oint \mathbf{A} \cdot d\boldsymbol{\ell}$ around the triangular contour shown in Fig. 2–36.
b) Evaluate $\int (\nabla \times \mathbf{A}) \cdot d\mathbf{s}$ over the triangular area.
c) Can $\mathbf{A}$ be expressed as the gradient of a scalar? Explain.

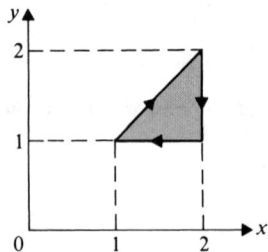

FIGURE 2–36
Graph for Problem P.2–34.

P.2–35 Use the definition in Eq. (2–126) to derive the expression of the $\mathbf{a}_R$-component of $\nabla \times \mathbf{A}$ in spherical coordinates for a vector field $\mathbf{A} = \mathbf{a}_R A_R + \mathbf{a}_\theta A_\theta + \mathbf{a}_\phi A_\phi$.

P.2–36 Given the vector function $\mathbf{A} = \mathbf{a}_\phi \sin (\phi/2)$, verify Stokes's theorem over the hemispherical surface and its circular contour that are shown in Fig. 2–37.

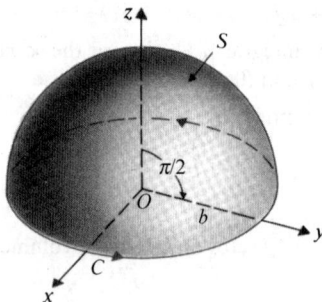

FIGURE 2–37
Graph for Problem P.2–36.

P.2–37 For a scalar function f and a vector function $\mathbf{G}$, prove that

$$\mathbf{V} \times (f\mathbf{G}) = f\mathbf{V} \times \mathbf{G} + (\mathbf{V}f) \times \mathbf{G}$$

in Cartesian coordinates.

P.2–38 Verify the null identities:

    a) $\mathbf{V} \times (\mathbf{V}V) \equiv 0$

    b) $\mathbf{V} \cdot (\mathbf{V} \times \mathbf{A}) \equiv 0$

by expansion in general orthogonal curvilinear coordinates.

P.2–39 Given a vector function $\mathbf{F} = \mathbf{a}_x(x + c_1 z) + \mathbf{a}_y(c_2 x - 3z) + \mathbf{a}_z(x + c_3 y + c_4 z)$.

    a) Determine the constants c_1, c_2, and c_3 if $\mathbf{F}$ is irrotational.

    b) Determine the constant c_4 if $\mathbf{F}$ is also solenoidal.

    c) Determine the scalar potential function V whose negative gradient equals $\mathbf{F}$.

3

Static
Electric Fields

3–1 Introduction

In Section 1–2 we mentioned that three essential steps are involved in constructing a deductive theory for the study of a scientific subject. They are: the definition of basic quantities, the development of rules of operation, and the postulation of fundamental relations. We have defined the source and field quantities for the electromagnetic model in Chapter 1 and developed the fundamentals of vector algebra and vector calculus in Chapter 2. We are now ready to introduce the fundamental postulates for the study of source-field relationships in electrostatics.

A *field* is a spatial distribution of a scalar or vector quantity, which may or may not be a function of time. An example of a scalar is the altitude of a location on a mountain relative to the sea level. It is a scalar, which is not a function of time if long-term erosion and earthquake effects are neglected. Various locations on the mountain have different altitudes, constituting an altitude field. The gradient of altitude is a vector that gives both the direction and the magnitude of the maximum rate of increase (the upward slope) of altitude. On a flat mountaintop or flat land the altitude is constant, and its gradient vanishes. The gravitational field of the earth, representing the force of gravity on a unit mass, is a vector field directed toward the center of the earth, having a magnitude depending on the altitude of the mass. Electric and magnetic field intensities are vector fields.

In electrostatics, electric charges (the sources) are at rest, and electric fields do not change with time. There are no magnetic fields; hence we deal with a relatively simple situation. After we have studied the behavior of static electric fields and mastered the techniques for solving electrostatic boundary-value problems, we will go on to the subject of magnetic fields and time-varying electromagnetic fields. Although electrostatics is relatively simple in the electromagnetics scheme of things, its mastery is fundamental to the understanding of more complicated electromagnetic models. Moreover, the explanation of many natural phenomena (such as lightning, corona, St. Elmo's fire, and grain explosion) and the principles of some important industrial applications (such as oscilloscope, ink-jet printer, xerography, and electret microphone) are based on electrostatics. Many articles on special applications of electrostatics appear in the literature, and a number of books on this subject have also been published.[†]

The development of electrostatics in elementary physics usually begins with the experimental Coulomb's law (formulated in 1785) for the force between two point charges. This law states that the force between two charged bodies, q_1 and q_2, that are very small in comparison with the distance of separation, R_{12}, is proportional to the product of the charges and inversely proportional to the square of the distance, the direction of the force being along the line connecting the charges. In addition, Coulomb found that unlike charges attract and like charges repel each other. Using vector notation, **Coulomb's law** can be written mathematically as

$$\mathbf{F}_{12} = \mathbf{a}_{R_{12}} k \frac{q_1 q_2}{R_{12}^2}, \qquad (3\text{–}1)$$

where $\mathbf{F}_{12}$ is the vector force exerted by q_1 on q_2, $\mathbf{a}_{R_{12}}$ is a unit vector in the direction from q_1 to q_2, and k is a proportionality constant depending on the medium and the system of units. Note that if q_1 and q_2 are of the same sign (both positive or both negative), $\mathbf{F}_{12}$ is positive (repulsive); and if q_1 and q_2 are of opposite signs, $\mathbf{F}_{12}$ is negative (attractive). Electrostatics can proceed from Coulomb's law to define electric field intensity **E**, electric scalar potential V, and electric flux density **D**, and then lead to Gauss's law and other relations. This approach has been accepted as "logical," perhaps because it begins with an experimental law observed in a laboratory and not with some abstract postulates.

We maintain, however, that Coulomb's law, though based on experimental evidence, is in fact also a postulate. Consider the two stipulations of Coulomb's law: that the charged bodies be very small in comparison with the distance of separation and that the force be inversely proportional to the square of the distance. The question arises regarding the first stipulation: How small must the charged bodies be in order to be considered "very small" in comparison with the distance? In practice the charged bodies cannot be of vanishing sizes (ideal point charges), and there is difficulty in determining the "true" distance between two bodies of finite dimensions. For given body sizes, the relative accuracy in distance measurements is better when the separation is larger. However, practical considerations (weakness of force, existence of extraneous charged bodies, etc.) restrict the usable distance of separation in the laboratory, and experimental inaccuracies cannot be entirely avoided. This leads to a more important question concerning the inverse-square relation of the second

---

[†] A. Klinkenberg and J. L. van der Minne, *Electrostatics in the Petroleum Industry*, Elsevier, Amsterdam, 1958. J. H. Dessauer and H. E. Clark, *Xerography and Related Processes*, Focal Press, London, 1965. A. D. Moore (Ed.), *Electrostatics and Its Applications*, John Wiley, New York, 1973. C. E. Jewett, *Electrostatics in the Electronics Environment*, John Wiley, New York, 1976. J.C. Crowley, *Fundamentals of Applied Electrostatics*, John Wiley, New York, 1986.

stipulation. Even if the charged bodies are of vanishing sizes, experimental measurements cannot be of infinite accuracy, no matter how skillful and careful an experimentor is. How then was it possible for Coulomb to know that the force was *exactly* inversely proportional to the *square* (not the 2.000001th or the 1.999999th power) of the distance of separation? This question cannot be answered from an experimental viewpoint because it is not likely that experiments could have been accurate to the seventh place during Coulomb's time.[†] We must therefore conclude that Coulomb's law is itself a postulate and that the exact relation stipulated by Eq. (3–1) is a law of nature discovered and assumed by Coulomb on the basis of his experiments of limited accuracy.

Instead of following the historical development of electrostatics, we introduce the subject by postulating both the divergence and the curl of the electric field intensity in free space. From Helmholtz's theorem in Section 2–12 we know that a vector field is determined if its divergence and curl are specified. We derive Gauss's law and Coulomb's law from the divergence and curl relations, and we do not present them as separate postulates. The concept of scalar potential follows naturally from a vector identity. Field behaviors in material media will be studied and expressions for electrostatic energy and forces will be developed.

3–2 Fundamental Postulates of Electrostatics in Free Space

We start the study of electromagnetism with the consideration of electric fields due to stationary (static) electric charges in free space. Electrostatics in free space is the simplest special case of electromagnetics. We need to consider only one of the four fundamental vector field quantities of the electromagnetic model discussed in Section 1–2, namely, the electric field intensity $\mathbf{E}$. Furthermore, only the permittivity of free space ϵ_0, of the three universal constants mentioned in Section 1–3 enters into our formulation.

Electric field intensity is defined as the force per unit charge that a very small stationary test charge experiences when it is placed in a region where an electric field exists. That is,

$$\mathbf{E} = \lim_{q \to 0} \frac{\mathbf{F}}{q} \qquad (\text{V/m}).$$

(3–2)

The electric field intensity $\mathbf{E}$ is, then, proportional to and in the direction of the force $\mathbf{F}$. If $\mathbf{F}$ is measured in newtons (N) and charge q in coulombs (C), then $\mathbf{E}$ is in newtons per coulomb (N/C), which is the same as volts per meter (V/m). The test charge

---

[†] The exponent on the distance in Coulomb's law has been verified by an indirect experiment to be 2 to within one part in 10^{15}. (See E. R. Williams, J. E. Faller, and H. A. Hall, *Phys. Rev. Letters*, vol. 26, 1971, p. 721.)

q, of course, cannot be zero in practice; as a matter of fact, it cannot be less than the charge on an electron. However, the finiteness of the test charge would not make the measured $\mathbf{E}$ differ appreciably from its calculated value if the test charge is small enough not to disturb the charge distribution of the source. An inverse relation of Eq. (3–2) gives the force $\mathbf{F}$ on a stationary charge q in an electric field $\mathbf{E}$:

$$\boxed{\mathbf{F} = q\mathbf{E} \qquad (\text{N}).}$$

(3–3)

The two fundamental postulates of electrostatics in free space specify the divergence and curl of $\mathbf{E}$. They are

$$\boxed{\nabla \cdot \mathbf{E} = \frac{\rho}{\epsilon_0}}$$

(3–4)

and

$$\boxed{\nabla \times \mathbf{E} = 0.}$$

(3–5)

In Eq. (3–4), ρ is the volume charge density of free charges (C/m^3), and ϵ_0 is the permittivity of free space, a universal constant.[†] Equation (3–5) asserts that *static electric fields are irrotational*, whereas Eq. (3–4) implies that a static electric field is not solenoidal unless $\rho = 0$. These two postulates are concise, simple, and independent of any coordinate system; and they can be used to derive all other relations, laws, and theorems in electrostatics! Such is the beauty of the deductive, axiomatic approach.

Equations (3–4) and (3–5) are point relations; that is, they hold at every point in space. They are referred to as the differential form of the postulates of electrostatics, since both divergence and curl operations involve spatial derivatives. In practical applications we are usually interested in the total field of an aggregate or a distribution of charges. This is more conveniently obtained by an integral form of Eq. (3–4). Taking the volume integral of both sides of Eq. (3–4) over an arbitrary volume V, we have

$$\int_V \nabla \cdot \mathbf{E} \, dv = \frac{1}{\epsilon_0} \int_V \rho \, dv.$$

(3–6)

In view of the divergence theorem in Eq. (2–115), Eq. (3–6) becomes

$$\boxed{\oint_S \mathbf{E} \cdot d\mathbf{s} = \frac{Q}{\epsilon_0},}$$

(3–7)

[†] The permittivity of free space $\epsilon_0 \cong \dfrac{1}{36\pi} \times 10^{-9}$ (F/m). See Eq. (1–11).

where Q is the total charge contained in volume V bounded by surface S. Equation (3–7) is a form of **Gauss's law**, which states that *the total outward flux of the electric field intensity over any closed surface in free space is equal to the total charge enclosed in the surface divided by* ϵ_0. Gauss's law is one of the most important relations in electrostatics. We will discuss it further in Section 3–4, along with illustrative examples.

An integral form can also be obtained for the curl relation in Eq. (3–5) by integrating $\nabla \times \mathbf{E}$ over an open surface and invoking Stokes's theorem as expressed in Eq. (2–143). We have

$$\oint_C \mathbf{E} \cdot d\boldsymbol{\ell} = 0. \tag{3–8}$$

The line integral is performed over a closed contour C bounding an arbitrary surface; hence C is itself arbitrary. As a matter of fact, the surface does not even enter into Eq. (3–8), which asserts that *the scalar line integral of the static electric field intensity around any closed path vanishes*. The scalar product $\mathbf{E} \cdot d\boldsymbol{\ell}$ integrated over any path is the voltage along that path. Thus Eq. (3–8) is an expression of **Kirchhoff's voltage law** in circuit theory that *the algebraic sum of voltage drops around any closed circuit is zero*. This will be discussed again in Section 5–3.

Equation (3–8) is another way of saying that $\mathbf{E}$ is irrotational (conservative). Referring to Fig. 3–1, we see that if the scalar line integral of $\mathbf{E}$ over the arbitrary closed contour $C_1 C_2$ is zero, then

$$\int_{C_1} \mathbf{E} \cdot d\boldsymbol{\ell} + \int_{C_2} \mathbf{E} \cdot d\boldsymbol{\ell} = 0 \tag{3–9}$$

or

$$\int_{P_1}^{P_2} \mathbf{E} \cdot d\boldsymbol{\ell} = - \int_{P_2}^{P_1} \mathbf{E} \cdot d\boldsymbol{\ell} \tag{3–10}$$
$$\text{Along } C_1 \qquad\qquad \text{Along } C_2$$

or

$$\int_{P_1}^{P_2} \mathbf{E} \cdot d\boldsymbol{\ell} = \int_{P_1}^{P_2} \mathbf{E} \cdot d\boldsymbol{\ell}. \tag{3–11}$$
$$\text{Along } C_1 \qquad\qquad \text{Along } C_2$$

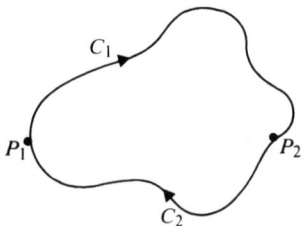

FIGURE 3–1
An arbitrary contour.

Equation (3–11) says that the scalar line integral of the irrotational **E** field is independent of the path; it depends only on the end points. As we shall see in Section 3–5, the integrals in Eq. (3–11) represent the work done *by* the electric field in moving a unit charge from point P_1 to point P_2; hence Eqs. (3–8) and (3–9) imply a statement of conservation of work or energy in an electrostatic field.

The two fundamental postulates of electrostatics in free space are repeated below because they form the foundation upon which we build the structure of electrostatics.

Postulates of Electrostatics in Free Space	
Differential Form	Integral Form
$\mathbf{V} \cdot \mathbf{E} = \dfrac{\rho}{\epsilon_0}$	$\displaystyle\oint_S \mathbf{E} \cdot d\mathbf{s} = \dfrac{Q}{\epsilon_0}$
$\mathbf{V} \times \mathbf{E} = 0$	$\displaystyle\oint_C \mathbf{E} \cdot d\ell = 0$

We consider these postulates, like the principle of conservation of charge, to be representations of laws of nature. In the following section we shall *derive* Coulomb's law.

3-3 Coulomb's Law

We consider the simplest possible electrostatic problem of a single point charge, q, at rest in a boundless free space. In order to find the electric field intensity due to q, we draw a hypothetical spherical surface of a radius R centered at q. Since a point charge has no preferred directions, its electric field must be everywhere radial and has the same intensity at all points on the spherical surface. Applying Eq. (3–7) to Fig. 3–2(a), we have

$$\oint_S \mathbf{E} \cdot d\mathbf{s} = \oint_S (\mathbf{a}_R E_R) \cdot \mathbf{a}_R \, ds = \frac{q}{\epsilon_0}$$

or

$$E_R \oint_S ds = E_R(4\pi R^2) = \frac{q}{\epsilon_0}.$$

Therefore,

$$\mathbf{E} = \mathbf{a}_R E_R = \mathbf{a}_R \frac{q}{4\pi\epsilon_0 R^2} \quad \text{(V/m)}. \qquad (3\text{--}12)$$

Equation (3–12) tells us that *the electric field intensity of a positive point charge is in the outward radial direction and has a magnitude proportional to the charge and inversely proportional to the square of the distance from the charge*. This is a very important basic formula in electrostatics. Using Eq. (2–139), we can verify that

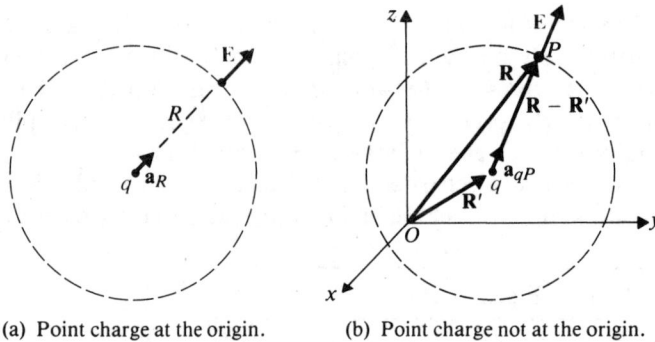

(a) Point charge at the origin.   (b) Point charge not at the origin.

FIGURE 3–2
Electric field iFIGUREdue to a point charge.

$\nabla \times \mathbf{E} = 0$ for the $\mathbf{E}$ given in Eq. (3–12). A flux-line graph for the electric field intensity of a positive point charge q will look like Fig. 2–25(b).

If the charge q is not located at the origin of a chosen coordinate system, suitable changes should be made to the unit vector $\mathbf{a}_R$ and the distance R to reflect the locations of the charge and of the point at which $\mathbf{E}$ is to be determined. Let the position vector of q be $\mathbf{R}'$ and that of a field point P be $\mathbf{R}$, as shown in Fig. 3–2(b). Then, from Eq. (3–12),

$$\mathbf{E}_P = \mathbf{a}_{qP} \frac{q}{4\pi\epsilon_0 |\mathbf{R} - \mathbf{R}'|^2}, \tag{3–13}$$

where $\mathbf{a}_{qP}$ is the unit vector drawn from q to P. Since

$$\mathbf{a}_{qP} = \frac{\mathbf{R} - \mathbf{R}'}{|\mathbf{R} - \mathbf{R}'|} \tag{3–14}$$

we have

$$\boxed{\mathbf{E}_p = \frac{q(\mathbf{R} - \mathbf{R}')}{4\pi\epsilon_0 |\mathbf{R} - \mathbf{R}'|^3}} \quad \text{(V/m)}. \tag{3–15}$$

EXAMPLE 3–1   Determine the electric field intensity at $P(-0.2, 0, -2.3)$ due to a point charge of $+5$ (nC) at $Q(0.2, 0.1, -2.5)$ in air. All dimensions are in meters.

Solution   The position vector for the field point P

$$\mathbf{R} = \overrightarrow{OP} = -\mathbf{a}_x 0.2 - \mathbf{a}_z 2.3.$$

The position vector for the point charge Q is

$$\mathbf{R}' = \overrightarrow{OQ} = \mathbf{a}_x 0.2 + \mathbf{a}_y 0.1 - \mathbf{a}_z 2.5.$$

The difference is

$$\mathbf{R} - \mathbf{R}' = -\mathbf{a}_x 0.4 - \mathbf{a}_y 0.1 + \mathbf{a}_z 0.2,$$

which has a magnitude

$$|\mathbf{R} - \mathbf{R}'| = [(-0.4)^2 + (-0.1)^2 + (0.2)^2]^{1/2} = 0.458 \text{ (m)}.$$

Substituting in Eq. (3–15), we obtain

$$\mathbf{E}_P = \left(\frac{1}{4\pi\epsilon_0}\right) \frac{q(\mathbf{R} - \mathbf{R}')}{|\mathbf{R} - \mathbf{R}'|^3}$$

$$= (9 \times 10^9) \frac{5 \times 10^{-9}}{0.458^3} (-\mathbf{a}_x 0.4 - \mathbf{a}_y 0.1 + \mathbf{a}_z 0.2)$$

$$= 214.5(-\mathbf{a}_x 0.873 - \mathbf{a}_y 0.218 + \mathbf{a}_z 0.437) \quad \text{(V/m)}.$$

The quantity within the parentheses is the unit vector $\mathbf{a}_{QP} = (\mathbf{R} - \mathbf{R}')/|\mathbf{R} - \mathbf{R}'|$, and $\mathbf{E}_P$ has a magnitude of 214.5 (V/m).   ▬

Note: The permittivity of air is essentially the same as that of the free space. The factor $1/(4\pi\epsilon_0)$ appears very frequently in electrostatics. From Eq. (1–11) we know that $\epsilon_0 = 1/(c^2\mu_0)$. But $\mu_0 = 4\pi \times 10^{-7}$ (H/m) in SI units; so

$$\frac{1}{4\pi\epsilon_0} = \frac{\mu_0 c^2}{4\pi} = 10^{-7} c^2 \quad \text{(m/F)} \tag{3–16}$$

exactly. If we use the approximate value $c = 3 \times 10^8$ (m/s), then $1/(4\pi\epsilon_0) = 9 \times 10^9$ (m/F).

When a point charge q_2 is placed in the field of another point charge q_1 at the origin, a force $\mathbf{F}_{12}$ is experienced by q_2 due to electric field intensity $\mathbf{E}_{12}$ of q_1 at q_2. Combining Eqs. (3–3) and (3–12), we have

$$\boxed{\mathbf{F}_{12} = q_2\mathbf{E}_{12} = \mathbf{a}_R \frac{q_1 q_2}{4\pi\epsilon_0 R^2} \quad \text{(N)}.} \tag{3–17}$$

Equation (3–17) is a mathematical form of **Coulomb's law** already stated in Section 3–1 in conjunction with Eq. (3–1). Note that the exponent on R is *exactly* 2, which is a consequence of the fundamental postulate Eq. (3–4). In SI units the proportionality constant k equals $1/(4\pi\epsilon_0)$, and the force is in newtons (N).

▬▬▬ **EXAMPLE 3–2**   A total charge Q is put on a thin spherical shell of radius b. Determine the electric field intensity at an arbitrary point inside the shell.

Solution   We shall solve this problem in two ways.

a) At any point, such as P, inside the hollow shell shown in Fig. 3–3, an arbitrary hypothetical closed surface (a **Gaussian surface**) may be drawn, over which we apply Gauss's law, Eq. (3–7). Since no charge exists inside the shell and the surface is arbitrary, we conclude readily that $\mathbf{E} = 0$ everywhere inside the shell.

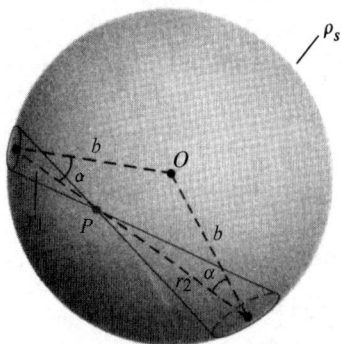

FIGURE 3–3
A charged shell (Example 3–2).

b) Let us now examine the problem in more detail. Draw a pair of elementary cones of solid angle $d\Omega$ with vertex at an arbitrary point P. The cones extend in both directions, intersecting the shell in areas ds_1 and ds_2 at distances r_1 and r_2, respectively, from the point P. Since charge Q distributes uniformly over the spherical shell, there is a uniform surface charge density

$$\rho_s = \frac{Q}{4\pi b^2}. \tag{3–18}$$

The magnitude of the electric field intensity at P due to charges on the elementary surfaces ds_1 and ds_2 is, from Eq. (3–12),

$$dE = \frac{\rho_s}{4\pi\epsilon_0}\left(\frac{ds_1}{r_1^2} - \frac{ds_2}{r_2^2}\right). \tag{3–19}$$

But the solid angle $d\Omega$ equals

$$d\Omega = \frac{ds_1}{r_1^2}\cos\alpha = \frac{ds_2}{r_2^2}\cos\alpha. \tag{3–20}$$

Combining the expressions of dE and $d\Omega$, we find that

$$dE = \frac{\rho_s}{4\pi\epsilon_0}\left(\frac{d\Omega}{\cos\alpha} - \frac{d\Omega}{\cos\alpha}\right) = 0. \tag{3–21}$$

Since the above result applies to every pair of elementary cones, we conclude that $\mathbf{E} = 0$ everywhere inside the conducting shell, as before. ◼

It will be noted that if Coulomb's law as expressed in Eq. (3–12) and used in Eq. (3–19) was slightly different from an inverse-square relation, the substitution of Eq. (3–20), which is a geometrical relation, in Eq. (3–19) would not yield the result $dE = 0$. Consequently, the electric field intensity inside the shell would not vanish; indeed, it would vary with the location of the point P. Coulomb originally used a torsion balance to conduct his experiments, which were necessarily of limited accuracy. Nevertheless, he was brilliant enough to *postulate* the inverse-square law. Many scientists subsequently made use of the vanishing field inside a spherical shell illustrated in this example to verify the inverse-square law. The field inside a charged shell, if it existed, could be detected to a very high accuracy by a probe through a small hole in the shell.

EXAMPLE 3–3 The electrostatic deflection system of a cathode-ray oscilloscope is depicted in Fig. 3–4. Electrons from a heated cathode are given an initial velocity $\mathbf{u}_0 = \mathbf{a}_z u_0$ by a positively charged anode (not shown). The electrons enter at $z = 0$ into a region of deflection plates where a uniform electric field $\mathbf{E}_d = -\mathbf{a}_y E_d$ is maintained over a width w. Ignoring gravitational effects, find the vertical deflection of the electrons on the fluorescent screen at $z = L$.

Solution Since there is no force in the z-direction in the $z > 0$ region, the horizontal velocity u_0 is maintained. The field $\mathbf{E}_d$ exerts a force on the electrons each carrying a charge $-e$, causing a deflection in the y-direction:

$$\mathbf{F} = (-e)\mathbf{E}_d = \mathbf{a}_y e E_d.$$

From Newton's second law of motion in the vertical direction we have

$$m\,\frac{du_y}{dt} = e E_d,$$

where m is the mass of an electron. Integrating both sides, we obtain

$$u_y = \frac{dy}{dt} = \frac{e}{m} E_d t,$$

where the constant of integration is set to zero because $u_y = 0$ at $t = 0$. Integrating again, we have

$$y = \frac{e}{2m} E_d t^2.$$

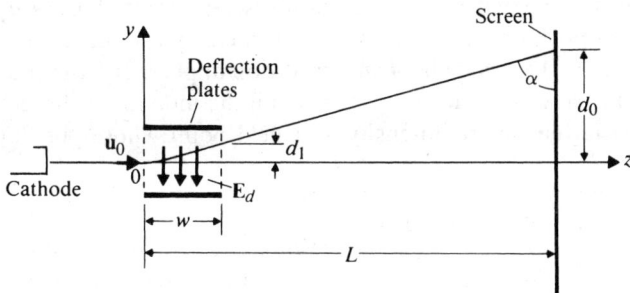

FIGURE 3–4
Electrostatic deflection system of a cathode-ray oscilloscope (Example 3–3)

The constant of integration is again zero because $y = 0$ at $t = 0$. Note that the electrons have a parabolic trajectory between the deflection plates. At the exit from the deflection plates, $t = w/u_0$,

$$d_1 = \frac{eE_d}{2m}\left(\frac{w}{u_0}\right)^2$$

and

$$u_{y1} = u_y\left(t = \frac{w}{u_0}\right) = \frac{eE_d}{m}\left(\frac{w}{u_0}\right).$$

When the electrons reach the screen, they have traveled a further horizontal distance of $(L - w)$ which takes $(L - w)/u_0$ seconds. During that time there is an additional vertical deflection

$$d_2 = u_{y1}\left(\frac{L - w}{u_0}\right) = \frac{eE_d}{m}\frac{w(L - w)}{u_0^2}.$$

Hence the deflection at the screen is

$$d_0 = d_1 + d_2 = \frac{eE_d}{mu_0^2}w\left(L - \frac{w}{2}\right).$$

    ■

Ink-jet printers used in computer output, like cathode-ray oscilloscopes, are devices based on the principle of electrostatic deflection of a stream of charged particles. Minute droplets of ink are forced through a vibrating nozzle controlled by a piezo-electric transducer. The output of the computer imparts variable amounts of charges on the ink droplets, which then pass through a pair of deflection plates where a uniform static electric field exists. The amount of droplet deflection depends on the charge it carries, causing the ink jet to strike the print surface and form an image as the print head moves in a horizontal direction.

3–3.1   ELECTRIC FIELD DUE TO A SYSTEM OF DISCRETE CHARGES

Suppose an electrostatic field is created by a group of n discrete point charges q_1, $q_2, \ldots, q_n$ located at different positions. Since electric field intensity is a linear function of (proportional to) $\mathbf{a}_R q/R^2$, the principle of superposition applies, and the total $\mathbf{E}$ field at a point is the *vector sum* of the fields caused by all the individual charges. From Eq. (3–15) we can write the electric intensity at a field point whose position vector is $\mathbf{R}$ as

$$\boxed{\mathbf{E} = \frac{1}{4\pi\epsilon_0}\sum_{k=1}^{n}\frac{q_k(\mathbf{R} - \mathbf{R}'_k)}{|\mathbf{R} - \mathbf{R}'_k|^3} \quad \text{(V/m)}.}\qquad (3\text{–}22)$$

Although Eq. (3–22) is a succinct expression, it is somewhat inconvenient to use because of the need to add vectors of different magnitudes and directions.

Let us consider the simple case of an **electric dipole** that consists of a pair of equal and opposite charges $+q$ and $-q$, separated by a small distance, d, as shown in Fig. 3–5. Let the center of the dipole coincide with the origin of a spherical coordinate system. Then the **E** field at the point P is the sum of the contributions due to $+q$ and $-q$. Thus,

$$\mathbf{E} = \frac{q}{4\pi\epsilon_0}\left\{\frac{\mathbf{R} - \dfrac{\mathbf{d}}{2}}{\left|\mathbf{R} - \dfrac{\mathbf{d}}{2}\right|^3} - \frac{\mathbf{R} + \dfrac{\mathbf{d}}{2}}{\left|\mathbf{R} + \dfrac{\mathbf{d}}{2}\right|^3}\right\}. \tag{3–23}$$

The first term on the right side of Eq. (3–23) can be simplified if $d \ll R$. We write

$$\begin{aligned}
\left|\mathbf{R} - \frac{\mathbf{d}}{2}\right|^{-3} &= \left[\left(\mathbf{R} - \frac{\mathbf{d}}{2}\right)\cdot\left(\mathbf{R} - \frac{\mathbf{d}}{2}\right)\right]^{-3/2}\\
&= \left[R^2 - \mathbf{R}\cdot\mathbf{d} + \frac{d^2}{4}\right]^{-3/2}\\
&\cong R^{-3}\left[1 - \frac{\mathbf{R}\cdot\mathbf{d}}{R^2}\right]^{-3/2}\\
&\cong R^{-3}\left[1 + \frac{3}{2}\frac{\mathbf{R}\cdot\mathbf{d}}{R^2}\right],
\end{aligned} \tag{3–24}$$

where the binomial expansion has been used and all terms containing the second and higher powers of (d/R) have been neglected. Similarly, for the second term on the right side of Eq. (3–23) we have

$$\left|\mathbf{R} + \frac{\mathbf{d}}{2}\right|^{-3} \cong R^{-3}\left[1 - \frac{3}{2}\frac{\mathbf{R}\cdot\mathbf{d}}{R^2}\right]. \tag{3–25}$$

Substitution of Eqs. (3–24) and (3–25) in Eq. (3–23) leads to

$$\mathbf{E} \cong \frac{q}{4\pi\epsilon_0 R^3}\left[3\frac{\mathbf{R}\cdot\mathbf{d}}{R^2}\mathbf{R} - \mathbf{d}\right]. \tag{3–26}$$

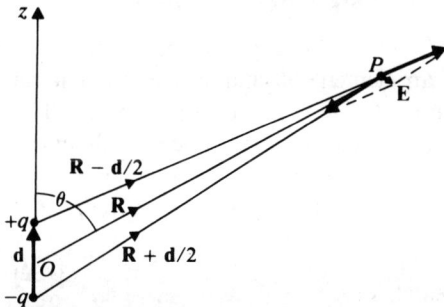

FIGURE 3–5
Electric field of a dipole.

The derivation and interpretation of Eq. (3–26) require the manipulation of vector quantities. We can appreciate that determining the electric field caused by three or more discrete charges will be even more tedious. In Section 3–5 we will introduce the concept of a scalar electric potential, with which the electric field intensity caused by a distribution of charges can be found more easily.

The electric dipole is an important entity in the study of the electric field in dielectric media. We define the product of the charge q and the vector $\mathbf{d}$ (going from $-q$ to $+q$) as the *electric dipole moment*, $\mathbf{p}$:

$$\mathbf{p} = q\mathbf{d}. \tag{3–27}$$

Equation (3–26) can then be rewritten as

$$\mathbf{E} = \frac{1}{4\pi\epsilon_0 R^3}\left[3\frac{\mathbf{R}\cdot\mathbf{p}}{R^2}\mathbf{R} - \mathbf{p}\right], \tag{3–28}$$

where the approximate sign ($\sim$) over the equal sign has been left out for simplicity. If the dipole lies along the z-axis as in Fig. 3–5, then (see Eq. 2–77)

$$\mathbf{p} = \mathbf{a}_z p = p(\mathbf{a}_R \cos\theta - \mathbf{a}_\theta \sin\theta), \tag{3–29}$$
$$\mathbf{R}\cdot\mathbf{p} = Rp\cos\theta, \tag{3–30}$$

and Eq. (3–28) becomes

$$\boxed{\mathbf{E} = \frac{p}{4\pi\epsilon_0 R^3}(\mathbf{a}_R 2\cos\theta + \mathbf{a}_\theta \sin\theta) \quad \text{(V/m)}.} \tag{3–31}$$

Equation (3–31) gives the electric field intensity of an electric dipole in spherical coordinates. We see that $\mathbf{E}$ of a dipole is inversely proportional to the cube of the distance R. This is reasonable because as R increases, the fields due to the closely spaced $+q$ and $-q$ tend to cancel each other more completely, thus decreasing more rapidly than that of a single point charge.

3–3.2 ELECTRIC FIELD DUE TO A CONTINUOUS DISTRIBUTION OF CHARGE

The electric field caused by a continuous distribution of charge can be obtained by integrating (superposing) the contribution of an element of charge over the charge distribution. Refer to Fig. 3–6, where a volume charge distribution is shown. The volume charge density ρ (C/m^3) is a function of the coordinates. Since a differential element of charge behaves like a point charge, the contribution of the charge $\rho\,dv'$ in a differential volume element dv' to the electric field intensity at the field point P is

$$d\mathbf{E} = \mathbf{a}_R \frac{\rho\,dv'}{4\pi\epsilon_0 R^2}. \tag{3–32}$$

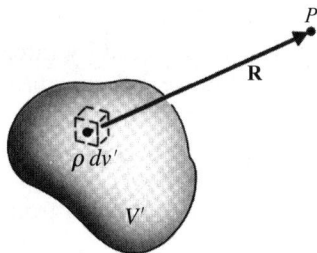

FIGURE 3-6
Electric field due to a continuous charge distribution.

We have

$$\mathbf{E} = \frac{1}{4\pi\epsilon_0} \int_{V'} \mathbf{a}_R \frac{\rho}{R^2} \, dv' \quad \text{(V/m)}, \tag{3-33}$$

or, since $\mathbf{a}_R = \mathbf{R}/R$,

$$\mathbf{E} = \frac{1}{4\pi\epsilon_0} \int_{V'} \rho \frac{\mathbf{R}}{R^3} \, dv' \quad \text{(V/m)}. \tag{3-34}$$

Except for some especially simple cases, the vector triple integral in Eq. (3-33) or Eq. (3-34) is difficult to carry out because, in general, all three quantities in the integrand ($\mathbf{a}_R$, ρ, and R) change with the location of the differential volume dv'.

If the charge is distributed on a surface with a surface charge density ρ_s (C/m^2), then the integration is to be carried out over the surface (not necessarily flat). Thus,

$$\mathbf{E} = \frac{1}{4\pi\epsilon_0} \int_{S'} \mathbf{a}_R \frac{\rho_s}{R^2} \, ds' \quad \text{(V/m)}. \tag{3-35}$$

For a line charge we have

$$\mathbf{E} = \frac{1}{4\pi\epsilon_0} \int_{L'} \mathbf{a}_R \frac{\rho_\ell}{R^2} \, d\ell' \quad \text{(V/m)}, \tag{3-36}$$

where ρ_ℓ (C/m) is the line charge density, and L' the line (not necessarily straight) along which the charge is distributed.

EXAMPLE 3-4 Determine the electric field intensity of an infinitely long, straight, line charge of a uniform density ρ_ℓ in air.

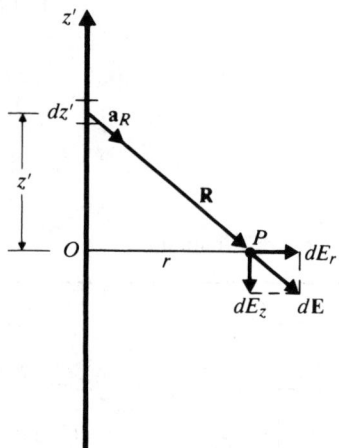

FIGURE 3–7
An infinitely long, straight, line charge.

Solution   Let us assume that the line charge lies along the z'-axis as shown in Fig. 3–7. (We are perfectly free to do this because the field obviously does not depend on how we designate the line. *It is an accepted convention to use primed coordinates for source points and unprimed coordinates for field points when there is a possibility of confusion.*) The problem asks us to find the electric field intensity at a point P, which is at a distance r from the line. Since the problem has a cylindrical symmetry (that is, the electric field is independent of the azimuth angle ϕ), it would be most convenient to work with cylindrical coordinates. We rewrite Eq. (3–36) as

$$
\boxed{\mathbf{E} = \frac{1}{4\pi\epsilon_0} \int_{L'} \rho_\ell \frac{\mathbf{R}}{R^3}\, d\ell' \quad \text{(V/m)}.}
\tag{3–37}
$$

For the problem at hand, ρ_ℓ is constant, and a line element $d\ell' = dz'$ is chosen to be at an arbitrary distance z' from the origin. It is most important to remember that $\mathbf{R}$ is the distance vector directed *from the source to the field point*, not the other way around. We have

$$
\mathbf{R} = \mathbf{a}_r r - \mathbf{a}_z z'.
\tag{3–38}
$$

The electric field, $d\mathbf{E}$, due to the differential line charge element $\rho_\ell\, d\ell' = \rho_\ell\, dz'$ is

$$
d\mathbf{E} = \frac{\rho_\ell\, dz'}{4\pi\epsilon_0} \frac{\mathbf{a}_r r - \mathbf{a}_z z'}{(r^2 + z'^2)^{3/2}}
\tag{3–39}
$$

$$
= \mathbf{a}_r\, dE_r + \mathbf{a}_z\, dE_z,
$$

where

$$
dE_r = \frac{\rho_\ell r\, dz'}{4\pi\epsilon_0 (r^2 + z'^2)^{3/2}}
\tag{3–39a}
$$

and

$$dE_z = \frac{-\rho_\ell z' \, dz'}{4\pi\epsilon_0 (r^2 + z'^2)^{3/2}}. \tag{3-39b}$$

In Eq. (3–39) we have decomposed $d\mathbf{E}$ into its components in the $\mathbf{a}_r$ and $\mathbf{a}_z$ directions. It is easy to see that for every $\rho_\ell \, dz'$ at $+z'$ there is a charge element $\rho_\ell \, dz'$ at $-z'$, which will produce a $d\mathbf{E}$ with components dE_r and $-dE_z$. Hence the $\mathbf{a}_z$ components will cancel in the integration process, and we only need to integrate the dE_r in Eq. (3–39a):

$$\mathbf{E} = \mathbf{a}_r E_r = \mathbf{a}_r \frac{\rho_\ell r}{4\pi\epsilon_0} \int_{-\infty}^{\infty} \frac{dz'}{(r^2 + z'^2)^{3/2}}$$

or

$$\boxed{\mathbf{E} = \mathbf{a}_r \frac{\rho_\ell}{2\pi\epsilon_0 r} \qquad (\text{V/m}).} \tag{3-40}$$

Equation (3–40) is an important result for an infinite line charge. Of course, no physical line charge is infinitely long; nevertheless, Eq. (3–40) gives the approximate $\mathbf{E}$ field of a long straight line charge at a point close to the line charge.

3–4   Gauss's Law and Applications

Gauss's law follows directly from the divergence postulate of electrostatics, Eq. (3–4), by the application of the divergence theorem. It was derived in Section 3–2 as Eq. (3–7) and is repeated here on account of its importance:

$$\boxed{\oint_S \mathbf{E} \cdot d\mathbf{s} = \frac{Q}{\epsilon_0}.} \tag{3-41}$$

Gauss's law asserts that the total outward flux of the E-field over any closed surface in free space is equal to the total charge enclosed in the surface divided by ϵ_0. We note that the surface S can be *any hypothetical (mathematical) closed surface chosen for convenience*; it does not have to be, and usually is not, a physical surface.

Gauss's law is particularly useful in determining the E-field of charge distributions with some symmetry conditions, such that *the normal component of the electric field intensity is constant over an enclosed surface.* In such cases the surface integral on the left side of Eq. (3–41) would be very easy to evaluate, and Gauss's law would be a much more efficient way for finding the electric field intensity than Eqs. (3–33) through (3–37). On the other hand, when symmetry conditions do not exist, Gauss's law would not be of much help. The essence of applying Gauss's law lies first in the recognition of symmetry conditions and second in the suitable choice of a surface over which the normal component of $\mathbf{E}$ resulting from a given charge distribution is a constant. Such a surface is referred to as a ***Gaussian surface***. This basic principle was used to obtain Eq. (3–12) for a point charge that possesses spherical symmetry; consequently, a proper Gaussian surface is the surface of a sphere centered at the point charge. Gauss's law could not help in the derivation of Eq. (3–26) or Eq. (3–31) for an electric dipole, since a surface about a separated pair of equal and opposite charges over which the normal component of **E** remains constant was not known.

EXAMPLE 3–5   Use Gauss's law to determine the electric field intensity of an infinitely long, straight, line charge of a uniform density ρ_ℓ in air.

Solution   This problem was solved in Example 3–4 by using Eq. (3–36). Since the line charge is infinitely long, the resultant **E** field must be radial and perpendicular to the line charge ($\mathbf{E} = \mathbf{a}_r E_r$), and a component of **E** along the line cannot exist. With the obvious cylindrical symmetry we construct a cylindrical Gaussian surface of a radius r and an arbitrary length L with the line charge as its axis, as shown in Fig. 3–8. On this surface, E_r is constant, and $d\mathbf{s} = \mathbf{a}_r r\, d\phi\, dz$ (from Eq. 2–53a). We have

$$\oint_S \mathbf{E} \cdot d\mathbf{s} = \int_0^L \int_0^{2\pi} E_r r\, d\phi\, dz = 2\pi r L E_r.$$

There is no contribution from the top or the bottom face of the cylinder because on the top face $d\mathbf{s} = \mathbf{a}_z r\, dr\, d\phi$ but **E** has no z-component there, making $\mathbf{E} \cdot d\mathbf{s} = 0$. Similarly for the bottom face. The total charge enclosed in the cylinder is $Q = \rho_\ell L$. Substitution into Eq. (3–41) gives us immediately

$$2\pi r L E_r = \frac{\rho_\ell L}{\epsilon_0}$$

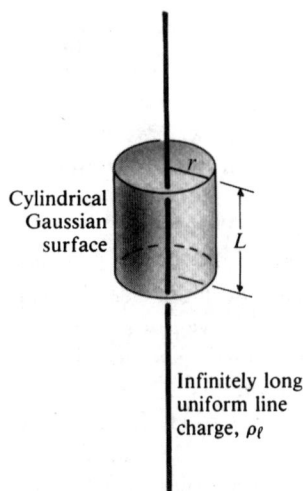

Cylindrical
Gaussian
surface

Infinitely long
uniform line
charge, ρ_ℓ

FIGURE 3–8
Applying Gauss's law to an infinitely long line charge (Example 3–5).

or

$$E = a_r E_r = a_r \frac{\rho_\ell}{2\pi\epsilon_0 r}.$$

This result is, of course, the same as that given in Eq. (3–40), but it is obtained here in a much simpler way. We note that the length L of the cylindrical Gaussian surface does not appear in the final expression; hence we could have chosen a cylinder of a unit length. ▬

EXAMPLE 3–6   Determine the electric field intensity of an infinite planar charge with a uniform surface charge density ρ_s.

Solution   It is clear that the **E** field caused by a charged sheet of an infinite extent is normal to the sheet. Equation (3–35) could be used to find **E**, but this would involve a double integration between infinite limits of a general expression of $1/R^2$. Gauss's law can be used to much advantage here.

We choose as the Gaussian surface a rectangular box with top and bottom faces of an arbitrary area A equidistant from the planar charge, as shown in Fig. 3–9. The sides of the box are perpendicular to the charged sheet. If the charged sheet coincides with the xy-plane, then on the top face,

$$\mathbf{E} \cdot d\mathbf{s} = (\mathbf{a}_z E_z) \cdot (\mathbf{a}_z \, ds) = E_z \, ds.$$

On the bottom face,

$$\mathbf{E} \cdot d\mathbf{s} = (-\mathbf{a}_z E_z) \cdot (-\mathbf{a}_z \, ds) = E_z \, ds.$$

Since there is no contribution from the side faces, we have

$$\oint_S \mathbf{E} \cdot d\mathbf{s} = 2E_z \int_A ds = 2E_z A.$$

The total charge enclosed in the box is $Q = \rho_s A$. Therefore,

$$2E_z A = \frac{\rho_s A}{\epsilon_0},$$

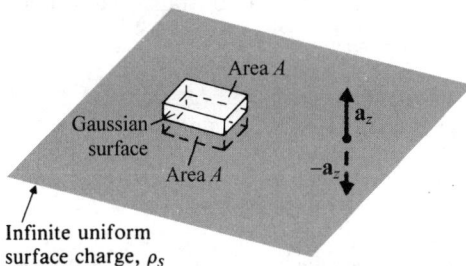

FIGURE 3–9
Applying Gauss's law to an infinite planar charge (Example 3–6).

from which we obtain

$$E = a_z E_z = a_z \frac{\rho_s}{2\epsilon_0}, \qquad z > 0, \qquad (3-42a)$$

and

$$E = -a_z E_z = -a_z \frac{\rho_s}{2\epsilon_0}, \qquad z < 0. \qquad (3-42b)$$

Of course, the charged sheet may not coincide with the *xy*-plane (in which case we do not speak in terms of above and below the plane), but the E field always points *away* from the sheet if ρ_s is *positive*. It is obvious that the Gaussian surface could have been a pillbox of any shape, not necessarily rectangular. ▬

The lighting scheme of an office or a classroom may consist of incandescent bulbs, long fluorescent tubes, or ceiling panel lights. These correspond roughly to point sources, line sources, and planar sources, respectively. From Eqs. (3–12), (3–40), and (3–42) we can estimate that light intensity will fall off rapidly as the square of the distance from the source in the case of incandescent bulbs, less rapidly as the first power of the distance for long fluorescent tubes, and not at all for ceiling panel lights.

▬▬▬ **EXAMPLE 3–7** Determine the E field caused by a spherical cloud of electrons with a volume charge density $\rho = -\rho_o$ for $0 \le R \le b$ (both ρ_o and b are positive) and $\rho = 0$ for $R > b$.

Solution First we recognize that the given source condition has spherical symmetry. The proper Gaussian surfaces must therefore be concentric spherical surfaces. We must find the E field in two regions. Refer to Fig. 3–10.

a) $0 \le R \le b$

A hypothetical spherical Gaussian surface S_i with $R < b$ is constructed within the electron cloud. On this surface, E is radial and has a constant magnitude:

$$E = a_R E_R, \qquad ds = a_R \, ds.$$

The total outward E flux is

$$\oint_{S_i} E \cdot ds = E_R \int_{S_i} ds = E_R 4\pi R^2.$$

The total charge enclosed within the Gaussian surface is

$$Q = \int_V \rho \, dv$$

$$= -\rho_o \int_V dv = -\rho_o \frac{4\pi}{3} R^3.$$

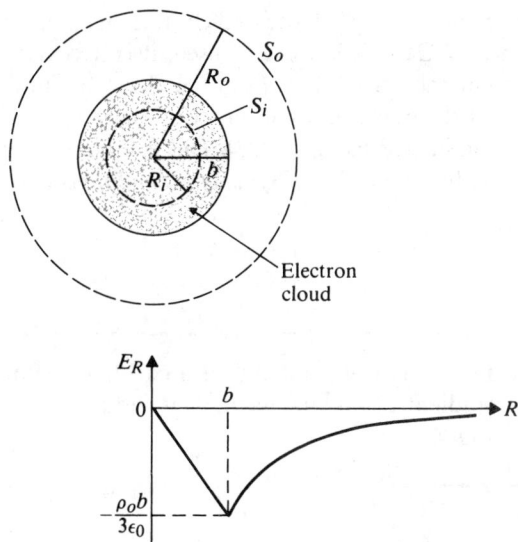

FIGURE 3–10
Electric field intensity of a spherical electron cloud (Example 3–7).

Substitution into Eq. (3–7) yields

$$\mathbf{E} = -\mathbf{a}_R \frac{\rho_o}{3\epsilon_0} R, \qquad 0 \le R \le b.$$

We see that within the uniform electron cloud the **E** field is directed toward the center and has a magnitude proportional to the distance from the center.

b) $R \ge b$

For this case we construct a spherical Gaussian surface S_o with $R > b$ outside the electron cloud. We obtain the same expression for $\oint_{S_o} \mathbf{E} \cdot d\mathbf{s}$ as in case (a). The total charge enclosed is

$$Q = -\rho_o \frac{4\pi}{3} b^3.$$

Consequently,

$$\mathbf{E} = -\mathbf{a}_R \frac{\rho_o b^3}{3\epsilon_0 R^2}, \qquad R \ge b,$$

which follows the inverse square law and could have been obtained directly from Eq. (3–12). We observe that *outside* the charged cloud the **E** field is exactly the same as though the total charge is concentrated on a single point charge at the center. This is true, in general, for a spherically symmetrical charged region even though ρ is a function of R. ▄

The variation of E_R versus R is plotted in Fig. 3–10. Note that the formal solution of this problem requires only a few lines. If Gauss's law is not used, it is necessary (1) to choose a differential volume element arbitrarily located in the electron cloud, (2) to express its vector distance $\mathbf{R}$ to a field point in a chosen coordinate system, and (3) to perform a triple integration as indicated in Eq. (3–33). This is a hopelessly involved process. The moral is: *Try to apply Gauss's law if symmetry conditions exist for the given charge distribution.*

3–5   Electric Potential

In connection with the null identity in Eq. (2–145) we noted that a curl-free vector field could always be expressed as the gradient of a scalar field. This induces us to *define* a scalar *electric potential V* such that

$$\boxed{\mathbf{E} = -\nabla V} \tag{3–43}$$

because scalar quantities are easier to handle than vector quantities. If we can determine V more easily, then $\mathbf{E}$ can be found by a gradient operation, which is a straightforward process in an orthogonal coordinate system. The reason for the inclusion of a negative sign in Eq. (3–43) will be explained presently.

Electric potential does have physical significance, and it is related to the work done in carrying a charge from one point to another. In Section 3–2 we defined the electric field intensity as the force acting on a unit test charge. Therefore in moving a unit charge from point P_1 to point P_2 in an electric field, work must be done *against the field* and is equal to

$$\frac{W}{q} = -\int_{P_1}^{P_2} \mathbf{E} \cdot d\boldsymbol{\ell} \qquad \text{(J/C or V)}. \tag{3–44}$$

Many paths may be followed in going from P_1 to P_2. Two such paths are drawn in Fig. 3–11. Since the path between P_1 and P_2 is not specified in Eq. (3–44), the

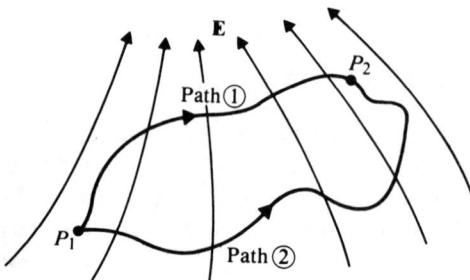

FIGURE 3–11
Two paths leading from P_1 to P_2 in an electric field.

question naturally arises, how does the work depend on the path taken? A little thought will lead us to conclude that W/q in Eq. (3-44) should not depend on the path; if it did, one would be able to go from P_1 to P_2 along a path for which W is smaller and then to come back to P_1 along another path, achieving a net gain in work or energy. This would be contrary to the principle of conservation of energy. We have already alluded to the path-independent nature of the scalar line integral of the irrotational (conservative) **E** field when we discussed Eq. (3-8).

Analogous to the concept of potential energy in mechanics, Eq. (3-44) represents the difference in electric potential energy of a unit charge between point P_2 and point P_1. Denoting the electric potential energy per unit charge by V, the **electric potential**, we have

$$V_2 - V_1 = -\int_{P_1}^{P_2} \mathbf{E} \cdot d\ell \qquad \text{(V)}. \qquad\qquad (3\text{-}45)$$

Mathematically, Eq. (3-45) can be obtained by substituting Eq. (3-43) in Eq. (3-44). Thus, in view of Eq. (2-88),

$$-\int_{P_1}^{P_2} \mathbf{E} \cdot d\ell = \int_{P_1}^{P_2} (\nabla V) \cdot (\mathbf{a}_\ell d\ell)$$

$$= \int_{P_1}^{P_2} dV = V_2 - V_1.$$

What we have defined in Eq. (3-45) is a **potential difference (electrostatic voltage)** between points P_2 and P_1. It makes no more sense to talk about the absolute potential of a point than about the absolute phase of a phasor or the absolute altitude of a geographical location; a reference zero-potential point, a reference zero phase (usually at $t = 0$), or a reference zero altitude (usually at sea level) must first be specified. In most (but not all) cases the zero-potential point is taken at infinity. When the reference zero-potential point is not at infinity, it should be specifically stated.

We want to make two more points about Eq. (3-43). First, the inclusion of the negative sign is necessary in order to conform with the convention that in going *against* the **E** field the electric potential V *increases*. For instance, when a d-c battery of a voltage V_0 is connected between two parallel conducting plates, as in Fig. 3-12, positive and negative charges cumulate on the top and bottom plates, respectively. The **E** field is directed from positive to negative charges, while the potential increases in the *opposite* direction. Second, we know from Section 2-6, when we defined the gradient of a scalar field, that the direction of ∇V is normal to the surfaces of constant

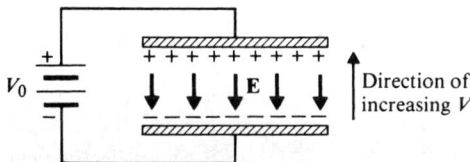

FIGURE 3-12
Relative directions of **E** and increasing V.

V. Hence if we use directed *field lines* or *streamlines* to indicate the direction of the **E** field, they are everywhere perpendicular to *equipotential lines* and *equipotential surfaces*.

3–5.1 ELECTRIC POTENTIAL DUE TO A CHARGE DISTRIBUTION

The electric potential of a point at a distance R from a point charge q referred to that at infinity can be obtained readily from Eq. (3–45):

$$V = - \int_{\infty}^{R} \left(\mathbf{a}_R \frac{q}{4\pi\epsilon_0 R^2} \right) \cdot (\mathbf{a}_R \, dR), \qquad (3\text{–}46)$$

which gives

$$\boxed{V = \frac{q}{4\pi\epsilon_0 R} \qquad \text{(V)}.} \qquad (3\text{–}47)$$

This is a scalar quantity and depends on, besides q, only the distance R. The potential difference between any two points P_2 and P_1 at distances R_2 and R_1, respectively, from q is

$$V_{21} = V_{P_2} - V_{P_1} = \frac{q}{4\pi\epsilon_0} \left(\frac{1}{R_2} - \frac{1}{R_1} \right). \qquad (3\text{–}48)$$

This result may appear a little surprising at first, since P_2 and P_1 may not lie on the same radial line through q, as illustrated in Fig. 3–13. However, the concentric circles (spheres) passing through P_2 and P_1 are equipotential lines (surfaces), and $V_{P_2} - V_{P_1}$ is the same as $V_{P_2} - V_{P_3}$. From the point of view of Eq. (3–45) we can choose the path of integration from P_1 to P_3 and then from P_3 to P_2. No work is done from P_1 to P_3 because **F** is perpendicular to $d\boldsymbol{\ell} = \mathbf{a}_\phi R_1 \, d\phi$ along the circular path ($\mathbf{E} \cdot d\boldsymbol{\ell} = 0$).

   The electric potential at **R** due to a system of n discrete point charges $q_1, q_2, \ldots, q_n$ located at $\mathbf{R}'_1, \mathbf{R}'_2, \ldots, \mathbf{R}'_n$ is, by superposition, the sum of the potentials due to

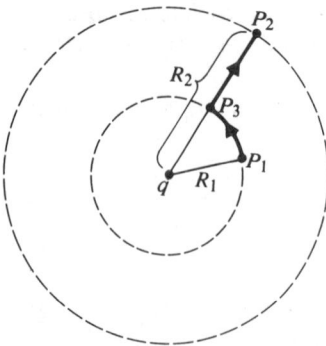

FIGURE 3–13
Path of integration about a point charge.

the individual charges:

$$V = \frac{1}{4\pi\epsilon_0} \sum_{k=1}^{n} \frac{q_k}{|\mathbf{R} - \mathbf{R}_k'|} \quad \text{(V).}$$

(3–49)

Since this is a scalar sum, it is, in general, easier to determine $\mathbf{E}$ by taking the negative gradient of V than from the vector sum in Eq. (3–22) directly.

As an example, let us again consider an electric dipole consisting of charges $+q$ and $-q$ with a small separation d. The distances from the charges to a field point P are designated R_+ and R_-, as shown in Fig. 3–14. The potential at P can be written down directly:

$$V = \frac{q}{4\pi\epsilon_0} \left(\frac{1}{R_+} - \frac{1}{R_-} \right).$$

(3–50)

If $d \ll R$, we have

$$\frac{1}{R_+} \cong \left(R - \frac{d}{2}\cos\theta \right)^{-1} \cong R^{-1} \left(1 + \frac{d}{2R}\cos\theta \right)$$

(3–51)

and

$$\frac{1}{R_-} \cong \left(R + \frac{d}{2}\cos\theta \right)^{-1} \cong R^{-1} \left(1 - \frac{d}{2R}\cos\theta \right).$$

(3–52)

Substitution of Eqs. (3–51) and (3–52) in Eq. (3–50) gives

$$V = \frac{qd\cos\theta}{4\pi\epsilon_0 R^2}$$

(3–53a)

or

$$V = \frac{\mathbf{p} \cdot \mathbf{a}_R}{4\pi\epsilon_0 R^2} \quad \text{(V),}$$

(3–53b)

where $\mathbf{p} = q\mathbf{d}$. (The "approximate" sign ($\sim$) has been dropped for simplicity.)

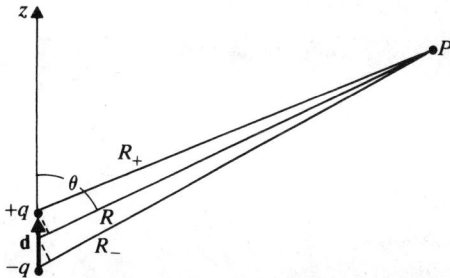

FIGURE 3–14
An electric dipole.

The **E** field can be obtained from $-\nabla V$. In spherical coordinates we have

$$\mathbf{E} = -\nabla V = -\mathbf{a}_R \frac{\partial V}{\partial R} - \mathbf{a}_\theta \frac{\partial V}{R \, \partial \theta}$$

$$= \frac{p}{4\pi\epsilon_0 R^3} (\mathbf{a}_R 2 \cos\theta + \mathbf{a}_\theta \sin\theta).$$

(3–54)

Equation (3–54) is the same as Eq. (3–31) but has been obtained by a simpler procedure without manipulating position vectors.

■■■■■ **EXAMPLE 3–8** Make a two-dimensional sketch of the equipotential lines and the electric field lines for an electric dipole.

Solution The equation of an equipotential surface of a charge distribution is obtained by setting the expression for V to equal a constant. Since q, d, and ϵ_0 in Eq. (3–53a) for an electric dipole are fixed quantities, a constant V requires a constant ratio $(\cos\theta/R^2)$. Hence the equation for an equipotential surface is

$$R = c_V \sqrt{\cos\theta},$$

(3–55)

where c_V is a constant. By plotting R versus θ for various values of c_V we draw the solid equipotential lines in Fig. 3–15. In the range $0 \le \theta \le \pi/2$, V is positive; R is maximum at $\theta = 0$ and zero at $\theta = 90°$. A mirror image is obtained in the range $\pi/2 \le \theta \le \pi$ where V is negative.

The electric field lines or streamlines represent the direction of the **E** field in space. We set

$$d\boldsymbol{\ell} = k\mathbf{E},$$

(3–56)

where k is a constant. In spherical coordinates, Eq. (3–56) becomes (see Eq. 2–66)

$$\mathbf{a}_R \, dR + \mathbf{a}_\theta R \, d\theta + \mathbf{a}_\phi R \sin\theta \, d\phi = k(\mathbf{a}_R E_R + \mathbf{a}_\theta E_\theta + \mathbf{a}_\phi E_\phi),$$

(3–57)

which can be written

$$\frac{dR}{E_R} = \frac{R \, d\theta}{E_\theta} = \frac{R \sin\theta \, d\phi}{E_\phi}.$$

(3–58)

For the electric dipole in Fig. 3–15 there is no E_ϕ component, and

$$\frac{dR}{2 \cos\theta} = \frac{R \, d\theta}{\sin\theta}$$

or

$$\frac{dR}{R} = \frac{2 \, d(\sin\theta)}{\sin\theta}.$$

(3–59)

Integrating Eq. (3–59), we obtain

$$R = c_E \sin^2\theta,$$

(3–60)

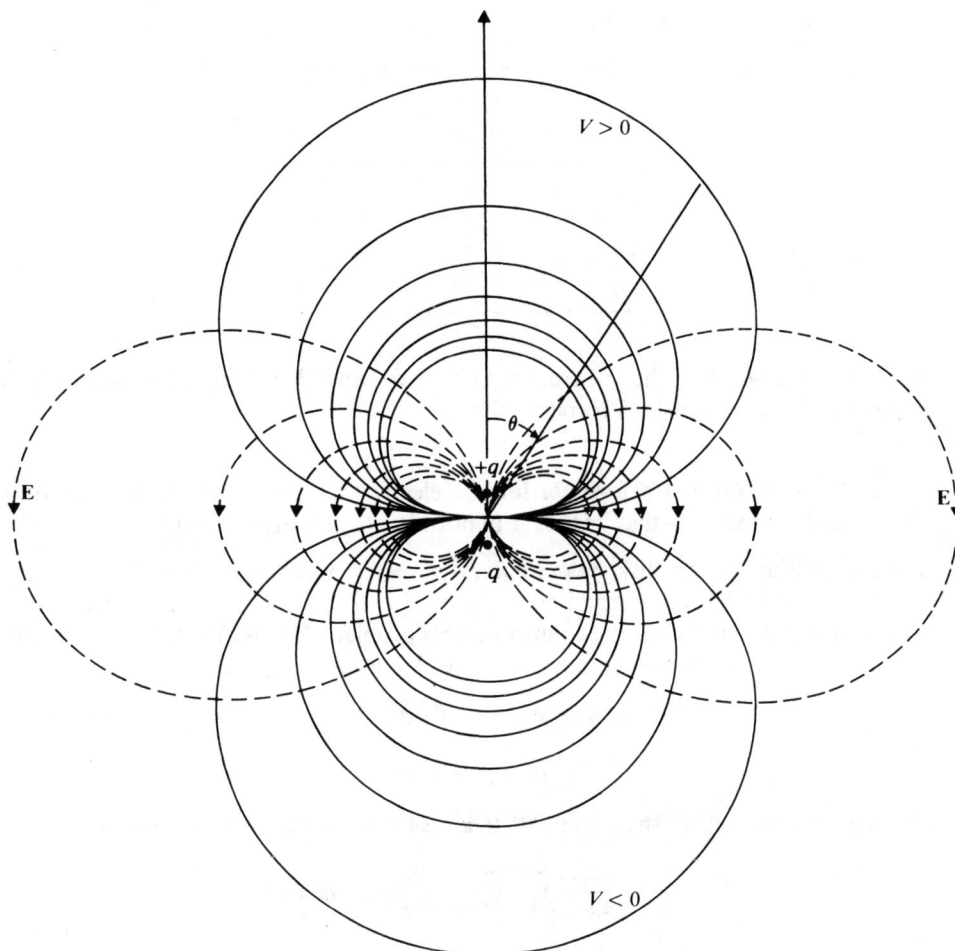

FIGURE 3–15
Equipotential and electric field lines of an electric dipole (Example 3–8).

where c_E is a constant. The electric field lines are drawn as dashed lines in Fig. 3–15. They are rotationally symmetrical about the z-axis (independent of ϕ) and are everywhere normal to the equipotential lines. ∎

The electric potential due to a continuous distribution of charge confined in a given region is obtained by integrating the contribution of an element of charge over the charged region. We have, for a volume charge distribution,

$$V = \frac{1}{4\pi\epsilon_0} \int_{V'} \frac{\rho}{R} \, dv' \quad \text{(V)}. \qquad (3\text{–}61)$$

For a surface charge distribution,

$$V = \frac{1}{4\pi\epsilon_0} \int_{S'} \frac{\rho_s}{R} \, ds' \quad \text{(V)};$$

(3–62)

and for a line charge,

$$V = \frac{1}{4\pi\epsilon_0} \int_{L'} \frac{\rho_\ell}{R} \, d\ell' \quad \text{(V)}.$$

(3–63)

We note here again that the integrals in Eqs. (3–61) and (3–62) represent integrations in three and two dimensions respectively.

EXAMPLE 3–9 Obtain a formula for the electric field intensity on the axis of a circular disk of radius b that carries a uniform surface charge density ρ_s.

Solution Although the disk has circular symmetry, we cannot visualize a surface around it over which the normal component of **E** has a constant magnitude; hence Gauss's law is not useful for the solution of this problem. We use Eq. (3–62). Working with cylindrical coordinates indicated in Fig. 3–16, we have

$$ds' = r' \, dr' \, d\phi'$$

and

$$R = \sqrt{z^2 + r'^2}.$$

The electric potential at the point $P(0, 0, z)$ referring to the point at infinity is

$$V = \frac{\rho_s}{4\pi\epsilon_0} \int_0^{2\pi} \int_0^b \frac{r'}{(z^2 + r'^2)^{1/2}} \, dr' \, d\phi'$$

$$= \frac{\rho_s}{2\epsilon_0} [(z^2 + b^2)^{1/2} - |z|].$$

(3–64)

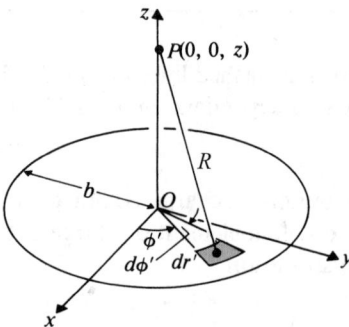

FIGURE 3–16
A uniformly charged disk (Example 3–9)

Therefore,

$$\mathbf{E} = -\nabla V = -\mathbf{a}_z \frac{\partial V}{\partial z}$$

$$= \begin{cases} \mathbf{a}_z \dfrac{\rho_s}{2\epsilon_0} [1 - z(z^2 + b^2)^{-1/2}], & z > 0 \qquad (3\text{–}65\text{a}) \\[4mm] -\mathbf{a}_z \dfrac{\rho_s}{2\epsilon_0} [1 + z(z^2 + b^2)^{-1/2}], & z < 0. \qquad (3\text{–}65\text{b}) \end{cases}$$

The determination of $\mathbf{E}$ field at an off-axis point would be a much more difficult problem. Do you know why?

For very large z, it is convenient to expand the second term in Eqs. (3–65a) and (3–65b) into a binomial series and neglect the second and all higher powers of the ratio (b^2/z^2). We have

$$z(z^2 + b^2)^{-1/2} = \left(1 + \frac{b^2}{z^2}\right)^{-1/2} \cong 1 - \frac{b^2}{2z^2}.$$

Substituting this into Eqs. (3–65a) and (3–65b), we obtain

$$\mathbf{E} = \mathbf{a}_z \frac{(\pi b^2 \rho_s)}{4\pi\epsilon_0 z^2}$$

$$= \begin{cases} \mathbf{a}_z \dfrac{Q}{4\pi\epsilon_0 z^2}, & z > 0 \qquad (3\text{–}66\text{a}) \\[4mm] -\mathbf{a}_z \dfrac{Q}{4\pi\epsilon_0 z^2}, & z < 0, \qquad (3\text{–}66\text{b}) \end{cases}$$

where Q is the total charge on the disk. Hence, when the point of observation is very far away from the charged disk, the $\mathbf{E}$ field approximately follows the inverse square law as if the total charge were concentrated at a point.   ▬

▬▬▬ **EXAMPLE 3–10**   Obtain a formula for the electric field intensity along the axis of a uniform line charge of length L. The uniform line-charge density is ρ_ℓ.

Solution   For an infinitely long line charge, the $\mathbf{E}$ field can be determined readily by applying Gauss's law, as in the solution to Example 3–5. However, for a line charge of finite length, as shown in Fig. 3–17, we cannot construct a Gaussian surface over which $\mathbf{E} \cdot d\mathbf{s}$ is constant. Gauss's law is therefore not useful here.

Instead, we use Eq. (3–63) by taking an element of charge $d\ell' = dz'$ at z'. The distance R from the charge element to the point $P(0, 0, z)$ along the axis of the line charge is

$$R = (z - z'), \qquad z > \frac{L}{2}.$$

Here it is extremely important to distinguish the position of the field point (un-primed coordinates) from the position of the source point (primed coordinates). We

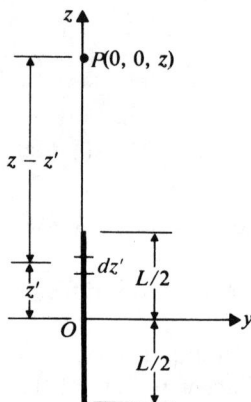

FIGURE 3-17
A finite line charge of a uniform line density ρ_ℓ (Example 3-10).

integrate over the source region:

$$V = \frac{\rho_\ell}{4\pi\epsilon_0} \int_{-L/2}^{L/2} \frac{dz'}{z - z'}$$

$$= \frac{\rho_\ell}{4\pi\epsilon_0} \ln\left[\frac{z + (L/2)}{z - (L/2)}\right], \qquad z > \frac{L}{2}. \tag{3-67}$$

The **E** field at P is the negative gradient of V with respect to the *unprimed* field coordinates. For this problem,

$$\mathbf{E} = -\mathbf{a}_z \frac{dV}{dz} = \mathbf{a}_z \frac{\rho_\ell L}{4\pi\epsilon_0[z^2 - (L/2)^2]}, \qquad z > \frac{L}{2}. \tag{3-68}$$

The preceding two examples illustrate the procedure for determining **E** by first finding V when Gauss's law cannot be conveniently applied. However, we emphasize that *if symmetry conditions exist such that a Gaussian surface can be constructed over which* **E** · *d***s** *is constant, it is always easier to determine* **E** *directly*. The potential V, if desired, may be obtained from **E** by integration.

3-6 Conductors in Static Electric Field

So far we have discussed only the electric field of stationary charge distributions in free space or air. We now examine the field behavior in material media. In general, we classify materials according to their electrical properties into three types: *conductors, semiconductors,* and *insulators* (or *dielectrics*). In terms of the crude atomic model of an atom consisting of a positively charged nucleus with orbiting electrons, the electrons in the outermost shells of the atoms of *conductors* are very loosely held and migrate easily from one atom to another. Most metals belong to this group. The electrons in the atoms of *insulators* or dielectrics, however, are confined to their orbits; they cannot be liberated in normal circumstances, even by the application of an external electric field. The electrical properties of *semiconductors* fall between those of conductors and insulators in that they possess a relatively small number of freely movable charges.

In terms of the band theory of solids we find that there are allowed energy bands for electrons, each band consisting of many closely spaced, discrete energy states. Between these energy bands there may be forbidden regions or gaps where no electrons of the solid's atom can reside. Conductors have an upper energy band partially filled with electrons or an upper pair of overlapping bands that are partially filled so that the electrons in these bands can move from one to another with only a small change in energy. Insulators or dielectrics are materials with a completely filled upper band, so conduction could not normally occur because of the existence of a large energy gap to the next higher band. If the energy gap of the forbidden region is relatively small, small amounts of external energy may be sufficient to excite the electrons in the filled upper band to jump into the next band, causing conduction. Such materials are semiconductors.

The macroscopic electrical property of a material medium is characterized by a constitutive parameter called *conductivity*, which we will define in Chapter 5. The definition of conductivity is not important in this chapter because we are not dealing with current flow and are now interested only in the behavior of static electric fields in material media. In this section we examine the electric field and charge distribution both inside the bulk and on the surface of a conductor.

Assume for the present that some positive (or negative) charges are introduced in the interior of a conductor. An electric field will be set up in the conductor, the field exerting a force on the charges and making them move away from one another. This movement will continue until *all* the charges reach the conductor surface and redistribute themselves in such a way that both the charge and the field inside vanish. Hence,

Inside a Conductor (Under Static Conditions)	
$\rho = 0$	(3–69)
$\mathbf{E} = 0$	(3–70)

When there is no charge in the interior of a conductor ($\rho = 0$), $\mathbf{E}$ must be zero because, according to Gauss's law, the total outward electric flux through *any* closed surface constructed inside the conductor must vanish.

The charge distribution on the surface of a conductor depends on the shape of the surface. Obviously, the charges would not be in a state of equilibrium if there were a tangential component of the electric field intensity that produces a tangential

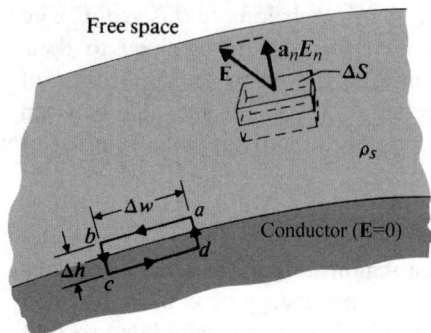

FIGURE 3–18
A conductor–free space interface.

force and moves the charges. Therefore, **under static conditions the E field on a conductor surface is everywhere normal to the surface**. In other words, **the surface of a conductor is an equipotential surface under static conditions**. As a matter of fact, since $\mathbf{E} = 0$ everywhere inside a conductor, the *whole* conductor has the same electrostatic potential. A finite time is required for the charges to redistribute on a conductor surface and reach the equilibrium state. This time depends on the conductivity of the material. For a good conductor such as copper this time is of the order of 10^{-19} (s), a very brief transient. (This point will be elaborated in Section 5–4.)

Figure 3–18 shows an interface between a conductor and free space. Consider the contour *abcda*, which has width $ab = cd = \Delta w$ and height $bc = da = \Delta h$. Sides *ab* and *cd* are parallel to the interface. Applying Eq. (3–8),[†] letting $\Delta h \rightarrow 0$, and noting that $\mathbf{E}$ in a conductor is zero, we obtain immediately

$$\oint_{abcda} \mathbf{E} \cdot d\ell = E_t \Delta w = 0$$

or

$$E_t = 0, \tag{3–71}$$

which says that **the tangential component of the E field on a conductor surface is zero**. In order to find E_n, the normal component of $\mathbf{E}$ at the surface of the conductor, we construct a Gaussian surface in the form of a thin pillbox with the top face in free space and the bottom face in the conductor where $\mathbf{E} = 0$. Using Eq. (3–7), we obtain

$$\oint_S \mathbf{E} \cdot d\mathbf{s} = E_n \Delta S = \frac{\rho_s \Delta S}{\epsilon_0}$$

or

$$E_n = \frac{\rho_s}{\epsilon_0}. \tag{3–72}$$

[†] We assume that Eqs. (3–7) and (3–8) are valid for regions containing discontinuous media.

Hence, *the normal component of the E field at a conductor/free space boundary is equal to the surface charge density on the conductor divided by the permittivity of free space*. Summarizing the *boundary conditions* at the conductor surface, we have

Boundary Conditions at a Conductor/Free Space Interface	
$E_t = 0$	(3–71)
$E_n = \dfrac{\rho_s}{\epsilon_0}$	(3–72)

When an uncharged conductor is placed in a static electric field, the external field will cause loosely held electrons inside the conductor to move in a direction opposite to that of the field and cause net positive charges to move in the direction of the field. These induced free charges will distribute on the conductor surface and create an *induced field* in such a way that they cancel the external field both inside the conductor and tangent to its surface. When the surface charge distribution reaches an equilibrium, all four relations, Eqs. (3–69) through (3–72), will hold; and the conductor is again an equipotential body.

EXAMPLE 3–11   A positive point charge Q is at the center of a spherical conducting shell of an inner radius R_i and an outer radius R_o. Determine **E** and V as functions of the radial distance R.

Solution   The geometry of the problem is shown in Fig. 3–19(a). Since there is spherical symmetry, it is simplest to use Gauss's law to determine **E** and then find V by integration. There are three distinct regions: (a) $R > R_o$, (b) $R_i < R < R_o$, and (c) $R < R_i$. Suitable spherical Gaussian surfaces will be constructed in these regions. Obviously, $\mathbf{E} = \mathbf{a}_R E_R$ in all three regions.

a) $R > R_o$ (Gaussian surface S_1):

$$\oint_S \mathbf{E} \cdot d\mathbf{s} = E_{R1} 4\pi R^2 = \frac{Q}{\epsilon_0}$$

or

$$E_{R1} = \frac{Q}{4\pi\epsilon_0 R^2}. \tag{3–73}$$

The **E** field is the same as that of a point charge Q without the presence of the shell. The potential referring to the point at infinity is

$$V_1 = -\int_\infty^R (E_{R1}) \, dR = \frac{Q}{4\pi\epsilon_0 R}. \tag{3–74}$$

b) $R_i < R < R_o$ (Gaussian surface S_2): Because of Eq. (3–70), we know that

$$E_{R2} = 0. \tag{3–75}$$

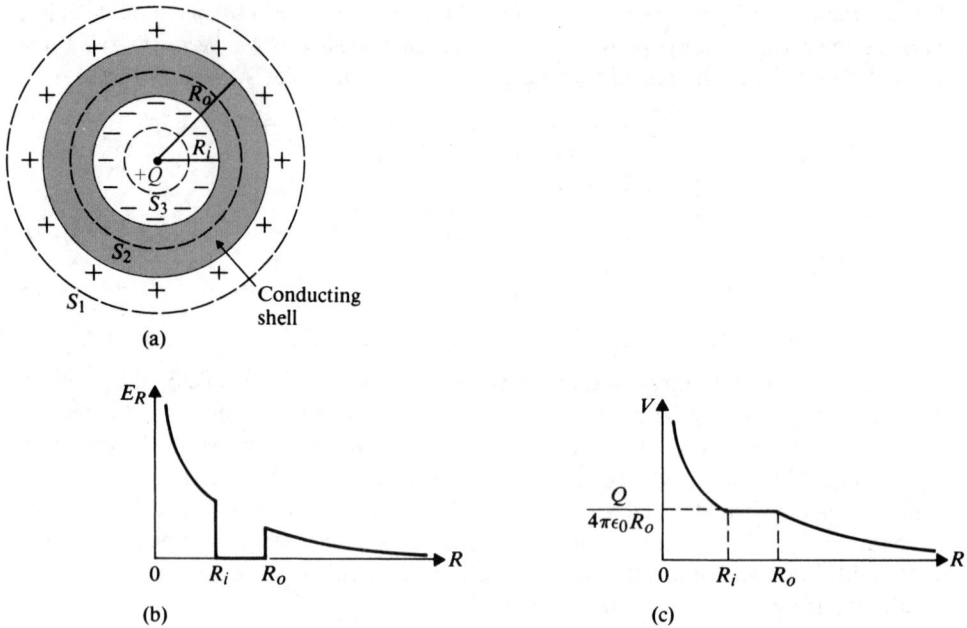

(a)

(b)

(c)

FIGURE 3-19
Electric field intensity and potential variations of a point charge $+Q$ at the center of a conducting shell (Example 3–11).

Since $\rho = 0$ in the conducting shell and since the total charge enclosed in surface S_2 must be zero, an amount of negative charge equal to $-Q$ must be induced on the inner shell surface at $R = R_i$. (This also means that an amount of positive charge equal to $+Q$ is induced on the outer shell surface at $R = R_o$.) The conducting shell is an equipotential body. Hence,

$$V_2 = V_1 \bigg|_{R = R_o} = \frac{Q}{4\pi\epsilon_0 R_o}. \tag{3–76}$$

c) $R < R_i$ (Gaussian surface S_3): Application of Gauss's law yields the same formula for E_{R3} as E_{R1} in Eq. (3–73) for the first region:

$$E_{R3} = \frac{Q}{4\pi\epsilon_0 R^2}. \tag{3–77}$$

The potential in this region is

$$V_3 = -\int E_{R3}\, dR + C = \frac{Q}{4\pi\epsilon_0 R} + C,$$

where the integration constant C is determined by requiring V_3 at $R = R_i$ to equal V_2 in Eq. (3–76). We have

$$C = \frac{Q}{4\pi\epsilon_0}\left(\frac{1}{R_o} - \frac{1}{R_i}\right)$$

and

$$V_3 = \frac{Q}{4\pi\epsilon_0}\left(\frac{1}{R} + \frac{1}{R_o} - \frac{1}{R_i}\right). \tag{3–78}$$

The variations of E_R and V versus R in all three regions are plotted in Figs. 3–19(b) and 3–19(c). Note that while the electric intensity has discontinuous jumps, the potential remains continuous. A discontinuous jump in potential would mean an infinite electric field intensity.   ▬

3–7   Dielectrics in Static Electric Field

Ideal dielectrics do not contain free charges. When a dielectric body is placed in an external electric field, there are no induced free charges that move to the surface and make the interior charge density and electric field vanish, as with conductors. However, since dielectrics contain *bound charges*, we cannot conclude that they have no effect on the electric field in which they are placed.

All material media are composed of atoms with a positively charged nucleus surrounded by negatively charged electrons. Although the molecules of dielectrics are macroscopically neutral, the presence of an external electric field causes a force to be exerted on each charged particle and results in small displacements of positive and negative charges in opposite directions. These displacements, though small in comparison to atomic dimensions, nevertheless *polarize* a dielectric material and create electric dipoles. The situation is depicted in Fig. 3–20. Inasmuch as electric dipoles do have nonvanishing electric potential and electric field intensity, we expect that the *induced electric dipoles* will modify the electric field both inside and outside the dielectric material.

The molecules of some dielectrics possess permanent dipole moments, even in the absence of an external polarizing field. Such molecules usually consist of two or

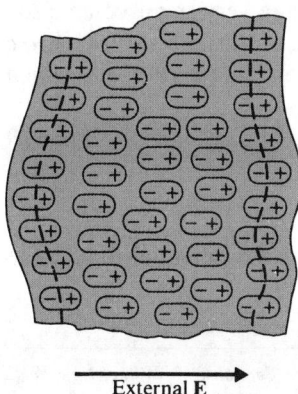

FIGURE 3–20
A cross section of a polarized dielectric medium.

External E

more dissimilar atoms and are called *polar molecules*, in contrast to **nonpolar molecules**, which do not have permanent dipole moments. An example is the water molecule H_2O, which consists of two hydrogen atoms and one oxygen atom. The atoms do not arrange themselves in a manner that makes the molecule have a zero dipole moment; that is, the hydrogen atoms do not lie exactly on diametrically opposite sides of the oxygen atom.

The dipole moments of polar molecules are of the order of 10^{-30} (C·m). When there is no external field, the individual dipoles in a polar dielectric are randomly oriented, producing no net dipole moment macroscopically. An applied electric field will exert a torque on the individual dipoles and tend to align them with the field in a manner similar to that shown in Fig. 3–20.

Some dielectric materials can exhibit a permanent dipole moment even in the absence of an externally applied electric field. Such materials are called *electrets*. Electrets can be made by heating (softening) certain waxes or plastics and placing them in an electric field. The polarized molecules in these materials tend to align with the applied field and to be frozen in their new positions after they return to normal temperatures. Permanent polarization remains without an external electric field. Electrets are the electrical equivalents of permanent magnets; they have found important applications in high fidelity electret microphones.[†]

3–7.1 EQUIVALENT CHARGE DISTRIBUTIONS OF POLARIZED DIELECTRICS

To analyze the macroscopic effect of induced dipoles we define a *polarization vector*, **P**, as

$$\mathbf{P} = \lim_{\Delta v \to 0} \frac{\sum_{k=1}^{n\Delta v} \mathbf{p}_k}{\Delta v} \quad (\text{C/m}^2), \qquad (3\text{–}79)$$

where n is the number of molecules per unit volume and the numerator represents the vector sum of the induced dipole moments contained in a very small volume Δv. The vector **P**, a smoothed point function, is the *volume density of electric dipole moment*. The dipole moment $d\mathbf{p}$ of an elemental volume dv' is $d\mathbf{p} = \mathbf{P}\, dv'$, which produces an electrostatic potential (see Eq. 3–53b):

$$dV = \frac{\mathbf{P} \cdot \mathbf{a}_R}{4\pi\epsilon_0 R^2}\, dv'. \qquad (3\text{–}80)$$

Integrating over the volume V' of the dielectric, we obtain the potential due to the polarized dielectric.

[†] See, for instance, J. M. Crowley, *Fundamentals of Applied Electrostatics, Section 8–3*, Wiley, New York, 1986.

$$V = \frac{1}{4\pi\epsilon_0} \int_{V'} \frac{\mathbf{P} \cdot \mathbf{a}_R}{R^2} \, dv', \tag{3-81}\dagger$$

where R is the distance from the elemental volume dv' to a fixed field point. In Cartesian coordinates,

$$R^2 = (x - x')^2 + (y - y')^2 + (z - z')^2, \tag{3-82}$$

and it is readily verified that the gradient of $1/R$ with respect to the *primed coordinates* is

$$\mathbf{V}'\left(\frac{1}{R}\right) = \frac{\mathbf{a}_R}{R^2}. \tag{3-83}$$

Hence Eq. (3–81) can be written as

$$V = \frac{1}{4\pi\epsilon_0} \int_{V'} \mathbf{P} \cdot \mathbf{V}'\left(\frac{1}{R}\right) dv'. \tag{3-84}$$

Recalling the vector identity (Problem 2–28),

$$\mathbf{V}' \cdot (f\mathbf{A}) = f\mathbf{V}' \cdot \mathbf{A} + \mathbf{A} \cdot \mathbf{V}'f, \tag{3-85}$$

and letting $\mathbf{A} = \mathbf{P}$ and $f = 1/R$, we can rewrite Eq. (3–84) as

$$V = \frac{1}{4\pi\epsilon_0} \left[\int_{V'} \mathbf{V}' \cdot \left(\frac{\mathbf{P}}{R}\right) dv' - \int_{V'} \frac{\mathbf{V}' \cdot \mathbf{P}}{R} \, dv' \right]. \tag{3-86}$$

The first volume integral on the right side of Eq. (3–86) can be converted into a closed surface integral by the divergence theorem. We have

$$V = \frac{1}{4\pi\epsilon_0} \oint_{S'} \frac{\mathbf{P} \cdot \mathbf{a}_n'}{R} \, ds' + \frac{1}{4\pi\epsilon_0} \int_{V'} \frac{(-\mathbf{V}' \cdot \mathbf{P})}{R} \, dv', \tag{3-87}$$

where $\mathbf{a}_n'$ is the outward normal from the surface element ds' of the dielectric. Comparison of the two integrals on the right side of Eq. (3–87) with Eqs. (3–62) and (3–61), respectively, reveals that the electric potential (and therefore the electric field intensity also) due to a polarized dielectric may be calculated from the contributions of surface and volume charge distributions having, respectively, densities

$$\boxed{\rho_{ps} = \mathbf{P} \cdot \mathbf{a}_n} \tag{3-88}\ddagger$$

and

$$\boxed{\rho_p = -\mathbf{V} \cdot \mathbf{P}.} \tag{3-89}\ddagger$$

---

† We note here that V on the left side of Eq. (3–81) represents the *electric potential* at a field point, and V' on the right side is the *volume* of the polarized dielectric.

‡ The prime sign on $\mathbf{a}_n$ and $\mathbf{V}$ has been dropped for simplicity, since Eqs. (3–88) and (3–89) involve only source coordinates and no confusion will result.

These are referred to as *polarization charge densities* or *bound-charge densities*. In other words, a *polarized dielectric may be replaced by an equivalent polarization surface charge density* ρ_{ps} *and an equivalent polarization volume charge density* ρ_p *for field calculations:*

$$V = \frac{1}{4\pi\epsilon_0} \oint_{S'} \frac{\rho_{ps}}{R}\, ds' + \frac{1}{4\pi\epsilon_0} \int_{V'} \frac{\rho_p}{R}\, dv'. \qquad (3\text{-}90)$$

Although Eqs. (3–88) and (3–89) were derived mathematically with the aid of a vector identity, a physical interpretation can be provided for the charge distributions. The sketch in Fig. 3–20 clearly indicates that charges from the ends of similarly oriented dipoles exist on surfaces not parallel to the direction of polarization. Consider an imaginary elemental surface Δs of a nonpolar dielectric. The application of an external electric field normal to Δs causes a separation d of the bound charges: positive charges $+q$ move a distance $d/2$ in the direction of the field, and negative charges $-q$ move an equal distance against the direction of the field. The net total charge ΔQ that crosses the surface Δs in the direction of the field is $nq\, d(\Delta s)$, where n is the number of molecules per unit volume. If the external field is not normal to Δs, the separation of the bound charges in the direction of $\mathbf{a}_n$ will be $\mathbf{d} \cdot \mathbf{a}_n$ and

$$\Delta Q = nq(\mathbf{d} \cdot \mathbf{a}_n)(\Delta s). \qquad (3\text{-}91)$$

But $nq\mathbf{d}$, the dipole moment per unit volume, is by definition the polarization vector $\mathbf{P}$. We have

$$\Delta Q = \mathbf{P} \cdot \mathbf{a}_n(\Delta s) \qquad (3\text{-}92)$$

and

$$\rho_{ps} = \frac{\Delta Q}{\Delta s} = \mathbf{P} \cdot \mathbf{a}_n,$$

as given in Eq. (3–88). Remember that $\mathbf{a}_n$ is always the *outward* normal. This relation correctly gives a positive surface charge on the right-hand surface in Fig. 3–20 and a negative surface charge on the left-hand surface.

For a surface S bounding a volume V, the net total charge flowing out of V as a result of polarization is obtained by integrating Eq. (3–92). The net charge *remaining* within the volume V is the *negative* of this integral:

$$\begin{aligned} Q &= -\oint_S \mathbf{P} \cdot \mathbf{a}_n\, ds \\ &= \int_V (-\nabla \cdot \mathbf{P})\, dv = \int_V \rho_p\, dv, \end{aligned} \qquad (3\text{-}93)$$

which leads to the expression for the volume charge density in Eq. (3–89). Hence, when the divergence of $\mathbf{P}$ does not vanish, the bulk of the polarized dielectric appears to be charged. However, since we started with an electrically neutral dielectric body, the total charge of the body after polarization must remain zero. This can be readily verified by noting that

$$\text{Total charge} = \oint_S \rho_{ps} \, ds + \int_V \rho_p \, dv$$
$$= \oint_S \mathbf{P} \cdot \mathbf{a}_n \, ds - \int_V \mathbf{\nabla} \cdot \mathbf{P} \, dv = 0, \tag{3-94}$$

where the divergence theorem has again been applied.

3-8 Electric Flux Density and Dielectric Constant

Because a polarized dielectric gives rise to an equivalent volume charge density ρ_p, we expect the electric field intensity due to a given source distribution in a dielectric to be different from that in free space. In particular, the divergence postulated in Eq. (3-4) must be modified to include the effect of ρ_p; that is,

$$\mathbf{\nabla} \cdot \mathbf{E} = \frac{1}{\epsilon_0} (\rho + \rho_p). \tag{3-95}$$

Using Eq. (3-89), we have

$$\mathbf{\nabla} \cdot (\epsilon_0 \mathbf{E} + \mathbf{P}) = \rho. \tag{3-96}$$

We now define a new fundamental field quantity, the *electric flux density*, or *electric displacement*, **D**, such that

$$\boxed{\mathbf{D} = \epsilon_0 \mathbf{E} + \mathbf{P} \qquad (\text{C/m}^2).} \tag{3-97}$$

The use of the vector **D** enables us to write a divergence relation between the electric field and the distribution of *free charges* in any medium without the necessity of dealing explicitly with the polarization vector **P** or the polarization charge density ρ_p. Combining Eqs. (3-96) and (3-97), we obtain the new equation

$$\boxed{\mathbf{\nabla} \cdot \mathbf{D} = \rho \qquad (\text{C/m}^3),} \tag{3-98}$$

where ρ is the volume density of *free charges*. Equations (3-98) and (3-5) are the two fundamental governing differential equations for electrostatics in any medium. Note that the permittivity of free space, ϵ_0, does not appear explicitly in these two equations.

The corresponding integral form of Eq. (3-98) is obtained by taking the volume integral of both sides. We have

$$\int_V \mathbf{\nabla} \cdot \mathbf{D} \, dv = \int_V \rho \, dv \tag{3-99}$$

or

$$\boxed{\oint_S \mathbf{D} \cdot d\mathbf{s} = Q \qquad (\text{C}).} \tag{3-100}$$

Equation (3–100), another form of **Gauss's law**, states that **the total outward flux of the electric displacement (or, simply, the total outward electric flux) over any closed surface is equal to the total free charge enclosed in the surface.** As was indicated in Section 3–4, Gauss's law is most useful in determining the electric field due to charge distributions under symmetry conditions.

When the dielectric properties of the medium are *linear* and *isotropic*, the polarization is directly proportional to the electric field intensity, and the proportionality constant is independent of the direction of the field. We write

$$\mathbf{P} = \epsilon_0 \chi_e \mathbf{E}, \tag{3–101}$$

where χ_e is a dimensionless quantity called **electric susceptibility**. A dielectric medium is linear if χ_e is independent of E and homogeneous if χ_e is independent of space coordinates. Substitution of Eq. (3–101) in Eq. (3–97) yields

$$\boxed{\begin{aligned} \mathbf{D} &= \epsilon_0(1 + \chi_e)\mathbf{E} \\ &= \epsilon_0\epsilon_r\mathbf{E} = \epsilon\mathbf{E} \qquad (\text{C/m}^2), \end{aligned}} \tag{3–102}$$

where

$$\epsilon_r = 1 + \chi_e = \frac{\epsilon}{\epsilon_0} \tag{3–103}$$

is a dimensionless quantity known as the **relative permittivity** or the **dielectric constant** of the medium. The coefficient $\epsilon = \epsilon_0\epsilon_r$ is the **absolute permittivity** (often called simply *permittivity*) of the medium and is measured in farads per meter (F/m). Air has a dielectric constant of 1.00059; hence its permittivity is usually taken as that of free space. The dielectric constants of some common materials are included in Table 3–1 on p. 114 and Appendix B–3.

Note that ϵ_r can be a function of space coordinates. If ϵ_r is independent of position, the medium is said to be **homogenous.** A linear, homogeneous, and isotropic medium is called a **simple medium.** The relative permittivity of a simple medium is a constant.

Later in the book we will learn that the effects of a lossy medium can be represented by a complex dielectric constant, whose imaginary part provides a measure of power loss in the medium and is, in general, frequency-dependent. For *anisotropic* materials the dielectric constant is different for different directions of the electric field, and **D** and **E** vectors generally have different directions; permittivity is a tensor. In matrix form we may write

$$\begin{bmatrix} D_x \\ D_y \\ D_z \end{bmatrix} = \begin{bmatrix} \epsilon_{11} & \epsilon_{12} & \epsilon_{13} \\ \epsilon_{21} & \epsilon_{22} & \epsilon_{23} \\ \epsilon_{31} & \epsilon_{32} & \epsilon_{33} \end{bmatrix} \begin{bmatrix} E_x \\ E_y \\ E_z \end{bmatrix}. \tag{3–104}$$

For crystals the reference coordinates can be chosen to be along the principal axes of the crystal so that the off-diagonal terms of the permittivity matrix in Eq. (3–104) are zero. We have

$$\begin{bmatrix} D_x \\ D_y \\ D_z \end{bmatrix} = \begin{bmatrix} \epsilon_1 & 0 & 0 \\ 0 & \epsilon_2 & 0 \\ 0 & 0 & \epsilon_3 \end{bmatrix} \begin{bmatrix} E_x \\ E_y \\ E_z \end{bmatrix}. \tag{3–105}$$

Media having the property represented by Eq. (3–105) are said to be **biaxial**. We may write

$$D_x = \epsilon_1 E_x, \tag{3–106a}$$
$$D_y = \epsilon_2 E_y, \tag{3–106b}$$
$$D_z = \epsilon_3 E_z. \tag{3–106c}$$

If further, $\epsilon_1 = \epsilon_2$, then the medium is said to be **uniaxial**. Of course, if $\epsilon_1 = \epsilon_2 = \epsilon_3$, we have an isotropic medium. We shall deal only with isotropic media in this book.

EXAMPLE 3–12   A positive point charge Q is at the center of a spherical dielectric shell of an inner radius R_i and an outer radius R_o. The dielectric constant of the shell is ϵ_r. Determine **E**, V, **D**, and **P** as functions of the radial distance R.

Solution   The geometry of this problem is the same as that of Example 3–11. The conducting shell has now been replaced by a dielectric shell, but the procedure of solution is similar. Because of the spherical symmetry, we apply Gauss's law to find **E** and **D** in three regions: (a) $R > R_o$; (b) $R_i < R < R_o$; and (c) $R < R_i$. Potential V is found from the negative line integral of **E**, and polarization **P** is determined by the relation

$$\mathbf{P} = \mathbf{D} - \epsilon_0 \mathbf{E} = \epsilon_0(\epsilon_r - 1)\mathbf{E}. \tag{3–107}$$

The **E**, **D**, and **P** vectors have only radial components. Refer to Fig. 3–21(a), where the Gaussian surfaces are not shown in order to avoid cluttering up the figure.

a) $R > R_o$

The situation in this region is exactly the same as that in Example 3–11. We have, from Eqs. (3–73) and (3–74),

$$E_{R1} = \frac{Q}{4\pi\epsilon_0 R^2}$$

$$V_1 = \frac{Q}{4\pi\epsilon_0 R}.$$

From Eqs. (3–102) and (3–107) we obtain

$$D_{R1} = \epsilon_0 E_{R1} = \frac{Q}{4\pi R^2} \tag{3–108}$$

and

$$P_{R1} = 0. \tag{3–109}$$

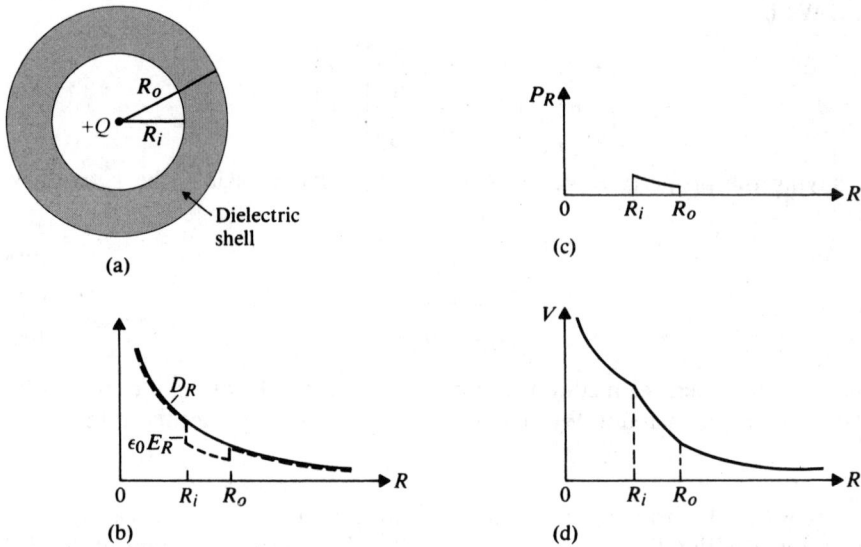

(a)

(b)

(c)

(d)

FIGURE 3–21
Field variations of a point charge $+Q$ at the center of a dielectric shell (Example 3–12).

b) $R_i < R < R_o$

The application of Gauss's law in this region gives us directly

$$E_{R2} = \frac{Q}{4\pi\epsilon_0\epsilon_r R^2} = \frac{Q}{4\pi\epsilon R^2},$$ (3–110)

$$D_{R2} = \frac{Q}{4\pi R^2},$$ (3–111)

$$P_{R2} = \left(1 - \frac{1}{\epsilon_r}\right)\frac{Q}{4\pi R^2}.$$ (3–112)

Note that D_{R2} has the same expression as D_{R1} and that both E_R and P_R have a discontinuity at $R = R_o$. In this region,

$$V_2 = -\int_{\infty}^{R_o} E_{R1}\, dR - \int_{R_o}^{R} E_{R2}\, dR$$

$$= V_1\big|_{R=R_o} - \frac{Q}{4\pi\epsilon}\int_{R_o}^{R} \frac{1}{R^2}\, dR$$ (3–113)

$$= \frac{Q}{4\pi\epsilon_0}\left[\left(1 - \frac{1}{\epsilon_r}\right)\frac{1}{R_o} + \frac{1}{\epsilon_r R}\right].$$

c) $R < R_i$

Since the medium in this region is the same as that in the region $R > R_o$, the application of Gauss's law yields the same expressions for E_R, D_R, and P_R in both regions:

$$E_{R3} = \frac{Q}{4\pi\epsilon_0 R^2},$$

$$D_{R3} = \frac{Q}{4\pi R^2},$$

$$P_{R3} = 0.$$

To find V_3, we must add to V_2 at $R = R_i$ the negative line integral of E_{R3}:

$$
\begin{aligned}
V_3 &= V_2\Big|_{R=R_i} - \int_{R_i}^{R} E_{R3}\, dR \\
&= \frac{Q}{4\pi\epsilon_0}\left[\left(1 - \frac{1}{\epsilon_r}\right)\frac{1}{R_o} - \left(1 - \frac{1}{\epsilon_r}\right)\frac{1}{R_i} + \frac{1}{R}\right].
\end{aligned}
\tag{3–114}
$$

The variations of $\epsilon_0 E_R$ and D_R versus R are plotted in Fig. 3–21(b). The difference $(D_R - \epsilon_0 E_R)$ is P_R and is shown in Fig. 3–21(c). The plot for V in Fig. 3–21(d) is a composite graph for V_1, V_2, and V_3 in the three regions. We note that D_R is a continuous curve exhibiting no sudden changes in going from one medium to another and that P_R exists only in the dielectric region. ▬

It is instructive to compare Figs. 3–21(b) and 3–21(d) with Figs. 3–19(b) and 3–19(c), respectively, of Example 3–11. From Eqs. (3–88) and (3–89) we find

$$
\begin{aligned}
\rho_{ps}\Big|_{R=R_i} &= \mathbf{P}\cdot(-\mathbf{a}_R)\Big|_{R=R_i} = -P_{R2}\Big|_{R=R_i} \\
&= -\left(1 - \frac{1}{\epsilon_r}\right)\frac{Q}{4\pi R_i^2}
\end{aligned}
\tag{3–115}
$$

on the inner shell surface,

$$
\begin{aligned}
\rho_{ps}\Big|_{R=R_o} &= \mathbf{P}\cdot\mathbf{a}_R\Big|_{R=R_o} = P_{R2}\Big|_{R=R_o} \\
&= \left(1 - \frac{1}{\epsilon_r}\right)\frac{Q}{4\pi R_o^2}
\end{aligned}
\tag{3–116}
$$

on the outer shell surface, and

$$
\begin{aligned}
\rho_p &= -\mathbf{\nabla}\cdot\mathbf{P} \\
&= -\frac{1}{R^2}\frac{\partial}{\partial R}(R^2 P_{R2}) = 0.
\end{aligned}
\tag{3–117}
$$

Equations (3–115), (3–116), and (3–117) indicate that there is no net polarization volume charge inside the dielectric shell. However, negative polarization surface charges exist on the inner surface and positive polarization surface charges on the outer surface. These surface charges produce an electric field intensity that is directed radially inward, thus reducing the **E** field in region 2 due to the point charge $+Q$ at the center.

TABLE 3-1
Dielectric Constants and Dielectric Strengths of Some Common Materials

Material	Dielectric Constant	Dielectric Strength (V/m)
Air (atmospheric pressure)	1.0	3×10^6
Mineral oil	2.3	15×10^6
Paper	2~4	15×10^6
Polystyrene	2.6	20×10^6
Rubber	2.3~4.0	25×10^6
Glass	4~10	30×10^6
Mica	6.0	200×10^6

3-8.1 DIELECTRIC STRENGTH

We have explained that an electric field causes small displacements of the bound charges in a dielectric material, resulting in polarization. If the electric field is very strong, it will pull electrons completely out of the molecules. The electrons will accelerate under the influence of the electric field, collide violently with the molecular lattice structure, and cause permanent dislocations and damage in the material. Avalanche effect of ionization due to collisions may occur. The material will become conducting, and large currents may result. This phenomenon is called a *dielectric breakdown*. The maximum electric field intensity that a dielectric material can withstand without breakdown is the *dielectric strength* of the material. The approximate dielectric strengths of some common substances are given in Table 3-1. The dielectric strength of a material must not be confused with its dielectric constant.

A convenient number to remember is that the dielectric strength of air at the atmospheric pressure is 3 kV/mm. When the electric field intensity exceeds this value, air breaks down. Massive ionization takes place, and sparking (corona discharge) follows. Charge tends to concentrate at sharp points. In view of Eq. (3–72), the electric field intensity in the immediate vicinity of sharp points is much higher than that at points on a relatively flat surface with a small curvature. This is the principle upon which a lightning arrester with a sharp metal lightning rod on top of tall buildings works. When a cloud containing an abundance of electric charges approaches a tall building equipped with a lightning rod connected to the ground, charges of an opposite sign are attracted from the ground to the tip of the rod, where the electric field intensity is the strongest. As the electric field intensity exceeds the dielectric strength of the wet air, breakdown occurs, and the air near the tip is ionized and becomes conducting. The electric charges in the cloud are then discharged safely to the ground through the conducting path.

The fact that the electric field intensity tends to be higher at a point near the surface of a charged conductor with a larger curvature is illustrated quantitatively in the following example.

EXAMPLE 3-13 Consider two spherical conductors with radii b_1 and b_2 ($b_2 > b_1$) that are connected by a conducting wire. The distance of separation between the conductors is assumed to be very large in comparison to b_2 so that the charges on the spherical conductors may be considered as uniformly distributed. A total charge Q is deposited on the spheres. Find (a) the charges on the two spheres, and (b) the electric field intensities at the sphere surfaces.

Solution

a) Refer to Fig. 3-22. Since the spherical conductors are at the same potential, we have

$$\frac{Q_1}{4\pi\epsilon_0 b_1} = \frac{Q_2}{4\pi\epsilon_0 b_2}$$

or

$$\frac{Q_1}{Q_2} = \frac{b_1}{b_2}.$$

Hence the charges on the spheres are directly proportional to their radii. But, since

$$Q_1 + Q_2 = Q,$$

we find that

$$Q_1 = \frac{b_1}{b_1 + b_2}Q \quad \text{and} \quad Q_2 = \frac{b_2}{b_1 + b_2}Q.$$

b) The electric field intensities at the surfaces of the two conducting spheres are

$$E_{1n} = \frac{Q_1}{4\pi\epsilon_0 b_1^2} \quad \text{and} \quad E_{2n} = \frac{Q_2}{4\pi\epsilon_0 b_2^2},$$

so

$$\frac{E_{1n}}{E_{2n}} = \left(\frac{b_2}{b_1}\right)^2 \frac{Q_1}{Q_2} = \frac{b_2}{b_1}.$$

The electric field intensities are therefore inversely proportional to the radii, being higher at the surface of the smaller sphere which has a larger curvature.

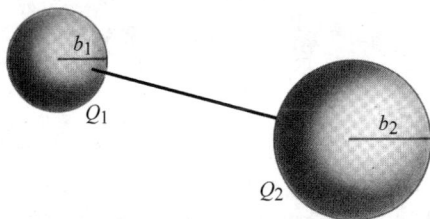

FIGURE 3-22
Two connected conducting spheres (Example 3-13).

3-9 Boundary Conditions for Electrostatic Fields

Electromagnetic problems often involve media with different physical properties and require the knowledge of the relations of the field quantities at an interface between two media. For instance, we may wish to determine how the **E** and **D** vectors change in crossing an interface. We already know the boundary conditions that must be satisfied at a conductor/free space interface. These conditions have been given in Eqs. (3–71) and (3–72). We now consider an interface between two general media shown in Fig. 3–23.

Let us construct a small path *abcda* with sides *ab* and *cd* in media 1 and 2, respectively, both being parallel to the interface and equal to Δw. Equation (3–8) is applied to this path. If we let sides $bc = da = \Delta h$ approach zero, their contributions to the line integral of **E** around the path can be neglected. We have

$$\oint_{abcda} \mathbf{E} \cdot d\ell = \mathbf{E}_1 \cdot \Delta\mathbf{w} + \mathbf{E}_2 \cdot (-\Delta\mathbf{w}) = E_{1t}\Delta w - E_{2t}\Delta w = 0.$$

Therefore

$$\boxed{E_{1t} = E_{2t} \qquad \text{(V/m),}} \qquad (3\text{–}118)$$

which states that **the tangential component of an E field is continuous across an interface**. Eq. (3–118) simplifies to Eq. (3–71) if one of the media is a conductor. When media 1 and 2 are dielectrics with permittivities ϵ_1 and ϵ_2, respectively, we have

$$\frac{D_{1t}}{\epsilon_1} = \frac{D_{2t}}{\epsilon_2}. \qquad (3\text{–}119)$$

In order to find a relation between the normal components of the fields at a boundary, we construct a small pillbox with its top face in medium 1 and bottom

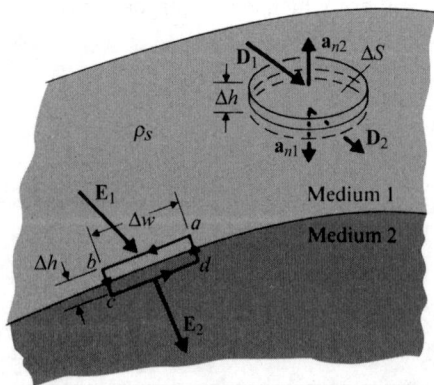

FIGURE 3–23
An interface between two media.

face in medium 2, as illustrated in Fig. 3–23. The faces have an area ΔS, and the height of the pillbox Δh is vanishingly small. Applying Gauss's law, Eq. (3–100), to the pillbox,[†] we have

$$\oint_S \mathbf{D} \cdot d\mathbf{s} = (\mathbf{D}_1 \cdot \mathbf{a}_{n2} + \mathbf{D}_2 \cdot \mathbf{a}_{n1}) \Delta S$$
$$= \mathbf{a}_{n2} \cdot (\mathbf{D}_1 - \mathbf{D}_2) \Delta S \qquad (3\text{–}120)$$
$$= \rho_s \Delta S,$$

where we have used the relation $\mathbf{a}_{n2} = -\mathbf{a}_{n1}$. Unit vectors $\mathbf{a}_{n1}$ and $\mathbf{a}_{n2}$ are, respectively, *outward* unit normals from media 1 and 2. From Eq. (3–120) we obtain

$$\boxed{\mathbf{a}_{n2} \cdot (\mathbf{D}_1 - \mathbf{D}_2) = \rho_s} \qquad (3\text{–}121\text{a})$$

or

$$\boxed{D_{1n} - D_{2n} = \rho_s \qquad (\text{C/m}^2),} \qquad (3\text{–}121\text{b})$$

where the reference unit normal is outward from medium 2.

Eq. (3–121b) states that *the normal component of **D** field is discontinuous across an interface where a surface charge exists—the amount of discontinuity being equal to the surface charge density.* If medium 2 is a conductor, $\mathbf{D}_2 = 0$ and Eq. (3–121b) becomes

$$D_{1n} = \epsilon_1 E_{1n} = \rho_s, \qquad (3\text{–}122)$$

which simplifies to Eq. (3–72) when medium 1 is free space.

When two dielectrics are in contact with *no free charges* at the interface, $\rho_s = 0$, we have

$$D_{1n} = D_{2n} \qquad (3\text{–}123)$$

or

$$\epsilon_1 E_{1n} = \epsilon_2 E_{2n}. \qquad (3\text{–}124)$$

Recapitulating, we find that the boundary conditions that must be satisfied for static electric fields are as follows:

$$\boxed{\begin{array}{ll} \text{Tangential components,} & E_{1t} = E_{2t}; \\ \text{Normal components,} & \mathbf{a}_{n2} \cdot (\mathbf{D}_1 - \mathbf{D}_2) = \rho_s. \end{array}} \qquad \begin{array}{l} (3\text{–}125) \\ (3\text{–}126) \end{array}$$

EXAMPLE 3–14   A lucite sheet ($\epsilon_r = 3.2$) is introduced perpendicularly in a uniform electric field $\mathbf{E}_o = \mathbf{a}_x E_o$ in free space. Determine $\mathbf{E}_i$, $\mathbf{D}_i$, and $\mathbf{P}_i$ inside the lucite.

[†] Equations (3–8) and (3–100) are assumed to hold for regions containing discontinuous media. See C. T. Tai, "On the presentation of Maxwell's theory," *Proceedings of the IEEE*, vol. 60, pp. 936–945, August 1972.

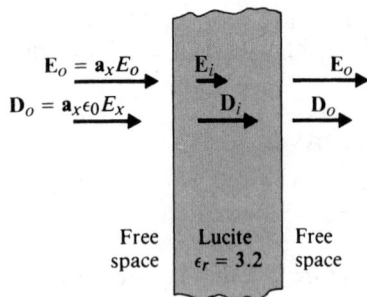

Free | Lucite | Free
space | $\epsilon_r = 3.2$ | space

FIGURE 3-24
A lucite sheet in a uniform electric field (Example 3-14).

Solution   We assume that the introduction of the lucite sheet does not disturb the original uniform electric field $\mathbf{E}_o$. The situation is depicted in Fig. 3-24. Since the interfaces are perpendicular to the electric field, only the normal field components need be considered. No free charges exist.

Boundary condition Eq. (3-123) at the left interface gives

$$\mathbf{D}_i = \mathbf{a}_x D_i = \mathbf{a}_x D_o$$

or

$$\mathbf{D}_i = \mathbf{a}_x \epsilon_0 E_o.$$

There is no change in electric flux density across the interface. The electric field intensity inside the lucite sheet is

$$\mathbf{E}_i = \frac{1}{\epsilon}\,\mathbf{D}_i = \frac{1}{\epsilon_0 \epsilon_r}\,\mathbf{D}_i = \mathbf{a}_x \frac{E_o}{3.2}.$$

The polarization vector is zero outside the lucite sheet ($\mathbf{P}_o = 0$). Inside the sheet,

$$\mathbf{P}_i = \mathbf{D}_i - \epsilon_0 \mathbf{E}_i = \mathbf{a}_x \left(1 - \frac{1}{3.2}\right)\epsilon_0 E_o$$

$$= \mathbf{a}_x 0.6875\epsilon_0 E_o \quad (\text{C/m}^2).$$ ▬

Clearly, a similar application of the boundary condition Eq. (3-123) on the right interface will yield the original $\mathbf{E}_o$ and $\mathbf{D}_o$ in the free space on the right of the lucite sheet. Does the solution of this problem change if the original electric field is not uniform; that is, if $\mathbf{E}_o = \mathbf{a}_x E(y)$?

▬▬▬ **EXAMPLE 3-15**   Two dielectric media with permittivities ϵ_1 and ϵ_2 are separated by a charge-free boundary as shown in Fig. 3-25. The electric field intensity in medium 1 at the point P_1 has a magnitude E_1 and makes an angle α_1 with the normal. Determine the magnitude and direction of the electric field intensity at point P_2 in medium 2.

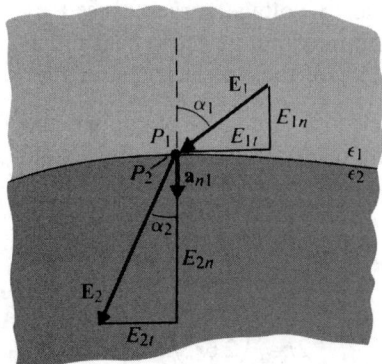

FIGURE 3–25
Boundary conditions at the interface between two dielectric media (Example 3–15).

Solution   Two equations are needed to solve for two unknowns E_{2t} and E_{2n}. After E_{2t} and E_{2n} have been found, E_2 and α_2 will follow directly. Using Eqs. (3–118) and (3–123), we have

$$E_2 \sin \alpha_2 = E_1 \sin \alpha_1 \tag{3–127}$$

and

$$\epsilon_2 E_2 \cos \alpha_2 = \epsilon_1 E_1 \cos \alpha_1. \tag{3–128}$$

Division of Eq. (3–127) by Eq. (3–128) gives

$$\boxed{\frac{\tan \alpha_2}{\tan \alpha_1} = \frac{\epsilon_2}{\epsilon_1}.} \tag{3–129}$$

The magnitude of $\mathbf{E}_2$ is

$$E_2 = \sqrt{E_{2t}^2 + E_{2n}^2} = \sqrt{(E_2 \sin \alpha_2)^2 + (E_2 \cos \alpha_2)^2}$$

$$= \left[(E_1 \sin \alpha_1)^2 + \left(\frac{\epsilon_1}{\epsilon_2} E_1 \cos \alpha_1 \right)^2 \right]^{1/2}$$

or

$$\boxed{E_2 = E_1 \left[\sin^2 \alpha_1 + \left(\frac{\epsilon_1}{\epsilon_2} \cos \alpha_1 \right)^2 \right]^{1/2}.} \tag{3–130}$$

By examining Fig. 3–25, can you tell whether ϵ_1 is larger or smaller than ϵ_2?   ■

■ EXAMPLE 3–16   When a coaxial cable is used to carry electric power, the radius of the inner conductor is determined by the load current, and the overall size by the voltage and the type of insulating material used. Assume that the radius of the inner conductor is 0.4 (cm) and that concentric layers of rubber ($\epsilon_{rr} = 3.2$) and polystyrene ($\epsilon_{rp} = 2.6$) are used as insulating materials. Design a cable that is to work at a voltage rating of 20 (kV). In order to avoid breakdown due to voltage surges caused by lightning and other abnormal external conditions, the maximum electric field intensities in the insulating materials are not to exceed 25% of their dielectric strengths.

Solution From Table 3–1, p. 114, we find the dielectric strengths of rubber and polystyrene to be 25×10^6 (V/m) and 20×10^6 (V/m), respectively. Using Eq. (3–40) for specified 25% of dielectric strengths, we have the following.

In rubber: $\quad$ Max $E_r = 0.25 \times 25 \times 10^6 = \dfrac{\rho_\ell}{2\pi\epsilon_0}\left(\dfrac{1}{3.2 r_i}\right).$ $\qquad$ (3–131a)

In polystyrene: $\quad$ Max $E_p = 0.25 \times 20 \times 10^6 = \dfrac{\rho_\ell}{2\pi\epsilon_0}\left(\dfrac{1}{2.6 r_p}\right).$ $\qquad$ (3–131b)

Combination of Eqs. (3–131a) and (3–131b) yields

$$r_p = 1.54 r_i = 0.616 \qquad \text{(cm)}. \qquad\qquad (3\text{–}132)$$

Equation (3–132) indicates that the insulating layer of polystyrene should be placed outside of that of rubber, as shown in Fig. 3–26(a). (It would be interesting to determine what would happen if the polystyrene layer were placed inside the rubber layer.)

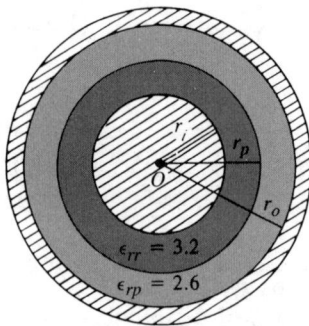

(a)

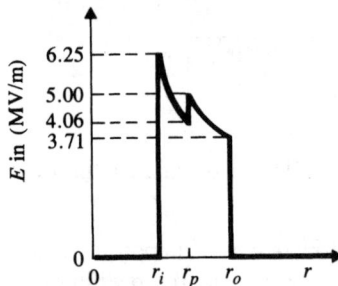

(b)

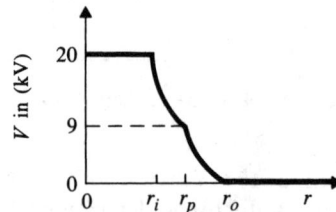

(c)

FIGURE 3–26
Cross section of coaxial cable with two different kinds of insulating material (Example 3–16).

The cable is to work at a potential difference of 20,000 (V) between the inner and outer conductors. We set

$$-\int_{r_o}^{r_p} E_p\, dr - \int_{r_p}^{r_i} E_r\, dr = 20{,}000,$$

where both E_p and E_r have the form given in Eq. (3–40). The above relation leads to

$$\frac{\rho_\ell}{2\pi\epsilon_0}\left(\frac{1}{\epsilon_{rp}}\ln\frac{r_o}{r_p} + \frac{1}{\epsilon_{rr}}\ln\frac{r_p}{r_i}\right) = 20{,}000$$

or

$$\frac{\rho_\ell}{2\pi\epsilon_0}\left(\frac{1}{2.6}\ln\frac{r_o}{1.54 r_i} + \frac{1}{3.2}\ln 1.54\right) = 20{,}000. \tag{3–133}$$

Since $r_i = 0.4$ (cm) is given, r_o can be determined by finding the factor $\rho_\ell/2\pi\epsilon_0$ from Eq. (3–131a) and then using it in Eq. (3–133). We obtain $\rho_\ell/2\pi\epsilon_0 = 8 \times 10^4$, and $r_o = 2.08 r_i = 0.832$ (cm).

In Figs. 3–26(b) and 3–26(c) are plotted the variations of the radial electric field intensity E and the potential V referred to that of the outer sheath. Note that E has discontinuous jumps, while the V curve is continuous. The reader should verify all the indicated numerical values. ▬

3–10 Capacitance and Capacitors

From Section 3–6 we understand that a conductor in a static electric field is an equipotential body and that charges deposited on a conductor will distribute themselves on its surface in such a way that the electric field inside vanishes. Suppose the potential due to a charge Q is V. Obviously, increasing the total charge by some factor k would merely increase the surface charge density ρ_s everywhere by the same factor without affecting the charge distribution because the conductor remains an equipotential body in a static situation. We may conclude from Eq. (3–62) that the potential of an isolated conductor is directly proportional to the total charge on it. This may also be seen from the fact that increasing V by a factor of k increases $\mathbf{E} = -\nabla V$ by a factor of k. But from Eq. (3–72), $\mathbf{E} = \mathbf{a}_n \rho_s/\epsilon_0$; it follows that ρ_s, and consequently the total charge Q will also increase by a factor of k. The ratio Q/V therefore remains unchanged. We write

$$\boxed{Q = CV,} \tag{3–134}$$

where the constant of proportionality C is called the **capacitance** of the isolated conducting body. The capacitance is the electric charge that must be added to the body per unit increase in its electric potential. Its SI unit is coulomb per volt, or farad (F).

Of considerable importance in practice is the **capacitor**, which consists of two conductors separated by free space or a dielectric medium. The conductors may be of arbitrary shapes as in Fig. 3–27. When a d-c voltage source is connected between the conductors, a charge transfer occurs, resulting in a charge $+Q$ on one conductor

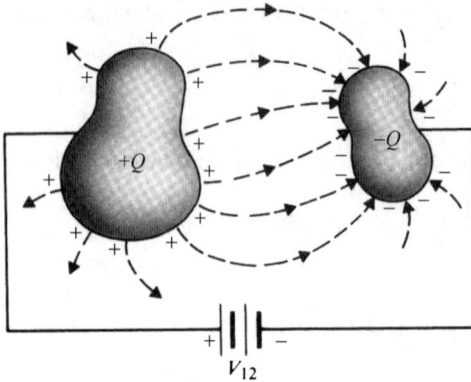

FIGURE 3–27
A two-conductor capacitor.

and $-Q$ on the other. Several electric field lines originating from positive charges and terminating on negative charges are shown in Fig. 3–27. Note that the field lines are perpendicular to the conductor surfaces, which are equipotential surfaces. Equation (3–134) applies here if V is taken to mean the potential difference between the two conductors, V_{12}. That is,

$$C = \frac{Q}{V_{12}} \quad \text{(F)}. \qquad\qquad (3\text{–}135)$$

The capacitance of a capacitor is a physical property of the two-conductor system. It depends on the geometry of the conductors and on the permittivity of the medium between them; it does *not* depend on either the charge Q or the potential difference V_{12}. A capacitor has a capacitance even when no voltage is applied to it and no free charges exist on its conductors. Capacitance C can be determined from Eq. (3–135) by either (1) assuming a V_{12} and determining Q in terms of V_{12}, or (2) assuming a Q and determining V_{12} in terms of Q. At this stage, since we have not yet studied the methods for solving boundary-value problems (which will be taken up in Chapter 4), we find C by the second method. The procedure is as follows:

1. Choose an appropriate coordinate system for the given geometry.
2. Assume charges $+Q$ and $-Q$ on the conductors.
3. Find **E** from Q by Eq. (3–122), Gauss's law, or other relations.
4. Find V_{12} by evaluating

$$V_{12} = -\int_2^1 \mathbf{E} \cdot d\ell$$

   from the conductor carrying $-Q$ to the other carrying $+Q$.
5. Find C by taking the ratio Q/V_{12}.

■■■■■ **EXAMPLE 3–17** A parallel-plate capacitor consists of two parallel conducting plates of area S separated by a uniform distance d. The space between the plates is filled with a dielectric of a constant permittivity ϵ. Determine the capacitance.

Solution A cross section of the capacitor is shown in Fig. 3–28. It is obvious that the appropriate coordinate system to use is the Cartesian coordinate system. Following the procedure outlined above, we put charges $+Q$ and $-Q$ on the upper and lower conducting plates, respectively. The charges are assumed to be uniformly distributed over the conducting plates with surface densities $+\rho_s$ and $-\rho_s$, where

$$\rho_s = \frac{Q}{S}.$$

From Eq. (3–122) we have

$$\mathbf{E} = -\mathbf{a}_y \frac{\rho_s}{\epsilon} = -\mathbf{a}_y \frac{Q}{\epsilon S},$$

which is constant within the dielectric if the fringing of the electric field at the edges of the plates is neglected. Now

$$V_{12} = -\int_{y=0}^{y=d} \mathbf{E} \cdot d\boldsymbol{\ell} = -\int_0^d \left(-\mathbf{a}_y \frac{Q}{\epsilon S}\right) \cdot (\mathbf{a}_y \, dy) = \frac{Q}{\epsilon S} d.$$

Therefore, *for a parallel-plate capacitor*,

$$\boxed{C = \frac{Q}{V_{12}} = \epsilon \frac{S}{d},} \qquad (3\text{–}136)$$

which is independent of Q or V_{12}. ■

For this problem we could have started by assuming a potential difference V_{12} between the upper and lower plates. The electric field intensity between the plates is uniform and equals

$$\mathbf{E} = -\mathbf{a}_y \frac{V_{12}}{d}.$$

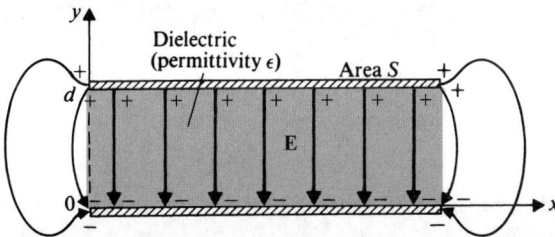

FIGURE 3–28
Cross section of a parallel-plate capacitor (Example 3–17).

The surface charge densities at the upper and lower conducting plates are $+\rho_s$ and $-\rho_s$, respectively, where, in view of Eq. (3–72),

$$\rho_s = \epsilon E_y = \epsilon \frac{V_{12}}{d}.$$

Therefore, $Q = \rho_s S = (\epsilon S/d)V_{12}$ and $C = Q/V_{12} = \epsilon S/d$, as before.

EXAMPLE 3–18   A cylindrical capacitor consists of an inner conductor of radius a and an outer conductor whose inner radius is b. The space between the conductors is filled with a dielectric of permittivity ϵ, and the length of the capacitor is L. Determine the capacitance of this capacitor.

Solution   We use cylindrical coordinates for this problem. First we assume charges $+Q$ and $-Q$ on the surface of the inner conductor and the inner surface of the outer conductor, respectively. The **E** field in the dielectric can be obtained by applying Gauss's law to a cylindrical Gaussian surface within the dielectric $a < r < b$. (Note that Eq. (3–122) gives only the *normal component* of the **E** field *at* a conductor surface. Since the conductor surfaces are not planes here, the **E** field is not constant in the dielectric and Eq. (3–122) cannot be used to find **E** in the $a < r < b$ region.) Referring to Fig. 3–29 and applying Gauss's law, we have

$$\mathbf{E} = \mathbf{a}_r E_r = \mathbf{a}_r \frac{Q}{2\pi\epsilon L r}. \tag{3–137}$$

Again we neglect the fringing effect of the field near the edges of the conductors. The potential difference between the inner and outer conductors is

$$
\begin{aligned}
V_{ab} &= -\int_{r=b}^{r=a} \mathbf{E} \cdot d\boldsymbol{\ell} = -\int_b^a \left(\mathbf{a}_r \frac{Q}{2\pi\epsilon L r} \right) \cdot (\mathbf{a}_r \, dr) \\
&= \frac{Q}{2\pi\epsilon L} \ln\left(\frac{b}{a} \right).
\end{aligned}
\tag{3–138}
$$

Dielectric, ϵ

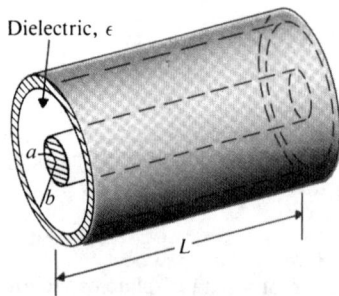

FIGURE 3–29
A cylindrical capacitor (Example 3–18).

Therefore, *for a cylindrical capacitor,*

$$C = \frac{Q}{V_{ab}} = \frac{2\pi\epsilon L}{\ln\left(\dfrac{b}{a}\right)}. \tag{3-139}$$

We could not solve this problem from an assumed V_{ab} because the electric field is not uniform between the inner and outer conductors. Thus we would not know how to express $\mathbf{E}$ and Q in terms of V_{ab} until we learned how to solve such a boundary-value problem. ■

■ **EXAMPLE 3–19**   A spherical capacitor consists of an inner conducting sphere of radius R_i and an outer conductor with a spherical inner wall of radius R_o. The space in between is filled with a dielectric of permittivity ϵ. Determine the capacitance.

Solution   Assume charges $+Q$ and $-Q$ on the inner and outer conductors, respectively, of the spherical capacitor in Fig. 3–30. Applying Gauss's law to a spherical Gaussian surface with radius $R(R_i < R < R_o)$, we have

$$\mathbf{E} = \mathbf{a}_R E_R = \mathbf{a}_R \frac{Q}{4\pi\epsilon R^2}$$

$$V = -\int_{R_o}^{R_i} \mathbf{E} \cdot (\mathbf{a}_R \, dR) = -\int_{R_o}^{R_i} \frac{Q}{4\pi\epsilon R^2} \, dR = \frac{Q}{4\pi\epsilon}\left(\frac{1}{R_i} - \frac{1}{R_o}\right).$$

Therefore, *for a spherical capacitor,*

$$C = \frac{Q}{V} = \frac{4\pi\epsilon}{\dfrac{1}{R_i} - \dfrac{1}{R_o}}. \tag{3-140}$$

■

For an isolated conducting sphere of a radius R_i, $R_o \to \infty$, $C = 4\pi\epsilon R_i$.

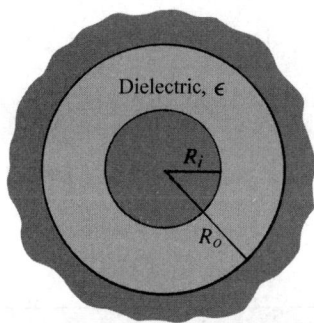

FIGURE 3–30
A spherical capacitor (Example 3–19).

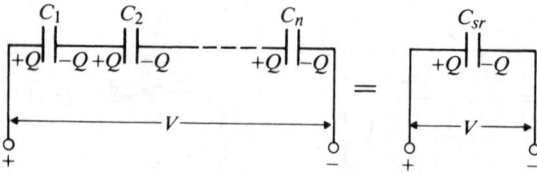

FIGURE 3–31
Series connection of capacitors.

3–10.1 SERIES AND PARALLEL CONNECTIONS OF CAPACITORS

Capacitors are often combined in various ways in electric circuits. The two basic ways are series and parallel connections. In the series, or head-to-tail, connection shown in Fig. 3–31,[†] the external terminals are from the first and last capacitors only. When a potential difference or electrostatic voltage V is applied, charge cumulations on the conductors connected to the external terminals are $+Q$ and $-Q$. Charges will be induced on the internally connected conductors such that $+Q$ and $-Q$ will appear on each capacitor independently of its capacitance. The potential differences across the individual capacitors are $Q/C_1, Q/C_2, \ldots, Q/C_n$, and

$$V = \frac{Q}{C_{sr}} = \frac{Q}{C_1} + \frac{Q}{C_2} + \cdots + \frac{Q}{C_n},$$

where C_{sr} is the equivalent capacitance of the series-connected capacitors. We have

$$\boxed{\frac{1}{C_{sr}} = \frac{1}{C_1} + \frac{1}{C_2} + \cdots + \frac{1}{C_n}.}$$
(3–141)

In the parallel connection of capacitors the external terminals are connected to the conductors of all the capacitors as in Fig. 3–32. When a potential difference V is applied to the terminals, the charge cumulated on a capacitor depends on its capacitance. The total charge is the sum of all the charges.

$$Q = Q_1 + Q_2 + \cdots + Q_n$$
$$= C_1 V + C_2 V + \cdots + C_n V = C_{||} V$$

Therefore, the equivalent capacitance of the parallel-connected capacitors is

$$\boxed{C_{||} = C_1 + C_2 + \cdots + C_n.}$$
(3–142)

[†] Capacitors, whatever their actual shape, are conventionally represented in circuits by pairs of parallel bars.

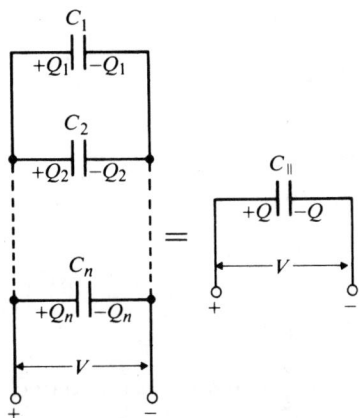

FIGURE 3–32
Parallel connection of capacitors.

We note that the formula for the equivalent capacitance of series-connected capacitors is similar to that for the equivalent resistance of parallel-connected resistors and that the formula for the equivalent capacitance of parallel-connected capacitors is similar to that for the equivalent resistance of series-connected resistors. Can you explain this?

EXAMPLE 3–20 Four capacitors $C_1 = 1$ (μF), $C_2 = 2$ (μF), $C_3 = 3$ (μF), and $C_4 = 4$ (μF) are connected as in Fig. 3–33. A d-c voltage of 100 (V) is applied to the external terminals $a–b$. Determine the following: (a) the total equivalent capacitance between terminals $a–b$, (b) the charge on each capacitor, and (c) the potential difference across each capacitor.

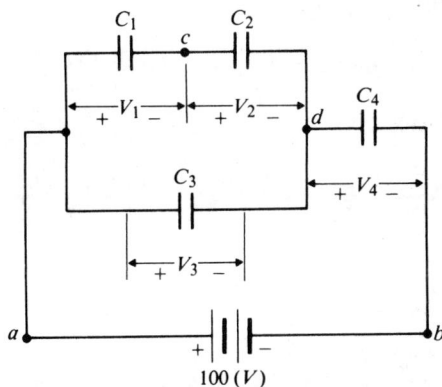

FIGURE 3–33
A combination of capacitors (Example 3–20).

Solution

a) The equivalent capacitance C_{12} of C_1 and C_2 in series is

$$C_{12} = \frac{1}{(1/C_1) + (1/C_2)} = \frac{C_1 C_2}{C_1 + C_2} = \frac{2}{3} \quad (\mu F).$$

The combination of C_{12} in parallel with C_3 gives

$$C_{123} = C_{12} + C_3 = \tfrac{11}{3} \quad (\mu F).$$

The total equivalent capacitance C_{ab} is then

$$C_{ab} = \frac{C_{123} C_4}{C_{123} + C_4} = \frac{44}{23} = 1.913 \quad (\mu F).$$

b) Since the capacitances are given, the voltages can be found as soon as the charges have been determined. We have four unknowns: $Q_1, Q_2, Q_3,$ and Q_4. Four equations are needed for their determination.

Series connection of C_1 and C_2: $\qquad\qquad\qquad Q_1 = Q_2.$

Kirchhoff's voltage law, $V_1 + V_2 = V_3$: $\qquad \dfrac{Q_1}{C_1} + \dfrac{Q_2}{C_2} = \dfrac{Q_3}{C_3}.$

Kirchhoff's voltage law, $V_3 + V_4 = 100$: $\qquad \dfrac{Q_3}{C_3} + \dfrac{Q_4}{C_4} = 100.$

Series connection at d: $\qquad\qquad\qquad\qquad Q_2 + Q_3 = Q_4.$

Using the given values of $C_1, C_2, C_3,$ and C_4 and solving the equations, we obtain

$$Q_1 = Q_2 = \frac{800}{23} = 34.8 \quad (\mu C),$$

$$Q_3 = \frac{3600}{23} = 156.5 \quad (\mu C),$$

$$Q_4 = \frac{4400}{23} = 191.3 \quad (\mu C).$$

c) Dividing the charges by the capacitances, we find

$$V_1 = \frac{Q_1}{C_1} = 34.8 \quad (V),$$

$$V_2 = \frac{Q_2}{C_2} = 17.4 \quad (V),$$

$$V_3 = \frac{Q_3}{C_3} = 52.2 \quad (V),$$

$$V_4 = \frac{Q_4}{C_4} = 47.8 \quad (V).$$

These results can be checked by verifying that $V_1 + V_2 = V_3$ and that $V_3 + V_4 = 100$ (V). ∎

3–10.2   CAPACITANCES IN MULTICONDUCTOR SYSTEMS

We now consider the situation of more than two conducting bodies in an isolated system, such as that shown in Fig. 3–34. The positions of the conductors are arbitrary, and one of the conductors may represent the ground. Obviously, the presence of a charge on any one of the conductors will affect the potential of all the others. Since the relation between potential and charge is linear, we may write the following set of N equations relating the potentials $V_1, V_2, \ldots, V_N$ of the N conductors to the charges $Q_1, Q_2, \ldots, Q_N$:

$$
\begin{aligned}
V_1 &= p_{11}Q_1 + p_{12}Q_2 + \cdots + p_{1N}Q_N, \\
V_2 &= p_{21}Q_1 + p_{22}Q_2 + \cdots + p_{2N}Q_N, \\
&\ \vdots \\
V_N &= p_{N1}Q_1 + p_{N2}Q_2 + \cdots + p_{NN}Q_N.
\end{aligned}
\tag{3–143}
$$

In Eqs. (3–143) the p_{ij}'s are called the **coefficients of potential**, which are constants whose values depend on the shape and position of the conductors as well as the permittivity of the surrounding medium. We note that in an isolated system,

$$
Q_1 + Q_2 + Q_3 + \cdots + Q_N = 0.
\tag{3–144}
$$

The N linear equations in (3–143) can be inverted to express the charges as functions of potentials as follows:

$$
\begin{aligned}
Q_1 &= c_{11}V_1 + c_{12}V_2 + \cdots + c_{1N}V_N, \\
Q_2 &= c_{21}V_1 + c_{22}V_2 + \cdots + c_{2N}V_N, \\
&\ \vdots \\
Q_N &= c_{N1}V_1 + c_{N2}V_2 + \cdots + c_{NN}V_N,
\end{aligned}
\tag{3–145}
$$

where the c_{ij}'s are constants whose values depend only on the p_{ij}'s in Eqs. (3–143). The coefficients c_{ii}'s are called the **coefficients of capacitance**, which equal the ratios of the charge Q_i on and the potential V_i of the ith conductor ($i = 1, 2, \ldots, N$) with all other conductors grounded. The c_{ij}'s ($i \neq j$) are called the **coefficients of induction**. If a positive Q_i exists on the ith conductor, V_i will be positive, but the charge Q_j

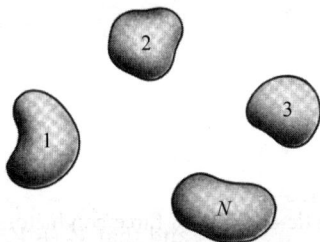

FIGURE 3–34
A multiconductor system.

induced on the jth $(j \neq i)$ conductor will be negative. Hence the coefficients of capacitance c_{ii} are positive, and the coefficients of induction c_{ij} are negative. The condition of reciprocity guarantees that $p_{ij} = p_{ji}$ and $c_{ij} = c_{ji}$.

To establish a physical meaning to the coefficients of capacitance and the coefficients of induction, let us consider a four-conductor system as depicted in Fig. 3–34 with the stipulation that the conductor labeled N is now the conducting earth at zero potential and is designated by the number 0. A schematic diagram of the four-conductor system is shown in Fig. 3–35, in which the conductors 1, 2, and 3 have been drawn as simple dots (nodes). Coupling capacitances have been shown between pairs of nodes and between the three nodes and the ground. If Q_1, Q_2, Q_3 and V_1, V_2, V_3 denote the charges and the potentials, respectively, of conductors 1, 2, and 3, the first three equations in (3–145) become

$$Q_1 = c_{11}V_1 + c_{12}V_2 + c_{13}V_3, \tag{3–146a}$$
$$Q_2 = c_{12}V_1 + c_{22}V_2 + c_{23}V_3, \tag{3–146b}$$
$$Q_3 = c_{13}V_1 + c_{23}V_2 + c_{33}V_3, \tag{3–146c}$$

where we have used the symmetry relation $c_{ij} = c_{ji}$. On the other hand, we can write another set of three $Q \sim V$ relations based on the schematic diagram in Fig. 3–35:

$$Q_1 = C_{10}V_1 + C_{12}(V_1 - V_2) + C_{13}(V_1 - V_3), \tag{3–147a}$$
$$Q_2 = C_{20}V_2 + C_{12}(V_2 - V_1) + C_{23}(V_2 - V_3), \tag{3–147b}$$
$$Q_3 = C_{30}V_3 + C_{13}(V_3 - V_1) + C_{23}(V_3 - V_2), \tag{3–147c}$$

where C_{10}, C_{20}, and C_{30} are self-partial capacitances and C_{ij} $(i \neq j)$ are mutual partial capacitances.

Equations (3–147a), (3–147b), and (3–147c) can be rearranged as

$$Q_1 = (C_{10} + C_{12} + C_{13})V_1 - C_{12}V_2 - C_{13}V_3, \tag{3–148a}$$
$$Q_2 = -C_{12}V_1 + (C_{20} + C_{12} + C_{23})V_2 - C_{23}V_3, \tag{3–148b}$$
$$Q_3 = -C_{13}V_1 - C_{23}V_2 + (C_{30} + C_{13} + C_{23})V_3. \tag{3–148c}$$

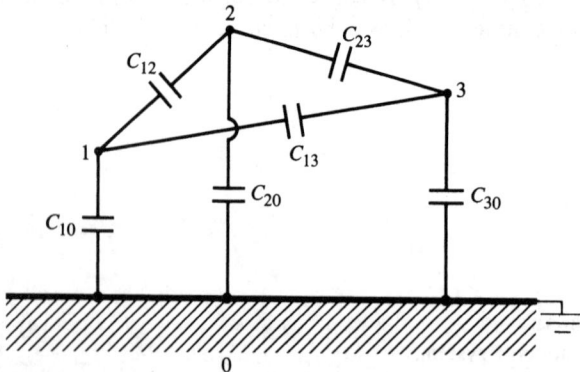

FIGURE 3–35
Schematic diagram of three conductors and the ground.

Comparing Eqs. (3–148) with Eqs. (3–146), we obtain

$$c_{11} = C_{10} + C_{12} + C_{13}, \qquad (3\text{--}149a)$$

$$c_{22} = C_{20} + C_{12} + C_{23}, \qquad (3\text{--}149b)$$

$$c_{33} = C_{30} + C_{13} + C_{23}, \qquad (3\text{--}149c)$$

and

$$c_{12} = -C_{12}, \qquad (3\text{--}150a)$$

$$c_{23} = -C_{23}, \qquad (3\text{--}150b)$$

$$c_{13} = -C_{13}. \qquad (3\text{--}150c)$$

On the basis of Eq. (3–149a) we can interpret the coefficient of capacitance c_{11} as the total capacitance between conductor 1 and all the other conductors connected together to ground; similarly for c_{22} and c_{33}. Equations (3–150) indicate that the coefficients of inductances are the negative of the mutual partial capacitances. Inverting Eqs. (3–149), we can express the conductor-to-ground capacitances in terms of the coefficients of capacitance and coefficients of induction:

$$C_{10} = c_{11} + c_{12} + c_{13}, \qquad (3\text{--}151a)$$

$$C_{20} = c_{22} + c_{12} + c_{23}, \qquad (3\text{--}151b)$$

$$C_{30} = c_{33} + c_{13} + c_{23}. \qquad (3\text{--}151c)$$

EXAMPLE 3–21 Three horizontal parallel conducting wires, each of radius a and isolated from the ground, are separated from one another as shown in Fig. 3–36. Assuming $d \gg a$, determine the partial capacitances per unit length between the wires.

Solution We designate the wires as conductors 0, 1, and 2, as indicated in Fig. 3–36. Choosing conductor 0 as the reference and using Eq. (3–138), we can write two equations for the potential differences V_{10} and V_{20} due to the three wires as follows:

$$V_{10} = \frac{\rho_{\ell 0}}{2\pi\epsilon_0} \ln \frac{a}{d} + \frac{\rho_{\ell 1}}{2\pi\epsilon_0} \ln \frac{d}{a} + \frac{\rho_{\ell 2}}{2\pi\epsilon_0} \ln \frac{3d}{2d}$$

or

$$2\pi\epsilon_0 V_{10} = \rho_{\ell 0} \ln \frac{a}{d} + \rho_{\ell 1} \ln \frac{d}{a} + \rho_{\ell 2} \ln \frac{3}{2}, \qquad (3\text{--}152a)$$

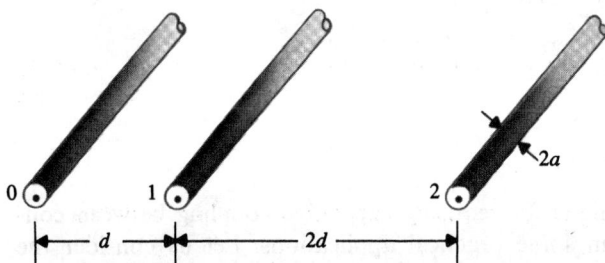

FIGURE 3–36
Three parallel wires (Example 3–21).

where $\rho_{\ell 0}$, $\rho_{\ell 1}$, and $\rho_{\ell 2}$ denote the charges per unit length on wires 0, 1, and 2 respectively. Similarly,

$$2\pi\epsilon_0 V_{20} = \rho_{\ell 0} \ln\frac{a}{3d} + \rho_{\ell 1} \ln\frac{d}{2d} + \rho_{\ell 2} \ln\frac{3d}{a}. \tag{3-152b}$$

For the isolated system of three conductors we have $\rho_{\ell 0} + \rho_{\ell 1} + \rho_{\ell 2} = 0$, or

$$\rho_{\ell 0} = -(\rho_{\ell 1} + \rho_{\ell 2}). \tag{3-153}$$

Combination of Eqs. (3–152a), (3–152b), and (3–153) yields

$$2\pi\epsilon_0 V_{10} = \rho_{\ell 1} 2 \ln\frac{d}{a} + \rho_{\ell 2} \ln\frac{3d}{2a}, \tag{3-154a}$$

$$2\pi\epsilon_0 V_{20} = \rho_{\ell 1} \ln\frac{3d}{2a} + \rho_{\ell 2} 2 \ln\frac{3d}{a}. \tag{3-154b}$$

Equations (3–154a) and (3–154b) can be used to solve for $\rho_{\ell 1}$ and $\rho_{\ell 2}$ as functions of V_{10} and V_{20}.

$$\rho_{\ell 1} = \Delta_0\left(V_{10} 2\ln\frac{3d}{a} - V_{20}\ln\frac{3d}{2a}\right), \tag{3-155a}$$

$$\rho_{\ell 2} = \Delta_0\left(-V_{10}\ln\frac{3d}{2a} + V_{20} 2\ln\frac{d}{a}\right), \tag{3-155b}$$

where

$$\Delta_0 = \frac{2\pi\epsilon_0}{4\ln\dfrac{d}{a}\ln\dfrac{3d}{a} - \left(\ln\dfrac{3d}{2a}\right)^2}. \tag{3-156}$$

Comparing Eqs. (3–155) with Eqs. (3–146), (3–148), and (3–151), we obtain the following partial capacitances per unit length for the given three-wire system:

$$C_{12} = -c_{12} = \Delta_0 \ln\frac{3d}{2a}, \tag{3-157a}$$

$$C_{10} = c_{11} + c_{12} = \Delta_0\left(2\ln\frac{3d}{a} - \ln\frac{3d}{2a}\right), \tag{3-157b}$$

$$C_{20} = c_{22} + c_{12} = \Delta_0\left(2\ln\frac{d}{a} - \ln\frac{3d}{2a}\right). \tag{3-157c}$$

3–10.3 ELECTROSTATIC SHIELDING

Electrostatic shielding, a technique for reducing capacitive coupling between conducting bodies, is important in some practical applications. Let us consider the situation shown in Fig. 3–37, in which a grounded conducting shell 2 completely

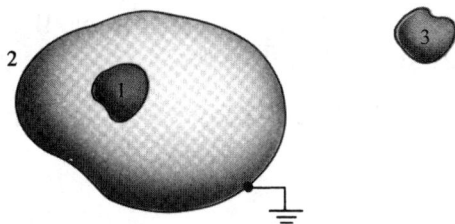

FIGURE 3-37
Illustrating electrostatic shielding.

encloses conducting body 1. Setting $V_2 = 0$ in Eq. (3–147a), we have

$$Q_1 = C_{10}V_1 + C_{12}V_1 + C_{13}(V_1 - V_3). \tag{3-158}$$

When $Q_1 = 0$, there is no field inside shell 2; hence body 1 and shell 2 have the same potential, $V_1 = V_2 = 0$. From Eq. (3–158) we see that the coupling capacitance C_{13} must vanish, since V_3 is arbitrary. This means that a change in V_3 will not affect Q_1, and vice versa. We then have electrostatic shielding between conducting bodies 1 and 3. Obviously, the same shielding effectiveness is obtained if the grounded conducting shell 2 encloses body 3 instead of body 1.

3-11 Electrostatic Energy and Forces

In Section 3–5 we indicated that electric potential at a point in an electric field is the work required to bring a unit positive charge from infinity (at reference zero-potential) to that point. To bring a charge Q_2 (slowly, so that kinetic energy and radiation effects may be neglected) from infinity *against* the field of a charge Q_1 in free space to a distance R_{12}, the amount of work required is

$$W_2 = Q_2 V_2 = Q_2 \frac{Q_1}{4\pi\epsilon_0 R_{12}}. \tag{3-159}$$

Because electrostatic fields are conservative, W_2 is independent of the path followed by Q_2. Another form of Eq. (3–159) is

$$W_2 = Q_1 \frac{Q_2}{4\pi\epsilon_0 R_{12}} = Q_1 V_1. \tag{3-160}$$

This work is stored in the assembly of the two charges as potential energy. Combining Eqs. (3–159) and (3–160), we can write

$$W_2 = \tfrac{1}{2}(Q_1 V_1 + Q_2 V_2). \tag{3-161}$$

Now suppose another charge Q_3 is brought from infinity to a point that is R_{13} from Q_1 and R_{23} from Q_2; an additional amount of work is required that equals

$$\Delta W = Q_3 V_3 = Q_3 \left(\frac{Q_1}{4\pi\epsilon_0 R_{13}} + \frac{Q_2}{4\pi\epsilon_0 R_{23}} \right). \tag{3-162}$$

The sum of ΔW in Eq. (3–162) and W_2 in Eq. (3–159) is the potential energy, W_3, stored in the assembly of the three charges $Q_1, Q_2,$ and Q_3. That is,

$$W_3 = W_2 + \Delta W = \frac{1}{4\pi\epsilon_0}\left(\frac{Q_1 Q_2}{R_{12}} + \frac{Q_1 Q_3}{R_{13}} + \frac{Q_2 Q_3}{R_{23}}\right). \qquad (3\text{–}163)$$

We can rewrite W_3 in the following form:

$$W_3 = \frac{1}{2}\left[Q_1\left(\frac{Q_2}{4\pi\epsilon_0 R_{12}} + \frac{Q_3}{4\pi\epsilon_0 R_{13}}\right) + Q_2\left(\frac{Q_1}{4\pi\epsilon_0 R_{12}} + \frac{Q_3}{4\pi\epsilon_0 R_{23}}\right)\right.$$
$$\left. + Q_3\left(\frac{Q_1}{4\pi\epsilon_0 R_{13}} + \frac{Q_2}{4\pi\epsilon_0 R_{23}}\right)\right] \qquad (3\text{–}164)$$
$$= \tfrac{1}{2}(Q_1 V_1 + Q_2 V_2 + Q_3 V_3).$$

In Eq. (3–164), V_1, the potential at the position of Q_1, is caused by charges Q_2 and Q_3; it is *different* from the V_1 in Eq. (3–160) in the two-charge case. Similarly, V_2 and V_3 are the potentials at Q_2 and Q_3, respectively, in the three-charge assembly.

Extending this procedure of bringing in additional charges, we arrive at the following general expression for the potential energy of a group of N discrete point charges at rest. (The purpose of the subscript e on W_e is to denote that the energy is of an electric nature.) We have

$$\boxed{\, W_e = \frac{1}{2}\sum_{k=1}^{N} Q_k V_k \quad \text{(J)}, \,} \qquad (3\text{–}165)$$

where V_k, the electric potential at Q_k, is caused by all the other charges and has the following expression:

$$V_k = \frac{1}{4\pi\epsilon_0}\sum_{\substack{j=1 \\ (j \neq k)}}^{N} \frac{Q_j}{R_{jk}}. \qquad (3\text{–}166)$$

Two remarks are in order here. First, W_e can be negative. For instance, W_2 in Eq. (3–159) will be negative if Q_1 and Q_2 are of opposite signs. In that case, work is done by the field (not against the field) established by Q_1 in moving Q_2 from infinity. Second, W_e in Eq. (3–165) represents only the interaction energy (mutual energy) and does not include the work required to assemble the individual point charges themselves (self-energy).

The SI unit for energy, *joule* (J), is too large a unit for work in physics of elementary particles, where energy is more conveniently measured in terms of a much smaller unit called *electron-volt* (eV). An electron-volt is the energy or work required to move an electron against a potential difference of one volt.

$$1 \quad \text{(eV)} = (1.60 \times 10^{-19}) \times 1 = 1.60 \times 10^{-19} \quad \text{(J)}. \qquad (3\text{–}167)$$

Energy in (eV) is essentially that in (J) per unit electronic charge. The proton beams of the world's most powerful high-energy particle accelerator collide with a kinetic energy of two trillion electron-volts (2 TeV), or $(2 \times 10^{12}) \times (1.60 \times 10^{-19}) = 3.20 \times 10^{-7}$ (J). A binding energy of $W = 5 \times 10^{-19}$ (J) in an ionic crystal is equal to $W/e = 5 \times 10^{-19}/1.60 \times 10^{-19} = 3.125$ (eV), which is a more convenient number to use than the one in terms of joules.

EXAMPLE 3-22   Find the energy required to assemble a uniform sphere of charge of radius b and volume charge density ρ.

Solution   Because of symmetry, it is simplest to assume that the sphere of charge is assembled by bringing up a succession of spherical layers of thickness dR. At a radius R shown in Fig. 3-38 the potential is

$$V_R = \frac{Q_R}{4\pi\epsilon_0 R},$$

where Q_R is the total charge contained in a sphere of radius R:

$$Q_R = \rho\tfrac{4}{3}\pi R^3.$$

The differential charge in a spherical layer of thickness dR is

$$dQ_R = \rho 4\pi R^2\, dR,$$

and the work or energy in bringing up dQ_R is

$$dW = V_R\, dQ_R = \frac{4\pi}{3\epsilon_0} \rho^2 R^4\, dR.$$

Hence the total work or energy required to assemble a uniform sphere of charge of radius b and charge density ρ is

$$W = \int dW = \frac{4\pi}{3\epsilon_0} \rho^2 \int_0^b R^4\, dR = \frac{4\pi\rho^2 b^5}{15\epsilon_0} \quad \text{(J)}. \qquad (3\text{-}168)$$

In terms of the total charge

$$Q = \rho \frac{4\pi}{3} b^3,$$

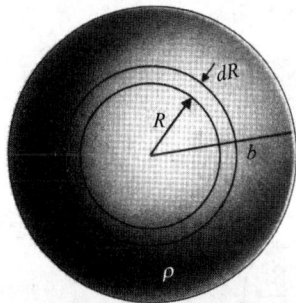

FIGURE 3-38
Assembling a uniform sphere of charge (Example 3-22).

we have

$$W = \frac{3Q^2}{20\pi\epsilon_0 b} \quad \text{(J).} \tag{3-169}$$

Equation (3–169) shows that the energy is directly proportional to the square of the total charge and inversely proportional to the radius. The sphere of charge in Fig. 3–38 could be a cloud of electrons, for instance. ■

For a continuous charge distribution of density ρ the formula for W_e in Eq. (3–165) for discrete charges must be modified. Without going through a separate proof we replace Q_k by $\rho\,dv$ and the summation by an integration and obtain

$$W_e = \tfrac{1}{2} \int_{V'} \rho V\,dv \quad \text{(J).} \tag{3-170}$$

In Eq. (3–170), V is the potential at the point where the volume charge density is ρ, and V' is the volume of the region where ρ exists.

■■■■ **EXAMPLE 3–23**   Solve the problem in Example 3–22 by using Eq. (3–170).

Solution   In Example 3–22 we solved the problem of assembling a sphere of charge by bringing up a succession of spherical layers of a differential thickness. Now we assume that the sphere of charge is already in place. Since ρ is a constant, it can be taken out of the integral sign. For a spherically symmetrical problem,

$$W_e = \frac{\rho}{2} \int_{V'} V\,dv = \frac{\rho}{2} \int_0^b V\,4\pi R^2\,dR, \tag{3-171}$$

where V is the potential at a point R from the center. To find V at R, we must find the negative of the line integral of $\mathbf{E}$ in two regions: (1) $\mathbf{E}_1 = \mathbf{a}_R E_{R1}$ from $R = \infty$ to $R = b$, and (2) $\mathbf{E}_2 = \mathbf{a}_R E_{R2}$ from $R = b$ to $R = R$. We have

$$E_{R1} = \mathbf{a}_R \frac{Q}{4\pi\epsilon_0 R^2} = \mathbf{a}_R \frac{\rho b^3}{3\epsilon_0 R^2}, \qquad R \geq b,$$

and

$$E_{R2} = \mathbf{a}_R \frac{QR}{4\pi\epsilon_0 R^2} = \mathbf{a}_R \frac{\rho R}{3\epsilon_0}, \qquad 0 < R \leq b.$$

Consequently, we obtain

$$V = -\int_\infty^R \mathbf{E} \cdot d\mathbf{R} = -\left[\int_\infty^b E_{R1}\,dR + \int_b^R E_{R2}\,dR \right]$$

$$= -\left[\int_\infty^b \frac{\rho b^3}{3\epsilon_0 R^2}\,dR + \int_b^R \frac{\rho R}{3\epsilon_0}\,dR \right] \tag{3-172}$$

$$= \frac{\rho}{3\epsilon_0}\left(b^2 + \frac{b^2}{2} - \frac{R^2}{2} \right) = \frac{\rho}{3\epsilon_0}\left(\frac{3}{2}b^2 - \frac{R^2}{2} \right).$$

Substituting Eq. (3–172) in Eq. (3–171), we get

$$W_e = \frac{\rho}{2} \int_0^b \frac{\rho}{3\epsilon_0} \left(\frac{3}{2} b^2 - \frac{R^2}{2} \right) 4\pi R^2 \, dR = \frac{4\pi \rho^2 b^5}{15\epsilon_0},$$

which is the same as the result in Eq. (3–168).   ■

Note that W_e in Eq. (3–170) includes the work (self-energy) required to assemble the distribution of macroscopic charges, because it is the energy of interaction of every infinitesimal charge element with all other infinitesimal charge elements. As a matter of fact, we have used Eq. (3–170) in Example 3–23 to find the self-energy of a uniform spherical charge. As the radius b approaches zero, the self-energy of a (mathematical) point charge of a given Q is infinite (see Eq. 3–169). The self-energies of point charges Q_k are not included in Eq. (3–165). Of course, there are, strictly, no point charges, inasmuch as the smallest charge unit, the electron, is itself a distribution of charge.

3–11.1   ELECTROSTATIC ENERGY IN TERMS OF FIELD QUANTITIES

In Eq. (3–170) the expression of electrostatic energy of a charge distribution contains the source charge density ρ and the potential function V. We frequently find it more convenient to have an expression of W_e in terms of field quantities $\mathbf{E}$ and/or $\mathbf{D}$, without knowing ρ explicitly. To this end, we substitute $\nabla \cdot \mathbf{D}$ for ρ in Eq. (3–170):

$$W_e = \tfrac{1}{2} \int_{V'} (\nabla \cdot \mathbf{D}) V \, dv. \tag{3–173}$$

Now, using the vector identity (from Problem P.2–28),

$$\nabla \cdot (V\mathbf{D}) = V\nabla \cdot \mathbf{D} + \mathbf{D} \cdot \nabla V, \tag{3–174}$$

we can write Eq. (3–173) as

$$\begin{aligned} W_e &= \tfrac{1}{2} \int_{V'} \nabla \cdot (V\mathbf{D}) \, dv - \tfrac{1}{2} \int_{V'} \mathbf{D} \cdot \nabla V \, dv \\ &= \tfrac{1}{2} \oint_{S'} V\mathbf{D} \cdot \mathbf{a}_n \, ds + \tfrac{1}{2} \int_{V'} \mathbf{D} \cdot \mathbf{E} \, dv, \end{aligned} \tag{3–175}$$

where the divergence theorem has been used to change the first volume integral into a closed surface integral and $\mathbf{E}$ been substituted for $-\nabla V$ in the second volume integral. Since V' can be any volume that includes all the charges, we may choose it to be a very large sphere with radius R. As we let $R \to \infty$, electric potential V and the magnitude of electric displacement D fall off at least as fast as $1/R$ and $1/R^2$, respectively.[†] The area of the bounding surface S' increases as R^2. Hence the surface integral in Eq. (3–175) decreases at least as fast as $1/R$ and will vanish as $R \to \infty$. We are then left with only the second integral on the right side of Eq. (3–175).

[†] For point charges $V \propto 1/R$ and $D \propto 1/R^2$; for dipoles $V \propto 1/R^2$ and $D \propto 1/R^3$.

$$W_e = \tfrac{1}{2} \int_{V'} \mathbf{D} \cdot \mathbf{E} \, dv \qquad \text{(J)}. \qquad\qquad (3\text{--}176a)$$

Using the relation $\mathbf{D} = \epsilon \mathbf{E}$ for a linear medium, Eq. (3–176a) can be written in two other forms:

$$W_e = \tfrac{1}{2} \int_{V'} \epsilon E^2 \, dv \qquad \text{(J)} \qquad\qquad (3\text{--}176b)$$

and

$$W_e = \tfrac{1}{2} \int_{V'} \frac{D^2}{\epsilon} \, dv \qquad \text{(J)}. \qquad\qquad (3\text{--}176c)$$

We can always define an *electrostatic energy density* w_e mathematically, such that its volume integral equals the total electrostatic energy:

$$W_e = \int_{V'} w_e \, dv. \qquad\qquad (3\text{--}177)$$

We can therefore write

$$w_e = \tfrac{1}{2}\mathbf{D} \cdot \mathbf{E} \qquad \text{(J/m}^3) \qquad\qquad (3\text{--}178a)$$

or

$$w_e = \tfrac{1}{2}\epsilon E^2 \qquad \text{(J/m}^3) \qquad\qquad (3\text{--}178b)$$

or

$$w_e = \frac{D^2}{2\epsilon} \qquad \text{(J/m}^3). \qquad\qquad (3\text{--}178c)$$

However, this definition of energy density is artificial because a physical justification has not been found to localize energy with an electric field; all we know is that the volume integrals in Eqs. (3–176a, b, c) give the correct total electrostatic energy.

EXAMPLE 3–24   In Fig. 3–39 a parallel-plate capacitor of area S and separation d is charged to a voltage V. The permittivity of the dielectric is ϵ. Find the stored electrostatic energy.

Solution   With the d-c source (batteries) connected as shown, the upper and lower plates are charged positive and negative, respectively. If the fringing of the field at

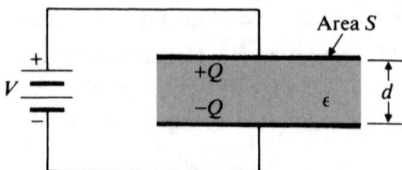

FIGURE 3–39
A charged parallel-plate capacitor (Example 3–24).

the edges is neglected, the electric field in the dielectric is uniform (over the plate) and constant (across the dielectric) and has a magnitude

$$E = \frac{V}{d}.$$

Using Eq. (3–176b), we have

$$W_e = \frac{1}{2} \int_{V'} \epsilon \left(\frac{V}{d}\right)^2 dv = \frac{1}{2} \epsilon \left(\frac{V}{d}\right)^2 (Sd) = \frac{1}{2} \left(\epsilon \frac{S}{d}\right) V^2. \tag{3–179}$$

The quantity in the parentheses of the last expression, $\epsilon S/d$, is the capacitance of the parallel-plate capacitor (see Eq. 3–136). So,

$$\boxed{W_e = \tfrac{1}{2} C V^2 \qquad \text{(J)}.} \tag{3–180a}$$

Since $Q = CV$, Eq. (3–180a) can be put in two other forms:

$$\boxed{W_e = \tfrac{1}{2} Q V \qquad \text{(J)}} \tag{3–180b}$$

and

$$\boxed{W_e = \frac{Q^2}{2C} \qquad \text{(J)}.} \tag{3–180c}$$

It so happens that Eqs. (3–180a, b, c) hold true for any two-conductor capacitor (see Problem P.3–43).

EXAMPLE 3–25   Use energy formulas (3–176) and (3–180) to find the capacitance of a cylindrical capacitor having a length L, an inner conductor of radius a, an outer conductor of inner radius b, and a dielectric of permittivity ϵ, as shown in Fig. 3–29.

Solution   By applying Gauss's law, we know that

$$\mathbf{E} = \mathbf{a}_r E_r = \mathbf{a}_r \frac{Q}{2\pi\epsilon L r}, \qquad a < r < b.$$

The electrostatic energy stored in the dielectric region is, from Eq. (3–176b),

$$\begin{aligned} W_e &= \frac{1}{2} \int_a^b \epsilon \left(\frac{Q}{2\pi\epsilon L r}\right)^2 (L 2\pi r \, dr) \\ &= \frac{Q^2}{4\pi\epsilon L} \int_a^b \frac{dr}{r} = \frac{Q^2}{4\pi\epsilon L} \ln \frac{b}{a}. \end{aligned} \tag{3–181}$$

On the other hand, W_e can also be expressed in the form of Eq. (3–180c). Equating (3–180c) and (3–181), we obtain

$$\frac{Q^2}{2C} = \frac{Q^2}{4\pi\epsilon L}\ln\frac{b}{a}$$

or

$$C = \frac{2\pi\epsilon L}{\ln\dfrac{b}{a}},$$

which is the same as that given in Eq. (3–139).   ▬

3–11.2   ELECTROSTATIC FORCES

Coulomb's law governs the force between two point charges. In a more complex system of charged bodies, using Coulomb's law to determine the force on one of the bodies that is caused by the charges on other bodies would be very tedious. This would be so even in the simple case of finding the force between the plates of a charged parallel-plate capacitor. We will now discuss a method for calculating the force on an object in a charged system from the electrostatic energy of the system. This method is based on the ***principle of virtual displacement***. We will consider two cases: (1) that of an isolated system of bodies with fixed charges, and (2) that of a system of conducting bodies with fixed potentials.

System of Bodies with Fixed Charges   We consider an isolated system of charged conducting, as well as dielectric, bodies separated from one another with no connection to the outside world. The charges on the bodies are constant. Imagine that the electric forces have displaced one of the bodies by a differential distance $d\boldsymbol{\ell}$ (a virtual displacement). The mechanical work done *by the system* would be

$$dW = \mathbf{F}_Q \cdot d\boldsymbol{\ell}, \tag{3–182}$$

where $\mathbf{F}_Q$ is the total electric force acting on the body under the condition of constant charges. Since we have an isolated system with no external supply of energy, this mechanical work must be done at the expense of the stored electrostatic energy; that is,

$$dW = -dW_e = \mathbf{F}_Q \cdot d\boldsymbol{\ell}. \tag{3–183}$$

Noting from Eq. (2–88) in Section 2–6 that the differential change of a scalar resulting from a position change $d\boldsymbol{\ell}$ is the dot product of the gradient of the scalar, and $d\boldsymbol{\ell}$, we write

$$dW_e = (\nabla W_e) \cdot d\boldsymbol{\ell}. \tag{3–184}$$

Since $d\boldsymbol{\ell}$ is arbitrary, comparison of Eqs. (3–183) and (3–184) leads to

$$\boxed{\mathbf{F}_Q = -\nabla W_e \quad \text{(N)}.} \tag{3–185}$$

Equation (3–185) is a very simple formula for the calculation of $\mathbf{F}_Q$ from the electrostatic energy of the system. In Cartesian coordinates the component forces are

$$(F_Q)_x = -\frac{\partial W_e}{\partial x}, \tag{3–186a}$$

$$(F_Q)_y = -\frac{\partial W_e}{\partial y}, \tag{3–186b}$$

$$(F_Q)_z = -\frac{\partial W_e}{\partial z}. \tag{3–186c}$$

If the body under consideration is constrained to rotate about an axis, say the z-axis, the mechanical work done by the system for a virtual angular displacement $d\phi$ would be

$$dW = (T_Q)_z \, d\phi, \tag{3–187}$$

where $(T_Q)_z$ is the z-component of the torque acting on the body under the condition of constant charges. The foregoing procedure will lead to

$$(T_Q)_z = -\frac{\partial W_e}{\partial \phi} \qquad (\text{N·m}). \tag{3–188}$$

System of Conducting Bodies with Fixed Potentials   Now consider a system in which conducting bodies are held at fixed potentials through connections to such external sources as batteries. Uncharged dielectric bodies may also be present. A displacement $d\ell$ by a conducting body would result in a change in total electrostatic energy and would require the sources to transfer charges to the conductors in order to keep them at their fixed potentials. If a charge dQ_k (which may be positive or negative) is added to the kth conductor that is maintained at potential V_k, the work done or energy supplied by the sources is $V_k \, dQ_k$. The total energy supplied by the sources to the system is

$$dW_s = \sum_k V_k \, dQ_k. \tag{3–189}$$

The mechanical work done by the system as a consequence of the virtual displacement is

$$dW = \mathbf{F}_V \cdot d\ell, \tag{3–190}$$

where $\mathbf{F}_V$ is the electric force on the conducting body under the condition of constant potentials. The charge transfers also change the electrostatic energy of the system by an amount dW_e, which, in view of Eq. (3–165), is

$$dW_e = \frac{1}{2} \sum_k V_k \, dQ_k = \frac{1}{2} dW_s. \tag{3–191}$$

Conservation of energy demands that

$$dW + dW_e = dW_s. \qquad (3-192)$$

Substitution of Eqs. (3–189), (3–190), and (3–191) in Eq. (3–192) gives

$$\mathbf{F}_V \cdot d\boldsymbol{\ell} = dW_e$$
$$= (\nabla W_e) \cdot d\boldsymbol{\ell}$$

or

$$\boxed{\mathbf{F}_V = \nabla W_e \qquad \text{(N)}.} \qquad (3-193)$$

Comparison of Eqs. (3–193) and (3–185) reveals that the only difference between the formulas for the electric forces in the two cases is in the sign. It is clear that if the conducting body is constrained to rotate about the z-axis, the z-component of the electric torque will be

$$\boxed{(T_V)_z = \frac{\partial W_e}{\partial \phi} \qquad \text{(N·m)},} \qquad (3-194)$$

which differs from Eq. (3–188) also only by a sign change.

EXAMPLE 3–26   Determine the force on the conducting plates of a charged parallel-plate capacitor. The plates have an area S and are separated in air by a distance x.

Soiution   We solve the problem in two ways: (a) by assuming fixed charges, and then (b) by assuming fixed potentials. The fringing of field around the edges of the plates will be neglected.

a) *Fixed charges.* With fixed charges $\pm Q$ on the plates, an electric field intensity $E_x = Q/(\epsilon_0 S) = V/x$ exists in the air between the plates regardless of their separation (unchanged by a virtual displacement). From Eq. (3–180b),

$$W_e = \tfrac{1}{2}QV = \tfrac{1}{2}QE_x x,$$

where Q and E_x are constants. Using Eq. (3–186a), we obtain

$$(F_Q)_x = -\frac{\partial}{\partial x}\left(\frac{1}{2}QE_x x\right) = -\frac{1}{2}QE_x = -\frac{Q^2}{2\epsilon_0 S}, \qquad (3-195)$$

where the negative signs indicate that the force is opposite to the direction of increasing x. It is an attractive force.

b) *Fixed potentials.* With fixed potentials it is more convenient to use the expression in Eq. (3–180a) for W_e. Capacitance C for the parallel-plate air capacitor is $\epsilon_0 S/x$. We have, from Eq. (3–193),

$$(F_V)_x = \frac{\partial W_e}{\partial x} = \frac{\partial}{\partial x}\left(\frac{1}{2}CV^2\right) = \frac{V^2}{2}\frac{\partial}{\partial x}\left(\frac{\epsilon_0 S}{x}\right) = -\frac{\epsilon_0 S V^2}{2x^2}. \qquad (3-196)$$

How different are $(F_Q)_x$ in Eq. (3–195) and $(F_V)_x$ in Eq. (3–196)? Recalling the relation

$$Q = CV = \frac{\epsilon_0 SV}{x},$$

we find

$$(F_Q)_x = (F_V)_x. \qquad (3\text{–}197)$$

The force is the same in both cases in spite of the apparent sign difference in the formulas as expressed by Eqs. (3–185) and (3–193). A little reflection on the physical problem will convince us that this must be true. Since the charged capacitor has fixed dimensions, a given Q will result in a fixed V, and vice versa. Therefore there is a unique force between the plates regardless of whether Q or V is given, and the force certainly does not depend on virtual displacements. A change in the conceptual constraint (fixed Q or fixed V) cannot change the unique force between the plates.

▬

The preceding discussion holds true for a general charged two-conductor capacitor with capacitance C. The electrostatic force F_ℓ in the direction of a virtual displacement $d\ell$ for fixed charges is

$$(F_Q)_\ell = -\frac{\partial W_e}{\partial \ell} = -\frac{\partial}{\partial \ell}\left(\frac{Q^2}{2C}\right) = \frac{Q^2}{2C^2}\frac{\partial C}{\partial \ell}. \qquad (3\text{–}198)$$

For fixed potentials,

$$(F_V)_\ell = \frac{\partial W_e}{\partial \ell} = \frac{\partial}{\partial \ell}\left(\frac{1}{2}CV^2\right) = \frac{V^2}{2}\frac{\partial C}{\partial \ell} = \frac{Q^2}{2C^2}\frac{\partial C}{\partial \ell}. \qquad (3\text{–}199)$$

It is clear that the forces calculated from the two procedures, which assumed different constraints imposed on the same charged capacitor, are equal.

Review Questions

R.3–1 Write the differential form of the fundamental postulates of electrostatics in free space.

R.3–2 Under what conditions will the electric field intensity be both solenoidal and irrotational?

R.3–3 Write the integral form of the fundamental postulates of electrostatics in free space, and state their meaning in words.

R.3–4 When the formula for the electric field intensity of a point charge, Eq. (3–12), was derived,
    a) why was it necessary to stipulate that q is in a boundless free space?
    b) why did we *not* construct a cubic or a cylindrical surface around q?

R.3–5 In what ways does the electric field intensity vary with distance for
    a) a point charge?   **b)** an electric dipole?

R.3–6 State *Coulomb's law.*

R.3–7 Explain the principle of operation of ink-jet printers.

R.3–8 State *Gauss's law.* Under what conditions is Gauss's law especially useful in determining the electric field intensity of a charge distribution?

R.3–9 Describe the ways in which the electric field intensity of an infinitely long, straight line charge of uniform density varies with distance.

R.3–10 Is Gauss's law useful in finding the **E** field of a finite line charge? Explain.

R.3–11 See Example 3–6, Fig. 3–9. Could a cylindrical pillbox with circular top and bottom faces be chosen as a Gaussian surface? Explain.

R.3–12 Make a two-dimensional sketch of the electric field lines and the equipotential lines of a point charge.

R.3–13 At what value of θ is the **E** field of a z-directed electric dipole pointed in the negative z-direction?

R.3–14 Refer to Eq. (3–64). Explain why the absolute sign around z is required.

R.3–15 If the electric potential at a point is zero, does it follow that the electrical field intensity is also zero at that point? Explain.

R.3–16 If the electric field intensity at a point is zero, does it follow that the electric potential is also zero at that point? Explain.

R.3–17 If an uncharged spherical conducting shell of a finite thickness is placed in an external electric field $\mathbf{E}_o$, what is the electric field intensity at the center of the shell? Describe the charge distributions on both the outer and the inner surfaces of the shell.

R.3–18 What are *electrets*? How can they be made?

R.3–19 Can $\nabla'(1/R)$ in Eq. (3–84) be replaced by $\nabla(1/R)$? Explain.

R.3–20 Define *polarization vector.* What is its SI unit?

R.3–21 What are *polarization charge densities*? What are the SI units for $\mathbf{P} \cdot \mathbf{a}_n$ and $\nabla \cdot \mathbf{P}$?

R.3–22 What do we mean by *simple medium*?

R.3–23 What properties do *anisotropic materials* have?

R.3–24 What characterizes a *uniaxial medium*?

R.3–25 Define *electric displacement vector.* What is its SI unit?

R.3–26 Define *electric susceptibility.* What is its unit?

R.3–27 What is the difference between the *permittivity* and the *dielectric constant* of a medium?

R.3–28 Does the electric flux density due to a given charge distribution depend on the properties of the medium? Does the electric field intensity? Explain.

R.3–29 What is the difference between the *dielectric constant* and the *dielectric strength* of a dielectric material?

R.3–30 Explain the principle of operation of lightning arresters.

R.3–31 What are the general boundary conditions for electrostatic fields at an interface between two different dielectric media?

R.3–32 What are the boundary conditions for electrostatic fields at an interface between a conductor and a dielectric with permittivity ϵ?

R.3–33 What is the boundary condition for electrostatic potential at an interface between two different dielectric media?

R.3–34 Does a force exist between a point charge and a dielectric body? Explain.

R.3–35 Define *capacitance* and *capacitor*.

R.3–36 Assume that the permittivity of the dielectric in a parallel-plate capacitor is not constant. Will Eq. (3–136) hold if the average value of permittivity is used for ϵ in the formula? Explain.

R.3–37 Given three 1-μF capacitors, explain how they should be connected in order to obtain a total capacitance of

    a) $\frac{1}{3}(\mu F)$,   **b)** $\frac{2}{3}(\mu F)$,   **c)** $\frac{3}{2}(\mu F)$,   **d)** $3(\mu F)$.

R.3–38 What are *coefficients of potential*, *coefficients of capacitance*, and *coefficients of induction*?

R.3–39 What are *partial capacitances*? How are they different from coefficients of capacitance?

R.3–40 Explain the principle of electrostatic shielding.

R.3–41 What is the definition of an *electron-volt*? How does it compare with a joule?

R.3–42 What is the expression for the electrostatic energy of an assembly of four discrete point charges?

R.3–43 What is the expression for the electrostatic energy of a continuous distribution of charge in a volume? on a surface? along a line?

R.3–44 Provide a mathematical expression for electrostatic energy in terms of **E** and/or **D**.

R.3–45 Discuss the meaning and use of the *principle of virtual displacement*.

R.3–46 What is the relation between the force and the stored energy in a system of stationary charged objects under the condition of constant charges? Under the condition of fixed potentials?

Problems

P.3–1 Refer to Fig. 3–4.
    a) Find the relation between the angle of arrival, α, of the electron beam at the screen and the deflecting electric field intensity E_d.
    b) Find the relation between w and L such that $d_1 = d_0/20$.

P.3–2 The cathode-ray oscilloscope (CRO) shown in Fig. 3–4 is used to measure the voltage applied to the parallel deflection plates.
    a) Assuming no breakdown in insulation, what is the maximum voltage that can be measured if the distance of separation between the plates is h?
    b) What is the restriction on L if the diameter of the screen is D?
    c) What can be done with a fixed geometry to double the CRO's maximum measurable voltage?

P.3–3 The deflection system of a cathode-ray oscilloscope usually consists of two pairs of parallel plates producing orthogonal electric fields. Assume the presence of another set of plates in Fig. 3–4 that establishes a uniform electric field $\mathbf{E}_x = \mathbf{a}_x E_x$ in the deflection region. Deflection voltages $v_x(t)$ and $v_y(t)$ are applied to produce $\mathbf{E}_x$ and $\mathbf{E}_y$, respectively. Determine

the types of waveforms that $v_x(t)$ and $v_y(t)$ should have if the electrons are to trace the following graphs on the fluorescent screen:
   a) a horizontal line,
   b) a straight line having a negative unity slope,
   c) a circle,
   d) two cycles of a sine wave.

P.3–4 Write a short article explaining the principle of operation of xerography. (Use library resources if needed.)

P.3–5 Two point charges, Q_1 and Q_2, are located at $(1, 2, 0)$ and $(2, 0, 0)$, respectively. Find the relation between Q_1 and Q_2 such that the total force on a test charge at the point $P(-1, 1, 0)$ will have
   a) no x-component,   **b)** no y-component.

P.3–6 Two very small conducting spheres, each of a mass 1.0×10^{-4} (kg), are suspended at a common point by very thin nonconducting threads of a length 0.2 (m). A charge Q is placed on each sphere. The electric force of repulsion separates the spheres, and an equilibrium is reached when the suspending threads make an angle of $10°$. Assuming a gravitational force of 9.80 (N/kg) and a negligible mass for the threads, find Q.

P.3–7 Find the force between a charged circular loop of radius b and uniform charge density ρ_ℓ and a point charge Q located on the loop axis at a distance h from the plane of the loop. What is the force when $h \gg b$, and when $h = 0$? Plot the force as a function of h.

P.3–8 A line charge of uniform density ρ_ℓ in free space forms a semicircle of radius b. Determine the magnitude and direction of the electric field intensity at the center of the semicircle.

P.3–9 Three uniform line charges—$\rho_{\ell 1}$, $\rho_{\ell 2}$, and $\rho_{\ell 3}$, each of length L—form an equilateral triangle. Assuming that $\rho_{\ell 1} = 2\rho_{\ell 2} = 2\rho_{\ell 3}$, determine the electric field intensity at the center of the triangle.

P.3–10 Assuming that the electric field intensity is $\mathbf{E} = \mathbf{a}_x 100x$ (V/m), find the total electric charge contained inside
   a) a cubical volume 100 (mm) on a side centered symmetrically at the origin,
   b) a cylindrical volume around the z-axis having a radius 50 (mm) and a height 100 (mm) centered at the origin.

P.3–11 A spherical distribution of charge $\rho = \rho_0[1 - (R^2/b^2)]$ exists in the region $0 \leq R \leq b$. This charge distribution is concentrically surrounded by a conducting shell with inner radius R_i $(>b)$ and outer radius R_o. Determine $\mathbf{E}$ everywhere.

P.3–12 Two infinitely long coaxial cylindrical surfaces, $r = a$ and $r = b$ $(b > a)$, carry surface charge densities ρ_{sa} and ρ_{sb}, respectively.
   a) Determine $\mathbf{E}$ everywhere.
   b) What must be the relation between a and b in order that $\mathbf{E}$ vanishes for $r > b$?

P.3–13 Determine the work done in carrying a -2 (μC) charge from $P_1(2, 1, -1)$ to $P_2(8, 2, -1)$ in the field $\mathbf{E} = \mathbf{a}_x y + \mathbf{a}_y x$
   a) along the parabola $x = 2y^2$,
   b) along the straight line joining P_1 and P_2.

P.3–14 At what values of θ does the electric field intensity of a z-directed dipole have no z-component?

P.3–15 Three charges $(+q, -2q,$ and $+q)$ are arranged along the z-axis at $z = d/2$, $z = 0$, and $z = -d/2$, respectively.

a) Determine V and $\mathbf{E}$ at a distant point $P(R, \theta, \phi)$.
b) Find the equations for equipotential surfaces and streamlines.
c) Sketch a family of equipotential lines and streamlines.

(Such an arrangement of three charges is called a *linear electrostatic quadrupole*.)

P.3–16 A finite line charge of length L carrying uniform line charge density ρ_ℓ is coincident with the x-axis.
a) Determine V in the plane bisecting the line charge.
b) Determine $\mathbf{E}$ from ρ_ℓ directly by applying Coulomb's law.
c) Check the answer in part (b) with $-\nabla V$.

P.3–17 In Example 3–5 we obtained the electric field intensity around an infinitely long line charge of a uniform charge density in a very simple manner by applying Gauss's law. Since $|\mathbf{E}|$ is a function of r only, any coaxial cylinder around the infinite line charge is an equipotential surface. In practice, all conductors are of finite length. A finite line charge carrying a constant charge density ρ_ℓ along the axis, however, does not produce a constant potential on a concentric cylindrical surface. Given the finite line charge ρ_ℓ of length L in Fig. 3–40, find the potential on the cylindrical surface of radius b as a function of x and plot it.

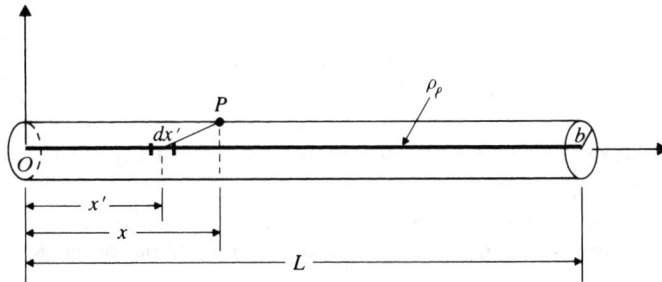

FIGURE 3–40
A finite line charge (Problem P.3–17).

(*Hint:* Find dV at P due to charge $\rho_\ell dx'$ and integrate.)

P.3–18 A charge Q is distributed uniformly over an $L \times L$ square plate. Determine V and $\mathbf{E}$ at a point on the axis perpendicular to the plate and through its center.

P.3–19 A charge Q is distributed uniformly over the wall of a circular tube of radius b and height h. Determine V and $\mathbf{E}$ on its axis
a) at a point outside the tube, then
b) at a point inside the tube.

P.3–20 An early model of the atomic structure of a chemical element was that the atom was a spherical cloud of uniformly distributed positive charge Ne, where N is the atomic number and e is the magnitude of electronic charge. Electrons, each carrying a negative charge $-e$, were considered to be imbedded in the cloud. Assuming the spherical charge cloud to have a radius R_0 and neglecting collision effects,
a) find the force experienced by an imbedded electron at a distance r from the center;
b) describe the motion of the electron;
c) explain why this atomic model is unsatisfactory.

P.3–21 A simple classical model of an atom consists of a nucleus of a positive charge Ne surrounded by a spherical electron cloud of the same total negative charge. (N is the atomic number and e is the magnitude of electronic charge.) An external electric field $\mathbf{E}_o$ will cause the nucleus to be displaced a distance r_o from the center of the electron cloud, thus polarizing the atom. Assuming a uniform charge distribution within the electron cloud of radius b, find r_o.

P.3–22 The polarization in a dielectric cube of side L centered at the origin is given by $\mathbf{P} = P_o(\mathbf{a}_x x + \mathbf{a}_y y + \mathbf{a}_z z)$.
  a) Determine the surface and volume bound-charge densities.
  b) Show that the total bound charge is zero.

P.3–23 Determine the electric field intensity at the center of a small spherical cavity cut out of a large block of dielectric in which a polarization $\mathbf{P}$ exists.

P.3–24 Solve the following problems:
  a) Find the breakdown voltage of a parallel-plate capacitor, assuming that conducting plates are 50 (mm) apart and the medium between them is air.
  b) Find the breakdown voltage if the entire space between the conducting plates is filled with plexiglass, which has a dielectric constant 3 and a dielectric strength 20 (kV/mm).
  c) If a 10-(mm) thick plexiglass is inserted between the plates, what is the maximum voltage that can be applied to the plates without a breakdown?

P.3–25 Assume that the $z = 0$ plane separates two lossless dielectric regions with $\epsilon_{r1} = 2$ and $\epsilon_{r2} = 3$. If we know that $\mathbf{E}_1$ in region 1 is $\mathbf{a}_x 2y - \mathbf{a}_y 3x + \mathbf{a}_z(5 + z)$, what do we also know about $\mathbf{E}_2$ and $\mathbf{D}_2$ in region 2? Can we determine $\mathbf{E}_2$ and $\mathbf{D}_2$ at any point in region 2? Explain.

P.3–26 Determine the boundary conditions for the tangential and the normal components of $\mathbf{P}$ at an interface between two perfect dielectric media with dielectric constants ϵ_{r1} and ϵ_{r2}.

P.3–27 What are the boundary conditions that must be satisfied by the electric potential at an interface between two perfect dielectrics with dielectric constants ϵ_{r1} and ϵ_{r2}?

P.3–28 Dielectric lenses can be used to collimate electromagnetic fields. In Fig. 3–41 the left surface of the lens is that of a circular cylinder, and the right surface is a plane. If $\mathbf{E}_1$ at point $P(r_o, 45°, z)$ in region 1 is $\mathbf{a}_r 5 - \mathbf{a}_\phi 3$, what must be the dielectric constant of the lens in order that $\mathbf{E}_3$ in region 3 is parallel to the x-axis?

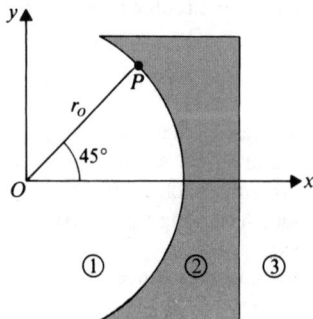

FIGURE 3–41
A dielectric lens (Problem P.3–28).

P.3–29 Refer to Example 3–16. Assuming the same r_i and r_o and requiring the maximum electric field intensities in the insulating materials not to exceed 25% of their dielectric strengths, determine the voltage rating of the coaxial cable

    a) if $r_p = 1.75r_i$;

    b) if $r_p = 1.35r_i$.

    c) Plot the variations of E_r and V versus r for both part (a) and part (b).

P.3–30 The space between a parallel-plate capacitor of area S is filled with a dielectric whose permittivity varies linearly from ϵ_1 at one plate $(y = 0)$ to ϵ_2 at the other plate $(y = d)$. Neglecting fringing effect, find the capacitance.

P.3–31 Assume that the outer conductor of the cylindrical capacitor in Example 3–18 is grounded and that the inner conductor is maintained at a potential V_0.

    a) Find the electric field intensity, $E(a)$, at the surface of the inner conductor.

    b) With the inner radius, b, of the outer conductor fixed, find a so that $E(a)$ is minimized.

    c) Find this minimum $E(a)$.

    d) Determine the capacitance under the conditions of part (b).

P.3–32 The radius of the core and the inner radius of the outer conductor of a very long coaxial transmission line are r_i and r_o, respectively. The space between the conductors is filled with two coaxial layers of dielectrics. The dielectric constants of the dielectrics are ϵ_{r1} for $r_i < r < b$ and ϵ_{r2} for $b < r < r_o$. Determine its capacitance per unit length.

P.3–33 A cylindrical capacitor of length L consists of coaxial conducting surfaces of radii r_i and r_o. Two dielectric media of different dielectric constants ϵ_{r1} and ϵ_{r2} fill the space between the conducting surfaces as shown in Fig. 3–42. Determine its capacitance.

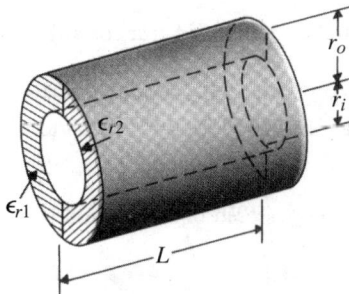

FIGURE 3–42
A cylindrical capacitor with two dielectric media
(Problem P.3–33).

P.3–34 A capacitor consists of two coaxial metallic cylindrical surfaces of a length 30 (mm) and radii 5 (mm) and 7 (mm). The dielectric material between the surfaces has a relative permittivity $\epsilon_r = 2 + (4/r)$, where r is measured in mm. Determine the capacitance of the capacitor.

P.3–35 Assuming the earth to be a large conducting sphere (radius $= 6.37 \times 10^3$ km) surrounded by air, find

    a) the capacitance of the earth;

    b) the maximum charge that can exist on the earth before the air breaks down.

P.3–36 Determine the capacitance of an isolated conducting sphere of radius b that is coated with a dielectric layer of uniform thickness d. The dielectric has an electric susceptibility χ_e.

P.3–37 A capacitor consists of two concentric spherical shells of radii R_i and R_o. The space between them is filled with a dielectric of relative permittivity ϵ_r from R_i to $b(R_i < b < R_o)$ and another dielectric of relative permittivity $2\epsilon_r$ from b to R_o.

a) Determine **E** and **D** everywhere in terms of an applied voltage V.

b) Determine the capacitance.

P.3–38 The two parallel conducting wires of a power transmission line have a radius a and are spaced at a distance d apart. The wires are at a height h above the ground. Assuming the ground to be perfectly conducting and both d and h to be much larger than a, find the expressions for the mutual and self-partial capacitances per unit length.

P.3–39 An isolated system consists of three very long parallel conducting wires. The axes of all three wires lie in a plane. The two outside wires are of a radius b and both are at a distance $d = 500b$ from a center wire of a radius $2b$. Determine the partial capacitances per unit length.

P.3–40 Calculate the amount of electrostatic energy of a uniform sphere of charge with radius b and volume charge density ρ stored in the following regions:

a) inside the sphere,

b) outside the sphere.

Check your results with those in Example 3–22.

P.3–41 Einstein's theory of relativity stipulates that the work required to assemble a charge is stored as energy in the mass and is equal to mc^2, where m is the mass and $c \cong 3 \times 10^8$ (m/s) is the velocity of light. Assuming the electron to be a perfect sphere, find its radius from its charge and mass (9.1×10^{-31} kg).

P.3–42 Find the electrostatic energy stored in the region of space $R > b$ around an electric dipole of moment **p**.

P.3–43 Prove that Eqs. (3–180) for stored electrostatic energy hold true for any two-conductor capacitor.

P.3–44 A parallel-plate capacitor of width w, length L, and separation d is partially filled with a dielectric medium of dielectric constant ϵ_r, as shown in Fig. 3–43. A battery of V_0 volts is connected between the plates.

a) Find **D**, **E**, and ρ_s in each region.

b) Find distance x such that the electrostatic energy stored in each region is the same.

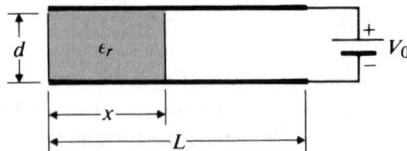

FIGURE 3–43
A parallel-plate capacitor (Problem P.3–44).

P.3–45 Using the principle of virtual displacement, derive an expression for the force between two point charges $+Q$ and $-Q$ separated by a distance x in free space.

P.3–46 A constant voltage V_0 is applied to a partially filled parallel-plate capacitor shown in Fig. 3–44. The permittivity of the dielectric is ϵ, and the area of the plates is S. Find the force on the upper plate.

P.3–47 The conductors of an isolated two-wire transmission line, each of radius b, are spaced at a distance D apart. Assuming $D \gg b$ and a voltage V_0 between the lines, find the force per unit length on the lines.

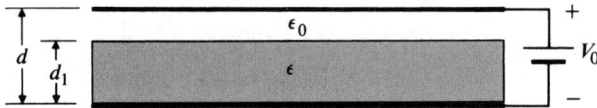

FIGURE 3–44
A parallel-plate capacitor (Problem P.3–46).

P.3–48 A parallel-plate capacitor of width w, length L, and separation d has a solid dielectric slab of permittivity ϵ in the space between the plates. The capacitor is charged to a voltage V_0 by a battery, as indicated in Fig. 3–45. Assuming that the dielectric slab is withdrawn to the position shown, determine the force acting on the slab

    a) with the switch closed,

    b) after the switch is first opened.

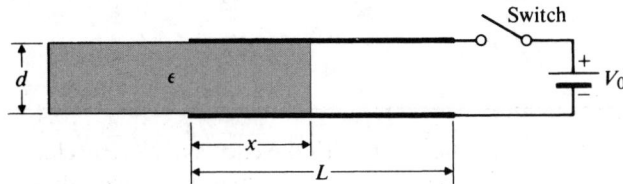

FIGURE 3–45
A partially filled parallel-plate capacitor (Problem P.3–48).

4

Solution of
Electrostatic Problems

4–1 Introduction

Electrostatic problems are those which deal with the effects of electric charges at rest. These problems can present themselves in several different ways according to what is initially known. The solution usually calls for the determination of electric potential, electric field intensity, and/or electric charge distribution. If the charge distribution is given, both the electric potential and the electric field intensity can be found by the formulas developed in Chapter 3. In many practical problems, however, the exact charge distribution is not known everywhere, and the formulas in Chapter 3 cannot be applied directly for finding the potential and field intensity. For instance, if the charges at certain discrete points in space and the potentials of some conducting bodies are given, it is rather difficult to find the distribution of surface charges on the conducting bodies and/or the electric field intensity in space. When the conducting bodies have boundaries of a simple geometry, the *method of images* may be used to great advantage. This method will be discussed in Section 4–4.

In another type of problem the potentials of all conducting bodies may be known, and we wish to find the potential and field intensity in the surrounding space as well as the distribution of surface charges on the conducting boundaries. Differential equations must be solved subject to the appropriate boundary conditions. These are *boundary-value problems*. The techniques for solving boundary-value problems in the various coordinate systems will be discussed in Sections 4–5 through 4–7.

4–2 Poisson's and Laplace's Equations

In Section 3–8 we pointed out that Eqs. (3–98) and (3–5) are the two fundamental governing differential equations for electrostatics in any medium. These equations are repeated below for convenience.

$$\text{Eq. (3–98):} \quad \mathbf{\nabla} \cdot \mathbf{D} = \rho. \tag{4–1}$$

$$\text{Eq. (3–5):} \quad \mathbf{\nabla} \times \mathbf{E} = 0. \tag{4–2}$$

The irrotational nature of $\mathbf{E}$ indicated by Eq. (4–2) enables us to define a scalar electric potential V, as in Eq. (3–43).

$$\text{Eq. (3–43):} \quad \mathbf{E} = -\mathbf{\nabla}V. \tag{4–3}$$

In a linear and isotropic medium $\mathbf{D} = \epsilon\mathbf{E}$, and Eq. (4–1) becomes

$$\mathbf{\nabla} \cdot \epsilon\mathbf{E} = \rho. \tag{4–4}$$

Substitution of Eq. (4–3) in Eq. (4–4) yields

$$\mathbf{\nabla} \cdot (\epsilon\mathbf{\nabla}V) = -\rho, \tag{4–5}$$

where ϵ can be a function of position. For a simple medium; that is, for a medium that is also homogeneous, ϵ is a constant and can then be taken out of the divergence operation. We have

$$\boxed{\nabla^2 V = -\frac{\rho}{\epsilon}.} \tag{4–6}$$

In Eq. (4–6) we have introduced a new operator, ∇^2 (del square), the **Laplacian operator**, which stands for "the divergence of the gradient of," or $\mathbf{\nabla} \cdot \mathbf{\nabla}$. Equation (4–6) is known as **Poisson's equation**; it states that the Laplacian (the divergence of the gradient) of V equals $-\rho/\epsilon$ *for a simple medium*, where ϵ is the permittivity of the medium (which is a constant) and ρ is the volume density of free charges (which may be a function of space coordinates).

Since both divergence and gradient operations involve first-order spatial derivatives, Poisson's equation is a second-order partial differential equation that holds at every point in space where the second-order derivatives exist. In Cartesian coordinates,

$$\nabla^2 V = \mathbf{\nabla} \cdot \mathbf{\nabla}V = \left(\mathbf{a}_x\frac{\partial}{\partial x} + \mathbf{a}_y\frac{\partial}{\partial y} + \mathbf{a}_z\frac{\partial}{\partial z}\right) \cdot \left(\mathbf{a}_x\frac{\partial V}{\partial x} + \mathbf{a}_y\frac{\partial V}{\partial y} + \mathbf{a}_z\frac{\partial V}{\partial z}\right);$$

and Eq. (4–6) becomes

$$\boxed{\frac{\partial^2 V}{\partial x^2} + \frac{\partial^2 V}{\partial y^2} + \frac{\partial^2 V}{\partial z^2} = -\frac{\rho}{\epsilon} \quad (\text{V/m}^2).} \tag{4–7}$$

Similarly, by using Eqs. (2–93) and (2–110) we can easily verify the following expressions for $\nabla^2 V$ in cylindrical and spherical coordinates.

Cylindrical coordinates:

$$\nabla^2 V = \frac{1}{r}\frac{\partial}{\partial r}\left(r\frac{\partial V}{\partial r}\right) + \frac{1}{r^2}\frac{\partial^2 V}{\partial \phi^2} + \frac{\partial^2 V}{\partial z^2}. \tag{4–8}$$

Spherical coordinates:

$$\mathbf{V}^2 V = \frac{1}{R^2} \frac{\partial}{\partial R} \left(R^2 \frac{\partial V}{\partial R} \right) + \frac{1}{R^2 \sin \theta} \frac{\partial}{\partial \theta} \left(\sin \theta \frac{\partial V}{\partial \theta} \right) + \frac{1}{R^2 \sin^2 \theta} \frac{\partial^2 V}{\partial \phi^2}. \qquad (4\text{-}9)$$

The solution of Poisson's equation in three dimensions subject to prescribed boundary conditions is, in general, not an easy task.

At points in a simple medium where there is no free charge, $\rho = 0$ and Eq. (4–6) reduces to

$$\boxed{\mathbf{V}^2 V = 0,} \qquad (4\text{-}10)$$

which is known as **Laplace's equation**. Laplace's equation occupies a very important position in electromagnetics. It is the governing equation for problems involving a set of conductors, such as capacitors, maintained at different potentials. Once V is found from Eq. (4–10), $\mathbf{E}$ can be determined from $-\nabla V$, and the charge distribution on the conductor surfaces can be determined from $\rho_s = \epsilon E_n$ (Eq. 3–72).

EXAMPLE 4–1   The two plates of a parallel-plate capacitor are separated by a distance d and maintained at potentials 0 and V_0, as shown in Fig. 4–1. Assuming negligible fringing effect at the edges, determine (a) the potential at any point between the plates, and (b) the surface charge densities on the plates.

Solution

a) Laplace's equation is the governing equation for the potential between the plates, since $\rho = 0$ there. Ignoring the fringing effect of the electric field is tantamount to assuming that the field distribution between the plates is the same as though the plates were infinitely large and that there is no variation of V in the x and z directions. Equation (4–7) then simplifies to

$$\frac{d^2 V}{dy^2} = 0, \qquad (4\text{-}11)$$

where d^2/dy^2 is used instead of $\partial^2/\partial y^2$, since y is the only space variable here.

Integration of Eq. (4–11) with respect to y gives

$$\frac{dV}{dy} = C_1,$$

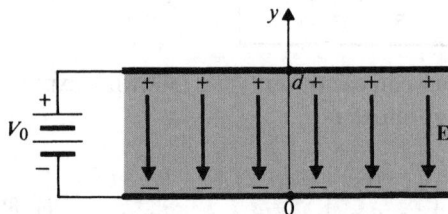

FIGURE 4–1
A parallel-plate capacitor (Example 4–1).

where the constant of integration C_1 is yet to be determined. Integrating again, we obtain

$$V = C_1 y + C_2.$$ (4–12)

Two boundary conditions are required for the determination of the two constants of integration, C_1 and C_2:

$$\text{At } y = 0, \qquad V = 0.$$ (4–13a)

$$\text{At } y = d, \qquad V = V_0.$$ (4–13b)

Substitution of Eqs. (4–13a) and (4–13b) in Eq. (4–12) yields immediately $C_1 = V_0/d$ and $C_2 = 0$. Hence the potential at any point y between the plates is, from Eq. (4–12),

$$V = \frac{V_0}{d} y.$$ (4–14)

The potential increases linearly from $y = 0$ to $y = d$.

b) In order to find the surface charge densities, we must first find **E** at the conducting plates at $y = 0$ and $y = d$. From Eqs. (4–3) and (4–14) we have

$$\mathbf{E} = -\mathbf{a}_y \frac{dV}{dy} = -\mathbf{a}_y \frac{V_0}{d},$$ (4–15)

which is a constant and is independent of y. Note that the direction of **E** is opposite to the direction of increasing V. The surface charge densities at the conducting plates are obtained by using Eq. (3–72),

$$E_n = \mathbf{a}_n \cdot \mathbf{E} = \frac{\rho_s}{\epsilon}.$$

At the lower plate,

$$\mathbf{a}_n = \mathbf{a}_y, \qquad E_{n\ell} = -\frac{V_0}{d}, \qquad \rho_{s\ell} = -\frac{\epsilon V_0}{d}.$$

At the upper plate,

$$\mathbf{a}_n = -\mathbf{a}_y, \qquad E_{nu} = \frac{V_0}{d}, \qquad \rho_{su} = \frac{\epsilon V_0}{d}.$$

Electric field lines in an electrostatic field begin from positive charges and end in negative charges. ▄

EXAMPLE 4–2 Determine the **E** field both inside and outside a spherical cloud of electrons with a uniform volume charge density $\rho = -\rho_0$ (where ρ_0 is a positive quantity) for $0 \le R \le b$ and $\rho = 0$ for $R > b$ by solving Poisson's and Laplace's equations for V.

Solution We recall that this problem was solved in Chapter 3 (Example 3–7) by applying Gauss's law. We now use the same problem to illustrate the solution of one-dimensional Poisson's and Laplace's equations. Since there are no variations in θ and ϕ directions, we are dealing only with functions of R in spherical coordinates.

a) Inside the cloud,

$$0 \leq R \leq b, \qquad \rho = -\rho_0.$$

In this region, Poisson's equation ($\nabla^2 V_i = -\rho/\epsilon_0$) holds. Dropping $\partial/\partial\theta$ and $\partial/\partial\phi$ terms from Eq. (4–9), we have

$$\frac{1}{R^2} \frac{d}{dR} \left(R^2 \frac{dV_i}{dR} \right) = \frac{\rho_0}{\epsilon_0},$$

which reduces to

$$\frac{d}{dR} \left(R^2 \frac{dV_i}{dR} \right) = \frac{\rho_0}{\epsilon_0} R^2. \tag{4–16}$$

Integration of Eq. (4–16) gives

$$\frac{dV_i}{dR} = \frac{\rho_0}{3\epsilon_0} R + \frac{C_1}{R^2}. \tag{4–17}$$

The electric field intensity inside the electron cloud is

$$\mathbf{E}_i = -\nabla V_i = -\mathbf{a}_R \left(\frac{dV_i}{dR} \right).$$

Since $\mathbf{E}_i$ cannot be infinite at $R = 0$, the integration constant C_1 in Eq. (4–17) must vanish. We obtain

$$\mathbf{E}_i = -\mathbf{a}_R \frac{\rho_0}{3\epsilon_0} R, \qquad 0 \leq R \leq b. \tag{4–18}$$

b) Outside the cloud,

$$R \geq b, \qquad \rho = 0.$$

Laplace's equation holds in this region. We have $\nabla^2 V_o = 0$ or

$$\frac{1}{R^2} \frac{\partial}{\partial R} \left(R^2 \frac{dV_o}{dR} \right) = 0. \tag{4–19}$$

Integrating Eq. (4–19), we obtain

$$\frac{dV_o}{dR} = \frac{C_2}{R^2} \tag{4–20}$$

or

$$\mathbf{E}_o = -\nabla V_o = -\mathbf{a}_R \frac{dV_o}{dR} = -\mathbf{a}_R \frac{C_2}{R^2}. \tag{4–21}$$

The integration constant C_2 can be found by equating $\mathbf{E}_o$ and $\mathbf{E}_i$ at $R = b$, where there is no discontinuity in medium characteristics.

$$\frac{C_2}{b^2} = \frac{\rho_0}{3\epsilon_0} b,$$

from which we find

$$C_2 = \frac{\rho_0 b^3}{3\epsilon_0} \tag{4–22}$$

and

$$\mathbf{E}_o = -\mathbf{a}_R \frac{\rho_0 b^3}{3\epsilon_0 R^2}, \qquad R \geq b. \tag{4–23}$$

Since the total charge contained in the electron cloud is

$$Q = -\rho_0 \frac{4\pi}{3} b^3,$$

Eq. (4–23) can be written as

$$\mathbf{E}_o = \mathbf{a}_R \frac{Q}{4\pi\epsilon_0 R^2}, \qquad (4-24)$$

which is the familiar expression for the electric field intensity at a point R from a point charge Q. ▬

Further insight to this problem can be gained by examining the potential as a function of R. Integrating Eq. (4–17), remembering that $C_1 = 0$, we have

$$V_i = \frac{\rho_0 R^2}{6\epsilon_0} + C_1'. \qquad (4-25)$$

It is important to note that C_1' is a new integration constant and is not the same as C_1. Substituting Eq. (4–22) in Eq. (4–20) and integrating, we obtain

$$V_o = -\frac{\rho_0 b^3}{3\epsilon_0 R} + C_2'. \qquad (4-26)$$

However, C_2' in Eq. (4–26) must vanish, since V_o is zero at infinity ($R \to \infty$). As electrostatic potential is continuous at a boundary, we determine C_1' by equating V_i and V_o at $R = b$:

$$\frac{\rho_0 b^2}{6\epsilon_0} + C_1' = -\frac{\rho_0 b^2}{3\epsilon_0}$$

or

$$C_1' = -\frac{\rho_0 b^2}{2\epsilon_0}; \qquad (4-27)$$

and from Eq. (4–25),

$$V_i = -\frac{\rho_0}{3\epsilon_0} \left(\frac{3b^2}{2} - \frac{R^2}{2} \right). \qquad (4-28)$$

We see that V_i in Eq. (4–28) is the same as V in Eq. (3–172), with $\rho = -\rho_0$.

4-3   Uniqueness of Electrostatic Solutions

In the two relatively simple examples in the last section we obtained the solutions by direct integration. In more complicated situations, other methods of solution must be used. Before these methods are discussed, it is important to know that *a solution of Poisson's equation* (of which Laplace's equation is a special case) *that satisfies the given boundary conditions is a unique solution.* This statement is called the *uniqueness theorem.* The implication of the uniqueness theorem is that a solution of an electrostatic problem satisfying its boundary conditions is *the only possible*

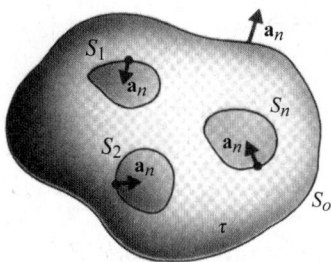

FIGURE 4–2
Surface S_o enclosing volume τ with conducting bodies.

solution, irrespective of the method by which the solution is obtained. A solution obtained even by intelligent guessing is the only correct solution. The importance of this theorem will be appreciated when we discuss the method of images in Section 4–4.

To prove the uniqueness theorem, suppose a volume τ is bounded outside by a surface S_o, which may be a surface at infinity. Inside the closed surface S_o there are a number of charged conducting bodies with surfaces $S_1, S_2, \ldots, S_n$ at specified potentials, as depicted in Fig. 4–2. Now assume that, contrary to the uniqueness theorem, there are two solutions, V_1 and V_2, to Poisson's equation in τ:

$$\nabla^2 V_1 = -\frac{\rho}{\epsilon}, \tag{4–29a}$$

$$\nabla^2 V_2 = -\frac{\rho}{\epsilon}. \tag{4–29b}$$

Also assume that both V_1 and V_2 satisfy the *same* boundary conditions on S_1, $S_2, \ldots, S_n$ and S_o. Let us try to define a new difference potential:

$$V_d = V_1 - V_2. \tag{4–30}$$

From Eqs. (4–29a) and (4–29b) we see that V_d satisfies Laplace's equation in τ:

$$\nabla^2 V_d = 0. \tag{4–31}$$

On conducting boundaries the potentials are specified and $V_d = 0$.

Recalling the vector identity (Problem P.2–28),

$$\nabla \cdot (f\mathbf{A}) = f\nabla \cdot \mathbf{A} + \mathbf{A} \cdot \nabla f; \tag{4–32}$$

and letting $f = V_d$ and $\mathbf{A} = \nabla V_d$; we have

$$\nabla \cdot (V_d \nabla V_d) = V_d \nabla^2 V_d + |\nabla V_d|^2, \tag{4–33}$$

where, because of Eq. (4–31), the first term on the right side vanishes. Integration of Eq. (4–33) over the volume τ yields

$$\oint_S (V_d \nabla V_d) \cdot \mathbf{a}_n \, ds = \int_\tau |\nabla V_d|^2 \, dv, \tag{4–34}$$

where $\mathbf{a}_n$ denotes the unit normal outward from τ. Surface S consists of S_o as well as $S_1, S_2, \ldots,$ and S_n. Over the conducting boundaries, $V_d = 0$. Over the large surface

S_o, which encloses the whole system, the surface integral on the left side of Eq. (4–34) can be evaluated by considering S_o as the surface of a very large sphere with radius R. As R increases, both V_1 and V_2 (and therefore also V_d) fall off as $1/R$; consequently, ∇V_d falls off as $1/R^2$, making the integrand $(V_d \nabla V_d)$ fall off as $1/R^3$. The surface area S_o, however, increases as R^2. Hence the surface integral on the left side of Eq. (4–34) decreases as $1/R$ and approaches zero at infinity. So must also the volume integral on the right side. We have

$$\int_\tau |\nabla V_d|^2 \, dv = 0. \tag{4–35}$$

Since the integrand $|\nabla V_d|^2$ is nonnegative everywhere, Eq. (4–35) can be satisfied only if $|\nabla V_d|$ is identically zero. A vanishing gradient everywhere means that V_d has the same value at all points in τ as it has on the bounding surfaces, $S_1, S_2, \ldots, S_n$, where $V_d = 0$. It follows that $V_d = 0$ throughout the volume τ. Therefore $V_1 = V_2$, and there is only one possible solution.

It is easy to see that the uniqueness theorem holds if the surface charge distributions ($\rho_s = \epsilon E_n = -\epsilon \, \partial V/\partial n$), rather than the potentials, of the conducting bodies are specified. In such a case, ∇V_d will be zero, which in turn, makes the left side of Eq. (4–34) vanish and leads to the same conclusion. In fact, the uniqueness theorem applies even if an inhomogeneous dielectric (one whose permittivity varies with position) is present. The proof, however, is more involved and will be omitted here.

4-4 Method of Images

There is a class of electrostatic problems with boundary conditions that appear to be difficult to satisfy if the governing Poisson's or Laplace's equation is to be solved directly, but the conditions on the bounding surfaces in these problems can be set up by appropriate *image* (equivalent) *charges*, and the potential distributions can then be determined in a straightforward manner. This method of replacing bounding surfaces by appropriate image charges in lieu of a formal solution of Poisson's or Laplace's equation is called the *method of images*.

Consider the case of a positive point charge, Q, located at a distance d above a large grounded (zero-potential) conducting plane, as shown in Fig. 4–3(a). The problem is to find the potential at every point above the conducting plane ($y > 0$). The formal procedure for doing so would be to solve Laplace's equation in Cartesian coordinates:

$$\nabla^2 V = \frac{\partial^2 V}{\partial x^2} + \frac{\partial^2 V}{\partial y^2} + \frac{\partial^2 V}{\partial z^2} = 0, \tag{4–36}$$

which must hold for $y > 0$ except at the point charge. The solution $V(x, y, z)$ should satisfy the following conditions:

1. At all points on the grounded conducting plane, the potential is zero; that is,

$$V(x, 0, z) = 0.$$

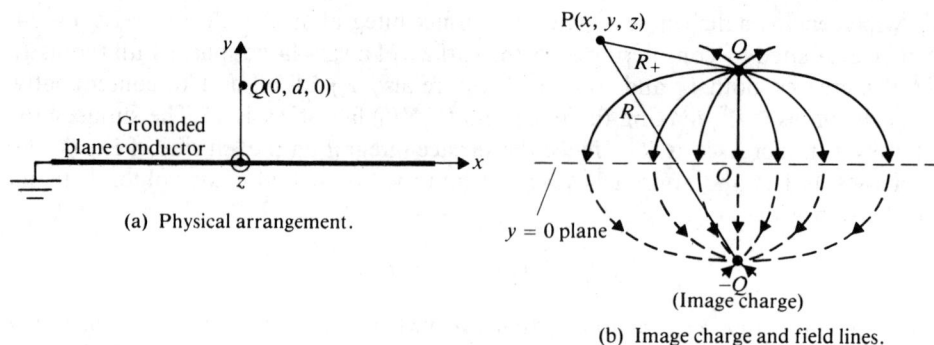

(a) Physical arrangement.

(b) Image charge and field lines.

FIGURE 4-3
Point charge and grounded plane conductor.

2. At points very close to Q the potential approaches that of the point charge alone; that is

$$V \to \frac{Q}{4\pi\epsilon_0 R}, \text{ as } R \to 0,$$

where R is the distance to Q.

3. At points very far from $Q(x \to \pm\infty, y \to +\infty, \text{ or } z \to \pm\infty)$ the potential approaches zero.

4. The potential function is even with respect to the x and z coordinates; that is,

$$V(x, y, z) = V(-x, y, z)$$

and

$$V(x, y, z) = V(x, y, -z).$$

It does appear difficult to construct a solution for V that will satisfy all of these conditions.

From another point of view, we may reason that the presence of a positive charge Q at $y = d$ would induce negative charges on the surface of the conducting plane, resulting in a surface charge density ρ_s. Hence the potential at points above the conducting plane would be

$$V(x, y, z) = \frac{Q}{4\pi\epsilon_0 \sqrt{x^2 + (y - d)^2 + z^2}} + \frac{1}{4\pi\epsilon_0} \int_S \frac{\rho_s}{R_1} \, ds,$$

where R_1 is the distance from ds to the point under consideration and S is the surface of the entire conducting plane. The trouble here is that ρ_s must first be determined from the boundary condition $V(x, 0, z) = 0$. Moreover, the indicated surface integral is difficult to evaluate even after ρ_s has been determined at every point on the conducting plane. In the following subsections we demonstrate how the method of images greatly simplifies these problems.

4–4.1 POINT CHARGE AND CONDUCTING PLANES

The problem in Fig. 4–3(a) is that of a positive point charge, Q, located at a distance d above a large plane conductor that is at zero potential. If we remove the conductor and replace it by an image point charge $-Q$ at $y = -d$, then the potential at a point $P(x, y, z)$ *in the* $y > 0$ *region* is

$$V(x, y, z) = \frac{Q}{4\pi\epsilon_0}\left(\frac{1}{R_+} - \frac{1}{R_-}\right), \qquad (4\text{–}37)$$

where R_+ and R_- are the distances from Q and $-Q$, respectively, to the point P.

$$R_+ = [x^2 + (y - d)^2 + z^2]^{1/2},$$
$$R_- = [x^2 + (y + d)^2 + z^2]^{1/2}.$$

It is easy to prove by direct substitution (Problem P.4–5a) that $V(x, y, z)$ in Eq. (4–37) satisfies the Laplace's equation in Eq. (4–36), and it is obvious that all four conditions listed after Eq. (4–36) are satisfied. Therefore Eq. (4–37) is a solution of this problem; and, in view of the uniqueness theorem, it is the only solution.

Electric field intensity $\mathbf{E}$ in the $y > 0$ region can be found easily from $-\nabla V$ with Eq. (4–37). It is exactly the same as that between two point charges, $+Q$ and $-Q$, spaced a distance $2d$ apart. A few of the field lines are shown in Fig. 4–3(b). The solution of this electrostatic problem by the method of images is extremely simple; but it must be emphasized that the image charge is located *outside* the region in which the field is to be determined. In this problem the point charges $+Q$ and $-Q$ *cannot* be used to calculate the V or $\mathbf{E}$ in the $y < 0$ region. As a matter of fact, both V and $\mathbf{E}$ are zero in the $y < 0$ region. Can you explain that?

It is readily seen that the electric field of a line charge ρ_ℓ above an infinite conducting plane can be found from ρ_ℓ and its image $-\rho_\ell$ (with the conducting plane removed).

EXAMPLE 4–3 A positive point charge Q is located at distances d_1 and d_2, respectively, from two grounded perpendicular conducting half-planes, as shown in Fig. 4–4(a). Determine the force on Q caused by the charges induced on the planes.

Solution A formal solution of Poisson's equation, subject to the zero-potential boundary condition at the conducting half-planes, would be quite difficult. Now an image charge $-Q$ in the fourth quadrant would make the potential of the horizontal half-plane (but not that of the vertical half-plane) zero. Similarly, an image charge $-Q$ in the second quadrant would make the potential of the vertical half-plane (but not that of the horizontal plane) zero. But if a third image charge $+Q$ is added in the third quadrant, we see from symmetry that the image-charge arrangement in Fig. 4–4(b) satisfies the zero-potential boundary condition on both half-planes and is electrically equivalent to the physical arrangement in Fig. 4–4(a).

Negative surface charges will be induced on the half-planes, but their effect on Q can be determined from that of the three image charges. Referring to Fig. 4–4(c),

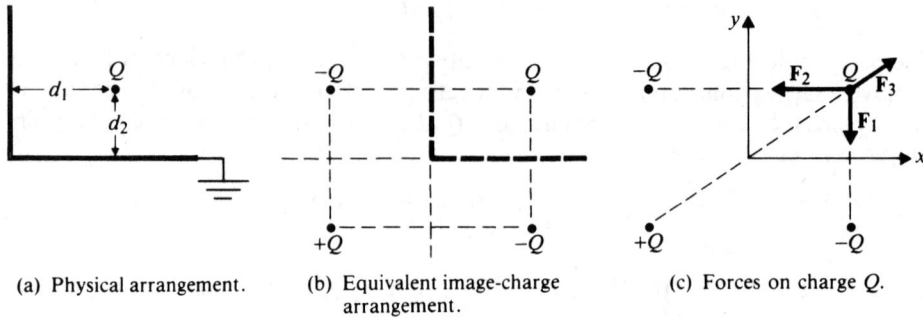

(a) Physical arrangement.   (b) Equivalent image-charge   (c) Forces on charge Q.
                                     arrangement.

FIGURE 4–4
Point charge and perpendicular conducting planes.

we have, for the net force on Q,

$$\mathbf{F} = \mathbf{F}_1 + \mathbf{F}_2 + \mathbf{F}_3,$$

where

$$\mathbf{F}_1 = -\mathbf{a}_y \frac{Q^2}{4\pi\epsilon_0 (2d_2)^2},$$

$$\mathbf{F}_2 = -\mathbf{a}_x \frac{Q^2}{4\pi\epsilon_0 (2d_1)^2},$$

$$\mathbf{F}_3 = \frac{Q^2}{4\pi\epsilon_0 [(2d_1)^2 + (2d_2)^2]^{3/2}} (\mathbf{a}_x 2d_1 + \mathbf{a}_y 2d_2).$$

Therefore,

$$\mathbf{F} = \frac{Q^2}{16\pi\epsilon_0} \left\{ \mathbf{a}_x \left[\frac{d_1}{(d_1^2 + d_2^2)^{3/2}} - \frac{1}{d_1^2} \right] + \mathbf{a}_y \left[\frac{d_2}{(d_1^2 + d_2^2)^{3/2}} - \frac{1}{d_2^2} \right] \right\}.$$

The electric potential and electric field intensity at points *in the first quadrant* and the surface charge density induced on the two half-planes can also be found from the system of four charges (Problem P.4–8).

4–4.2 LINE CHARGE AND PARALLEL CONDUCTING CYLINDER

We now consider the problem of a line charge ρ_ℓ (C/m) located at a distance d from the axis of a parallel, conducting, circular cylinder of radius a. Both the line charge and the conducting cylinder are assumed to be infinitely long. Figure 4–5(a) shows a cross section of this arrangement. Preparatory to the solution of this problem by the method of images, we note the following: (1) The image must be a parallel line charge inside the cylinder in order to make the cylindrical surface at $r = a$ an equipotential surface. Let us call this image line charge ρ_i. (2) Because of symmetry with respect to the line OP, the image line charge must lie somewhere along OP, say at point P_i, which is at a distance d_i from the axis (Fig. 4–5b). We need to determine the two unknowns, ρ_i and d_i.

As a first approach, let us assume that

$$\rho_i = -\rho_\ell. \tag{4–38}$$

At this stage, Eq. (4–38) is just a trial solution (an intelligent guess), and we are not sure that it will hold true. We will, on one hand, proceed with this trial solution until we find that it fails to satisfy the boundary conditions. On the other hand, if Eq. (4–38) leads to a solution that does satisfy all boundary conditions, then by the uniqueness theorem it is the only solution. Our next job will be to see whether we can determine d_i.

The electric potential at a distance r from a line charge of density ρ_ℓ can be obtained by integrating the electric field intensity $\mathbf{E}$ given in Eq. (3–40):

$$V = -\int_{r_0}^{r} E_r \, dr = -\frac{\rho_\ell}{2\pi\epsilon_0} \int_{r_0}^{r} \frac{1}{r} \, dr$$

$$= \frac{\rho_\ell}{2\pi\epsilon_0} \ln \frac{r_0}{r}. \tag{4–39}$$

Note that the reference point for zero potential, r_0, cannot be at infinity because setting $r_0 = \infty$ in Eq. (4–39) would make V infinite everywhere else. Let us leave r_0 unspecified for the time being. The potential at a point on or outside the cylindrical surface is obtained by adding the contributions of ρ_ℓ and ρ_i. In particular, at a point M on the cylindrical surface shown in Fig. 4–5(b) we have

$$V_M = \frac{\rho_\ell}{2\pi\epsilon_0} \ln \frac{r_0}{r} - \frac{\rho_\ell}{2\pi\epsilon_0} \ln \frac{r_0}{r_i}$$

$$= \frac{\rho_\ell}{2\pi\epsilon_0} \ln \frac{r_i}{r}. \tag{4–40}$$

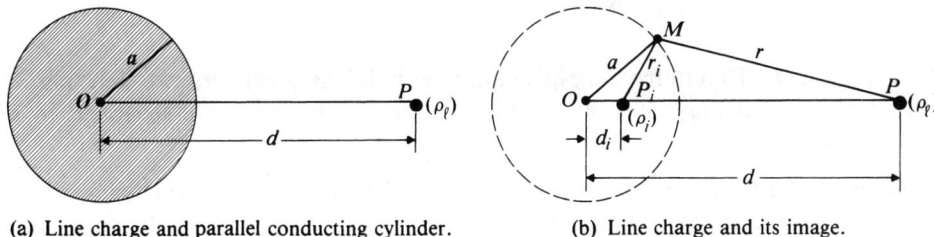

(a) Line charge and parallel conducting cylinder.

(b) Line charge and its image.

FIGURE 4–5
Cross section of line charge and its image in a parallel, conducting, circular cylinder.

In Eq. (4–40) we have chosen, for simplicity, a point equidistant from ρ_ℓ and ρ_i as the reference point for zero potential so that the $\ln r_0$ terms cancel. Otherwise, a constant term should be included in the right side of Eq. (4–40), but it would not affect what follows. Equipotential surfaces are specified by

$$\frac{r_i}{r} = \text{Constant.} \tag{4–41}$$

If an equipotential surface is to coincide with the cylindrical surface ($\overline{OM} = a$), the point P_i must be located in such a way as to make triangles OMP_i and OPM similar. Note that these two triangles already have one common angle, $\angle MOP_i$. Point P_i should be chosen to make $\angle OMP_i = \angle OPM$. We have

$$\frac{\overline{P_iM}}{\overline{PM}} = \frac{\overline{OP_i}}{\overline{OM}} = \frac{\overline{OM}}{\overline{OP}}$$

or

$$\frac{r_i}{r} = \frac{d_i}{a} = \frac{a}{d} = \text{Constant.} \tag{4–42}$$

From Eq. (4–42) we see that if

$$\boxed{d_i = \frac{a^2}{d}} \tag{4–43}$$

the image line charge $-\rho_\ell$, together with ρ_ℓ, will make the dashed cylindrical surface in Fig. 4–5(b) equipotential. As the point M changes its location on the dashed circle, both r_i and r will change; but their ratio remains a constant that equals a/d. Point P_i is called the **inverse point** of P with respect to a circle of radius a.

The image line charge $-\rho_\ell$ can then replace the cylindrical conducting surface, and V and $\mathbf{E}$ at any point outside the surface can be determined from the line charges ρ_ℓ and $-\rho_\ell$. By symmetry we find that the parallel cylindrical surface surrounding the original line charge ρ_ℓ with radius a and its axis at a distance d_i to the right of P is also an equipotential surface. This observation enables us to calculate the capacitance per unit length of an open-wire transmission line consisting of two parallel conductors of circular cross section.

EXAMPLE 4–4 Determine the capacitance per unit length between two long, parallel, circular conducting wires of radius a. The axes of the wires are separated by a distance D.

Solution Refer to the cross section of the two-wire transmission line shown in Fig. 4–6. The equipotential surfaces of the two wires can be considered to have been generated by a pair of line charges $+\rho_\ell$ and $-\rho_\ell$ separated by a distance $(D - 2d_i) = d - d_i$. The potential difference between the two wires is that between any two points

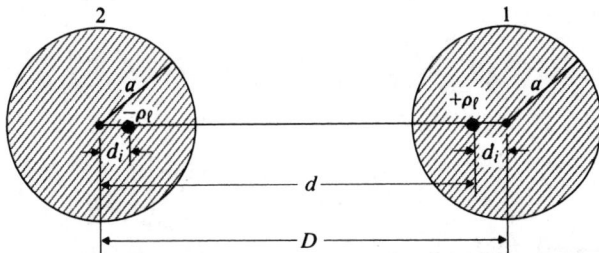

FIGURE 4–6
Cross section of two-wire transmission
line and equivalent line charges
(Example 4–4).

on the respective wires. Let subscripts 1 and 2 denote the wires surrounding the equivalent line charges $+\rho_\ell$ and $-\rho_\ell$, respectively. We have, from Eqs. (4–40) and (4–42).

$$V_2 = \frac{\rho_\ell}{2\pi\epsilon_0} \ln \frac{a}{d}$$

and, similarly,

$$V_1 = -\frac{\rho_\ell}{2\pi\epsilon_0} \ln \frac{a}{d}.$$

We note that V_1 is a positive quantity, whereas V_2 is negative because $a < d$. The capacitance per unit length is

$$C = \frac{\rho_\ell}{V_1 - V_2} = \frac{\pi\epsilon_0}{\ln (d/a)}, \qquad (4\text{–}44)$$

where

$$d = D - d_i = D - \frac{a^2}{d},$$

from which we obtain[†]

$$d = \tfrac{1}{2}(D + \sqrt{D^2 - 4a^2}). \qquad (4\text{–}45)$$

Using Eq. (4–45) in Eq. (4–44), we have

$$\boxed{C = \frac{\pi\epsilon_0}{\ln \left[(D/2a) + \sqrt{(D/2a)^2 - 1}\right]}} \quad \text{(F/m)}. \qquad (4\text{–}46)$$

Since

$$\ln \left[x + \sqrt{x^2 - 1}\right] = \cosh^{-1} x$$

[†] The other solution, $d = \tfrac{1}{2}(D - \sqrt{D^2 - 4a^2})$, is discarded because both D and d are usually much larger than a.

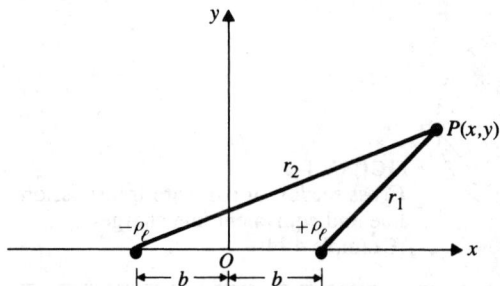

FIGURE 4–7
Cross section of a pair of line charges.

for $x > 1$, Eq. (4–46) can be written alternatively as

$$C = \frac{\pi\epsilon_0}{\cosh^{-1}(D/2a)} \quad \text{(F/m)}. \qquad (4\text{–}47)$$

The potential distribution and electric field intensity around the two-wire line in Fig. 4–6 can also be determined easily from the equivalent line charges.

We now consider the more general case of a two-wire line of different radii. We know that our problem would be solved if we could find the location of the equivalent line charges that make the wire surfaces equipotential. Let us then first study the potential distribution around a pair of positive and negative line charges, a cross section of which is given in Fig. 4–7. The potential at any point $P(x, y)$ due to $+\rho_\ell$ and $-\rho_\ell$ is, from Eq. (4–40),

$$V_P = \frac{\rho_\ell}{2\pi\epsilon_0} \ln \frac{r_2}{r_1}. \qquad (4\text{–}48)$$

In the xy-plane the equipotential lines are defined by $r_2/r_1 = k$ (constant). We have

$$\frac{r_2}{r_1} = \frac{\sqrt{(x + b)^2 + y^2}}{\sqrt{(x - b)^2 + y^2}} = k, \qquad (4\text{–}49)$$

which reduces to

$$\left(x - \frac{k^2 + 1}{k^2 - 1} b\right)^2 + y^2 = \left(\frac{2k}{k^2 - 1} b\right)^2 \qquad (4\text{–}50)$$

Eq. (4–49) represents a family of circles in the xy-plane with radii

$$a = \left|\frac{2kb}{k^2 - 1}\right|, \qquad (4\text{–}51)$$

where the absolute-value sign is necessary because k in Eq. (4–49) can be less than unity and a must be positive. The centers of the circles are displaced from the origin by a distance

$$c = \frac{k^2 + 1}{k^2 - 1} b.$$

(4–52)

A particularly simple relation exists among a, b, and c:

$$c^2 = a^2 + b^2,$$

(4–53)

or

$$b = \sqrt{c^2 - a^2}.$$

(4–54)

Two families of the displaced circular equipotential lines are shown in Fig. 4–8: one family around $+\rho_\ell$ for $k > 1$ and another around $-\rho_\ell$ for $k < 1$. The y-axis is the zero-potential line (a circle of infinite radius) corresponding to $k = 1$. The dashed lines in Fig. 4–8 are circles representing electric field lines, which are everywhere perpendicular to the equipotential lines (Problem P.4–12). Thus the electrostatic

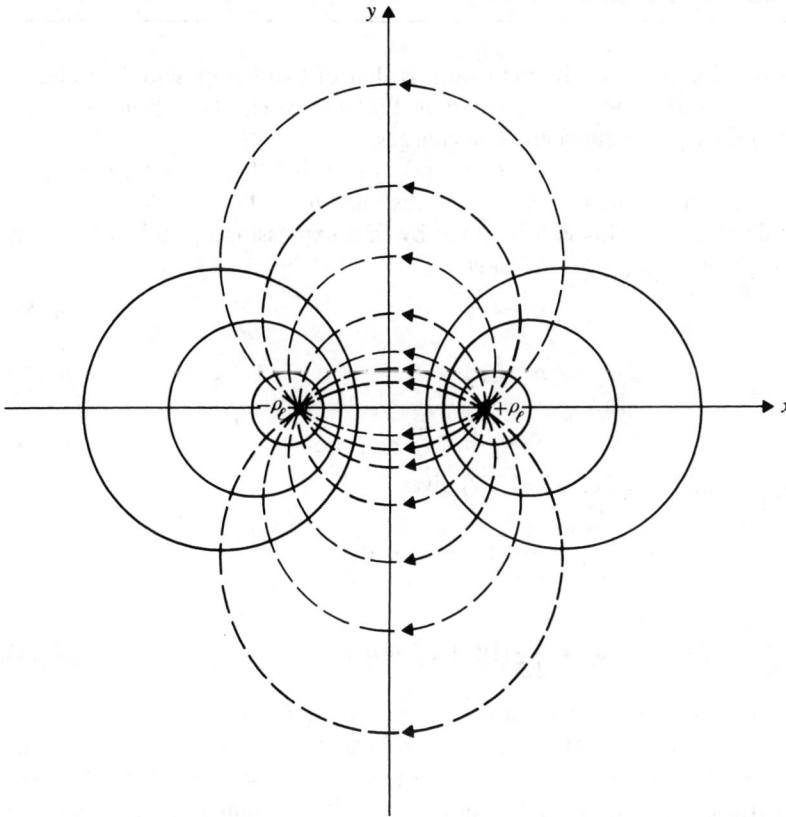

FIGURE 4–8
Equipotential (solid) and electric field (dashed) lines around a pair of line charges.

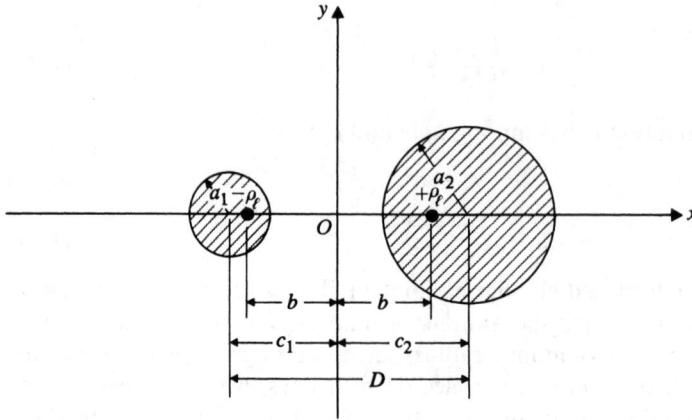

FIGURE 4–9
Cross section of two parallel wires with different radii.

problem of a two-wire line with different radii is that of two equipotential circles of unequal radii, one on each side of the y-axis in Fig. 4–8; it can be solved by determining the locations of the equivalent line charges.

Assume that the radii of the wires are a_1 and a_2 and that their axes are separated by a distance D, as shown in Fig. 4–9. The distance b of the line charges to the origin is to be determined. This can be done by first expressing c_1 and c_2 in terms of a_1, a_2, and D. From Eq. (4–54) we have

$$b^2 = c_1^2 - a_1^2 \tag{4-55}$$

and

$$b^2 = c_2^2 - a_2^2. \tag{4-56}$$

But

$$c_1 + c_2 = D. \tag{4-57}$$

Solution of Eqs. (4–55), (4–56), and (4–57) gives

$$c_1 = \frac{1}{2D}(D^2 + a_1^2 - a_2^2) \tag{4-58}$$

and

$$c_2 = \frac{1}{2D}(D^2 + a_2^2 - a_1^2). \tag{4-59}$$

The distance b can then be found from Eq. (4–55) or Eq. (4–56).

An interesting variation of the two-wire problem is that of an off-center conductor inside a conducting cylindrical tunnel shown in Fig. 4–10(a). Here the two equipotential surfaces are on the same side of a pair of equal and opposite line charges. This is depicted in Fig. 4–10(b). We have, in addition to Eqs. (4–55) and (4–56),

$$c_2 - c_1 = D. \tag{4-60}$$

Combination of Eqs. (4–55), (4–56), and (4–60) yields

$$c_1 = \frac{1}{2D}(a_2^2 - a_1^2 - D^2) \tag{4-61}$$

and

$$c_2 = \frac{1}{2D}(a_2^2 - a_1^2 + D^2). \tag{4-62}$$

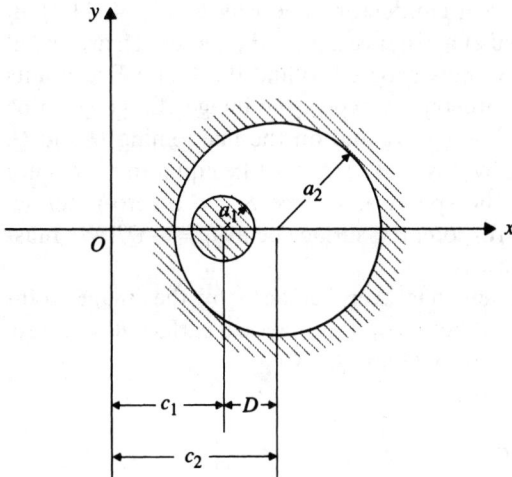

(a) A cross-sectional view.

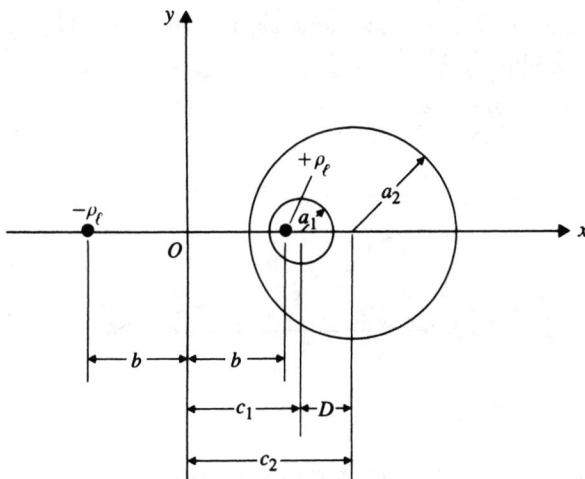

(b) Equivalent line charges.

FIGURE 4–10
An off-center conductor inside a
cylindrical tunnel.

The distance b can be found from Eq. (4–55) or Eq. (4–56). With the locations of the equivalent line charges known, the determination of the potential and electric field distributions and of the capacitance between the conductors per unit length becomes straightforward (Problems P.4–13 and P.4–14).

4–4.3 POINT CHARGE AND CONDUCTING SPHERE

The method of images can also be applied to solve the electrostatic problem of a point charge in the presence of a spherical conductor. Referring to Fig. 4–11(a), in which a positive point charge Q is located at a distance d from the center of a grounded conducting sphere of radius a ($a < d$), we now proceed to find the V and $\mathbf{E}$ at points external to the sphere. By reason of symmetry we expect the image charge Q_i to be a negative point charge situated inside the sphere and on the line joining O and Q. Let it be at a distance d_i from O. It is obvious that Q_i cannot be equal to $-Q$, since $-Q$ and the original Q do not make the spherical surface $R = a$ a zero-potential surface as required. (What would the zero-potential surface be if $Q_i = -Q$?) We must therefore treat both d_i and Q_i as unknowns.

In Fig. 4–11(b) the conducting sphere has been replaced by the image point charge Q_i, which makes the potential at all points on the spherical surface $R = a$ zero. At a typical point M, the potential caused by Q and Q_i is

$$V_M = \frac{1}{4\pi\epsilon_0} \left(\frac{Q}{r} + \frac{Q_i}{r_i} \right) = 0, \tag{4–63}$$

which requires

$$\frac{r_i}{r} = -\frac{Q_i}{Q} = \text{Constant.} \tag{4–64}$$

Noting that the requirement on the ratio r_i/r is the same as that in Eq. (4–41), we conclude from Eqs. (4–42), (4–43), and (4–64) that

$$-\frac{Q_i}{Q} = \frac{a}{d}$$

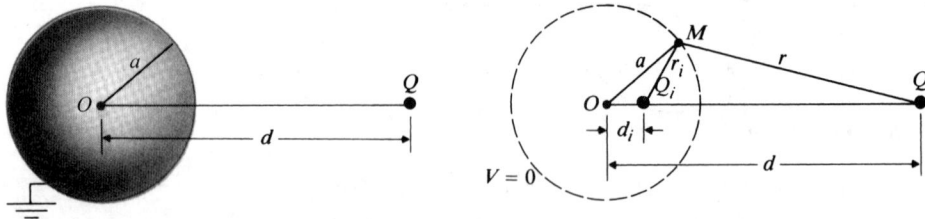

(a) Point charge and grounded conducting sphere.　　　　(b) Point charge and its image.

FIGURE 4–11
Point charge and its image in a grounded sphere.

or

$$Q_i = -\frac{a}{d} Q \tag{4-65}$$

and

$$d_i = \frac{a^2}{d}. \tag{4-66}$$

The point Q_i is thus the **inverse point** of Q with respect to a sphere of radius a. The V and $\mathbf{E}$ of all points external to the grounded sphere can now be calculated from the V and $\mathbf{E}$ caused by the two point charges Q and $-aQ/d$.

■■■■ **EXAMPLE 4–5**   A point charge Q is at a distance d from the center of a grounded conducting sphere of radius a $(a < d)$. Determine (a) the charge distribution induced on the surface of the sphere, and (b) the total charge induced on the sphere.

Solution   The physical problem is that shown in Fig. 4–11(a). We solve the problem by the method of images and replace the grounded sphere by the image charge $Q_i = -aQ/d$ at a distance $d_i = a^2/d$ from the center of the sphere, as shown in Fig. 4–12. The electric potential V at an arbitrary point $P(R, \theta)$ is

$$V(R, \theta) = \frac{Q}{4\pi\epsilon_0} \left(\frac{1}{R_Q} - \frac{a}{dR_{Q_i}} \right), \tag{4-67}$$

where, by the law of cosines,

$$R_Q = [R^2 + d^2 - 2Rd \cos \theta]^{1/2} \tag{4-68}$$

and

$$R_{Q_i} = \left[R^2 + \left(\frac{a^2}{d} \right)^2 - 2R \left(\frac{a^2}{d} \right) \cos \theta \right]^{1/2}. \tag{4-69}$$

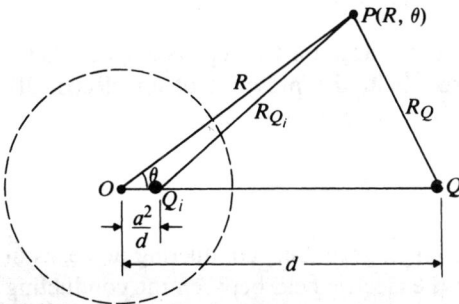

FIGURE 4–12
Diagram for computing induced charge distribution (Example 4–5).

Note that θ is measured from the line OQ. The R-component of the electric field intensity, E_R, is

$$E_R(R, \theta) = -\frac{\partial V(R, \theta)}{\partial R}. \qquad (4\text{--}70)$$

Using Eq. (4–67) in Eq. (4–70) and noting Eqs. (4–68) and (4–69), we have

$$E_R(R, \theta) = \frac{Q}{4\pi\epsilon_0} \left\{ \frac{R - d\cos\theta}{(R^2 + d^2 - 2Rd\cos\theta)^{3/2}} - \frac{a[R - (a^2/d)\cos\theta]}{d[R^2 + (a^2/d)^2 - 2R(a^2/d)\cos\theta]^{3/2}} \right\}. \qquad (4\text{--}71)$$

a) To find the induced surface charge on the sphere, we set $R = a$ in Eq. (4–71) and evaluate

$$\rho_s = \epsilon_0 E_R(a, \theta), \qquad (4\text{--}72)$$

which yields the following after simplification:

$$\rho_s = -\frac{Q(d^2 - a^2)}{4\pi a(a^2 + d^2 - 2ad\cos\theta)^{3/2}}. \qquad (4\text{--}73)$$

Eq. (4–73) tells us that the induced surface charge is negative and that its magnitude is maximum at $\theta = 0$ and minimum at $\theta = \pi$, as expected.

b) The total charge induced on the sphere is obtained by integrating ρ_s over the surface of the sphere. We have

$$\text{Total induced charge} = \oint \rho_s \, ds = \int_0^{2\pi} \int_0^{\pi} \rho_s a^2 \sin\theta \, d\theta \, d\phi$$

$$= -\frac{a}{d} Q = Q_i. \qquad (4\text{--}74)$$

We note that the total induced charge is exactly equal to the image charge Q_i that replaced the sphere. Can you explain this? ∎

   If the conducting sphere is electrically neutral and is not grounded, the image of a point charge Q at a distance d from the center of the sphere would still be Q_i at d_i given by Eqs. (4–65) and (4–66), respectively, in order to make the spherical surface $R = a$ equipotential. However, an additional point charge

$$Q' = -Q_i = \frac{a}{d} Q \qquad (4\text{--}75)$$

at the center would be needed to make the net charge on the replaced sphere zero. The electrostatic problem of a point charge Q in the presence of an electrically neutral sphere can then be solved as a problem with three point charges : Q' at $R = 0$, Q_i at $R = a^2/d$, and Q at $R = d$.

4–4.4 CHARGED SPHERE AND GROUNDED PLANE

When a charged conducting sphere is near a large, grounded, conducting plane, as in Fig. 4–13(a), the charge distribution on and the electric field between the conducting bodies are obviously nonuniform. Since the geometry contains a mixture of spherical and Cartesian coordinates, field determination and capacitance calculation through a solution of Laplace's equation is a rather difficult problem. We shall now show how the repeated application of the method of images can be used to solve this problem.

Assume that a charge Q_0 is put at the center of the sphere. We wish to find a system of image charges that, together with Q_0, will make both the sphere and the plane equipotential surfaces. The problem of a charged sphere near a grounded plane can then be replaced by that of the much simpler system of point charges. A cross section in the xy-plane is shown in Fig. 4–13(b). The presence of Q_0 at $(-c, 0)$ requires an image charge $-Q_0$ at $(c, 0)$ to make the yz-plane equipotential; but the pair of charges Q_0 and $-Q_0$ destroy the equipotential property of the sphere unless, according to Eqs. (4–65) and (4–66), an image charge $Q_1 = (a/2c)Q_0$ is placed at $(-c + a^2/2c, 0)$ inside the dashed circle. This, in turn, requires an image charge $-Q_1$ to make the yz-plane equipotential. This process of successive application of the method of images is continued, and we obtain two groups of image point charges: one group $(-Q_0, -Q_1, -Q_2, \ldots)$ on the right side of the y-axis, and another group $(Q_1, Q_2, \ldots)$ inside the sphere. We have

$$Q_1 = \left(\frac{a}{2c}\right)Q_0 = \alpha Q_0, \tag{4-76a}$$

$$Q_2 = \frac{a}{\left(2c - \dfrac{a^2}{2c}\right)} Q_1 = \frac{\alpha^2}{1 - \alpha^2} Q_0, \tag{4-76b}$$

$$Q_3 = \frac{a}{2c - \dfrac{a^2}{\left(2c - \dfrac{a^2}{2c}\right)}} Q_2 = \frac{\alpha^3}{(1 - \alpha^2)\left(1 - \dfrac{\alpha^3}{1 - \alpha^3}\right)} Q_0, \tag{4-76c}$$

$$\vdots$$

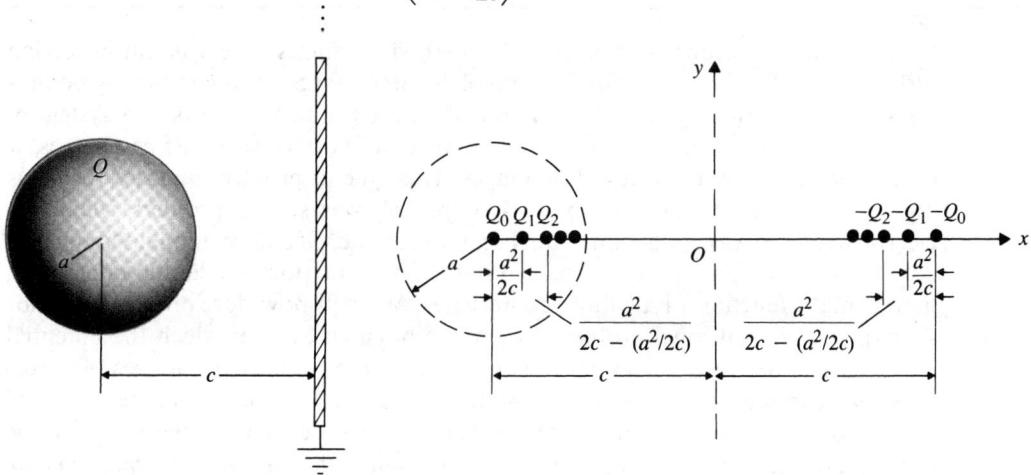

(a) Physical arrangement.

(b) Two groups of image point charges.

FIGURE 4–13
Charged sphere and grounded conducting plane.

where

$$\alpha = \frac{a}{2c}. \tag{4-77}$$

The total charge on the sphere is

$$Q = Q_0 + Q_1 + Q_2 + \cdots$$
$$= Q_0\left(1 + \alpha + \frac{\alpha^2}{1 - \alpha^2} + \cdots\right). \tag{4-78}$$

The series in Eq. (4–78) usually converges rapidly ($\alpha < 1/2$). Now since the charge pairs $(-Q_0, Q_1), (-Q_1, Q_2), \ldots$ yield a zero potential on the sphere, only the original Q_0 contributes to the potential of the sphere, which is

$$V_0 = \frac{Q_0}{4\pi\epsilon_0 a}. \tag{4-79}$$

Hence the capacitance between the sphere and the conducting plane is, from Eqs. (4–78) and (4–79),

$$C = \frac{Q}{V_0} = 4\pi\epsilon_0 a\left(1 + \alpha + \frac{\alpha^2}{1 - \alpha^2} + \cdots\right), \tag{4-80}$$

which is larger than the capacitance of an isolated sphere of radius a, as expected. The potential and electric field distributions between the sphere and the conducting plane can also be obtained from the image point charges.

4–5 Boundary-Value Problems in Cartesian Coordinates

We saw in the preceding section that the method of images is very useful in solving certain types of electrostatic problems involving free charges near conducting boundaries that are geometrically simple. However, if the problem consists of a system of conductors maintained at specified potentials and with no isolated free charges, it cannot be solved by the method of images. This type of problem requires the solution of Laplace's equation. Example 4–1 (p. 154) was such a problem where the electric potential was a function of only one coordinate. Of course, Laplace's equation applied to three dimensions is a partial differential equation, where the potential is, in general, a function of all three coordinates. We will now develop a method for solving three-dimensional problems where the boundaries, over which the potential or its normal derivative is specified, coincide with the coordinate surfaces of an orthogonal, curvilinear coordinate system. In such cases, the solution can be expressed as a product of three one-dimensional functions, each depending separately on one coordinate variable only. The procedure is called the ***method of separation of variables***.

Problems (electromagnetic or otherwise) governed by partial differential equations with prescribed boundary conditions are called ***boundary-value problems***.

Boundary-value problems for potential functions can be classified into three types: (1) **Dirichlet problems**, in which the value of the potential is specified everywhere on the boundaries; (2) **Neumann problems**, in which the normal derivative of the potential is specified everywhere on the boundaries; (3) **Mixed boundary-value problems**, in which the potential is specified over some boundaries and the normal derivative of the potential is specified over the remaining ones. Different specified boundary conditions will require the choice of different potential functions, but the procedure of solving these types of problems—that is, by the method of separation of variables— for the three types of problems is the same. The solutions of Laplace's equation are often called **harmonic functions**.

Laplace's equation for scalar electric potential V in Cartesian coordinates is

$$\frac{\partial^2 V}{\partial x^2} + \frac{\partial^2 V}{\partial y^2} + \frac{\partial^2 V}{\partial z^2} = 0. \tag{4–81}$$

To apply the method of separation of variables, we assume that the solution $V(x, y, z)$ can be expressed as a product in the following form:

$$V(x, y, z) = X(x)Y(y)Z(z), \tag{4–82}$$

where $X(x)$, $Y(y)$, and $Z(z)$ are functions of only x, y, and z, respectively. Substituting Eq. (4–82) in Eq. (4–81), we have

$$Y(y)Z(z)\frac{d^2 X(x)}{dx^2} + X(x)Z(z)\frac{d^2 Y(y)}{dy^2} + X(x)Y(y)\frac{d^2 Z(z)}{dz^2} = 0,$$

which, when divided through by the product $X(x)Y(y)Z(z)$, yields

$$\frac{1}{X(x)}\frac{d^2 X(x)}{dx^2} + \frac{1}{Y(y)}\frac{d^2 Y(y)}{dy^2} + \frac{1}{Z(z)}\frac{d^2 Z(z)}{dz^2} = 0. \tag{4–83}$$

Note that each of the three terms on the left side of Eq. (4–83) is a function of only one coordinate variable and that only ordinary derivatives are involved. In order for Eq. (4–83) to be satisfied for *all values* of x, y, z, each of the three terms must be a constant. For instance, if we differentiate Eq. (4–83) with respect to x, we have

$$\frac{d}{dx}\left[\frac{1}{X(x)}\frac{d^2 X(x)}{dx^2}\right] = 0, \tag{4–84}$$

since the other two terms are independent of x. Equation (4–84) requires that

$$\frac{1}{X(x)}\frac{d^2 X(x)}{dx^2} = -k_x^2, \tag{4–85}$$

where k_x^2 is a constant of integration to be determined from the boundary conditions of the problem. The negative sign on the right side of Eq. (4–85) is arbitrary, just as the square sign on k_x is arbitrary. The **separation constant** k_x can be a real or an imaginary number. If k_x is imaginary, k_x^2 is a negative real number, making $-k_x^2$ a

TABLE 4–1
Possible Solutions of $X''(x) + k_x^2 X(x) = 0$

k_x^2	k_x	$X(x)$	Exponential forms[†] of $X(x)$
0	0	$A_0 x + B_0$	
+	k	$A_1 \sin kx + B_1 \cos kx$	$C_1 e^{jkx} + D_1 e^{-jkx}$
−	jk	$A_2 \sinh kx + B_2 \cosh kx$	$C_2 e^{kx} + D_2 e^{-kx}$

[†] The exponential forms of $X(x)$ are related to the trigonometric and hyperbolic forms listed in the third column by the following formulas:

$$e^{\pm jkx} = \cos kx \pm j \sin kx, \quad \cos kx = \tfrac{1}{2}(e^{jkx} + e^{-jkx}), \quad \sin kx = \frac{1}{2j}(e^{jkx} - e^{-jkx});$$

$$e^{\pm kx} = \cosh kx \pm \sinh kx, \quad \cosh kx = \tfrac{1}{2}(e^{kx} + e^{-kx}), \quad \sinh kx = \tfrac{1}{2}(e^{kx} - e^{-kx}).$$

positive real number. It is convenient to rewrite Eq. (4–85) as

$$\frac{d^2 X(x)}{dx^2} + k_x^2 X(x) = 0. \tag{4–86}$$

In a similar manner, we have

$$\frac{d^2 Y(y)}{dy^2} + k_y^2 Y(y) = 0 \tag{4–87}$$

and

$$\frac{d^2 Z(z)}{dz^2} + k_z^2 Z(z) = 0, \tag{4–88}$$

where the separation constants k_y and k_z will, in general, be different from k_x; but, because of Eq. (4–83), the following condition must be satisfied:

$$k_x^2 + k_y^2 + k_z^2 = 0. \tag{4–89}$$

Our problem has now been reduced to finding the appropriate solutions—$X(x)$, $Y(y)$, and $Z(z)$—from the second-order *ordinary* differential equations Eqs. (4–86), (4–87), and (4–88), respectively. The possible solutions of Eq. (4–86) are known from our study of ordinary differential equations with constant coefficients. They are listed in Table 4–1. That the listed solutions satisfy Eq. (4–86) is easily verified by direct substitution.

Of the listed solutions in Table 4–1, the first one, $A_0 x + B_0$ for $k_x = 0$, is a straight line with a slope A_0 and an intercept B_0 at $x = 0$. When $A_0 = 0$, $X(x) = B_0$, which means that V, the solution of Laplace's equation, is independent of the dimension x.

We are, of course, familiar with the sine and cosine functions, both of which are periodic with a period 2π. If plotted versus x, $\sin kx$ and $\cos kx$ have a period $2\pi/k$. Frequently, a careful examination of a given problem enables us to decide whether a sine or a cosine function is the proper choice. For example, if the solution is to vanish at $x = 0$, $\sin kx$ must be chosen; on the other hand, if the solution is expected to be symmetrical with respect to $x = 0$, then $\cos kx$ is the right choice. In

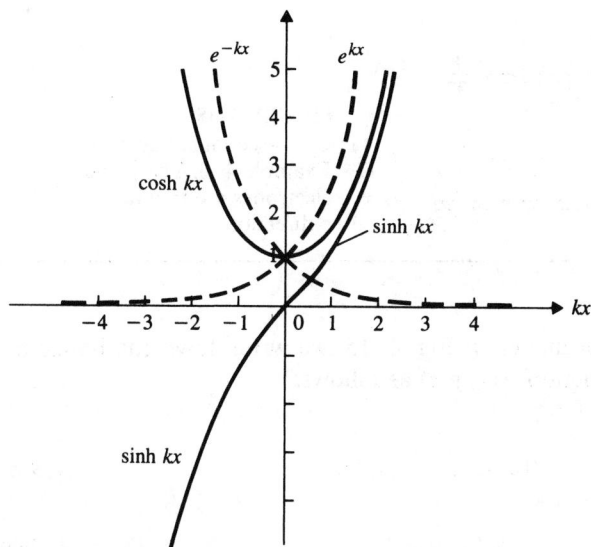

FIGURE 4–14
Hyperbolic and exponential functions.

the general case, both terms are required. Sometimes it may be desirable to write $A_1 \sin kx + B_1 \cos kx$ as $A_s \sin(kx + \psi_s)$ or $A_c \cos(kx + \psi_c)$.[†]

For $k_x = jk$ the solution converts to hyperbolic functions:

$$\sin jkx = -j \sinh kx$$

and

$$\cos jkx = \cosh kx.$$

Hyperbolic functions are combinations of exponential functions with real exponents, and are nonperiodic. They are plotted in Fig. 4–14 for easy reference. The important characteristics of $\sinh kx$ are that it is an odd function of x and that its value approaches $\pm\infty$ as x goes to $\pm\infty$. The function $\cosh kx$ is an even function of x, equals unity at $x = 0$, and approaches $+\infty$ as x goes to $+\infty$ or $-\infty$.

The specified boundary conditions will determine the choice of the proper form of the solution and of the constants A and B or C and D. The solutions of Eqs. (4–87) and (4–88) for $Y(y)$ and $Z(z)$ are entirely similar.

EXAMPLE 4–6 Two grounded, semi-infinite, parallel-plane electrodes are separated by a distance b. A third electrode perpendicular to and insulated from both is maintained at a constant potential V_0 (see Fig. 4–15). Determine the potential distribution in the region enclosed by the electrodes.

[†] $A_s \sin(kx + \psi_s) = (A_s \cos \psi_s) \sin kx + (A_s \sin \psi_s) \cos kx$; $A_1 = A_s \cos \psi_s$, $B_1 = A_s \sin \psi_s$; $A_s = (A_1^2 + B_1^2)^{1/2}$, $\psi_s = \tan^{-1}(B_1/A_1)$. $A_c \cos(kx + \psi_c) = (-A_c \sin \psi_c) \sin kx + (A_c \cos \psi_c) \cos kx$; $A_1 = -A_c \sin \psi_c$, $B_1 = A_c \cos \psi_c$; $A_c = (A_1^2 + B_1^2)^{1/2}$, $\psi_c = \tan^{-1}(-A_1/B_1)$.

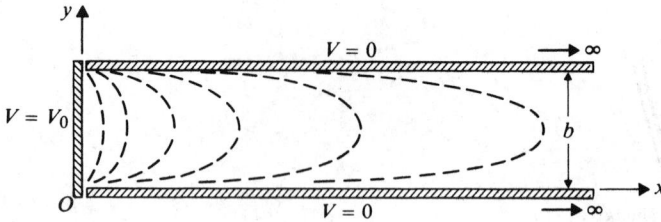

FIGURE 4-15
Cross-sectional figure for
Example 4-6. The plane
electrodes are infinite in
z-direction.

Solution Referring to the coordinates in Fig. 4–15, we write down the boundary conditions for the potential function $V(x, y, z)$ as follows.

With V independent of z:

$$V(x, y, z) = V(x, y). \tag{4-90a}$$

In the x-direction:

$$V(0, y) = V_0 \tag{4-90b}$$

$$V(\infty, y) = 0. \tag{4-90c}$$

In the y-direction:

$$V(x, 0) = 0 \tag{4-90d}$$

$$V(x, b) = 0. \tag{4-90e}$$

Condition (4–90a) implies $k_z = 0$, and from Table 4–1,

$$Z(z) = B_0. \tag{4-91}$$

The constant A_0 vanishes because Z is independent of z. From Eq. (4–89) we have

$$k_y^2 = -k_x^2 = k^2, \tag{4-92}$$

where k is a real number. This choice of k implies that k_x is imaginary and that k_y is real. The use of $k_x = jk$, together with the condition of Eq. (4–90c), requires us to choose the exponentially decreasing form for $X(x)$, which is

$$X(x) = D_2 e^{-kx}. \tag{4-93}$$

In the y-direction, $k_y = k$. Condition (4–90d) indicates that the proper choice for $Y(y)$ from Table 4–1 is

$$Y(y) = A_1 \sin ky. \tag{4-94}$$

Combining the solutions given by Eqs. (4–91), (4–93), and (4–94) in Eq. (4–82), we obtain an appropriate solution of the following form:

$$V_n(x, y) = (B_0 D_2 A_1) e^{-kx} \sin ky$$
$$= C_n e^{-kx} \sin ky, \tag{4-95}$$

where the arbitrary constant C_n has been written for the product $B_0 D_2 A_1$.

Now, of the five boundary conditions listed in Eqs. (4–90a) through (4–90e) we have used conditions (4–90a), (4–90c), and (4–90d). To meet condition (4–90e), we require

$$V_n(x, b) = C_n e^{-kx} \sin kb = 0, \tag{4–96}$$

which can be satisfied, for all values of x, only if

$$\sin kb = 0$$

or

$$kb = n\pi$$

or

$$k = \frac{n\pi}{b}, \qquad n = 1, 2, 3, \ldots. \tag{4–97}$$

Therefore, Eq. (4–95) becomes

$$V_n(x\ y) = C_n e^{-n\pi x/b} \sin \frac{n\pi}{b} y. \tag{4–98}$$

Question: Why are 0 and negative integral values of n not included in Eq. (4–97)?

We can readily verify by direct substitution that $V_n(x, y)$ in Eq. (4–98) satisfies the Laplace's equation (4–81). However, $V_n(x, y)$ alone cannot satisfy the remaining boundary condition (4–90b) at $x = 0$ for all values of y from 0 to b. Since Laplace's equation is a *linear* partial differential equation, a sum (superposition) of $V_n(x, y)$ of the form in Eq. (4–98) with different values of n is also a solution. At $x = 0$, we write

$$V(0, y) = \sum_{n=1}^{\infty} V_n(0, y) = \sum_{n=1}^{\infty} C_n \sin \frac{n\pi}{b} y$$
$$= V_0, \qquad 0 < y < b. \tag{4–99}$$

Equation (4–99) is essentially a Fourier-series expansion of the periodic rectangular wave at $x = 0$ shown in Fig. 4–16, which has a constant value V_0 in the interval $0 < y < b$.

In order to evaluate the coefficients C_n, we multiply both sides of Eq. (4–99) by $\sin \frac{m\pi}{b} y$ and integrate the products from $y = 0$ to $y = b$:

$$\sum_{n=1}^{\infty} \int_0^b C_n \sin \frac{n\pi}{b} y \sin \frac{m\pi}{b} y\, dy = \int_0^b V_0 \sin \frac{m\pi}{b} y\, dy. \tag{4–100}$$

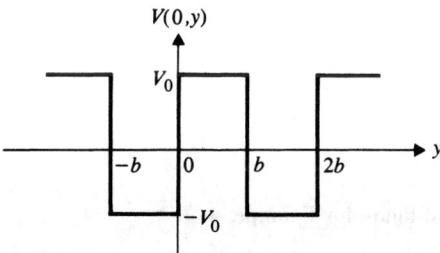

FIGURE 4–16
For Fourier-series expansion of boundary condition at $x = 0$ (Example 4–6).

The integral on the right side of Eq. (4–100) is easily evaluated:

$$\int_0^b V_0 \sin \frac{m\pi}{b} y \, dy = \begin{cases} \dfrac{2bV_0}{m\pi} & \text{if } m \text{ is odd,} \\ 0 & \text{if } m \text{ is even.} \end{cases} \tag{4–101}$$

Each integral on the left side of Eq. (4–100) is

$$\int_0^b C_n \sin \frac{n\pi}{b} y \sin \frac{m\pi}{b} y \, dy = \frac{C_n}{2} \int_0^b \left[\cos \frac{(n-m)\pi}{b} y - \cos \frac{(n+m)\pi}{b} y \right] dy$$

$$= \begin{cases} \dfrac{C_n}{2} b & \text{if } m = n, \\ 0 & \text{if } m \neq n. \end{cases} \tag{4–102}$$

Substituting Eqs. (4–101) and (4–102) in Eq. (4–100), we obtain

$$C_n = \begin{cases} \dfrac{4V_0}{n\pi} & \text{if } n \text{ is odd,} \\ 0 & \text{if } n \text{ is even.} \end{cases} \tag{4–103}$$

The desired potential distribution is, then, a superposition of $V_n(x, y)$ in Eq. (4–98).

$$V(x, y) = \sum_{n=1}^{\infty} C_n e^{-n\pi x/b} \sin \frac{n\pi}{b} y$$

$$= \frac{4V_0}{\pi} \sum_{\substack{n=\text{odd}}}^{\infty} \frac{1}{n} e^{-n\pi x/b} \sin \frac{n\pi}{b} y, \tag{4–104}$$

$$n = 1, 3, 5, \ldots,$$

$$x > 0 \quad \text{and} \quad 0 < y < b.$$

Equation (4–104) is a rather complicated expression to plot in two dimensions; but since the amplitude of the sine terms in the series decreases very rapidly as n increases, only the first few terms are needed to obtain a good approximation. Several equipotential lines are sketched in Fig. 4–15. ■

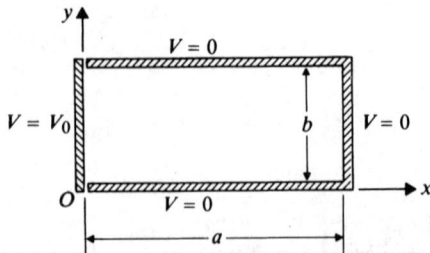

FIGURE 4–17
Cross-sectional figure for Example 4–7.

■■■■ **EXAMPLE 4–7**   Consider the region enclosed on three sides by grounded conducting planes shown in Fig. 4–17. The end plate on the left is insulated from the grounded sides and has a constant potential V_0. All planes are assumed to be infinite in extent in the z-direction. Determine the potential distribution within this region.

Solution   The boundary conditions for the potential function $V(x, y, z)$ are as follows.

With V independent of z:

$$V(x, y, z) = V(x, y).\tag{4–105a}$$

In the x-direction:

$$V(0, y) = V_0,\tag{4–105b}$$

$$V(a, y) = 0.\tag{4–105c}$$

In the y-direction:

$$V(x, 0) = 0,\tag{4–105d}$$

$$V(x, b) = 0.\tag{4–105e}$$

Condition (4–105a) implies that $k_z = 0$, and, from Table 4–1,

$$Z(z) = B_0.\tag{4–106}$$

As a consequence, Eq. (4–89) reduces to

$$k_y^2 = -k_x^2 = k^2,\tag{4–107}$$

which is the same as Eq. (4–92) in Example 4–6.

The boundary conditions in the y-direction, Eqs. (4–105d) and Eq. (4–105e), are the same as those specified by Eqs. (4–90d) and (4–90e). To make $V(x, 0) = 0$ for all values of x between 0 and a, $Y(0)$ must be zero, and we have

$$Y(y) = A_1 \sin ky,\tag{4–108}$$

as in Eq. (4–94). However, $X(x)$ given by Eq. (4–93) is obviously not a solution here because it does not satisfy the boundary condition (4–105c). In this case it is convenient to use the general form for $k_x = jk$ given in the third column of Table 4–1. (The exponential solution form given in the last column could be used as well, but it would not be as convenient because it is not as easy to see the condition under which the sum of two exponential terms vanishes at $x = a$ as it is to make a sinh term zero. This will be clear presently.) We have

$$X(x) = A_2 \sinh kx + B_2 \cosh kx.\tag{4–109}$$

A relation exists between the arbitrary constants A_2 and B_2 because of the boundary condition in Eq. (4–105c), which demands that $X(a) = 0$; that is,

$$0 = A_2 \sinh ka + B_2 \cosh ka$$

or

$$B_2 = -A_2 \frac{\sinh ka}{\cosh ka}.$$

From Eq. (4–109) we have

$$X(x) = A_2 \left[\sinh kx - \frac{\sinh ka}{\cosh ka} \cosh kx \right]$$

$$= \frac{A_2}{\cosh ka} [\cosh ka \sinh kx - \sinh ka \cosh kx] \qquad (4–110)$$

$$= A_3 \sinh k(x - a),$$

where A_3 has been written for $A_2/\cosh ka$. It is evident that Eq. (4–110) satisfies the condition $X(a) = 0$. With experience we should be able to write the solution given in Eq. (4–110) directly, without the steps leading to it, as only a shift in the argument of the sinh function is needed to make it vanish at $x = a$.

Collecting Eqs. (4–106), (4–108) and (4–110), we obtain the appropriate product solution

$$V_n(x, y) = B_0 A_1 A_3 \sinh k(x - a) \sin ky$$

$$= C'_n \sinh \frac{n\pi}{b} (x - a) \sin \frac{n\pi}{b} y, \qquad n = 1, 2, 3, \ldots, \qquad (4–111)$$

where $C'_n = B_0 A_1 A_3$, and k has been set to equal $n\pi/b$ in order to satisfy boundary condition (4–105e).

We have now used all of the boundary conditions except Eq. (4–105b), which may be satisfied by a Fourier-series expansion of $V(0, y) = V_0$ over the interval from $y = 0$ to $y = b$. We have

$$V_0 = \sum_{n=1}^{\infty} V_n(0, y) = -\sum_{n=1}^{\infty} C'_n \sinh \frac{n\pi}{b} a \sin \frac{n\pi}{b} y, \qquad 0 < y < b. \qquad (4–112)$$

We note that Eq. (4–112) is of the same form as Eq. (4–99), except that C_n is replaced by $-C'_n \sinh (n\pi a/b)$. The values for the coefficient C'_n can then be written down from Eq. (4–103):

$$C'_n = \begin{cases} -\dfrac{4V_0}{n\pi \sinh (n\pi a/b)} & \text{if } n \text{ is odd,} \\ 0 & \text{if } n \text{ is even.} \end{cases} \qquad (4–113)$$

The desired potential distribution within the enclosed region in Fig. 4–17 is a summation of $V_n(x, y)$ in Eq. (4–111):

$$V(x, y) = \sum_{n=1}^{\infty} C'_n \sinh \frac{n\pi}{b} (x - a) \sin \frac{n\pi}{b} y$$

$$= \frac{4V_0}{\pi} \sum_{n=\text{odd}}^{\infty} \frac{\sinh [n\pi(a - x)/b]}{n \sinh (n\pi a/b)} \sin \frac{n\pi}{b} y, \qquad (4–114)$$

$$n = 1, 3, 5, \ldots,$$

$$0 < x < a \qquad \text{and} \qquad 0 < y < b.$$

The electric field distribution within the enclosure is obtained by the relation

$$\mathbf{E}(x, y) = -\nabla V(x, y).$$

4-6 Boundary-Value Problems in Cylindrical Coordinates

For problems with circular cylindrical boundaries we write the governing equations in the cyclindrical coordinate system. Laplace's equation for scalar electric potential V in cylindrical coordinates is, from Eq. (4-8),

$$\frac{1}{r}\frac{\partial}{\partial r}\left(r\frac{\partial V}{\partial r}\right) + \frac{1}{r^2}\frac{\partial^2 V}{\partial \phi^2} + \frac{\partial^2 V}{\partial z^2} = 0. \tag{4-115}$$

A general solution of Eq. (4-115) requires the knowlege of **Bessel functions**, the discussion of which will be deferred until Chapter 10. In situations in which the lengthwise dimension of the cylindrical geometry is large in comparison to its radius, the associated field quantities may be considered to be approximately independent of z. In such cases, $\partial^2 V/\partial z^2 = 0$ and Eq. (4-115) becomes the governing equation of a two-dimensional problem:

$$\frac{1}{r}\frac{\partial}{\partial r}\left(r\frac{\partial V}{\partial r}\right) + \frac{1}{r^2}\frac{\partial^2 V}{\partial \phi^2} = 0. \tag{4-116}$$

Applying the method of separation of variables, we assume a product solution

$$V(r, \phi) = R(r)\Phi(\phi), \tag{4-117}$$

where $R(r)$ and $\Phi(\phi)$ are, respectively, functions of r and ϕ only. Substituting solution (4-117) in Eq. (4-116) and dividing by $R(r)\Phi(\phi)$, we have

$$\frac{r}{R(r)}\frac{d}{dr}\left[r\frac{dR(r)}{dr}\right] + \frac{1}{\Phi(\phi)}\frac{d^2\Phi(\phi)}{d\phi^2} = 0. \tag{4-118}$$

In Eq. (4-118) the first term on the left side is a function of r only, and the second term is a function of ϕ only. (Note that ordinary derivatives have replaced partial derivatives.) For Eq. (4-118) to hold for all values of r and ϕ, each term must be a constant and be the negative of the other. We have

$$\frac{r}{R(r)}\frac{d}{dr}\left[r\frac{dR(r)}{dr}\right] = k^2 \tag{4-119}$$

and

$$\frac{1}{\Phi(\phi)}\frac{d^2\Phi(\phi)}{d\phi^2} = -k^2, \tag{4-120}$$

where k is a separation constant.

Equation (4-120) can be rewritten as

$$\frac{d^2\Phi(\phi)}{d\phi^2} + k^2\Phi(\phi) = 0. \tag{4-121}$$

This is of the same form as Eq. (4-86), and its solution can be any one of those listed in Table 4-1. For circular cylindrical configurations, potential functions and therefore $\Phi(\phi)$ are periodic in ϕ, and the hyperbolic functions do not apply. In fact, if the range of ϕ is unrestricted, k must be an integer. Let k equal n. The appropriate solution is

$$\Phi(\phi) = A_\phi \sin n\phi + B_\phi \cos n\phi, \qquad (4-122)$$

where A_ϕ and B_ϕ are arbitrary constants.

We now turn our attention to Eq. (4–119), which can be rearranged as

$$r^2 \frac{d^2 R(r)}{dr^2} + r \frac{dR(r)}{dr} - n^2 R(r) = 0, \qquad (4-123)$$

where integer n has been written for k, implying a 2π range for ϕ. The solution of Eq. (4–123) is

$$R(r) = A_r r^n + B_r r^{-n}. \qquad (4-124)$$

This can be verified by direct substitution. Taking the product of the solutions in (4–122) and (4–124), we obtain a general solution of z-independent Laplace's equation (4–116) for circular cylindrical regions with an unrestricted range for ϕ:

$$V_n(r, \phi) = r^n(A_n \sin n\phi + B_n \cos n\phi) + r^{-n}(A'_n \sin n\phi + B'_n \cos n\phi), \qquad n \neq 0. \qquad (4-125)$$

Depending on the boundary conditions the complete solution of a problem may be a summation of the terms in Eq. (4–125). It is useful to note that, when the region of interest includes the cylindrical axis where $r = 0$, the terms containing the r^{-n} factor cannot exist. On the other hand, if the region of interest includes the point at infinity, the terms containing the r^n factor cannot exist, since the potential must be zero as $r \to \infty$.

Eq. (4–121) has the simplest form when $k = 0$. We have

$$\frac{d^2 \Phi(\phi)}{d\phi^2} = 0. \qquad (4-126)$$

The general solution of Eq. (4–126) is $\Phi(\phi) = A_0 \phi + B_0$. If there is no circumferential variation, A_0 vanishes,[†] and we have

$$\Phi(\phi) = B_0, \qquad k = 0. \qquad (4-127)$$

The equation for $R(r)$ also becomes simpler when $k = 0$. We obtain from Eq. (4–119)

$$\frac{d}{dr}\left[r \frac{dR(r)}{dr} \right] = 0, \qquad (4-128)$$

which has a solution

$$R(r) = C_0 \ln r + D_0, \qquad k = 0. \qquad (4-129)$$

The product of Eqs. (4–127) and (4–129) gives a solution that is independent of either z or ϕ:

$$V(r) = C_1 \ln r + C_2, \qquad (4-130)$$

where the arbitrary constants C_1 and C_2 are determined from boundary conditions.

[†] The term $A_0 \phi$ should be retained if there is circumferential variation, such as in problems involving a wedge. (See Problem P.4–23.)

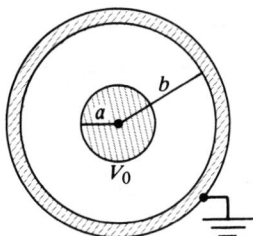

FIGURE 4–18
Cross section of a coaxial cable (Example 4–8).

We shall now illustrate the above procedures with two examples. One (Example 4–8) deals with a situation that is circularly symmetrical, and the other (Example 4–9) solves a problem with circumferential variation.

EXAMPLE 4–8 Consider a very long coaxial cable. The inner conductor has a radius a and is maintained at a potential V_0. The outer conductor has an inner radius b and is grounded. Determine the potential distribution in the space between the conductors.

Solution Figure 4–18 shows a cross section of the coaxial cable. We assume no z-dependence and, by symmetry, also no ϕ-dependence ($k = 0$). Therefore, the electric potential is a function of r only and is given by Eq. (4–130).

The boundary conditions are

$$V(b) = 0, \tag{4–131a}$$

$$V(a) = V_0. \tag{4–131b}$$

Substitution of Eqs. (4–131a) and (4–131b) in Eq. (4–130) leads to two relations:

$$C_1 \ln b + C_2 = 0, \tag{4–132a}$$

$$C_1 \ln a + C_2 = V_0. \tag{4–132b}$$

From Eqs. (4–132a) and (4–132b), C_1 and C_2 are readily determined:

$$C_1 = -\frac{V_0}{\ln (b/a)}, \qquad C_2 = \frac{V_0 \ln b}{\ln (b/a)}.$$

Therefore, the potential distribution in the space $a \leq r \leq b$ is

$$V(r) = \frac{V_0}{\ln (b/a)} \ln \left(\frac{b}{r}\right). \tag{4–133}$$

Obviously, equipotential surfaces are coaxial cylindrical surfaces. ∎

EXAMPLE 4–9 An infinitely long, thin, conducting circular tube of radius b is split in two halves. The upper half is kept at a potential $V = V_0$ and the lower half at $V = -V_0$. Determine the potential distribution both inside and outside the tube.

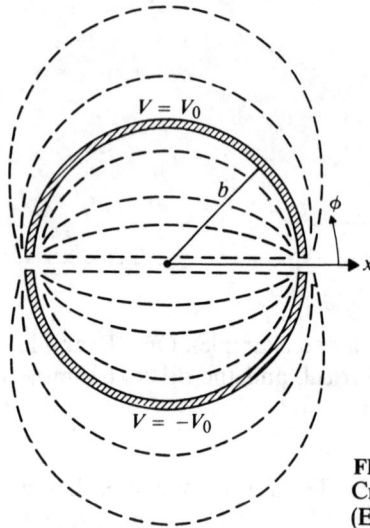

FIGURE 4–19
Cross section of split circular cylinder and equipotential lines
(Example 4–9).

Solution A cross section of the split circular tube is shown in Fig. 4–19. Since the tube is assumed to be infinitely long, the potential is independent of z and the two-dimensional Laplace's equation (4–116) applies. The boundary conditions are

$$V(b, \phi) = \begin{cases} V_0 & \text{for } 0 < \phi < \pi, \\ -V_0 & \text{for } \pi < \phi < 2\pi. \end{cases} \tag{4–134}$$

These conditions are plotted in Fig. 4–20. Obviously, $V(r, \phi)$ is an odd function of ϕ. We shall determine $V(r, \phi)$ inside and outside the tube separately.

a) Inside the tube,

$$r < b.$$

Because this region includes $r = 0$, terms containing the r^{-n} factor cannot exist. Moreover, since $V(r, \phi)$ is an odd function of ϕ, the appropriate form of solution

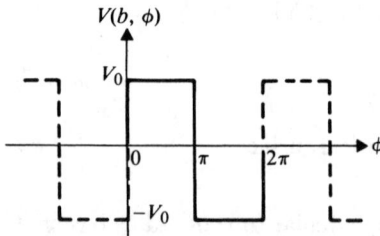

FIGURE 4–20
Boundary condition for Example 4–9.

is, from Eq. (4–125),

$$V_n(r, \phi) = A_n r^n \sin n\phi. \tag{4–135}$$

However, a single such term does not satisfy the boundary conditions specified in Eq. (4–134). We form a series solution

$$V(r, \phi) = \sum_{n=1}^{\infty} V_n(r, \phi)$$

$$= \sum_{n=1}^{\infty} A_n r^n \sin n\phi, \tag{4–136}$$

and require that Eq. (4–134) be satisfied at $r = b$. This amounts to expanding the rectangular wave (period $= 2\pi$) shown in Fig. 4–20 into a Fourier sine series.

$$\sum_{n=1}^{\infty} A_n b^n \sin n\phi = \begin{cases} V_0 & \text{for } 0 < \phi < \pi, \\ -V_0 & \text{for } \pi < \phi < 2\pi. \end{cases} \tag{4–137}$$

The coefficients A_n can be found by the method illustrated in Example 4–6. As a matter of fact, because we already have the result in Eq. (4–103), we can directly write

$$A_n = \begin{cases} \dfrac{4V_0}{n\pi b^n} & \text{if } n \text{ is odd,} \\ 0 & \text{if } n \text{ is even.} \end{cases} \tag{4–138}$$

The potential distribution inside the tube is obtained by substituting Eq. (4–138) in Eq. (4–136):

$$V(r, \phi) = \frac{4V_0}{\pi} \sum_{n=\text{odd}}^{\infty} \frac{1}{n} \left(\frac{r}{b}\right)^n \sin n\phi, \qquad r < b. \tag{4–139}$$

b) Outside the tube,

$$r > b.$$

In this region the potential must decrease to zero as $r \to \infty$. Terms containing the factor r^n cannot exist, and the appropriate form of solution is, from Eq. (4–125),

$$V(r, \phi) = \sum_{n=1}^{\infty} V_n(r, \phi)$$

$$= \sum_{n=1}^{\infty} B'_n r^{-n} \sin n\phi. \tag{4–140}$$

At $r = b$,

$$V(b, \phi) = \sum_{n=1}^{\infty} B'_n b^{-n} \sin n\phi$$

$$= \begin{cases} V_0 & \text{for } 0 < \phi < \pi, \\ -V_0 & \text{for } \pi < \phi < 2\pi. \end{cases} \tag{4–141}$$

The coefficients B'_n in Eq. (4–141) are analogous to A_n in Eq. (4–137). From Eq. (4–138) we obtain

$$B'_n = \begin{cases} \dfrac{4V_0 b^n}{n\pi} & \text{if } n \text{ is odd,} \\[2mm] 0 & \text{if } n \text{ is even.} \end{cases} \tag{4–142}$$

Therefore, the potential distribution outside the tube is, from Eq. (4–140),

$$V(r, \phi) = \frac{4V_0}{\pi} \sum_{n=\text{odd}}^{\infty} \frac{1}{n} \left(\frac{b}{r}\right)^n \sin n\phi, \qquad r > b. \tag{4–143}$$

Several equipotential lines both inside and outside the tube have been sketched in Fig. 4–19.   ▬

4–7   Boundary-Value Problems in Spherical Coordinates

The general solution of Laplace's equation in spherical coordinates is a very involved procedure, so we will limit our discussion to cases in which the electric potential is independent of the azimuthal angle ϕ. Even with this limitation we will need to introduce some new functions. From Eq. (4–9) we have

$$\frac{1}{R^2} \frac{\partial}{\partial R} \left(R^2 \frac{\partial V}{\partial R} \right) + \frac{1}{R^2 \sin\theta} \frac{\partial}{\partial\theta} \left(\sin\theta \frac{\partial V}{\partial\theta} \right) = 0. \tag{4–144}$$

Applying the method of separation of variables, we assume a product solution

$$V(R, \theta) = \Gamma(R)\Theta(\theta). \tag{4–145}$$

Substitution of this solution in Eq. (4–144) yields, after rearrangement,

$$\frac{1}{\Gamma(R)} \frac{d}{dR} \left[R^2 \frac{d\Gamma(R)}{dR} \right] + \frac{1}{\Theta(\theta) \sin\theta} \frac{d}{d\theta} \left[\sin\theta \frac{d\Theta(\theta)}{d\theta} \right] = 0. \tag{4–146}$$

In Eq. (4–146) the first term on the left side is a function of R only, and the second term is a function of θ only. If the equation is to hold for all values of R and θ, each term must be a constant and be the negative of the other. We write

$$\frac{1}{\Gamma(R)} \frac{d}{dR} \left[R^2 \frac{d\Gamma(R)}{dR} \right] = k^2 \tag{4–147}$$

and

$$\frac{1}{\Theta(\theta) \sin\theta} \frac{d}{d\theta} \left[\sin\theta \frac{d\Theta(\theta)}{d\theta} \right] = -k^2, \tag{4–148}$$

where k is a separation constant. We must now solve the two second-order, ordinary differential equations, Eqs. (4–147) and (4–148).

TABLE 4–2
Several Legendre
Polynomials

n	$P_n(\cos \theta)$
0	1
1	$\cos \theta$
2	$\frac{1}{2}(3 \cos^2 \theta - 1)$
3	$\frac{1}{2}(5 \cos^3 \theta - 3 \cos \theta)$

Equation (4–147) can be rewritten as

$$R^2 \frac{d^2\Gamma(R)}{dR^2} + 2R \frac{d\Gamma(R)}{dR} - k^2\Gamma(R) = 0, \tag{4–149}$$

which has a solution of the form

$$\Gamma_n(R) = A_n R^n + B_n R^{-(n+1)}. \tag{4–150}$$

In Eq. (4–150), A_n and B_n are arbitrary constants, and the following relation between n and k can be verified by substitution:

$$n(n + 1) = k^2, \tag{4–151}$$

where $n = 0, 1, 2, \ldots$ is a positive integer.

With the value of k^2 given in Eq. (4–151), we have, from Eq. (4–148),

$$\frac{d}{d\theta}\left[\sin \theta \frac{d\Theta(\theta)}{d\theta}\right] + n(n + 1)\Theta(\theta) \sin \theta = 0, \tag{4–152}$$

which is a form of *Legendre's equation.* For problems involving the full range of θ, from 0 to π, the solutions to Legendre's equation (4–152) are called *Legendre functions*, usually denoted by $P(\cos \theta)$. Since Legendre functions for integral values of n are polynomials in $\cos \theta$, they are also called *Legendre polynomials*. We write

$$\Theta_n(\theta) = P_n(\cos \theta). \tag{4–153}$$

Table 4–2 lists the expressions for Legendre polynomials[†] for several values of n.

Combining solutions (4–150) and (4–153) in Eq. (4–145), we have, for spherical boundary-value problems with no azimuthal variation,

$$V_n(R, \theta) = [A_n R^n + B_n R^{-(n+1)}]P_n(\cos \theta). \tag{4–154}$$

Depending on the boundary conditions of the given problem, the complete solution may be a summation of the terms in Eq. (4–154). We illustrate the application of

---

[†] Actually, Legendre polynomials are Legendre functions of the first kind. There is another set of solutions to Legendre's equation, called Legendre functions of the second kind; but they have singularities at $\theta = 0$ and π and must therefore be excluded if the polar axis is a region of interest.

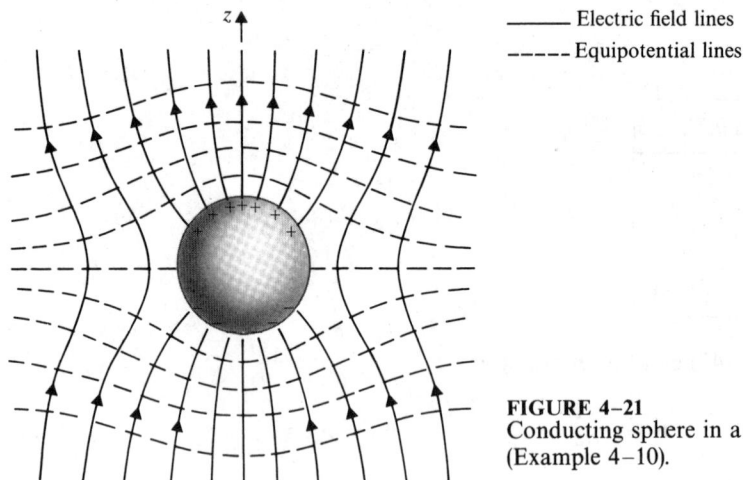

——— Electric field lines
————— Equipotential lines

FIGURE 4–21
Conducting sphere in a uniform electric field
(Example 4–10).

Legendre polynomials in the solution of a simple boundary-value problem in the following example.

EXAMPLE 4–10 An uncharged conducting sphere of radius b is placed in an initially uniform electric field $\mathbf{E}_0 = \mathbf{a}_z E_0$. Determine (a) the potential distribution $V(R, \theta)$, and (b) the electric field intensity $\mathbf{E}(R, \theta)$ after the introduction of the sphere.

Solution After the conducting sphere is introduced into the electric field, a separation and redistribution of charges will take place in such a way that the surface of the sphere is maintained equipotential. The electric field intensity within the sphere is zero. Outside the sphere the field lines will intersect the surface normally, and the field intensity at points very far away from the sphere will not be affected appreciably. The geometry of this problem is depicted in Fig. 4–21. The potential is, obviously, independent of the azimuthal angle ϕ, and the solution obtained in this section applies.

a) To determine the potential distribution $V(R, \theta)$ for $R \geq b$, we note the following boundary conditions:

$$V(b, \theta) = 0^{\dagger} \tag{4–155a}$$

$$V(R, \theta) = -E_0 z = -E_0 R \cos \theta, \qquad \text{for } R \gg b. \tag{4–155b}$$

Equation (4–155b) is a statement that the original $\mathbf{E}_0$ is not disturbed at points very far away from the sphere. By using Eq. (4–154) we write the general solution

---

† For this problem it is convenient to assume that $V = 0$ in the equatorial plane ($\theta = \pi/2$), which leads to $V(b, \theta) = 0$, since the surface of the conducting sphere is equipotential. (See Problem P.4–28 for $V(b, \theta) = V_0$.)

as

$$V(R, \theta) = \sum_{n=0}^{\infty} [A_n R^n + B_n R^{-(n+1)}] P_n(\cos \theta), \qquad R \geq b. \qquad (4\text{–}156)$$

However, in view of Eq. (4–155b), all A_n except A_1 must vanish, and $A_1 = -E_0$. We have, from Eq. (4–156) and Table 4–2,

$$V(R, \theta) = -E_0 R P_1(\cos \theta) + \sum_{n=0}^{\infty} B_n R^{-(n+1)} P_n(\cos \theta)$$

$$= B_0 R^{-1} + (B_1 R^{-2} - E_0 R) \cos \theta + \sum_{n=2}^{\infty} B_n R^{-(n+1)} P_n(\cos \theta), \qquad R \geq b. \qquad (4\text{–}157)$$

Actually, the first term on the right side of Eq. (4–157) corresponds to the potential of a charged sphere. Since the sphere is uncharged, $B_0 = 0$, and Eq. (4–157) becomes

$$V(R, \theta) = \left(\frac{B_1}{R^2} - E_0 R\right) \cos \theta + \sum_{n=2}^{\infty} B_n R^{-(n+1)} P_n(\cos \theta), \qquad R \geq b. \qquad (4\text{–}158)$$

Now applying boundary condition (4–155a) at $R = b$, we require

$$0 = \left(\frac{B_1}{b^2} - E_0 b\right) \cos \theta + \sum_{n=2}^{\infty} B_n b^{-(n+1)} P_n(\cos \theta),$$

from which we obtain

$$B_1 = E_0 b^3$$

and

$$B_n = 0, \qquad n \geq 2.$$

We have, finally, from Eq. (4–158),

$$V(R, \theta) = -E_0 \left[1 - \left(\frac{b}{R}\right)^3\right] R \cos \theta, \qquad R \geq b. \qquad (4\text{–}159)$$

b) The electric field intensity $\mathbf{E}(R, \theta)$ for $R \geq b$ can be easily determined from $-\nabla V(R, \theta)$:

$$\mathbf{E}(R, \theta) = \mathbf{a}_R E_R + \mathbf{a}_\theta E_\theta, \qquad (4\text{–}160)$$

where

$$E_R = -\frac{\partial V}{\partial R} = E_0 \left[1 + 2\left(\frac{b}{R}\right)^3\right] \cos \theta, \qquad R \geq b \qquad (4\text{–}160a)$$

and

$$E_\theta = -\frac{\partial V}{R \, \partial \theta} = -E_0 \left[1 - \left(\frac{b}{R}\right)^3\right] \sin \theta, \qquad R \geq b. \qquad (4\text{–}160b)$$

The surface charge density on the sphere can be found by noting that

$$\rho_s(\theta) = \epsilon_0 E_R \big|_{R=b} = 3\epsilon_0 E_0 \cos \theta, \qquad (4\text{–}161)$$

which is proportional to $\cos \theta$, being zero at $\theta = \pi/2$. Some equipotential and field lines are sketched in Fig. 4–21.

It is interesting to note from Eq. (4–159) that the potential is the sum of two terms: $-E_0 R \cos \theta$ due to the applied uniform electric field; and $(E_0 b^3 \cos \theta)/R^2$ due to an electric dipole of a dipole moment:

$$\mathbf{p} = \mathbf{a}_z 4\pi\epsilon_0 b^3 E_0 \tag{4–162}$$

at the center of the sphere. The contribution of the equivalent dipole can be verified by referring to Eq. (3–53). The expressions in Eqs. (4–160a) and (4–160b) for the resultant electric field intensity, being derived from the potential, obviously also represent the combination of the applied uniform field and that of the equivalent dipole, given in Eq. (3–54).

In this chapter we have discussed the analytical solution of electrostatic problems by the method of images and by direct solution of Laplace's equation. The method of images is useful when charges exist near conducting bodies of a simple and compatible geometry: a point charge near a conducting sphere or an infinite conducting plane; and a line charge near a parallel conducting cylinder or a parallel conducting plane. The solution of Laplace's equation by the method of separation of variables requires that the boundaries coincide with coordinate surfaces. These requirements restrict the usefulness of both methods. In practical problems we are often faced with more complicated boundaries, which are not amenable to neat analytical solutions. In such cases we must resort to approximate graphical or numerical methods. These methods are beyond the scope of this book.[†]

Review Questions

R.4–1 Write Poisson's equation in vector notation
   a) for a simple medium,
   b) for a linear and isotropic but inhomogeneous medium.

R.4–2 Repeat in Cartesian coordinates both parts of Question R.4–1.

R.4–3 Write Laplace's equation for a simple medium
   a) in vector notation,   **b)** in Cartesian coordinates.

R.4–4 If $\nabla^2 U = 0$, why does it not follow that U is identically zero?

R.4–5 A fixed voltage is connected across a parallel-plate capacitor.
   a) Does the electric field intensity in the space between the plates depend on the permittivity of the medium?
   b) Does the electric flux density depend on the permittivity of the medium?
Explain.

R.4–6 Assume that fixed charges $+Q$ and $-Q$ are deposited on the plates of an isolated parallel-plate capacitor.
   a) Does the electric field intensity in the space between the plates depend on the permittivity of the medium?
   b) Does the electric flux density depend on the permittivity of the medium?
Explain.

[†] See, for instance, B. D. Popović, *Introductory Engineering Electromagnetics*, Chapter 5, Addison-Wesley Publishing Co., Reading, Mass., 1971.

R.4–7 Why is the electrostatic potential continuous at a boundary?

R.4–8 State in words the uniqueness theorem of electrostatics.

R.4–9 What is the image of a spherical cloud of electrons with respect to an infinite conducting plane?

R.4–10 Why cannot the point at infinity be used as the point for the zero reference potential for an infinite line charge as it is for a point charge? What is the physical reason for this difference?

R.4–11 What is the image of an infinitely long line charge of density ρ_ℓ with respect to a parallel conducting circular cylinder?

R.4–12 Where is the zero-potential surface of the two-wire transmission line in Fig. 4–6?

R.4–13 In finding the surface charge induced on a grounded sphere by a point charge, can we set $R = a$ in Eq. (4–67) and then evaluate ρ_s by $-\epsilon_0 \, \partial V(a, \theta)/\partial R$? Explain.

R.4–14 What is the method of separation of variables? Under what conditions is it useful in solving Laplace's equation?

R.4–15 What are boundary-value problems?

R.4–16 Can all three separation constants (k_x, k_y, and k_z) in Cartesian coordinates be real? Can they all be imaginary? Explain.

R.4–17 Can the separation constant k in the solution of the two-dimensional Laplace's equation (4–120) be imaginary? Explain.

R.4–18 What should we do to modify the solution in Eq. (4–133) for Example 4–8 if the inner conductor of the coaxial cable is grounded and the outer conductor is kept at a potential V_0?

R.4–19 What should we do to modify the solution in Eq. (4–139) for Example 4–9 if the conducting circular cylinder is split vertically in two halves, with $V = V_0$ for $-\pi/2 < \phi < \pi/2$ and $V = -V_0$ for $\pi/2 < \phi < 3\pi/2$?

R.4–20 Can functions $V_1(R, \theta) = C_1 R \cos \theta$ and $V_2(R, \theta) = C_2 R^{-2} \cos \theta$, where C_1 and C_2 are arbitrary constants, be solutions of Laplace's equation in spherical coordinates? Explain.

Problems

P.4–1 The upper and lower conducting plates of a large parallel-plate capacitor are separated by a distance d and maintained at potentials V_0 and 0, respectively. A dielectric slab of dielectric constant 6.0 and uniform thickness $0.8d$ is placed over the lower plate. Assuming negligible fringing effect, determine
   a) the potential and electric field distribution in the dielectric slab,
   b) the potential and electric field distribution in the air space between the dielectric slab and the upper plate,
   c) the surface charge densities on the upper and lower plates.
   d) Compare the results in part (b) with those without the dielectric slab.

P.4–2 Prove that the scalar potential V in Eq. (3–61) satisfies Poisson's equation, Eq. (4–6).

P.4–3 Prove that a potential function satisfying Laplace's equation in a given region possesses no maximum or minimum within the region.

P.4–4 Verify that

$$V_1 = C_1/R \qquad \text{and} \qquad V_2 = C_2 z/(x^2 + y^2 + z^2)^{3/2},$$

where C_1 and C_2 are arbitrary constants, are solutions of Laplace's equation.

P.4–5 Assume a point charge Q above an infinite conducting plane at $y = 0$.

  a) Prove that $V(x, y, z)$ in Eq. (4–37) satisfies Laplace's equation if the conducting plane is maintained at zero potential.
  b) What should the expression for $V(x, y, z)$ be if the conducting plane has a nonzero potential V_0?
  c) What is the electrostatic force of attraction between the charge Q and the conducting plane?

P.4–6 Assume that the space between the inner and outer conductors of a long coaxial cylindrical structure is filled with an electron cloud having a volume density of charge $\rho = A/r$ for $a < r < b$, where a and b are, the radii of the inner and outer conductors, respectively. The inner conductor is maintained at a potential V_0, and the outer conductor is grounded. Determine the potential distribution in the region $a < r < b$ by solving Poisson's equation.

P.4–7 A point charge Q exists at a distance d above a large grounded conducting plane. Determine

  a) the surface charge density ρ_s,
  b) the total charge induced on the conducting plane.

P.4–8 For a positive point charge Q located at distances d_1 and d_2, respectively, from two grounded perpendicular conducting half-planes shown in Fig. 4–4(a), find the expressions for

  a) the potential and the electric field intensity at an arbitrary point $P(x, y)$ in the first quadrant,
  b) the surface charge densities induced on the two half-planes. Sketch the variations of the surface charge densities in the xy-plane.

P.4–9 Determine the systems of image charges that will replace the conducting boundaries that are maintained at zero potential for

  a) a point charge Q located between two large, grounded, parallel conducting planes as shown in Fig. 4–22(a),
  b) an infinite line charge ρ_ℓ located midway between two large, intersecting conducting planes forming a 60-degree angle, as shown in Fig. 4–22(b).

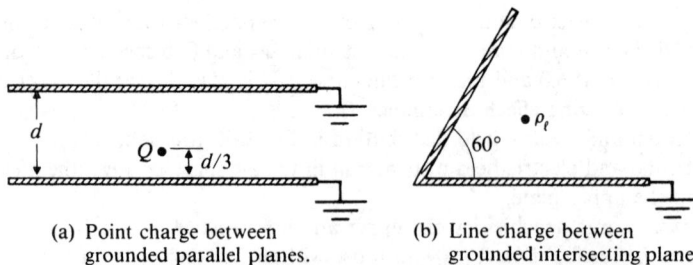

(a) Point charge between grounded parallel planes.

(b) Line charge between grounded intersecting plane.

FIGURE 4–22
Diagrams for Problem P.4–9.

P.4–10 A straight conducting wire of radius a is parallel to and at height h from the surface of the earth. Assuming that the earth is perfectly conducting, determine the capacitance and the force per unit length between the wire and the earth.

P.4–11 A very long two-wire transmission line, each wire of radius a and separated by a distance d, is supported at a height h above a flat conducting ground. Assuming both d and h to be much larger than a, find the capacitance per unit length of the line.

P.4–12 For the pair of equal and opposite line charges shown in Fig. 4–7,
   a) write the expression for electric field intensity **E** at point $P(x, y)$ in Cartesian coordinates,
   b) find the equation of the electric field lines sketched in Fig. 4–8.

P.4–13 Determine the capacitance per unit length of a two-wire transmission line with parallel conducting cylinders of different radii a_1 and a_2, their axes being separated by a distance D (where $D > a_1 + a_2$).

P.4–14 A long wire of radius a_1 lies inside a conducting circular tunnel of radius a_2, as shown in Fig. 4–10(a). The distance between their axes is D.
   a) Find the capacitance per unit length.
   b) Determine the force per unit length on the wire if the wire and the tunnel carry equal and opposite line charges of magnitude ρ_ℓ.

P.4–15 A point charge Q is located inside and at distance d from the center of a grounded spherical conducting shell of radius b (where $b > d$). Use the method of images to determine
   a) the potential distribution inside the shell,
   b) the charge density ρ_s induced on the inner surface of the shell.

P.4–16 Two conducting spheres of equal radius a are maintained at potentials V_0 and 0, respectively. Their centers are separated by a distance D.
   a) Find the image charges and their locations that can electrically replace the two spheres.
   b) Find the capacitance between the two spheres.

P.4–17 Two dielectric media with dielectric constants ϵ_1 and ϵ_2 are separated by a plane boundary at $x = 0$, as shown in Fig. 4–23. A point charge Q exists in medium 1 at distance d from the boundary.

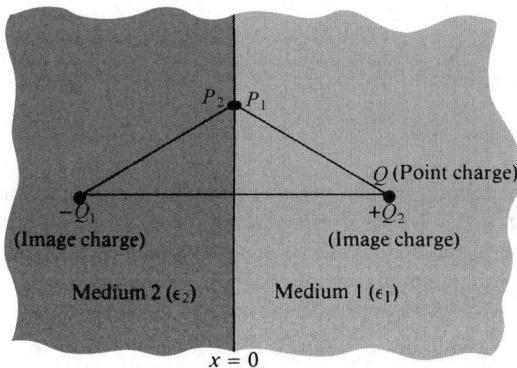

FIGURE 4–23
Image charges in dielectric media (Problem P.4–17).

a) Verify that the field in medium 1 can be obtained from Q and an image charge $-Q_1$, both acting in medium 1.

b) Verify that the field in medium 2 can be obtained from Q and an image charge $+Q_2$ coinciding with Q, both acting in medium 2.

c) Determine Q_1 and Q_2. (*Hint:* Consider neighboring points P_1 and P_2 in media 1 and 2, respectively, and require the continuity of the tangential component of the E-field and of the normal component of the D-field.)

P.4–18 Describe the geometry of the region in which the potential function can be represented by a single term as follows:

a) $V(x, y) = c_1 xy$,

b) $V(x, y) = c_2 \sin kx \sinh ky$.

Find c_1, c_2, and k in terms of the dimensions and a fixed potential V_0.

P.4–19 In what way should we modify the solution in Eq. (4–114) for Example 4–7 if the boundary conditions on the top, bottom, and right planes in Fig. 4–17 are $\partial V/\partial n = 0$?

P.4–20 In what way should we modify the solution in Eq. (4–114) for Example 4–7 if the top, bottom, and left planes in Fig. 4–17 are grounded ($V = 0$) and an end plate on the right is maintained at a constant potential V_0?

P.4–21 Consider the rectangular region shown in Fig. 4–17 as the cross section of an enclosure formed by four conducting plates. The left and right plates are grounded, and the top and bottom plates are maintained at constant potentials V_1 and V_2, respectively. Determine the potential distribution inside the enclosure.

P.4–22 Consider a metallic rectangular box with sides a and b and height c. The side walls and the bottom surface are grounded. The top surface is isolated and kept at a constant potential V_0. Determine the potential distribution inside the box.

P.4–23 Two infinite insulated conducting planes maintained at potentials 0 and V_0 form a wedge-shaped configuration, as shown in Fig. 4–24. Determine the potential distributions for the regions: (a) $0 < \phi < \alpha$, and (b) $\alpha < \phi < 2\pi$.

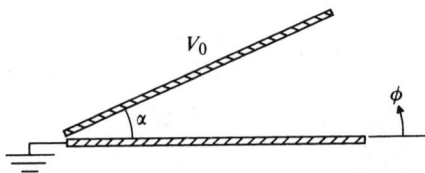

FIGURE 4–24
Two infinite insulated conducting planes maintained at constant potentials (Problem P.4–23).

P.4–24 An infinitely long, thin conducting circular cylinder of radius b is split in four quarter-cylinders, as shown in Fig. 4–25. The quarter-cylinders in the second and fourth quadrants are grounded, and those in the first and third quadrants are kept at potentials V_0 and $-V_0$, respectively. Determine the potential distribution both inside and outside the cylinder.

P.4–25 A long, grounded conducting cylinder of radius b is placed along the z-axis in an initially uniform electric field $\mathbf{E}_0 = \mathbf{a}_x E_0$. Determine potential distribution $V(r, \phi)$ and electric field intensity $\mathbf{E}(r, \phi)$ outside the cylinder. Show that the electric field intensity at the surface of the cylinder may be twice as high as that in the distance, which may cause a

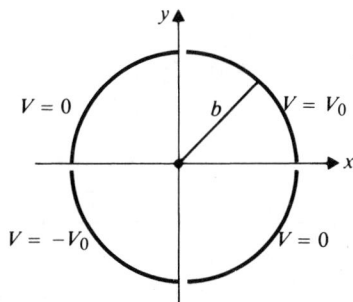

FIGURE 4–25
Cross section of long circular cylinder split in four quarters (Problem P.4–24).

local breakdown or corona. (This phenomenon of corona discharge along the rigging and spars of ships and on airplanes near storms is known as *St. Elmo's fire.*[†])

P.4–26 A long dielectric cylinder of radius b and dielectric constant ϵ_r is placed along the z-axis in an initially uniform electric field $\mathbf{E}_0 = \mathbf{a}_x E_0$. Determine $V(r, \phi)$ and $\mathbf{E}(r, \phi)$ both inside and outside the dielectric cylinder.

P.4–27 An infinite conducting cone of half-angle α is maintained at potential V_0 and insulated from a grounded conducting plane, as illustrated in Fig. 4–26. Determine

   a) the potential distribution $V(\theta)$ in the region $\alpha < \theta < \pi/2$,
   b) the electric field intensity in the region $\alpha < \theta < \pi/2$,
   c) the charge densities on the cone surface and on the grounded plane.

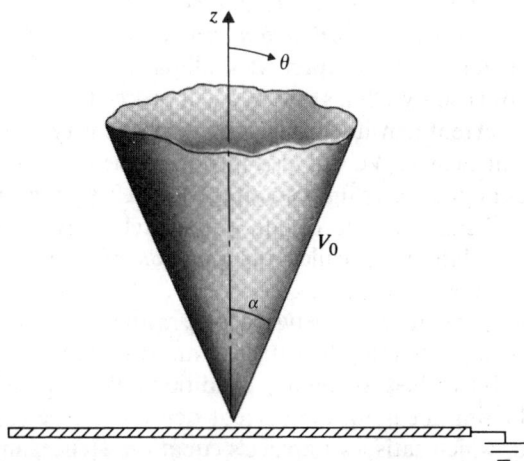

FIGURE 4–26
An infinite conducting cone and a grounded conducting plane (Problem P.4–27).

P.4–28 Rework Example 4–10, assuming that $V(b, \theta) = V_0$ in Eq. (4–155a).

P.4–29 A dielectric sphere of radius b and dielectric constant ϵ_r is placed in an initially uniform electric field, $\mathbf{E}_0 = \mathbf{a}_z E_0$, in air. Determine $V(R, \theta)$ and $\mathbf{E}(R, \theta)$ both inside and outside the dielectric sphere.

[†] R. H. Golde (Ed.), *Lightning*, Academic Press, New York, 1977, vol. 2, Chap. 21.

5

Steady
Electric Currents

5–1 Introduction

In Chapters 3 and 4 we dealt with electrostatic problems, field problems associated with electric charges at rest. We now consider the charges in motion that constitute current flow. There are several types of electric currents caused by the *motion of free charges*.[†] *Conduction currents* in conductors and semiconductors are caused by drift motion of conduction electrons and/or holes; *electrolytic currents* are the result of migration of positive and negative ions; and *convection currents* result from motion of electrons and/or ions in a vacuum. In this chapter we shall pay special attention to conduction currents that are governed by Ohm's law. We will proceed from the point form of Ohm's law that relates current density and electric field intensity and obtain the $V = IR$ relationship in circuit theory. We will also introduce the concept of electromotive force and derive the familiar Kirchhoff's voltage law. Using the principle of *conservation of charge*, we will show how to obtain a point relationship between current and charge densities, a relationship called the *equation of continuity* from which Kirchhoff's current law follows.

When a current flows across the interface between two media of different conductivities, certain boundary conditions must be satisfied, and the direction of current flow is changed. We will discuss these boundary conditions. We will also show that for a homogeneous conducting medium, the current density can be expressed as the gradient of a scalar field, which satisfies Laplace's equation. Hence, an analogous situation exists between steady-current and electrostatic fields that is the basis for mapping the potential distribution of an electrostatic problem in an *electrolytic tank*.

The electrolyte in an electrolytic tank is essentially a liquid medium with a low conductivity, usually a diluted salt solution. Highly conducting metallic electrodes

---

[†] In a time-varying situation there is another type of current caused by bound charges. The time-rate of change of electric displacement leads to a *displacement current*. This will be discussed in Chapter 7.

are inserted in the solution. When a voltage or potential difference is applied to the electrodes, an electric field is established within the solution, and the molecules of the electrolyte are decomposed into oppositely charged ions by a chemical process called *electrolysis*. Positive ions move in the direction of the electric field, and negative ions move in a direction opposite to the field, both contributing to a current flow in the direction of the field. An experimental model can be set up in an electrolytic tank, with electrodes of proper geometrical shapes simulating the boundaries in electrostatic problems. The measured potential distribution in the electrolyte is then the solution to Laplace's equation for difficult-to-solve analytic problems having complex boundaries in a homogeneous medium.

Convection currents are the result of the motion of positively or negatively charged particles in a vacuum or rarefied gas. Familiar examples are electron beams in a cathode-ray tube and the violent motions of charged particles in a thunderstorm. Convection currents, the result of hydrodynamic motion involving a mass transport, are not governed by Ohm's law.

The mechanism of conduction currents is different from that of both electrolytic currents and convection currents. In their normal state the atoms of a conductor occupy regular positions in a crystalline structure. The atoms consist of positively charged nuclei surrounded by electrons in a shell-like arrangement. The electrons in the inner shells are tightly bound to the nuclei and are not free to move away. The electrons in the outermost shells of a conductor atom do not completely fill the shells; they are valence or conduction electrons and are only very loosely bound to the nuclei. These latter electrons may wander from one atom to another in a random manner. The atoms, on the average, remain electrically neutral, and there is no net drift motion of electrons. When an external electric field is applied on a conductor, an organized motion of the conduction electrons will result, producing an electric current. The average drift velocity of the electrons is very low (on the order of 10^{-4} or 10^{-3} m/s) even for very good conductors because they collide with the atoms in the course of their motion, dissipating part of their kinetic energy as heat. Even with the drift motion of conduction electrons, a conductor remains electrically neutral. Electric forces prevent excess electrons from accumulating at any point in a conductor. We will show analytically that the charge density in a conductor decreases exponentially with time. In a good conductor the charge density diminishes extremely rapidly toward zero as the state of equilibrium is approached.

5-2 Current Density and Ohm's Law

Consider the steady motion of one kind of charge carriers, each of charge q (which is negative for electrons), across an element of surface Δs with a velocity $\mathbf{u}$, as shown in Fig. 5-1. If N is the number of charge carriers per unit volume, then in time Δt each charge carrier moves a distance $\mathbf{u}\,\Delta t$, and the amount of charge passing through the surface Δs is

$$\Delta Q = Nq\mathbf{u} \cdot \mathbf{a}_n \Delta s \, \Delta t \qquad \text{(C)}. \qquad (5-1)$$

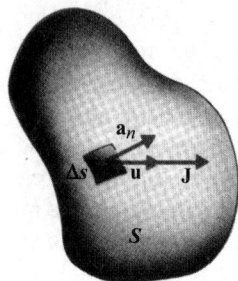

FIGURE 5–1
Conduction current due to drift motion of charge carriers across a surface.

Since current is the time rate of change of charge, we have

$$\Delta I = \frac{\Delta Q}{\Delta t} = Nq\mathbf{u} \cdot \mathbf{a}_n \Delta s = Nq\mathbf{u} \cdot \Delta \mathbf{s} \qquad \text{(A)}. \qquad (5\text{–}2)$$

In Eq. (5–2) we have written $\Delta \mathbf{s} = \mathbf{a}_n \Delta s$ as a vector quantity. It is convenient to define a vector point function, *volume current density*, or simply *current density*, **J**, in amperes per *square* meter,

$$\mathbf{J} = Nq\mathbf{u} \qquad \text{(A/m}^2\text{)}, \qquad (5\text{–}3)$$

so that Eq. (5–2) can be written as

$$\Delta I = \mathbf{J} \cdot \Delta \mathbf{s}. \qquad (5\text{–}4)$$

The total current I flowing through an arbitrary surface S is then the flux of the **J** vector through S:

$$\boxed{I = \int_S \mathbf{J} \cdot d\mathbf{s} \qquad \text{(A)}.} \qquad (5\text{–}5)$$

Noting that the product Nq is in fact free charge per unit volume, we may rewrite Eq. (5–3) as

$$\boxed{\mathbf{J} = \rho\mathbf{u} \qquad \text{(A/m}^2\text{)},} \qquad (5\text{–}6)$$

which is the relation between the *convection current density* and the velocity of the charge carrier.

EXAMPLE 5–1   In vacuum-tube diodes, electrons are emitted from a hot cathode at zero potential and collected by an anode maintained at a potential V_0, resulting in a convection current flow. Assuming that the cathode and the anode are parallel conducting plates and that the electrons leave the cathode with a zero initial velocity (space-charge limited condition), find the relation between the current density J and V_0.

Solution The region between the cathode and the anode is shown in Fig. 5–2, where a cloud of electrons (negative space charge) exists such that the force of repulsion makes the electrons boiled off the hot cathode leave essentially with a zero velocity. In other words, the net electric field at the cathode is zero. Neglecting fringing effects, we have

$$\mathbf{E}(0) = \mathbf{a}_y E_y(0) = -\mathbf{a}_y \left.\frac{dV(y)}{dy}\right|_{y=0} = 0. \tag{5-7}$$

In the steady state the current density is constant, independent of y:

$$\mathbf{J} = -\mathbf{a}_y J = \mathbf{a}_y \rho(y) u(y), \tag{5-8}$$

where the charge density $\rho(y)$ is a negative quantity. The velocity $\mathbf{u} = \mathbf{a}_y u(y)$ is related to the electric field intensity $\mathbf{E}(y) = \mathbf{a}_y E(y)$ by Newton's law of motion:

$$m\frac{du(y)}{dt} = -eE(y) = e\frac{dV(y)}{dy}, \tag{5-9}$$

where $m = 9.11 \times 10^{-31}$ (kg) and $-e = -1.60 \times 10^{-19}$ (C) are the mass and charge, respectively, of an electron. Noting that

$$m\frac{du}{dt} = m\frac{du}{dy}\frac{dy}{dt} = mu\frac{du}{dy}$$

$$= \frac{d}{dy}\left(\frac{1}{2}mu^2\right),$$

we can rewrite Eq. (5–9) as

$$\frac{d}{dy}\left(\frac{1}{2}mu^2\right) = e\frac{dV}{dy}. \tag{5-10}$$

Integration of Eq. (5–10) gives

$$\tfrac{1}{2}mu^2 = eV, \tag{5-11}$$

where the constant of integration has been set to zero because at $y = 0$, $u(0) = V(0) = 0$. From Eq. (5–11) we obtain

$$u = \left(\frac{2e}{m}V\right)^{1/2}. \tag{5-12}$$

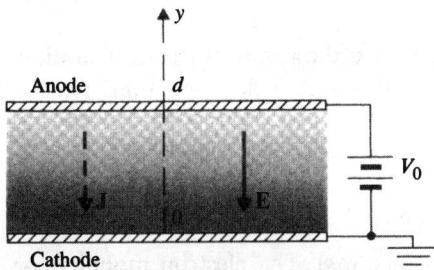

FIGURE 5–2
Space-charge-limited vacuum diode (Example 5–1).

In order to find $V(y)$ in the interelectrode region we must solve Poisson's equation with ρ expressed in terms of $V(y)$ from Eq. (5-8):

$$\rho = -\frac{J}{u} = -J \sqrt{\frac{m}{2e}} \, V^{-1/2}.$$ (5-13)

We have, from Eq. (4-6),

$$\frac{d^2 V}{dy^2} = -\frac{\rho}{\epsilon_0} = \frac{J}{\epsilon_0} \sqrt{\frac{m}{2e}} \, V^{-1/2}.$$ (5-14)

Equation (5-14) can be integrated if both sides are first multiplied by $2\,dV/dy$. The result is

$$\left(\frac{dV}{dy}\right)^2 = \frac{4J}{\epsilon_0} \sqrt{\frac{m}{2e}} \, V^{1/2} + c.$$ (5-15)

At $y = 0$, $V = 0$, and $dV/dy = 0$ from Eq. (5-7), so $c = 0$. Equation (5-15) becomes

$$V^{-1/4} \, dV = 2 \sqrt{\frac{J}{\epsilon_0}} \left(\frac{m}{2e}\right)^{1/4} dy.$$ (5-16)

Integrating the left side of Eq. (5-16) from $V = 0$ to V_0 and the right side from $y = 0$ to d, we obtain

$$\frac{4}{3} V_0^{3/4} = 2 \sqrt{\frac{J}{\epsilon_0}} \left(\frac{m}{2e}\right)^{1/4} d,$$

or

$$J = \frac{4\epsilon_0}{9d^2} \sqrt{\frac{2e}{m}} \, V_0^{3/2} \qquad \text{(A/m}^2\text{)}.$$ (5-17)

Equation (5-17) states that the convection current density in a space-charge limited vacuum diode is proportional to the three-halves power of the potential difference between the anode and the cathode. This nonlinear relation is known as the ***Child-Langmuir law***. ▬

In the case of conduction currents there may be more than one kind of charge carriers (electrons, holes, and ions) drifting with different velocities. Equation (5-3) should be generalized to read

$$\mathbf{J} = \sum_i N_i q_i \mathbf{u}_i \qquad \text{(A/m}^2\text{)}.$$ (5-18)

As indicated in Section 5-1, conduction currents are the result of the drift motion of charge carriers under the influence of an applied electric field. The atoms remain neutral ($\rho = 0$). It can be justified analytically that for most conducting materials the average drift velocity is directly proportional to the electric field intensity. For metallic conductors we write

$$\mathbf{u} = -\mu_e \mathbf{E} \qquad \text{(m/s)},$$ (5-19)

where μ_e is the electron ***mobility*** measured in $(\text{m}^2/\text{V}\cdot\text{s})$. The electron mobility for copper is $3.2 \times 10^{-3} \, (\text{m}^2/\text{V}\cdot\text{s})$. It is $1.4 \times 10^{-4} \, (\text{m}^2/\text{V}\cdot\text{s})$ for aluminum and 5.2×10^{-3}

$(m^2/V \cdot s)$ for silver. From Eqs. (5–3) and (5–19) we have

$$\mathbf{J} = -\rho_e \mu_e \mathbf{E}, \tag{5-20}$$

where $\rho_e = -Ne$ is the charge density of the drifting electrons and is a negative quantity. Equation (5–20) can be rewritten as

$$\boxed{\mathbf{J} = \sigma \mathbf{E} \qquad (A/m^2),} \tag{5-21}$$

where the proportionality constant, $\sigma = -\rho_e \mu_e$, is a macroscopic constitutive parameter of the medium called **conductivity**.

For semiconductors, conductivity depends on the concentration and mobility of both electrons and holes:

$$\sigma = -\rho_e \mu_e + \rho_h \mu_h, \tag{5-22}$$

where the subscript h denotes hole. In general, $\mu_e \neq \mu_h$. For germanium, typical values are $\mu_e = 0.38$, $\mu_h = 0.18$; for silicon, $\mu_e = 0.12$, $\mu_h = 0.03$ $(m^2/V \cdot s)$.

Equation (5–21) is a constitutive relation of a conducting medium. Isotropic materials for which the linear relation Eq. (5–21) holds are called **ohmic media**. The unit for σ is ampere per volt-meter $(A/V \cdot m)$ or siemens per meter (S/m). Copper, the most commonly used conductor, has a conductivity 5.80×10^7 (S/m). On the other hand, the conductivity of germanium is around 2.2 (S/m), and that of silicon is 1.6×10^{-3} (S/m). The conductivity of semiconductors is highly dependent of (increases with) temperature. Hard rubber, a good insulator, has a conductivity of only 10^{-15} (S/m). Appendix B–4 lists the conductivities of some other frequently used materials. However, note that, unlike the dielectric constant, the conductivity of materials varies over an extremely wide range. The reciprocal of conductivity is called **resistivity**, in ohm-meters $(\Omega \cdot m)$. We prefer to use conductivity; there is really no compelling need to use both conductivity and resistivity.

We recall **Ohm's law** from circuit theory that the voltage V_{12} across a resistance R, in which a current I flows from point 1 to point 2, is equal to RI; that is,

$$V_{12} = RI. \tag{5-23}$$

Here R is usually a piece of conducting material of a given length; V_{12} is the voltage between two terminals 1 and 2; and I is the total current flowing from terminal 1 to terminal 2 through a finite cross section.

Equation (5–23) is *not* a point relation. Although there is little resemblance between Eq. (5–21) and Eq. (5–23), the former is generally referred to as the **point form of Ohm's law**. It holds at all points in space, and σ can be a function of space coordinates.

Let us use the point form of Ohm's law to derive the voltage-current relationship of a piece of homogeneous material of conductivity σ, length ℓ, and uniform cross section S, as shown in Fig. 5–3. Within the conducting material, $\mathbf{J} = \sigma \mathbf{E}$, where both $\mathbf{J}$ and $\mathbf{E}$ are in the direction of current flow. The potential difference or voltage

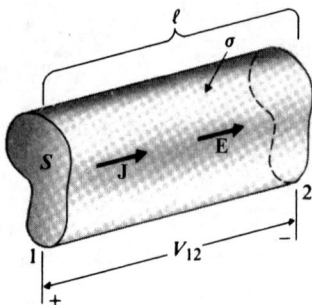

FIGURE 5–3
Homogeneous conductor with a constant cross section.

between terminals 1 and 2 is[†]

$$V_{12} = E\ell$$

or

$$E = \frac{V_{12}}{\ell}. \qquad (5\text{–}24)$$

The total current is

$$I = \int_S \mathbf{J} \cdot d\mathbf{s} = JS$$

or

$$J = \frac{I}{S}. \qquad (5\text{–}25)$$

Using Eqs. (5–24) and (5–25) in Eq. (5–21), we obtain

$$\frac{I}{S} = \sigma \frac{V_{12}}{\ell}$$

or

$$V_{12} = \left(\frac{\ell}{\sigma S}\right) I = RI, \qquad (5\text{–}26)$$

which is the same as Eq. (5–23). From Eq. (5–26) we have the formula for the *resistance* of a straight piece of homogeneous material of a uniform cross section for steady current (d.c.):

$$\boxed{R = \frac{\ell}{\sigma S} \quad (\Omega).} \qquad (5\text{–}27)$$

We could have started with Eq. (5–23) as the experimental Ohm's law and applied it to a homogeneous conductor of length ℓ and uniform cross-section S. Using the formula in Eq. (5–27), we could derive the point relationship in Eq. (5–21).

[†] We will discuss the significance of V_{12} and E more in detail in Section 5–3.

■■■■■ **EXAMPLE 5–2**   Determine the d-c resistance of 1-(km) of wire having a 1-(mm) radius (a) if the wire is made of copper, and (b) if the wire is made of aluminum.

Solution   Since we are dealing with conductors of a uniform cross section, Eq. (5–27) applies.

a) For copper wire, $\sigma_{cu} = 5.80 \times 10^7$ (S/m):

$$\ell = 10^3 \text{ (m)}, \qquad S = \pi(10^{-3})^2 = 10^{-6}\pi \quad (\text{m}^2).$$

We have

$$R_{cu} = \frac{\ell}{\sigma_{cu}S} = \frac{10^3}{5.80 \times 10^7 \times 10^{-6}\pi} = 5.49 \quad (\Omega).$$

b) For aluminum wire, $\sigma_{al} = 3.54 \times 10^7$ (S/m):

$$R_{al} = \frac{\ell}{\sigma_{al}S} = \frac{\sigma_{cu}}{\sigma_{al}} R_{cu} = \frac{5.80}{3.54} \times 5.49 = 8.99 \quad (\Omega). \qquad ■$$

The **conductance**, G, or the reciprocal of resistance, is useful in combining resistances in parallel. The unit for conductance is (Ω^{-1}), or siemens (S).

$$G = \frac{1}{R} = \sigma \frac{S}{\ell} \quad (\text{S}). \tag{5–28}$$

From circuit theory we know the following:

a) When resistances R_1 and R_2 are connected in series (same current), the total resistance R is

$$\boxed{R_{sr} = R_1 + R_2.} \tag{5–29}$$

b) When resistances R_1 and R_2 are connected in parallel (same voltage), we have

$$\boxed{\frac{1}{R_{\parallel}} = \frac{1}{R_1} + \frac{1}{R_2}} \tag{5–30a}$$

or

$$\boxed{G_{\parallel} = G_1 + G_2.} \tag{5–30b}$$

5–3   Electromotive Force and Kirchhoff's Voltage Law

In Section 3–2 we pointed out that static electric field is conservative and that the scalar line integral of static electric intensity around any closed path is zero; that is,

$$\oint_C \mathbf{E} \cdot d\ell = 0. \tag{5–31}$$

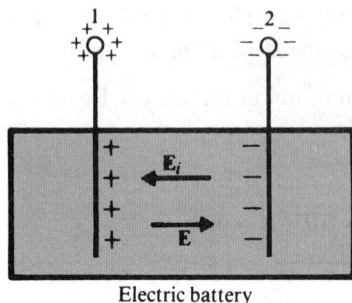

FIGURE 5-4

Electric battery

Electric fields inside an electric battery.

For an ohmic material $\mathbf{J} = \sigma\mathbf{E}$, Eq. (5-31) becomes

$$\oint_C \frac{1}{\sigma} \mathbf{J} \cdot d\boldsymbol{\ell} = 0. \qquad (5\text{-}32)$$

Equation (5-32) tells us that *a steady current cannot be maintained in the same direction in a closed circuit by an electrostatic field*. A steady current in a circuit is the result of the motion of charge carriers, which, in their paths, collide with atoms and dissipate energy in the circuit. This energy must come from a nonconservative field, since a charge carrier completing a closed circuit in a conservative field neither gains nor loses energy. The source of the nonconservative field may be electric batteries (conversion of chemical energy to electric energy), electric generators (conversion of mechanical energy to electric energy), thermocouples (conversion of thermal energy to electric energy), photovoltaic cells (conversion of light energy to electric energy), or other devices. These electrical energy sources, when connected in an electric circuit, provide a driving force for the charge carriers. This force manifests itself as an equivalent *impressed electric field intensity* $\mathbf{E}_i$.

Consider an electric battery with electrodes 1 and 2, shown schematically in Fig. 5-4. Chemical action creates a cumulation of positive and negative charges at electrodes 1 and 2, respectively. These charges give rise to an electrostatic field intensity $\mathbf{E}$ both outside and inside the battery. Inside the battery, $\mathbf{E}$ must be equal in magnitude and opposite in direction to the nonconservative $\mathbf{E}_i$ produced by chemical action, since no current flows in the open-circuited battery and the net force acting on the charge carriers must vanish. The line integral of the impressed field intensity $\mathbf{E}_i$ from the negative to the positive electrode (from electrode 2 to electrode 1 in Fig. 5-4) inside the battery is customarily called the *electromotive force*[†] (emf) of the battery. The SI unit for emf is volt, and an emf is *not* a force in newtons. Denoted by $\mathscr{V}$, the electromotive force is a measure of the strength of the nonconservative source. We have

$$\mathscr{V} = \int_2^1 \mathbf{E}_i \cdot d\boldsymbol{\ell} = -\int_2^1 \mathbf{E} \cdot d\boldsymbol{\ell}. \qquad (5\text{-}33)$$
$$\text{Inside}$$
$$\text{the source}$$

[†] Also called *electromotance*.

The conservative electrostatic field intensity $\mathbf{E}$ satisfies Eq. (5–31):

$$\oint_C \mathbf{E} \cdot d\ell = \underbrace{\int_1^2 \mathbf{E} \cdot d\ell}_{\substack{\text{Outside} \\ \text{the source}}} + \underbrace{\int_2^1 \mathbf{E} \cdot d\ell}_{\substack{\text{Inside} \\ \text{the source}}} = 0. \tag{5–34}$$

Combining Eqs. (5–33) and (5–34), we have

$$\mathscr{V} = \underbrace{\int_1^2 \mathbf{E} \cdot d\ell}_{\substack{\text{Outside} \\ \text{the source}}} \tag{5–35}$$

or

$$\mathscr{V} = V_{12} = V_1 - V_2. \tag{5–36}$$

In Eqs. (5–35) and (5–36) we have expressed the emf of the source as a line integral of the conservative $\mathbf{E}$ and interpreted it as a *voltage rise*. In spite of the nonconservative nature of $\mathbf{E}_i$, the emf can be expressed as a potential difference between the positive and negative terminals. This was what we did in arriving at Eq. (5–24).

When a resistor in the form of Fig. 5–3 is connected between terminals 1 and 2 of the battery, completing the circuit, the *total* electric field intensity (electrostatic $\mathbf{E}$ caused by charge cumulation, as well as impressed $\mathbf{E}_i$ caused by chemical action), must be used in the point form of Ohm's law. We have, instead of Eq. (5–21),

$$\mathbf{J} = \sigma(\mathbf{E} + \mathbf{E}_i), \tag{5–37}$$

where $\mathbf{E}_i$ exists inside the battery only, while $\mathbf{E}$ has a nonzero value both inside and outside the source. From Eq. (5–37) we obtain

$$\mathbf{E} + \mathbf{E}_i = \frac{\mathbf{J}}{\sigma}. \tag{5–38}$$

The scalar line integral of Eq. (5–38) around the closed circuit yields, in view of Eqs. (5–31) and (5–33),

$$\mathscr{V} = \oint_C (\mathbf{E} + \mathbf{E}_i) \cdot d\ell = \oint_C \frac{1}{\sigma} \mathbf{J} \cdot d\ell. \tag{5–39}$$

Equation (5–39) should be compared to Eq. (5–32), which holds when there is no source of nonconservative field. If the resistor has a conductivity σ, length ℓ, and uniform cross section S, $J = I/S$ and the right side of Eq. (5–39) becomes RI. We have[†]

$$\mathscr{V} = RI. \tag{5–40}$$

If there are more than one source of electromotive force and more than one resistor (including the internal resistances of the sources) in the closed path, we generalize

---

[†] We assume the battery to have a negligible internal resistance; otherwise, its effect must be included in Eq. (5–40). An *ideal voltage source* is one whose terminal voltage is equal to its emf and is independent of the current flowing through it. This implies that an ideal voltage source has a zero internal resistance.

Eq. (5–40) to

$$\sum_j \mathscr{V}_j = \sum_k R_k I_k \qquad \text{(V)}.$$

(5–41)

Equation (5–41) is an expression of **Kirchhoff's voltage law**. It states that, **around a closed path in an electric circuit, the algebraic sum of the emf's (voltage rises) is equal to the algebraic sum of the voltage drops across the resistances.** It applies to *any closed path* in a network. The direction of tracing the path can be arbitrarily assigned, and the currents in the different resistances need not be the same. Kirchhoff's voltage law is the basis for loop analysis in circuit theory.

5–4   Equation of Continuity and Kirchhoff's Current Law

The **principle of conservation of charge** is one of the fundamental postulates of physics. Electric charges may not be created or destroyed; all charges either at rest or in motion must be accounted for at all times. Consider an arbitrary volume V bounded by surface S. A net charge Q exists within this region. If a net current I flows across the surface *out* of this region, the charge in the volume must *decrease* at a rate that equals the current. Conversely, if a net current flows across the surface *into* the region, the charge in the volume must *increase* at a rate equal to the current. The current leaving the region is the total outward flux of the current density vector through the surface S. We have

$$I = \oint_S \mathbf{J} \cdot d\mathbf{s} = -\frac{dQ}{dt} = -\frac{d}{dt} \int_V \rho \, dv.$$

(5–42)

Divergence theorem, Eq. (2–115), may be invoked to convert the surface integral of $\mathbf{J}$ to the volume integral of $\nabla \cdot \mathbf{J}$. We obtain, for a stationary volume,

$$\int_V \nabla \cdot \mathbf{J} \, dv = -\int_V \frac{\partial \rho}{\partial t} \, dv.$$

(5–43)

In moving the time derivative of ρ inside the volume integral, it is necessary to use partial differentiation because ρ may be a function of time as well as of space coordinates. Since Eq. (5–43) must hold regardless of the choice of V, the integrands must be equal. Thus we have

$$\nabla \cdot \mathbf{J} = -\frac{\partial \rho}{\partial t} \qquad \text{(A/m}^3\text{)}.$$

(5–44)

This point relationship derived from the principle of conservation of charge is called the **equation of continuity**.

For steady currents, charge density does not vary with time, $\partial \rho / \partial t = 0$. Equation (5–44) becomes

$$\mathbf{V} \cdot \mathbf{J} = 0. \tag{5-45}$$

Thus, steady electric currents are divergenceless or solenoidal. Equation (5–45) is a point relationship and holds also at points where $\rho = 0$ (no flow source). It means that the field lines or streamlines of steady currents close upon themselves, unlike those of electrostatic field intensity that originate and end on charges. Over any enclosed surface, Eq. (5–45) leads to the following integral form:

$$\oint_S \mathbf{J} \cdot d\mathbf{s} = 0, \tag{5-46}$$

which can be written as

$$\boxed{\sum_j I_j = 0 \qquad \text{(A)}.} \tag{5-47}$$

Equation (5–47) is an expression of ***Kirchhoff's current law***. It states that *the algebraic sum of all the currents flowing out of a junction in an electric circuit is zero.*[†] Kirchhoff's current law is the basis for node analysis in circuit theory.

In Section 3–6, we stated that charges introduced in the interior of a conductor will move to the conductor surface and redistribute themselves in such a way as to make $\rho = 0$ and $\mathbf{E} = 0$ inside under equilibrium conditions. We are now in a position to prove this statement and to calculate the time it takes to reach an equilibrium. Combining Ohm's law, Eq. (5–21), with the equation of continuity and assuming a constant σ, we have

$$\sigma \mathbf{V} \cdot \mathbf{E} = -\frac{\partial \rho}{\partial t}. \tag{5-48}$$

In a simple medium, $\mathbf{V} \cdot \mathbf{E} = \rho/\epsilon$, and Eq. (5–48) becomes

$$\frac{\partial \rho}{\partial t} + \frac{\sigma}{\epsilon} \rho = 0. \tag{5-49}$$

The solution of Eq. (5–49) is

$$\boxed{\rho = \rho_0 e^{-(\sigma/\epsilon)t} \qquad \text{(C/m}^3\text{),}} \tag{5-50}$$

where ρ_0 is the initial charge density at $t = 0$. Both ρ and ρ_0 can be functions of the space coordinates, and Eq. (5–50) says that the charge density at a given location will decrease with time exponentially. An initial charge density ρ_0 will decay to $1/e$

[†] This includes the currents of current generators at the junction, if any. An ***ideal current generator*** is one whose current is independent of its terminal voltage. This implies that an ideal current source has an infinite internal resistance.

or 36.8% of its value in a time equal to

$$\tau = \frac{\epsilon}{\sigma} \quad \text{(s)}. \qquad (5\text{–}51)$$

The time constant τ is called the **relaxation time**. For a good conductor such as copper—$\sigma = 5.80 \times 10^7$ (S/m), $\epsilon \cong \epsilon_0 = 8.85 \times 10^{-12}$ (F/m)—τ equals 1.52×10^{-19} (s), a very short time indeed. The transient time is so brief that, for all practical purposes, ρ can be considered zero in the interior of a conductor—see Eq. (3–69) in Section 3–6. The relaxation time for a good insulator is not infinite but can be hours or days.

5–5   Power Dissipation and Joule's Law

In Section 5–1 we indicated that under the influence of an electric field, conduction electrons in a conductor undergo a drift motion macroscopically. Microscopically, these electrons collide with atoms on lattice sites. Energy is thus transmitted from the electric field to the atoms in thermal vibration. The work Δw done by an electric field $\mathbf{E}$ in moving a charge q a distance $\Delta \ell$ is $q\mathbf{E} \cdot (\Delta \ell)$, which corresponds to a power

$$p = \lim_{\Delta t \to 0} \frac{\Delta w}{\Delta t} = q\mathbf{E} \cdot \mathbf{u}, \qquad (5\text{–}52)$$

where $\mathbf{u}$ is the drift velocity. The total power delivered to all the charge carriers in a volume dv is

$$dP = \sum_i p_i = \mathbf{E} \cdot \left(\sum_i N_i q_i \mathbf{u}_i \right) dv,$$

which, by virtue of Eq. (5–18), is

$$dP = \mathbf{E} \cdot \mathbf{J}\, dv$$

or

$$\frac{dP}{dv} = \mathbf{E} \cdot \mathbf{J} \quad \text{(W/m}^3\text{)}. \qquad (5\text{–}53)$$

Thus the point function $\mathbf{E} \cdot \mathbf{J}$ is a **power density** under steady-current conditions. For a given volume V the total electric power converted into heat is

$$\boxed{P = \int_V \mathbf{E} \cdot \mathbf{J}\, dv \quad \text{(W)}.} \qquad (5\text{–}54)$$

This is known as **Joule's law**. (Note that the SI unit for P is watt, not joule, which is the unit for energy or work.) Equation (5–53) is the corresponding point relationship.

In a conductor of a constant cross section, $dv = ds\, d\ell$, with $d\ell$ measured in the direction $\mathbf{J}$. Equation (5–54) can be written as

$$P = \int_L E\, d\ell \int_S J\, ds = VI,$$

where I is the current in the conductor. Since $V = RI$, we have

$$\boxed{P = I^2R \quad \text{(W)}.}$$

(5–55)

Equation (5–55) is, of course, the familiar expression for ohmic power representing the heat dissipated in resistance R per unit time.

5-6 Boundary Conditions for Current Density

When current obliquely crosses an interface between two media with different conductivities, the current density vector changes both in direction and in magnitude. A set of boundary conditions can be derived for **J** in a way similar to that used in Section 3–9 for obtaining the boundary conditions for **D** and **E**. The governing equations for steady current density **J** in the absence of nonconservative energy sources are

Governing Equations for Steady Current Density		
Differential Form	Integral Form	
$\mathbf{V} \cdot \mathbf{J} = 0$	$\oint_S \mathbf{J} \cdot d\mathbf{s} = 0$	(5–56)
$\mathbf{V} \times \left(\dfrac{\mathbf{J}}{\sigma}\right) = 0$	$\oint_C \dfrac{1}{\sigma} \mathbf{J} \cdot d\ell = 0$	(5–57)

The divergence equation is the same as Eq. (5–45), and the curl equation is obtained by combining Ohm's law ($\mathbf{J} = \sigma\mathbf{E}$) with $\mathbf{V} \times \mathbf{E} = 0$. By applying Eqs. (5–56) and (5–57) at the interface between two ohmic media with conductivities σ_1 and σ_2, we obtain the boundary conditions for the normal and tangential components of **J**.

Without actually constructing a pillbox at the interface as was done in Fig. 3–23, we know from Section 3–9 that *the normal component of a divergenceless vector field is continuous*. Hence from $\mathbf{V} \cdot \mathbf{J} = 0$ we have

$$\boxed{J_{1n} = J_{2n} \quad \text{(A/m}^2\text{)}.}$$

(5–58)

Similarly, *the tangential component of a curl-free vector field is continuous across an interface*. We conclude from $\mathbf{V} \times (\mathbf{J}/\sigma) = 0$ that

$$\boxed{\dfrac{J_{1t}}{J_{2t}} = \dfrac{\sigma_1}{\sigma_2}.}$$

(5–59)

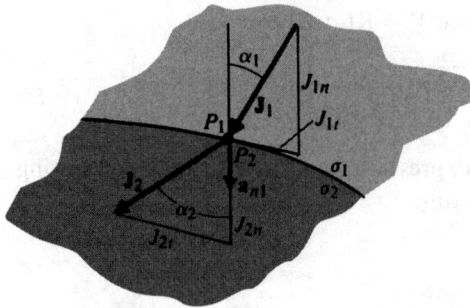

FIGURE 5-5
Boundary conditions at interface between two
conducting media (Example 5-3).

Equation (5–59) states that *the ratio of the tangential components of* **J** *at two sides of an interface is equal to the ratio of the conductivities*. Comparing the boundary conditions Eqs. (5–58) and (5–59) for steady current density in ohmic media with the boundary conditions Eqs. (3–123) and (3–119), respectively, for electrostatic flux density at an interface of dielectric media where there are no free charges, we note an exact analogy of **J** and σ with **D** and ϵ.

EXAMPLE 5-3   Two conducting media with conductivities σ_1 and σ_2 are separated by an interface, as shown in Fig. 5–5. The steady current density in medium 1 at point P_1 has a magnitude J_1 and makes an angle α_1 with the normal. Determine the magnitude and direction of the current density at point P_2 in medium 2.

Solution   Using Eqs. (5–58) and (5–59), we have

$$J_1 \cos \alpha_1 = J_2 \cos \alpha_2 \tag{5-60}$$

and

$$\sigma_2 J_1 \sin \alpha_1 = \sigma_1 J_2 \sin \alpha_2. \tag{5-61}$$

Division of Eq. (5–61) by Eq. (5–60) yields

$$\boxed{\frac{\tan \alpha_2}{\tan \alpha_1} = \frac{\sigma_2}{\sigma_1}.} \tag{5-62}$$

If medium 1 is a much better conductor than medium 2 ($\sigma_1 \gg \sigma_2$ or $\sigma_2/\sigma_1 \to 0$), α_2 approaches zero, and $\mathbf{J}_2$ emerges almost perpendicularly to the interface (normal to the surface of the good conductor). The magnitude of $\mathbf{J}_2$ is

$$J_2 = \sqrt{J_{2t}^2 + J_{2n}^2} = \sqrt{(J_2 \sin \alpha_2)^2 + (J_2 \cos \alpha_2)^2}$$
$$= \left[\left(\frac{\sigma_2}{\sigma_1} J_1 \sin \alpha_1 \right)^2 + (J_1 \cos \alpha_1)^2 \right]^{1/2}$$

or

$$J_2 = J_1 \left[\left(\frac{\sigma_2}{\sigma_1} \sin \alpha_1 \right)^2 + \cos^2 \alpha_1 \right]^{1/2}.$$ (5-63)

By examining Fig. 5-5, can you tell whether medium 1 or medium 2 is the better conductor? ■

For a homogeneous conducting medium the differential form of Eq. (5-57) simplifies to

$$\mathbf{V} \times \mathbf{J} = 0.$$ (5-64)

From Section 2-11 we know that a curl-free vector field can be expressed as the gradient of a scalar potential field. Let us write

$$\mathbf{J} = -\mathbf{V}\psi.$$ (5-65)

Substitution of Eq. (5-65) into $\mathbf{V} \cdot \mathbf{J} = 0$ yields a Laplace's equation in ψ; that is,

$$\mathbf{V}^2\psi = 0.$$ (5-66)

A problem in steady-current flow can therefore be solved by determining ψ (A/m) from Eq. (5-66), subject to appropriate boundary conditions and then by finding $\mathbf{J}$ from its negative gradient in exactly the same way as a problem in electrostatics is solved. As a matter of fact, ψ and electrostatic potential are simply related: $\psi = \sigma V$. As indicated in Section 5-1, this similarity between electrostatic and steady-current fields is the basis for using an electrolytic tank to map the potential distribution of difficult-to-solve electrostatic boundary-value problems.[†]

When a steady current flows across the boundary between two different lossy dielectrics (dielectrics with permittivities ϵ_1 and ϵ_2 and finite conductivities σ_1 and σ_2), the tangential component of the electric field is continuous across the interface as usual; that is, $E_{2t} = E_{1t}$, which is equivalent to Eq. (5-59). The normal component of the electric field, however, must simultaneously satisfy both Eq. (5-58) and Eq. (3-121b). We require

$$J_{1n} = J_{2n} \rightarrow \sigma_1 E_{1n} = \sigma_2 E_{2n},$$ (5-67)

$$D_{1n} - D_{2n} = \rho_s \rightarrow \epsilon_1 E_{1n} - \epsilon_2 E_{2n} = \rho_s,$$ (5-68)

where the reference unit normal is *outward from medium 2*. Hence, unless $\sigma_2/\sigma_1 = \epsilon_2/\epsilon_1$, a surface charge must exist at the interface. From Eqs. (5-67) and (5-68) we find

$$\rho_s = \left(\epsilon_1 \frac{\sigma_2}{\sigma_1} - \epsilon_2 \right) E_{2n} = \left(\epsilon_1 - \epsilon_2 \frac{\sigma_1}{\sigma_2} \right) E_{1n}.$$ (5-69)

[†] See, for instance, E. Weber, *Electromagnetic Fields*, John Wiley and Sons, 1950, Vol. I: *Mapping of Fields*, pp. 187–193.

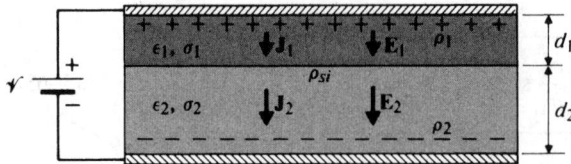

FIGURE 5-6
Parallel-plate capacitor with two lossy dielectrics (Example 5-4).

Again, if medium 2 is a much better conductor than medium 1 ($\sigma_2 \gg \sigma_1$ or $\sigma_1/\sigma_2 \to 0$), Eq. (5-69) becomes approximately

$$\rho_s = \epsilon_1 E_{1n} = D_{1n}, \qquad (5\text{-}70)$$

which is the same as Eq. (3-122).

EXAMPLE 5-4 An emf $\mathscr{V}$ is applied across a parallel-plate capacitor of area S. The space between the conducting plates is filled with two different lossy dielectrics of thicknesses d_1 and d_2, permittivities ϵ_1 and ϵ_2, and conductivities σ_1 and σ_2, respectively. Determine (a) the current density between the plates, (b) the electric field intensities in both dielectrics, and (c) the surface charge densities on the plates and at the interface.

Solution   Refer to Fig. 5-6.

a) The continuity of the normal component of **J** assures that the current densities and therefore the currents in both media are the same. By Kirchhoff's voltage law we have

$$\mathscr{V} = (R_1 + R_2)I = \left(\frac{d_1}{\sigma_1 S} + \frac{d_2}{\sigma_2 S}\right)I.$$

Hence,

$$J = \frac{I}{S} = \frac{\mathscr{V}}{(d_1/\sigma_1) + (d_2/\sigma_2)} = \frac{\sigma_1\sigma_2\mathscr{V}}{\sigma_2 d_1 + \sigma_1 d_2} \qquad (\text{A/m}^2). \qquad (5\text{-}71)$$

b) To determine the electric field intensities E_1 and E_2 in both media, two equations are needed. Neglecting fringing effect at the edges of the plates, we have

$$\mathscr{V} = E_1 d_1 + E_2 d_2 \qquad (5\text{-}72)$$

and

$$\sigma_1 E_1 = \sigma_2 E_2. \qquad (5\text{-}73)$$

Equation (5-73) comes from $J_1 = J_2$. Solving Eqs. (5-72) and (5-73), we obtain

$$E_1 = \frac{\sigma_2\mathscr{V}}{\sigma_2 d_1 + \sigma_1 d_2} \qquad (\text{V/m}) \qquad (5\text{-}74)$$

and

$$E_2 = \frac{\sigma_1\mathscr{V}}{\sigma_2 d_1 + \sigma_1 d_2} \qquad (\text{V/m}). \qquad (5\text{-}75)$$

c) The surface charge densities on the upper and lower plates can be determined by using Eq. (5–70):

$$\rho_{s1} = \epsilon_1 E_1 = \frac{\epsilon_1 \sigma_2 \mathscr{V}}{\sigma_2 d_1 + \sigma_1 d_2} \qquad (\text{C/m}^2) \tag{5–76}$$

$$\rho_{s2} = -\epsilon_2 E_2 = -\frac{\epsilon_2 \sigma_1 \mathscr{V}}{\sigma_2 d_1 + \sigma_1 d_2} \qquad (\text{C/m}^2). \tag{5–77}$$

The negative sign in Eq. (5–77) comes about because $\mathbf{E}_2$ and the *outward* normal at the lower plate are in opposite directions.

Equation (5–69) can be used to find the surface charge density at the interface of the dielectrics. We have

$$\begin{aligned}
\rho_{si} &= \left(\epsilon_2 \frac{\sigma_1}{\sigma_2} - \epsilon_1 \right) \frac{\sigma_2 \mathscr{V}}{\sigma_2 d_1 + \sigma_1 d_2} \\
&= \frac{(\epsilon_2 \sigma_1 - \epsilon_1 \sigma_2) \mathscr{V}}{\sigma_2 d_1 + \sigma_1 d_2} \qquad (\text{C/m}^2).
\end{aligned} \tag{5–78}$$

From these results we see that $\rho_{s2} \neq -\rho_{s1}$ but that $\rho_{s1} + \rho_{s2} + \rho_{si} = 0$. ∎

In Example 5–4 we encounter a situation in which both static charges and a steady current exist. As we shall see in Chapter 6, a steady current gives rise to a steady magnetic field. We have, then, both a static electric field and a steady magnetic field. They constitute an **electromagnetostatic field**. The electric and magnetic fields of an electromagnetostatic field are coupled through the constitutive relation $\mathbf{J} = \sigma \mathbf{E}$ of the conducting medium.

5–7 Resistance Calculations

In Section 3–10 we discussed the procedure for finding the capacitance between two conductors separated by a dielectric medium. These conductors may be of arbitrary shapes, as was shown in Fig. 3–27, which is reproduced here as Fig. 5–7. In terms of electric field quantities the basic formula for capacitance can be written as

$$C = \frac{Q}{V} = \frac{\oint_S \mathbf{D} \cdot d\mathbf{s}}{-\int_L \mathbf{E} \cdot d\boldsymbol{\ell}} = \frac{\oint_S \epsilon \mathbf{E} \cdot d\mathbf{s}}{-\int_L \mathbf{E} \cdot d\boldsymbol{\ell}}, \tag{5–79}$$

where the surface integral in the numerator is carried out over a surface enclosing the positive conductor and the line integral in the denominator is from the negative (lower-potential) conductor to the positive (higher-potential) conductor (see Eq. 5–35).

When the dielectric medium is lossy (having a small but nonzero conductivity), a current will flow from the positive to the negative conductor, and a current-density field will be established in the medium. Ohm's law, $\mathbf{J} = \sigma \mathbf{E}$, ensures that the streamlines for $\mathbf{J}$ and $\mathbf{E}$ will be the same in an isotropic medium. The resistance between

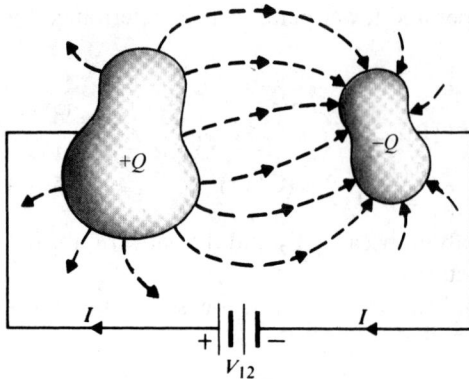

FIGURE 5–7
Two conductors in a lossy dielectric medium.

the conductors is

$$R = \frac{V}{I} = \frac{-\int_L \mathbf{E} \cdot d\boldsymbol{\ell}}{\oint_S \mathbf{J} \cdot d\mathbf{s}} = \frac{-\int_L \mathbf{E} \cdot d\boldsymbol{\ell}}{\oint_S \sigma \mathbf{E} \cdot d\mathbf{s}}, \tag{5–80}$$

where the line and surface integrals are taken over the same L and S as those in Eq. (5–79). Comparison of Eqs. (5–79) and (5–80) shows the following interesting relationship:

$$RC = \frac{C}{G} = \frac{\epsilon}{\sigma}. \tag{5–81}$$

Equation (5–81) holds if ϵ and σ of the medium have the same space dependence or if the medium is homogeneous (independent of space coordinates). In these cases, if the capacitance between two conductors is known, the resistance (or conductance) can be obtained directly from the ϵ/σ ratio without recomputation.

EXAMPLE 5–5   Find the leakage resistance per unit length (a) between the inner and outer conductors of a coaxial cable that has an inner conductor of radius a, an outer conductor of inner radius b, and a medium with conductivity σ, and (b) of a parallel-wire transmission line consisting of wires of radius a separated by a distance D in a medium with conductivity σ.

Solution

a) The capacitance per unit length of a coaxial cable has been obtained as Eq. (3–139) in Example 3–18:

$$C_1 = \frac{2\pi\epsilon}{\ln(b/a)} \qquad \text{(F/m)}.$$

Hence the leakage resistance per unit length is, from Eq. (5–81),

$$R_1 = \frac{\epsilon}{\sigma}\left(\frac{1}{C_1}\right) = \frac{1}{2\pi\sigma}\ln\left(\frac{b}{a}\right) \quad (\Omega\cdot m). \qquad (5\text{–}82)$$

The conductance per unit length is $G_1 = 1/R_1$.

b) For the parallel-wire transmission line, Eq. (4–47) in Example 4–4 gives the capacitance per unit length:

$$C_1' = \frac{\pi\epsilon}{\cosh^{-1}\left(\dfrac{D}{2a}\right)} \quad (\text{F/m}).$$

Therefore the leakage resistance per unit length is, without further ado,

$$\begin{aligned}
R_1' &= \frac{\epsilon}{\sigma}\left(\frac{1}{C_1'}\right) = \frac{1}{\pi\sigma}\cosh^{-1}\left(\frac{D}{2a}\right) \\
&= \frac{1}{\pi\sigma}\ln\left[\frac{D}{2a} + \sqrt{\left(\frac{D}{2a}\right)^2 - 1}\right] \quad (\Omega\cdot m).
\end{aligned} \qquad (5\text{–}83)$$

The conductance per unit length is $G_1' = 1/R_1'$. ▬

It must be emphasized here that the resistance *between* the conductors for a length ℓ of the coaxial cable is R_1/ℓ, not ℓR_1; similarly, the leakage resistance of a length ℓ of the parallel-wire transmission line is R_1'/ℓ, not $\ell R_1'$. Do you know why?

In certain situations, electrostatic and steady-current problems are not exactly analogous, even when the geometrical configurations are the same. This is because current flow can be confined strictly within a conductor (which has a *very large* σ in comparison to that of the surrounding medium), whereas electric flux usually cannot be contained within a dielectric slab of finite dimensions. The range of the dielectric constant of available materials is very limited (see Appendix B–3), and the flux-fringing around conductor edges makes the computation of capacitance less accurate.

The procedure for computing the resistance of a piece of conducting material between specified equipotential surfaces (or terminals) is as follows:

1. Choose an appropriate coordinate system for the given geometry.

2. Assume a potential difference V_0 between conductor terminals.

3. Find electric field intensity $\mathbf{E}$ within the conductor. (If the material is homogeneous, having a *constant* conductivity, the general method is to solve Laplace's equation $\nabla^2 V = 0$ for V in the chosen coordinate system, and then obtain $\mathbf{E} = -\nabla V$.)

4. Find the total current

$$I = \int_S \mathbf{J}\cdot d\mathbf{s} = \int_S \sigma\mathbf{E}\cdot d\mathbf{s},$$

where S is the cross-sectional area over which I flows.

5. Find resistance R by taking the ratio V_0/I.

It is important to note that if the conducting material is inhomogeneous and if the conductivity is a function of space coordinates, Laplace's equation for V does not hold. Can you explain why and indicate how $\mathbf{E}$ can be determined under these circumstances?

When the given geometry is such that $\mathbf{J}$ can be determined easily from a total current I, we may start the solution by assuming an I. From I, $\mathbf{J}$ and $\mathbf{E} = \mathbf{J}/\sigma$ are found. Then the potential difference V_0 is determined from the relation

$$V_0 = -\int \mathbf{E} \cdot d\boldsymbol{\ell},$$

where the integration is from the low-potential terminal to the high-potential terminal. The resistance $R = V_0/I$ is independent of the assumed I, which will be canceled in the process.

EXAMPLE 5–6   A conducting material of uniform thickness h and conductivity σ has the shape of a quarter of a flat circular washer, with inner radius a and outer radius b, as shown in Fig. 5–8. Determine the resistance between the end faces.

Solution   Obviously, the appropriate coordinate system to use for this problem is the cylindrical coordinate system. Following the foregoing procedure, we first assume a potential difference V_0 between the end faces, say $V = 0$ on the end face at $y = 0$ ($\phi = 0$) and $V = V_0$ on the end face at $x = 0$ ($\phi = \pi/2$). We are to solve Laplace's equation in V subject to the following boundary conditions:

$$V = 0 \qquad \text{at} \qquad \phi = 0, \tag{5–84a}$$
$$V = V_0 \qquad \text{at} \qquad \phi = \pi/2. \tag{5–84b}$$

Since potential V is a function of ϕ only, Laplace's equation in cylindrical coordinates simplifies to

$$\frac{d^2 V}{d\phi^2} = 0. \tag{5–85}$$

The general solution of Eq. (5–85) is

$$V = c_1 \phi + c_2,$$

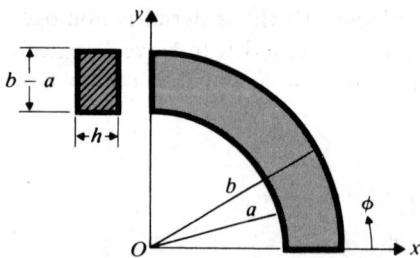

FIGURE 5–8
A quarter of a flat conducting circular washer (Example 5–6).

which, upon using the boundary conditions in Eqs. (5–84a) and (5–84b), becomes

$$V = \frac{2V_0}{\pi} \phi. \tag{5–86}$$

The current density is

$$\mathbf{J} = \sigma\mathbf{E} = -\sigma\nabla V$$

$$= -\mathbf{a}_\phi \sigma \frac{\partial V}{r\partial\phi} = -\mathbf{a}_\phi \frac{2\sigma V_0}{\pi r}. \tag{5–87}$$

The total current I can be found by integrating $\mathbf{J}$ over the $\phi = \pi/2$ surface at which $d\mathbf{s} = -\mathbf{a}_\phi h\, dr$. We have

$$I = \int_S \mathbf{J} \cdot d\mathbf{s} = \frac{2\sigma V_0}{\pi} h \int_a^b \frac{dr}{r}$$

$$= \frac{2\sigma h V_0}{\pi} \ln\frac{b}{a}. \tag{5–88}$$

Therefore,

$$R = \frac{V_0}{I} = \frac{\pi}{2\sigma h \ln(b/a)}. \tag{5–89}$$

Note that, for this problem, it is not convenient to begin by assuming a total current I because it is not obvious how $\mathbf{J}$ varies with r for a given I. Without $\mathbf{J}$, $\mathbf{E}$ and V_0 cannot be determined. ▬

Review Questions

R.5–1 Explain the difference between conduction and convection currents.

R.5–2 Explain the operation of an electrolytic tank. In what ways do electrolytic currents differ from conduction and convection currents?

R.5–3 Define *mobility* of the electron in a conductor. What is its SI unit?

R.5–4 What is the *Child-Langmuir law*?

R.5–5 What is the point form for *Ohm's law*?

R.5–6 Define *conductivity*. What is its SI unit?

R.5–7 Why does the resistance formula in Eq. (5–27) require that the material be homogeneous and straight and that it have a uniform cross section?

R.5–8 Prove Eqs. (5–29) and (5–30b).

R.5–9 Define *electromotive force* in words.

R.5–10 What is the difference between impressed and electrostatic field intensities?

R.5–11 State *Kirchhoff's voltage law* in words.

R.5–12 What are the characteristics of an ideal voltage source?

R.5–13 Can the currents in different branches (resistors) of a closed loop in an electric network flow in opposite directions? Explain.

R.5–14 What is the physical significance of the *equation of continuity*?

R.5–15 State *Kirchhoff's current law* in words.

R.5–16 What are the characteristics of an ideal current source?

R.5–17 Define *relaxation time*. What is the order of magnitude of the relaxation time in copper?

R.5–18 In what ways should Eq. (5–48) be modified when σ is a function of space coordinates?

R.5–19 State Joule's law. Express the power dissipated in a volume
    a) in terms of $\mathbf{E}$ and σ,
    b) in terms of $\mathbf{J}$ and σ.

R.5–20 Does the relation $\mathbf{V} \times \mathbf{J} = 0$ hold in a medium whose conductivity is not constant? Explain.

R.5–21 What are the boundary conditions of the normal and tangential components of steady current at the interface of two media with different conductivities?

R.5–22 What quantities in electrostatics are analogous to the steady current density vector and conductivity in an ohmic medium?

R.5–23 What is the basis of using an electrolytic tank to map the potential distribution of electrostatic boundary-value problems?

R.5–24 What is the relation between the resistance and the capacitance formed by two conductors immersed in a lossy dielectric medium that has permittivity ϵ and conductivity σ?

R.5–25 Under what situations will the relation between R and C in R.5–24 be only approximately correct? Give a specific example.

Problems

P.5–1 Assuming S to be the area of the electrodes in the space-charge-limited vacuum diode in Fig. 5–2, find
    a) $V(y)$ and $E(y)$ within the interelectrode region,
    b) the total amount of charge in the interelectrode region,
    c) the total surface charge on the cathode and on the anode,
    d) the transit time of an electron from the cathode to the anode with $V_0 = 200$ (V) and
        $d = 1$ (cm).

P.5–2 Starting with Ohm's law as expressed in Eq. (5–26) applied to a resistor of length ℓ, conductivity σ, and uniform cross-section S, verify the point form of Ohm's law represented by Eq. (5–21).

P.5–3 A long, round wire of radius a and conductivity σ is coated with a material of conductivity 0.1σ.
    a) What must be the thickness of the coating so that the resistance per unit length of the uncoated wire is reduced by 50%?
    b) Assuming a total current I in the coated wire, find $\mathbf{J}$ and $\mathbf{E}$ in both the core and the coating material.

P.5–4 Find the current and the heat dissipated in each of the five resistors in the network shown in Fig. 5–9 if

$$R_1 = \tfrac{1}{3}\,(\Omega), \quad R_2 = 20\,(\Omega), \quad R_3 = 30\,(\Omega), \quad R_4 = 8\,(\Omega), \quad R_5 = 10\,(\Omega),$$

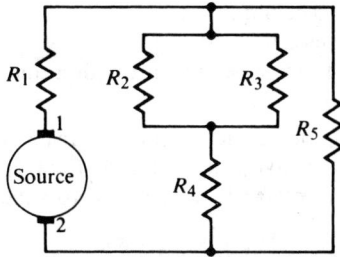

FIGURE 5–9
A network problem (Problem P.5–4).

and if the source is an ideal d-c voltage generator of 0.7 (V) with its positive polarity at terminal 1. What is the total resistance seen by the source at terminal pair 1–2?

P.5–5 Solve Problem P.5–4, assuming that the source is an ideal current generator that supplies a direct current of 0.7 (A) out of terminal 1.

P.5–6 Lightning strikes a lossy dielectric sphere—$\epsilon = 1.2\,\epsilon_0$, $\sigma = 10$ (S/m)—of radius 0.1 (m) at time $t = 0$, depositing uniformly in the sphere a total charge 1 (mC). Determine, for all t,
   a) the electric field intensity both inside and outside the sphere,
   b) the current density in the sphere.

P.5–7 Refer to Problem P.5–6.
   a) Calculate the time it takes for the charge density in the sphere to diminish to 1% of its initial value.
   b) Calculate the change in the electrostatic energy stored in the sphere as the charge density diminishes from the initial value to 1% of its value. What happens to this energy?
   c) Determine the electrostatic energy stored in the space outside the sphere. Does this energy change with time?

P.5–8 A d-c voltage of 6 (V) applied to the ends of 1 (km) of a conducting wire of 0.5 (mm) radius results in a current of 1/6 (A). Find
   a) the conductivity of the wire,
   b) the electric field intensity in the wire,
   c) the power dissipated in the wire,
   d) the electron drift velocity, assuming electron mobility in the wire to be 1.4×10^{-3} (m^2/V·s).

P.5–9 Two lossy dielectric media with permittivities and conductivities (ϵ_1, σ_1) and (ϵ_2, σ_2) are in contact. An electric field with a magnitude E_1 is incident from medium 1 upon the interface at an angle α_1 measured from the common normal, as in Fig. 5–10.

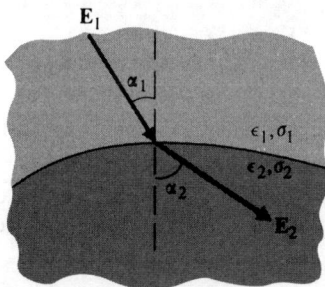

FIGURE 5–10
Boundary between two lossy dielectric media (Problem P.5–9).

a) Find the magnitude and direction of $\mathbf{E}_2$ in medium 2.

b) Find the surface charge density at the interface.

c) Compare the results in parts (a) and (b) with the case in which both media are perfect dielectrics.

P.5–10 The space between two parallel conducting plates each having an area S is filled with an inhomogeneous ohmic medium whose conductivity varies linearly from σ_1 at one plate ($y = 0$) to σ_2 at the other plate ($y = d$). A d-c voltage V_0 is applied across the plates as in Fig. 5–11. Determine

a) the total resistance between the plates,

b) the surface charge densities on the plates,

c) the volume charge density and the total amount of charge between the plates.

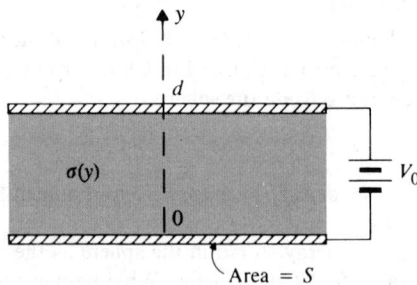

FIGURE 5–11
Inhomogeneous ohmic medium with conductivity $\sigma(y)$ (Problem P.5–10).

P.5–11 Refer to Example 5–4.

a) Draw the equivalent circuit of the two-layer, parallel-plate capacitor with lossy dielectrics, and identify the magnitude of each component.

b) Determine the power dissipated in the capacitor.

P.5–12 Refer again to Example 5–4. Assuming that a voltage V_0 is applied across the parallel-plate capacitor with the two layers of different lossy dielectrics at $t = 0$,

a) express the surface charge density ρ_{si} at the dielectric interface as a function of t,

b) express the electric field intensities $\mathbf{E}_1$ and $\mathbf{E}_2$ as functions of t.

P.5–13 A d-c voltage V_0 is applied across a cylindrical capacitor of length L. The radii of the inner and outer conductors are a and b, respectively. The space between the conductors is filled with two different lossy dielectrics having, respectively, permittivity ϵ_1 and conductivity σ_1 in the region $a < r < c$, and permittivity ϵ_2 and conductivity σ_2 in the region $c < r < b$. Determine

a) the current density in each region,

b) the surface charge densities on the inner and outer conductors and at the interface between the two dielectrics.

P.5–14 Refer to the flat conducting quarter-circular washer in Example 5–6 and Fig. 5–8. Find the resistance between the curved sides.

P.5–15 Find the resistance between two concentric spherical surfaces of radii R_1 and R_2 ($R_1 < R_2$) if the space between the surfaces is filled with a homogeneous and isotropic material having a conductivity σ.

P.5–16 Determine the resistance between two concentric spherical surfaces of radii R_1 and R_2 $(R_1 < R_2)$, assuming that a material of conductivity $\sigma = \sigma_0(1 + k/R)$ fills the space between them. (*Note:* Laplace's equation for V does not apply here.)

P.5–17 A homogeneous material of uniform conductivity σ is shaped like a truncated conical block and defined in spherical coordinates by

$$R_1 \leq R \leq R_2 \quad \text{and} \quad 0 \leq \theta \leq \theta_0.$$

Determine the resistance between the $R = R_1$ and $R = R_2$ surfaces.

P.5–18 Redo Problem P.5–17, assuming that the truncated conical block is composed of an inhomogeneous material with a nonuniform conductivity $\sigma(R) = \sigma_0 R_1/R$, where $R_1 \leq R \leq R_2$.

P.5–19 Two conducting spheres of radii b_1 and b_2 that have a very high conductivity are immersed in a poorly conducting medium (for example, they are buried very deep in the ground) of conductivity σ and permittivity ϵ. The distance, d, between the spheres is very large in comparison with the radii. Determine the resistance between the conducting spheres. (*Hint:* Find the capacitance between the spheres by following the procedure in Section 3–10 and using Eq. (5–81).)

P.5–20 Justify the statement that the steady-current problem associated with a conductor buried in a poorly conducting medium near a plane boundary with air, as shown in Fig. 5–12(a), can be replaced by that of the conductor and its image, both immersed in the poorly conducting medium as shown in Fig. 5–12(b).

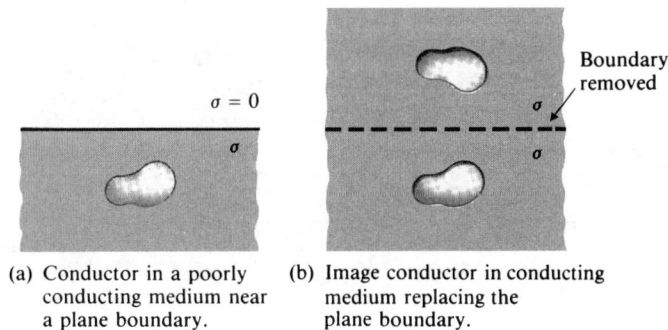

(a) Conductor in a poorly conducting medium near a plane boundary.

(b) Image conductor in conducting medium replacing the plane boundary.

FIGURE 5–12
Steady-current problem with a plane boundary (Problem P.5–20).

P.5–21 A ground connection is made by burying a hemispherical conductor of radius 25 (mm) in the earth with its base up, as shown in Fig. 5–13. Assuming the earth conductivity to be 10^{-6} S/m, find the resistance of the conductor to far-away points in the ground. (*Hint:* Use the image method in P.5–20.)

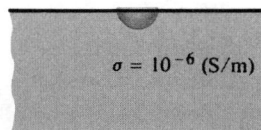

$\sigma = 10^{-6}$ (S/m)

FIGURE 5–13
Hemispherical conductor in ground (Problem P.5–21).

P.5–22 Assume a rectangular conducting sheet of conductivity σ, width a, and height b. A potential difference V_0 is applied to the side edges, as shown in Fig. 5–14. Find

    a) the potential distribution,

    b) the current density everywhere within the sheet. (*Hint:* Solve Laplace's equation in Cartesian coordinates subject to appropriate boundary conditions.)

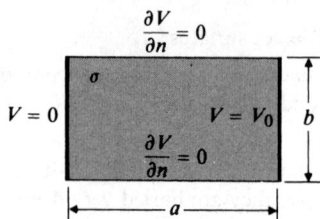

FIGURE 5–14
A conducting sheet (Problem P.5–22).

P.5–23 A uniform current density $\mathbf{J} = \mathbf{a}_x J_0$ flows in a very large rectangular block of homogeneous material of a uniform thickness having a conductivity σ. A hole of radius b is drilled in the material. Find the new current density $\mathbf{J}'$ in the conducting material. (*Hint:* Solve Laplace's equation in cylindrical coordinates and note that V approaches $-(J_0 r/\sigma)\cos\phi$ as $r \to \infty$, where ϕ is the angle measured from the x-axis.)

6

Static Magnetic Fields

6–1 Introduction

In Chapter 3 we dealt with static electric fields caused by electric charges at rest. We saw that electric field intensity **E** is the only fundamental vector field quantity required for the study of electrostatics in free space. In a material medium it is convenient to define a second vector field quantity, the electric flux density (or electric displacement) **D**, to account for the effect of polarization. The following two equations form the basis of the electrostatic model:

$$\nabla \cdot \mathbf{D} = \rho, \tag{6-1}$$

$$\nabla \times \mathbf{E} = 0. \tag{6-2}$$

The electrical property of the medium determines the relation between **D** and **E**. If the medium is linear and isotropic, we have the simple *constitutive relation* $\mathbf{D} = \epsilon \mathbf{E}$, where the permittivity ϵ is a scalar.

When a small test charge q is placed in an electric field **E**, it experiences an *electric force* $\mathbf{F}_e$, which is a function of the position of q. We have

$$\boxed{\mathbf{F}_e = q\mathbf{E} \qquad (\text{N}).} \tag{6-3}$$

When the test charge is in motion in a magnetic field (to be defined presently), experiments show that it experiences another force, $\mathbf{F}_m$, which has the following characteristics: (1) The magnitude of $\mathbf{F}_m$ is proportional to q; (2) the direction of $\mathbf{F}_m$ at any point is at right angles to the velocity vector of the test charge as well as to a fixed direction at that point; and (3) the magnitude of $\mathbf{F}_m$ is also proportional to the component of the velocity at right angles to this fixed direction. The force $\mathbf{F}_m$ is a *magnetic force*; it cannot be expressed in terms of **E** or **D**. The characteristics of $\mathbf{F}_m$ can be described by defining a new vector field quantity, the *magnetic flux density* **B**, that specifies both the fixed direction and the constant of proportionality. In SI units the magnetic force can be expressed as

$$\mathbf{F}_m = q\mathbf{u} \times \mathbf{B} \qquad \text{(N)},$$

(6–4)

where $\mathbf{u}$ (m/s) is the velocity vector, and $\mathbf{B}$ is measured in webers per square meter (Wb/m^2) or teslas (T).[†] The total *electromagnetic force* on a charge q is, then, $\mathbf{F} = \mathbf{F}_e + \mathbf{F}_m$; that is,

$$\mathbf{F} = q(\mathbf{E} + \mathbf{u} \times \mathbf{B}) \qquad \text{(N)},$$

(6–5)

which is called *Lorentz's force equation*. Its validity has been unquestionably established by experiments. We may consider $\mathbf{F}_e/q$ for a small q as the definition for electric field intensity $\mathbf{E}$ (as we did in Eq. 3–2), and $\mathbf{F}_m/q = \mathbf{u} \times \mathbf{B}$ as the defining relation for magnetic flux density $\mathbf{B}$. Alternatively, we may consider Lorentz's force equation as a fundamental postulate of our electromagnetic model; it cannot be derived from other postulates.

We begin the study of static magnetic fields in free space by two postulates specifying the divergence and the curl of $\mathbf{B}$. From the solenoidal character of $\mathbf{B}$ a vector magnetic potential is defined, which is shown to obey a vector Poisson's equation. Next we derive the Biot-Savart law, which can be used to determine the magnetic field of a current-carrying circuit. The postulated curl relation leads directly to Ampère's circuital law, which is particularly useful when symmetry exists.

The macroscopic effect of magnetic materials in a magnetic field can be studied by defining a magnetization vector. Here we introduce a fourth vector field quantity, the magnetic field intensity $\mathbf{H}$. From the relation between $\mathbf{B}$ and $\mathbf{H}$ we define the permeability of the material, following which we discuss magnetic circuits and the microscopic behavior of magnetic materials. We then examine the boundary conditions of $\mathbf{B}$ and $\mathbf{H}$ at the interface of two different magnetic media; self- and mutual inductances; and magnetic energy, forces, and torques.

6–2 Fundamental Postulates of Magnetostatics in Free Space

To study magnetostatics (steady magnetic fields) in free space, we need only consider the magnetic flux density vector, $\mathbf{B}$. The two fundamental postulates of magnetostatics that specify the divergence and the curl of $\mathbf{B}$ *in free space* are

$$\nabla \cdot \mathbf{B} = 0,$$

(6–6)

$$\nabla \times \mathbf{B} = \mu_0 \mathbf{J}.$$

(6–7)

[†] One weber per square meter or one tesla equals 10^4 gauss in CGS units. The earth magnetic field is about $\frac{1}{2}$ gauss or 0.5×10^{-4} T. (A weber is the same as a volt-second.)

In Eq. (6–7), μ_0 is the permeability of free space:

$$\mu_0 = 4\pi \times 10^{-7} \quad (H/m)$$

(see Eq. 1–9), and **J** is the current density. Since the divergence of the curl of any vector field is zero (see Eq. 2–149), we obtain from Eq. (6–7)

$$\mathbf{V} \cdot \mathbf{J} = 0, \tag{6–8}$$

which is consistent with Eq. (5–44) for steady currents.

Comparison of Eq. (6–6) with the analogous equation for electrostatics in free space, $\mathbf{V} \cdot \mathbf{E} = \rho/\epsilon_0$ (Eq. 3–4), leads us to conclude that there is no magnetic analogue for electric charge density ρ. Taking the volume integral of Eq. (6–6) and applying the divergence theorem, we have

$$\oint_S \mathbf{B} \cdot d\mathbf{s} = 0, \tag{6–9}$$

where the surface integral is carried out over the bounding surface of an arbitrary volume. Comparing Eq. (6–9) with Eq. (3–7), we again deny the existence of isolated magnetic charges. *There are no magnetic flow sources, and the magnetic flux lines always close upon themselves.* Equation (6–9) is also referred to as an expression for *the law of conservation of magnetic flux* because it states that the total outward magnetic flux through any closed surface is zero.

The traditional designation of north and south poles in a permanent bar magnet does not imply that an isolated positive magnetic charge exists at the north pole and a corresponding amount of isolated negative magnetic charge exists at the south pole. Consider the bar magnet with north and south poles in Fig. 6–1(a). If this magnet is cut into two segments, new south and north poles appear, and we have two shorter magnets as in Fig. 6–1(b). If each of the two shorter magnets is cut again into two segments, we have four magnets, each with a north pole and a south pole as in Fig. 6–1(c). This process could be continued until the magnets are of atomic dimensions; but each infinitesimally small magnet would still have a north pole and a south pole. Obviously, then, magnetic poles cannot be isolated. The magnetic flux lines follow closed paths from one end of a magnet to the other end outside the magnet and then

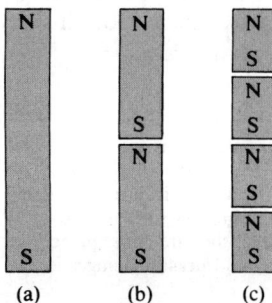

FIGURE 6–1
Successive division of a bar magnet.

continue inside the magnet back to the first end. The designation of north and south poles is in accordance with the fact that the respective ends of a bar magnet freely suspended in the earth's magnetic field will seek the north and south directions.[†]

The integral form of the curl relation in Eq. (6–7) can be obtained by integrating both sides over an open surface and applying Stokes's theorem. We have

$$\int_S (\nabla \times \mathbf{B}) \cdot d\mathbf{s} = \mu_0 \int_S \mathbf{J} \cdot d\mathbf{s}$$

or

$$\oint_C \mathbf{B} \cdot d\ell = \mu_0 I, \qquad (6\text{–}10)$$

where the path C for the line integral is the contour bounding the surface S, and I is the total current through S. The sense of tracing C and the direction of current flow follow the right-hand rule. Equation (6–10) is a form of *Ampère's circuital law*, which states that *the circulation of the magnetic flux density in free space around any closed path is equal to μ_0 times the total current flowing through the surface bounded by the path*. Ampère's circuital law is very useful in determining the magnetic flux density **B** caused by a current I when there is a closed path C around the current such that the magnitude of **B** is constant over the path.

The following is a summary of the two fundamental postulates of magnetostatics in free space:

Postulates of Magnetostatics in Free Space	
Differential Form	Integral Form
$\nabla \cdot \mathbf{B} = 0$	$\oint_S \mathbf{B} \cdot d\mathbf{s} = 0$
$\nabla \times \mathbf{B} = \mu_0 \mathbf{J}$	$\oint_C \mathbf{B} \cdot d\ell = \mu_0 I$

EXAMPLE 6–1 An infinitely long, straight conductor with a circular cross section of radius b carries a steady current I. Determine the magnetic flux density both inside and outside the conductor.

[†] We note here parenthetically that examination of some prehistoric rock formations has led to the belief that there have been dramatic reversals of the earth's magnetic field every ten million years or so. The earth's magnetic field is thought to be produced by the rolling motions of the molten iron in the earth's outer core, but the exact reasons for the field reversals are still not well understood. The next such reversal is predicted to be only about 2000 years from now. One cannot conjecture all the dire consequences of such a reversal, but among them would be disruptions in global navigation and drastic changes in the migratory patterns of birds.

Solution First we note that this is a problem with cylindrical symmetry and that Ampère's circuital law can be used to advantage. If we align the conductor along the z-axis, the magnetic flux density $\mathbf{B}$ will be ϕ-directed and will be constant along any circular path around the z-axis. Figure 6–2(a) shows a cross section of the conductor and the two circular paths of integration, C_1 and C_2, inside and outside, respectively, the current-carrying conductor. Note again that the directions of C_1 and C_2 and the direction of I follow the right-hand rule. (When the fingers of the right hand follow the directions of C_1 and C_2, the thumb of the right hand points to the direction of I.)

a) *Inside the conductor:*

$$\mathbf{B}_1 = \mathbf{a}_\phi B_{\phi 1}, \qquad d\boldsymbol{\ell} = \mathbf{a}_\phi r_1 d\phi$$

$$\oint_{C_1} \mathbf{B}_1 \cdot d\boldsymbol{\ell} = \int_0^{2\pi} B_{\phi 1} r_1 \, d\phi = 2\pi r_1 B_{\phi 1}.$$

The current through the area enclosed by C_1 is

$$I_1 = \frac{\pi r_1^2}{\pi b^2} I = \left(\frac{r_1}{b}\right)^2 I.$$

Therefore, from Ampère's circuital law,

$$\mathbf{B}_1 = \mathbf{a}_\phi B_{\phi 1} = \mathbf{a}_\phi \frac{\mu_0 r_1 I}{2\pi b^2}, \qquad r_1 \leq b. \tag{6–11a}$$

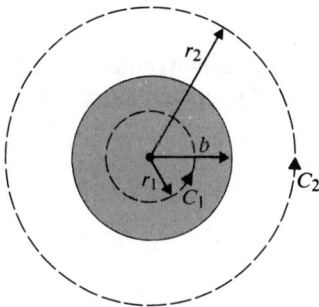

(a)

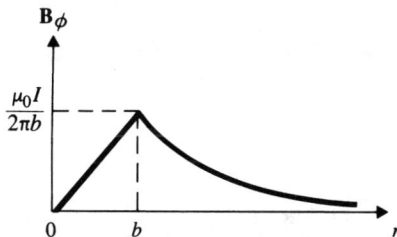

(b)

FIGURE 6–2
Magnetic flux density of an infinitely long circular conductor carrying a current I out of paper (Example 6–1).

b) *Outside the conductor:*

$$\mathbf{B}_2 = \mathbf{a}_\phi B_{\phi 2}, \qquad d\ell = \mathbf{a}_\phi r_2 \, d\phi$$

$$\oint_{C_2} \mathbf{B}_2 \cdot d\ell = 2\pi r_2 B_{\phi 2}.$$

Path C_2 outside the conductor encloses the total current I. Hence

$$\mathbf{B}_2 = \mathbf{a}_\phi B_{\phi 2} = \mathbf{a}_\phi \frac{\mu_0 I}{2\pi r_2}, \qquad r_2 \geq b. \qquad (6\text{--}11b)$$

Examination of Eqs. (6–11a) and (6–11b) reveals that the magnitude of **B** increases linearly with r_1 from 0 until $r_1 = b$, after which it decreases inversely with r_2. The variation of B_ϕ versus r is sketched in Fig. 6–2(b).   ■

If the problem is not that of a solid cylindrical conductor carrying a total steady current I, but that of a very thin circular tube carrying a surface current, then it is obvious from Ampère's circuital law that $\mathbf{B} = 0$ inside the tube. Outside the tube, Eq. (6–11b) still applies with I = total current flowing in the tube. Thus, for an infinitely long, hollow cylinder carrying a surface current density $\mathbf{J}_s = \mathbf{a}_z J_s$ (A/m), $I = 2\pi b J_s$, we have

$$B = \begin{cases} 0, & r < b, \\[2mm] \mathbf{a}_\phi \dfrac{\mu_0 b}{r} J_s, & r > b. \end{cases} \qquad (6\text{--}12)$$

■ **EXAMPLE 6–2**   Determine the magnetic flux density inside a closely wound toroidal coil with an air core having N turns and carrying a current I. The toroid has a mean radius b, and the radius of each turn is a.

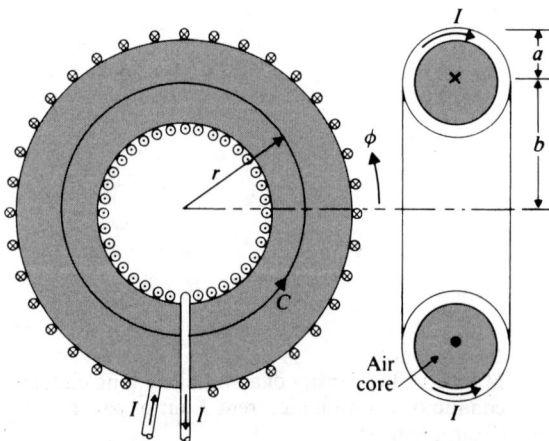

FIGURE 6–3
A current-carrying toroidal oil
(Example 6–2).

Solution Figure 6–3 depicts the geometry of this problem. Cylindrical symmetry ensures that **B** has only a ϕ-component and is constant along any circular path about the axis of the toroid. We construct a circular contour C with radius r as shown. For $(b - a) < r < b + a$, Eq. (6–10) leads directly to

$$\oint \mathbf{B} \cdot d\ell = 2\pi r B_\phi = \mu_0 N I,$$

where we have assumed that the toroid has an air core with permeability μ_0. Therefore,

$$\mathbf{B} = \mathbf{a}_\phi B_\phi = \mathbf{a}_\phi \frac{\mu_0 N I}{2\pi r}, \qquad (b - a) < r < (b + a). \tag{6–13}$$

It is apparent that $\mathbf{B} = 0$ for $r < (b - a)$ and $r > (b + a)$, since the net total current enclosed by a contour constructed in these two regions is zero. ▬

▬▬ **EXAMPLE 6–3** Determine the magnetic flux density inside an infinitely long solenoid with air core having n closely wound turns per unit length and carrying a current I as shown in Fig. 6–4.

Solution This problem can be solved in two ways.

a) *As a direct application of Ampère's circuital law.* It is clear that there is no magnetic field outside of the solenoid. To determine the **B**-field inside, we construct a rectangular contour C of length L that is partially inside and partially outside the solenoid. By reason of symmetry the **B**-field inside must be parallel to the axis. Applying Ampère's circuital law, we have

$$BL = \mu_0 n L I$$

or

$$B = \mu_0 n I. \tag{6–14}$$

The direction of **B** goes from right to left, conforming to the right-hand rule with respect to the direction of the current I in the solenoid, as indicated in Fig. 6–4.

b) *As a special case of toroid.* The straight solenoid may be regarded as a special case of the toroidal coil in Example 6–2 with an infinite radius ($b \to \infty$). In such

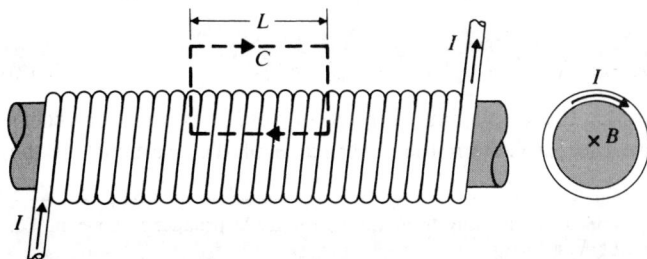

FIGURE 6–4
A current-carrying long solenoid
(Example 6–3).

a case the dimensions of the cross section of the core are very small in comparison with b, and the magnetic flux density inside the core is approximately constant. We have, from Eq. (6–13),

$$B = \mu_0 \left(\frac{N}{2\pi b} \right) I = \mu_0 n I,$$

which is the same as Eq. (6–14).

6–3 Vector Magnetic Potential

The divergence-free postulate of $\mathbf{B}$ in Eq. (6–6), $\mathbf{V} \cdot \mathbf{B} = 0$, assures that $\mathbf{B}$ is solenoidal. As a consequence, $\mathbf{B}$ can be expressed as the curl of another vector field, say $\mathbf{A}$, such that (see Identity II, Eq. (2–149), in Section 2–11)

$$\boxed{\mathbf{B} = \mathbf{V} \times \mathbf{A} \qquad (\text{T}).} \qquad (6\text{–}15)$$

The vector field A so defined is called the *vector magnetic potential*. Its SI unit is weber per meter (Wb/m). Thus if we can find $\mathbf{A}$ of a current distribution, $\mathbf{B}$ can be obtained from $\mathbf{A}$ by a differential (curl) operation. This is quite similar to the introduction of the scalar electric potential V for the curl-free $\mathbf{E}$ in electrostatics (Section 3–5) and the obtaining of $\mathbf{E}$ from the relation $\mathbf{E} = -\mathbf{V}V$. However, the definition of a vector requires the specification of both its curl and its divergence. Hence Eq. (6–15) alone is not sufficient to define $\mathbf{A}$; we must still specify its divergence.

How do we choose $\mathbf{V} \cdot \mathbf{A}$? Before we answer this question, let us take the curl of $\mathbf{B}$ in Eq. (6–15) and substitute it in Eq. (6–7). We have

$$\mathbf{V} \times \mathbf{V} \times \mathbf{A} = \mu_0 \mathbf{J}. \qquad (6\text{–}16)$$

Here we digress to introduce a formula for the curl curl of a vector:

$$\mathbf{V} \times \mathbf{V} \times \mathbf{A} = \mathbf{V}(\mathbf{V} \cdot \mathbf{A}) - \mathbf{V}^2 \mathbf{A} \qquad (6\text{–}17\text{a})$$

or

$$\mathbf{V}^2 \mathbf{A} = \mathbf{V}(\mathbf{V} \cdot \mathbf{A}) - \mathbf{V} \times \mathbf{V} \times \mathbf{A}. \qquad (6\text{–}17\text{b})$$

Equation (6–17a)[†] or (6–17b) can be regarded as the definition of $\mathbf{V}^2 \mathbf{A}$, the Laplacian of $\mathbf{A}$. For Cartesian coordinates it can be readily verified by direct substitution (Problem P.6–16) that

$$\mathbf{V}^2 \mathbf{A} = \mathbf{a}_x \mathbf{V}^2 A_x + \mathbf{a}_y \mathbf{V}^2 A_y + \mathbf{a}_z \mathbf{V}^2 A_z. \qquad (6\text{–}18)$$

Thus, for Cartesian coordinates the Laplacian of a vector field $\mathbf{A}$ is another vector field whose components are the Laplacian (the divergence of the gradient) of the

---

[†] Equation (6–17a) can also be obtained heuristically from the vector triple product formula in Eq. (2–20) by considering the del operator, $\mathbf{V}$, a vector:

$$\mathbf{V} \times (\mathbf{V} \times \mathbf{A}) = \mathbf{V}(\mathbf{V} \cdot \mathbf{A}) - (\mathbf{V} \cdot \mathbf{V})\mathbf{A} = \mathbf{V}(\mathbf{V} \cdot \mathbf{A}) - \mathbf{V}^2 \mathbf{A}.$$

corresponding components of **A**. This, however, is not true for other coordinate systems.

We now expand $\mathbf{V} \times \mathbf{V} \times \mathbf{A}$ in Eq. (6–16) according to Eq. (6–17a) and obtain

$$\mathbf{V}(\mathbf{V} \cdot \mathbf{A}) - \mathbf{V}^2\mathbf{A} = \mu_0\mathbf{J}. \qquad (6\text{–}19)$$

With the purpose of simplifying Eq. (6–19) to the greatest extent possible we choose

$$\boxed{\mathbf{V} \cdot \mathbf{A} = 0,} \qquad (6\text{–}20)^\dagger$$

and Eq. (6–19) becomes

$$\boxed{\mathbf{V}^2\mathbf{A} = -\mu_0\mathbf{J}.} \qquad (6\text{–}21)$$

This is a **vector Poisson's equation**. In Cartesian coordinates, Eq. (6–21) is equivalent to three scalar Poisson's equations:

$$\mathbf{V}^2 A_x = -\mu_0 J_x, \qquad (6\text{–}22\text{a})$$
$$\mathbf{V}^2 A_y = -\mu_0 J_y, \qquad (6\text{–}22\text{b})$$
$$\mathbf{V}^2 A_z = -\mu_0 J_z. \qquad (6\text{–}22\text{c})$$

Each of these three equations is mathematically the same as the scalar Poisson's equation, Eq. (4–6), in electrostatics. In free space the equation

$$\mathbf{V}^2 V = -\frac{\rho}{\epsilon_0}$$

has a particular solution (see Eq. (3–61)),

$$V = \frac{1}{4\pi\epsilon_0} \int_{V'} \frac{\rho}{R} \, dv'.$$

Hence the solution for Eq. (6–22a) is

$$A_x = \frac{\mu_0}{4\pi} \int_{V'} \frac{J_x}{R} \, dv'.$$

We can write similar solutions for A_y and A_z. Combining the three components, we have the solution for Eq. (6–21):

$$\boxed{\mathbf{A} = \frac{\mu_0}{4\pi} \int_{V'} \frac{\mathbf{J}}{R} \, dv' \qquad \text{(Wb/m)}.} \qquad (6\text{–}23)$$

---

† This relation is called **Coulomb condition** or **Coulomb gauge**.

Equation (6–23) enables us to find the vector magnetic potential **A** from the volume current density **J**. The magnetic flux density **B** can then be obtained from $\mathbf{V} \times \mathbf{A}$ by differentiation, in a way similar to that of obtaining the static electric field **E** from $-\mathbf{V}V$.

Vector potential **A** relates to the magnetic flux Φ through a given area S that is bounded by contour C in a simple way:

$$\Phi = \int_S \mathbf{B} \cdot d\mathbf{s}. \qquad (6\text{–}24)$$

The SI unit for magnetic flux is weber (Wb), which is equivalent to tesla-square meter $(\text{T} \cdot \text{m}^2)$. Using Eq. (6–15) and Stokes's theorem, we have

$$\Phi = \int_S (\mathbf{V} \times \mathbf{A}) \cdot d\mathbf{s} = \oint_C \mathbf{A} \cdot d\boldsymbol{\ell} \qquad \text{(Wb)}. \qquad (6\text{–}25)$$

Thus, vector magnetic potential **A** does have physical significance in that its line integral around any closed path equals the total magnetic flux passing through the area enclosed by the path.

6–4 The Biot-Savart Law and Applications

In many applications we are interested in determining the magnetic field due to a current-carrying circuit. For a thin wire with cross-sectional area S, dv' equals $S\,d\ell'$, and the current flow is entirely along the wire. We have

$$\mathbf{J}\,dv' = JS\,d\ell' = I\,d\boldsymbol{\ell}', \qquad (6\text{–}26)$$

and Eq. (6–23) becomes

$$\boxed{\mathbf{A} = \frac{\mu_0 I}{4\pi} \oint_{C'} \frac{d\boldsymbol{\ell}'}{R} \qquad \text{(Wb/m)},} \qquad (6\text{–}27)$$

where a circle has been put on the integral sign because the current I must flow in a closed path,[†] which is designated C'. The magnetic flux density is then

$$\begin{aligned}
\mathbf{B} = \mathbf{V} \times \mathbf{A} &= \mathbf{V} \times \left[\frac{\mu_0 I}{4\pi} \oint_{C'} \frac{d\boldsymbol{\ell}'}{R} \right] \\
&= \frac{\mu_0 I}{4\pi} \oint_{C'} \mathbf{V} \times \left(\frac{d\boldsymbol{\ell}'}{R} \right).
\end{aligned} \qquad (6\text{–}28)$$

[†] We are now dealing with direct (non-time-varying) currents that give rise to steady magnetic fields. Circuits containing time-varying sources may send time-varying currents along an open wire and deposit charges at its ends. Antennas are examples.

It is very important to note in Eq. (6–28) that the *unprimed* curl operation implies differentiations with respect to the space coordinates of the *field point*, and that the integral operation is with respect to the *primed source coordinates*. The integrand in Eq. (6–28) can be expanded into two terms by using the following identity (see Problem P.2–37):

$$\mathbf{V} \times (f\mathbf{G}) = f\mathbf{V} \times \mathbf{G} + (\nabla f) \times \mathbf{G}. \tag{6-29}$$

We have, with $f = 1/R$ and $\mathbf{G} = d\ell'$,

$$\mathbf{B} = \frac{\mu_0 I}{4\pi} \oint_{C'} \left[\frac{1}{R} \mathbf{V} \times d\ell' + \left(\mathbf{V}\frac{1}{R} \right) \times d\ell' \right]. \tag{6-30}$$

Now, since the unprimed and primed coordinates are independent, $\mathbf{V} \times d\ell'$ equals 0, and the first term on the right side of Eq. (6–30) vanishes. The distance R is measured from $d\ell'$ at (x', y', z') to the field point at (x, y, z). Thus we have

$$\frac{1}{R} = [(x - x')^2 + (y - y')^2 + (z - z')^2]^{-1/2};$$

$$\mathbf{V}\left(\frac{1}{R}\right) = \mathbf{a}_x \frac{\partial}{\partial x}\left(\frac{1}{R}\right) + \mathbf{a}_y \frac{\partial}{\partial y}\left(\frac{1}{R}\right) + \mathbf{a}_z \frac{\partial}{\partial z}\left(\frac{1}{R}\right)$$

$$= -\frac{\mathbf{a}_x(x - x') + \mathbf{a}_y(y - y') + \mathbf{a}_z(z - z')}{[(x - x')^2 + (y - y')^2 + (z - z')^2]^{3/2}} \tag{6-31}$$

$$= -\frac{\mathbf{R}}{R^3} = -\mathbf{a}_R \frac{1}{R^2},$$

where $\mathbf{a}_R$ is the unit vector directed *from the source point to the field point*. Substituting Eq. (6–31) in Eq. (6–30), we get

$$\mathbf{B} = \frac{\mu_0 I}{4\pi} \oint_{C'} \frac{d\ell' \times \mathbf{a}_R}{R^2} \quad \text{(T).} \tag{6-32}$$

Equation (6–32) is known as **Biot-Savart law**. It is a formula for determining $\mathbf{B}$ caused by a current I in a closed path C' and is obtained by taking the curl of $\mathbf{A}$ in Eq. (6–27). Sometimes it is convenient to write Eq. (6–32) in two steps:

$$\mathbf{B} = \oint_{C'} d\mathbf{B} \quad \text{(T),} \tag{6-33a}$$

with

$$d\mathbf{B} = \frac{\mu_0 I}{4\pi} \left(\frac{d\ell' \times \mathbf{a}_R}{R^2} \right) \quad \text{(T),} \tag{6-33b}$$

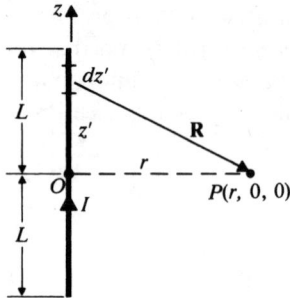

FIGURE 6–5
A current-carrying straight wire (Example 6–4).

which is the magnetic flux density due to a current element $I \, d\ell'$. An alternative and sometimes more convenient form for Eq. (6–33b) is

$$d\mathbf{B} = \frac{\mu_0 I}{4\pi} \left(\frac{d\ell' \times \mathbf{R}}{R^3} \right) \quad \text{(T)}. \qquad (6\text{–}33c)$$

Comparison of Eq. (6–32) with Eq. (6–10) will reveal that Biot-Savart law is, in general, more difficult to apply than Ampère's circuital law. However, Ampère's circuital law is not useful for determining $\mathbf{B}$ from I in a circuit if a closed path cannot be found over which $\mathbf{B}$ has a constant magnitude.

EXAMPLE 6–4   A direct current I flows in a straight wire of length $2L$. Find the magnetic flux density $\mathbf{B}$ at a point located at a distance r from the wire in the bisecting plane: (a) by determining the vector magnetic potential $\mathbf{A}$ first, and (b) by applying Biot-Savart law.

Solution   Currents exist only in closed circuits. Hence the wire in the present problem must be a part of a current-carrying loop with several straight sides. Since we do not know the rest of the circuit, Ampère's circuital law cannot be used to advantage. Refer to Fig. 6–5. The current-carrying line segment is aligned with the z-axis. A typical element on the wire is

$$d\ell' = \mathbf{a}_z \, dz'.$$

The cylindrical coordinates of the field point P are $(r, 0, 0)$.

a) *By finding $\mathbf{B}$ from $\nabla \times \mathbf{A}$.* Substituting $R = \sqrt{z'^2 + r^2}$ into Eq. (6–27), we have

$$\begin{aligned}
\mathbf{A} &= \mathbf{a}_z \frac{\mu_0 I}{4\pi} \int_{-L}^{L} \frac{dz'}{\sqrt{z'^2 + r^2}} \\
&= \mathbf{a}_z \frac{\mu_0 I}{4\pi} \left[\ln \left(z' + \sqrt{z'^2 + r^2} \right) \right] \Big|_{-L}^{L} \qquad (6\text{–}34) \\
&= \mathbf{a}_z \frac{\mu_0 I}{4\pi} \ln \frac{\sqrt{L^2 + r^2} + L}{\sqrt{L^2 + r^2} - L}.
\end{aligned}$$

Therefore,

$$\mathbf{B} = \nabla \times \mathbf{A} = \nabla \times (\mathbf{a}_z A_z) = \mathbf{a}_r \frac{1}{r} \frac{\partial A_z}{\partial \phi} - \mathbf{a}_\phi \frac{\partial A_z}{\partial r}.$$

Cylindrical symmetry around the wire assures that $\partial A_z / \partial \phi = 0$. Thus,

$$\mathbf{B} = -\mathbf{a}_\phi \frac{\partial}{\partial r} \left[\frac{\mu_0 I}{4\pi} \ln \frac{\sqrt{L^2 + r^2} + L}{\sqrt{L^2 + r^2} - L} \right]$$

$$= \mathbf{a}_\phi \frac{\mu_0 I L}{2\pi r \sqrt{L^2 + r^2}}. \tag{6–35}$$

When $r \ll L$, Eq. (6–35) reduces to

$$\mathbf{B}_\phi = \mathbf{a}_\phi \frac{\mu_0 I}{2\pi r}, \tag{6–36}$$

which is the expression for $\mathbf{B}$ at a point located at a distance r from an infinitely long, straight wire carrying current I, as given in Eq. (6–11b).

b) *By applying Biot-Savart law.* From Fig. 6–5 we see that the distance vector *from* the source element dz' *to* the field point P is

$$\mathbf{R} = \mathbf{a}_r r - \mathbf{a}_z z'$$

$$d\boldsymbol{\ell}' \times \mathbf{R} = \mathbf{a}_z \, dz' \times (\mathbf{a}_r r - \mathbf{a}_z z') = \mathbf{a}_\phi r \, dz'.$$

Substitution in Eq. (6–33c) gives

$$\mathbf{B} = \int d\mathbf{B} = \mathbf{a}_\phi \frac{\mu_0 I}{4\pi} \int_{-L}^{L} \frac{r \, dz'}{(z'^2 + r^2)^{3/2}}$$

$$= \mathbf{a}_\phi \frac{\mu_0 I L}{2\pi r \sqrt{L^2 + r^2}},$$

which is the same as Eq. (6–35). ▄

EXAMPLE 6–5 Find the magnetic flux density at the center of a square loop, with side w carrying a direct current I.

Solution Assume that the loop lies in the xy-plane, as shown in Fig. 6–6. The magnetic flux density at the center of the square loop is equal to four times that caused

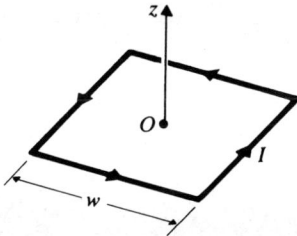

FIGURE 6–6
A square loop carrying current I (Example 6–5).

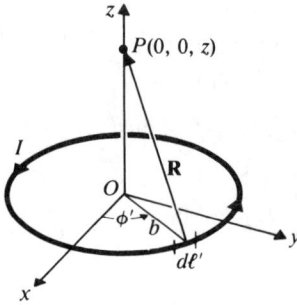

FIGURE 6–7
A circular loop carrying current I (Example 6–6).

by a single side of length w. We have, by setting $L = r = w/2$ in Eq. (6–35),

$$\mathbf{B} = \mathbf{a}_z \frac{\mu_0 I}{\sqrt{2}\pi w} \times 4 = \mathbf{a}_z \frac{2\sqrt{2}\mu_0 I}{\pi w}, \qquad (6\text{--}37)$$

where the direction of **B** and that of the current in the loop follow the right-hand rule. ■

EXAMPLE 6–6   Find the magnetic flux density at a point on the axis of a circular loop of radius b that carries a direct current I.

Solution   We apply Biot-Savart law to the circular loop shown in Fig. 6–7:

$$d\boldsymbol{\ell}' = \mathbf{a}_\phi b\, d\phi',$$
$$\mathbf{R} = \mathbf{a}_z z - \mathbf{a}_r b,$$
$$R = (z^2 + b^2)^{1/2}.$$

Again it is important to remember that **R** is the vector *from* the source element $d\boldsymbol{\ell}'$ *to* the field point P. We have

$$d\boldsymbol{\ell}' \times \mathbf{R} = \mathbf{a}_\phi b\, d\phi' \times (\mathbf{a}_z z - \mathbf{a}_r b)$$
$$= \mathbf{a}_r bz\, d\phi' + \mathbf{a}_z b^2\, d\phi'.$$

Because of cylindrical symmetry, it is easy to see that the $\mathbf{a}_r$-component is canceled by the contribution of the element located diametrically opposite to $d\boldsymbol{\ell}'$, so we need only consider the $\mathbf{a}_z$-component of this cross product.

We write, from Eqs. (6–33a) and (6–33c),

$$\mathbf{B} = \frac{\mu_0 I}{4\pi} \int_0^{2\pi} \mathbf{a}_z \frac{b^2\, d\phi'}{(z^2 + b^2)^{3/2}}$$

or

$$\boxed{\mathbf{B} = \mathbf{a}_z \frac{\mu_0 I b^2}{2(z^2 + b^2)^{3/2}} \qquad (\text{T}).} \qquad (6\text{--}38)$$

6–5   The Magnetic Dipole

We begin this section with an example.

■■■■■ **EXAMPLE 6-7**   Find the magnetic flux density at a distant point of a small circular loop of radius b that carries current I (a **magnetic dipole**).

Solution   It is apparent from the statement of the problem that we are interested in determining **B** at a point whose distance, R, from the center of the loop satisfies the relation $R \gg b$; that being the case, we may make certain simplifying approximations.

We select the center of the loop to be the origin of spherical coordinates, as shown in Fig. 6–8. The source coordinates are primed. We first find the vector magnetic potential **A** and then determine **B** by $\nabla \times \mathbf{A}$:

$$\mathbf{A} = \frac{\mu_0 I}{4\pi} \oint_{C'} \frac{d\boldsymbol{\ell}'}{R_1}. \tag{6-39}$$

Equation (6–39) is the same as Eq. (6–27), except for one important point: R in Eq. (6–27) denotes the distance between the source element $d\boldsymbol{\ell}'$ at P' and the field point P; but it must be replaced by R_1 in accordance with the notation in Fig. 6–8. Because of symmetry, the magnetic field is obviously independent of the angle ϕ of the field point. We pick $P(R, \theta, \pi/2)$ in the yz-plane for convenience.

Another point of importance is that $\mathbf{a}_{\phi'}$ at $d\boldsymbol{\ell}'$ is *not* the same as $\mathbf{a}_\phi$ at point P. In fact, $\mathbf{a}_\phi$ at P, shown in Fig. 6–8 is $-\mathbf{a}_x$, and

$$d\boldsymbol{\ell}' = (-\mathbf{a}_x \sin \phi' + \mathbf{a}_y \cos \phi')b \, d\phi'. \tag{6-40}$$

For every $I \, d\boldsymbol{\ell}'$ there is another symmetrically located differential current element on the other side of the y-axis that will contribute an equal amount to **A** in the $-\mathbf{a}_x$ direction but will cancel the contribution of $I \, d\boldsymbol{\ell}'$ in the $\mathbf{a}_y$ direction. Eq. (6–39)

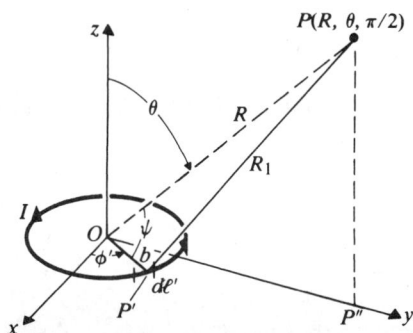

FIGURE 6–8
A small circular loop carrying current I (Example 6–7).

can be written as

$$\mathbf{A} = -\mathbf{a}_x \frac{\mu_0 I}{4\pi} \int_0^{2\pi} \frac{b \sin \phi'}{R_1} \, d\phi'$$

or

$$\mathbf{A} = \mathbf{a}_\phi \frac{\mu_0 I b}{2\pi} \int_{-\pi/2}^{\pi/2} \frac{\sin \phi'}{R_1} \, d\phi'. \tag{6–41}$$

The law of cosines applied to the triangle OPP' gives

$$R_1^2 = R^2 + b^2 - 2bR \cos \psi,$$

where $R \cos \psi$ is the projection of R on the radius OP', which is the same as the projection of OP'' ($OP'' = R \sin \theta$) on OP'. Hence,

$$R_1^2 = R^2 + b^2 - 2bR \sin \theta \sin \phi',$$

and

$$\frac{1}{R_1} = \frac{1}{R} \left(1 + \frac{b^2}{R^2} - \frac{2b}{R} \sin \theta \sin \phi' \right)^{-1/2}.$$

When $R^2 \gg b^2$, b^2/R^2 can be neglected in comparison with 1:

$$\frac{1}{R_1} \cong \frac{1}{R} \left(1 - \frac{2b}{R} \sin \theta \sin \phi' \right)^{-1/2}$$

$$\cong \frac{1}{R} \left(1 + \frac{b}{R} \sin \theta \sin \phi' \right). \tag{6–42}$$

Substitution of Eq. (6–42) in Eq. (6–41) yields

$$\mathbf{A} = \mathbf{a}_\phi \frac{\mu_0 I b}{2\pi R} \int_{-\pi/2}^{\pi/2} \left(1 + \frac{b}{R} \sin \theta \sin \phi' \right) \sin \phi' \, d\phi'$$

or

$$\mathbf{A} = \mathbf{a}_\phi \frac{\mu_0 I b^2}{4R^2} \sin \theta. \tag{6–43}$$

The magnetic flux density is $\mathbf{B} = \nabla \times \mathbf{A}$. Equation (2–139) can be used to find

$$\mathbf{B} = \frac{\mu_0 I b^2}{4R^3} (\mathbf{a}_R \, 2 \cos \theta + \mathbf{a}_\theta \sin \theta), \tag{6–44}$$

which is our answer.   ▄▄

At this point we recognize the similarity between Eq. (6–44) and the expression for the electric field intensity *in the far field* of an electrostatic dipole as given in Eq. (3–54). Hence, at distant points the magnetic flux lines of a magnetic dipole (placed in the xy-plane) such as that in Fig. 6–8 will have the same form as the dashed electric field lines of an electric dipole (lying in the z-direction) given in Fig. 3–15. In the vicinity of the dipoles, however, the flux lines of a magnetic dipole are continuous, whereas the field lines of an electric dipole terminate on the charges, always going from the positive to the negative charge. This is illustrated in Fig. 6–9.

Let us now rearrange the expression of the vector magnetic potential in Eq. (6–43) as

$$\mathbf{A} = \mathbf{a}_\phi \frac{\mu_0(I\pi b^2)}{4\pi R^2} \sin \theta$$

or

$$\mathbf{A} = \frac{\mu_0 \mathbf{m} \times \mathbf{a}_R}{4\pi R^2} \quad \text{(Wb/m)}, \tag{6–45}$$

where

$$\mathbf{m} = \mathbf{a}_z I\pi b^2 = \mathbf{a}_z IS = \mathbf{a}_z m \quad \text{(A·m}^2) \tag{6–46}$$

is defined as the **magnetic dipole moment**, which is a vector whose magnitude is the product of the current in and the area of the loop and whose direction is the direction of the thumb as the fingers of the right hand follow the direction of the current. Comparison of Eq. (6–45) with the expression for the scalar electric potential of an electric dipole in Eq. (3–53b),

$$V = \frac{\mathbf{p} \cdot \mathbf{a}_R}{4\pi\epsilon_0 R^2} \quad \text{(V)}, \tag{6–47}$$

reveals that, for the two cases, $\mathbf{A}$ is analogous to V. We call a small current-carrying loop a **magnetic dipole**.

In a similar manner we can also rewrite Eq. (6–44) as

$$\mathbf{B} = \frac{\mu_0 m}{4\pi R^3} (\mathbf{a}_R 2 \cos \theta + \mathbf{a}_\theta \sin \theta) \quad \text{(T)}. \tag{6–48}$$

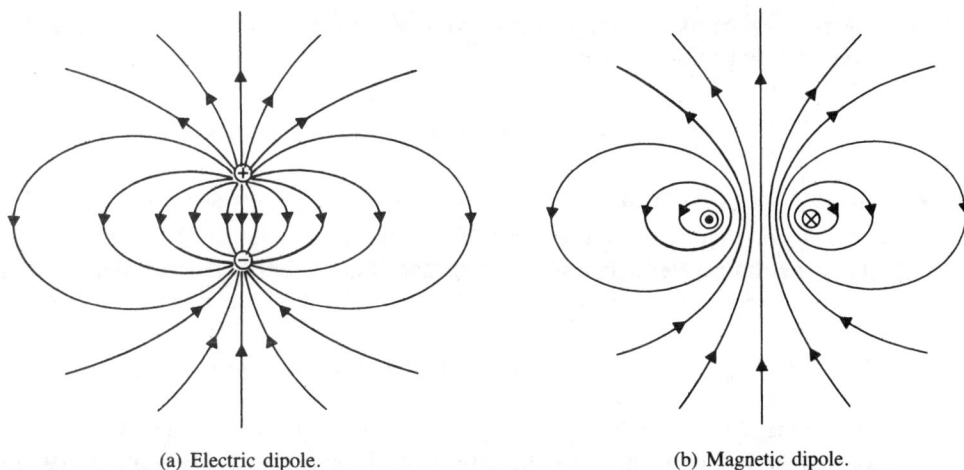

(a) Electric dipole.   (b) Magnetic dipole.

FIGURE 6–9
Electric field lines of an electric dipole and magnetic flux lines of a magnetic dipole.

Except for the change of p to m and ϵ_0 to $1/\mu_0$, Eq. (6–48) has the same form as Eq. (3–54) does for the expression for $\mathbf{E}$ at a distant point of an electric dipole. Hence the magnetic flux lines of a magnetic dipole lying in the xy-plane will have the same form as that of the electric field lines of an electric dipole positioned along the z-axis, as noted before.

Although the magnetic dipole in Example 6–7 was taken to be a circular loop, it can be shown (Problem P.6–19) that the same expressions—Eqs. (6–45) and (6–48)—are obtained when the loop has a rectangular shape, with $m = IS$, as given in Eq. (6–46).

6–5.1 SCALAR MAGNETIC POTENTIAL

In a current-free region $\mathbf{J} = 0$, Eq. (6–7) becomes

$$\nabla \times \mathbf{B} = 0. \tag{6–49}$$

The magnetic flux density $\mathbf{B}$ is then curl-free and can be expressed as the gradient of a scalar field. Let

$$\mathbf{B} = -\mu_0 \nabla V_m, \tag{6–50}$$

where V_m is called the *scalar magnetic potential* (expressed in amperes). The negative sign in Eq. (6–50) is conventional (see the definition of the scalar electric potential V in Eq. 3–43), and the permeability of free space μ_0 is simply a proportionality constant. Analogous to Eq. (3–45), we can write the scalar magnetic potential difference between two points, P_2 and P_1, in free space as

$$V_{m2} - V_{m1} = -\int_{P_1}^{P_2} \frac{1}{\mu_0} \mathbf{B} \cdot d\boldsymbol{\ell}. \tag{6–51}$$

If there *were* magnetic charges with a volume density ρ_m (A/m^2) in a volume V', we *would* be able to find V_m from

$$V_m = \frac{1}{4\pi} \int_{V'} \frac{\rho_m}{R} \, dv' \qquad \text{(A).} \tag{6–52}$$

The magnetic flux density $\mathbf{B}$ could then be determined from Eq. (6–50). However, isolated magnetic charges have never been observed experimentally; they must be considered fictitious. Nevertheless, the consideration of fictitious magnetic charges in a mathematical (not physical) model is expedient both to the discussion of some magnetostatic relations in terms of our knowledge of electrostatics and to the establishment of a bridge between the traditional magnetic-pole viewpoint of magnetism and the concept of microscopic circulating currents as sources of magnetism.

The magnetic field of a small bar magnet is the same as that of a magnetic dipole. This can be verified experimentally by observing the contours of iron filings around a magnet. The traditional understanding is that the ends (the north and south poles) of a permanent magnet are the location of positive and negative magnetic charges, respectively. For a bar magnet the fictitious magnetic charges $+q_m$ and $-q_m$ are assumed to be separated by a distance d and to form an equivalent magnetic dipole of moment

$$\mathbf{m} = q_m\mathbf{d} = \mathbf{a}_n IS. \tag{6–53}$$

The scalar magnetic potential V_m caused by this magnetic dipole can then be found by following the procedure used in Subsection 3–5.1 for finding the scalar electric potential that is caused by an electric dipole. We obtain, as in Eq. (3–53b),

$$V_m = \frac{\mathbf{m} \cdot \mathbf{a}_R}{4\pi R^2} \qquad \text{(A).} \tag{6–54}$$

Substitution of Eq. (6–54) in Eq. (6–50) yields the same $\mathbf{B}$ as is given in Eq. (6–48).

We note that the expression of the scalar magnetic potential V_m in Eq. (6–54) for a magnetic dipole is exactly analogous to that of the scalar electric potential V in Eq. (6–47) for an electric dipole. The likeness between the vector magnetic potential $\mathbf{A}$ in Eq. (6–45) and V in Eq. (6–47) is, however, not as exact. It is noted that the curl-free nature of $\mathbf{B}$ indicated in Eq. (6–49), from which the scalar magnetic potential V_m is defined, holds only at points with no currents. In a region where currents exist, the magnetic field is *not conservative*, and the scalar magnetic potential is not a single-valued function; hence the magnetic potential difference evaluated by Eq. (6–51) depends on the path of integration. For these reasons we will use the circulating-current-and-vector-potential approach, instead of the fictitious magnetic-charge-and-scalar-potential approach, for the study of magnetic fields in magnetic materials. We ascribe the macroscopic properties of a bar magnet to circulating atomic currents (Ampèrian currents) caused by orbiting and spinning electrons. Some aspects of equivalent (fictitious) magnetic charge densities will be discussed in Subsection 6–6.1.

6–6 Magnetization and Equivalent Current Densities

According to the elementary atomic model of matter, all materials are composed of atoms, each with a positively charged nucleus and a number of orbiting negatively charged electrons. The orbiting electrons cause circulating currents and form microscopic magnetic dipoles. In addition, both the electrons and the nucleus of an atom rotate (spin) on their own axes with certain magnetic dipole moments. The magnetic dipole moment of a spinning nucleus is usually negligible in comparison to that of an orbiting or spinning electron because of the much larger mass and lower angular velocity of the nucleus. A complete understanding of the magnetic effects of materials requires a knowledge of quantum mechanics. (We give a qualitative description of the behavior of different kinds of magnetic materials in Section 6–9.)

In the absence of an external magnetic field the magnetic dipoles of the atoms of most materials (except permanent magnets) have random orientations, resulting in no net magnetic moment. The application of an external magnetic field causes both an alignment of the magnetic moments of the spinning electrons and an induced magnetic moment due to a change in the orbital motion of electrons. To obtain a formula for determining the quantitative change in the magnetic flux density caused by the presence of a magnetic material, we let $\mathbf{m}_k$ be the magnetic dipole moment of an atom. If there are n atoms per unit volume, we define a *magnetization vector*, $\mathbf{M}$, as

$$\mathbf{M} = \lim_{\Delta v \to 0} \frac{\sum\limits_{k=1}^{n\,\Delta v} \mathbf{m}_k}{\Delta v} \quad \text{(A/m)}, \tag{6-55}$$

which is the volume density of magnetic dipole moment. The magnetic dipole moment $d\mathbf{m}$ of an elemental volume dv' is $d\mathbf{m} = \mathbf{M}\,dv'$ that, according to Eq. (6–45), will produce a vector magnetic potential

$$d\mathbf{A} = \frac{\mu_0 \mathbf{M} \times \mathbf{a}_R}{4\pi R^2}\,dv'. \tag{6-56}$$

Using Eq. (3–83), we can write Eq. (6–56) as

$$d\mathbf{A} = \frac{\mu_0}{4\pi}\,\mathbf{M} \times \nabla'\!\left(\frac{1}{R}\right) dv'.$$

Thus,

$$\mathbf{A} = \int_{V'} d\mathbf{A} = \frac{\mu_0}{4\pi} \int_{V'} \mathbf{M} \times \nabla'\!\left(\frac{1}{R}\right) dv', \tag{6-57}$$

where V' is the volume of the magnetized material.

We now use the vector identity in Eq. (6–29) to write

$$\mathbf{M} \times \nabla'\!\left(\frac{1}{R}\right) = \frac{1}{R}\,\nabla' \times \mathbf{M} - \nabla' \times \left(\frac{\mathbf{M}}{R}\right) \tag{6-58}$$

and expand the right side of Eq. (6–57) into two terms:

$$\mathbf{A} = \frac{\mu_0}{4\pi} \int_{V'} \frac{\nabla' \times \mathbf{M}}{R}\,dv' - \frac{\mu_0}{4\pi} \int_{V'} \nabla' \times \left(\frac{\mathbf{M}}{R}\right) dv'. \tag{6-59}$$

The following vector identity (see Problem P.6–20) enables us to change the volume integral of the curl of a vector into a surface integral:

$$\int_{V'} \nabla' \times \mathbf{F}\,dv' = -\oint_{S'} \mathbf{F} \times d\mathbf{s}', \tag{6-60}$$

where $\mathbf{F}$ is any vector with continuous first derivatives. We have, from Eq. (6–59),

$$\mathbf{A} = \frac{\mu_0}{4\pi} \int_{V'} \frac{\nabla' \times \mathbf{M}}{R}\,dv' + \frac{\mu_0}{4\pi} \oint_{S'} \frac{\mathbf{M} \times \mathbf{a}_n'}{R}\,ds', \tag{6-61}$$

where $\mathbf{a}_n'$ is the unit outward normal vector from ds' and S' is the surface bounding the volume V'.

A comparison of the expressions on the right side of Eq. (6–61) with the form of **A** in Eq. (6–23) expressed in terms of volume current density **J** suggests that the effect of the magnetization vector is equivalent to both a volume current density

$$\boxed{\mathbf{J}_m = \nabla \times \mathbf{M} \qquad (\text{A/m}^2)}$$

(6–62)

and a surface current density

$$\boxed{\mathbf{J}_{ms} = \mathbf{M} \times \mathbf{a}_n \qquad (\text{A/m}).}$$

(6–63)

In Eqs. (6–62) and (6–63) we have omitted the primes on ∇ and $\mathbf{a}_n$ for simplicity, since it is clear that both refer to the coordinates of the source point where the magnetization vector **M** exists. However, the primes should be retained when there is a possibility of confusing the coordinates of the source and field points.

The problem of finding the magnetic flux density **B** caused by a given volume density of magnetic dipole moment **M** is then reduced to finding the equivalent ***magnetization current densities*** $\mathbf{J}_m$ and $\mathbf{J}_{ms}$ by using Eqs. (6–62) and (6–63), determining **A** from Eq. (6–61), and then obtaining **B** from the curl of **A**. The externally applied magnetic field, if it also exists, must be accounted for separately.

The mathematical derivation of Eqs. (6–62) and (6–63) is straightforward. The equivalence of a volume density of magnetic dipole moment to a volume current density and a surface current density can be appreciated qualitatively by referring to Fig. 6–10, in which a cross section of a magnetized material is shown. It is assumed that an externally applied magnetic field has caused the atomic circulating currents to align with it, thereby magnetizing the material. The strength of this magnetizing effect is measured by the magnetization vector **M**. On the surface of the material there will be a surface current density $\mathbf{J}_{ms}$, whose direction is correctly given

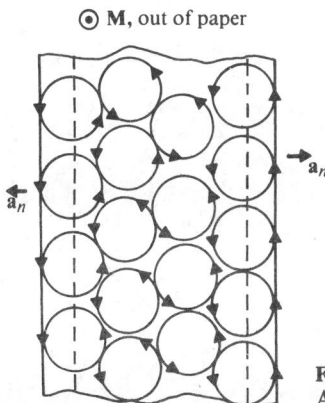

FIGURE 6–10
A cross section of a magnetized material.

by that of the cross product $\mathbf{M} \times \mathbf{a}_n$. If $\mathbf{M}$ is uniform inside the material, the currents of the neighboring atomic dipoles that flow in opposite directions will cancel everywhere, leaving no net currents in the interior. This is predicted by Eq. (6–62), since the space derivatives (and therefore the curl) of a constant $\mathbf{M}$ vanish. However, if $\mathbf{M}$ has space variations, the internal atomic currents do not completely cancel, resulting in a net volume current density $\mathbf{J}_m$. It is possible to justify the quantitative relationships between $\mathbf{M}$ and the current densities by deriving the atomic currents on the surface *and* in the interior. But since this additional derivation is really not necessary and tends to be tedious, we will not attempt it here.

EXAMPLE 6–8   Determine the magnetic flux density on the axis of a uniformly magnetized circular cylinder of a magnetic material. The cylinder has a radius b, length L, and axial magnetization $\mathbf{M} = \mathbf{a}_z M_0$.

Solution   In this problem concerning a cylindrical bar magnet, let the axis of the magnetized cylinder coincide with the z-axis of a cylindrical coordinate system, as shown in Fig. 6–11. Since the magnetization $\mathbf{M}$ is a constant within the magnet, $\mathbf{J}_m = \mathbf{\nabla}' \times \mathbf{M} = 0$, and there is no equivalent volume current density. The equivalent magnetization surface current density on the side wall is

$$\begin{aligned}
\mathbf{J}_{ms} &= \mathbf{M} \times \mathbf{a}_n' = (\mathbf{a}_z M_0) \times \mathbf{a}_r \\
&= \mathbf{a}_\phi M_0.
\end{aligned} \tag{6–64}$$

The magnet is then like a cylindrical sheet with a lineal current density of M_0 (A/m). There is no surface current on the top and bottom faces. To find $\mathbf{B}$ at $P(0, 0, z)$, we consider a differential length dz' with a current $\mathbf{a}_\phi M_0 \, dz'$ and use Eq. (6–38) to obtain

$$d\mathbf{B} = \mathbf{a}_z \frac{\mu_0 M_0 b^2 \, dz'}{2[(z - z')^2 + b^2]^{3/2}}$$

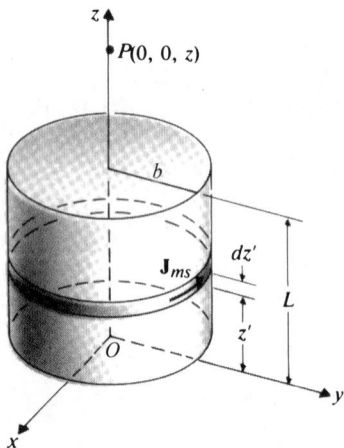

FIGURE 6–11
A uniformly magnetized circular cylinder (Example 6–8).

and

$$\mathbf{B} = \int d\mathbf{B} = \mathbf{a}_z \int_0^L \frac{\mu_0 M_0 b^2 \, dz'}{2[(z - z')^2 + b^2]^{3/2}}$$

$$= \mathbf{a}_z \frac{\mu_0 M_0}{2} \left[\frac{z}{\sqrt{z^2 + b^2}} - \frac{z - L}{\sqrt{(z - L)^2 + b^2}} \right]. \tag{6-65}$$

6–6.1 EQUIVALENT MAGNETIZATION CHARGE DENSITIES

In subsection 6–5.1 we noted that in a current-free region we may define a scalar magnetic potential V_m, from which the magnetic flux density $\mathbf{B}$ can be found by differentiation, as in Eq. (6–50). In terms of magnetization vector $\mathbf{M}$ (volume density of magnetic dipole moment) we may write, in lieu of Eq. (6–54),

$$dV_m = \frac{\mathbf{M} \cdot \mathbf{a}_R}{4\pi R^2}. \tag{6-66}$$

Integrating Eq. (6–66) over a magnetized body (a magnet) carrying no current, we have

$$V_m = \frac{1}{4\pi} \int_{V'} \frac{\mathbf{M} \cdot \mathbf{a}_R}{R^2} \, dv'. \tag{6-67}$$

Equation (6–67) is of exactly the same form as Eq. (3–81) for the scalar electric potential of a polarized dielectric. Following the steps leading to Eq. (3–87), we obtain

$$V_m = \frac{1}{4\pi} \oint_{S'} \frac{\mathbf{M} \cdot \mathbf{a}_n'}{R} \, ds' + \frac{1}{4\pi} \int_{V'} \frac{-(\mathbf{\nabla}' \cdot \mathbf{M})}{R} \, dv', \tag{6-68}$$

where $\mathbf{a}_n'$ is the outward normal to the surface element ds' of the magnetized body. We saw in Section 3–7 that, for field calculations, a polarized dielectric may be replaced by an equivalent polarization surface charge density, given in Eq. (3–88), and an equivalent polarization volume charge density, given in Eq. (3–89). Similarly, we can conclude that, for field calculations, a magnetized body may be replaced by an equivalent (fictitious) magnetization surface charge density ρ_{ms} and an equivalent (fictitious) magnetization volume charge density ρ_m such that

$$\boxed{\rho_{ms} = \mathbf{M} \cdot \mathbf{a}_n \qquad (\text{A/m})} \tag{6-69}$$

and

$$\boxed{\rho_m = -\mathbf{\nabla} \cdot \mathbf{M} \qquad (\text{A/m}^2).} \tag{6-70}$$

The use of the equivalent magnetization charge density concept for determining the magnetic flux density of a magnetized body will be illustrated in the following example.

EXAMPLE 6–9   A cylindrical bar magnet of radius b and length L has a uniform magnetization $\mathbf{M} = \mathbf{a}_z M_0$ along its axis. Use the equivalent magnetization charge density concept to determine the magnetic flux density at an arbitrary distant point.

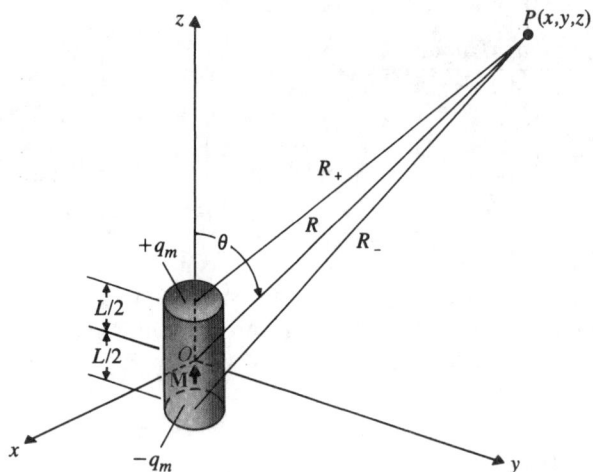

FIGURE 6–12
A cylindrical bar magnet (Example 6–9).

Solution   Refer to Fig. 6–12. The equivalent magnetization charge densities for $\mathbf{M} = \mathbf{a}_z M_0$ are, according to Eqs. (6–69) and (6–70):

$$\rho_{ms} = \begin{cases} M_0 & \text{on top face,} \\ -M_0 & \text{on bottom face,} \\ 0 & \text{on side wall;} \end{cases}$$

$$\rho_m = 0 \quad \text{in the interior.}$$

At a distant point the total equivalent magnetic charges on the top and bottom faces appear as point charges: $q_m = \pi b^2 \rho_{ms} = \pi b^2 M_0$. We have at $P(x, y, z)$

$$V_m = \frac{q_m}{4\pi}\left(\frac{1}{R_+} - \frac{1}{R_-}\right) \quad \text{(A)}, \tag{6–71}$$

which is similar to Eq. (3–50) for an electric dipole. If $R \gg b$, Eq. (6–71) can be reduced to (see Eq. 3–53a)

$$V_m = \frac{q_m L \cos\theta}{4\pi R^2} = \frac{(\pi b^2 M_0)L \cos\theta}{4\pi R^2}$$

$$= \frac{M_T \cos\theta}{4\pi R^2}, \tag{6–72}$$

where $M_T = \pi b^2 L M_0$ is the total dipole moment of the cylindrical magnet. The magnetic flux density $\mathbf{B}$ can then be found by applying Eq. (6–50):

$$\mathbf{B} = -\mu_0 \nabla V_m = \frac{\mu_0 M_T}{4\pi R^3}(\mathbf{a}_R 2\cos\theta + \mathbf{a}_\theta \sin\theta) \quad \text{(T)}, \tag{6–73}$$

which is of the same form as the expression in Eq. (6–44) for **B** at a distant point due to a single magnetic dipole having a moment $I\pi b^2$. ▬

This problem can be solved just as easily by using the equivalent magnetization current density concept. (See Problem P.6–25.)

6–7 Magnetic Field Intensity and Relative Permeability

Because the application of an external magnetic field causes both an alignment of the internal dipole moments and an induced magnetic moment in a magnetic material, we expect that the resultant magnetic flux density in the presence of a magnetic material will be different from its value in free space. The macroscopic effect of magnetization can be studied by incorporating the equivalent volume current density, $\mathbf{J}_m$ in Eq. (6–62), into the basic curl equation, Eq. (6–7). We have

$$\frac{1}{\mu_0} \nabla \times \mathbf{B} = \mathbf{J} + \mathbf{J}_m = \mathbf{J} + \nabla \times \mathbf{M}$$

or

$$\nabla \times \left(\frac{\mathbf{B}}{\mu_0} - \mathbf{M} \right) = \mathbf{J}. \tag{6–74}$$

We now define a new fundamental field quantity, the *magnetic field intensity* **H**, such that

$$\boxed{\mathbf{H} - \frac{\mathbf{B}}{\mu_0} - \mathbf{M} \quad \text{(A/m)}.} \tag{6–75}$$

The use of the vector **H** enables us to write a curl equation relating the magnetic field and the distribution of free currents in any medium. There is no need to deal explicitly with the magnetization vector **M** or the equivalent volume current density $\mathbf{J}_m$. Combining Eqs. (6–74) and (6–75), we obtain the new equation

$$\boxed{\nabla \times \mathbf{H} = \mathbf{J} \quad \text{(A/m}^2\text{)},} \tag{6–76}$$

where **J** (A/M²) is the volume density of *free current*. Equations (6–6) and (6–76) are the two fundamental governing differential equations for magnetostatics. The permeability of the medium does not appear explicitly in these two equations.

The corresponding integral form of Eq. (6–76) is obtained by taking the scalar surface integral of both sides:

$$\int_S (\nabla \times \mathbf{H}) \cdot d\mathbf{s} = \int_S \mathbf{J} \cdot d\mathbf{s} \tag{6–77}$$

or, according to Stokes's theorem,

$$\oint_C \mathbf{H} \cdot d\boldsymbol{\ell} = I \qquad \text{(A)}, \qquad (6\text{–}78)$$

where C is the contour (closed path) bounding the surface S and I is the total free current passing through S. The relative directions of C and current flow I follow the right-hand rule. Equation (6–78) is another form of *Ampère's circuital law*: It states that *the circulation of the magnetic field intensity around any closed path is equal to the free current flowing through the surface bounded by the path*. As we indicated in Section 6–2, Ampère's circuital law is most useful in determining the magnetic field caused by a current when cylindrical symmetry exists—that is, when there is a closed path around the current over which the magnetic field is constant.

When the magnetic properties of the medium are *linear* and *isotropic*, the magnetization is directly proportional to the magnetic field intensity:

$$\mathbf{M} = \chi_m \mathbf{H}, \qquad (6\text{–}79)$$

where χ_m is a dimensionless quantity called *magnetic susceptibility*. Substitution of Eq. (6–79) in Eq. (6–75) yields

$$\begin{aligned} \mathbf{B} &= \mu_0(1 + \chi_m)\mathbf{H} \\ &= \mu_0\mu_r\mathbf{H} = \mu\mathbf{H} \qquad (\text{Wb/m}^2) \end{aligned} \qquad (6\text{–}80\text{a})$$

or

$$\mathbf{H} = \frac{1}{\mu}\,\mathbf{B} \qquad (\text{A/m}), \qquad (6\text{–}80\text{b})$$

where

$$\mu_r = 1 + \chi_m = \frac{\mu}{\mu_0} \qquad (6\text{–}81)$$

is another dimensionless quantity known as the *relative permeability* of the medium. The parameter $\mu = \mu_0\mu_r$ is the *absolute permeability* (or sometimes just *permeability*) of the medium and is measured in H/m; χ_m, and therefore μ_r, can be a function of space coordinates. For a simple medium—linear, isotropic, and homogeneous—χ_m and μ_r are constants.

The permeability of most materials is very close to that of free space (μ_0). For ferromagnetic materials such as iron, nickel, and cobalt, μ_r could be very large (50–5000 and up to 10^6 or more for special alloys); the permeability depends not only on the magnitude of $\mathbf{H}$ but also on the previous history of the material. Section 6–9 contains some qualitative discussions of the macroscopic behavior of magnetic materials.

   At this point we note a number of analogous relations between the quantities
in electrostatics and those in magnetostatics as follows:

Electrostatics	Magnetostatics
$\mathbf{E}$	$\mathbf{B}$
$\mathbf{D}$	$\mathbf{H}$
ϵ	$\dfrac{1}{\mu}$
$\mathbf{P}$	$-\mathbf{M}$
ρ	$\mathbf{J}$
V	$\mathbf{A}$
$\cdot$	$\times$
$\times$	$\cdot$

With the above table, most of the equations relating the basic quantities in electrostatics can be converted into corresponding analogous ones in magnetostatics.

6–8 Magnetic Circuits

In electric-circuit problems we are required to find the voltages across and the currents in various branches and elements of an electric network that are excited by
voltage and/or current sources. There is an analogous class of problems dealing with
magnetic circuits. In a magnetic circuit we are generally concerned with the determination of the magnetic fluxes and magnetic field intensities in various parts of a
circuit caused by windings carrying currents around ferromagnetic cores. Magnetic
circuit problems arise in transformers, generators, motors, relays, magnetic recording
devices, and so on.

   Analysis of magnetic circuits is based on the two basic equations for magnetostatics, (6–6) and (6–76), which are repeated below for convenience:

$$\nabla \cdot \mathbf{B} = 0, \tag{6–82}$$

$$\nabla \times \mathbf{H} = \mathbf{J}. \tag{6–83}$$

We have seen in Eq. (6–78) that Eq. (6–83) converts to Ampère's circuital law. If
the closed path C is chosen to enclose N turns of a winding carrying a current I
that excites a magnetic circuit, we have

$$\oint_C \mathbf{H} \cdot d\ell = NI = \mathscr{V}_m. \tag{6–84}$$

The quantity $\mathscr{V}_m \; (=NI)$ here plays a role that is analogous to electromotive force
(emf) in an electric circuit and is therefore called a ***magnetomotive force*** (mmf). Its
SI unit is ampere (A); but, because of Eq. (6–84), mmf is frequently measured in
ampere-turns (A·t). An mmf is *not* a force measured in newtons.

EXAMPLE 6–10 Assume that N turns of wire are wound around a toroidal core of a ferromagnetic material with permeability μ. The core has a mean radius r_o, a circular cross section of radius a ($a \ll r_o$), and a narrow air gap of length ℓ_g, as shown in Fig. 6–13. A steady current I_o flows in the wire. Determine (a) the magnetic flux density, $\mathbf{B}_f$, in the ferromagnetic core; (b) the magnetic field intensity, $\mathbf{H}_f$, in the core; and (c) the magnetic field intensity, $\mathbf{H}_g$, in the air gap.

Solution

a) Applying Ampère's circuital law, Eq. (6–84), around the circular contour C in Fig. 6–13, which has a mean radius r_o, we have

$$\oint_C \mathbf{H} \cdot d\boldsymbol{\ell} = NI_o. \tag{6–85}$$

If flux leakage is neglected, the same total flux will flow in both the ferromagnetic core and in the air gap. If the fringing effect of the flux in the air gap is also neglected, the magnetic flux density $\mathbf{B}$ in both the core and the air gap will also be the same. However, because of the different permeabilities, the magnetic field intensities in both parts will be different. We have

$$\mathbf{B}_f = \mathbf{B}_g = \mathbf{a}_\phi B_f, \tag{6–86}$$

where the subscripts f and g denote ferromagnetic and gap, respectively. In the ferromagnetic core,

$$\mathbf{H}_f = \mathbf{a}_\phi \frac{B_f}{\mu}; \tag{6–87}$$

and, in the air gap,

$$\mathbf{H}_g = \mathbf{a}_\phi \frac{B_f}{\mu_0}. \tag{6–88}$$

Substituting Eqs. (6–87) and (6–88) in Eq. (6–85), we obtain

$$\frac{B_f}{\mu}(2\pi r_o - \ell_g) + \frac{B_f}{\mu_0}\ell_g = NI_o$$

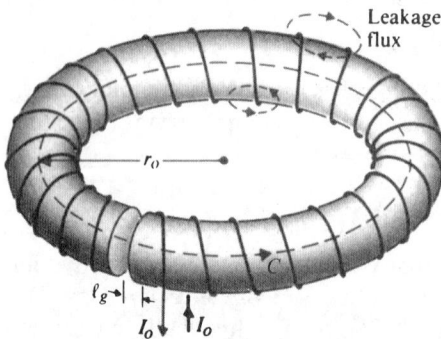

FIGURE 6–13
Coil on ferromagnetic toroid with air gap (Example 6–10).

and

$$\mathbf{B}_f = \mathbf{a}_\phi \frac{\mu_0 \mu N I_o}{\mu_0 (2\pi r_o - \ell_g) + \mu \ell_g}. \tag{6–89}$$

b) From Eqs. (6–87) and (6–89) we get

$$\mathbf{H}_f = \mathbf{a}_\phi \frac{\mu_0 N I_o}{\mu_0 (2\pi r_o - \ell_g) + \mu \ell_g}. \tag{6–90}$$

c) Similarly, from Eqs. (6–88) and (6–89) we have

$$\mathbf{H}_g = \mathbf{a}_\phi \frac{\mu N I_o}{\mu_0 (2\pi r_o - \ell_g) + \mu \ell_g}. \tag{6–91}$$

Since $H_g/H_f = \mu/\mu_0$, the magnetic field intensity in the air gap is much stronger than that in the ferromagnetic core.   ▄▄▄

If the radius of the cross section of the core is much smaller than the mean radius of the toroid, the magnetic flux density **B** in the core is approximately constant, and the magnetic flux in the circuit is

$$\Phi \cong BS, \tag{6–92}$$

where S is the cross-sectional area of the core. Combination of Eqs. (6–92) and (6–89) yields

$$\Phi = \frac{NI_o}{(2\pi r_o - \ell_g)/\mu S + \ell_g/\mu_0 S}. \tag{6–93}$$

Equation (6–93) can be rewritten as

$$\Phi = \frac{\mathcal{V}_m}{\mathcal{R}_f + \mathcal{R}_g}, \tag{6–94}$$

with

$$\mathcal{R}_f = \frac{2\pi r_o - \ell_g}{\mu S} = \frac{\ell_f}{\mu S}, \tag{6–95}$$

where $\ell_f = 2\pi r_o - \ell_g$ is the length of the ferromagnetic core, and

$$\mathcal{R}_g = \frac{\ell_g}{\mu_0 S}. \tag{6–96}$$

Both $\mathcal{R}_f$ and $\mathcal{R}_g$ have the same form as the formula, Eq. (5–27), for the d-c resistance of a straight piece of homogeneous material with a uniform cross section S. Both are called **reluctance**: $\mathcal{R}_f$, of the ferromagnetic core; and $\mathcal{R}_g$, of the air gap. The SI unit for reluctance is reciprocal henry (H^{-1}). The fact that Eqs. (6–95) and (6–96) are as they are, even though the core is not straight, is a consequence of assuming that **B** is approximately constant over the core cross section.

Equation (6–94) is analogous to the expression for the current I in an electric circuit, in which an ideal voltage source of emf $\mathcal{V}$ is connected in series with two

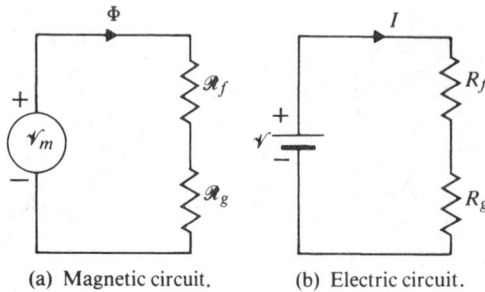

(a) Magnetic circuit.   (b) Electric circuit.

FIGURE 6–14
Equivalent magnetic circuit and analogous electric
circuit for toroidal coil with air gap in Fig. 6–13.

resistances R_f and R_g:

$$I = \frac{\mathscr{V}}{R_f + R_g}. \tag{6-97}$$

The analogous magnetic and electric circuits are shown in Figs. 6–14(a) and 6–14(b),
respectively. Magnetic circuits can, by analogy, be analyzed by the same techniques
we have used in analyzing electric circuits. The analogous quantities are as follows:

Magnetic Circuits	Electric Circuits
mmf, $\mathscr{V}_m\,(=NI)$	emf, $\mathscr{V}$
magnetic flux, Φ	electric current, I
reluctance, $\mathscr{R}$	resistance, R
permeability, μ	conductivity, σ

In spite of this convenient likeness an exact analysis of magnetic circuits is
inherently very difficult to achieve.

First, it is very difficult to account for leakage fluxes, fluxes that stray or leak
from the main flux paths of a magnetic circuit. For the toroidal coil in Fig. 6–13,
leakage flux paths encircle every turn of the winding; they partially transverse the
space around the core, as illustrated, because the permeability of air is not zero.
(There is little need for considering leakage currents outside the conducting paths of
electric circuits that carry direct currents. The reason is that the conductivity of air
is practically zero compared to that of a good conductor.)

A second difficulty is the fringing effect that causes the magnetic flux lines at the
air gap to spread and bulge.[†] (The purpose of specifying the "narrow air gap" in
Example 6–10 was to minimize this fringing effect.)

[†] To obtain a more accurate numerical result, it is customary to consider the effective area of the air gap
as slightly larger than the cross-sectional area of the ferromagnetic core, with each of the lineal dimensions
of the core cross section increased by the length of the air gap. If we were to make a correction like this
in Eq. (6–86), B_g would become

$$B_g = \frac{a^2 B_f}{(a + \ell_g)^2} < B_f.$$

A third difficulty is that the permeability of ferromagnetic materials depends on the magnetic field intensity; that is, **B** and **H** have a nonlinear relationship. (They might not even be in the same direction). The problem of Example 6–10, which assumes a given μ before either $\mathbf{B}_c$ or $\mathbf{H}_c$ is known, is therefore not a realistic one.

In a practical problem the B–H curve of the ferromagnetic material, such as that shown later in Fig. 6–17, should be given. The ratio of B to H is obviously not a constant, and B_f can be known only when H_f is known. So how does one solve the problem? Two conditions must be satisfied. First, the sum of $H_g\ell_g$ and $H_f\ell_f$ must equal the total mmf NI_o:

$$H_g\ell_g + H_f\ell_f = NI_o. \tag{6–98}$$

Second, if we assume no leakage flux, the total flux Φ in the ferromagnetic core and in the air gap must be the same, or $B_f = B_g$:[†]

$$B_f = \mu_0 H_g. \tag{6–99}$$

Substitution of Eq. (6–99) in Eq. (6–98) yields an equation relating B_f and H_f in the core:

$$B_f + \mu_0 \frac{\ell_f}{\ell_g} H_f = \frac{\mu_0}{\ell_g} NI_o. \tag{6–100}$$

This is an equation for a straight line in the B–H plane with a negative slope $(-\mu_0\ell_f/\ell_g)$. The intersection of this line and the given B–H curve determines the operating point. Once the operating point has been found, μ and H_f and all other quantities can be obtained.

The similarity between Eqs. (6–94) and (6–97) can be extended to the writing of two basic equations for magnetic circuits that correspond to Kirchhoff's voltage and current laws for electric circuits. Similar to Kirchhoff's voltage law in Eq. (5–41), we may write, for any closed path in a magnetic circuit,

$$\boxed{\sum_j N_j I_j = \sum_k \mathscr{R}_k \Phi_k.} \tag{6–101}$$

Equation (6–101) states that *around a closed path in a magnetic circuit the algebraic sum of ampere-turns is equal to the algebraic sum of the products of the reluctances and fluxes.*

Kirchhoff's current law for a junction in an electric circuit, Eq. (5–47), is a consequence of $\nabla \cdot \mathbf{J} = 0$. Similarly, the fundamental postulate $\nabla \cdot \mathbf{B} = 0$ in Eq. (6–82) leads to Eq. (6–9). Thus, we have

$$\boxed{\sum_j \Phi_j = 0,} \tag{6–102}$$

[†] This assumes an equal cross-sectional area for the core and the gap. If the core were to be constructed of insulated laminations of ferromagnetic material, the effective area for flux passage in the core would be smaller than the geometrical cross-sectional area, and B_c would be larger than B_g by a factor. This factor can be determined from the data on the insulated laminations.

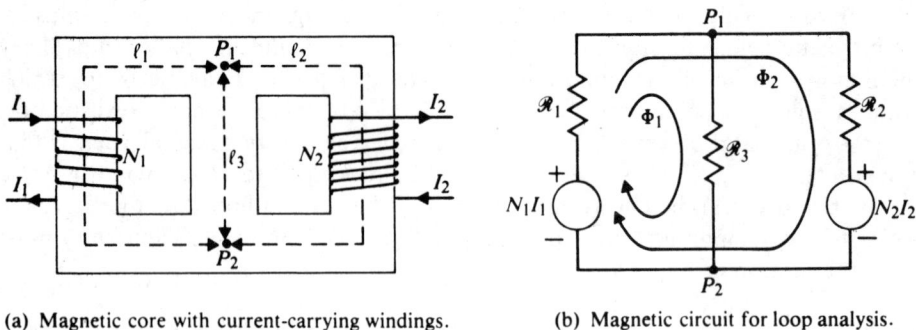

(a) Magnetic core with current-carrying windings.   (b) Magnetic circuit for loop analysis.

FIGURE 6–15
A magnetic circuit (Example 6–11).

which states that *the algebraic sum of all the magnetic fluxes flowing out of a junction in a magnetic circuit is zero*. Equations (6–101) and (6–102) form the bases for the loop and node analysis, respectively, of magnetic circuits.

EXAMPLE 6–11   Consider the magnetic circuit in Fig. 6–15(a). Steady currents I_1 and I_2 flow in windings of N_1 and N_2 turns, respectively, on the outside legs of the ferromagnetic core. The core has a cross-sectional area S_c and a permeability μ. Determine the magnetic flux in the center leg.

Solution   The equivalent magnetic circuit for loop analysis is shown in Fig. 6–15(b). Two sources of mmf's, N_1I_1 and N_2I_2, are shown with proper polarities in series with reluctances $\mathscr{R}_1$ and $\mathscr{R}_2$, respectively. This is obviously a two-loop network. Since we are determining magnetic flux in the center leg P_1P_2, it is expedient to choose the two loops in such a way that only one loop flux (Φ_1) flows through the center leg. The reluctances are computed on the basis of average path lengths. These are, of course, approximations. We have

$$\mathscr{R}_1 = \frac{\ell_1}{\mu S_c}, \tag{6–103a}$$

$$\mathscr{R}_2 = \frac{\ell_2}{\mu S_c}, \tag{6–103b}$$

$$\mathscr{R}_3 = \frac{\ell_3}{\mu S_c}. \tag{6–103c}$$

The two loop equations are, from Eq. (6–101),

Loop 1: $N_1I_1 = (\mathscr{R}_1 + \mathscr{R}_3)\Phi_1 + \mathscr{R}_1\Phi_2;$ \hfill (6–104)

Loop 2: $N_1I_1 - N_2I_2 = \mathscr{R}_1\Phi_1 + (\mathscr{R}_1 + \mathscr{R}_2)\Phi_2.$ \hfill (6–105)

Solving these simultaneous equations, we obtain

$$\Phi_1 = \frac{\mathscr{R}_2 N_1 I_1 + \mathscr{R}_1 N_2 I_2}{\mathscr{R}_1 \mathscr{R}_2 + \mathscr{R}_1 \mathscr{R}_3 + \mathscr{R}_2 \mathscr{R}_3},\qquad (6\text{–}106)$$

which is the desired answer.   ∎

Actually, since the magnetic fluxes and therefore the magnetic flux densities in the three legs are different, different permeabilities should be used in computing the reluctances in Eqs. (6–103a), (6–103b), and (6–103c). But the value of permeability, in turn, depends on the magnetic flux density. The only way to improve the accuracy of the solution, provided that the B–H curve of the core material is given, is to use a procedure of successive approximation. For instance, Φ_1, Φ_2, and Φ_3 (and therefore B_1, B_2, and B_3) are first solved with an assumed μ and reluctances computed from the three parts of Eq. (6–103). From B_1, B_2, and B_3 the corresponding μ_1, μ_2, and μ_3 can be found from the B–H curve. These will modify the reluctances. A second approximation for B_1, B_2, and B_3 is then obtained with the modified reluctances. From the new flux densities, new permeabilities and new reluctances are determined. This procedure is repeated until further iterations bring little change in the computed values.

We remark here that the currents in the windings in Fig. 6–15(a) are independent of time and that Example 6–11 is strictly a d-c magnetic circuit problem. If the currents vary with time, we must deal with the effects of electromagnetic induction, and we will have a transformer problem. Other fundamental laws are involved, which we shall discuss in Chapter 7.

6–9   Behavior of Magnetic Materials

In Eq. (6–79), Section 6–7, we described the macroscopic magnetic property of a linear, isotropic medium by defining the magnetic susceptibility χ_m, a dimensionless coefficient of proportionality between magnetization **M** and magnetic field intensity **H**. The relative permeability μ_r is simply $1 + \chi_m$. Magnetic materials can be roughly classified into three main groups in accordance with their μ_r values. A material is said to be

Diamagnetic, if $\mu_r \lesssim 1$ (χ_m is a very small negative number).

Paramagnetic, if $\mu_r \gtrsim 1$ (χ_m is a very small positive number).

Ferromagnetic, if $\mu_r \gg 1$ (χ_m is a large positive number).

As mentioned before, a thorough understanding of microscopic magnetic phenomena requires a knowledge of quantum mechanics. In the following we give a qualitative description of the behavior of the various types of magnetic materials based on the classical atomic model.

In a *diamagnetic* material the net magnetic moment due to the orbital and spin-ning motions of the electrons in any particular atom is zero in the absence of an externally applied magnetic field. As predicted by Eq. (6–4), the application of an external magnetic field to this material produces a force on the orbiting electrons, causing a perturbation in the angular velocities. As a consequence, a net magnetic moment is created. This is a process of induced magnetization. According to **Lenz's law** of electromagnetic induction (Section 7–2), the induced magnetic moment always *opposes* the applied field, thus reducing the magnetic flux density. The macroscopic effect of this process is equivalent to that of a negative magnetization that can be described by a negative magnetic susceptibility. This effect is usually very small, and χ_m for most known diamagnetic materials (bismuth, copper, lead, mercury, germanium, silver, gold, diamond) is of the order of -10^{-5}.

Diamagnetism arises mainly from the orbital motion of the electrons within an atom and is present in all materials. In most materials it is too weak to be of any practical importance. The diamagnetic effect is masked in paramagnetic and ferromagnetic materials. Diamagnetic materials exhibit no permanent magnetism, and the induced magnetic moment disappears when the applied field is withdrawn.

In some materials the magnetic moments due to the orbiting and spinning electrons do not cancel completely, and the atoms and molecules have a net average magnetic moment. An externally applied magnetic field, in addition to causing a very weak diamagnetic effect, tends to align the molecular magnetic moments *in the direction of* the applied field, thus increasing the magnetic flux density. The macroscopic effect is, then, equivalent to that of a positive magnetization that is described by a positive magnetic susceptibility. The alignment process is, however, impeded by the forces of random thermal vibrations. There is little coherent interaction, and the increase in magnetic flux density is quite small. Materials with this behavior are said to be *paramagnetic*. Paramagnetic materials generally have very small positive values of magnetic susceptibility, of the order of 10^{-5} for aluminum, magnesium, titanium, and tungsten.

Paramagnetism arises mainly from the magnetic dipole moments of the spinning electrons. The alignment forces, acting upon molecular dipoles by the applied field, are counteracted by the deranging effects of thermal agitation. Unlike diamagnetism, which is essentially independent of temperature, the paramagnetic effect is temperature dependent, being stronger at lower temperatures where there is less thermal collision.

The magnetization of *ferromagnetic* materials can be many orders of magnitude larger than that of paramagnetic substances. (See Appendix B–5 for typical values of relative permeability.) **Ferromagnetism** can be explained in terms of magnetized **domains**. According to this model, which has been experimentally confirmed, a ferromagnetic material (such as cobalt, nickel, and iron) is composed of many small domains, their linear dimensions ranging from a few microns to about 1 mm. These domains, each containing about 10^{15} or 10^{16} atoms, are fully magnetized in the sense that they contain aligned magnetic dipoles resulting from spinning electrons even in the absence of an applied magnetic field. Quantum theory asserts that strong coupling forces exist between the magnetic dipole moments of the atoms in a domain, holding the dipole moments in parallel. Between adjacent domains there is a transition region

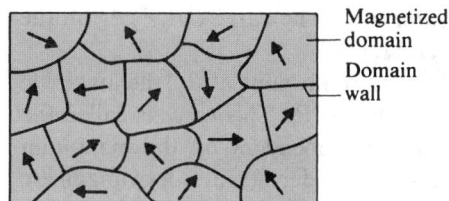

FIGURE 6–16
Domain structure of a polycrystalline ferromagnetic specimen.

about 100 atoms thick called a ***domain wall***. In an unmagnetized state the magnetic moments of the adjacent domains in a ferromagnetic material have different directions, as exemplified in Fig. 6–16 by the polycrystalline specimen shown. Viewed as a whole, the random nature of the orientations in the various domains results in no net magnetization.

When an external magnetic field is applied to a ferromagnetic material, the walls of those domains having magnetic moments aligned with the applied field move in such a way as to make the volumes of those domains grow at the expense of other domains. As a result, magnetic flux density is increased. For weak applied fields, say up to point P_1 in Fig. 6–17, domain-wall movements are reversible. But when an applied field becomes stronger (past P_1), domain-wall movements are no longer reversible, and domain rotation toward the direction of the applied field will also occur. For example, if an applied field is reduced to zero at point P_2, the B–H relationship will not follow the solid curve $P_2 P_1 O$, but will go down from P_2 to P_2', along the lines of the broken curve in the figure. This phenomenon of magnetization lagging behind the field producing it is called ***hysteresis***, which is derived from a Greek word meaning "to lag." As the applied field becomes even much stronger (past P_2 to P_3), domain-wall motion and domain rotation will cause essentially a total alignment of the microscopic magnetic moments with the applied field, at which point

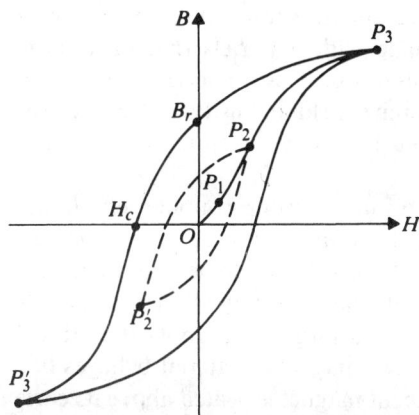

FIGURE 6–17
Hysteresis loops in the B–H plane for ferromagnetic material.

the magnetic material is said to have reached *saturation*. The curve $OP_1P_2P_3$ on the *B–H* plane is called the **normal magnetization curve**.

If the applied magnetic field is reduced to zero from the value at P_3, the magnetic flux density does not go to zero but assumes the value at B_r. This value is called the **residual** or **remanent flux density** (in Wb/m^2) and is dependent on the maximum applied field intensity. The existence of a remanent flux density in a ferromagnetic material makes permanent magnets possible.

To make the magnetic flux density of a specimen zero, it is necessary to apply a magnetic field intensity H_c in the opposite direction. This required H_c is called *coercive force*, but a more appropriate name is **coercive field intensity** (in A/m). Like B_r, H_c also depends on the maximum value of the applied magnetic field intensity.

It is evident from Fig. 6–17 that the *B–H* relationship for a ferromagnetic material is nonlinear. Hence, if we write $\mathbf{B} = \mu\mathbf{H}$ as in Eq. (6–80a), the permeability μ itself is a function of the magnitude of $\mathbf{H}$. Permeability μ also depends on the history of the material's magnetization, since—even for the same $\mathbf{H}$—we must know the location of the operating point on a particular branch of a particular hysteresis loop in order to determine the value of μ exactly. In some applications a small alternating current may be superimposed on a large steady magnetizing current. The steady magnetizing field intensity locates the operating point, and the local slope of the hysteresis curve at the operating point determines the **incremental permeability**.

Ferromagnetic materials for use in electric generators, motors, and transformers should have a large magnetization for a very small applied field; they should have tall, narrow hysteresis loops. As the applied magnetic field intensity varies periodically between $\pm H_{\max}$, the hysteresis loop is traced once per cycle. The area of the hysteresis loop corresponds to energy loss (**hysteresis loss**) per unit volume per cycle (Problem P.6–29). Hysteresis loss is the energy lost in the form of heat in overcoming the friction encountered during domain-wall motion and domain rotation. Ferromagnetic materials, which have tall, narrow hysteresis loops with small loop areas, are referred to as "soft" materials; they are usually well-annealed materials with very few dislocations and impurities so that the domain walls can move easily.

Good permanent magnets, on the other hand, should show a high resistance to demagnetization. This requires that they be made with materials that have large coercive field intensities H_c and hence fat hysteresis loops. These materials are referred to as "hard" ferromagnetic materials. The coercive field intensity of hard ferromagnetic materials (such as Alnico alloys) can be 10^5 (A/m) or more, whereas that for soft materials is usually 50 (A/m) or less.

As indicated before, ferromagnetism is the result of strong coupling effects between the magnetic dipole moments of the atoms in a domain. Figure 6–18(a) depicts the atomic spin structure of a ferromagnetic material. When the temperature of a ferromagnetic material is raised to such an extent that the thermal energy exceeds the coupling energy, the magnetized domains become disorganized. Above this critical temperature, known as the **curie temperature**, a ferromagnetic material behaves like a paramagnetic substance. Hence, when a permanent magnet is heated above its curie temperature it loses its magnetization. The curie temperature of most ferromagnetic materials lies between a few hundred to a thousand degrees Celsius, that of iron being 770°C.

Some elements, such as chromium and manganese, which are close to ferromagnetic elements in atomic number and are neighbors of iron in the periodic table, also have strong coupling forces between the atomic magnetic dipole moments; but their coupling forces produce antiparallel alignments of electron spins, as illustrated in Fig. 6–18(b). The spins alternate in direction from atom to atom and result in no net magnetic moment. A material possessing this property is said to be **antiferromagnetic**. Antiferromagnetism is also temperature dependent. When an antiferromagnetic material is heated above its curie temperature, the spin directions suddenly become random, and the material becomes paramagnetic.

There is another class of magnetic materials that exhibit a behavior between ferromagnetism and antiferromagnetism. Here quantum mechanical effects make the directions of the magnetic moments in the ordered spin structure alternate and the magnitudes unequal, resulting in a net nonzero magnetic moment, as depicted in Fig. 6–18(c). These materials are said to be **ferrimagnetic**. Because of the partial cancellation, the maximum magnetic flux density attained in a ferrimagnetic substance is substantially lower than that in a ferromagnetic specimen. Typically, it is about 0.3 Wb/m^2, approximately one-tenth that for ferromagnetic substances.

Ferrites are a subgroup of ferrimagnetic material. One type of ferrites, called **magnetic spinels**, crystallize in a complicated spinel structure and have the formula $XO \cdot Fe_2O_3$, where X denotes a divalent metallic ion such as Fe, Co, Ni, Mn, Mg, Zn, Cd, etc. These are ceramiclike compounds with very low conductivities (for instance, 10^{-4} to 1 (S/m) compared with 10^7 (S/m) for iron). Low conductivity limits eddy-current losses at high frequencies. Hence ferrites find extensive uses in such high-frequency and microwave applications as cores for FM antennas, high-frequency transformers, and phase shifters. Ferrite material also has broad applications in

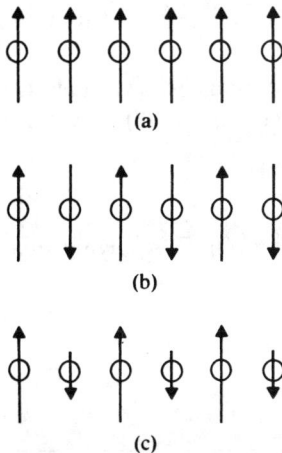

FIGURE 6–18
Schematic atomic spin structures for (a) ferromagnetic, (b) antiferromagnetic, and (c) ferrimagnetic materials.

computer magnetic-core and magnetic-disk memory devices. Other ferrites include magnetic-oxide garnets, of which yttrium-iron-garnet ("YIG," $Y_3Fe_5O_{12}$) is typical. Garnets are used in microwave multiport junctions.

Ferrites are anisotropic in the presence of a magnetic field. This means that **H** and **B** vectors in ferrites generally have different directions, and permeability is a tensor. The relation between the components of **H** and **B** can be represented in a matrix form similar to that between the components of **D** and **E** in an anisotropic dielectric medium, as given in Eq. (3–104) or Eq. (3–105). Analysis of problems containing anisotropic and/or nonlinear media is beyond the scope of this book.

6–10 Boundary Conditions for Magnetostatic Fields

In order to solve problems concerning magnetic fields in regions having media with different physical properties, it is necessary to study the conditions (boundary conditions) that **B** and **H** vectors must satisfy at the interfaces of different media. Using techniques similar to those employed in Section 3–9 to obtain the boundary conditions for electrostatic fields, we derive magnetostatic boundary conditions by applying the two fundamental governing equations, Eqs. (6–82) and (6–83), to a small pillbox and a small closed path, respectively, which include the interface. From the divergenceless nature of the **B** field in Eq. (6–82) we may conclude directly, in light of past experience, that *the normal component of* **B** *is continuous across an interface*; that is,

$$\boxed{B_{1n} = B_{2n} \quad \text{(T)}.}$$

(6–107)

For linear media, $\mathbf{B}_1 = \mu_1\mathbf{H}_1$ and $\mathbf{B}_2 = \mu_2\mathbf{H}_2$, Eq. (6–107) becomes

$$\boxed{\mu_1 H_{1n} = \mu_2 H_{2n}.}$$

(6–108)

The boundary condition for the tangential components of magnetostatic field is obtained from the integral form of the curl equation for **H**, Eq. (6–78), which is repeated here for convenience:

$$\oint_C \mathbf{H} \cdot d\ell = I.$$

(6–109)

We now choose the closed path *abcda* in Fig. 6–19 as the contour C. Applying Eq. (6–109) and letting $bc = da = \Delta h$ approach zero, we have[†]

$$\oint_{abcda} \mathbf{H} \cdot d\ell = \mathbf{H}_1 \cdot \Delta\mathbf{w} + \mathbf{H}_2 \cdot (-\Delta\mathbf{w}) = J_{sn}\Delta w$$

or

$$H_{1t} - H_{2t} = J_{sn} \quad \text{(A/m)},$$

(6–110)

[†] Equation (6–109) is assumed to be valid for regions containing discontinuous media.

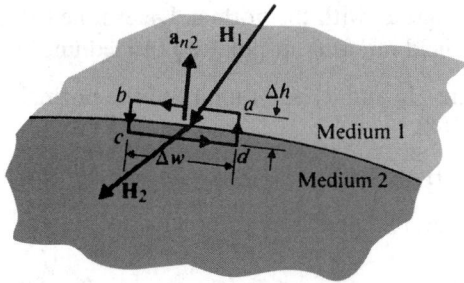

FIGURE 6–19
Closed path about the interface of two media for determining the boundary condition of H_t.

where J_{sn} is the surface current density on the interface normal to the contour C. The direction of J_{sn} is that of the thumb when the fingers of the right hand follow the direction of the path. In Fig. 6–19 the positive direction of J_{sn} for the chosen path is out of the paper. The following is a more concise expression of the boundary condition for the tangential components of **H**, which includes both magnitude and direction relations (Problem P.6–30).

$$\mathbf{a}_{n2} \times (\mathbf{H}_1 - \mathbf{H}_2) = \mathbf{J}_s \qquad (\text{A/m}), \tag{6–111}$$

where $\mathbf{a}_{n2}$ is the *outward unit normal from medium* 2 at the interface. Thus, *the tangential component of the* **H** *field is discontinuous across an interface where a free surface current exists*, the amount of discontinuity being determined by Eq. (6–111).

When the conductivities of both media are finite, currents are defined by volume current densities and free surface currents do not exist on the interface. Hence $\mathbf{J}_s$ equals zero, and *the tangential component of* **H** *is continuous across the boundary of almost all physical media; it is discontinuous only when an interface with an ideal perfect conductor or a superconductor is assumed.*

EXAMPLE 6–12 Two magnetic media with permeabilities μ_1 and μ_2 have a common boundary, as shown in Fig. 6–20. The magnetic field intensity in medium 1 at the

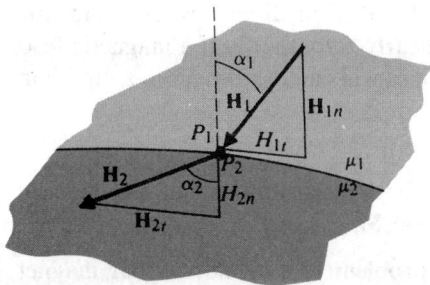

FIGURE 6–20
Boundary conditions for magnetostatic field at an interface (Example 6–12).

point P_1 has a magnitude H_1 and makes an angle α_1 with the normal. Determine the magnitude and the direction of the magnetic field intensity at point P_2 in medium 2.

Solution The desired unknown quantities are H_2 and α_2. Continuity of the normal component of **B** field requires, from Eq. (6–108),

$$\mu_2 H_2 \cos \alpha_2 = \mu_1 H_1 \cos \alpha_1. \tag{6–112}$$

Since neither of the media is a perfect conductor, the tangential component of **H** field is continuous. We have

$$H_2 \sin \alpha_2 = H_1 \sin \alpha_1. \tag{6–113}$$

Division of Eq. (6–113) by Eq. (6–112) gives

$$\boxed{\frac{\tan \alpha_2}{\tan \alpha_1} = \frac{\mu_2}{\mu_1}} \tag{6–114}$$

or

$$\alpha_2 = \arctan^{-1}\left(\frac{\mu_2}{\mu_1} \tan \alpha_1\right), \tag{6–115}$$

which describes the refraction property of the magnetic field. The magnitude of $\mathbf{H_2}$ is

$$H_2 = \sqrt{H_{2t}^2 + H_{2n}^2} = \sqrt{(H_2 \sin \alpha_2)^2 + (H_2 \cos \alpha_2)^2}.$$

From Eqs. (6–112) and (6–113) we obtain

$$\boxed{H_2 = H_1\left[\sin^2 \alpha_1 + \left(\frac{\mu_1}{\mu_2} \cos \alpha_1\right)^2\right]^{1/2}.} \tag{6–116}$$

We make three remarks here. First, Eqs. (6–114) and (6–116) are entirely similar to Eqs. (3–129) and (3–130), respectively, for the electric fields in dielectric media—except for the use of permeabilities (instead of permittivities) in the case of magnetic fields. Second, if medium 1 is nonmagnetic (like air) and medium 2 is ferromagnetic (like iron), then $\mu_2 \gg \mu_1$, and, from Eq. (6–114), α_2 will be nearly 90°. This means that for any arbitrary angle α_1 that is not close to zero, the magnetic field in a ferromagnetic medium runs almost parallel to the interface. Third, if medium 1 is ferromagnetic and medium 2 is air ($\mu_1 \gg \mu_2$), then α_2 will be nearly zero; that is, if a magnetic field originates in a ferromagnetic medium, the flux lines will emerge into air in a direction almost normal to the interface.

EXAMPLE 6–13 Sketch the magnetic flux lines both inside and outside a cylindrical bar magnet having a uniform axial magnetization $\mathbf{M} = \mathbf{a}_z M_0$.

Solution In Example 6–8 we noted that the problem of a cylindrical bar magnet could be replaced by that of a magnetization current sheet having a surface current density $\mathbf{J}_{ms} = \mathbf{a}_\phi M_0$ (the equivalent volume current density being zero). The determination of $\mathbf{B}$ at an arbitrary point inside and outside the magnet involves integrals that are difficult to evaluate. We shall use the result in Example 6–8 for a point on the magnet axis to obtain a rough sketch of the $\mathbf{B}$ lines.

A cross section of a cylindrical bar magnet having a radius b and length L is shown in Fig. 6–21. From Eq. (6–65) we get

$$\mathbf{B}_{P_0} = \mathbf{a}_z \frac{\mu_0 M_0}{2} \left[\frac{L}{\sqrt{(L/2)^2 + b^2}} \right] \tag{6-117}$$

$$\mathbf{B}_{P_1} = \mathbf{a}_z \frac{\mu_0 M_0}{2} \left[\frac{L}{\sqrt{L^2 + b^2}} \right] = \mathbf{B}_{P_1'}. \tag{6-118}$$

It is obvious from Eqs. (6–117) and (6–118) that $\mathbf{B}_{P_1} = \mathbf{B}_{P_1'} < \mathbf{B}_{P_0}$; that is, the magnetic flux density along the axis at the end faces of the magnet is less than that at the center. This is because the flux lines tend to flare out at the end faces. We know that, at points off the axis, $\mathbf{B}$ has a radial component. We also know that $\mathbf{B}$ lines are not refracted at the end faces and that they close upon themselves.

On the side of the magnet there is a surface current given by Eq. (6–64):

$$\mathbf{J}_{ms} = \mathbf{a}_\phi M_0. \tag{6-119}$$

Hence according to Eq. (6–111), the axial component of $\mathbf{B}$ changes by an amount equal to $\mu_0 M_0$. From Eqs. (6–117) and (6–118) we see that B_z inside the magnet is less

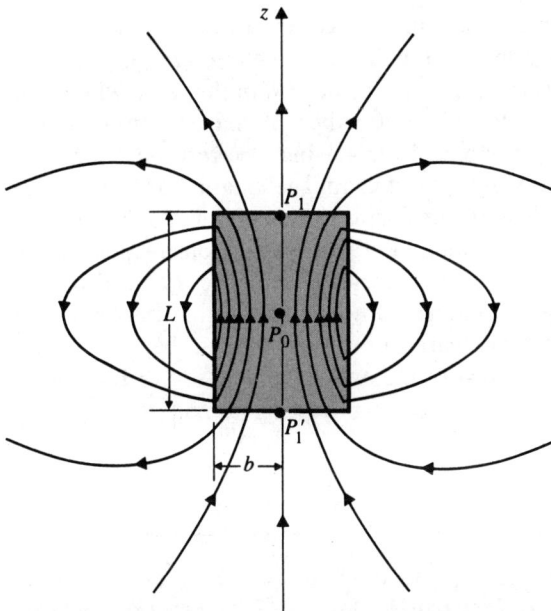

FIGURE 6–21
Magnetic flux lines around a cylindrical bar magnet (Example 6–13).

than $\mu_0 M_0$. Consequently, there is a change in both the magnitude and the direction for B_z as it crosses the side wall. The magnetic flux lines will then assume the form sketched in Fig. 6–21.

It must be remarked here that while $\mathbf{H} = \mathbf{B}/\mu_0$ outside the magnet, $\mathbf{H}$ and $\mathbf{B}$ inside the magnet are far from being proportional vectors in the same direction. From Eq. (6–75),

$$\mathbf{H} = \frac{\mathbf{B}}{\mu_0} - \mathbf{M}, \tag{6–120}$$

and the fact that B/μ_0 along the axis inside is less than M_0, we observe that $\mathbf{H}$ and $\mathbf{B}$ are in opposite directions along the axis inside. For a long, thin magnet, $L \gg b$, Eq. (6–117) gives approximately $B_{P_0} = \mu_0 M_0$. From Eq. (6–120) we obtain $H_{P_0} \cong 0$. Hence $\mathbf{H}$ nearly vanishes at the center of a long, thin magnet, where $\mathbf{B}$ is maximum. By hypothesis the magnetization vector $\mathbf{M}$ is zero outside and is a constant vector everywhere inside the magnet.   ▬▬

In current-free regions the magnetic flux density $\mathbf{B}$ is irrotational and can be expressed as the gradient of a scalar magnetic potential V_m, as indicated in Section 6–5.1.

$$\mathbf{B} = -\mu \nabla V_m. \tag{6–121}$$

Assuming a constant μ, substitution of Eq. (6–121) in $\nabla \cdot \mathbf{B} = 0$ (Eq. 6–6) yields a Laplace's equation in V_m:

$$\nabla^2 V_m = 0. \tag{6–122}$$

Equation (6–122) is entirely similar to the Laplace's equation, Eq. (4–10), for the scalar electric potential V in a charge-free region. That the solution for Eq. (6–122) satisfying given boundary conditions is unique can be proved in the same way as for Eq. (4–10)—see Section 4–3. Thus the techniques (method of images and method of separation of variables) discussed in Chapter 4 for solving electrostatic boundary-value problems can be adapted to solving analogous magnetostatic boundary-value problems. However, although electrostatic problems with conducting boundaries maintained at fixed potentials occur quite often in practice, analogous magnetostatic problems with constant magnetic-potential boundaries are of little practical importance. (We recall that isolated magnetic charges do not exist and that magnetic flux lines always form closed paths.) The nonlinearity in the relationship between $\mathbf{B}$ and $\mathbf{H}$ in ferromagnetic materials also complicates the analytical solution of boundary-value problems in magnetostatics.

6–11 Inductances and Inductors

Consider two neighboring closed loops, C_1 and C_2 bounding surfaces S_1 and S_2, respectively, as shown in Fig. 6–22. If a current I_1 flows in C_1, a magnetic field $\mathbf{B}_1$ will be created. Some of the magnetic flux due to $\mathbf{B}_1$ will link with C_2—that is, will pass through the surface S_2 bounded by C_2. Let us designate this mutual flux Φ_{12}. We have

$$\Phi_{12} = \int_{S_2} \mathbf{B}_1 \cdot d\mathbf{s}_2 \qquad \text{(Wb)}. \qquad (6\text{-}123)$$

From physics we know that a time-varying I_1 (and therefore a time-varying Φ_{12}) will produce an induced electromotive force or voltage in C_2 as a result of Faraday's law of electromagnetic induction. (We defer the discussion of Faraday's law until the next chapter.) However, Φ_{12} exists even if I_1 is a steady d-c current.

From the Biot-Savart law, Eq. (6-32), we see that B_1 is directly proportional to I_1; hence Φ_{12} is also proportional to I_1. We write

$$\Phi_{12} = L_{12}I_1, \qquad (6\text{-}124)$$

where the proportionality constant L_{12} is called the **mutual inductance** between loops C_1 and C_2, with SI unit henry (H). In case C_2 has N_2 turns, the **flux linkage** Λ_{12} due to Φ_{12} is

$$\Lambda_{12} = N_2\Phi_{12} \qquad \text{(Wb)}, \qquad (6\text{-}125)$$

and Eq. (6-124) generalizes to

$$\Lambda_{12} = L_{12}I_1 \qquad \text{(Wb)} \qquad (6\text{-}126)$$

or

$$\boxed{L_{12} = \frac{\Lambda_{12}}{I_1} \qquad \text{(H)}.} \qquad (6\text{-}127)$$

The **mutual inductance between two circuits** is then the magnetic flux linkage with one circuit per unit current in the other. In Eq. (6-124) it is implied that the permeability of the medium does not change with I_1. In other words, Eq. (6-124) and hence Eq.

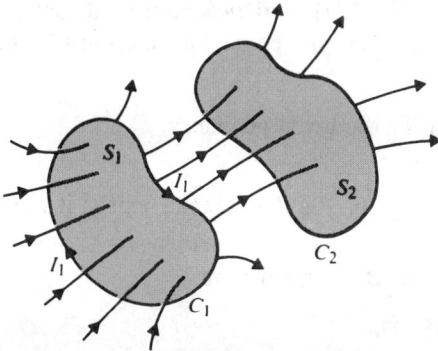

FIGURE 6-22
Two magnetically coupled loops.

(6–127) apply only to *linear* media. A more general definition for L_{12} is

$$L_{12} = \frac{d\Lambda_{12}}{dI_1} \quad (H).$$

(6–128)

Some of the magnetic flux produced by I_1 links only with C_1 itself, and not with C_2. The total flux linkage with C_1 caused by I_1 is

$$\Lambda_{11} = N_1\Phi_{11} > N_1\Phi_{12}.$$

(6–129)

*The **self-inductance** of loop C_1 is defined as the magnetic flux linkage per unit current in the loop itself; that is,*

$$L_{11} = \frac{\Lambda_{11}}{I_1} \quad (H),$$

(6–130)

for a linear medium. In general,

$$L_{11} = \frac{d\Lambda_{11}}{dI_1} \quad (H).$$

(6–131)

The self-inductance of a loop or circuit depends on the geometrical shape and the physical arrangement of the conductor constituting the loop or circuit, as well as on the permeability of the medium. With a linear medium, self-inductance does not depend on the current in the loop or circuit. As a matter of fact, it exists regardless of whether the loop or circuit is open or closed, or whether it is near another loop or circuit.

A conductor arranged in an appropriate shape (such as a conducting wire wound as a coil) to supply a certain amount of self-inductance is called an **inductor**. Just as a capacitor can store electric energy, an inductor can storage magnetic energy, as we shall see in Section 6–12. When we deal with only one loop or coil, there is no need to carry the subscripts in Eq. (6–130) or Eq. (6–131), and **inductance** without an adjective will be taken to mean self-inductance. The procedure for determining the self-inductance of an inductor is as follows:

1. Choose an appropriate coordinate system for the given geometry.
2. Assume a current I in the conducting wire.
3. Find **B** from I by Ampère's circuital law, Eq. (6–10), if symmetry exists; if not, Biot-Savart law, Eq. (6–32), must be used.
4. Find the flux linking with each turn, Φ, from **B** by integration:

$$\Phi = \int_S \mathbf{B} \cdot d\mathbf{s},$$

where S is the area over which **B** exists and links with the assumed current.

5. Find the flux linkage Λ by multiplying Φ by the number of turns.

6. Find L by taking the ratio $L = \Lambda/I$.

Only a slight modification of this procedure is needed to determine the mutual inductance L_{12} between two circuits. After choosing an appropriate coordinate system, proceed as follows: Assume $I_1 \rightarrow$ Find $\mathbf{B}_1 \rightarrow$ Find Φ_{12} by integrating $\mathbf{B}_1$ over surface $S_2 \rightarrow$ Find flux linkage $\Lambda_{12} = N_2\Phi_{12} \rightarrow$ Find $L_{12} = \Lambda_{12}/I_1$.

EXAMPLE 6–14   Assume that N turns of wire are tightly wound on a toroidal frame of a rectangular cross section with dimensions as shown in Fig. 6–23. Then, assuming the permeability of the medium to be μ_0, find the self-inductance of the toroidal coil.

Solution   It is clear that the cylindrical coordinate system is appropriate for this problem because the toroid is symmetrical about its axis. Assuming a current I in the conducting wire, we find, by applying Eq. (6–10) to a circular path with radius r $(a < r < b)$:

$$\mathbf{B} = \mathbf{a}_\phi B_\phi,$$

$$d\boldsymbol{\ell} = \mathbf{a}_\phi r\,d\phi,$$

$$\oint_C \mathbf{B} \cdot d\boldsymbol{\ell} = \int_0^{2\pi} B_\phi r\,d\phi = 2\pi r B_\phi.$$

This result is obtained because both B_ϕ and r are constant around the circular path C. Since the path encircles a total current NI, we have

$$2\pi r B_\phi = \mu_0 NI$$

and

$$B_\phi = \frac{\mu_0 NI}{2\pi r}.$$

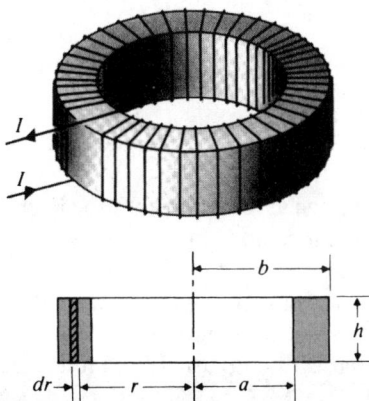

FIGURE 6–23
A closely wound toroidal coil (Example 6–14).

Next we find

$$\Phi = \int_S \mathbf{B} \cdot d\mathbf{s} = \int_S \left(\mathbf{a}_\phi \frac{\mu_0 N I}{2\pi r} \right) \cdot (\mathbf{a}_\phi h\, dr)$$

$$= \frac{\mu_0 N I h}{2\pi} \int_a^b \frac{dr}{r} = \frac{\mu_0 N I h}{2\pi} \ln \frac{b}{a}.$$

The flux linkage Λ is $N\Phi$ or

$$\Lambda = \frac{\mu_0 N^2 I h}{2\pi} \ln \frac{b}{a}.$$

Finally, we obtain

$$L = \frac{\Lambda}{I} = \frac{\mu_0 N^2 h}{2\pi} \ln \frac{b}{a} \quad \text{(H)}. \tag{6-132}$$

We note that the self-inductance is not a function of I (for a constant medium permeability). The qualification that the coil be closely wound on the toroid is to minimize the linkage flux around the individual turns of the wire.   ■

■■■■ **EXAMPLE 6-15**   Find the inductance per unit length of a very long solenoid with air core having n turns per unit length.

Solution   The magnetic flux density inside an infinitely long solenoid has been found in Example 6-3. For current I we have, from Eq. (6-14),

$$B = \mu_0 n I,$$

which is constant inside the solenoid. Hence,

$$\Phi = BS = \mu_0 n S I, \tag{6-133}$$

where S is the cross-sectional area of the solenoid. The flux linkage per unit length is

$$\Lambda' = n\Phi = \mu_0 n^2 S I. \tag{6-134}$$

Therefore the inductance per unit length is

$$L' = \mu_0 n^2 S \quad \text{(H/m)}. \tag{6-135}$$

Equation (6-135) is an approximate formula, based on the assumption that the length of the solenoid is very much greater than the linear dimensions of its cross section. A more accurate derivation for the magnetic flux density and flux linkage per unit length near the ends of a finite solenoid will show that they are less than the values given, respectively, by Eqs. (6-14) and (6-134). Hence the total inductance of a finite solenoid is somewhat less than the values of L', as given in Eq. (6-135), multiplied by the length.   ■

The following is a significant observation about the results of the previous two examples: The self-inductance of wire-wound inductors is proportional to the *square* of the number of turns.

EXAMPLE 6-16   An air coaxial transmission line has a solid inner conductor of radius a and a very thin outer conductor of inner radius b. Determine the inductance per unit length of the line.

Solution   Refer to Fig. 6-24. Assume that a current I flows in the inner conductor and returns via the outer conductor in the other direction. Because of the cylindrical symmetry, **B** has only a ϕ-component with different expressions in the two regions: (a) inside the inner conductor, and (b) between the inner and outer conductors. Also assume that the current I is uniformly distributed over the cross section of the inner conductor.

a) *Inside the inner conductor,*

$$0 \leq r \leq a.$$

From Eq. (6-11a),

$$\mathbf{B}_1 = \mathbf{a}_\phi B_{\phi 1} = \mathbf{a}_\phi \frac{\mu_0 r I}{2\pi a^2}. \tag{6-136}$$

b) *Between the inner and outer conductors,*

$$a \leq r \leq b.$$

From Eq. (6-11b),

$$\mathbf{B}_2 = \mathbf{a}_\phi B_{\phi 2} = \mathbf{a}_\phi \frac{\mu_0 I}{2\pi r}. \tag{6-137}$$

Now consider an annular ring in the inner conductor between radii r and $r + dr$. The current in a unit length of this annular ring is linked by the flux that can be obtained by integrating Eqs. (6-136) and (6-137). We have

$$
\begin{aligned}
d\Phi' &= \int_r^a B_{\phi 1} \, dr + \int_a^b B_{\phi 2} \, dr \\
&= \frac{\mu_0 I}{2\pi a^2} \int_r^a r \, dr + \frac{\mu_0 I}{2\pi} \int_a^b \frac{dr}{r} \\
&= \frac{\mu_0 I}{4\pi a^2} (a^2 - r^2) + \frac{\mu_0 I}{2\pi} \ln \frac{b}{a}.
\end{aligned}
\tag{6-138}
$$

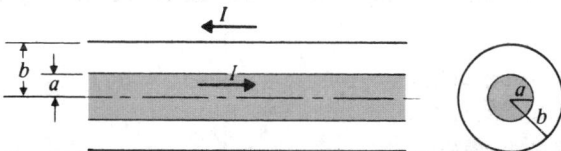

FIGURE 6-24
Two views of a coaxial transmission line (Example 6-16).

But the current in the annular ring is only a fraction $(2\pi r\, dr/\pi a^2 = 2r\, dr/a^2)$ of the total current I.[†] Hence the flux linkage for this annular ring is

$$d\Lambda' = \frac{2r\, dr}{a^2}\, d\Phi'. \tag{6-139}$$

The total flux linkage per unit length is

$$\begin{aligned} \Lambda' &= \int_{r=0}^{r=a} d\Lambda' \\ &= \frac{\mu_0 I}{\pi a^2}\left[\frac{1}{2a^2}\int_0^a (a^2 - r^2)r\, dr + \left(\ln\frac{b}{a}\right)\int_0^a r\, dr\right] \\ &= \frac{\mu_0 I}{2\pi}\left(\frac{1}{4} + \ln\frac{b}{a}\right). \end{aligned}$$

The inductance of a unit length of the coaxial transmission line is therefore

$$\boxed{L' = \frac{\Lambda'}{I} = \frac{\mu_0}{8\pi} + \frac{\mu_0}{2\pi}\ln\frac{b}{a} \qquad \text{(H/m)}.} \tag{6-140}$$

The first term $\mu_0/8\pi$ arises from the flux linkage internal to the solid inner conductor; it is known as the **internal inductance** per unit length of the inner conductor. The second term comes from the linkage of the flux that exists between the inner and the outer conductors; this term is known as the **external inductance** per unit length of the coaxial line.   ▄

In high-frequency applications the current in a good conductor tends to shift to the surface of the conductor (due to **skin effect**, as we shall see in Chapter 8), resulting in an uneven current distribution in the inner conductor and thereby changing the value of the internal inductances. In the extreme case the current may essentially concentrate in the "skin" of the inner conductor as a surface current, and the internal self-inductance is reduced to zero.

▬▬▬▬ **EXAMPLE 6-17**   Calculate the internal and external inductances per unit length of a transmission line consisting of two long parallel conducting wires of radius a that carry currents in opposite directions. The axes of the wires are separated by a distance d, which is much larger than a.

Solution   The internal self-inductance per unit length of each wire is, from Eq. (6-140), $\mu_0/8\pi$. So for two wires we have

$$L_i' = 2 \times \frac{\mu_0}{8\pi} = \frac{\mu_0}{4\pi} \qquad \text{(H/m)}. \tag{6-141}$$

[†] It is assumed that the current is distributed uniformly in the inner conductor. This assumption does not hold for high-frequency a-c currents.

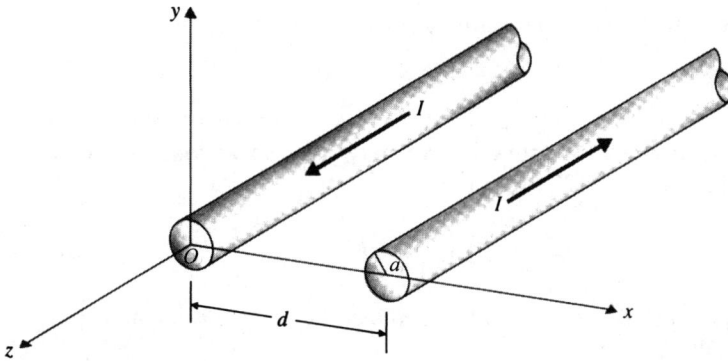

FIGURE 6-25
A two-wire transmission line (Example 6-17).

To find the external self-inductance per unit length, we first calculate the magnetic flux linking with a unit length of the transmission line for an assumed current I in the wires. In the xz-plane where the two wires lie, as in Fig. 6-25, the contributing **B** vectors due to the equal and opposite currents in the two wires have only a y-component:

$$B_{y1} = \frac{\mu_0 I}{2\pi x} \tag{6-142}$$

$$B_{y2} = \frac{\mu_0 I}{2\pi(d - x)}. \tag{6-143}$$

The flux linkage per unit length is then

$$\Phi' = \int_a^{d-a} (B_{y1} + B_{y2})\, dx$$

$$= \int_a^{d-a} \frac{\mu_0 I}{2\pi} \left[\frac{1}{x} + \frac{1}{d - x}\right] dx$$

$$= \frac{\mu_0 I}{\pi} \ln\left(\frac{d - a}{a}\right) \cong \frac{\mu_0 I}{\pi} \ln \frac{d}{a} \quad \text{(Wb/m)}.$$

Therefore,

$$L'_e = \frac{\Phi'}{I} = \frac{\mu_0}{\pi} \ln \frac{d}{a} \quad \text{(H/m)}, \tag{6-144}$$

and the total self-inductance per unit length of the two-wire line is

$$L' = L'_i + L'_e = \frac{\mu_0}{\pi}\left(\frac{1}{4} + \ln \frac{d}{a}\right) \quad \text{(H/m)}. \tag{6-145}$$

Before we present some examples showing how to determine the mutual inductance between two circuits, we pose the following question about Fig. 6-22 and Eq. (6-127): Is the flux linkage with loop C_2 caused by a unit current in loop C_1 equal to the flux linkage with C_1 caused by a unit current in C_2? That is, is it true that

$$L_{12} = L_{21}? \tag{6-146}$$

We may vaguely and intuitively expect that the answer is in the affirmative "because of reciprocity." But how do we prove it? We may proceed as follows. Combining Eqs. (6–123), (6–125) and (6–127), we obtain

$$L_{12} = \frac{N_2}{I_1} \int_{S_2} \mathbf{B}_1 \cdot d\mathbf{s}_2. \tag{6-147}$$

But in view of Eq. (6–15), $\mathbf{B}_1$ can be written as the curl of a vector magnetic potential $\mathbf{A}_1$, $\mathbf{B}_1 = \nabla \times \mathbf{A}_1$. We have

$$L_{12} = \frac{N_2}{I_1} \int_{S_2} (\nabla \times \mathbf{A}_1) \cdot d\mathbf{s}_2$$
$$= \frac{N_2}{I_1} \oint_{C_2} \mathbf{A}_1 \cdot d\boldsymbol{\ell}_2. \tag{6-148}$$

Now, from Eq. (6–27),

$$\mathbf{A}_1 = \frac{\mu_0 N_1 I_1}{4\pi} \oint_{C_1} \frac{d\boldsymbol{\ell}_1}{R}. \tag{6-149}$$

In Eqs. (6–148) and (6–149) the contour integrals are evaluated only *once* over the periphery of the *loops* C_2 and C_1, respectively—the effects of multiple turns having been taken care of separately by the factors N_2 and N_1. Substitution of Eq. (6–149) in Eq. (6–148) yields

$$L_{12} = \frac{\mu_0 N_1 N_2}{4\pi} \oint_{C_1} \oint_{C_2} \frac{d\boldsymbol{\ell}_1 \cdot d\boldsymbol{\ell}_2}{R}, \tag{6-150a}$$

where R is the distance between the differential lengths $d\boldsymbol{\ell}_1$ and $d\boldsymbol{\ell}_2$. It is customary to write Eq. (6–150a) as

$$\boxed{L_{12} = \frac{\mu_0}{4\pi} \oint_{C_1} \oint_{C_2} \frac{d\boldsymbol{\ell}_1 \cdot d\boldsymbol{\ell}_2}{R} \qquad \text{(H)},} \tag{6-150b}$$

where N_1 and N_2 have been absorbed in the contour integrals over the *circuits* C_1 and C_2 from one end to the other. Equation (6–150b) is the **Neumann formula** for mutual inductance. It is a general formula requiring the evaluation of a double line integral. For any given problem we always first look for symmetry conditions that may simplify the determination of flux linkage and mutual inductance without resorting to Eq. (6–150b) directly.

It is clear from Eq. (6–150b) that mutual inductance is a property of the geometrical shape and the physical arrangement of coupled circuits. For a *linear medium*, mutual inductance is proportional to the medium's permeability and is independent of the currents in the circuits. It is obvious that interchanging the subscripts 1 and

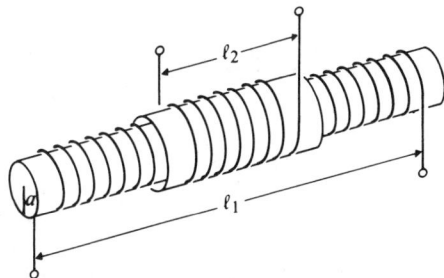

FIGURE 6-26
A solenoid with two windings (Example 6-18).

2 does not change the value of the double integral; hence an affirmative answer to the question posed in Eq. (6-146) follows. This is an important conclusion because it allows us to use the simpler of the two ways (finding L_{12} or L_{21}) to determine the mutual inductance.[†]

EXAMPLE 6-18 Two coils of N_1 and N_2 turns are wound concentrically on a straight cylindrical core of radius a and permeability μ. The windings have lengths ℓ_1 and ℓ_2, respectively. Find the mutual inductance between the coils.

Solution Figure 6-26 shows such a solenoid with two concentric windings. Assume that current I_1 flows in the inner coil. From Eq. (6-133) we find that the flux Φ_{12} in the solenoid core that links with the outer coil is

$$\Phi_{12} = \mu \left(\frac{N_1}{\ell_1} \right) (\pi a^2) I_1.$$

Since the outer coil has N_2 turns, we have

$$\Lambda_{12} = N_2 \Phi_{12} = \frac{\mu}{\ell_1} N_1 N_2 \pi a^2 I_1.$$

Hence the mutual inductance is

$$L_{12} = \frac{\Lambda_{12}}{I_1} = \frac{\mu}{\ell_1} N_1 N_2 \pi a^2 \quad \text{(H)}. \tag{6-151}$$

Leakage flux has been neglected.

EXAMPLE 6-19 Determine the mutual inductance between a conducting triangular loop and a very long straight wire as shown in Fig. 6-27.

Solution Let us designate the triangular loop as circuit 1 and the long wire as circuit 2. If we assume a current I_1 in the triangular loop, it is difficult to find the magnetic flux density $\mathbf{B}_1$ everywhere. Consequently, it is difficult to determine the mutual

---

[†] In circuit theory books the symbol M is frequently used to denote mutual inductance.

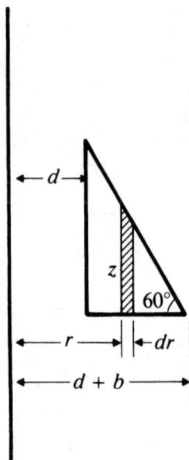

FIGURE 6–27
A conducting triangular loop and a long straight wire (Example 6–19).

inductance L_{12} from Λ_{12}/I_1 in Eq. (6–127). We can, however, apply Ampère's circuital law and readily write the expression for $\mathbf{B}_2$ that is caused by a current I_2 in the long straight wire:

$$\mathbf{B}_2 = \mathbf{a}_\phi \frac{\mu_0 I_2}{2\pi r}. \qquad (6\text{–}152)$$

The flux linkage $\Lambda_{21} = \Phi_{21}$ is

$$\Lambda_{21} = \int_{S_1} \mathbf{B}_2 \cdot d\mathbf{s}_1, \qquad (6\text{–}153)$$

where

$$d\mathbf{s}_1 = \mathbf{a}_\phi z \, dr. \qquad (6\text{–}154)$$

The relation between z and r is given by the equation of the hypotenuse of the triangle:

$$\begin{aligned} z &= -[r - (d + b)] \tan 60^\circ \\ &= -\sqrt{3}[r - (d + b)]. \end{aligned} \qquad (6\text{–}155)$$

Substituting Eqs. (6–152), (6–154), and (6–155) in Eq. (6–153), we have

$$\begin{aligned} \Lambda_{21} &= -\frac{\sqrt{3}\,\mu_0 I_2}{2\pi} \int_d^{d+b} \frac{1}{r}[r - (d + b)]\, dr \\ &= \frac{\sqrt{3}\,\mu_0 I_2}{2\pi}\left[(d + b)\ln\left(1 + \frac{b}{d}\right) - b\right]. \end{aligned}$$

Therefore, the mutual inductance is

$$L_{21} = \frac{\Lambda_{21}}{I_2} = \frac{\sqrt{3}\,\mu_0}{2\pi}\left[(d + b)\ln\left(1 + \frac{b}{d}\right) - b\right] \qquad \text{(H)}. \qquad (6\text{–}156)$$

6–12 Magnetic Energy

So far we have discussed self- and mutual inductances in static terms. Because inductances depend on the geometrical shape and the physical arrangement of the conductors constituting the circuits, and, for a linear medium, are independent of the currents, we were not concerned with nonsteady currents in the defining of inductances. However, we know that resistanceless inductors appear as short-circuits to steady (d-c) currents; it is obviously necessary that we consider alternating currents when the effects of inductances on circuits and magnetic fields are of interest. A general consideration of time-varying electromagnetic fields (electrodynamics) will be deferred until the next chapter. For now we assume *quasi-static conditions*, which imply that the currents vary very slowly in time (are low of frequency) and that the dimensions of the circuits are very small in comparison to the wavelength. These conditions are tantamount to ignoring retardation and radiation effects, as we shall see when electromagnetic waves are discussed in Chapter 8.

In Section 3–11 we discussed the fact that work is required to assemble a group of charges and that the work is stored as electric energy. We certainly expect that work also needs to be expended in sending currents into conducting loops and that it will be stored as magnetic energy. Consider a single closed loop with a self-inductance L_1 in which the current is initially zero. A current generator is connected to the loop, which increases the current i_1 from zero to I_1. From physics we know that an electromotive force (emf) will be induced in the loop that opposes the current change.[†] An amount of work must be done to overcome this induced emf. Let $v_1 = L_1 \, di_1/dt$ be the voltage across the inductance. The work required is

$$W_1 = \int v_1 i_1 \, dt = L_1 \int_0^{I_1} i_1 \, di_1 = \tfrac{1}{2} L_1 I_1^2. \tag{6–157}$$

Since $L_1 = \Phi_1/I_1$ for *linear* media, Eq. (6–157) can be written alternatively in terms of flux linkage as

$$W_1 = \tfrac{1}{2} I_1 \Phi_1, \tag{6–158}$$

which is stored as *magnetic energy*.

Now consider two closed loops C_1 and C_2 carrying currents i_1 and i_2, respectively. The currents are initially zero and are to be increased to I_1 and I_2, respectively. To find the amount of work required, we first keep $i_2 = 0$ and increase i_1 from zero to I_1. This requires a work W_1 in loop C_1, as given in Eq. (6–157) or (6–158); no work is done in loop C_2, since $i_2 = 0$. Next we keep i_1 at I_1 and increase i_2 from zero to I_2. Because of mutual coupling, some of the magnetic flux due to i_2 will link with loop C_1, giving rise to an induced emf that must be overcome by a voltage $v_{21} = L_{21} \, di_2/dt$ in order to keep i_1 constant at its value I_1. The work involved is

$$W_{21} = \int v_{21} I_1 \, dt = L_{21} I_1 \int_0^{I_2} di_2 = L_{21} I_1 I_2. \tag{6–159}$$

[†] The subject of electromagnetic induction will be discussed in Chapter 7.

At the same time a work W_{22} must be done in loop C_2 in order to counteract the induced emf and increase i_2 to I_2.

$$W_{22} = \tfrac{1}{2}L_2I_2^2. \tag{6-160}$$

The total amount of work done in raising the currents in loops C_1 and C_2 from zero to I_1 and I_2, respectively, is then the sum of W_1, W_{21}, and W_{22}:

$$\begin{aligned} W_2 &= \tfrac{1}{2}L_1I_1^2 + L_{21}I_1I_2 + \tfrac{1}{2}L_2I_2^2 \\ &= \frac{1}{2}\sum_{j=1}^{2}\sum_{k=1}^{2} L_{jk}I_jI_k. \end{aligned} \tag{6-161}$$

Generalizing this result to a system of N loops carrying currents $I_1, I_2, \ldots, I_n$, we obtain

$$\boxed{W_m = \frac{1}{2}\sum_{j=1}^{N}\sum_{k=1}^{N} L_{jk}I_jI_k \quad \text{(J)},} \tag{6-162}$$

which is the energy stored in the magnetic field. For a current I flowing in a single inductor with inductance L, the stored magnetic energy is

$$\boxed{W_m = \tfrac{1}{2}LI^2 \quad \text{(J)}.} \tag{6-163}$$

It is instructive to derive Eq. (6–162) in an alternative way. Consider a typical kth loop of N magnetically coupled loops. Let v_k and i_k be the voltage across and the current in the loop, respectively. The work done to the kth loop in time dt is

$$dW_k = v_k i_k \, dt = i_k \, d\phi_k, \tag{6-164}$$

where we have used the relation $v_k = d\phi_k/dt$. Note that the change, $d\phi_k$, in the flux ϕ_k linking with the kth loop is the result of the changes of the currents in all the coupled loops. The differential work done to, or the differential magnetic energy stored in, the system is then

$$dW_m = \sum_{k=1}^{N} dW_k = \sum_{k=1}^{N} i_k \, d\phi_k. \tag{6-165}$$

The total stored energy is the integration of dW_m and is independent of the manner in which the final values of the currents and fluxes are reached. Let us assume that all the currents and fluxes are brought to their final values in concert by an equal fraction α that increases from 0 to 1; that is, $i_k = \alpha I_k$, and $\phi_k = \alpha \Phi_k$ at any instant of time. We obtain the total *magnetic energy*:

$$W_m = \int dW_m = \sum_{k=1}^{N} I_k \Phi_k \int_0^1 \alpha \, d\alpha$$

or

$$W_m = \frac{1}{2} \sum_{k=1}^{N} I_k \Phi_k \qquad \text{(J)}, \qquad\qquad (6\text{--}166)$$

which simplifies to Eq. (6–158) for $N = 1$, as expected. Noting that, for *linear* media,

$$\Phi_k = \sum_{j=1}^{N} L_{jk} I_j,$$

we obtain Eq. (6–162) immediately.

6–12.1   MAGNETIC ENERGY IN TERMS OF FIELD QUANTITIES

Equation (6–166) can be generalized to determine the magnetic energy of a continuous distribution of current within a volume. A single current-carrying loop can be considered as consisting of a large number, N, of contiguous filamentary current elements of closed paths C_k, each with a current ΔI_k flowing in an infinitesimal cross-sectional area $\Delta a_k'$ and linking with magnetic flux Φ_k.

$$\Phi_k = \int_{S_k} \mathbf{B} \cdot \mathbf{a}_n \, ds_k' = \oint_{C_k} \mathbf{A} \cdot d\boldsymbol{\ell}_k', \qquad\qquad (6\text{--}167)$$

where S_k is the surface bounded by C_k. Substituting Eq. (6–167) in Eq. (6–166), we have

$$W_m = \frac{1}{2} \sum_{k=1}^{N} \Delta I_k \oint_{C_k} \mathbf{A} \cdot d\boldsymbol{\ell}_k'. \qquad\qquad (6\text{--}168)$$

Now,

$$\Delta I_k \, d\boldsymbol{\ell}_k' = J(\Delta a_k') \, d\boldsymbol{\ell}_k' = \mathbf{J} \, \Delta v_k'.$$

As $N \to \infty$, $\Delta v_k'$ becomes dv', and the summation in Eq. (6–168) can be written as an integral. We have

$$W_m = \frac{1}{2} \int_{V'} \mathbf{A} \cdot \mathbf{J} \, dv' \qquad \text{(J)}, \qquad\qquad (6\text{--}169)$$

where V' is the volume of the loop or the *linear medium* in which $\mathbf{J}$ exists. This volume can be extended to include all space, since the inclusion of a region where $\mathbf{J} = 0$ does not change W_m. Equation (6–169) should be compared with the expression for the electric energy W_e in Eq. (3–170).

It is often desirable to express the magnetic energy in terms of field quantities $\mathbf{B}$ and $\mathbf{H}$ instead of current density $\mathbf{J}$ and vector potential $\mathbf{A}$. Making use of the vector identity,

$$\nabla \cdot (\mathbf{A} \times \mathbf{H}) = \mathbf{H} \cdot (\nabla \times \mathbf{A}) - \mathbf{A} \cdot (\nabla \times \mathbf{H}),$$

(see Problem P.2–33 or the list at the end of book), we have

$$\mathbf{A} \cdot (\nabla \times \mathbf{H}) = \mathbf{H} \cdot (\nabla \times \mathbf{A}) - \nabla \cdot (\mathbf{A} \times \mathbf{H})$$

or

$$A \cdot J = H \cdot B - \nabla \cdot (A \times H). \tag{6-170}$$

Substituting Eq. (6–170) in Eq. (6–169), we obtain

$$W_m = \tfrac{1}{2} \int_{V'} H \cdot B \, dv' - \tfrac{1}{2} \oint_{S'} (A \times H) \cdot a_n \, ds'. \tag{6-171}$$

In Eq. (6–171) we have applied the divergence theorem, and S' is the surface bounding V'. If V' is taken to be sufficiently large, the points on its surface S' will be very far from the currents. At those far-away points the contribution of the surface integral in Eq. (6–171) tends to zero because $|A|$ falls off as $1/R$ and $|H|$ falls off as $1/R^2$, as can be seen from Eqs. (6–23) and (6–32), respectively. Thus, the magnitude of $(A \times H)$ decreases as $1/R^3$, whereas at the same time the surface S' increases only as R^2. When R approaches infinity, the surface integral in Eq. (6–171) vanishes. We have then

$$\boxed{W_m = \tfrac{1}{2} \int_{V'} H \cdot B \, dv' \quad \text{(J)}.} \tag{6-172a}$$

Noting that $H = B/\mu$, we can write Eq. (6–172a) in two alternative forms:

$$\boxed{W_m = \frac{1}{2} \int_{V'} \frac{B^2}{\mu} \, dv' \quad \text{(J)}} \tag{6-172b}$$

and

$$\boxed{W_m = \tfrac{1}{2} \int_{V'} \mu H^2 \, dv' \quad \text{(J)}.} \tag{6-172c}$$

The expressions in Eqs. (6–172a), (6–172b), and (6–172c) for the magnetic energy W_m in a linear medium are analogous to those of electrostatic energy W_e in Eqs. (3–176a), (3–176b), and (3–176c), respectively.

If we define a *magnetic energy density*, w_m, such that its volume integral equals the total magnetic energy

$$W_m = \int_{V'} w_m \, dv', \tag{6-173}$$

we can write w_m in three different forms:

$$w_m = \tfrac{1}{2} H \cdot B \quad \text{(J/m}^3\text{)} \tag{6-174a}$$

or

$$w_m = \frac{B^2}{2\mu} \quad \text{(J/m}^3\text{)} \tag{6-174b}$$

or

$$w_m = \tfrac{1}{2}\mu H^2 \quad \text{(J/m}^3\text{)}. \tag{6-174c}$$

By using Eq. (6–163) we can often determine self-inductance more easily from stored magnetic energy calculated in terms of B and/or H, than from flux linkage.

We have

$$L = \frac{2W_m}{I^2} \quad \text{(H)}. \tag{6-175}$$

EXAMPLE 6–20 By using stored magnetic energy, determine the inductance per unit length of an air coaxial transmission line that has a solid inner conductor of radius a and a very thin outer conductor of inner radius b.

Solution This is the same problem as that in Example 6–16, in which the self-inductance was determined through a consideration of flux linkages. Refer again to Fig. 6–24. Assume that a uniform current I flows in the inner conductor and returns in the outer conductor. The magnetic energy per unit length stored in the inner conductor is, from Eqs. (6–136) and (6–172b),

$$\begin{aligned}
W'_{m1} &= \frac{1}{2\mu_0} \int_0^a B_{\phi 1}^2 2\pi r \, dr \\
&= \frac{\mu_0 I^2}{4\pi a^4} \int_0^a r^3 \, dr = \frac{\mu_0 I^2}{16\pi} \quad \text{(J/m)}.
\end{aligned} \tag{6-176a}$$

The magnetic energy per unit length stored in the region between the inner and outer conductors is, from Eq. (6–137) and (6–172b),

$$\begin{aligned}
W'_{m2} &= \frac{1}{2\mu_0} \int_a^b B_{\phi 2}^2 2\pi r \, dr \\
&= \frac{\mu_0 I^2}{4\pi} \int_a^b \frac{1}{r} \, dr = \frac{\mu_0 I^2}{4\pi} \ln \frac{b}{a} \quad \text{(J/m)}.
\end{aligned} \tag{6-176b}$$

Therefore, from Eq. (6–175) we have

$$\begin{aligned}
L' &= \frac{2}{I^2} (W'_{m1} + W'_{m2}) \\
&= \frac{\mu_0}{8\pi} + \frac{\mu_0}{2\pi} \ln \frac{b}{a} \quad \text{(H/m)},
\end{aligned}$$

which is the same as Eq. (6–140). The procedure used in this solution is comparatively simpler than that used in Example 6–16, especially the part leading to the internal inductance $\mu_0/8\pi$.

6–13 Magnetic Forces and Torques

In Section 6–1 we noted that a charge q moving with a velocity $\mathbf{u}$ in a magnetic field with flux density $\mathbf{B}$ experiences a magnetic force $\mathbf{F}_m$ given by Eq. (6–4), which is repeated below:

$$\mathbf{F}_m = q\mathbf{u} \times \mathbf{B} \quad \text{(N)}. \tag{6-177}$$

In this section we will discuss various aspects of forces and torques in static magnetic fields.

6–13.1   HALL EFFECT

Consider a conducting material of a $d \times b$ rectangular cross section in a uniform magnetic field $\mathbf{B} = \mathbf{a}_z B_0$, as shown in Fig. 6–28. A uniform direct current flows in the y-direction:

$$\mathbf{J} = \mathbf{a}_y J_0 = Nq\mathbf{u}, \tag{6-178}$$

where N is the number of charge carriers per unit volume, moving with a velocity $\mathbf{u}$, and q is the charge on each charge carrier. Because of Eq. (6–177), the charge carriers experience a force perpendicular to both $\mathbf{B}$ and $\mathbf{u}$. If the material is a conductor or an n-type semiconductor, the charge carriers are electrons, and q is negative. The magnetic force tends to move the electrons in the positive x-direction, creating a transverse electric field. This will continue until the transverse field is sufficient to stop the drift of the charge carriers. In the steady state the net force on the charge carriers is zero:

$$\mathbf{E}_h + \mathbf{u} \times \mathbf{B} = 0 \tag{6-179a}$$

or

$$\mathbf{E}_h = -\mathbf{u} \times \mathbf{B}. \tag{6-179b}$$

This is known as the *Hall effect*, and $\mathbf{E}_h$ is called the *Hall field*. For conductors and n-type semiconductors and a positive J_0, $\mathbf{u} = -\mathbf{a}_y u_0$, and

$$\begin{aligned} \mathbf{E}_h &= -(-\mathbf{a}_y u_0) \times \mathbf{a}_z B_0 \\ &= \mathbf{a}_x u_0 B_0. \end{aligned} \tag{6-180}$$

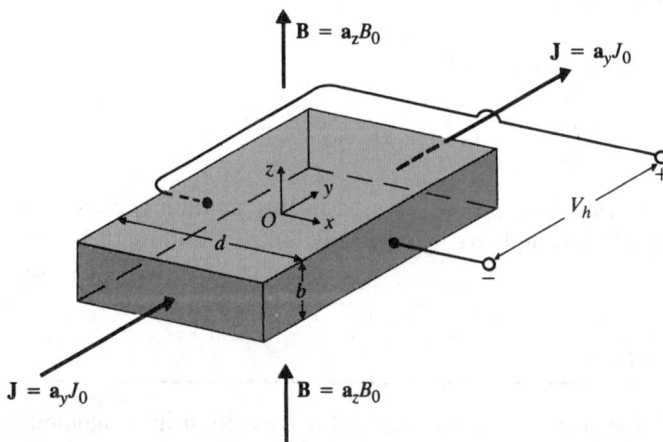

FIGURE 6–28
Illustrating the Hall effect.

A transverse potential appears across the sides of the material. Thus, we have

$$V_h = - \int_0^d E_h \, dx = u_0 B_0 d, \tag{6–181}$$

for electron carriers. In Eq. (6–181), V_h is called the **Hall voltage**. The ratio $E_x/J_y B_z = 1/Nq$ is a characteristic of the material and is known as the **Hall coefficient**.

If the charge carriers are holes, such as in a p-type semiconductor, the Hall field will be reversed, and the Hall voltage in Eq. (6–181) will be negative with the reference polarities shown in Fig. 6–28.

The Hall effect can be used for measuring the magnetic field and determining the sign of the predominant charge carriers (distinguishing an n-type from a p-type semiconductor). We have given here a simplified version of the Hall effect. In actuality it is a complex affair involving quantum theory concepts.

6–13.2 FORCES AND TORQUES ON CURRENT-CARRYING CONDUCTORS

Let us consider an element of conductor $d\ell$ with a cross-sectional area S. If there are N charge carriers (electrons) per unit volume moving with a velocity $\mathbf{u}$ in the direction of $d\ell$, then the magnetic force on the differential element is

$$\begin{aligned} d\mathbf{F}_m &= -NeS|d\ell|\mathbf{u} \times \mathbf{B} \\ &= -NeS|\mathbf{u}|\, d\ell \times \mathbf{B}, \end{aligned} \tag{6–182}$$

where e is the electronic charge. The two expressions in Eq. (6–182) are equivalent, since $\mathbf{u}$ and $d\ell$ have the same direction. Now, since $-NeS|\mathbf{u}|$ equals the current in the conductor, we can write Eq. (6–182) as

$$\boxed{d\mathbf{F}_m = I \, d\ell \times \mathbf{B} \quad \text{(N).}} \tag{6–183}$$

The magnetic force on a complete (closed) circuit of contour C that carries a current I in a magnetic field $\mathbf{B}$ is then

$$\boxed{\mathbf{F}_m = I \oint_C d\ell \times \mathbf{B} \quad \text{(N).}} \tag{6–184}$$

When we have two circuits carrying currents I_1 and I_2, respectively, the situation is that of one current-carrying circuit in the magnetic field of the other. In the presence of the magnetic field $\mathbf{B}_{21}$, which was caused by the current I_2 in C_2, the force $\mathbf{F}_{21}$ on circuit C_1 can be written as

$$\mathbf{F}_{21} = I_1 \oint_{C_1} d\ell_1 \times \mathbf{B}_{21}, \tag{6–185a}$$

where $\mathbf{B}_{21}$ is, from the Biot-Savart law in Eq. (6–32),

$$\mathbf{B}_{21} = \frac{\mu_0 I_2}{4\pi} \oint_{C_2} \frac{d\ell_2 \times \mathbf{a}_{R_{21}}}{R_{21}^2}. \tag{6–185b}$$

Combining Eqs. (6–185a) and (6–185b), we obtain

$$
\boxed{\mathbf{F}_{21} = \frac{\mu_0}{4\pi} I_1 I_2 \oint_{C_1} \oint_{C_2} \frac{d\boldsymbol{\ell}_1 \times (d\boldsymbol{\ell}_2 \times \mathbf{a}_{R_{21}})}{R_{21}^2}} \quad (\text{N}),
\qquad (6\text{–}186\text{a})
$$

which is **Ampere's law of force** between two current-carrying circuits. It is an inverse-square relationship and should be compared with Coulomb's law of force in Eq. (3–17) between two stationary charges, the latter being much the simpler.

The force $\mathbf{F}_{12}$ on circuit C_2, in the presence of the magnetic field set up by the current I_1 in circuit C_1, can be written from Eq. (6–186a) by interchanging the subscripts 1 and 2:

$$
\mathbf{F}_{12} = \frac{\mu_0}{4\pi} I_2 I_1 \oint_{C_2} \oint_{C_1} \frac{d\boldsymbol{\ell}_2 \times (d\boldsymbol{\ell}_1 \times \mathbf{a}_{R_{12}})}{R_{12}^2}.
\qquad (6\text{–}186\text{b})
$$

However, since $d\boldsymbol{\ell}_2 \times (d\boldsymbol{\ell}_1 \times \mathbf{a}_{R_{12}}) \neq -d\boldsymbol{\ell}_1 \times (d\boldsymbol{\ell}_2 \times \mathbf{a}_{R_{21}})$, we inquire whether this means $\mathbf{F}_{21} \neq -\mathbf{F}_{12}$—that is, whether Newton's third law governing the forces of action and reaction fails here. Let us expand the vector triple product in the integrand of Eq. (6–186a) by the back-cab rule, Eq. (2–20):

$$
\frac{d\boldsymbol{\ell}_1 \times (d\boldsymbol{\ell}_2 \times \mathbf{a}_{R_{21}})}{R_{21}^2} = \frac{d\boldsymbol{\ell}_2 (d\boldsymbol{\ell}_1 \cdot \mathbf{a}_{R_{21}})}{R_{21}^2} - \frac{\mathbf{a}_{R_{21}}(d\boldsymbol{\ell}_1 \cdot d\boldsymbol{\ell}_2)}{R_{21}^2}.
\qquad (6\text{–}187)
$$

Now the double closed line integral of the first term on the right side of Eq. (6–187) is

$$
\oint_{C_1} \oint_{C_2} \frac{d\boldsymbol{\ell}_2(d\boldsymbol{\ell}_1 \cdot \mathbf{a}_{R_{21}})}{R_{21}^2} = \oint_{C_2} d\boldsymbol{\ell}_2 \oint_{C_1} \frac{d\boldsymbol{\ell}_1 \cdot \mathbf{a}_{R_{21}}}{R_{21}^2}
$$

$$
= \oint_{C_2} d\boldsymbol{\ell}_2 \oint_{C_1} d\boldsymbol{\ell}_1 \cdot \left(-\nabla_1 \frac{1}{R_{21}}\right)
\qquad (6\text{–}188)
$$

$$
= -\oint_{C_2} d\boldsymbol{\ell}_2 \oint_{C_1} d\left(\frac{1}{R_{21}}\right) = 0.
$$

In Eq. (6–188) we have made use of Eq. (2–88) and the relation $\nabla_1(1/R_{21}) = -\mathbf{a}_{R_{21}}/R_{21}^2$. The closed line integral (with identical upper and lower limits) of $d(1/R_{21})$ around circuit C_1 vanishes. Substituting Eq. (6–187) in Eq. (6–186a) and using Eq. (6–188), we get

$$
\mathbf{F}_{21} = -\frac{\mu_0}{4\pi} I_1 I_2 \oint_{C_1} \oint_{C_2} \frac{\mathbf{a}_{R_{21}}(d\boldsymbol{\ell}_1 \cdot d\boldsymbol{\ell}_2)}{R_{21}^2},
\qquad (6\text{–}189)
$$

which obviously equals $-\mathbf{F}_{12}$, inasmuch as $\mathbf{a}_{R_{12}} = -\mathbf{a}_{R_{21}}$. It follows that Newton's third law holds here, as expected.

━━━ **EXAMPLE 6–21**   Determine the force per unit length between two infinitely long parallel conducting wires carrying currents I_1 and I_2 in the same direction. The wires are separated by a distance d.

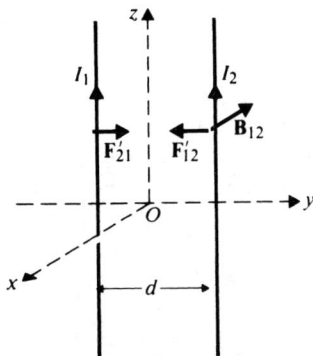

FIGURE 6–29
Force between two parallel current-carrying wires (Example 6–21).

Solution Let the wires lie in the yz-plane and be parallel to the z-axis, as shown in Fig. 6–29. This problem is a straightforward application of Eq. (6–185a). Using $\mathbf{F}'_{12}$ to denote the force per unit length on wire 2, we have

$$\mathbf{F}'_{12} = I_2(\mathbf{a}_z \times \mathbf{B}_{12}), \qquad (6\text{–}190)$$

where $\mathbf{B}_{12}$, the magnetic flux density at wire 2, set up by the current I_1 in wire 1, is constant over wire 2. Because the wires are assumed to be infinitely long and cylindrical symmetry exists, it is not necessary to use Eq. (6–185b) for the determination of $\mathbf{B}_{12}$. We apply Ampère's circuital law and write, from Eq. (6–11b),

$$\mathbf{B}_{12} = -\mathbf{a}_x \frac{\mu_0 I_1}{2\pi d}. \qquad (6\text{–}191)$$

Substitution of Eq. (6–191) in Eq. (6–190) yields

$$\mathbf{F}'_{12} = -\mathbf{a}_y \frac{\mu_0 I_1 I_2}{2\pi d} \qquad \text{(N/m)}. \qquad (6\text{–}192)$$

We see that the force on wire 2 pulls it toward wire 1. Hence the force between two wires carrying *currents in the same direction* is one of *attraction* (unlike the force between two charges of the same polarity, which is one of repulsion). It is trivial to prove that $\mathbf{F}'_{21} = -\mathbf{F}'_{12} = \mathbf{a}_y(\mu_0 I_1 I_2/2\pi d)$ and that the force between two wires carrying *currents in opposite directions* is one of *repulsion*. ∎

Let us now consider a small circular loop of radius b and carrying a current I in a uniform magnetic field of flux density $\mathbf{B}$. It is convenient to resolve $\mathbf{B}$ into two components, $\mathbf{B} = \mathbf{B}_\perp + \mathbf{B}_{\parallel}$, where $\mathbf{B}_\perp$ and $\mathbf{B}_{\parallel}$ are perpendicular and parallel, respectively, to the plane of the loop. As illustrated in Fig. 6–30(a), the perpendicular component $\mathbf{B}_\perp$ tends to expand the loop (or contract it if the direction of I is reversed), but $\mathbf{B}_\perp$ exerts no net force to move the loop. The parallel component $\mathbf{B}_{\parallel}$ produces an upward force $d\mathbf{F}_1$ (out from the paper) on element $d\ell_1$ and a downward force (into the paper) $d\mathbf{F}_2 = -d\mathbf{F}_1$ on the symmetrically located element $d\ell_2$, as shown in

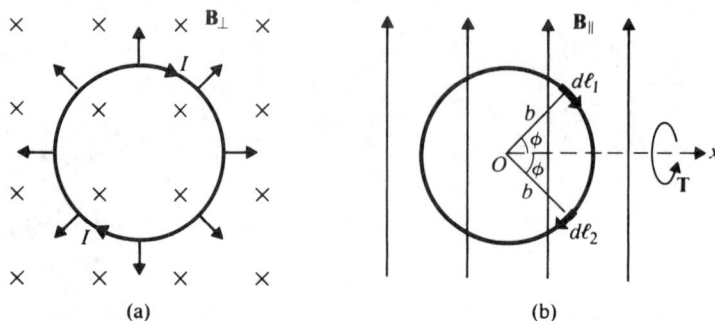

FIGURE 6–30
A circular loop in a uniform magnetic field $\mathbf{B} = \mathbf{B}_\perp + \mathbf{B}_\parallel$.

Fig. 6–30(b). Although the net force on the entire loop caused by $\mathbf{B}_\parallel$ is also zero, a torque exists that tends to rotate the loop about the x-axis in such a way as to *align* the magnetic field (due to I) with the external $\mathbf{B}_\parallel$ field. The differential torque produced by $d\mathbf{F}_1$ and $d\mathbf{F}_2$ is

$$
\begin{aligned}
d\mathbf{T} &= \mathbf{a}_x(dF)2b \sin \phi \\
&= \mathbf{a}_x(I \, d\ell \, B_\parallel \sin \phi)2b \sin \phi \qquad\qquad (6\text{--}193)\\
&= \mathbf{a}_x 2Ib^2 B_\parallel \sin^2 \phi \, d\phi,
\end{aligned}
$$

where $dF = |d\mathbf{F}_1| = |d\mathbf{F}_2|$ and $d\ell = |d\ell_1| = |d\ell_2| = b \, d\phi$. The total torque acting on the loop is then

$$
\begin{aligned}
\mathbf{T} = \int d\mathbf{T} &= \mathbf{a}_x 2Ib^2 B_\parallel \int_0^\pi \sin^2 \phi \, d\phi \\
&= \mathbf{a}_x I(\pi b^2) B_\parallel.
\end{aligned} \qquad\qquad (6\text{--}194)
$$

If the definition of the magnetic dipole moment in Eq. (6–46) is used,

$$
\mathbf{m} = \mathbf{a}_n I(\pi b^2) = \mathbf{a}_n IS,
$$

where $\mathbf{a}_n$ is a unit vector in the direction of the right thumb (normal to the plane of the loop) as the fingers of the right hand follow the direction of the current, we can write Eq. (6–194) as

$$
\boxed{\mathbf{T} = \mathbf{m} \times \mathbf{B} \qquad (\text{N·m}).} \qquad\qquad (6\text{--}195)
$$

The vector $\mathbf{B}$ (instead of $\mathbf{B}_\parallel$) is used in Eq. (6–195) because $\mathbf{m} \times (\mathbf{B}_\perp + \mathbf{B}_\parallel) = \mathbf{m} \times \mathbf{B}_\parallel$. This is the torque that aligns the microscopic magnetic dipoles in magnetic materials and causes the materials to be magnetized by an applied magnetic field. It should be remembered that Eq. (6–195) does not hold if $\mathbf{B}$ is not uniform over the current-carrying loop.

▬▬▬▬ **EXAMPLE 6–22**   A rectangular loop in the xy-plane with sides b_1 and b_2 carrying a current I lies in a *uniform* magnetic field $\mathbf{B} = \mathbf{a}_xB_x + \mathbf{a}_yB_y + \mathbf{a}_zB_z$. Determine the force and torque on the loop.

Solution   Resolving $\mathbf{B}$ into perpendicular and parallel components $\mathbf{B}_\perp$ and $\mathbf{B}_{\parallel}$, we have

$$\mathbf{B}_\perp = \mathbf{a}_zB_z \tag{6–196a}$$
$$\mathbf{B}_{\parallel} = \mathbf{a}_xB_x + \mathbf{a}_yB_y. \tag{6–196b}$$

Assuming that the current flows in a clockwise direction, as shown in Fig. 6–31, we find that the perpendicular component $\mathbf{a}_zB_z$ results in forces Ib_1B_z on sides (1) and (3) and forces Ib_2B_z on sides (2) and (4), all directed toward the center of the loop. The vector sum of these four contracting forces is zero, and no torque is produced.

The parallel component of the magnetic flux density, $\mathbf{B}_{\parallel}$, produces the following forces on the four sides:

$$\mathbf{F}_1 = Ib_1\mathbf{a}_x \times (\mathbf{a}_xB_x + \mathbf{a}_yB_y)$$
$$= \mathbf{a}_zIb_1B_y = -\mathbf{F}_3; \tag{6–197a}$$
$$\mathbf{F}_2 = Ib_2(-\mathbf{a}_y) \times (\mathbf{a}_xB_x + \mathbf{a}_yB_y)$$
$$= \mathbf{a}_zIb_2B_x = -\mathbf{F}_4. \tag{6–197b}$$

Again, the net force on the loop, $\mathbf{F}_1 + \mathbf{F}_2 + \mathbf{F}_3 + \mathbf{F}_4$, is zero. However, these forces result in a net torque that can be computed as follows. The torque $\mathbf{T}_{13}$, due to forces $\mathbf{F}_1$ and $\mathbf{F}_3$ on sides (1) and (3), is

$$\mathbf{T}_{13} = \mathbf{a}_xIb_1b_2B_y; \tag{6–198a}$$

the torque $\mathbf{T}_{24}$, due to forces $\mathbf{F}_2$ and $\mathbf{F}_4$ on sides (2) and (4), is

$$\mathbf{T}_{24} = -\mathbf{a}_yIb_1b_2B_x. \tag{6–198b}$$

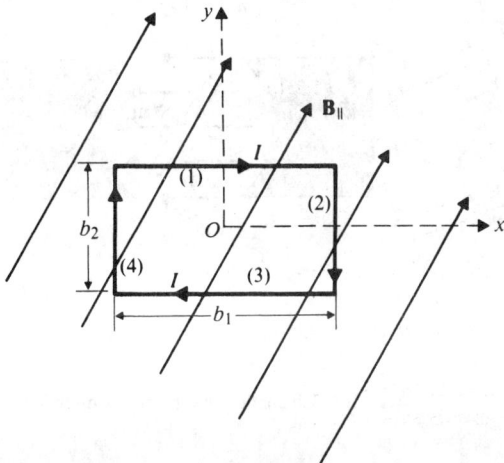

FIGURE 6–31
A rectangular loop in a uniform magnetic field (Example 6–22).

The total torque on the rectangular loop is then

$$\mathbf{T} = \mathbf{T}_{13} + \mathbf{T}_{24} = Ib_1b_2(\mathbf{a}_xB_y - \mathbf{a}_yB_x) \quad (\text{N·m}).$$ (6–199)

Since the magnetic moment of the loop is $\mathbf{m} = -\mathbf{a}_zIb_1b_2$, the result in Eq. (6–199) is exactly $\mathbf{T} = \mathbf{m} \times (\mathbf{a}_xB_x + \mathbf{a}_yB_y) = \mathbf{m} \times \mathbf{B}$. Hence in spite of the fact that Eq. (6–195) was derived for a circular loop, the torque formula holds also for a rectangular loop. As a matter of fact, it can be proved that Eq. (6–195) holds for a planar loop of any shape as long as it is located in a uniform magnetic field. Can you suggest a proof for the last statement?

The principle of operation of direct-current (d-c) motor is based on Eq. (6–195). Figure 6–32(a) shows a schematic diagram of such a motor. The magnetic field $\mathbf{B}$ is produced by a field current I_f in a winding around the pole pieces. When a current I is sent through the rectangular loop, a torque results that makes the loop rotate in a clockwise direction as viewed from the $+x$-direction. This is illustrated in Fig. 6–32(b). A split ring with brushes is necessary so that the currents in the two legs of the coil reverse their directions every half of a turn in order to maintain the torque $\mathbf{T}$ always in the same direction; the magnetic moment $\mathbf{m}$ of the loop must have a positive z-component.

To obtain a smooth operation, an actual d-c motor has many such rectangular loops wound and distributed around an armature. The ends of each loop are attached to a pair of conducting bars arranged on a small cylindrical drum called a *commutator*.

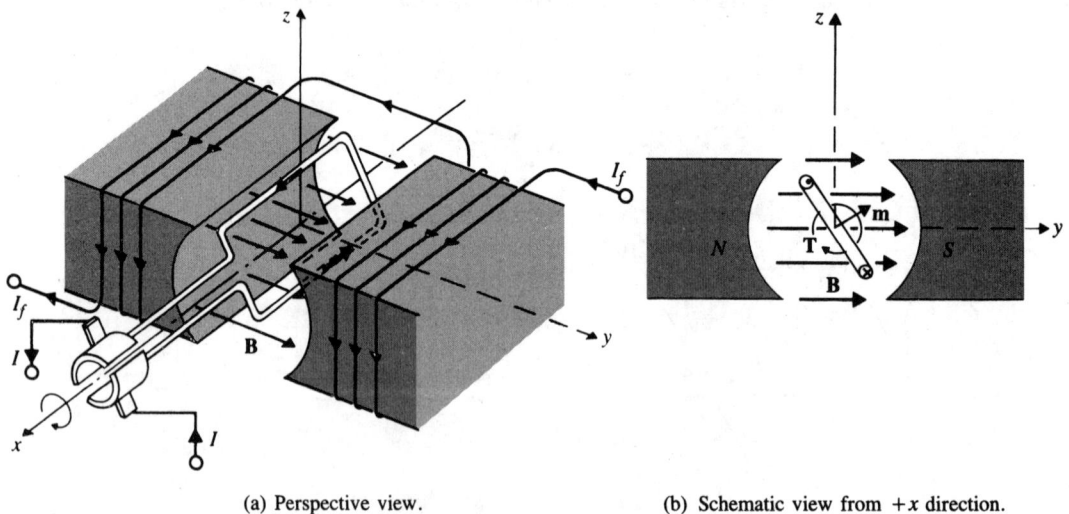

(a) Perspective view.   (b) Schematic view from $+x$ direction.

FIGURE 6–32
Illustrating the principle of operation of d-c motor.

The commutator has twice as many parallel conducting bars insulated from one another as there are loops.

6–13.3   FORCES AND TORQUES IN TERMS OF STORED MAGNETIC ENERGY

All current-carrying conductors and circuits experience magnetic forces when situated in a magnetic field. They are held in place only if mechanical forces, equal and opposite to the magnetic forces, exist. Except for special symmetrical cases (such as the case of the two infinitely long, current-carrying, parallel conducting wires in Example 6–21), determining the magnetic forces between current-carrying circuits by Ampère's law of force is a tedious task. We now examine an alternative method of finding magnetic forces and torques based on the **principle of virtual displacement**. This principle was used in Section 3–11.2 to determine electrostatic forces between charged conductors. We consider two cases: first, a system of circuits with constant magnetic flux linkages, and second, a system of circuits with constant currents.

System of Circuits with Constant Flux Linkages   If we assume that no changes in flux linkages result from a virtual differential displacement $d\ell$ of one of the current-carrying circuits, there will be no induced emf's, and the sources will supply no energy to the system. The mechanical work, $\mathbf{F}_\Phi \cdot d\ell$, done *by the system* is at the expense of a decrease in the stored magnetic energy, where $\mathbf{F}_\Phi$ denotes the force under the constant-flux condition. Thus,

$$\mathbf{F}_\Phi \cdot d\ell = -dW_m = -(\nabla W_m) \cdot d\ell, \tag{6–200}$$

from which we obtain

$$\boxed{\mathbf{F}_\Phi = -\nabla W_m \quad \text{(N)}.} \tag{6–201}$$

In Cartesian coordinates the component forces are

$$(F_\Phi)_x = -\frac{\partial W_m}{\partial x}, \tag{6–202a}$$

$$(F_\Phi)_y = -\frac{\partial W_m}{\partial y}, \tag{6–202b}$$

$$(F_\Phi)_z = -\frac{\partial W_m}{\partial z}. \tag{6–202c}$$

If the circuit is constrained to rotate about an axis, say the z-axis, the mechanical work done by the system will be $(T_\Phi)_z \, d\phi$, and

$$\boxed{(T_\Phi)_z = -\frac{\partial W_m}{\partial \phi} \quad \text{(N·m)},} \tag{6–203}$$

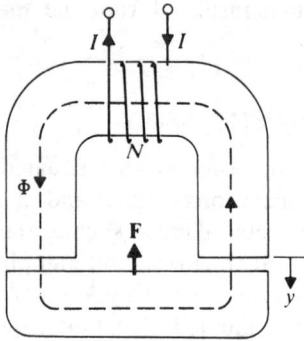

FIGURE 6–33
An electromagnet (Example 6–23).

which is the z-component of the torque acting on the circuit under the condition of constant flux linkages.

EXAMPLE 6–23   Consider the electromagnet in Fig. 6–33, in which a current I in an N-turn coil produces a flux Φ in the magnetic circuit. The cross-sectional area of the core is S. Determine the lifting force on the armature.

Solution   Let the armature take a virtual displacement dy (a differential increase in y) and the source be adjusted to keep the flux Φ constant. A displacement of the armature changes only the length of the air gaps; consequently, the displacement changes only the magnetic energy stored in the two air gaps. We have, from Eq. (6–172b),

$$dW_m = d(W_m)_{\text{air} \atop \text{gap}} = 2\left(\frac{B^2}{2\mu_0} S\,dy\right)$$
$$= \frac{\Phi^2}{\mu_0 S}\,dy. \tag{6–204}$$

An increase in the air-gap length (a positive dy) increases the stored magnetic energy if Φ is constant. Using Eq. (6–202b), we obtain

$$\mathbf{F}_\Phi = \mathbf{a}_y (F_\Phi)_y = -\mathbf{a}_y \frac{\Phi^2}{\mu_0 S} \qquad \text{(N)}. \tag{6–205}$$

Here the negative sign indicates that the force tends to reduce the air-gap length; that is, it is a force of attraction.   ▬

System of Circuits with Constant Currents   In this case the circuits are connected to current sources that counteract the induced emf's resulting from changes in flux linkages that are caused by a virtual displacement $d\ell$. The work done or energy supplied by the sources is (see Eq. (6–165))

$$dW_s = \sum_k I_k \, d\Phi_k. \tag{6–206}$$

This energy must be equal to the sum of the mechanical work done by the system dW ($dW = \mathbf{F}_I \cdot d\boldsymbol{\ell}$, where $\mathbf{F}_I$ denotes the force on the displaced circuit under the constant-current condition) and the increase in the stored magnetic energy, dW_m. That is,

$$dW_s = dW + dW_m. \tag{6–207}$$

From Eq. (6–166) we have

$$dW_m = \frac{1}{2} \sum_k I_k \, d\Phi_k = \frac{1}{2} dW_s. \tag{6–208}$$

Equations (6–207) and (6–208) combine to give

$$dW = \mathbf{F}_I \cdot d\boldsymbol{\ell} = dW_m$$
$$= (\nabla W_m) \cdot d\boldsymbol{\ell}$$

or

$$\boxed{\mathbf{F}_I = \nabla W_m \qquad (\text{N}),} \tag{6–209}$$

which differs from the expression for $\mathbf{F}_\Phi$ in Eq. (6–201) only by a sign change. If the circuit is constrained to rotate about the z-axis, the z-component of the torque acting on the circuit is

$$\boxed{(T_I)_z = \frac{\partial W_m}{\partial \phi} \qquad (\text{N·m}).} \tag{6–210}$$

The difference between the expression above and $(T_\Phi)_z$ in Eq. (6–203) is, again, only in the sign. It must be understood that, despite the difference in the sign, Eqs. (6–201) and (6–203) should yield the same answers to a given problem as do Eqs. (6–209) and (6–210), respectively. The formulations using the method of virtual displacement under constant-flux-linkage and constant-current conditions are simply two means of solving the same problem.

Let us solve the electromagnet problem in Example 6–23 assuming a virtual displacement under the constant-current condition. For this purpose we express W_m in terms of the current I:

$$W_m = \tfrac{1}{2}LI^2, \tag{6–211}$$

where L is the self-inductance of the coil. The flux, Φ, in the electromagnet is obtained by dividing the applied magnetomotive force (NI) by the sum of the reluctance of the core ($\mathcal{R}_c$) and that of the two air gaps ($2y/\mu_0 S$). Thus,

$$\Phi = \frac{NI}{\mathcal{R}_c + 2y/\mu_0 S}. \tag{6–212}$$

Inductance L is equal to flux linkage per unit current:

$$L = \frac{N\Phi}{I} = \frac{N^2}{\mathcal{R}_c + 2y/\mu_0 S}. \qquad (6\text{-}213)$$

Combining Eqs. (6–209) and (6–211) and using Eq. (6–213), we obtain

$$\begin{aligned}
\mathbf{F}_I &= \mathbf{a}_y \frac{I^2}{2} \frac{dL}{dy} = -\mathbf{a}_y \frac{1}{\mu_0 S} \left(\frac{NI}{\mathcal{R}_c + 2y/\mu_0 S} \right)^2 \\
&= -\mathbf{a}_y \frac{\Phi^2}{\mu_0 S} \qquad (\text{N}),
\end{aligned} \qquad (6\text{-}214)$$

which is exactly the same as the $\mathbf{F}_\Phi$ in Eq. (6–205).

6–13.4 FORCES AND TORQUES IN TERMS OF MUTUAL INDUCTANCE

The method of virtual displacement for constant currents provides a powerful technique for determining the forces and torques between rigid current-carrying circuits. For two circuits with currents I_1 and I_2, self-inductances L_1 and L_2, and mutual inductance L_{12}, the magnetic energy is, from Eq. (6–161),

$$W_m = \tfrac{1}{2} L_1 I_1^2 + L_{12} I_1 I_2 + \tfrac{1}{2} L_2 I_2^2. \qquad (6\text{-}215)$$

If one of the circuits is given a virtual displacement under the condition of constant currents, there would be a change in W_m, and Eq. (6–209) applies. Substitution of Eq. (6–215) in Eq. (6–209) yields,

$$\boxed{\mathbf{F}_I = I_1 I_2 (\nabla L_{12}) \qquad (\text{N}).} \qquad (6\text{-}216)$$

Similarly, we obtain, from Eq. (6–210),

$$\boxed{(T_I)_z = I_1 I_2 \frac{\partial L_{12}}{\partial \phi} \qquad (\text{N·m}).} \qquad (6\text{-}217)$$

■■■■■ **EXAMPLE 6–24** Determine the force between two coaxial circular coils of radii b_1 and b_2 separated by a distance d that is much larger than the radii ($d \gg b_1, b_2$). The coils consist of N_1 and N_2 closely wound turns and carry currents I_1 and I_2, respectively.

Solution This problem is rather a difficult one if we try to solve it with Ampère's law of force, as expressed in Eq. (6–185a). Therefore we will base our solution on Eq. (6–216). First, we determine the mutual inductance between the two coils. In Example 6–7 we found, in Eq. (6–43), the vector potential at a distant point, which was caused by a single-turn circular loop carrying a current I. Referring to Fig. 6–34 for this problem, at the point P on coil 2 we have $\mathbf{A}_{12}$ due to current I_1 in coil 1 with N_1 turns as follows:

$$\mathbf{A}_{12} = \mathbf{a}_\phi \frac{\mu_0 N_1 I_1 b_1^2}{4R^2} \sin\theta$$

$$= \mathbf{a}_\phi \frac{\mu_0 N_1 I_1 b_1^2}{4R^2} \left(\frac{b_2}{R}\right) \qquad (6\text{–}218)$$

$$= \mathbf{a}_\phi \frac{\mu_0 N_1 I_1 b_1^2 b_2}{4(z^2 + b_2^2)^{3/2}}.$$

In Eq. (6–218), z, instead of d, is used because we anticipate a virtual displacement, and z is to be kept as a variable for the time being. Using Eq. (6–218) in Eq. (6–25), we find the mutual flux.

$$\Phi_{12} = \oint_{C_2} \mathbf{A}_{12} \cdot d\boldsymbol{\ell}_2 = \int_0^{2\pi} A_{12} b_2 \, d\phi$$

$$= \frac{\mu_0 N_1 I_1 b_1^2 b_2^2 \pi}{2(z^2 + b_2^2)^{3/2}}. \qquad (6\text{–}219)$$

The mutual inductance is then, from Eq. (6–127),

$$L_{12} = \frac{N_2 \Phi_{12}}{I_1} = \frac{\mu_0 N_1 N_2 \pi b_1^2 b_2^2}{2(z^2 + b_2^2)^{3/2}} \qquad (\text{H}). \qquad (6\text{–}220)$$

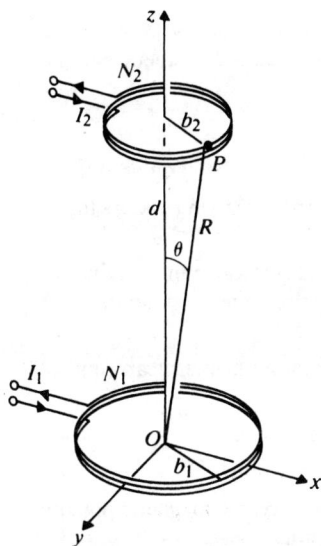

FIGURE 6–34
Coaxial current-carrying circular loops (Example 6–24).

On coil 2 the force due to the magnetic field of coil 1 can now be obtained directly by substituting Eq. (6–220) in Eq. (6–216):

$$\mathbf{F}_{12} = \mathbf{a}_z I_1 I_2 \frac{dL_{12}}{dz}\bigg|_{z=d} = -\mathbf{a}_z I_1 I_2 \frac{3\mu_0 N_1 N_2 \pi b_1^2 b_2^2 d}{2(d^2 + b_2^2)^{5/2}},$$

which can be written as

$$F_{12} \cong -\mathbf{a}_z \frac{3\mu_0 m_1 m_2}{2\pi d^4} \quad \text{(N)}, \tag{6–221}$$

where $(d^2 + b_2^2)$ has been replaced approximately by d^2, and m_1 and m_2 are the magnitudes of the magnetic moments of coils 1 and 2, respectively:

$$m_1 = N_1 I_1 \pi b_1^2, \qquad m_2 = N_2 I_2 \pi b_2^2.$$

The negative sign in Eq. (6–221) indicates that $\mathbf{F}_{12}$ is a force of attraction for currents flowing in the same direction. This force diminishes very rapidly as the inverse fourth power of the distance of separation. ▬

Review Questions

R.6–1 What is the expression for the force on a test charge q that moves with velocity **u** in a magnetic field of flux density **B**?

R.6–2 Verify that tesla (T), the unit for magnetic flux density, is the same as volt-second per square meter (V·s/m²).

R.6–3 Write Lorentz's force equation.

R.6–4 Which postulate of magnetostatics denies the existence of isolated magnetic charges?

R.6–5 State the law of conservation of magnetic flux.

R.6–6 State Ampère's circuital law.

R.6–7 In applying Ampère's circuital law, must the path of integration be circular? Explain.

R.6–8 Why cannot the **B**-field of an infinitely long, straight, current-carrying conductor have a component in the direction of the current?

R.6–9 Do the formulas for **B**, as derived in Eqs. (6–11) and (6–12) for a round conductor, apply to a conductor having a square cross section of the same area and carrying the same current? Explain.

R.6–10 In what manner does the **B**-field of an infinitely long straight filament carrying a direct current I vary with distance?

R.6–11 Can a static magnetic field exist in a good conductor? Explain.

R.6–12 Define in words *vector magnetic potential* **A**. What is its SI unit?

R.6–13 What is the relation between magnetic flux density **B** and vector magnetic potential **A**? Give an example of a situation in which **B** is zero and **A** is not.

R.6–14 What is the relation between vector magnetic potential **A** and the magnetic flux through a given area?

R.6–15 State Biot-Savart law.

R.6–16 Compare the usefulness of Ampère's circuital law and Biot-Savart law in determining **B** of a current-carrying circuit.

R.6–17 What is a magnetic dipole? Define *magnetic dipole moment*. What is its SI unit?

R.6–18 Define *scalar magnetic potential* V_m. What is its SI unit?

R.6–19 Discuss the relative merits of using the vector and scalar magnetic potentials in magnetostatics.

R.6–20 Define *magnetization vector*. What is its SI unit?

R.6–21 What is meant by "equivalent magnetization current densities"? What are the SI units for $\mathbf{V} \times \mathbf{M}$ and $\mathbf{M} \times \mathbf{a}_n$?

R.6–22 Define *magnetic field intensity vector*. What is its SI unit?

R.6–23 What are *magnetization charge densities*? What are the SI units for $\mathbf{M} \cdot \mathbf{a}_n$ and $-\mathbf{V} \cdot \mathbf{M}$?

R.6–24 Describe a procedure for finding the external magnetic field of a bar magnet having a known volume density of dipole moment.

R.6–25 Define *magnetic susceptibility* and *relative permeability*. What are their SI units?

R.6–26 Does the magnetic field intensity due to a current distribution depend on the properties of the medium? Does the magnetic flux density?

R.6–27 Define *magnetomotive force*. What is its SI unit?

R.6–28 What is the reluctance of a piece of magnetic material of permeability μ, length ℓ, and a constant cross section S? What is its SI unit?

R.6–29 An air gap is cut in a ferromagnetic toroidal core. The core is excited with an mmf of NI ampere-turns. Is the magnetic field intensity in the air gap higher or lower than that in the core?

R.6–30 Define *diamagnetic*, *paramagnetic*, and *ferromagnetic* materials.

R.6–31 What is a magnetic domain?

R.6–32 Define *remanent flux density* and *coercive field intensity*.

R.6–33 Discuss the difference between soft and hard ferromagnetic materials.

R.6–34 What is *curie temperature*?

R.6–35 What are the characteristics of ferrites?

R.6–36 What are the boundary conditions for magnetostatic fields at an interface between two different magnetic media?

R.6–37 Explain why magnetic flux lines leave the surface of a ferromagnetic medium perpendicularly.

R.6–38 Explain qualitatively the statement that **H** and **B** along the axis of a cylindrical bar magnet are in opposite directions.

R.6–39 Define (a) the mutual inductance between two circuits, and (b) the self-inductance of a single coil.

R.6–40 Explain how the self-inductance of a wire-wound inductor depends on its number of turns.

R.6–41 In Example 6–16, would the answer be the same if the outer conductor were not "very thin"? Explain.

R.6–42 What is implied by "quasi-static conditions" in electromagnetics?

R.6–43 Give an expression of magnetic energy in terms of **B** and/or **H**.

R.6–44 Explain the *Hall effect*.

R.6–45 Give the integral expression for the force on a closed circuit that carries a current I in a magnetic field **B**.

R.6–46 Discuss first the net force and then the net torque acting on a current-carrying circuit situated in a uniform magnetic field.

R.6–47 Explain the principle of operation of d-c motors.

R.6–48 What is the relation between the force and the stored magnetic energy in a system of current-carrying circuits under the condition of constant flux linkages? Under the condition of constant currents?

Problems

P.6–1 A positive point charge q of mass m is injected with a velocity $\mathbf{u}_0 = \mathbf{a}_y u_0$ into the $y > 0$ region where a uniform magnetic field $\mathbf{B} = \mathbf{a}_x B_0$ exists. Obtain the equation of motion of the charge, and describe the path that the charge follows.

P.6–2 An electron is injected with a velocity $\mathbf{u}_0 = \mathbf{a}_y u_0$ into a region where both an electric field **E** and a magnetic field **B** exist. Describe the motion of the electron if
   a) $\mathbf{E} = \mathbf{a}_z E_0$ and $\mathbf{B} = \mathbf{a}_x B_0$,
   b) $\mathbf{E} = -\mathbf{a}_z E_0$ and $\mathbf{B} = -\mathbf{a}_x B_0$.
Discuss the effect of the relative magnitudes of E_0 and B_0 on the electron paths in parts (a) and (b).

P.6–3 A current I flows in the inner conductor of an infinitely long coaxial line and returns via the outer conductor. The radius of the inner conductor is a, and the inner and outer radii of the outer conductor are b and c, respectively. Find the magnetic flux density **B** for all regions and plot $|\mathbf{B}|$ versus r.

P.6–4 A current I flows lengthwise in a very long, thin conducting sheet of width w, as shown in Fig. 6–35.
   a) Assuming that the current flows into the paper, determine the magnetic flux density $\mathbf{B}_1$ at point $P_1(0, d)$.
   b) Use the result in part (a) to find the magnetic flux density $\mathbf{B}_2$ at point $P_2(2w/3, d)$.

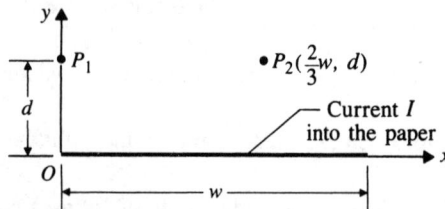

FIGURE 6–35
A thin conducting sheet carrying a current I (Problem P.6–4).

P.6–5 A current I flows in a $w \times w$ square loop as in Fig. 6–36. Find the magnetic flux density at the off-center point $P(w/4, w/2)$.

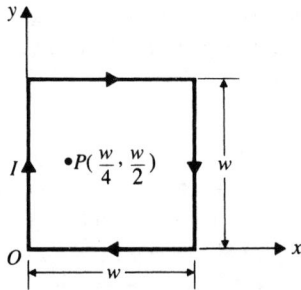

FIGURE 6-36
A square loop carrying a current I (Problem P.6-5).

P.6-6 Figure 6-37 shows an infinitely long solenoid with air core having a radius b and n closely wound turns per unit length. The windings are slanted at an angle α and carry a current I. Determine the magnetic flux density both inside and outside the solenoid.

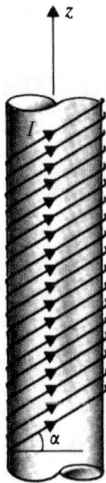

FIGURE 6-37
A long solenoid with closely wound windings carrying a current I (Problem P.6-6).

P.6-7 Determine the magnetic flux density at a point on the axis of a solenoid with radius b and length L, and with a current I in its N turns of closely wound coil. Show that the result reduces to that given in Eq. (6-14) when L approaches infinity.

P.6-8 Starting from the expression for vector magnetic potential **A** in Eq. (6-23), prove that

$$\mathbf{B} = \frac{\mu_0}{4\pi} \int_{V'} \frac{\mathbf{J} \times \mathbf{a}_R}{R^2} \, dv'. \tag{6-222}$$

Furthermore, prove that **B** in Eq. (6-222) satisfies the fundamental postulates of magnetostatics in free space, Eqs. (6-6) and (6-7).

P.6-9 Combine Eqs. (6-4) and (6-33) to obtain a formula for the magnetic force $\mathbf{F}_{12}$ exerted by a charge q_1 moving with a velocity $\mathbf{u}_1$ on a charge q_2 moving with a velocity $\mathbf{u}_2$.

P.6-10 A very long, thin conducting strip of width w lies in the xz-plane between $x = \pm w/2$. A surface current $\mathbf{J}_s = \mathbf{a}_z J_{s0}$ flows in the strip. Find the magnetic flux density at an arbitrary point outside the strip.

P.6–11 A long wire carrying a current I folds back with a semicircular bend of radius b as in Fig. 6–38. Determine the magnetic flux density at the center point P of the bend.

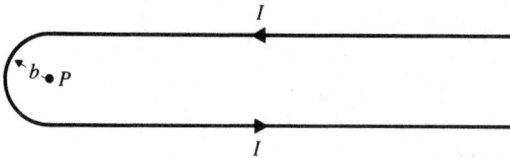

FIGURE 6–38
A very long wire with a semicircular bend (Problem P.6–11).

P.6–12 Two identical coaxial coils, each of N turns and radius b, are separated by a distance d, as depicted in Fig. 6–39. A current I flows in each coil in the same direction.
   a) Find the magnetic flux density $\mathbf{B} = \mathbf{a}_x B_x$ at a point midway between the coils.
   b) Show that dB_x/dx vanishes at the midpoint.
   c) Find the relation between b and d such that $d^2 B_x/dx^2$ also vanishes at the midpoint. Such a pair of coils are used to obtain an approximately uniform magnetic field in the midpoint region. They are known as **Helmholtz coils**.

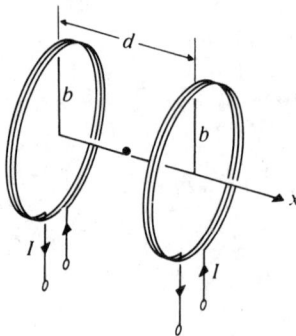

FIGURE 6–39
Helmholtz coils (Problems P.6–12).

P.6–13 A thin conducting wire is bent into the shape of a regular polygon of N sides. A current I flows in the wire. Show that the magnetic flux density at the center is

$$\mathbf{B} = \mathbf{a}_n \frac{\mu_0 NI}{2\pi b} \tan \frac{\pi}{N},$$

where b is the radius of the circle circumscribing the polygon and $\mathbf{a}_n$ is a unit vector normal to the plane of the polygon. Show also that, as N becomes very large, this result reduces to that given in Eq. (6–38) with $z = 0$.

P.6–14 Find the total magnetic flux through a circular toroid with a rectangular cross section of height h. The inner and outer radii of the toroid are a and b, respectively. A current I flows in N turns of closely wound wire around the toroid. Determine the percentage of error if the flux is found by multiplying the cross-sectional area by the flux density at the mean radius.

P.6–15 In certain experiments it is desirable to have a region of constant magnetic flux density. This can be created in an off-center cylindrical cavity that is cut in a very long cylindrical conductor carrying a uniform current density. Refer to the cross section in

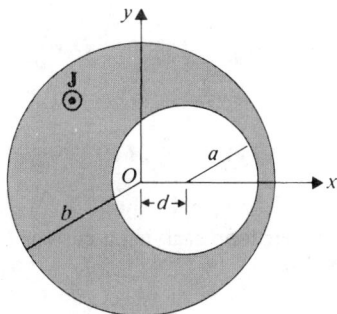

FIGURE 6–40
Cross section of a long cylindrical conductor with cavity (Problem P.6–15).

Fig. 6–40. The uniform axial current density is $\mathbf{J} = \mathbf{a}_z J$. Find the magnitude and direction of $\mathbf{B}$ in the cylindrical cavity whose axis is displaced from that of the conducting part by a distance d. (*Hint:* Use principle of superposition and consider $\mathbf{B}$ in the cavity as that due to two long cylindrical conductors with radii b and a and current densities $\mathbf{J}$ and $-\mathbf{J}$, respectively.)

P.6–16 Prove the following:
   a) If Cartesian coordinates are used, Eq. (6–18) for the Laplacian of a vector field holds.
   b) If cylindrical coordinates are used, $\nabla^2 \mathbf{A} \neq \mathbf{a}_r \nabla^2 A_r + \mathbf{a}_\phi \nabla^2 A_\phi + \mathbf{a}_z \nabla^2 A_z$.

P.6–17 The magnetic flux density $\mathbf{B}$ for an infinitely long cylindrical conductor has been found in Example 6–1. Determine the vector magnetic potential $\mathbf{A}$ both inside and outside the conductor from the relation $\mathbf{B} = \nabla \times \mathbf{A}$.

P.6–18 Starting from the expression of $\mathbf{A}$ in Eq. (6–34) for the vector magnetic potential at a point in the bisecting plane of a straight wire of length $2L$ that carries a current I:
   a) Find $\mathbf{A}$ at point $P(x, y, 0)$ in the bisecting plane of two parallel wires each of length $2L$, located at $y - \pm d/2$ and carrying equal and opposite currents, as shown in Fig. 6–41.
   b) Find $\mathbf{A}$ due to equal and opposite currents in a very long two-wire transmission line.
   c) Find $\mathbf{B}$ from $\mathbf{A}$ in part (b), and check your answer against the result obtained by applying Ampère's circuital law.
   d) Find the equation for the magnetic flux lines in the xy-plane.

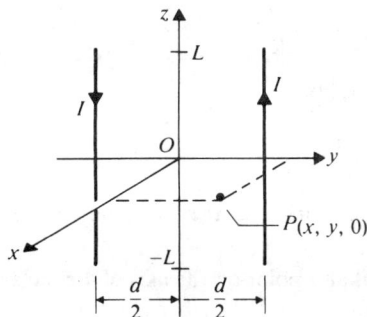

FIGURE 6–41
Parallel wires carrying equal and opposite currents (Problem P.6–18).

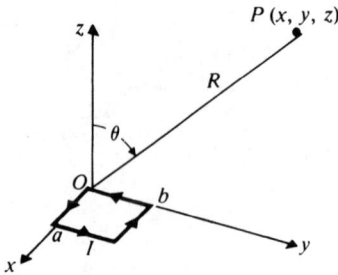

FIGURE 6–42
A small rectangular loop carrying a current I
(Problem P.6–19).

P.6–19 For the small rectangular loop with sides a and b that carries a current I, shown in Fig. 6–42:
   a) Find the vector magnetic potential $\mathbf{A}$ at a distant point, $P(x, y, z)$. Show that it can be put in the form of Eq. (6–45).
   b) Determine the magnetic flux density $\mathbf{B}$ from $\mathbf{A}$, and show that it is the same as that given in Eq. (6–48).

P.6–20 For a vector field $\mathbf{F}$ with continuous first derivatives, prove that

$$\int_V (\nabla \times \mathbf{F})\, dv = -\oint_S \mathbf{F} \times d\mathbf{s},$$

where S is the surface enclosing the volume V. (*Hint:* Apply the divergence theorem to $(\mathbf{F} \times \mathbf{C})$, where $\mathbf{C}$ is a constant vector.)

P.6–21 A very large slab of material of thickness d lies perpendicularly to a uniform magnetic field of intensity $\mathbf{H}_0 = \mathbf{a}_z H_0$. Ignoring edge effect, determine the magnetic field intensity in the slab:
   a) if the slab material has a permeability μ,
   b) if the slab is a permanent magnet having a magnetization vector $\mathbf{M}_i = \mathbf{a}_z M_i$.

P.6–22 A circular rod of magnetic material with permeability μ is inserted coaxially in the long solenoid of Fig. 6–4. The radius of the rod, a, is less than the inner radius, b, of the solenoid. The solenoid's winding has n turns per unit length and carries a current I.
   a) Find the values of $\mathbf{B}$, $\mathbf{H}$, and $\mathbf{M}$ inside the solenoid for $r < a$ and for $a < r < b$.
   b) What are the equivalent magnetization current densities $\mathbf{J}_m$ and $\mathbf{J}_{ms}$ for the magnetized rod?

P.6–23 The scalar magnetic potential, V_m, due to a current loop can be obtained by first dividing the loop area into many small loops and then summing up the contribution of these small loops (magnetic dipoles); that is,

$$V_m = \int dV_m = \int \frac{d\mathbf{m} \cdot \mathbf{a}_R}{4\pi R^2}, \tag{6–223a}$$

where

$$d\mathbf{m} = \mathbf{a}_n I \, ds. \tag{6–223b}$$

Prove that

$$V_m = -\frac{I}{4\pi} \Omega, \tag{6–224}$$

where Ω is the solid angle subtended by the loop surface at the field point P (see Fig. 6–43).

P.6–24 Do the following by using Eq. (6–224):
   a) Determine the scalar magnetic potential at a point on the axis of a circular loop having a radius b and carrying a current I.

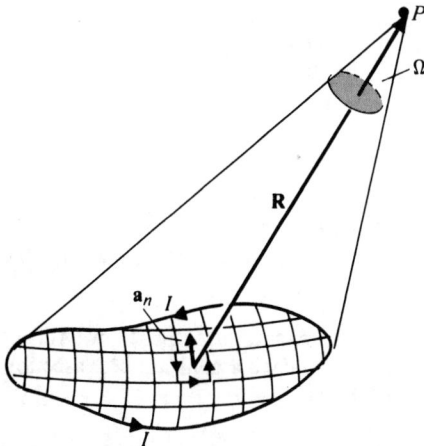

FIGURE 6–43
Subdivided current loop for determination of scalar magnetic potential (Problem P.6–23).

b) Obtain the magnetic flux density $\mathbf{B}$ from $-\mu_0 \nabla V_m$, and compare the result with Eq. (6–38).

P.6–25 Solve the cylindrical bar magnet problem in Example 6–9, using the equivalent magnetization current density concept.

P.6–26 A ferromagnetic sphere of radius b is magnetized uniformly with a magnetization $\mathbf{M} = \mathbf{a}_z M_0$.
    a) Determine the equivalent magnetization current densities $\mathbf{J}_m$ and $\mathbf{J}_{ms}$.
    b) Determine the magnetic flux density at the center of the sphere.

P.6–27 A toroidal iron core of relative permeability 3000 has a mean radius $R = 80$ (mm) and a circular cross section with radius $b = 25$ (mm). An air gap $\ell_g = 3$ (mm) exists, and a current I flows in a 500-turn winding to produce a magnetic flux of 10^{-5} (Wb). (See Fig. 6–44.) Neglecting flux leakage and using mean path length, find
    a) the reluctances of the air gap and of the iron core,
    b) B_g and H_g in the air gap, and B_c and H_c in the iron core,
    c) the required current I.

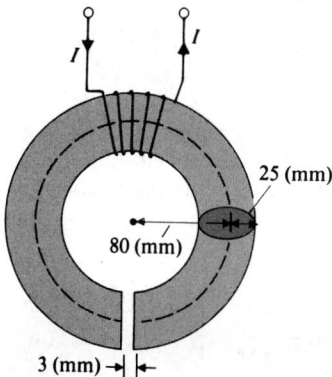

FIGURE 6–44
A toroidal iron core with air gap (Problem P.6–27).

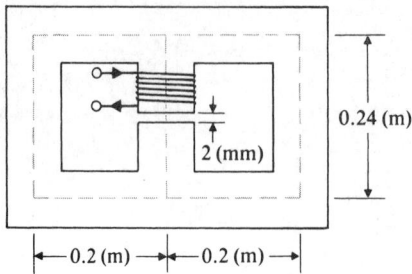

FIGURE 6–45
A magnetic circuit with air gap (Problem P.6–28).

P.6–28 Consider the magnetic circuit in Fig. 6–45. A current of 3 (A) flows through 200 turns of wire on the center leg. Assuming the core to have a constant cross-sectional area of 10^{-3} (m^2) and a relative permeability of 5000:
   a) Determine the magnetic flux in each leg.
   b) Determine the magnetic field intensity in each leg of the core and in the air gap.

P.6–29 Consider an infinitely long solenoid with n turns per unit length around a ferromagnetic core of cross-sectional area S. When a current is sent through the coil to create a magnetic field, a voltage $v_1 = -n\,d\Phi/dt$ is induced per unit length, which opposes the current change. Power $P_1 = -v_1 I$ per unit length must be supplied to overcome this induced voltage in order to increase the current to I.
   a) Prove that the work per unit volume required to produce a final magnetic flux density B_f is
$$W_1 = \int_0^{B_f} H\,dB. \qquad (6\text{–}225)$$
   b) Assuming that the current is changed in a periodic manner such that B is reduced from B_f to $-B_f$ and then is increased again to B_f, prove that the work done per unit volume for such a cycle of change in the ferromagnetic core is represented by the area of the hysteresis loop of the core material.

P.6–30 Prove that the relation $\nabla \times \mathbf{H} = \mathbf{J}$ leads to Eq. (6–111) at an interface between two media.

P.6–31 What boundary conditions must the scalar magnetic potential V_m satisfy at an interface between two different magnetic media?

P.6–32 Consider a plane boundary ($y = 0$) between air (region 1, $\mu_{r1} = 1$) and iron (region 2, $\mu_{r2} = 5000$).
   a) Assuming $\mathbf{B}_1 = \mathbf{a}_x 0.5 - \mathbf{a}_y 10$ (mT), find $\mathbf{B}_2$ and the angle that $\mathbf{B}_2$ makes with the interface.
   b) Assuming $\mathbf{B}_2 = \mathbf{a}_x 10 + \mathbf{a}_y 0.5$ (mT), find $\mathbf{B}_1$ and the angle that $\mathbf{B}_1$ makes with the normal to the interface.

P.6–33 The *method of images* can also be applied to certain magnetostatic problems. Consider a straight, thin conductor in air parallel to and at a distance d above the plane interface of a magnetic material of relative permeability μ_r. A current I flows in the conductor.
   a) Show that all boundary conditions are satisfied if
      i) the magnetic field in the air is calculated from I and an image current I_i,
$$I_i = \left(\frac{\mu_r - 1}{\mu_r + 1}\right)I,$$
      and these currents are equidistant from the interface and situated in air;

ii) the magnetic field below the boundary plane is calculated from I and $-I_i$, both at the same location. These currents are situated in an infinite magnetic material of relative permeability μ_r.

b) For a long conductor carrying a current I and for $\mu_r \gg 1$, determine the magnetic flux density **B** at the point P in Fig. 6–46.

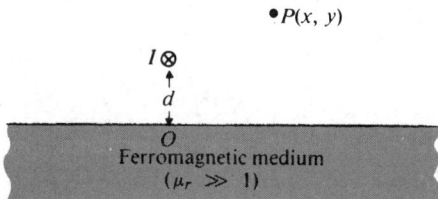

FIGURE 6–46
A current-carrying conductor near a ferromagnetic medium (Problem P.6–33).

P.6–34 A very long conductor in free space carrying a current I is parallel to, and at a distance d from, an infinite plane interface with a medium.

a) Discuss the behavior of the normal and tangential components of **B** and **H** at the interface:
i) if the medium is infinitely conducting;
ii) if the medium is infinitely permeable.
b) Find and compare the magnetic field intensities **H** at an arbitrary point in the free space for the two cases in part (a).
c) Determine the surface current densities at the interface, if any, for the two cases.

P.6–35 Determine the self-inductance of a toroidal coil of N turns of wire wound on an air frame with mean radius r_o and a circular cross section of radius b. Obtain an approximate expression assuming $b \ll r_o$.

P.6–36 Refer to Example 6–16. Determine the inductance per unit length of the air coaxial transmission line assuming that its outer conductor is not very thin but is of a thickness d.

P.6–37 Calculate the mutual inductance per unit length between two parallel two-wire transmission lines A–A' and B–B' separated by a distance D, as shown in Fig. 6–47. Assume the wire radius to be much smaller than D and the wire spacing d.

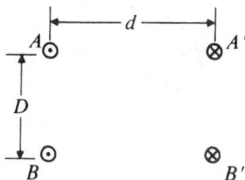

FIGURE 6–47
Coupled two-wire transmission lines (Problem P.6–37).

P.6–38 Determine the mutual inductance between a very long, straight wire and a conducting equilateral triangular loop, as shown in Fig. 6–48.

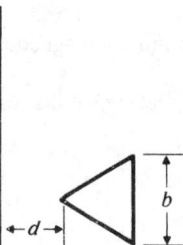

FIGURE 6–48
A long, straight wire and a conducting equilateral triangular loop
(Problem P.6–38).

P.6–39 Determine the mutual inductance between a very long, straight wire and a
conducting circular loop, as shown in Fig. 6–49.

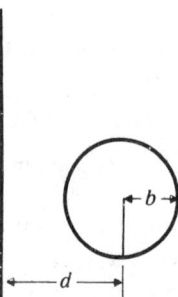

FIGURE 6–49
A long, straight wire and a conducting circular loop (Problem P.6–39).

P.6–40 Find the mutual inductance between two coplanar rectangular loops with parallel
sides, as shown in Fig. 6–50. Assume that $h_1 \gg h_2$ ($h_2 > w_2 > d$).

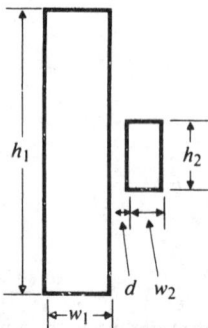

FIGURE 6–50
Two coplanar rectangular loops, $h_1 \gg h_2$ (Problem P.6–40).

P.6–41 Consider two coupled circuits, having self-inductances L_1 and L_2, that carry currents I_1 and I_2, respectively. The mutual inductance between the circuits is M.

    a) Using Eq. (6–161), find the ratio I_1/I_2 that makes the stored magnetic energy W_2 a minimum.

    b) Show that $M \le \sqrt{L_1 L_2}$.

P.6–42 Calculate the force per unit length on each of three equidistant, infinitely long, parallel wires 0.15 (m) apart, each carrying a current of 25 (A) in the same direction. Specify the direction of the force.

P.6–43 The cross section of a long thin metal strip and a parallel wire is shown in Fig. 6–51. Equal and opposite currents I flow in the conductors. Find the force per unit length on the conductors.

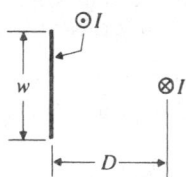

FIGURE 6–51
Cross section of parallel strip and wire conductor (Problem P.6–43).

P.6–44 Determine the force per unit length between two parallel, long, thin conducting strips of equal width w. The strips are at a distance d apart and carry currents I_1 and I_2 in opposite directions as in Fig. 6–52.

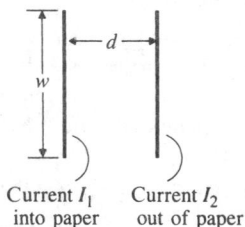

Current I_1 Current I_2
into paper   out of paper

FIGURE 6–52
Cross section of two parallel strips carrying opposite currents (Problem P.6–44).

P.6–45 Refer to Problem 6–39 and Fig. 6–49. Find the force on the circular loop that is exerted by the magnetic field due to an upward current I_1 in the long straight wire. The circular loop carries a current I_2 in the counterclockwise direction.

P.6–46 The bar AA' in Fig. 6–53 serves as a conducting path (such as the blade of a circuit breaker) for the current I in two very long parallel lines. The lines have a radius b and are spaced at a distance d apart. Find the direction and the magnitude of the magnetic force on the bar.

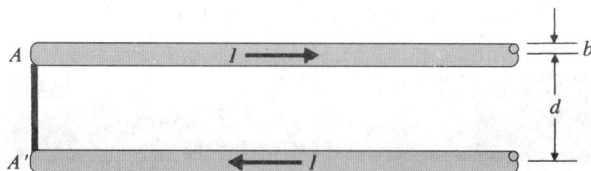

FIGURE 6–53
Force on end conducting bar (Problem P.6–46).

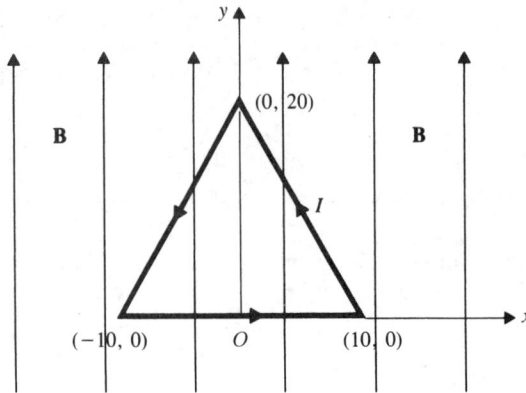

FIGURE 6–54
A triangular loop in a uniform
magnetic field (Problem P.6–47).

P.6–47 A d-c current $I = 10$ (A) flows in a triangular loop in the xy-plane as in Fig. 6–54. Assuming a uniform magnetic flux density $\mathbf{B} = \mathbf{a}_y 0.5$ (T) in the region, find the forces and torque on the loop. The dimensions are in (cm).

P.6–48 One end of a long air-core coaxial transmission line having an inner conductor of radius a and an outer conductor of inner radius b is short-circuited by a thin, tight-fitting conducting washer. Find the magnitude and the direction of the magnetic force on the washer when a current I flows in the line.

P.6–49 Assuming that the circular loop in Problem P.6–45 is rotated about its horizontal axis by an angle α, find the torque exerted on the circular loop.

P.6–50 A small circular turn of wire of radius r_1 that carries a steady current I_1 is placed at the center of a much larger turn of wire of radius r_2 $(r_2 \gg r_1)$ that carries a steady current I_2 in the same direction. The angle between the normals of the two circuits is θ and the small circular wire is free to turn about its diameter. Determine the magnitude and the direction of the torque on the small circular wire.

P.6–51 A magnetized compass needle will line up with the earth's magnetic field. A small bar magnet (a magnetic dipole) with a magnetic moment 2 (A·m²) is placed at a distance 0.15 (m) from the center of a compass needle. Assuming the earth's magnetic flux density at the needle to be 0.1 (mT), find the maximum angle at which the bar magnet can cause the needle to deviate from the north–south direction. How should the bar magnet be oriented?

P.6–52 The total mean length of the flux path in iron for the electromagnet in Fig. 6–33 is 3 (m), and the yoke-bar contact areas measure 0.01 (m²). Assuming the permeability of iron to be $4000\mu_0$ and each of air gaps to be 2 (mm), calculate the mmf needed to lift a total mass of 100 (kg).

P.6–53 A current I flows in a long solenoid with n closely wound coil-turns per unit length. The cross-sectional area of its iron core, which has permeability μ, is S. Determine the force acting on the core if it is withdrawn to the position shown in Fig. 6–55.

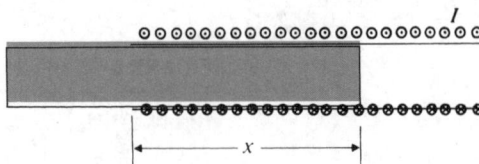

FIGURE 6–55
A long solenoid with iron core partially withdrawn (Problem P.6–53).

7

Time-Varying Fields and Maxwell's Equations

7-1 Introduction

In constructing the electrostatic model we defined an electric field intensity vector, **E**, and an electric flux density (electric displacement) vector, **D**. The fundamental governing differential equations are

$$\nabla \times \mathbf{E} = 0, \tag{3-5}$$

$$\nabla \cdot \mathbf{D} = \rho. \tag{3-98}$$

For linear and isotropic (not necessarily homogeneous) media, **E** and **D** are related by the constitutive relation

$$\mathbf{D} = \epsilon \mathbf{E}. \tag{3-102}$$

For the magnetostatic model we defined a magnetic flux density vector, **B**, and a magnetic field intensity vector, **H**. The fundamental governing differential equations are

$$\nabla \cdot \mathbf{B} = 0, \tag{6-6}$$

$$\nabla \times \mathbf{H} = \mathbf{J}. \tag{6-76}$$

The constitutive relation for **B** and **H** in linear and isotropic media is

$$\mathbf{H} = \frac{1}{\mu} \mathbf{B}. \tag{6-80b}$$

These fundamental relations are summarized in Table 7–1.

We observe that, in the static (non-time-varying) case, electric field vectors **E** and **D** and magnetic field vectors **B** and **H** form separate and independent pairs. In other words, **E** and **D** in the electrostatic model are not related to **B** and **H** in the magnetostatic model. In a conducting medium, static electric and magnetic fields may both exist and form an ***electromagnetostatic field*** (see the statement following Example 5–4 on p. 215). A static electric field in a conducting medium causes a steady current

307

TABLE 7–1
Fundamental Relations for Electrostatic and Magnetostatic Models

Fundamental Relations	Electrostatic Model	Magnetostatic Model
Governing equations	$\nabla \times \mathbf{E} = 0$ $\nabla \cdot \mathbf{D} = \rho$	$\nabla \cdot \mathbf{B} = 0$ $\nabla \times \mathbf{H} = \mathbf{J}$
Constitutive relations (linear and isotropic media)	$\mathbf{D} = \epsilon \mathbf{E}$	$\mathbf{H} = \dfrac{1}{\mu} \mathbf{B}$

to flow that, in turn, gives rise to a static magnetic field. However, the electric field can be completely determined from the static electric charges or potential distributions. The magnetic field is a consequence; it does not enter into the calculation of the electric field.

In this chapter we will see that a changing magnetic field gives rise to an electric field, and vice versa. To explain electromagnetic phenomena under time-varying conditions, it is necessary to construct an electromagnetic model in which the electric field vectors $\mathbf{E}$ and $\mathbf{D}$ are properly related to the magnetic field vectors $\mathbf{B}$ and $\mathbf{H}$. The two pairs of the governing equations in Table 7–1 must therefore be modified to show a mutual dependence between the electric and magnetic field vectors in the time-varying case.

We will begin with a fundamental postulate that modifies the $\nabla \times \mathbf{E}$ equation in Table 7–1 and leads to Faraday's law of electromagnetic induction. The concepts of transformer emf and motional emf will be discussed. With the new postulate we will also need to modify the $\nabla \times \mathbf{H}$ equation in order to make the governing equations consistent with the equation of continuity (law of conservation of charge). The two modified curl equations together with the two divergence equations in Table 7–1 are known as Maxwell's equations and form the foundation of electromagnetic theory. The governing equations for electrostatics and magnetostatics are special forms of Maxwell's equations when all quantities are independent of time. Maxwell's equations can be combined to yield wave equations that predict the existence of electromagnetic waves propagating with the velocity of light. The solutions of the wave equations, especially for time-harmonic fields, will be discussed in this chapter.

7–2   Faraday's Law of Electromagnetic Induction

A major advance in electromagnetic theory was made by Michael Faraday, who, in 1831, discovered experimentally that a current was induced in a conducting loop when the magnetic flux linking the loop changed. The quantitative relationship between the induced emf and the rate of change of flux linkage, based on experimental observation, is known as ***Faraday's law***. It is an experimental law and can be considered as a postulate. However, we do not take the experimental relation concerning a finite loop as the starting point for developing the theory of electromagnetic induction. Instead, we follow our approach in Chapter 3 for electrostatics and in Chapter 6 for magnetostatics by putting forth the following fundamental postulate and developing from it the integral forms of Faraday's law.

Fundamental Postulate for Electromagnetic Induction

$$\boxed{\nabla \times \mathbf{E} = -\frac{\partial \mathbf{B}}{\partial t}.}\tag{7-1}$$

Equation (7–1) expresses a point-function relationship; that is, it applies to every point in space, whether it be in free space or in a material medium. *The electric field intensity in a region of time-varying magnetic flux density is therefore nonconservative and cannot be expressed as the gradient of a scalar potential.*

Taking the surface integral of both sides of Eq. (7–1) over an open surface and applying Stokes's theorem, we obtain

$$\oint_C \mathbf{E} \cdot d\boldsymbol{\ell} = -\int_S \frac{\partial \mathbf{B}}{\partial t} \cdot d\mathbf{s}.\tag{7-2}$$

Equation (7–2) is valid for any surface S with a bounding contour C, whether or not a physical circuit exists around C. Of course, in a field with no time variation, $\partial \mathbf{B}/\partial t = 0$, Eqs. (7–1) and (7–2) reduce to Eqs. (3–5) and (3–8), respectively, for electrostatics.

In the following subsections we discuss separately the cases of a stationary circuit in a time-varying magnetic field, a moving conductor in a static magnetic field, and a moving circuit in a time-varying magnetic field.

7–2.1   A STATIONARY CIRCUIT IN A TIME-VARYING MAGNETIC FIELD

For a stationary circuit with a contour C and surface S, Eq. (7–2) can be written as

$$\oint_C \mathbf{E} \cdot d\boldsymbol{\ell} = -\frac{d}{dt}\int_S \mathbf{B} \cdot d\mathbf{s}.\tag{7-3}$$

If we define

$$\mathscr{V} = \oint_C \mathbf{E} \cdot d\boldsymbol{\ell} = \text{emf induced in circuit with contour } C \quad \text{(V)}\tag{7-4}$$

and

$$\Phi = \int_S \mathbf{B} \cdot d\mathbf{s} = \text{magnetic flux crossing surface } S \quad \text{(Wb),}\tag{7-5}$$

then Eq. (7–3) becomes

$$\mathscr{V} = -\frac{d\Phi}{dt} \quad \text{(V)}. \qquad (7\text{--}6)$$

Equation (7–6) states that *the electromotive force induced in a stationary closed circuit is equal to the negative rate of increase of the magnetic flux linking the circuit*. This is a statement of *Faraday's law of electromagnetic induction*. A time-rate of change of magnetic flux induces an electric field according to Eq. (7–3), even in the absence of a physical closed circuit. The negative sign in Eq. (7–6) is an assertion that the induced emf will cause a current to flow in the closed loop in such a direction as to oppose the change in the linking magnetic flux. This assertion is known as *Lenz's law*. The emf induced in a stationary loop caused by a time-varying magnetic field is a *transformer emf*.

EXAMPLE 7–1   A circular loop of N turns of conducting wire lies in the xy-plane with its center at the origin of a magnetic field specified by $\mathbf{B} = \mathbf{a}_z B_0 \cos(\pi r/2b) \sin \omega t$, where b is the radius of the loop and ω is the angular frequency. Find the emf induced in the loop.

Solution   The problem specifies a stationary loop in a time-varying magnetic field; hence Eq. (7–6) can be used directly to find the induced emf, $\mathscr{V}$. The magnetic flux linking each turn of the circular loop is

$$\Phi = \int_S \mathbf{B} \cdot d\mathbf{s}$$

$$= \int_0^b \left[\mathbf{a}_z B_0 \cos \frac{\pi r}{2b} \sin \omega t \right] \cdot (\mathbf{a}_z 2\pi r \, dr)$$

$$= \frac{8b^2}{\pi} \left(\frac{\pi}{2} - 1 \right) B_0 \sin \omega t.$$

Since there are N turns, the total flux linkage is $N\Phi$, and we obtain

$$\mathscr{V} = -N \frac{d\Phi}{dt}$$

$$= -\frac{8N}{\pi} b^2 \left(\frac{\pi}{2} - 1 \right) B_0 \omega \cos \omega t \quad \text{(V)}.$$

The induced emf is seen to be 90° out of time phase with the magnetic flux.   ■

7–2.2   TRANSFORMERS

A transformer is an alternating-current (a-c) device that transforms voltages, currents, and impedances. It usually consists of two or more coils coupled magnetically through a common ferromagnetic core, such as that sketched in Fig. 7–1. Faraday's law of electromagnetic induction is the principle of operation of transformers.

For the closed path in the magnetic circuit in Fig. 7–1(a) traced by magnetic flux Φ, we have, from Eq. (6–101),

$$N_1 i_1 - N_2 i_2 = \mathcal{R}\Phi, \qquad (7\text{–}7)$$

where N_1, N_2 and i_1, i_2 are the numbers of turns and the currents in the primary and secondary circuits, respectively, and $\mathcal{R}$ denotes the reluctance of the magnetic circuit. In Eq. (7–7) we have noted, in accordance with Lenz's law, that the induced mmf in the secondary circuit, $N_2 i_2$, opposes the flow of the magnetic flux Φ created by the mmf in the primary circuit, $N_1 i_1$. From Section 6–8 we know that the reluctance of the ferromagnetic core of length ℓ, cross-sectional area S, and permeability μ is

$$\mathcal{R} = \frac{\ell}{\mu S}. \qquad (7\text{–}8)$$

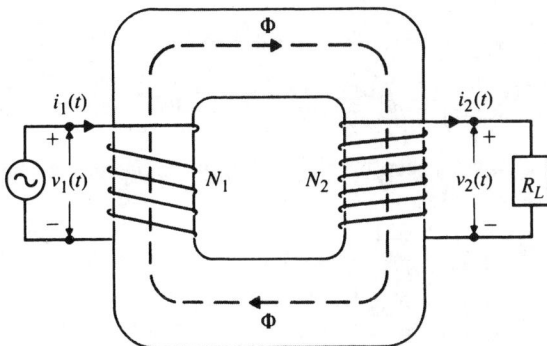

(a) Schematic diagram of a transformer.

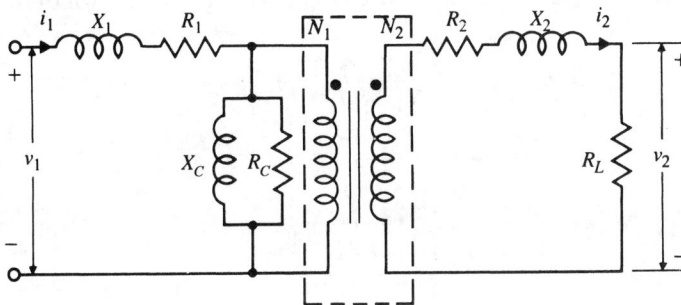

Ideal transformer

(b) An equivalent circuit.

FIGURE 7–1
Schematic diagram and equivalent circuit of a transformer.

Substituting Eq. (7–8) in Eq. (7–7), we obtain

$$N_1 i_1 - N_2 i_2 = \frac{\ell}{\mu S} \Phi. \tag{7–9}$$

a) *Ideal transformer.* For an ideal transformer we assume that $\mu \to \infty$, and Eq. (7–9) becomes

$$\boxed{\frac{i_1}{i_2} = \frac{N_2}{N_1}.} \tag{7–10}$$

Equation (7–10) states that *the ratio of the currents in the primary and secondary windings of an ideal transformer is equal to the inverse ratio of the numbers of turns.* Faraday's law tells us that

$$v_1 = N_1 \frac{d\Phi}{dt} \tag{7–11}$$

and

$$v_2 = N_2 \frac{d\Phi}{dt}, \tag{7–12}$$

the proper signs for v_1 and v_2 having been taken care of by the designated polarities in Fig. 7–1(a). From Eqs. (7–11) and (7–12) we have

$$\boxed{\frac{v_1}{v_2} = \frac{N_1}{N_2}.} \tag{7–13}$$

Thus, *the ratio of the voltages across the primary and secondary windings of an an ideal transformer is equal to the turns ratio.*

When the secondary winding is terminated in a load resistance R_L, as shown in Fig. 7–1(a), the effective load seen by the source connected to primary winding is

$$(R_1)_{\text{eff}} = \frac{v_1}{i_1} = \frac{(N_1/N_2)v_2}{(N_2/N_1)i_2},$$

or

$$\boxed{(R_1)_{\text{eff}} = \left(\frac{N_1}{N_2}\right)^2 R_L,} \tag{7–14a}$$

which is the load resistance multiplied by the square of the turns ratio. For a sinusoidal source $v_1(t)$ and a load impedance Z_L, it is obvious that the effective load seen by the source is $(N_1/N_2)^2 Z_L$, an impedance transformation. We have

$$\boxed{(Z_1)_{\text{eff}} = \left(\frac{N_1}{N_2}\right)^2 Z_L.} \tag{7–14b}$$

b) *Real transformer.* Referring back to Eq. (7–9), we can write the magnetic flux linkages of the primary and secondary windings as

$$\Lambda_1 = N_1\Phi = \frac{\mu S}{\ell}(N_1^2 i_1 - N_1 N_2 i_2), \tag{7–15}$$

$$\Lambda_2 = N_2\Phi = \frac{\mu S}{\ell}(N_1 N_2 i_1 - N_2^2 i_2). \tag{7–16}$$

Using Eqs. (7–15) and (7–16) in Eqs. (7–11) and (7–12), we obtain

$$v_1 = L_1\frac{di_1}{dt} - L_{12}\frac{di_2}{dt}, \tag{7–17}$$

$$v_2 = L_{12}\frac{di_1}{dt} - L_2\frac{di_2}{dt}, \tag{7–18}$$

where

$$L_1 = \frac{\mu S}{\ell}N_1^2, \tag{7–19}$$

$$L_2 = \frac{\mu S}{\ell}N_2^2, \tag{7–20}$$

$$L_{12} = \frac{\mu S}{\ell}N_1 N_2 \tag{7–21}$$

are the self-inductance of the primary winding, the self-inductance of the secondary winding, and the mutual inductance between the primary and secondary windings, respectively. For an ideal transformer there is no leakage flux, and $L_{12} = \sqrt{L_1 L_2}$. For real transformers,

$$L_{12} = k\sqrt{L_1 L_2}, \qquad k < 1, \tag{7–22}$$

where k is called the **coefficient of coupling**. We see that the expressions in Eqs. (7–19), (7–20), and (7–21) are consistent with the inductance per unit length formula, Eq. (6–135), for a long solenoid. In both cases we assume no leakage flux. Note that the assumption of an infinite μ for an ideal transformer also implies infinite inductances.

For real transformers we have the following real-life conditions: the existence of leakage flux ($k < 1$), noninfinite inductances, nonzero winding resistances, and the presence of hysteresis and eddy-current losses. (Eddy-current losses will be discussed presently.) The nonlinear nature of the ferromagnetic core (the dependence of permeability on magnetic field intensity) further compounds the difficulty of an exact analysis of real transformers. Figure 7–1(b) is an approximate equivalent circuit for the transformer in Fig. 7–1(a). In Fig. 7–1(b). R_1 and R_2 are winding resistances, X_1 and X_2 are leakage inductive reactances, R_c represents the power loss due to hysteresis and eddy-current effects, and X_c is a nonlinear inductive reactance representing the nonlinear magnetization behavior of the ferromagnetic core. Analytical determination of these quantities is an exceedingly

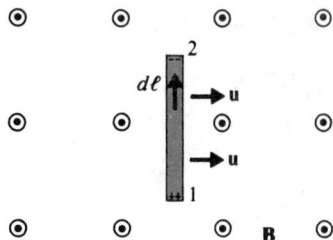

FIGURE 7–2
A conducting bar moving in a magnetic field.

difficult task. The two dots appearing on the winding terminals of the ideal trans-former indicate that the potentials of these terminals rise and fall together due to electromagnetic induction. This dot convention is a simple way of showing the relative sense of the windings on the transformer core.[†]

When time-varying magnetic flux flows in the ferromagnetic core, an induced emf will result in accordance with Faraday's law. This induced emf will produce local currents in the conducting core normal to the magnetic flux. These cur-rents are called *eddy currents*. Eddy currents produce ohmic power loss and cause local heating. As a matter of fact, this is the principle of induction heating. Induction furnaces have been built to produce high enough temperatures to melt metals. In transformers this eddy-current power loss is undesirable and can be reduced by using core materials that have high permeability but low conductivity (high μ and low σ). Ferrites are such materials. For low-frequency, high-power applications an economical way for reducing eddy-current power loss is to use laminated cores; that is, to make transformer cores out of stacked ferromagnetic (iron) sheets, each electrically insulated from its neighbors by thin varnish or oxide coatings. The insulating coatings are parallel to the direction of the magnetic flux so that eddy currents normal to the flux are restricted to the laminated sheets. It can be proved that the total eddy-current power loss decreases as the number of laminations increases. (See Problem P.7–6.) The amount of power-loss reduction depends on the shape and size of the cross section as well as on the method of lamination. For instance, the circular core in Fig. 7–12(a) could also be laminated into stacked insulated sheets, instead of the filamentary parts shown in Fig. 7–12(b).

7–2.3 A MOVING CONDUCTOR IN A STATIC MAGNETIC FIELD

When a conductor moves with a velocity **u** in a static (non-time-varying) magnetic field **B** as shown in Fig. 7–2, a force $\mathbf{F}_m = q\mathbf{u} \times \mathbf{B}$ will cause the freely movable

---

[†] See, for instance, D. K. Cheng, *Analysis of Linear Systems*, Addison-Wesley, Reading, Mass. 1959, p. 50.

electrons in the conductor to drift toward one end of the conductor and leave the other end positively charged. This separation of the positive and negative charges creates a Coulombian force of attraction. The charge-separation process continues until the electric and magnetic forces balance each other and a state of equilibrium is reached. At equilibrium, which is reached very rapidly, the net force on the free charges in the moving conductor is zero.

To an observer moving with the conductor there is no apparent motion, and the magnetic force per unit charge $\mathbf{F}_m/q = \mathbf{u} \times \mathbf{B}$ can be interpreted as an induced electric field acting along the conductor and producing a voltage

$$V_{21} = \int_1^2 (\mathbf{u} \times \mathbf{B}) \cdot d\boldsymbol{\ell}. \tag{7-23}$$

If the moving conductor is a part of a closed circuit C, then the emf generated around the circuit is

$$\boxed{\mathscr{V}' = \oint_C (\mathbf{u} \times \mathbf{B}) \cdot d\boldsymbol{\ell} \qquad (\text{V}).} \tag{7-24}$$

This is referred to as a *flux cutting emf* or a *motional emf*. Obviously, only the part of the circuit that moves in a direction not parallel to (and hence, figuratively, "cutting") the magnetic flux will contribute to $\mathscr{V}'$ in Eq. (7-24).

EXAMPLE 7-2 A metal bar slides over a pair of conducting rails in a uniform magnetic field $\mathbf{B} = \mathbf{a}_z B_0$ with a constant velocity $\mathbf{u}$, as shown in Fig. 7-3.

a) Determine the open-circuit voltage V_0 that appears across terminals 1 and 2.

b) Assuming that a resistance R is connected between the terminals, find the electric power dissipated in R.

c) Show that this electric power is equal to the mechanical power required to move the sliding bar with a velocity $\mathbf{u}$. Neglect the electric resistance of the metal bar and of the conducting rails. Neglect also the mechanical friction at the contact points.

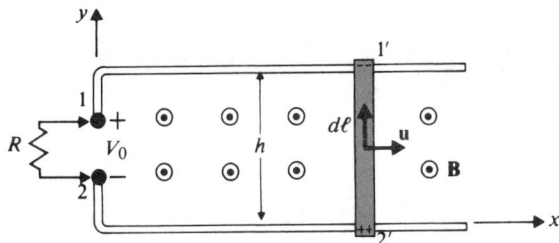

FIGURE 7-3
A metal bar sliding over conducting rails
(Example 7-2).

Solution

a) The moving bar generates a flux-cutting emf. We use Eq. (7–24) to find the open-circuit voltage V_0:

$$V_0 = V_1 - V_2 = \oint_C (\mathbf{u} \times \mathbf{B}) \cdot d\ell$$
$$= \int_{2'}^{1'} (\mathbf{a}_x u \times \mathbf{a}_z B_0) \cdot (\mathbf{a}_y \, d\ell) \qquad (7\text{–}25)$$
$$= -u B_0 h \qquad (\text{V}).$$

b) When a resistance R is connected between terminals 1 and 2, a current $I = u B_0 h / R$ will flow from terminal 2 to terminal 1, so the electric power, P_e, dissipated in R is

$$P_e = I^2 R = \frac{(u B_0 h)^2}{R} \qquad (\text{W}). \qquad (7\text{–}26)$$

c) The mechanical power, P_m, required to move the sliding bar is

$$P_m = \mathbf{F} \cdot \mathbf{u} \qquad (\text{W}), \qquad (7\text{–}27)$$

where $\mathbf{F}$ is the mechanical force required to counteract the magnetic force, $\mathbf{F}_m$, which the magnetic field exerts on the current-carrying metal bar. From Eq. (6–184) we have

$$\mathbf{F}_m = I \int_{2'}^{1'} d\ell \times \mathbf{B} = -\mathbf{a}_x I B_0 h \qquad (\text{N}). \qquad (7\text{–}28)$$

The negative sign in Eq. (7–28) arises because current I flows in a direction opposite to that of $d\ell$. Hence,

$$\mathbf{F} = -\mathbf{F}_m = \mathbf{a}_x I B_0 h = \mathbf{a}_x u B_0^2 h^2 / R \qquad (\text{N}). \qquad (7\text{–}29)$$

Substitution of Eq. (7–29) in Eq. (7–27) proves that $P_m = P_e$, which upholds the principle of conservation of energy.   ▬

EXAMPLE 7–3   The *Faraday disk generator* consists of a circular metal disk rotating with a constant angular velocity ω in a uniform and constant magnetic field of

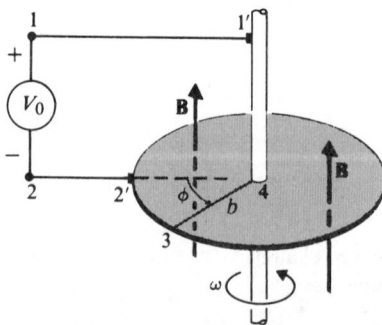

FIGURE 7–4
Faraday disk generator (Example 7–3).

flux density $\mathbf{B} = \mathbf{a}_z B_0$ that is parallel to the axis of rotation. Brush contacts are provided at the axis and on the rim of the disk, as depicted in Fig. 7–4. Determine the open-circuit voltage of the generator if the radius of the disk is b.

Solution   Let us consider the circuit 122′341′1. Of the part 2′34 that moves with the disk, only the straight portion 34 "cuts" the magnetic flux. We have, from Eq. (7–24),

$$
\begin{aligned}
V_0 &= \oint (\mathbf{u} \times \mathbf{B}) \cdot d\ell \\
&= \int_3^4 \left[(\mathbf{a}_\phi r\omega) \times \mathbf{a}_z B_0 \right] \cdot (\mathbf{a}_r \, dr) \\
&= \omega B_0 \int_b^0 r \, dr = -\frac{\omega B_0 b^2}{2} \quad \text{(V)},
\end{aligned}
\tag{7–30}
$$

which is the emf of the Faraday disk generator. To measure V_0, we must use a voltmeter of a very high resistance so that no appreciable current flows in the circuit to modify the externally applied magnetic field.   ▬

7–2.4   A MOVING CIRCUIT IN A TIME-VARYING MAGNETIC FIELD

When a charge q moves with a velocity $\mathbf{u}$ in a region where both an electric field $\mathbf{E}$ and a magnetic field $\mathbf{B}$ exist, the electromagnetic force $\mathbf{F}$ on q, as measured by a laboratory observer, is given by Lorentz's force equation, Eq. (6–5), which is repeated below:

$$
\mathbf{F} = q(\mathbf{E} + \mathbf{u} \times \mathbf{B}). \tag{7–31}
$$

To an observer moving with q, there is no apparent motion, and the force on q can be interpreted as caused by an electric field $\mathbf{E}'$, where

$$
\mathbf{E}' = \mathbf{E} + \mathbf{u} \times \mathbf{B} \tag{7–32}
$$

or

$$
\mathbf{E} = \mathbf{E}' - \mathbf{u} \times \mathbf{B}. \tag{7–33}
$$

Hence, when a conducting circuit with contour C and surface S moves with a velocity $\mathbf{u}$ in a field $(\mathbf{E}, \mathbf{B})$, we use Eq. (7–33) in Eq. (7–2) to obtain

$$
\boxed{\oint_C \mathbf{E}' \cdot d\ell = -\int_S \frac{\partial \mathbf{B}}{\partial t} \cdot d\mathbf{s} + \oint_C (\mathbf{u} \times \mathbf{B}) \cdot d\ell \quad \text{(V)}.} \tag{7–34}
$$

Equation (7–34) is the general form of *Faraday's law* for a moving circuit in a time-varying magnetic field. The line integral on the left side is the emf induced in the moving frame of reference. The first term on the right side represents the transformer emf due to the time variation of $\mathbf{B}$; and the second term represents the motional emf due to the motion of the circuit in $\mathbf{B}$. The division of the induced emf between the transformer and the motional parts depends on the chosen frame of reference.

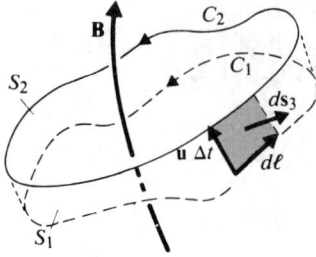

FIGURE 7–5
A moving circuit in a time-varying magnetic field.

Let us consider a circuit with contour C that moves from C_1 at time t to C_2 at time $t + \Delta t$ in a changing magnetic field $\mathbf{B}$. The motion may include translation, rotation, and distortion in an arbitrary manner. Figure 7–5 illustrates the situation. The time-rate of change of magnetic flux through the contour is

$$
\begin{aligned}
\frac{d\Phi}{dt} &= \frac{d}{dt} \int_S \mathbf{B} \cdot d\mathbf{s} \\
&= \lim_{\Delta t \to 0} \frac{1}{\Delta t} \left[\int_{S_2} \mathbf{B}(t + \Delta t) \cdot d\mathbf{s}_2 - \int_{S_1} \mathbf{B}(t) \cdot d\mathbf{s}_1 \right].
\end{aligned}
\tag{7–35}
$$

$\mathbf{B}(t + \Delta t)$ in Eq. (7–35) can be expanded as a Taylor's series:

$$
\mathbf{B}(t + \Delta t) = \mathbf{B}(t) + \frac{\partial \mathbf{B}(t)}{\partial t} \Delta t + \text{H.O.T.},
\tag{7–36}
$$

where the higher-order terms (H.O.T.) contain the second and higher powers of (Δt). Substitution of Eq. (7–36) in Eq. (7–35) yields

$$
\frac{d}{dt} \int_S \mathbf{B} \cdot d\mathbf{s} = \int_S \frac{\partial \mathbf{B}}{\partial t} \cdot d\mathbf{s} + \lim_{\Delta t \to 0} \frac{1}{\Delta t} \left[\int_{S_2} \mathbf{B} \cdot d\mathbf{s}_2 - \int_{S_1} \mathbf{B} \cdot d\mathbf{s}_1 + \text{H.O.T.} \right],
\tag{7–37}
$$

where $\mathbf{B}$ has been written for $\mathbf{B}(t)$ for simplicity. In going from C_1 to C_2 the circuit covers a region that is bounded by S_1, S_2, and S_3. Side surface S_3 is the area swept out by the contour in time Δt. An element of the side surface is

$$
d\mathbf{s}_3 = d\boldsymbol{\ell} \times \mathbf{u}\, \Delta t.
\tag{7–38}
$$

We now apply the divergence theorem for $\mathbf{B}$ at time t to the region sketched in Fig. 7–5:

$$
\int_V \boldsymbol{\nabla} \cdot \mathbf{B}\, dv = \int_{S_2} \mathbf{B} \cdot d\mathbf{s}_2 - \int_{S_1} \mathbf{B} \cdot d\mathbf{s}_1 + \int_{S_3} \mathbf{B} \cdot d\mathbf{s}_3,
\tag{7–39}
$$

where a negative sign is included in the term involving $d\mathbf{s}_1$ because *outward* normals must be used in the divergence theorem. Using Eq. (7–38) in Eq. (7–39) and noting that $\boldsymbol{\nabla} \cdot \mathbf{B} = 0$, we have

$$
\int_{S_2} \mathbf{B} \cdot d\mathbf{s}_2 - \int_{S_1} \mathbf{B} \cdot d\mathbf{s}_1 = -\Delta t \oint_C (\mathbf{u} \times \mathbf{B}) \cdot d\boldsymbol{\ell}.
\tag{7–40}
$$

Combining Eqs. (7–37) and (7–40), we obtain

$$\frac{d}{dt}\int_S \mathbf{B} \cdot d\mathbf{s} = \int_S \frac{\partial \mathbf{B}}{\partial t} \cdot d\mathbf{s} - \oint_C (\mathbf{u} \times \mathbf{B}) \cdot d\boldsymbol{\ell}, \qquad (7\text{–}41)$$

which can be identified as the negative of the right side of Eq. (7–34).

If we designate

$$\mathscr{V}' = \oint_C \mathbf{E}' \cdot d\boldsymbol{\ell}$$

$$= \text{emf induced in circuit } C \text{ measured in the moving frame}, \qquad (7\text{–}42)$$

Eq. (7–34) can be written simply as

$$\boxed{\begin{aligned} \mathscr{V}' &= -\frac{d}{dt}\int_S \mathbf{B} \cdot d\mathbf{s} \\ &= -\frac{d\Phi}{dt} \quad (\text{V}), \end{aligned}} \qquad (7\text{–}43)$$

which is of the same form as Eq. (7–6). Of course, if a circuit is not in motion, $\mathscr{V}'$ reduces to $\mathscr{V}$, and Eqs. (7–43) and (7–6) are exactly the same. Hence, Faraday's law that the emf induced in a closed circuit equals the negative time-rate of increase of the magnetic flux linking a circuit applies to a stationary circuit as well as a moving one. Either Eq. (7–34) or Eq. (7–43) can be used to evaluate the induced emf in the general case. If a high-impedance voltmeter is inserted in a conducting circuit, it will read the open-circuit voltage due to electromagnetic induction whether the circuit is stationary or moving. We have mentioned that the division of the induced emf in Eq. (7–34) into transformer and motional emf's is not unique, but their sum is always equal to that computed by using Eq. (7–43).

In Example 7–2 (Fig. 7–3) we determined the open-circuit voltage V_0 by using Eq. (7–24). If we use Eq. (7–43), we have

$$\Phi = \int_S \mathbf{B} \cdot d\mathbf{s} = B_0(hut)$$

and

$$V_0 = -\frac{d\Phi}{dt} = -uB_0h \qquad (\text{V}),$$

which is the same as Eq. (7–25).

Similarly, for the Faraday disk generator in Example 7–3 the magnetic flux linking the circuit 122′341′1 is that which passes through the wedge-shaped area 2′342′:

$$\Phi = \int_S \mathbf{B} \cdot d\mathbf{s} = B_0 \int_0^b \int_0^{\omega t} r \, d\phi \, dr$$

$$= B_0(\omega t)\frac{b^2}{2}$$

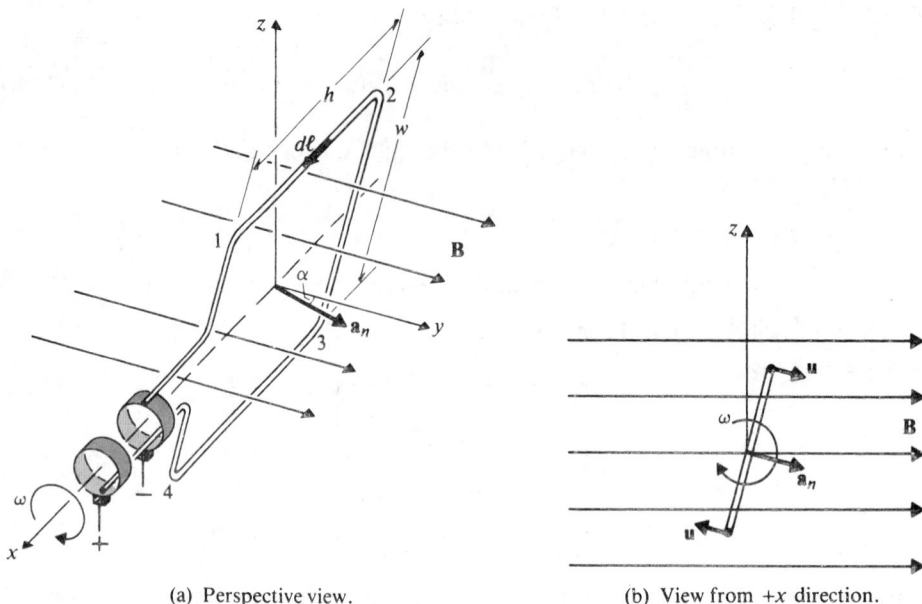

(a) Perspective view.   (b) View from +x direction.

FIGURE 7–6
A rectangular conducting loop rotating in a changing magnetic field (Example 7–4).

and

$$V_0 = -\frac{d\Phi}{dt} = -\frac{\omega B_0 b^2}{2},$$

which is the same as Eq. (7–30).

EXAMPLE 7–4   An h by w rectangular conducting loop is situated in a changing magnetic field $\mathbf{B} = \mathbf{a}_y B_0 \sin \omega t$. The normal of the loop initially makes an angle α with $\mathbf{a}_y$, as shown in Fig. 7–6. Find the induced emf in the loop: (a) when the loop is at rest, and (b) when the loop rotates with an angular velocity ω about the x-axis.

Solution

a) When the loop is at rest, we use Eq. (7–6):

$$\Phi = \int \mathbf{B} \cdot d\mathbf{s}$$
$$= (\mathbf{a}_y B_0 \sin \omega t) \cdot (\mathbf{a}_n hw)$$
$$= B_0 hw \sin \omega t \cos \alpha.$$

Therefore,

$$\mathcal{V}_a = -\frac{d\Phi}{dt} = -B_0 S\omega \cos \omega t \cos \alpha, \tag{7–44}$$

where $S = hw$ is the area of the loop. The relative polarities of the terminals are as indicated. If the circuit is completed through an external load, $\mathscr{V}_a$ will produce a current that will oppose the change in Φ.

b) When the loop rotates about the x-axis, both terms in Eq. (7–34) contribute: the first term contributes the transformer emf $\mathscr{V}_a$ in Eq. (7–44), and the second term contributes a motional emf $\mathscr{V}'_a$ where

$$\mathscr{V}'_a = \oint_C (\mathbf{u} \times \mathbf{B}) \cdot d\boldsymbol{\ell}$$

$$= \int_2^1 \left[\left(\mathbf{a}_n \frac{w}{2} \omega \right) \times (\mathbf{a}_y B_0 \sin \omega t) \right] \cdot (\mathbf{a}_x\, dx)$$

$$+ \int_4^3 \left[\left(-\mathbf{a}_n \frac{w}{2} \omega \right) \times (\mathbf{a}_y B_0 \sin \omega t) \right] \cdot (\mathbf{a}_x\, dx)$$

$$= 2\left(\frac{w}{2} \omega B_0 \sin \omega t \sin \alpha \right) h.$$

Note that the sides 23 and 41 do not contribute to $\mathscr{V}'_a$ and that the contributions of sides 12 and 34 are of equal magnitude and in the same direction. If $\alpha = 0$ at $t = 0$, then $\alpha = \omega t$, and we can write

$$\mathscr{V}'_a = B_0 S \omega \sin \omega t \sin \omega t. \tag{7–45}$$

The total emf induced or generated in the rotating loop is the sum of $\mathscr{V}_a$ in Eq. (7–44) and $\mathscr{V}'_a$ in Eq. (7–45):

$$\mathscr{V}'_t = -B_0 S \omega (\cos^2 \omega t - \sin^2 \omega t) = -B_0 S \omega \cos 2\omega t, \tag{7–46}$$

which has an angular frequency 2ω.

We can determine the total induced emf $\mathscr{V}'_t$ by applying Eq. (7–43) directly. At any time t, the magnetic flux linking the loop is

$$\Phi(t) = \mathbf{B}(t) \cdot [\mathbf{a}_n(t)S] = B_0 S \sin \omega t \cos \alpha$$

$$= B_0 S \sin \omega t \cos \omega t = \tfrac{1}{2} B_0 S \sin 2\omega t.$$

Hence,

$$\mathscr{V}'_t = -\frac{d\Phi}{dt} = -\frac{d}{dt}\left(\frac{1}{2} B_0 S \sin 2\omega t \right)$$

$$= -B_0 S \omega \cos 2\omega t$$

as before.

7-3 Maxwell's Equations

The fundamental postulate for electromagnetic induction assures us that a time-varying magnetic field gives rise to an electric field. This assurance has been amply verified by numerous experiments. The $\nabla \times \mathbf{E} = 0$ equation in Table 7–1 must therefore be replaced by Eq. (7–1) in the time-varying case. Following are the revised set of two curl and two divergence equations from Table 7–1:

$$\mathbf{V} \times \mathbf{E} = -\frac{\partial \mathbf{B}}{\partial t},$$ (7–47a)

$$\mathbf{V} \times \mathbf{H} = \mathbf{J},$$ (7–47b)

$$\mathbf{V} \cdot \mathbf{D} = \rho,$$ (7–47c)

$$\mathbf{V} \cdot \mathbf{B} = 0.$$ (7–47d)

In addition, we know that the principle of conservation of charge must be satisfied at all times. The mathematical expression of charge conservation is the equation of continuity, Eq. (5–44), which is repeated below:

$$\mathbf{V} \cdot \mathbf{J} = -\frac{\partial \rho}{\partial t}.$$ (7–48)

The crucial question here is whether the set of four equations in (7–47a, b, c, and d) are now consistent with the requirement specified by Eq. (7–48) in a time-varying situation. That the answer is in the negative is immediately obvious by simply taking the divergence of Eq. (7–47b),

$$\mathbf{V} \cdot (\mathbf{V} \times \mathbf{H}) = 0 = \mathbf{V} \cdot \mathbf{J},$$ (7–49)

which follows from the null identity, Eq. (2–149). We are reminded that the divergence of the curl of any well-behaved vector field is zero. Since Eq. (7–48) asserts that $\mathbf{V} \cdot \mathbf{J}$ does not vanish in a time-varying situation, Eq. (7–49) is, in general, not true.

How should Eqs. (7–47a, b, c, and d) be modified so that they are consistent with Eq. (7–48)? First of all, a term $\partial \rho/\partial t$ must be added to the right side of Eq. (7–49):

$$\mathbf{V} \cdot (\mathbf{V} \times \mathbf{H}) = 0 = \mathbf{V} \cdot \mathbf{J} + \frac{\partial \rho}{\partial t}.$$ (7–50)

Using Eq. (7–47c) in Eq. (7–50), we have

$$\mathbf{V} \cdot (\mathbf{V} \times \mathbf{H}) = \mathbf{V} \cdot \left(\mathbf{J} + \frac{\partial \mathbf{D}}{\partial t} \right),$$ (7–51)

which implies that

$$\boxed{\mathbf{V} \times \mathbf{H} = \mathbf{J} + \frac{\partial \mathbf{D}}{\partial t}.}$$ (7–52)[†]

Equation (7–52) indicates that a time-varying electric field will give rise to a magnetic field, even in the absence of a current flow. The additional term $\partial \mathbf{D}/\partial t$ is necessary to make Eq. (7–52) consistent with the principle of conservation of charge.

[†] An integration constant could be added to Eq. (7–52) without violating Eq. (7–51), but this constant must be zero in order that Eq. (7–52) reduces to Eq. (7–47b) in the static case.

It is easy to verify that $\partial \mathbf{D}/\partial t$ has the dimension of a current density (SI unit: A/m^2). The term $\partial \mathbf{D}/\partial t$ is called **displacement current density**, and its introduction in the $\nabla \times \mathbf{H}$ equation was one of the major contributions of James Clerk Maxwell (1831–1879). In order to be consistent with the equation of continuity in a time-varying situation, both of the curl equations in Table 7–1 must be generalized. The set of four consistent equations to replace the inconsistent equations, Eqs. (7–47a, b, c, and d), are

$$\nabla \times \mathbf{E} = -\frac{\partial \mathbf{B}}{\partial t}, \tag{7–53a}$$

$$\nabla \times \mathbf{H} = \mathbf{J} + \frac{\partial \mathbf{D}}{\partial t}, \tag{7–53b}$$

$$\nabla \cdot \mathbf{D} = \rho, \tag{7–53c}$$

$$\nabla \cdot \mathbf{B} = 0. \tag{7–53d}$$

They are known as **Maxwell's equations**. Note that ρ in Eq. (7–53c) is the volume density of *free charges*, and $\mathbf{J}$ in Eq. (7–53b) is the density of *free currents*, which may comprise both convection current ($\rho \mathbf{u}$) and conduction current ($\sigma \mathbf{E}$). These four equations, together with the equation of continuity in Eq. (7–48) and Lorentz's force equation in Eq. (6–5), form the foundation of electromagnetic theory. These equations can be used to explain and predict *all* macroscopic electromagnetic phenomena.

Although the four Maxwell's equations in Eqs. (7–53a, b, c, and d) are consistent, they are not all independent. As a matter of fact, the two divergence equations, Eqs. (5–53c and d), can be derived from the two curl equations, Eqs. (7–53a and b), by making use of the equation of continuity, Eq. (7–48) (see Problem P.7–11). The four fundamental field vectors $\mathbf{E}$, $\mathbf{D}$, $\mathbf{B}$, $\mathbf{H}$ (each having three components) represent twelve unknowns. Twelve scalar equations are required for the determination of these twelve unknowns. The required equations are supplied by the two vector curl equations and the two vector constitutive relations $\mathbf{D} = \epsilon \mathbf{E}$ and $\mathbf{H} = \mathbf{B}/\mu$, each vector equation being equivalent to three scalar equations.

7–3.1   INTEGRAL FORM OF MAXWELL'S EQUATIONS

The four Maxwell's equations in (7–53a, b, c, and d) are differential equations that are valid at every point in space. In explaining electromagnetic phenomena in a physical environment we must deal with finite objects of specified shapes and boundaries. It is convenient to convert the differential forms into their integral-form equivalents. We take the surface integral of both sides of the curl equations in Eqs. (7–53a) and (7–53b) over an open surface S with a contour C and apply Stokes's theorem to obtain

$$\oint_C \mathbf{E} \cdot d\boldsymbol{\ell} = -\int_S \frac{\partial \mathbf{B}}{\partial t} \cdot d\mathbf{s} \tag{7–54a}$$

and

$$\oint_C \mathbf{H} \cdot d\boldsymbol{\ell} = \int_S \left(\mathbf{J} + \frac{\partial \mathbf{D}}{\partial t} \right) \cdot d\mathbf{s}. \tag{7–54b}$$

Taking the volume integral of both sides of the divergence equations in Eqs. (7–53c) and (7–53d) over a volume V with a *closed* surface S and using divergence theorem, we have

$$\oint_S \mathbf{D} \cdot d\mathbf{s} = \int_V \rho \, dv \tag{7–54c}$$

and

$$\oint_S \mathbf{B} \cdot d\mathbf{s} = 0. \tag{7–54d}$$

The set of four equations in (7–54a, b, c, and d) are the integral form of Maxwell's equations. We see that Eq. (7–54a) is the same as Eq. (7–2), which is an expression of Faraday's law of electromagnetic induction. Equation (7–54b) is a generalization of Ampère's circuital law given in Eq. (6–78), the latter applying only to static magnetic fields. Note that the current density $\mathbf{J}$ may consist of a convection current density $\rho\mathbf{u}$ due to the motion of a free-charge distribution, as well as a conduction current density $\sigma\mathbf{E}$ caused by the presence of an electric field in a conducting medium. The surface integral of $\mathbf{J}$ is the current I flowing through the open surface S.

Equation (7–54c) can be recognized as Gauss's law, which we used extensively in electrostatics and which remains the same in the time-varying case. The volume integral of ρ equals the total charge Q that is enclosed in surface S. No particular law is associated with Eq. (7–54d); but, in comparing it with Eq. (7–54c) we conclude that there are no isolated magnetic charges and that the total outward magnetic flux through any closed surface is zero. Both the differential and the integral forms

TABLE 7–2
Maxwell's Equations

Differential Form	Integral Form	Significance
$\nabla \times \mathbf{E} = -\dfrac{\partial \mathbf{B}}{\partial t}$	$\oint_C \mathbf{E} \cdot d\boldsymbol{\ell} = -\dfrac{d\Phi}{dt}$	Faraday's law
$\nabla \times \mathbf{H} = \mathbf{J} + \dfrac{\partial \mathbf{D}}{\partial t}$	$\oint_C \mathbf{H} \cdot d\boldsymbol{\ell} = I + \int_S \dfrac{\partial \mathbf{D}}{\partial t} \cdot d\mathbf{s}$	Ampère's circuital law
$\nabla \cdot \mathbf{D} = \rho$	$\oint_S \mathbf{D} \cdot d\mathbf{s} = Q$	Gauss's law
$\nabla \cdot \mathbf{B} = 0$	$\oint_S \mathbf{B} \cdot d\mathbf{s} = 0$	No isolated magnetic charge

of Maxwell's equations are collected in Table 7–2 for easy reference. It is obvious that in non-time-varying cases these equations simplify to the fundamental relations in Table 7–1 for electrostatic and magnetostatic models.

EXAMPLE 7–5   An a-c voltage source of amplitude V_0 and angular frequency ω, $v_c = V_0 \sin \omega t$, is connected across a parallel-plate capacitor C_1, as shown in Fig. 7–7. (a) Verify that the displacement current in the capacitor is the same as the conduction current in the wires. (b) Determine the magnetic field intensity at a distance r from the wire.

Solution

a) The conduction current in the connecting wire is

$$i_C = C_1 \frac{dv_C}{dt} = C_1 V_0 \omega \cos \omega t \qquad \text{(A)}.$$

For a parallel-plate capacitor with an area A, plate separation d, and a dielectric medium of permittivity ϵ the capacitance is

$$C_1 = \epsilon \frac{A}{d}.$$

With a voltage v_C appearing between the plates, the uniform electric field intensity E in the dielectric is equal to (neglecting fringing effects) $E = v_C/d$, whence

$$D = \epsilon E = \epsilon \frac{V_0}{d} \sin \omega t.$$

The displacement current is then

$$i_D = \int_A \frac{\partial \mathbf{D}}{\partial t} \cdot d\mathbf{s} = \left(\epsilon \frac{A}{d} \right) V_0 \omega \cos \omega t$$

$$= C_1 V_0 \omega \cos \omega t = i_C. \qquad \text{Q.E.D.}$$

b) The magnetic field intensity at a distance r from the conducting wire can be found by applying the generalized Ampère's circuital law, Eq. (7–54b), to contour

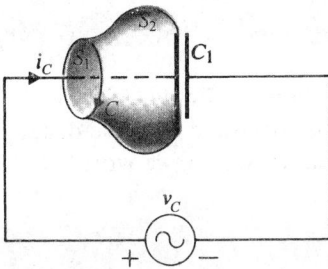

FIGURE 7–7
A parallel-plate capacitor connected to an a-c voltage source (Example 7–5).

C in Fig. 7–7. Two typical open surfaces with rim C may be chosen: (1) a planar disk surface S_1, or (2) a curved surface S_2 passing through the dielectric medium. Symmetry around the wire ensures a constant H_ϕ along the contour C. The line integral on the left side of Eq. (7–54b) is

$$\oint_C \mathbf{H} \cdot d\ell = 2\pi r H_\phi.$$

For the surface S_1, only the first term on the right side of Eq. (7–54b) is nonzero because no charges are deposited along the wire and, consequently, $\mathbf{D} = 0$.

$$\int_{S_1} \mathbf{J} \cdot d\mathbf{s} = i_C = C_1 V_0 \omega \cos \omega t.$$

Since the surface S_2 passes through the dielectric medium, no conduction current flows through S_2. If the second surface integral were not there, the right side of Eq. (7–54b) would be zero. This would result in a contradiction. The inclusion of the displacement-current term by Maxwell eliminates this contradiction. As we have shown in part (a), $i_D = i_C$. Hence we obtain the same result whether surface S_1 or surface S_2 is chosen. Equating the two previous integrals, we find that

$$H_\phi = \frac{C_1 V_0}{2\pi r} \omega \cos \omega t \qquad \text{(A/m)}.$$

7–4 Potential Functions

In Section 6–3 the concept of the vector magnetic potential $\mathbf{A}$ was introduced because of the solenoidal nature of $\mathbf{B}$ ($\nabla \cdot \mathbf{B} = 0$):

$$\boxed{\mathbf{B} = \nabla \times \mathbf{A} \qquad \text{(T).}} \qquad (7\text{–}55)$$

If Eq. (7–55) is substituted in the differential form of Faraday's law, Eq. (7–1), we get

$$\nabla \times \mathbf{E} = -\frac{\partial}{\partial t}(\nabla \times \mathbf{A})$$

or

$$\nabla \times \left(\mathbf{E} + \frac{\partial \mathbf{A}}{\partial t}\right) = 0. \qquad (7\text{–}56)$$

Since the sum of the two vector quantities in the parentheses of Eq. (7–56) is curl-free, it can be expressed as the gradient of a scalar. To be consistent with the definition of the scalar electric potential V in Eq. (3–43) for electrostatics, we write

$$\mathbf{E} + \frac{\partial \mathbf{A}}{\partial t} = -\nabla V,$$

from which we obtain

$$\boxed{\mathbf{E} = -\nabla V - \frac{\partial \mathbf{A}}{\partial t} \quad \text{(V/m)}.}$$

(7–57)

In the static case, $\partial \mathbf{A}/\partial t = 0$, and Eq. (7–57) reduces to $\mathbf{E} = -\nabla V$. Hence $\mathbf{E}$ can be determined from V alone, and $\mathbf{B}$ from $\mathbf{A}$ by Eq. (7–55). For time-varying fields, $\mathbf{E}$ depends on both V and $\mathbf{A}$; that is, an electric field intensity can result both from accumulations of charge through the $-\nabla V$ term and from time-varying magnetic fields through the $-\partial \mathbf{A}/\partial t$ term. Inasmuch as $\mathbf{B}$ also depends on $\mathbf{A}$, $\mathbf{E}$ and $\mathbf{B}$ are coupled.

The electric field in Eq. (7–57) can be viewed as composed of two parts: the first part, $-\nabla V$, is due to charge distribution ρ; and the second part, $-\partial \mathbf{A}/\partial t$, is due to time-varying current $\mathbf{J}$. We are tempted to find V from ρ by Eq. (3–61):

$$V = \frac{1}{4\pi\epsilon_0} \int_{V'} \frac{\rho}{R}\, dv',$$

(7–58)

and to find $\mathbf{A}$ by Eq. (6–23):

$$\mathbf{A} = \frac{\mu_0}{4\pi} \int_{V'} \frac{\mathbf{J}}{R}\, dv'.$$

(7–59)

However, the preceding two equations were obtained under static conditions, and V and $\mathbf{A}$ as given were, in fact, solutions of Poisson's equations, Eqs. (4–6) and (6–21), respectively. These solutions may themselves be time-dependent because ρ and $\mathbf{J}$ may be functions of time, but they neglect the time-retardation effects associated with the finite velocity of propagation of time-varying electromagnetic fields. When ρ and $\mathbf{J}$ vary slowly with time (at a very low frequency) and the range of interest R is small in comparison with the wavelength, it is allowable to use Eqs. (7–58) and (7–59) in Eqs. (7–55) and (7–57) to find *quasi-static fields*. We will discuss this again in Subsection 7–7.2.

Quasi-static fields are approximations. Their consideration leads from field theory to circuit theory. However, when the source frequency is high and the range of interest is no longer small in comparison to the wavelength, quasi-static solutions will not suffice. Time-retardation effects must then be included, as in the case of electromagnetic radiation from antennas. These points will be discussed more fully when we study solutions to wave equations.

Let us substitute Eqs. (7–55) and (7–57) into Eq. (7–53b) and make use of the constitutive relations $\mathbf{H} = \mathbf{B}/\mu$ and $\mathbf{D} = \epsilon\mathbf{E}$. We have

$$\nabla \times \nabla \times \mathbf{A} = \mu\mathbf{J} + \mu\epsilon \frac{\partial}{\partial t}\left(-\nabla V - \frac{\partial \mathbf{A}}{\partial t}\right),$$

(7–60)

where a homogeneous medium has been assumed. Recalling the vector identity for $\mathbf{V} \times \mathbf{V} \times \mathbf{A}$ in Eq. (6–17a), we can write Eq. (7–60) as

$$\mathbf{V}(\mathbf{V} \cdot \mathbf{A}) - \mathbf{V}^2\mathbf{A} = \mu\mathbf{J} - \mathbf{V}\left(\mu\epsilon\frac{\partial V}{\partial t}\right) - \mu\epsilon\frac{\partial^2\mathbf{A}}{\partial t^2}$$

or

$$\mathbf{V}^2\mathbf{A} - \mu\epsilon\frac{\partial^2\mathbf{A}}{\partial t^2} = -\mu\mathbf{J} + \mathbf{V}\left(\mathbf{V} \cdot \mathbf{A} + \mu\epsilon\frac{\partial V}{\partial t}\right). \tag{7–61}$$

Now, the definition of a vector requires the specification of both its curl and its divergence. Although the curl of $\mathbf{A}$ is designated $\mathbf{B}$ in Eq. (7–55), we are still at liberty to choose the divergence of $\mathbf{A}$. We let

$$\boxed{\mathbf{V} \cdot \mathbf{A} + \mu\epsilon\frac{\partial V}{\partial t} = 0,} \tag{7–62}$$

which makes the second term on the right side of Eq. (7–61) vanish, so we obtain

$$\boxed{\mathbf{V}^2\mathbf{A} - \mu\epsilon\frac{\partial^2\mathbf{A}}{\partial t^2} = -\mu\mathbf{J}.} \tag{7–63}$$

Equation (7–63) is the *nonhomogeneous wave equation for vector potential* $\mathbf{A}$. It is called a wave equation because its solutions represent waves traveling with a velocity equal to $1/\sqrt{\mu\epsilon}$. This will become clear in Section 7–6 when the solution of wave equations is discussed. The relation between $\mathbf{A}$ and V in Eq. (7–62) is called the *Lorentz condition (or Lorentz gauge) for potentials*. It reduces to the condition $\mathbf{V} \cdot \mathbf{A} = 0$ in Eq. (6–20) for static fields. The Lorentz condition can be shown to be consistent with the equation of continuity (Problem P.7–12).

A corresponding wave equation for the scalar potential V can be obtained by substituting Eq. (7–57) in Eq. (7–53c). We have

$$-\mathbf{V} \cdot \epsilon\left(\mathbf{V}V + \frac{\partial\mathbf{A}}{\partial t}\right) = \rho,$$

which, for a constant ϵ, leads to

$$\mathbf{V}^2V + \frac{\partial}{\partial t}(\mathbf{V} \cdot \mathbf{A}) = -\frac{\rho}{\epsilon}. \tag{7–64}$$

Using Eq. (7–62), we get

$$\boxed{\mathbf{V}^2V - \mu\epsilon\frac{\partial^2 V}{\partial t^2} = -\frac{\rho}{\epsilon},} \tag{7–65}$$

which is the *nonhomogeneous wave equation for scalar potential V*. Hence the Lorentz condition in Eq. (7–62) uncouples the wave equations for $\mathbf{A}$ and for V. The nonhomogeneous wave equations in (7–63) and (7–65) reduce to Poisson's equations in static cases. Since the potential functions given in Eqs. (7–58) and (7–59) are solutions of Poisson's equations, they cannot be expected to be the solutions of nonhomogeneous wave equations in time-varying situations without modification.

7–5   Electromagnetic Boundary Conditions

In order to solve electromagnetic problems involving contiguous regions of different constitutive parameters, it is necessary to know the boundary conditions that the field vectors **E, D, B,** and **H** must satisfy at the interfaces. Boundary conditions are derived by applying the integral form of Maxwell's equations (7–54a, b, c, and d) to a small region at an interface of two media in manners similar to those used in obtaining the boundary conditions for static electric and magnetic fields. The integral equations are assumed to hold for regions containing discontinuous media. The reader should review the procedures followed in Sections 3–9 and 6–10. In general, the application of the integral form of a curl equation to a flat closed path at a boundary with top and bottom sides in the two touching media yields the boundary condition for the tangential components; and the application of the integral form of a divergence equation to a shallow pillbox at an interface with top and bottom faces in the two contiguous media gives the boundary condition for the normal components.

The boundary conditions for the tangential components of **E** and **H** are obtained from Eqs. (7–54a) and (7–54b), respectively:

$$\boxed{E_{1t} = E_{2t} \quad \text{(V/m)};} \tag{7–66a}$$

$$\boxed{\mathbf{a}_{n2} \times (\mathbf{H}_1 - \mathbf{H}_2) = \mathbf{J}_s \quad \text{(A/m)}.} \tag{7–66b}$$

We note that Eqs. (7–66a) and (7–66b) for the time-varying case are exactly the same as Eq. (3–118) for static electric fields and Eq. (6–111) for static magnetic fields, respectively, in spite of the existence of the time-varying terms in Eqs. (7–54a) and (7–54b). The reason is that, in letting the height of the flat closed path (*abcda* in Figs. 3–23 and 6–19) approach zero, the area bounded by the path approaches zero, causing the surface integrals of $\partial \mathbf{B}/\partial t$ and $\partial \mathbf{D}/\partial t$ to vanish.

Similarly, the boundary conditions for the normal components of **D** and **B** are obtained from Eqs. (7–54c) and (7–54d):

$$\boxed{\mathbf{a}_{n2} \cdot (\mathbf{D}_1 - \mathbf{D}_2) = \rho_s \quad \text{(C/m}^2\text{)};} \tag{7–66c}$$

$$\boxed{B_{1n} = B_{2n} \quad \text{(T)}.} \tag{7–66d}$$

These are the same as, respectively, Eq. (3–121a) for static electric fields and Eq. (6–107) for static magnetic fields because we start from the same divergence equations.

We can make the following general statements about electromagnetic boundary conditions:

1. *The tangential component of an* **E** *field is continuous across an interface.*

2. *The tangential component of an* **H** *field is discontinuous across an interface where a surface current exists, the amount of discontinuity being determined by Eq. (7–66b).*

3. *The normal component of a* **D** *field is discontinuous across an interface where a surface charge exists, the amount of discontinuity being determined by Eq. (7–66c).*

4. *The normal component of a* **B** *field is continuous across an interface.*

As we have noted previously, the two divergence equations can be derived from the two curl equations and the equation of continuity; hence, the boundary conditions in Eqs. (7–66c) and (7–66d), which are obtained from the divergence equations, cannot be independent from those in Eqs. (7–66a) and (7–66b), which are obtained from the curl equations. As a matter of fact, in the time-varying case the boundary condition for the tangential component of **E** in Eq. (7–66a) is equivalent to that for the normal component of **B** in Eq. (7–66d), and the boundary condition for the tangential component of **H** in Eq. (7–66b) is equivalent to that for the normal component of **D** in Eq. (7–66c). The simultaneous specification of the tangential component of **E** and the normal component of **B** at a boundary surface in a time-varying situation, for example, would be redundant and, if we are not careful, could result in contradictions.

We now examine the important special cases of (1) a boundary between two lossless linear media, and (2) a boundary between a good dielectric and a good conductor.

7–5.1 INTERFACE BETWEEN TWO LOSSLESS LINEAR MEDIA

A lossless linear medium can be specified by a permittivity ϵ and a permeability μ, with $\sigma = 0$. There are usually no free charges and no surface currents at the interface between two lossless media. We set $\rho_s = 0$ and $\mathbf{J}_s = 0$ in Eqs. (7–66a, b, c, and d) and obtain the boundary conditions listed in Table 7–3.

TABLE 7–3
Boundary Conditions between
Two Lossless Media

$$E_{1t} = E_{2t} \rightarrow \frac{D_{1t}}{D_{2t}} = \frac{\epsilon_1}{\epsilon_2} \qquad (7\text{–}67a)$$

$$H_{1t} = H_{2t} \rightarrow \frac{B_{1t}}{B_{2t}} = \frac{\mu_1}{\mu_2} \qquad (7\text{–}67b)$$

$$D_{1n} = D_{2n} \rightarrow \epsilon_1 E_{1n} = \epsilon_2 E_{2n} \qquad (7\text{–}67c)$$

$$B_{1n} = B_{2n} \rightarrow \mu_1 H_{1n} = \mu_2 H_{2n} \qquad (7\text{–}67d)$$

TABLE 7–4

**Boundary Conditions between a Dielectric (Medium 1) and
a Perfect Conductor (Medium 2) (Time-varying Case)**

On the Side of Medium 1	On the Side of Medium 2	
$E_{1t} = 0$	$E_{2t} = 0$	(7–68a)
$\mathbf{a}_{n2} \times \mathbf{H}_1 = \mathbf{J}_s$	$H_{2t} = 0$	(7–68b)
$\mathbf{a}_{n2} \cdot \mathbf{D}_1 = \rho_s$	$D_{2n} = 0$	(7–68c)
$B_{1n} = 0$	$B_{2n} = 0$	(7–68d)

7–5.2 INTERFACE BETWEEN A DIELECTRIC AND A PERFECT CONDUCTOR

A perfect conductor is one with an infinite conductivity. In the physical world we have an abundance of "good conductors" such as silver, copper, gold, and aluminum, whose conductivities are of the order of 10^7 (S/m). (See the table in Appendix B–4). There are superconducting materials whose conductivities are essentially infinite (in excess of 10^{20} S/m) at cryogenic temperatures. They are called *superconductors*. Because of the requirement of extremely low temperatures, they have not found much practical use. (The apparent upper limit for transition temperature in 1973 was 23 K. Cooling by expensive liquid helium was required.) However, this situation is expected to change in the near future, since scientists have recently found ceramic materials that show superconducting properties at much higher transition temperatures (20–30 degrees above the 77 K boiling point of nitrogen, raising the possibility of using inexpensive liquid nitrogen as a coolant). At the present time the brittleness of the ceramic materials and limitations on usable current density and magnetic field intensity remain obstacles to industrial applications.[†] Room-temperature superconductivity is still a dream.

In order to simplify the analytical solution of field problems, good conductors are often considered perfect conductors in regard to boundary conditions. In the *interior* of a perfect conductor the electric field is zero (otherwise, it would produce an infinite current density), and any charges the conductor will have will reside on the surface only. The interrelationship between (**E, D**) and (**B, H**) through Maxwell's equations ensures that **B** and **H** are also zero in the *interior* of a conductor in a *time-varying situation.*[‡] Consider an interface between a lossless dielectric (medium 1) and a perfect conductor (medium 2). In medium 2, $\mathbf{E}_2 = 0$, $\mathbf{H}_2 = 0$, $\mathbf{D}_2 = 0$, and $\mathbf{B}_2 = 0$. The general boundary conditions in Eqs. (7–66a, b, c, and d) reduce to those listed in Table 7–4. When we apply Eqs. (7–68b) and (7–68c), it is important to note that the reference unit normal is an *outward normal from medium 2* in order to avoid an error in sign. As mentioned in Section 6–10, currents in media with finite conductivities are expressed in terms of volume current densities, and surface current densities defined for currents flowing through an infinitesimal thickness is zero. In this case, Eq.

[†] R. K. Jurgen, "Technology '88—The main event," *IEEE Spectrum*, vol. 25, pp. 27–28, January 1988.

[‡] In the *static* case a steady current in a conductor produces a static magnetic field that does not affect the electric field. Hence, **E** and **D** within a good conductor may be zero, but **B** and **H** may not be zero.

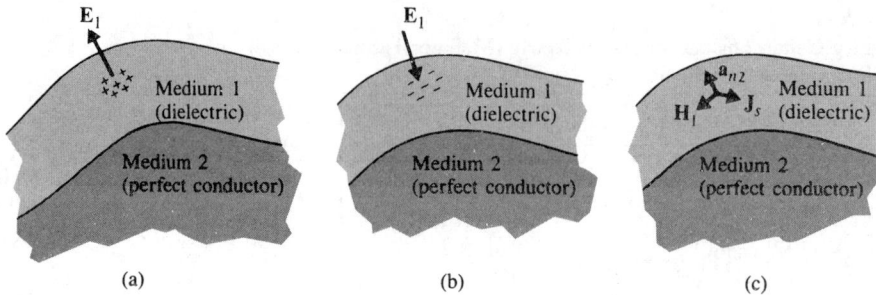

FIGURE 7–8
Boundary conditions at an interface between a dielectric (medium 1) and a perfect
conductor (medium 2).

(7–68b) leads to the condition that the tangential component of **H** is continuous
across an interface with a conductor having a finite conductivity.

At an interface between a dielectric and a perfect conductor, it is possible to
conclude from Eqs. (7–68a) and (7–68c) that the electric field intensity **E** is normal
to and points away from (into) the conductor surface when the surface charges are
positive (negative), as illustrated in Figs. 7–8(a) and 7–8(b). The magnitude E_{1n} of $\mathbf{E}_1$
at the interface is related to ρ_s by the equation

$$|\mathbf{E}_1| = E_{1n} = \frac{\rho_s}{\epsilon_1}. \qquad (7\text{–}69)$$

Similarly, Eqs. (7–68b) and (7–68d) show that the magnetic field intensity $\mathbf{H}_1$ is tangential to the interface with a magnitude equal to that of the surface current density:

$$|\mathbf{H}_1| = |\mathbf{H}_{1t}| = |\mathbf{J}_s|. \qquad (7\text{–}70)$$

The direction of $\mathbf{H}_{1t}$ is determined from Eq. (7–68b). This is illustrated in Fig. 7–8(c).
Equations (7–69) and (7–70) are analytically quite simple relations.

In this section we have discussed the relations that field vectors must satisfy at
an interface between different media. Boundary conditions are of basic importance
in the solution of electromagnetic problems because general solutions of Maxwell's
equations carry little meaning until they are adapted to physical problems each with
a given region and associated boundary conditions. Maxwell's equations are partial
differential equations. Their solutions will contain integration constants that are determined from the additional information supplied by boundary conditions so that
each solution will be unique for each given problem.

7–6 Wave Equations and Their Solutions

At this point we are in possession of the essentials of the fundamental structure of
electromagnetic theory. Maxwell's equations give a complete description of the relation between electromagnetic fields and charge and current distributions. Their solu-

tions provide the answers to all electromagnetic problems, albeit in some cases the solutions are difficult to obtain. Special analytical and numerical techniques may be devised to aid in the solution procedure; but they do not add to or refine the fundamental structure. Such is the importance of Maxwell's equations.

For given charge and current distributions, ρ and $\mathbf{J}$, we first solve the nonhomogeneous wave equations, Eqs. (7–63) and (7–65), for potentials $\mathbf{A}$ and V. With $\mathbf{A}$ and V determined, $\mathbf{E}$ and $\mathbf{B}$ can be found from Eqs. (7–57) and (7–55), respectively, by differentiation.

7–6.1   SOLUTION OF WAVE EQUATIONS FOR POTENTIALS

We now consider the solution of the nonhomogeneous wave equation, Eq. (7–65), for scalar electric potential V. We can do this by first finding the solution for an elemental point charge at time t, $\rho(t)\,\Delta v'$, located at the origin of the coordinates and then by summing the effects of all the charge elements in a given region. For a point charge at the origin it is most convenient to use spherical coordinates. Because of spherical symmetry, V depends only on R and t (not on θ or ϕ). Except at the origin, V satisfies the following homogeneous equation:

$$\frac{1}{R^2}\frac{\partial}{\partial R}\left(R^2\frac{\partial V}{\partial R}\right) - \mu\epsilon\frac{\partial^2 V}{\partial t^2} = 0. \tag{7–71}$$

We introduce a new variable

$$V(R, t) = \frac{1}{R}U(R, t), \tag{7–72}$$

which converts Eq. (7–71) to

$$\frac{\partial^2 U}{\partial R^2} - \mu\epsilon\frac{\partial^2 U}{\partial t^2} = 0. \tag{7–73}$$

Equation (7–73) is a one-dimensional homogeneous wave equation. It can be verified by direct substitution (see Problem P.7–20) that *any* twice-differentiable function of $(t - R\sqrt{\mu\epsilon})$ or of $(t + R\sqrt{\mu\epsilon})$ is a solution of Eq. (7–73). Later in this section we will see that a function of $(t + R\sqrt{\mu\epsilon})$ does not correspond to a physically useful solution. Hence we have

$$U(R, t) = f(t - R\sqrt{\mu\epsilon}). \tag{7–74}$$

Equation (7–74) represents a wave traveling in the positive R direction with a velocity $1/\sqrt{\mu\epsilon}$. As we see, the function at $R + \Delta R$ at a later time $t + \Delta t$ is

$$U(R + \Delta R, t + \Delta t) = f[t + \Delta t - (R + \Delta R)\sqrt{\mu\epsilon}] = f(t - R\sqrt{\mu\epsilon}).$$

Thus the function retains its form if $\Delta t = \Delta R\sqrt{\mu\epsilon} = \Delta R/u$, where $u = 1/\sqrt{\mu\epsilon}$ is the *velocity of propagation*, a characteristic of the medium. From Eq. (7–72) we get

$$V(R, t) = \frac{1}{R}f(t - R/u). \tag{7–75}$$

To determine what the specific function $f(t - R/u)$ must be, we note from Eq. (3–47) that for a static point charge $\rho(t)\,\Delta v'$ at the origin,

$$\Delta V(R) = \frac{\rho(t)\,\Delta v'}{4\pi\epsilon R}. \tag{7–76}$$

Comparison of Eqs. (7–75) and (7–76) enables us to identify

$$\Delta f(t - R/u) = \frac{\rho(t - R/u)\,\Delta v'}{4\pi\epsilon}.$$

The potential due to a charge distribution over a volume V' is then

$$\boxed{V(R, t) = \frac{1}{4\pi\epsilon} \int_{V'} \frac{\rho(t - R/u)}{R}\,dv' \qquad (\text{V}).} \tag{7–77}$$

Equation (7–77) indicates that the scalar potential at a distance R from the source at time t depends on the value of the charge density at an *earlier* time $(t - R/u)$. It takes time R/u for the effect of ρ to be felt at distance R. For this reason, $V(R, t)$ in Eq. (7–77) is called the ***retarded scalar potential***. It is now clear that a function of $(t + R/u)$ cannot be a physically useful solution, since it would lead to the impossible situation that the effect of ρ would be felt at a distant point before it occurs at the source.

The solution of the nonhomogeneous wave equation, Eq. (7–63), for vector magnetic potential $\mathbf{A}$ can proceed in exactly the same way as that for V. The vector equation, Eq. (7–63), can be decomposed into three scalar equations, each similar to Eq. (7–65) for V. The ***retarded vector potential*** is thus given by

$$\boxed{\mathbf{A}(R, t) = \frac{\mu}{4\pi} \int_{V'} \frac{\mathbf{J}(t - R/u)}{R}\,dv' \qquad (\text{Wb/m}).} \tag{7–78}$$

The electric and magnetic fields derived from $\mathbf{A}$ and V by differentiation will obviously also be functions of $(t - R/u)$ and therefore retarded in time. It takes time for electromagnetic waves to travel and for the effects of time-varying charges and currents to be felt at distant points. In the quasi-static approximation we ignore this time-retardation effect and assume instant response. This assumption is implicit in dealing with circuit problems.

7-6.2   SOURCE-FREE WAVE EQUATIONS

In problems of wave propagation we are concerned with the behavior of an electromagnetic wave in a source-free region where ρ and $\mathbf{J}$ are both zero. In other words, we are often interested not so much in how an electromagnetic wave is originated, but in how it propagates. If the wave is in a simple (linear, isotropic, and homogeneous) nonconducting medium characterized by ϵ and μ ($\sigma = 0$), Maxwell's equations (7–53a, b, c, and d) reduce to

$$\mathbf{\nabla} \times \mathbf{E} = -\mu \frac{\partial \mathbf{H}}{\partial t}, \tag{7-79a}$$

$$\mathbf{\nabla} \times \mathbf{H} = \epsilon \frac{\partial \mathbf{E}}{\partial t}, \tag{7-79b}$$

$$\mathbf{\nabla} \cdot \mathbf{E} = 0, \tag{7-79c}$$

$$\mathbf{\nabla} \cdot \mathbf{H} = 0. \tag{7-79d}$$

Equations (7–79a, b, c, and d) are first-order differential equations in the two variables $\mathbf{E}$ and $\mathbf{H}$. They can be combined to give a second-order equation in $\mathbf{E}$ or $\mathbf{H}$ alone. To do this, we take the curl of Eq. (7–79a) and use Eq. (7–79b):

$$\mathbf{\nabla} \times \mathbf{\nabla} \times \mathbf{E} = -\mu \frac{\partial}{\partial t} (\mathbf{\nabla} \times \mathbf{H}) = -\mu\epsilon \frac{\partial^2 \mathbf{E}}{\partial t^2}.$$

Now $\mathbf{\nabla} \times \mathbf{\nabla} \times \mathbf{E} = \mathbf{\nabla}(\mathbf{\nabla} \cdot \mathbf{E}) - \nabla^2 \mathbf{E} = -\nabla^2 \mathbf{E}$ because of Eq. (7–79c). Hence we have

$$\nabla^2 \mathbf{E} - \mu\epsilon \frac{\partial^2 \mathbf{E}}{\partial t^2} = 0; \tag{7-80}$$

or, since $u = 1/\sqrt{\mu\epsilon}$,

$$\boxed{\nabla^2 \mathbf{E} - \frac{1}{u^2} \frac{\partial^2 \mathbf{E}}{\partial t^2} = 0.} \tag{7-81}$$

In an entirely similar way we can also obtain an equation in $\mathbf{H}$:

$$\boxed{\nabla^2 \mathbf{H} - \frac{1}{u^2} \frac{\partial^2 \mathbf{H}}{\partial t^2} = 0.} \tag{7-82}$$

Equations (7–81) and (7–82) are *homogeneous vector wave equations*.

   We can see that in Cartesian coordinates Eqs. (7–81) and (7–82) can each be decomposed into three one-dimensional, homogeneous, scalar wave equations. Each component of $\mathbf{E}$ and of $\mathbf{H}$ will satisfy an equation exactly like Eq. (7–73), whose solutions represent waves. We will extensively discuss wave behavior in various environments in the next two chapters.

7-7   Time-Harmonic Fields

Maxwell's equations and all the equations derived from them so far in this chapter hold for electromagnetic quantities with an arbitrary time-dependence. The actual type of time functions that the field quantities assume depends on the source functions

ρ and **J**. In engineering, sinusoidal time functions occupy a unique position. They are easy to generate; arbitrary periodic time functions can be expanded into Fourier series of harmonic sinusoidal components; and transient nonperiodic functions can be expressed as Fourier integrals.[†] Since Maxwell's equations are *linear* differential equations, sinusoidal time variations of source functions of a given frequency will produce sinusoidal variations of **E** and **H** with the *same frequency* in the steady state. For source functions with an arbitrary time dependence, electrodynamic fields can be determined in terms of those caused by the various frequency components of the source functions. The application of the principle of superposition will give us the total fields. In this section we examine *time-harmonic* (steady-state sinusoidal) field relationships.

7–7.1   THE USE OF PHASORS—A REVIEW

For time-harmonic fields it is convenient to use a phasor notation. At this time we digress briefly to review the use of phasors. Conceptually, it is simpler to discuss a scalar phasor. The instantaneous (time-dependent) expression of a sinusoidal scalar quantity, such as a current i, can be written as either a cosine or a sine function. *If we choose a cosine function as the reference* (which is usually dictated by the functional form of the excitation), then all derived results will refer to the cosine function. The specification of a sinusoidal quantity requires the knowledge of three parameters: amplitude, frequency, and phase. For example,

$$i(t) = I \cos (\omega t + \phi), \qquad (7–83)$$

where I is the amplitude; ω is the angular frequency (rad/s)—ω is always equal to $2\pi f$, f being the frequency in hertz; and ϕ is the phase referred to the cosine function. We could write $i(t)$ in Eq. (7–83) as a sine function if we wish: $i(t) = I \sin (\omega t + \phi')$, with $\phi' = \phi + \pi/2$. Thus it is important to decide at the outset whether our reference is a cosine or a sine function, then to stick to that decision throughout a problem.

To work directly with an instantaneous expression such as the cosine function is inconvenient when differentiations or integrations of $i(t)$ are involved because they lead to both sine (first-order differentiation or integration) and cosine (second-order differentiation or integration) functions and because it is tedious to combine sine and cosine functions. For instance, the loop equation for a series RLC circuit with an applied voltage $e(t) = E \cos \omega t$ is

$$L \frac{di}{dt} + Ri + \frac{1}{C} \int i \, dt = e(t). \qquad (7–84)$$

If we write $i(t)$ as in Eq. (7–83), Eq. (7–84) yields

$$I \left[-\omega L \sin (\omega t + \phi) + R \cos (\omega t + \phi) + \frac{1}{\omega C} \sin (\omega t + \phi) \right] = E \cos \omega t. \qquad (7–85)$$

[†] D. K. Cheng, *op. cit.*, Chapter 5.

Complicated mathematical manipulations are required in order to determine the unknown I and ϕ from Eq. (7–85).

It is much simpler to use exponential functions by writing the applied voltage as

$$
\begin{aligned}
e(t) = E \cos \omega t &= \mathscr{R}e\big[(Ee^{j0})e^{j\omega t}\big] \\
&= \mathscr{R}e(E_s e^{j\omega t})
\end{aligned}
\tag{7–86}
$$

and $i(t)$ in Eq. (7–83) as

$$
\begin{aligned}
i(t) &= \mathscr{R}e\big[(Ie^{j\phi})e^{j\omega t}\big] \\
&= \mathscr{R}e(I_s e^{j\omega t}),
\end{aligned}
\tag{7–87}
$$

where $\mathscr{R}e$ means "the real part of." In Eqs. (7–86) and (7–87),

$$
E_s = Ee^{j0} = E
\tag{7–88a}
$$

$$
I_s = Ie^{j\phi}
\tag{7–88b}
$$

are (scalar) **phasors** that contain amplitude and phase information but are independent of t. The phasor E_s in Eq. (7–88a) with zero phase angle is the reference phasor. Now,

$$
\frac{di}{dt} = \mathscr{R}e(j\omega I_s e^{j\omega t}),
\tag{7–89}
$$

$$
\int i\,dt = \mathscr{R}e\left(\frac{I_s}{j\omega}\,e^{j\omega t}\right).
\tag{7–90}
$$

Substitution of Eqs. (7–86) through (7–90) in Eq. (7–84) yields

$$
\left[R + j\left(\omega L - \frac{1}{\omega C}\right)\right]I_s = E_s,
\tag{7–91}
$$

from which the current phasor I_s can be solved easily. Note that the time-dependent factor $e^{j\omega t}$ disappears from Eq. (7–91) because it is present in every term in Eq. (7–84) after the substitution and is therefore canceled. This is the essence of the usefulness of phasors in the analysis of linear systems with time-harmonic excitations. After I_s has been determined, the instantaneous current response $i(t)$ can be found from Eq. (7–87) by (1) multiplying I_s by $e^{j\omega t}$, and (2) taking the real part of the product.

If the applied voltage had been given as a *sine function* such as $e(t) = E \sin \omega t$, the series RLC-circuit problem would be solved in terms of phasors in exactly the same way; only the instantaneous expressions would be obtained by taking the *imaginary part* of the product of the phasors with $e^{j\omega t}$. The complex phasors represent the magnitudes and the phase shifts of the quantities in the solution of time-harmonic problems.

━━━ **EXAMPLE 7–6** Express $3 \cos \omega t - 4 \sin \omega t$ as first (a) $A_1 \cos(\omega t + \theta_1)$, and then (b) $A_2 \sin(\omega t + \theta_2)$. Determine A_1, θ_1, A_2, and θ_2.

Solution We can conveniently use phasors to solve this problem.

a) To express $3 \cos \omega t - 4 \sin \omega t$ as $A_1 \cos (\omega t + \theta_1)$, we *use* $\cos \omega t$ *as the reference* and consider the sum of the two phasors 3 and $-4e^{-j\pi/2}$ $(=j4)$, since $\sin \omega t = \cos (\omega t - \pi/2)$ lags behind $\cos \omega t$ by $\pi/2$ rad:

$$3 + j4 = 5e^{j \tan^{-1}(4/3)} = 5e^{j53.1°}.$$

Taking the *real part* of the product of this phasor and $e^{j\omega t}$, we have

$$3 \cos \omega t - 4 \sin \omega t = \mathscr{R}e[(5e^{j53.1°})e^{j\omega t}]$$
$$= 5 \cos (\omega t + 53.1°). \qquad (7\text{-}92\text{a})$$

So, $A_1 = 5$, and $\theta_1 = 53.1° = 0.927$ (rad).

b) To express $3 \cos \omega t - 4 \sin \omega t$ as $A_2 \sin (\omega t + \theta_2)$, we *use* $\sin \omega t$ *as the reference* and consider the sum of the two phasors $3e^{j\pi/2}$ $(=j3)$ and -4:

$$j3 - 4 = 5e^{j \tan^{-1} 3/(-4)} = 5e^{j143.1°}.$$

(The reader should note that the angle above is 143.1°, *not* $-36.9°$.) Now we take the *imaginary part* of the product of the phasor above and $e^{j\omega t}$ to obtain the desired answer:

$$3 \cos \omega t - 4 \sin \omega t = \mathscr{I}m[(5e^{j143.1°})e^{j\omega t}]$$
$$= 5 \sin (\omega t + 143.1°). \qquad (7\text{-}92\text{b})$$

Hence, $A_2 = 5$ and $\theta_2 = 143.1° = 2.50$ (rad).

The reader should recognize that the results in Eqs. (7–92a) and (7–92b) are identical. ∎

7-7.2 TIME-HARMONIC ELECTROMAGNETICS

Field vectors that vary with space coordinates and are sinusoidal functions of time can similarly be represented by vector phasors that depend on space coordinates but not on time. As an example, we can write a time-harmonic E field *referring to* $\cos \omega t$[†] as

$$\mathbf{E}(x, y, z, t) = \mathscr{R}e[\mathbf{E}(x, y, z)e^{j\omega t}], \qquad (7\text{-}93)$$

where $\mathbf{E}(x, y, z)$ is a **vector phasor** that contains information on direction, magnitude, and phase. Phasors are, in general, complex quantities. From Eqs. (7–93), (7–87), (7–89), and (7–90) we see that, if $\mathbf{E}(x, y, z, t)$ is to be represented by the vector phasor $\mathbf{E}(x, y, z)$, then $\partial \mathbf{E}(x, y, z, t)/\partial t$ and $\int \mathbf{E}(x, y, z, t) \, dt$ would be represented by vector phasors $j\omega \mathbf{E}(x, y, z)$ and $\mathbf{E}(x, y, z)/j\omega$, respectively. Higher-order differentiations and integrations with respect to t would be represented by multiplications and divisions, respectively, of the phasor $\mathbf{E}(x, y, z)$ by higher powers of $j\omega$.

We now write time-harmonic Maxwell's equations (7–53a, b, c, and d) in terms of vector field phasors $(\mathbf{E}, \mathbf{H})$ and source phasors $(\rho, \mathbf{J})$ in a simple (linear, isotropic, and homogeneous) medium as follows.

[†] If the time reference is not explicitly specified, it is customarily taken as $\cos \omega t$.

$$\mathbf{V} \times \mathbf{E} = -j\omega\mu\mathbf{H}, \tag{7–94a}$$

$$\mathbf{V} \times \mathbf{H} = \mathbf{J} + j\omega\epsilon\mathbf{E}, \tag{7–94b}$$

$$\mathbf{V} \cdot \mathbf{E} = \rho/\epsilon, \tag{7–94c}$$

$$\mathbf{V} \cdot \mathbf{H} = 0. \tag{7–94d}$$

The space-coordinate arguments have been omitted for simplicity. The fact that the same notations are used for the phasors as are used for their corresponding time-dependent quantities should create little confusion because we will deal almost exclusively with time-harmonic fields (and therefore with phasors) in the rest of this book. When there is a need to distinguish an instantaneous quantity from a phasor, the time dependence of the instantaneous quantity will be indicated explicitly by the inclusion of a t in its argument. Phasor quantities are not functions of t. It is useful to note that any quantity containing j must necessarily be a phasor.

The time-harmonic wave equations for scalar potential V and vector potential $\mathbf{A}$—Eqs. (7–65) and (7–63)—become, respectively,

$$\mathbf{V}^2 V + k^2 V = -\frac{\rho}{\epsilon} \tag{7–95}$$

and

$$\mathbf{V}^2 \mathbf{A} + k^2 \mathbf{A} = -\mu\mathbf{J}, \tag{7–96}$$

where

$$k = \omega\sqrt{\mu\epsilon} = \frac{\omega}{u} \tag{7–97}$$

is called the **wavenumber**. Equations (7–95) and (7–96) are referred to as **nonhomogeneous Helmholtz's equations**. The Lorentz condition for potentials, Eq. (7–62), is now

$$\mathbf{V} \cdot \mathbf{A} + j\omega\mu\epsilon V = 0. \tag{7–98}$$

The phasor solutions of Eqs. (7–95) and (7–96) are obtained from Eqs. (7–77) and (7–78), respectively:

$$V(R) = \frac{1}{4\pi\epsilon} \int_{V'} \frac{\rho e^{-jkR}}{R} \, dv' \quad (V), \tag{7–99}$$

$$\mathbf{A}(R) = \frac{\mu}{4\pi} \int_{V'} \frac{\mathbf{J} e^{-jkR}}{R} \, dv' \quad (\text{Wb/m}). \tag{7–100}$$

These are the expressions for the retarded scalar and vector potentials due to time-harmonic sources. Now the Taylor-series expansion for the exponential factor e^{-jkR} is

$$e^{-jkR} = 1 - jkR + \frac{k^2 R^2}{2} + \cdots, \tag{7–101}$$

where k, defined in Eq. (7–97), can be expressed in terms of the wavelength $\lambda = u/f$ in the medium. We have

$$k = \frac{2\pi f}{u} = \frac{2\pi}{\lambda}.$$ (7–102)

Thus, if

$$kR = 2\pi \frac{R}{\lambda} \ll 1,$$ (7–103)

or if the distance R is very small in comparison to the wavelength λ, e^{-jkR} can be approximated by 1. Equations (7–99) and (7–100) then simplify to the static expressions in Eqs. (7–58) and (7–59), which are used in Eqs. (7–55) and (7–57) to find quasi-static fields.

The formal procedure for determining the electric and magnetic fields due to time-harmonic charge and current distributions is as follows:[†]

1. Find phasors $V(R)$ and $\mathbf{A}(R)$ from Eqs. (7–99) and (7–100).
2. Find phasors $\mathbf{E}(R) = -\nabla V - j\omega\mathbf{A}$ and $\mathbf{B}(R) = \nabla \times \mathbf{A}$.
3. Find instantaneous $\mathbf{E}(R, t) = \mathcal{R}e[\mathbf{E}(R)e^{j\omega t}]$ and $\mathbf{B}(R, t) = \mathcal{R}e[\mathbf{B}(R)e^{j\omega t}]$ for a cosine reference.

The degree of difficulty of a problem depends on how difficult it is to perform the integrations in Step 1.

7–7.3 SOURCE-FREE FIELDS IN SIMPLE MEDIA

In a simple, nonconducting source-free medium characterized by $\rho = 0$, $\mathbf{J} = 0$, $\sigma = 0$, the time-harmonic Maxwell's equations (7–94a, b, c, and d) become

$$\nabla \times \mathbf{E} = -j\omega\mu\mathbf{H},$$ (7–104a)

$$\nabla \times \mathbf{H} = j\omega\epsilon\mathbf{E},$$ (7–104b)

$$\nabla \cdot \mathbf{E} = 0,$$ (7–104c)

$$\nabla \cdot \mathbf{H} = 0.$$ (7–104d)

Equations (7–104a, b, c, and d) can be combined to yield second-order partial differential equations in $\mathbf{E}$ and $\mathbf{H}$. From Eqs. (7–81) and (7–82) we obtain

$$\boxed{\nabla^2\mathbf{E} + k^2\mathbf{E} = 0}$$ (7–105)

and

$$\boxed{\nabla^2\mathbf{H} + k^2\mathbf{H} = 0,}$$ (7–106)

[†] Alternatively, Steps 1 and 2 can be replaced by the following: (1′) Find phasor $A(R)$ from Eq. (7–100). (2′) Find $\mathbf{H}(R) = \dfrac{1}{\mu}\nabla \times \mathbf{A}$, and $\mathbf{E}(R) = \dfrac{1}{j\omega\epsilon}(\nabla \times \mathbf{H} - \mathbf{J})$ from Eq. (7–94b).

which are **homogeneous vector Helmholtz's equations**. Solutions of homogeneous Helmholtz's equations with various boundary conditions is the main concern of Chapters 8 and 10.

EXAMPLE 7-7 Show that if $(\mathbf{E}, \mathbf{H})$ are solutions of source-free Maxwell's equations in a simple medium characterized by ϵ and μ, then so also are $(\mathbf{E}', \mathbf{H}')$, where

$$\mathbf{E}' = \eta\mathbf{H} \tag{7-107a}$$

$$\mathbf{H}' = -\frac{\mathbf{E}}{\eta}. \tag{7-107b}$$

In the above equations, $\eta = \sqrt{\mu/\epsilon}$ is called the **intrinsic impedance** of the medium.

Solution We prove the statement by taking the curl and the divergence of $\mathbf{E}'$ and $\mathbf{H}'$ and using Eqs. (7-104a, b, c, and d):

$$\nabla \times \mathbf{E}' = \eta(\nabla \times \mathbf{H}) = \eta(j\omega\epsilon\mathbf{E})$$
$$= -j\omega\epsilon\eta^2\left(-\frac{\mathbf{E}}{\eta}\right) = -j\omega\mu\mathbf{H}' \tag{7-108a}$$

$$\nabla \times \mathbf{H}' = -\frac{1}{\eta}(\nabla \times \mathbf{E}) = -\frac{1}{\eta}(-j\omega\mu\mathbf{H})$$
$$= j\omega\mu\frac{1}{\eta^2}(\eta\mathbf{H}) = j\omega\epsilon\mathbf{E}' \tag{7-108b}$$

$$\nabla \cdot \mathbf{E}' = \eta(\nabla \cdot \mathbf{H}) = 0 \tag{7-108c}$$

$$\nabla \cdot \mathbf{H}' = -\frac{1}{\eta}(\nabla \cdot \mathbf{E}) = 0. \tag{7-108d}$$

Equations (7-108a, b, c, and d) are source-free Maxwell's equations in $\mathbf{E}'$ and $\mathbf{H}'$. (Q.E.D.)

This example shows that source-free Maxwell's equations in a simple medium are invariant under the linear transformation specified by Eqs. (7-107a) and (7-107b). This is a statment of the **principle of duality**. This principle is a consequence of the symmetry of source-free Maxwell's equations. An illustration of the principle of duality and dual devices can be found in Subsection 11-2.2.

If the simple medium is conducting ($\sigma \neq 0$), a current $\mathbf{J} = \sigma\mathbf{E}$ will flow, and Eq. (7-104b) should be changed to

$$\nabla \times \mathbf{H} = (\sigma + j\omega\epsilon)\mathbf{E} = j\omega\left(\epsilon + \frac{\sigma}{j\omega}\right)\mathbf{E} \tag{7-109}$$
$$= j\omega\epsilon_c\mathbf{E}$$

with

$$\epsilon_c = \epsilon - j\frac{\sigma}{\omega} \quad \text{(F/m)}. \tag{7-110}$$

The other three equations, Eqs. (7–104a, c, and d), are unchanged. Hence, all the previous equations for nonconducting media will apply to conducting media if ϵ is replaced by the *complex permittivity* ϵ_c.

As we discussed in Section 3–7, when an external time-varying electric field is applied to material bodies, small displacements of bound charges result, giving rise to a volume density of polarization. This polarization vector will vary with the same frequency as that of the applied field. As the frequency increases, the inertia of the charged particles tends to prevent the particle displacements from keeping in phase with the field changes, leading to a frictional damping mechanism that causes power loss because work must be done to overcome the damping forces. This phenomenon of out-of-phase polarization can be characterized by a complex electric susceptibility and hence a complex permittivity. If, in addition, the material body or medium has an appreciable amount of free charge carriers such as the electrons in a conductor, the electrons and holes in a semiconductor, or the ions in an electrolyte, there will also be ohmic losses. In treating such media it is customary to include the effects of both the damping and the ohmic losses in the imaginary part of a complex permittvity ϵ_c:

$$\epsilon_c = \epsilon' - j\epsilon'' \qquad \text{(F/m)}, \qquad (7\text{--}111)$$

where both ϵ' and ϵ'' may be functions of frequency. Alternatively, we may define an equivalent conductivity representing all losses and write

$$\sigma = \omega\epsilon'' \qquad \text{(S/m)}. \qquad (7\text{--}112)$$

Combination of Eqs. (7–111) and (7–112) gives Eq. (7–110). In low-loss media, damping losses are very small, and the real part of ϵ_c in Eq. (7–110) is usually written as ϵ without a prime.

Similar loss arguments apply to the existence of an out-of-phase component of magnetization under the influence of an external time-varying magnetic field. We expect the permeability also to be complex at high frequencies:

$$\mu = \mu' - j\mu''. \qquad (7\text{--}113)$$

For ferromagnetic materials the real part, μ', is many orders of magnitude larger than the imaginary part, μ'', and the effect of the latter is normally neglected. In view of the above, the real wavenumber k in the Helmholtz's equations, Eqs. (7–105) and (7–106), should be changed to a complex wavenumber:

$$\begin{aligned} k_c &= \omega\sqrt{\mu\epsilon_c} \\ &= \omega\sqrt{\mu(\epsilon' - j\epsilon'')} \end{aligned} \qquad (7\text{--}114)$$

in a lossy dielectric medium.

The ratio ϵ''/ϵ' is called a *loss tangent* because it is a measure of the power loss in the medium:

$$\tan \delta_c = \frac{\epsilon''}{\epsilon'} \cong \frac{\sigma}{\omega\epsilon}. \qquad (7\text{--}115)$$

The quantity δ_c in Eq. (7–115) may be called the *loss angle*.

On the basis of Eq. (7–110) a medium is said to be a *good conductor* if $\sigma \gg \omega\epsilon$, and a *good insulator* if $\omega\epsilon \gg \sigma$. Thus, a material may be a good conductor at low frequencies but may have the properties of a lossy dielectric at very high frequencies. For example, a moist ground has a dielectric constant ϵ_r and a conductivity σ that are in the neighborhood of 10 and 10^{-2} (S/m), respectively. The loss tangent $\sigma/\omega\epsilon$ of the moist ground then equals 1.8×10^4 at 1 (kHz), making it a relatively good conductor. At 10 (GHz), $\sigma/\omega\epsilon$ becomes 1.8×10^{-3}, and the moist ground behaves more like an insulator.[†]

EXAMPLE 7–8   A sinusoidal electric intensity of amplitude 250 (V/m) and frequency 1 (GHz) exists in a lossy dielectric medium that has a relative permittivity of 2.5 and a loss tangent of 0.001. Find the average power dissipated in the medium per cubic meter.

Solution   First we must find the effective conductivity of the lossy medium:

$$\tan \delta_c = 0.001 = \frac{\sigma}{\omega\epsilon_0\epsilon_r},$$

$$\sigma = 0.001(2\pi10^9)\left(\frac{10^{-9}}{36\pi}\right)(2.5)$$

$$= 1.39 \times 10^{-4} \text{ (S/m)}.$$

The average power dissipated per unit volume is

$$p = \tfrac{1}{2}JE = \tfrac{1}{2}\sigma E^2$$
$$= \tfrac{1}{2} \times (1.39 \times 10^{-4}) \times 250^2 = 4.34 \quad \text{(W/m}^3\text{)}.$$

A microwave oven cooks food by irradiating the food with microwave power generated by a magnetron. The operating frequency is usually set at 2.45 GHz $(2.45 \times 10^9$ Hz). For a beef steak that has approximately a dielectric constant of 40 and a loss tangent of 0.35 at 2.45 (GHz), calculations following those in Example 7–8 will yield $\sigma = 1.91$ (S/m) and $p = 59.6$ (kW/m^3). However, since high-frequency currents in a conducting body tend to concentrate near the surface layer (due to *skin effect*—see Subsection 8–3.2), the value of p obtained here is only a rough estimate.

7–7.4   THE ELECTROMAGNETIC SPECTRUM

We have seen that **E** and **H** in source-free regions satisfy homogeneous wave equations (7–81) and (7–82), respectively. If the sources of the fields are time-harmonic, these equations reduce to homogeneous Helmholtz's equations (7–105) and (7–106). That the solutions of Eqs. (7–105) and (7–106) represent propagating waves will become clear in the beginning of the next chapter. For the moment we note two

---

[†] Actually, the loss mechanism of a dielectric material is a very complicated process, and the assumption of a constant conductivity is only a rough approximation.

important points. First, Maxwell's equations, and therefore the wave and Helmholtz's equations, impose *no limit* on the frequency of the waves. The electromagnetic spectrum that has been investigated experimentally extends from very low power frequencies through radio, television, microwave, infrared, visible light, ultraviolet, X-ray, and gamma (γ)-ray frequencies exceeding 10^{24} (Hz). Second, all electromagnetic

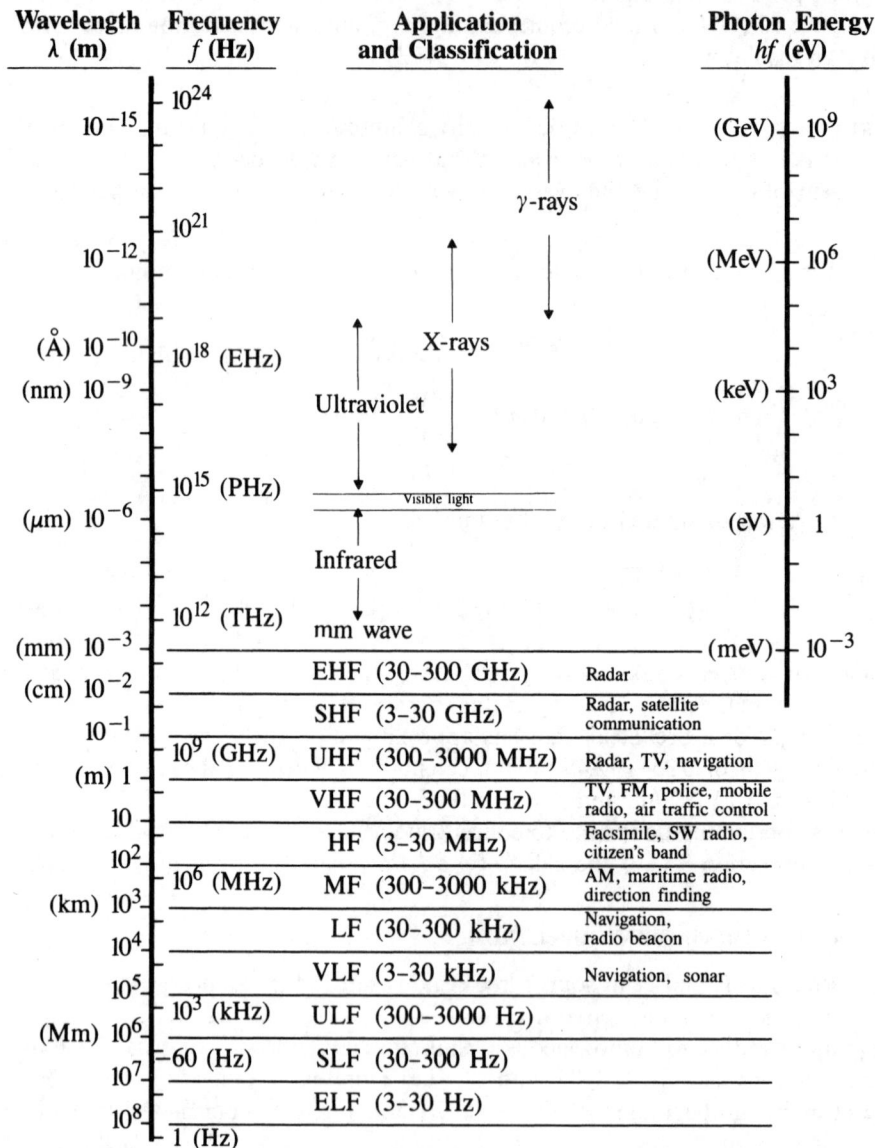

Wavelength λ (m)	Frequency f (Hz)	Application and Classification		Photon Energy hf (eV)
10^{-15}	10^{24}			(GeV) 10^9
			γ-rays	
	10^{21}			(MeV) 10^6
10^{-12}				
(Å) 10^{-10}	10^{18} (EHz)		X-rays	(keV) 10^3
(nm) 10^{-9}		Ultraviolet		
	10^{15} (PHz)			
(μm) 10^{-6}		Visible light		(eV) 1
		Infrared		
	10^{12} (THz)	mm wave		
(mm) 10^{-3}				(meV) 10^{-3}
(cm) 10^{-2}		EHF (30–300 GHz)	Radar	
10^{-1}		SHF (3–30 GHz)	Radar, satellite communication	
	10^9 (GHz)	UHF (300–3000 MHz)	Radar, TV, navigation	
(m) 1		VHF (30–300 MHz)	TV, FM, police, mobile radio, air traffic control	
10		HF (3–30 MHz)	Facsimile, SW radio, citizen's band	
10^2	10^6 (MHz)	MF (300–3000 kHz)	AM, maritime radio, direction finding	
(km) 10^3		LF (30–300 kHz)	Navigation, radio beacon	
10^4		VLF (3–30 kHz)	Navigation, sonar	
10^5	10^3 (kHz)	ULF (300–3000 Hz)		
(Mm) 10^6	60 (Hz)	SLF (30–300 Hz)		
10^7		ELF (3–30 Hz)		
10^8	1 (Hz)			

FIGURE 7–9
Spectrum of electromagnetic waves.

TABLE 7–5
Band Designations for Microwave Frequency Ranges

Old[†]	New	Frequency Ranges (GHz)
Ka	K	26.5–40
K	K	20–26.5
K	J	18–20
Ku	J	12.4–18
X	J	10–12.4
X	I	8–10
C	H	6–8
C	G	4–6
S	F	3–4
S	E	2–3
L	D	1–2
UHF	C	0.5–1

[†] Because the old band designations have been in wide use since the early days of radar, they are still in common use because of habit.

waves in whatever frequency range propagate in a medium with the *same velocity*, $u = 1/\sqrt{\mu\epsilon}$ ($c \cong 3 \times 10^8$ m/s in air).

Figure 7–9 shows the electromagnetic spectrum divided into frequency and wavelength ranges on logarithmic scales according to application and natural occurrence. The term "microwave" is somewhat nebulous and imprecise; it could mean electromagnetic waves above a frequency of 1 (GHz) and all the way up to the lower limit of the infrared band, encompassing UHF, SHF, EHF, and mm-wave regions. Beyond the frequency range of visible light it is also customary to show the energy level of a photon (quantum of radiation), hf in electron-volts (eV), where h = Planck's constant = 6.63×10^{-34} (J·s). This is included in Fig. 7–9.[†] The wavelength range of visible light is from deep red at 720 (nm) to violet at 380 (nm), or from 0.72 (μm) to 0.38 (μm), corresponding to a frequency range of from 4.2×10^{14} (Hz) to 7.9×10^{14} (Hz). The bands used for radar, satellite communication, navigation aids, television (TV), FM and AM radio, citizen's band radio (CB), sonar, and others are also noted. Frequencies below the VLF range are seldom used for wireless transmission because huge antennas would be needed for efficient radiation of electromagnetic waves and because of the very low data rate at these low frequencies. There have been proposals to use these frequencies for strategic global communication with submarines submerged in conducting seawater. In radar work it has been found convenient to assign alphabet names to the different microwave frequency bands. They are listed in Table 7–5.

In the next chapter we shall discuss the characteristics of plane electromagnetic waves and examine their behavior as they propagate across discontinuous boundaries.

[†] The conversion relations are: 1 (Hz) $\leftrightarrow$ 4.14×10^{-15} (eV) $\leftrightarrow$ 3×10^8 (m), or 2.42×10^{14} (Hz) $\leftrightarrow$ 1 (eV) $\leftrightarrow$ 1.24×10^{-6} (m).

Review Questions

R.7–1 What constitutes an *electromagnetostatic field*? In what ways are **E** and **B** related in a conducting medium under static conditions?

R.7–2 Write the fundamental postulate for electromagnetic induction, and explain how it leads to Faraday's law.

R.7–3 State Lenz's law.

R.7–4 Write the expression for transformer emf.

R.7–5 What are the characteristics of an ideal transformer?

R.7–6 What is the definition of *coefficient of coupling* in inductive circuits?

R.7–7 What are *eddy currents*?

R.7–8 What are *superconductors*?

R.7–9 Why are materials having high permeability and low conductivity preferred as transformer cores?

R.7–10 Why are the cores of power transformers laminated?

R.7–11 Write the expression for flux-cutting emf.

R.7–12 Write the expression for the induced emf in a closed circuit that moves in a changing magnetic field.

R.7–13 What is a Faraday disk generator?

R.7–14 Write the differential form of Maxwell's equations.

R.7–15 Are all four Maxwell's equations independent? Explain.

R.7–16 Write the integral form of Maxwell's equations, and identify each equation with the proper experimental law.

R.7–17 Explain the significance of *displacement current*.

R.7–18 Why are potential functions used in electromagnetics?

R.7–19 Express **E** and **B** in terms of potential functions V and **A**.

R.7–20 What do we mean by *quasi-static fields*? Are they exact solutions of Maxwell's equations? Explain.

R.7–21 What is the Lorentz condition for potentials? What is its physical significance?

R.7–22 Write the nonhomogeneous wave equation for scalar potential V and for vector potential **A**.

R.7–23 State the boundary conditions for the tangential component of **E** and for the normal component of **B**.

R.7–24 Write the boundary conditions for the tangential component of **H** and for the normal component of **D**.

R.7–25 Why is the **E** field immediately outside of a perfect conductor perpendicular to the conductor surface?

R.7–26 Why is the **H** field immediately outside of a perfect conductor tangential to the conductor surface?

R.7–27 Can a static magnetic field exist in the interior of a perfect conductor? Explain. Can a time-varying magnetic field? Explain.

R.7–28 What do we mean by a *retarded potential?*

R.7–29 In what ways do the retardation time and the velocity of wave propagation depend on the constitutive parameters of the medium?

R.7–30 Write the source-free wave equations for **E** and **H** in free space.

R.7–31 What is a *phasor?* Is a phasor a function of t? A function of ω?

R.7–32 What is the difference between a phasor and a vector?

R.7–33 Discuss the advantages of using phasors in electromagnetics.

R.7–34 Are conduction and displacement currents in phase for time-harmonic fields? Explain.

R.7–35 Write in terms of phasors the time-harmonic Maxwell's equations for a simple medium.

R.7–36 Define *wavenumber.*

R.7–37 Write the expressions for time-harmonic retarded scalar and vector potentials in terms of charge and current distributions.

R.7–38 Write the homogeneous vector Helmholtz's equation for **E** in a simple, non-conducting, source-free medium.

R.7–39 Write the expression for the wavenumber of a lossy medium in terms of its permittivity and permeability.

R.7–40 What is meant by the *loss tangent* of a medium?

R.7–41 In a time-varying situation how do we define a *good conductor?* A *lossy dielectric?*

R.7–42 What is the velocity of propagation of electromagnetic waves? Is it the same in air as in vacuum? Explain.

R.7–43 What is the wavelength range of visible light?

R.7–44 Why are frequencies below the VLF range rarely used for wireless transmission?

Problems

P.7–1 Express the transformer emf induced in a stationary loop in terms of time-varying vector potential **A**.

P.7–2 The circuit in Fig. 7–10 is situated in a magnetic field

$$\mathbf{B} = \mathbf{a}_z 3 \cos \left(5\pi 10^7 t - \tfrac{2}{3}\pi x\right) \qquad (\mu\text{T}).$$

Assuming $R = 15\ (\Omega)$, find the current i.

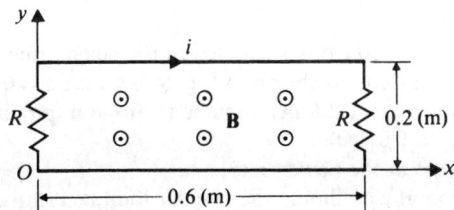

FIGURE 7–10
A circuit in a time-varying magnetic field (Problem P.7–2).

(a)

0   T

(b)

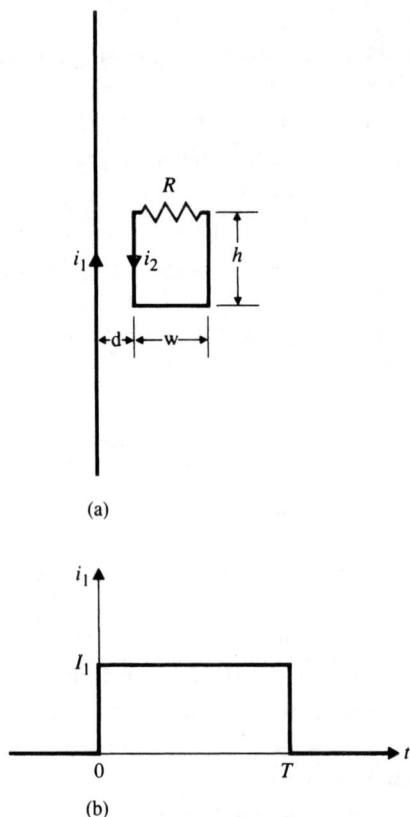

FIGURE 7–11
A rectangular loop near a long current-carrying wire (Problem P.7–3).

P.7–3 A rectangular loop of width w and height h is situated near a very long wire carrying a current i_1 as in Fig. 7–11(a). Assume i_1 to be a rectangular pulse as shown in Fig. 7–11(b).
   a) Find the induced current i_2 in the rectangular loop whose self-inductance is L.
   b) Find the energy dissipated in the resistance R if $T \gg L/R$.

P.7–4 A conducting equilateral triangular loop is placed near a very long straight wire, shown in Fig. 6–48, with $d = b/2$. A current $i(t) = I \sin \omega t$ flows in the straight wire.
   a) Determine the voltage registered by a high-impedance rms voltmeter inserted in the loop.
   b) Determine the voltmeter reading when the triangular loop is rotated by 60° about a perpendicular axis through its center.

P.7–5 A conducting circular loop of a radius 0.1 (m) is situated in the neighborhood of a very long power line carrying a 60-(Hz) current, as shown in Fig. 6–49, with $d = 0.15$ (m). An a-c milliammeter inserted in the loop reads 0.3 (mA). Assume the total impedance of the loop including the milliammeter to be 0.01 (Ω).
   a) Find the magnitude of the current in the power line.
   b) To what angle about the horizontal axis should the circular loop be rotated in order to reduce the milliammeter reading to 0.2 (mA)?

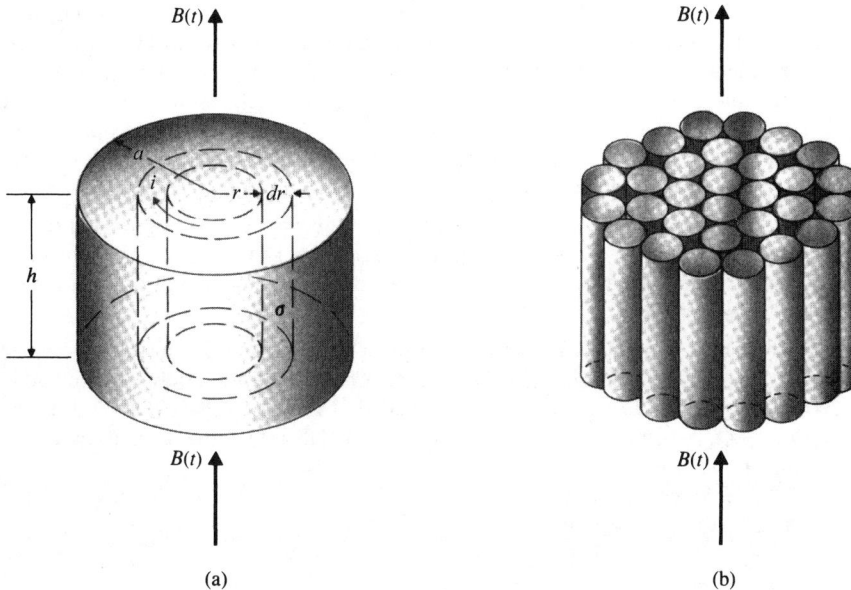

FIGURE 7–12
Suggested eddy-current power-loss reduction scheme (Problem P.7–6).

P.7–6 A suggested scheme for reducing eddy-current power loss in transformer cores with a circular cross section is to divide the cores into a large number of small insulated filamentary parts. As illustrated in Fig. 7–12, the section shown in part (a) is replaced by that in part (b). Assuming that $B(t) = B_0 \sin \omega t$ and that N filamentary areas fill 95% of the original cross-sectional area, find

    a) the average eddy-current power loss in the section of core of height h in Fig. 7–12(a),

    b) the total average eddy-current power loss in the N filamentary sections in Fig. 7–12(b).

The magnetic field due to eddy currents is assumed to be negligible. (*Hint:* First find the current and power dissipated in the differential circular ring section of height h and width dr at radius r.)

P.7–7 A conducting sliding bar oscillates over two parallel conducting rails in a sinusoidally varying magnetic field

$$\mathbf{B} = \mathbf{a}_z 5 \cos \omega t \qquad \text{(mT)},$$

as shown in Fig. 7–13. The position of the sliding bar is given by $x = 0.35(1 - \cos \omega t)$ (m), and the rails are terminated in a resistance $R = 0.2$ (Ω). Find i.

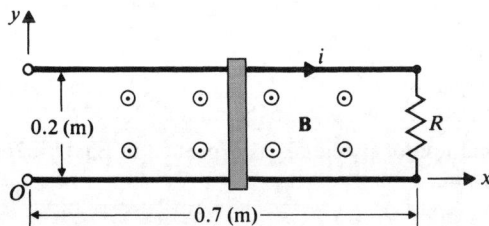

FIGURE 7–13
A conducting bar sliding over parallel rails in a time-varying magnetic field (Problem P.7–7).

P.7–8 In the d-c motor illustrated in Fig. 6–32 we noted that a current I sent through the loop in a magnetic field **B** produces a torque that makes the loop rotate. As the loop rotates, the amount of the magnetic flux linking with the loop changes, giving rise to an induced emf. Energy must be expended by an external electric source to counter this emf and establish the current in the loop. Prove that this electric energy is equal to the mechanical work done by the rotating loop. (*Hint:* Consider the normal of the loop at an arbitary angle α with **B**, and let it rotate by an angle $\Delta\alpha$.)

P.7–9 Assuming that a resistance R is connected across the slip rings of the rectangular conducting loop that rotates in a constant magnetic field $\mathbf{B} = \mathbf{a}_y B_0$, shown in Fig. 7–6, prove that the power dissipated in R is equal to the power required to rotate the loop at an angular frequency ω.

P.7–10 A hollow cylindrical magnet with inner radius a and outer radius b rotates about its axis at an angular frequency ω. The magnet has a uniform axial magnetization $\mathbf{M} = \mathbf{a}_z M_0$. Sliding brush contacts are provided at the inner and outer surfaces as shown in Fig. 7–14. Assuming that $\mu_r = 5000$ and $\sigma = 10^7$ (S/m) for the magnet, find
   a) **H** and **B** in the magnet,
   b) open-circuit voltage V_0,
   c) short-circuit current.

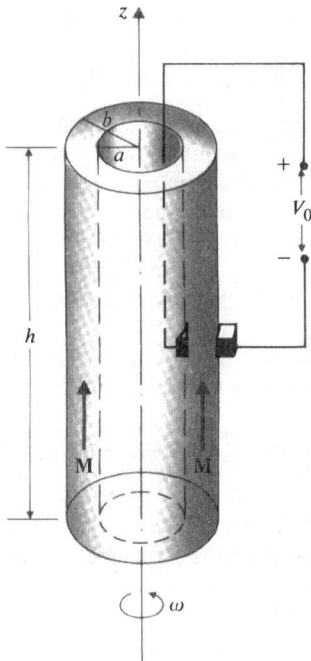

FIGURE 7–14
A rotating hollow cylindrical magnet (Problem P.7–10).

P.7–11 Derive the two divergence equations, Eqs. (7–53c) and (7–53d), from the two curl equations, Eqs. (7–53a) and (7–53b), and the equation of continuity, Eq. (7–48).

P.7–12 Prove that the Lorentz condition for potentials as expressed in Eq. (7–62) is consistent with the equation of continuity.

P.7–13 The vector magnetic potential $\mathbf{A}$ and scalar electric potential V defined in Section 7–4 are not unique in that it is possible to add to $\mathbf{A}$ the gradient of a scalar ψ, $\nabla\psi$, with no change in $\mathbf{B}$ from Eq. (7–55).

$$\mathbf{A}' = \mathbf{A} + \nabla\psi. \tag{7–116}$$

In order not to change $\mathbf{E}$ in using Eq. (7–57), V must be modified to V'.
   a) Find the relation between V' and V.
   b) Discuss the condition that ψ must satisfy so that the new potentials $\mathbf{A}'$ and V'
      remain governed by the uncoupled wave equations (7–63) and (7–65).

P.7–14 Substitute Eqs. (7–55) and (7–57) in Maxwell's equations to obtain wave equations for scalar potential V and vector potential $\mathbf{A}$ for a linear, isotropic but inhomogeneous medium. Show that these wave equations reduce to Eqs. (7–65) and (7–63) for simple media. (*Hint:* Use the following gauge condition for potentials in an inhomogeneous medium:

$$\nabla \cdot (\epsilon\mathbf{A}) + \mu\epsilon^2 \frac{\partial V}{\partial t} = 0. \tag{7–117}$$

P.7–15 Write the set of four Maxwell's equations, Eqs. (7–53a, b, c and d), as eight scalar equations
   a) in Cartesian coordinates,
   b) in cylindrical coordinates,
   c) in spherical coordinates.

P.7–16 Supply the detailed steps for the derivation of the electromagnetic boundary conditions, Eqs. (7–66a, b, c, and d).

P.7–17 Discuss the relations
   a) between the boundary conditions for the tangential components of $\mathbf{E}$ and those for
      the normal components of $\mathbf{B}$,
   b) between the boundary conditions for the normal components of $\mathbf{D}$ and those for
      the tangential components of $\mathbf{H}$.

P.7–18 In Eqs. (3–88) and (3–89) it was shown that for field calculations a polarized dielectric may be replaced by an equivalent polarization surface charge density ρ_{ps} and an equivalent polarization volume charge density ρ_p. Find the boundary conditions at the interface of two different media for
   a) the normal component of $\mathbf{P}$,
   b) the normal components of $\mathbf{E}$
in terms of free and equivalent polarization surface charge densities ρ_s and ρ_{ps}.

P.7–19 Write the boundary conditions that exist at the interface of free space and a magnetic material of infinite (an approximation) permeability.

P.7–20 Prove by direct substitution that any twice differentiable function of $(t - R\sqrt{\mu\epsilon})$ or of $(t + R\sqrt{\mu\epsilon})$ is a solution of the homogeneous wave equation, Eq. (7–73).

P.7–21 Prove that the retarded potential in Eq. (7–77) satisfies the nonhomogeneous wave equation, Eq. (7–65).

P.7–22 For the assumed $f(t)$ at $R = 0$ in Fig. 7–15, sketch
   a) $f(t - R/u)$ versus t,
   b) $f(t - R/u)$ versus R for $t > T$.

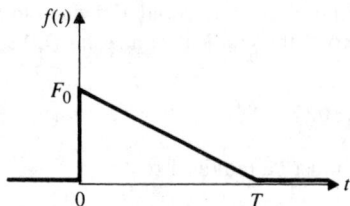

FIGURE 7–15
A triangular time function (Problem P.7–22).

P.7–23 The electric field of an electromagnetic wave

$$\mathbf{E} = \mathbf{a}_x E_0 \cos\left[10^8 \pi \left(t - \frac{z}{c}\right) + \theta\right]$$

is the sum of

$$\mathbf{E}_1 = \mathbf{a}_x 0.03 \sin 10^8 \pi \left(t - \frac{z}{c}\right)$$

and

$$\mathbf{E}_2 = \mathbf{a}_x 0.04 \cos\left[10^8 \pi \left(t - \frac{z}{c}\right) - \frac{\pi}{3}\right].$$

Find E_0 and θ.

P.7–24 Derive the general wave equations for **E** and **H** in a nonconducting simple medium where a charge distribution ρ and a current distribution **J** exist. Convert the wave equations to Helmholtz's equations for sinusoidal time dependence. Write the general solutions for $\mathbf{E}(R, t)$ and $\mathbf{H}(R, t)$ in terms of ρ and **J**.

P.7–25 Given that
$$\mathbf{E} = \mathbf{a}_y 0.1 \sin (10\pi x) \cos (6\pi 10^9 t - \beta z) \qquad \text{(V/m)}$$
in air, find **H** and β.

P.7–26 Given that
$$\mathbf{H} = \mathbf{a}_y 2 \cos (15\pi x) \sin (6\pi 10^9 t - \beta z) \qquad \text{(A/m)}$$
in air, find **E** and β.

P.7–27 It is known that the electric field intensity of a spherical wave in free space is

$$\mathbf{E} = \mathbf{a}_\theta \frac{E_0}{R} \sin \theta \cos (\omega t - kR).$$

Determine the magnetic field intensity **H** and the value of k.

P.7–28 In Section 7–4 we indicated that **E** and **B** can be determined from the potentials V and **A**, which are related by the Lorentz condition, Eq. (7–98), in the time-harmonic case. The vector potential **A** was introduced through the relation $\mathbf{B} = \nabla \times \mathbf{A}$ because of the solenoidal nature of **B**. In a source-free region, $\nabla \cdot \mathbf{E} = 0$, we can define another type of vector potential $\mathbf{A}_e$, such that $\mathbf{E} = \nabla \times \mathbf{A}_e$. Assuming harmonic time dependence:
   a) Express **H** in terms of $\mathbf{A}_e$.
   b) Show that $\mathbf{A}_e$ is a solution of a homogeneous Helmholtz's equation.

P.7–29 For a source-free polarized medium where $\rho = 0$, $\mathbf{J} = 0$, $\mu = \mu_0$, but where there is a volume density of polarization **P**, a single vector potential π_e may be defined such that
$$\mathbf{H} = j\omega\epsilon_0 \nabla \times \pi_e. \qquad (7\text{–}118)$$
   a) Express electric field intensity **E** in terms of π_e and **P**.

b) Show that π_e satisfies the nonhomogeneous Helmholtz's equation

$$\nabla^2 \pi_e + k_0^2 \pi_e = -\frac{\mathbf{P}}{\epsilon_0}. \qquad (7\text{--}119)$$

The quantity π_e is known as the *electric Hertz potential*.

P.7–30 Calculations concerning the electromagnetic effect of currents in a good conductor usually neglect the displacement current even at microwave frequencies.

  a) Assuming $\epsilon_r = 1$ and $\sigma = 5.70 \times 10^7$ (S/m) for copper, compare the magnitude of the displacement current density with that of the conduction current density at 100 (GHz).

  b) Write the governing differential equation for magnetic field intensity $\mathbf{H}$ in a source-free good conductor.

8

Plane Electromagnetic Waves

8–1 Introduction

In Chapter 7 we showed that in a source-free nonconducting simple medium, Maxwell's equations (Eqs. 7–79a, b, c, and d) can be combined to yield homogeneous vector wave equations in **E** and in **H**. These two equations, Eqs. (7–81) and (7–82), have exactly the same form. In free space the source-free wave equation for **E** is

$$\nabla^2 \mathbf{E} - \frac{1}{c^2}\frac{\partial^2 \mathbf{E}}{\partial t^2} = 0, \qquad (8\text{–}1)$$

where

$$c = \frac{1}{\sqrt{\mu_0 \epsilon_0}} \cong 3 \times 10^8 \ (\text{m/s}) = 300 \quad (\text{Mm/s}) \qquad (8\text{–}2)$$

is the velocity of wave propagation (the speed of light) in free space. The solutions of Eq. (8–1) represent waves. The study of the behavior of waves that have a one-dimensional spatial dependence (*plane waves*) is the main concern of this chapter.

We begin the chapter with a study of the propagation of time-harmonic plane-wave fields in an unbounded homogeneous medium. Medium parameters such as intrinsic impedance, attenuation constant, and phase constant will be introduced. The meaning of *skin depth*, the depth of wave penetration into a good conductor, will be explained. Electromagnetic waves carry with them electromagnetic power. The concept of *Poynting vector*, a power flux density, will be discussed.

We will examine the behavior of a plane wave incident normally on a plane boundary. The laws governing the reflection and refraction of plane waves incident obliquely on a plane boundary will then be discussed, and the conditions for no reflection and for total reflection will be examined.

A *uniform plane wave* is a particular solution of Maxwell's equations with **E** assuming the same direction, same magnitude, and same phase in infinite planes perpendicular to the direction of propagation (similarly for **H**). Strictly speaking, a uniform plane wave does not exist in practice because a source infinite in extent would be required to creat it, and practical wave sources are always finite in extent. But, if we are far enough away from a source, the *wavefront* (surface of constant phase) becomes almost spherical; and a very small portion of the surface of a giant sphere is very nearly a plane. The characteristics of uniform plane waves are particularly simple, and their study is of fundamental theoretical, as well as practical, importance.

8–2 Plane Waves in Lossless Media

In this and future chapters we focus our attention on wave behavior in the sinusoidal steady state, using phasors to great advantage. The source-free wave equation, Eq. (8–1), for free space becomes a homogeneous vector Helmholtz's equation (see Eq. 7–105):

$$\mathbf{V}^2\mathbf{E} + k_0^2\mathbf{E} = 0, \tag{8–3}$$

where k_0 is the *free-space wavenumber*

$$k_0 = \omega\sqrt{\mu_0\epsilon_0} = \frac{\omega}{c} \quad \text{(rad/m)}. \tag{8–4}$$

In Cartesian coordinates, Eq. (8–3) is equivalent to three scalar Helmholtz's equations, one each in the components E_x, E_y, and E_z. Writing it for the component E_x, we have

$$\left(\frac{\partial^2}{\partial x^2} + \frac{\partial^2}{\partial y^2} + \frac{\partial^2}{\partial z^2} + k_0^2\right)E_x = 0. \tag{8–5}$$

Consider a uniform plane wave characterized by a uniform E_x (uniform magnitude and constant phase) over plane surfaces perpendicular to z; that is,

$$\partial^2 E_x/\partial x^2 = 0 \quad \text{and} \quad \partial^2 E_x/\partial y^2 = 0.$$

Equation (8–5) simplifies to

$$\frac{d^2 E_x}{dz^2} + k_0^2 E_x = 0, \tag{8–6}$$

which is an ordinary differential equation because E_x, a phasor, depends only on z. The solution of Eq. (8–6) is readily seen to be

$$E_x(z) = E_x^+(z) + E_x^-(z)$$
$$= E_0^+ e^{-jk_0 z} + E_0^- e^{jk_0 z}, \tag{8–7}$$

where E_0^+ and E_0^- are arbitrary (and, in general, complex) constants that must be determined by boundary conditions. Note that since Eq. (8–6) is a second-order equation, its general solution in Eq. (8–7) contains two integration constants.

$E_x^+(z)$

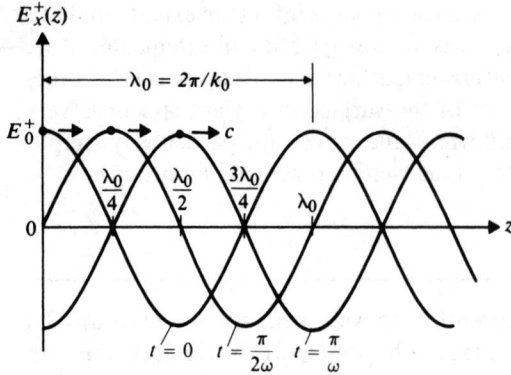

FIGURE 8–1
Wave traveling in positive z direction
$E_x^+(z, t) = E_0^+ \cos (\omega t - k_0 z)$, for several values
of t.

Now let us examine what the first phasor term on the right side of Eq. (8–7) represents in real time. Using $\cos \omega t$ as the reference and assuming E_0^+ to be a real constant (zero reference phase at $z = 0$), we have

$$E_x^+(z, t) = \mathcal{R}e\big[E_x^+(z)e^{j\omega t}\big]$$
$$= \mathcal{R}e\big[E_0^+ e^{j(\omega t - k_0 z)}\big] \qquad (8\text{--}8)$$
$$= E_0^+ \cos (\omega t - k_0 z) \qquad (\text{V/m}).$$

Equation (8–8) has been plotted in Fig. 8–1 for several values of t. At $t = 0$, $E_x^+(z, 0) = E_0^+ \cos k_0 z$ is a cosine curve with an amplitude E_0^+. At successive times the curve effectively travels in the positive z direction. We have, then, a **traveling wave**. If we fix our attention on a particular point (a point of a particular phase) on the wave, we set $\cos (\omega t - k_0 z) = $ a constant or

$$\omega t - k_0 z = \text{A constant phase},$$

from which we obtain

$$u_p = \frac{dz}{dt} = \frac{\omega}{k_0} = \frac{1}{\sqrt{\mu_0 \epsilon_0}} = c. \qquad (8\text{--}9)$$

Equation (8–9) assures us that the velocity of propagation of an equiphase front (the **phase velocity**) in free space is equal to the velocity of light, which is approximately 3×10^8 (m/s) in free space.

Wavenumber k_0 bears a definite relation to the wavelength. From Eq. (8–4), $k_0 = 2\pi f/c$ or

$$\boxed{k_0 = \frac{2\pi}{\lambda_0} \qquad (\text{rad/m}),} \qquad (8\text{--}10)$$

which measures the number of wavelengths in a complete cycle, hence its name. An inverse relation of Eq. (8–10) is

$$\lambda_0 = \frac{2\pi}{k_0} \qquad \text{(m)}. \tag{8–11}$$

Equations (8–10) and (8–11) are valid without the subscript 0 if the medium is a lossless material such as a perfect dielectric, instead of free space.

It is obvious without replotting that the second phasor term on the right side of Eq. (8–7), $E_0^- e^{jk_0 z}$, represents a cosinusoidal wave traveling in the $-z$ direction with the same velocity c. If we are concerned only with the wave traveling in the $+z$ direction, $E_0^- = 0$. However, if there are discontinuities in the medium, reflected waves traveling in the opposite direction must also be considered, as we will see later in this chapter.

The associated magnetic field $\mathbf{H}$ can be found from Eq. (7–104a)

$$\nabla \times \mathbf{E} = \begin{vmatrix} \mathbf{a}_x & \mathbf{a}_y & \mathbf{a}_z \\ 0 & 0 & \dfrac{\partial}{\partial z} \\ E_x^+(z) & 0 & 0 \end{vmatrix} = -j\omega\mu_0(\mathbf{a}_x H_x^+ + \mathbf{a}_y H_y^+ + \mathbf{a}_z H_z^+),$$

which leads to

$$H_x^+ = 0, \tag{8–12a}$$

$$H_y^+ = \frac{1}{-j\omega\mu_0} \frac{\partial E_x^+(z)}{\partial z}, \tag{8–12b}$$

$$H_z^+ = 0. \tag{8–12c}$$

Thus H_y^+ is the only nonzero component of $\mathbf{H}$; and since

$$\frac{\partial E_x^+(z)}{\partial z} = \frac{\partial}{\partial z}(E_0^+ e^{-jk_0 z}) = -jk_0 E_x^+(z),$$

Eq. (8–12b) yields

$$H_y^+(z) = \frac{k_0}{\omega\mu_0} E_x^+(z) = \frac{1}{\eta_0} E_x^+(z) \qquad \text{(A/m)}. \tag{8–13}^\dagger$$

We have introduced a new quantity, η_0, in Eq. (8–13):

$$\eta_0 = \sqrt{\frac{\mu_0}{\epsilon_0}} \cong 120\pi \cong 377 \qquad (\Omega), \tag{8–14}$$

$\dagger$ If we had started with $E_x^-(z) = E_0^- e^{jk_0 z}$, we would obtain $H_y^-(z) = -\dfrac{1}{\eta_0} E_x^-(z)$.

which is called the **intrinsic impedance of the free space**. Because η_0 is a real number, $H_y^+(z)$ is in phase with $E_x^+(z)$, and we can write the instantaneous expression for **H** as

$$\mathbf{H}(z, t) = \mathbf{a}_y H_y^+(z, t) = \mathbf{a}_y \mathscr{R}e\left[H_y^+(z)e^{j\omega t}\right]$$

$$= \mathbf{a}_y \frac{E_0^+}{\eta_0} \cos(\omega t - k_0 z) \qquad \text{(A/m)}. \tag{8-15}$$

Hence, for a uniform plane wave the ratio of the magnitudes of **E** and **H** is the intrinsic impedance of the medium. We also note that **H** is perpendicular to **E** and that both are normal to the direction of propagation. The fact that we specified $\mathbf{E} = \mathbf{a}_x E_x$ is not as restrictive as it appears, inasmuch as we are free to designate the direction of **E** as the $+x$-direction, which is normal to the direction of propagation $\mathbf{a}_z$.

EXAMPLE 8-1   A uniform plane wave with $\mathbf{E} = \mathbf{a}_x E_x$ propagates in a lossless simple medium ($\epsilon_r = 4$, $\mu_r = 1$, $\sigma = 0$) in the $+z$-direction. Assume that E_x is sinusoidal with a frequency 100 (MHz) and has a maximum value of $+10^{-4}$ (V/m) at $t = 0$ and $z = \frac{1}{8}$ (m).

a) Write the instantaneous expression for **E** for any t and z.

b) Write the instantaneous expression for **H**.

c) Determine the locations where E_x is a positive maximum when $t = 10^{-8}$ (s).

Solution   First we find k:

$$k = \omega\sqrt{\mu\epsilon} = \frac{\omega}{c}\sqrt{\mu_r\epsilon_r}$$

$$= \frac{2\pi 10^8}{3 \times 10^8}\sqrt{4} = \frac{4\pi}{3} \quad \text{(rad/m)}.$$

a) Using $\cos\omega t$ as the reference, we find the instantaneous expression for **E** to be

$$\mathbf{E}(z, t) = \mathbf{a}_x E_x = \mathbf{a}_x 10^{-4}\cos(2\pi 10^8 t - kz + \psi).$$

Since E_x equals $+10^{-4}$ when the argument of the cosine function equals zero—that is, when

$$2\pi 10^8 t - kz + \psi = 0,$$

we have, at $t = 0$ and $z = \frac{1}{8}$,

$$\psi = kz = \left(\frac{4\pi}{3}\right)\left(\frac{1}{8}\right) = \frac{\pi}{6} \quad \text{(rad)}.$$

Thus,

$$\mathbf{E}(z, t) = \mathbf{a}_x 10^{-4}\cos\left(2\pi 10^8 t - \frac{4\pi}{3}z + \frac{\pi}{6}\right)$$

$$= \mathbf{a}_x 10^{-4}\cos\left[2\pi 10^8 t - \frac{4\pi}{3}\left(z - \frac{1}{8}\right)\right] \quad \text{(V/m)}.$$

This expression shows a shift of $\frac{1}{8}$ (m) in the $+z$-direction and could have been written down directly from the statement of the problem.

b) The phasor expression for **H** is

$$\mathbf{H} = \mathbf{a}_y H_y = \mathbf{a}_y \frac{E_x}{\eta},$$

where

$$\eta = \sqrt{\frac{\mu}{\epsilon}} = \frac{\eta_0}{\sqrt{\epsilon_r}} = 60\pi \quad (\Omega).$$

Hence,

$$\mathbf{H}(z, t) = \mathbf{a}_y \frac{10^{-4}}{60\pi} \cos\left[2\pi10^8 t - \frac{4\pi}{3}\left(z - \frac{1}{8} \right) \right] \quad (\text{A/m}).$$

c) At $t = 10^{-8}$, we equate the argument of the cosine function to $+2n\pi$ in order to make E_x a positive maximum:

$$2\pi10^8(10^{-8}) - \frac{4\pi}{3}\left(z_m - \frac{1}{8} \right) = \pm 2n\pi,$$

from which we get

$$z_m = \frac{13}{8} \pm \frac{3}{2}n \quad (\text{m}), \qquad n = 0, 1, 2, \ldots; \qquad z_m > 0.$$

Examining this result more closely, we note that the wavelength in the given medium is

$$\lambda = \frac{2\pi}{k} = \frac{3}{2} \quad (\text{m}).$$

Hence the positive maximum value of E_x occurs at

$$z_m = \frac{13}{8} \pm n\lambda \quad (\text{m}).$$

The **E** and **H** fields are shown in Fig. 8–2 as functions of z for the reference time $t = 0$. ▬

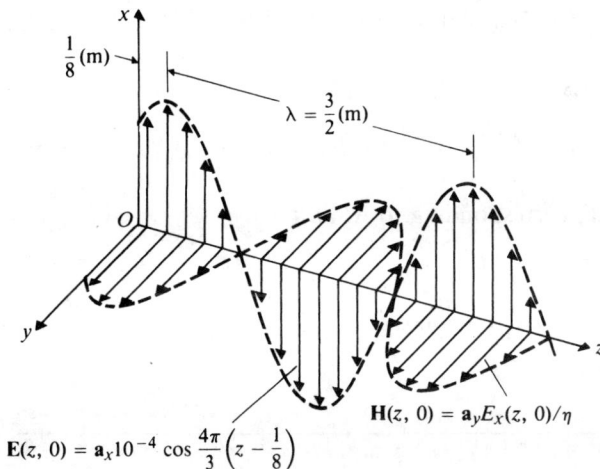

$$\mathbf{E}(z, 0) = \mathbf{a}_x 10^{-4} \cos\frac{4\pi}{3}\left(z - \frac{1}{8} \right)$$

$$\mathbf{H}(z, 0) = \mathbf{a}_y E_x(z, 0)/\eta$$

FIGURE 8–2
E and H fields of a uniform plane wave at $t = 0$ (Example 8–1).

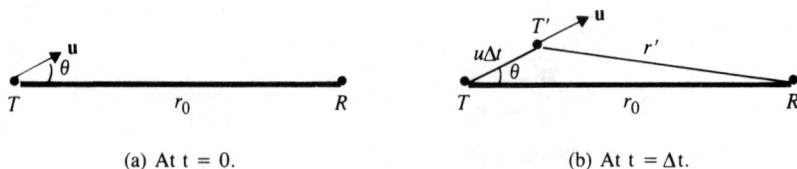

(a) At t = 0.   (b) At t = Δt.

FIGURE 8–3
Illustrating the Doppler effect.

8–2.1   DOPPLER EFFECT

When there is relative motion between a time-harmonic source and a receiver, the frequency of the wave detected by the receiver tends to be different from that emitted by the source. This phenomenon is known as the **Doppler effect**.[†] The Doppler effect manifests itself in acoustics as well as in electromagnetics. Perhaps you have experienced the changes in the pitch of a fast-moving locomotive whistle. In the following we give an explanation of the Doppler effect.

Let us assume that the source (transmitter) T of a time-harmonic wave of a frequency f moves with a velocity $\mathbf{u}$ at an angle θ relative to the direct line to a stationary receiver R, as illustrated in Fig. 8–3(a). The electromagnetic wave emitted by T at a reference time $t = 0$ will reach R at

$$t_1 = \frac{r_0}{c}. \tag{8–16}$$

At a later time $t = \Delta t$, T has moved to the new position T', and the wave emitted by T' at that time will reach R at

$$t_2 = \Delta t + \frac{r'}{c}$$
$$= \Delta t + \frac{1}{c}\left[r_0^2 - 2r_0(u\,\Delta t)\cos\theta + (u\,\Delta t)^2\right]^{1/2}. \tag{8–17}$$

If $(u\,\Delta t)^2 \ll r_0^2$, Eq. (8–17) becomes

$$t_2 \cong \Delta t + \frac{r_0}{c}\left(1 - \frac{u\,\Delta t}{r_0}\cos\theta\right). \tag{8–18}$$

Hence the time elapsed at R, $\Delta t'$, corresponding to Δt at T is

$$\Delta t' = t_2 - t_1$$
$$= \Delta t\left(1 - \frac{u}{c}\cos\theta\right), \tag{8–19}$$

which is not equal to Δt.

[†] C. Doppler (1803–1853).

If Δt represents a period of the time-harmonic source—that is, if $\Delta t = 1/f$—then the frequency of the received wave at R is

$$f' = \frac{1}{\Delta t'} = \frac{f}{\left(1 - \dfrac{u}{c}\cos\theta\right)}$$

$$\cong f\left(1 + \frac{u}{c}\cos\theta\right) \tag{8–20}$$

for the usual case of $(u/c)^2 \ll 1$. Equation (8–20) is an approximate formula and does not hold when θ is close to $\pi/2$. (Do you know why?) For $\theta = 0$, Eq. (8–20) clearly indicates that the frequency perceived at R is higher than the transmitted frequency when T moves toward R. Conversely, the perceived frequency is lower than the transmitted frequency when T moves away from R ($\theta = \pi$). It is obvious that similar results are obtained if R moves and T is stationary.

The Doppler effect is the basis of operation of the (Doppler) radar used by police to check the speed of a moving vehicle. The frequency shift of the received wave reflected by a moving vehicle is proportional to the speed of the vehicle and can be detected and displayed on a hand-held unit. (See Problem P.8–3). The Doppler effect is also the cause of the so-called *red shift* of the light spectrum emitted by a receding distant star in astronomy. As the star *moves away* at a high speed from an observer on earth, the received frequency shifts toward the *lower frequency* (*red*) end of the spectrum.

8–2.2 TRANSVERSE ELECTROMAGNETIC WAVES

We have seen that a uniform plane wave characterized by $\mathbf{E} = \mathbf{a}_x E_x$ propagating in the $+z$-direction has associated with it a magnetic field $\mathbf{H} = \mathbf{a}_y H_y$. Thus $\mathbf{E}$ and $\mathbf{H}$ are perpendicular to each other, and both are transverse to the direction of propagation. It is a particular case of a *transverse electromagnetic (*TEM*) wave*. The phasor field quantities are functions of only the distance z along a single coordinate axis. We now consider the propagation of a uniform plane wave along an arbitrary direction that does not necessarily coincide with a coordinate axis.

The phasor electric field intensity for a uniform plane wave propagating in the $+z$-direction is

$$\mathbf{E}(z) = \mathbf{E}_0 e^{-jkz}, \tag{8–21}$$

where $\mathbf{E}_0$ is a constant vector. A more general form of Eq. (8–21) is

$$\mathbf{E}(x, y, z) = \mathbf{E}_0 e^{-jk_x x - jk_y y - jk_z z}. \tag{8–22}$$

It can be easily proved by direct substitution that this expression satisfies the homogeneous Helmholtz's equation, provided that

$$k_x^2 + k_y^2 + k_z^2 = \omega^2 \mu\epsilon. \tag{8–23}$$

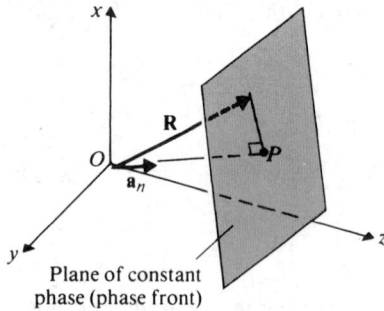

FIGURE 8-4
Radius vector and wave normal to a phase front of a uniform plane wave.

Plane of constant
phase (phase front)

If we define a *wavenumber vector* as

$$\mathbf{k} = \mathbf{a}_x k_x + \mathbf{a}_y k_y + \mathbf{a}_z k_z = k\mathbf{a}_n \qquad (8-24)$$

and a radius vector from the origin

$$\mathbf{R} = \mathbf{a}_x x + \mathbf{a}_y y + \mathbf{a}_z z, \qquad (8-25)$$

then Eq. (8-22) can be written compactly as

$$\boxed{\mathbf{E}(\mathbf{R}) = \mathbf{E}_0 e^{-j\mathbf{k}\cdot\mathbf{R}} = \mathbf{E}_0 e^{-jk\mathbf{a}_n\cdot\mathbf{R}} \qquad (V/m),} \qquad (8-26)$$

where $\mathbf{a}_n$ is a unit vector in the direction of propagation. From Eq. (8-24) it is clear that

$$k_x = \mathbf{k}\cdot\mathbf{a}_x = k\mathbf{a}_n\cdot\mathbf{a}_x, \qquad (8-27a)$$

$$k_y = \mathbf{k}\cdot\mathbf{a}_y = k\mathbf{a}_n\cdot\mathbf{a}_y, \qquad (8-27b)$$

$$k_z = \mathbf{k}\cdot\mathbf{a}_z = k\mathbf{a}_n\cdot\mathbf{a}_z, \qquad (8-27c)$$

and that $\mathbf{a}_n\cdot\mathbf{a}_x$, $\mathbf{a}_n\cdot\mathbf{a}_y$ and $\mathbf{a}_n\cdot\mathbf{a}_z$ are direction cosines of $\mathbf{a}_n$.

The geometrical relations of $\mathbf{a}_n$ and $\mathbf{R}$ are illustrated in Fig. 8-4, from which we see that

$$\mathbf{a}_n\cdot\mathbf{R} = \text{Length } \overline{OP} \quad \text{(a constant)}$$

is the equation of a plane normal to $\mathbf{a}_n$, the direction of propagation. Just as $z = $ Constant denotes a plane of constant phase and uniform amplitude for the wave in Eq. (8-21), $\mathbf{a}_n\cdot\mathbf{R} = $ Constant is a plane of constant phase and uniform amplitude for the wave in Eq. (8-26). In a charge-free region, $\mathbf{V}\cdot\mathbf{E} = 0$. As a result,

$$\mathbf{E}_0\cdot\mathbf{V}(e^{-jk\mathbf{a}_n\cdot\mathbf{R}}) = 0. \qquad (8-28a)^\dagger$$

---

† This is a consequence of the fact that $\mathbf{V}\cdot\mathbf{E}_0 = 0$, where $\mathbf{E}_0$ is a constant vector (see Problem P.2-28).

But

$$\mathbf{V}(e^{-jk\mathbf{a}_n \cdot \mathbf{R}}) = \left(\mathbf{a}_x \frac{\partial}{\partial x} + \mathbf{a}_y \frac{\partial}{\partial y} + \mathbf{a}_z \frac{\partial}{\partial z}\right) e^{-j(k_x x + k_y y + k_z z)}$$

$$= -j(\mathbf{a}_x k_x + \mathbf{a}_y k_y + \mathbf{a}_z k_z)e^{-j(k_x x + k_y y + k_z z)}$$

$$= -jk\mathbf{a}_n e^{-jk\mathbf{a}_n \cdot \mathbf{R}},$$

hence Eq. (8–28a) can be written as

$$-jk(\mathbf{E}_0 \cdot \mathbf{a}_n)e^{-jk\mathbf{a}_n \cdot \mathbf{R}} = 0,$$

which requires

$$\mathbf{a}_n \cdot \mathbf{E}_0 = 0. \tag{8–28b}$$

Thus the plane-wave solution in Eq. (8–26) imples that $\mathbf{E}_0$ is transverse to the direction of propagation.

The magnetic field associated with $\mathbf{E}(\mathbf{R})$ in Eq. (8–26) may be obtained from Eq. (7–104a) as

$$\mathbf{H}(\mathbf{R}) = -\frac{1}{j\omega\mu} \mathbf{V} \times \mathbf{E}(\mathbf{R})$$

or

$$\boxed{\mathbf{H}(\mathbf{R}) = \frac{1}{\eta} \mathbf{a}_n \times \mathbf{E}(\mathbf{R}) \qquad \text{(A/m)},} \tag{8–29}$$

where

$$\boxed{\eta = \frac{\omega\mu}{k} = \sqrt{\frac{\mu}{\epsilon}} \qquad (\Omega)} \tag{8–30}$$

is the *intrinsic impedance*[†] of the medium. Substitution of Eq. (8–26) in Eq. (8–29) yields

$$\boxed{\mathbf{H}(\mathbf{R}) = \frac{1}{\eta} (\mathbf{a}_n \times \mathbf{E}_0)e^{-jk\mathbf{a}_n \cdot \mathbf{R}} \qquad \text{(A/m)}.} \tag{8–31}$$

It is now clear that a uniform plane wave propagating in an arbitrary direction, $\mathbf{a}_n$, is a TEM wave with $\mathbf{E} \perp \mathbf{H}$ and that both $\mathbf{E}$ and $\mathbf{H}$ are normal to $\mathbf{a}_n$.

EXAMPLE 8–2   If $\mathbf{E}(\mathbf{R})$ of a TEM wave is given, as in Eq. (8–26), $\mathbf{H}(\mathbf{R})$ can be found by using Eq. (8–29). Obtain a relation expressing $\mathbf{E}(\mathbf{R})$ in terms of $\mathbf{H}(\mathbf{R})$.

Solution   Assuming $\mathbf{H}(\mathbf{R})$ to have the form

$$\mathbf{H}(\mathbf{R}) = \mathbf{H}_0 e^{-jk\mathbf{a}_n \cdot \mathbf{R}}, \tag{8–32}$$

[†] Also called *wave impedance*.

we obtain from Eq. (7–104b)

$$\mathbf{E(R)} = \frac{1}{j\omega\epsilon} \nabla \times \mathbf{H(R)}$$

$$= \frac{1}{j\omega\epsilon} (-jk)\mathbf{a}_n \times \mathbf{H(R)}$$

or

$$\boxed{\mathbf{E(R)} = -\eta\mathbf{a}_n \times \mathbf{H(R)} \quad \text{(V/m).}} \tag{8–33}$$

Alternatively, we can obtain the same result by cross-multiplying both sides of Eq. (8–29) by $\mathbf{a}_n$ and using the back-cab rule in Eq. (2–20). ■

8–2.3 POLARIZATION OF PLANE WAVES

The *polarization* of a uniform plane wave describes the time-varying behavior of the electric field intensity vector at a given point in space. Since the $\mathbf{E}$ vector of the plane wave in Example 8–1 is fixed in the x direction ($\mathbf{E} = \mathbf{a}_x E_x$, where E_x may be positive or negative), the wave is said to be *linearly polarized* in the x-direction. A separate description of magnetic-field behavior is not necessary, inasmuch as the direction of $\mathbf{H}$ is definitely related to that of $\mathbf{E}$.

In some cases the direction of $\mathbf{E}$ of a plane wave at a given point may change with time. Consider the superposition of two linearly polarized waves: one polarized in the x-direction, and the other polarized in the y-direction and lagging 90° (or $\pi/2$ rad) in time phase. In phasor notation we have

$$\begin{aligned}\mathbf{E}(z) &= \mathbf{a}_x E_1(z) + \mathbf{a}_y E_2(z) \\ &= \mathbf{a}_x E_{10} e^{-jkz} - \mathbf{a}_y j E_{20} e^{-jkz},\end{aligned} \tag{8–34}$$

where E_{10} and E_{20} are real numbers denoting the amplitudes of the two linearly polarized waves.

The instantaneous expression for $\mathbf{E}$ is

$$\begin{aligned}\mathbf{E}(z, t) &= \mathscr{R}e\{[\mathbf{a}_x E_1(z) + \mathbf{a}_y E_2(z)]e^{j\omega t}\} \\ &= \mathbf{a}_x E_{10} \cos(\omega t - kz) + \mathbf{a}_y E_{20} \cos\left(\omega t - kz - \frac{\pi}{2}\right).\end{aligned}$$

In examining the direction change of $\mathbf{E}$ at a given point as t changes, it is convenient to set $z = 0$. We have

$$\begin{aligned}\mathbf{E}(0, t) &= \mathbf{a}_x E_1(0, t) + \mathbf{a}_y E_2(0, t) \\ &= \mathbf{a}_x E_{10} \cos \omega t + \mathbf{a}_y E_{20} \sin \omega t.\end{aligned} \tag{8–35}$$

As ωt increases from 0 through $\pi/2$, π, and $3\pi/2$—completing the cycle at 2π—the tip of the vector $\mathbf{E}(0, t)$ will traverse an elliptical locus in the counterclockwise

direction. Analytically, we have

$$\cos \omega t = \frac{E_1(0, t)}{E_{10}}$$

and

$$\sin \omega t = \frac{E_2(0, t)}{E_{20}}$$

$$= \sqrt{1 - \cos^2 \omega t} = \sqrt{1 - \left[\frac{E_1(0, t)}{E_{10}}\right]^2},$$

which leads to the following equation for an ellipse:

$$\left[\frac{E_2(0, t)}{E_{20}}\right]^2 + \left[\frac{E_1(0, t)}{E_{10}}\right]^2 = 1. \tag{8-36}$$

Hence **E**, which is the sum of two linearly polarized waves in both space and time quadrature, is *elliptically polarized* if $E_{20} \neq E_{10}$, and is *circularly polarized* if $E_{20} = E_{10}$. A typical polarization circle is shown in Fig. 8-5(a).

When $E_{20} = E_{10}$, the instantaneous angle α that **E** makes with the x-axis at $z = 0$ is

$$\alpha = \tan^{-1} \frac{E_2(0, t)}{E_1(0, t)} = \omega t, \tag{8-37}$$

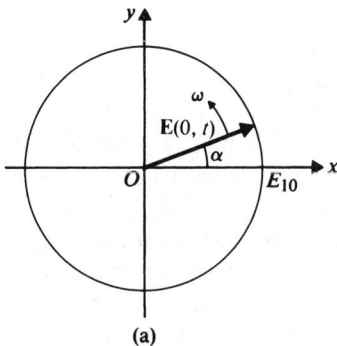

(a)

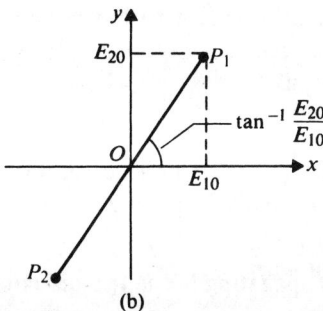

(b)

FIGURE 8-5
Polarization diagrams for sum of two linearly polarized waves in space quadrature at $z = 0$: (a) circular polarization, $E(0, t) = E_{10}(\mathbf{a}_x \cos \omega t + \mathbf{a}_y \sin \omega t)$; (b) linear polarization, $\mathbf{E}(0, t) = (\mathbf{a}_x E_{10} + \mathbf{a}_y E_{20}) \cos \omega t$.

which indicates that **E** rotates at a uniform rate with an angular velocity ω in a *counterclockwise* direction. When the fingers of the right hand follow the direction of the rotation of **E**, the thumb points to the direction of propagation of the wave. This is a ***right-hand*** or ***positive circularly polarized wave***.

If we start with an $E_2(z)$, which *leads* $E_1(z)$ by 90° ($\pi/2$ rad) in time phase, Eqs. (8–34) and (8–35) will be, respectively,

$$\mathbf{E}(z) = \mathbf{a}_x E_{10} e^{-jkz} + \mathbf{a}_y j E_{20} e^{-jkz} \tag{8–38}$$

and

$$\mathbf{E}(0, t) = \mathbf{a}_x E_{10} \cos \omega t - \mathbf{a}_y E_{20} \sin \omega t. \tag{8–39}$$

Comparing Eq. (8–39) with Eq. (8–35), we see that **E** will still be elliptically polarized. If $E_{20} = E_{10}$, **E** will be circularly polarized, and its angle measured from the x-axis at $z = 0$ will now be $-\omega t$, indicating that **E** will rotate with an angular velocity ω in a *clockwise* direction; this is a ***left-hand*** or ***negative circularly polarized wave***.

If $E_2(z)$ and $E_1(z)$ are in space quadrature but in time phase, their sum **E** will be linearly polarized along a line that makes an angle $\tan^{-1}(E_{20}/E_{10})$ with the x-axis, as depicted in Fig. 8–5(b). The instantaneous expression for **E** at $z = 0$ is

$$\mathbf{E}(0, t) = (\mathbf{a}_x E_{10} + \mathbf{a}_y E_{20}) \cos \omega t. \tag{8–40}$$

The tip of the $\mathbf{E}(0, t)$ will be at the point P_1 when $\omega t = 0$. Its magnitude will decrease toward zero as ωt increases toward $\pi/2$. After that, $\mathbf{E}(0, t)$ starts to increase again, in the opposite direction, toward the point P_2 where $\omega t = \pi$.

In the general case, $E_2(z)$ and $E_1(z)$, which are in space quadrature, can have unequal amplitudes ($E_{20} \neq E_{10}$) and can differ in phase by an arbitrary amount (not zero or an integral multiple of $\pi/2$). Their sum **E** will be elliptically polarized, and the principal axes of the polarization ellipse will not coincide with the axes of the coordinates (see Problem P.8–7).

We note here that the electromagnetic waves radiated by AM broadcast stations from their antenna towers are linearly polarized with the **E**-field perpendicular to the ground. For maximum reception the receiving antenna should be parallel to the **E**-field—that is, vertical. Television signals, on the other hand, are linearly polarized in the horizontal direction. This is why the wires of rooftop TV receiving antennas are horizontal. The waves radiated by FM broadcast stations are generally circularly polarized; hence the orientation of an FM receiving antenna is not critical as long as it lies in a plane normal to the direction of the signal.

EXAMPLE 8–3 Prove that a linearly polarized plane wave can be resolved into a right-hand circularly polarized wave and a left-hand circularly polarized wave of equal amplitude.

Solution Consider a linearly polarized plane wave propagating in the $+z$-direction. We can assume, with no loss of generality, that **E** is polarized in the x-direction. In phasor notation we have

$$\mathbf{E}(z) = \mathbf{a}_x E_0 e^{-jkz}.$$

But this can be written as

$$\mathbf{E}(z) = \mathbf{E}_{rc}(z) + \mathbf{E}_{lc}(z),$$

where

$$\mathbf{E}_{rc}(z) = \frac{E_0}{2}(\mathbf{a}_x - j\mathbf{a}_y)e^{-jkz} \tag{8–41a}$$

and

$$\mathbf{E}_{lc}(z) = \frac{E_0}{2}(\mathbf{a}_x + j\mathbf{a}_y)e^{-jkz}. \tag{8–41b}$$

From previous discussions we recognize that $\mathbf{E}_{rc}(z)$ in Eq. (8–41a) and $\mathbf{E}_{lc}(z)$ in Eq. (8–41b) represent right-hand and left-hand circularly polarized waves, respectively, each having an amplitude $E_0/2$. The statement of this problem is therefore proved. The converse statement that the sum of two oppositely rotating circularly polarized waves of equal amplitude is a linearly polarized wave is, of course, also true. ■

8–3 Plane Waves in Lossy Media

In a source-free lossy medium the homogeneous vector Helmholtz's equation to be solved is

$$\nabla^2 \mathbf{E} + k_c^2 \mathbf{E} = 0, \tag{8–42}$$

where the wavenumber $k_c = \omega\sqrt{\mu\epsilon_c}$ is a complex number, as given in Eq. (7–114). The derivations and discussions pertaining to plane waves in a lossless medium in Section 8–2 can be modified to apply to wave propagation in a lossy medium by simply replacing k with k_c. However, in an effort to conform with the conventional notation used in transmission-line theory, it is customary to define a propagation constant, γ, such that

$$\boxed{\gamma = jk_c = j\omega\sqrt{\mu\epsilon_c} \qquad (\text{m}^{-1}).} \tag{8–43}$$

Since γ is complex, we write, with the help of Eq. (7–110),

$$\gamma = \alpha + j\beta = j\omega\sqrt{\mu\epsilon}\left(1 + \frac{\sigma}{j\omega\epsilon}\right)^{1/2}, \tag{8–44}$$

or, from Eq. (7–114),

$$\gamma = \alpha + j\beta = j\omega\sqrt{\mu\epsilon'}\left(1 - j\frac{\epsilon''}{\epsilon'}\right)^{1/2}, \tag{8–45}$$

where α and β are the real and imaginary parts of γ, respectively. Their physical significance will be explained presently. For a lossless medium, $\sigma = 0$ ($\epsilon'' = 0$, $\epsilon = \epsilon'$), $\alpha = 0$, and $\beta = k = \omega\sqrt{\mu\epsilon}$.

The Helmholtz's equation, Eq. (8–42), becomes

$$\nabla^2 \mathbf{E} - \gamma^2 \mathbf{E} = 0. \tag{8–46}$$

The solution of Eq. (8–46), representing a uniform plane wave propagating in the $+z$-direction, is

$$\mathbf{E} = \mathbf{a}_x E_x = \mathbf{a}_x E_0 e^{-\gamma z}, \tag{8–47}$$

where we have assumed that the wave is linearly polarized in the x-direction. The propagation factor $e^{-\gamma z}$ can be written as a product of two factors:

$$E_x = E_0 e^{-\alpha z} e^{-j\beta z}.$$

As we shall see, both α and β are positive quantities. The first factor, $e^{-\alpha z}$, decreases as z increases and thus is an attenuation factor, and α is called an **attenuation constant**. The SI unit of the attenuation constant is neper per meter (Np/m).[†] The second factor, $e^{-j\beta z}$, is a phase factor; β is called a **phase constant** and is expressed in radians per meter (rad/m). The phase constant expresses the amount of phase shift that occurs as the wave travels one meter.

General expressions of α and β in terms of ω and the constitutive parameters—ϵ, μ, and σ—of the medium are rather involved (see Problem P.8–9). In the following paragraphs we examine the approximate expressions for low-loss dielectrics, good conductors, and ionized gases.

8–3.1   LOW-LOSS DIELECTRICS

A low-loss dielectric is a good but imperfect insulator with a nonzero equivalent conductivity, such that $\epsilon'' \ll \epsilon'$ or $\sigma/\omega\epsilon \ll 1$. Under this condition, γ in Eq. (8–45) can be approximated by using the binomial expansion:

$$\gamma = \alpha + j\beta \cong j\omega\sqrt{\mu\epsilon'}\left[1 - j\frac{\epsilon''}{2\epsilon'} + \frac{1}{8}\left(\frac{\epsilon''}{\epsilon'}\right)^2\right],$$

from which we obtain the attenuation constant

$$\alpha \cong \frac{\omega\epsilon''}{2}\sqrt{\frac{\mu}{\epsilon'}} \qquad (\text{Np/m}) \tag{8–48}$$

and the phase constant

$$\beta \cong \omega\sqrt{\mu\epsilon'}\left[1 + \frac{1}{8}\left(\frac{\epsilon''}{\epsilon'}\right)^2\right] \qquad (\text{rad/m}). \tag{8–49}$$

It is seen from Eq. (8–48) that the attenuation constant of a low-loss dielectric is a positive quantity and is approximately directly proportional to the frequency. The phase constant in Eq. (8–49) deviates only very slightly from the value $\omega\sqrt{\mu\epsilon}$ for a perfect (lossless) dielectric.

[†] Neper is a dimensionless quantity. If $\alpha = 1$ (Np/m), then a unit wave amplitude decreases to a magnitude e^{-1} ($=0.368$) as it travels a distance of 1 (m). An attenuation of 1 (Np/m) equals $20\log_{10}e = 8.69$ (dB/m).

The intrinsic impedance of a low-loss dielectric is a complex quantity.

$$\eta_c = \sqrt{\frac{\mu}{\epsilon'}} \left(1 - j\frac{\epsilon''}{\epsilon'}\right)^{-1/2}$$

$$\cong \sqrt{\frac{\mu}{\epsilon'}} \left(1 + j\frac{\epsilon''}{2\epsilon'}\right) \quad (\Omega). \tag{8–50}$$

Since the intrinsic impedance is the ratio of E_x and H_y for a uniform plane wave, the electric and magnetic field intensities in a lossy dielectric are thus not in time phase, as they are in a lossless medium.

The phase velocity u_p is obtained from the ratio ω/β in a manner similar to that in Eq. (8–9). Using Eq. (8–49), we have

$$u_p = \frac{\omega}{\beta} \cong \frac{1}{\sqrt{\mu\epsilon'}} \left[1 - \frac{1}{8}\left(\frac{\epsilon''}{\epsilon'}\right)^2\right] \quad (\text{m/s}). \tag{8–51}$$

8–3.2   GOOD CONDUCTORS

A good conductor is a medium for which $\sigma/\omega\epsilon \gg 1$. Under this condition it is convenient to use Eq. (8–44) and neglect 1 in comparison with the term $\sigma/j\omega\epsilon$. We write

$$\gamma \cong j\omega\sqrt{\mu\epsilon} \sqrt{\frac{\sigma}{j\omega\epsilon}} = \sqrt{j}\sqrt{\omega\mu\sigma} = \frac{1+j}{\sqrt{2}}\sqrt{\omega\mu\sigma}$$

or

$$\gamma = \alpha + j\beta \cong (1 + j)\sqrt{\pi f\mu\sigma}, \tag{8–52}$$

where we have used the relations

$$\sqrt{j} = (e^{j\pi/2})^{1/2} = e^{j\pi/4} = (1 + j)/\sqrt{2}$$

and $\omega = 2\pi f$. Equation (8–52) indicates that α and β for a good conductor are approximately equal and both increase as $\sqrt{f}$ and $\sqrt{\sigma}$. For a good conductor,

$$\boxed{\alpha = \beta = \sqrt{\pi f\mu\sigma}.} \tag{8–53}$$

The intrinsic impedance of a good conductor is

$$\eta_c = \sqrt{\frac{\mu}{\epsilon_c}} \cong \sqrt{\frac{j\omega\mu}{\sigma}} = (1 + j)\sqrt{\frac{\pi f\mu}{\sigma}} = (1 + j)\frac{\alpha}{\sigma} \quad (\Omega), \tag{8–54}$$

which has a phase angle of 45°. Hence the magnetic field intensity lags behind the electric field intensity by 45°.

The phase velocity in a good conductor is

$$u_p = \frac{\omega}{\beta} \cong \sqrt{\frac{2\omega}{\mu\sigma}} \quad (\text{m/s}), \tag{8–55}$$

which is proportional to $\sqrt{f}$ and $1/\sqrt{\sigma}$. Consider copper as an example:

$$\sigma = 5.80 \times 10^7 \quad \text{(S/m)},$$
$$\mu = 4\pi \times 10^{-7} \quad \text{(H/m)},$$
$$u_p = 720 \text{ (m/s)} \quad \text{at} \quad 3 \text{ (MHz)},$$

which is about twice the velocity of sound in air and is many orders of magnitude slower than the velocity of light in air. The wavelength of a plane wave in a good conductor is

$$\lambda = \frac{2\pi}{\beta} = \frac{u_p}{f} = 2\sqrt{\frac{\pi}{f\mu\sigma}} \quad \text{(m)}. \tag{8-56}$$

For copper at 3 (MHz), $\lambda = 0.24$ (mm). As a comparison, a 3 (MHz) electromagnetic wave in air has a wavelength of 100 (m).

At very high frequencies the attenuation constant α for a good conductor, as given by Eq. (8–53), tends to be very large. For copper at 3 (MHz),

$$\alpha = \sqrt{\pi(3 \times 10^6)(4\pi \times 10^{-7})(5.80 \times 10^7)} = 2.62 \times 10^4 \quad \text{(Np/m)}.$$

Since the attenuation factor is $e^{-\alpha z}$, the amplitude of a wave will be attenuated by a factor of $e^{-1} = 0.368$ when it travels a distance $\delta = 1/\alpha$. For copper at 3 (MHz) this distance is $(1/2.62) \times 10^{-4}$ (m), or 0.038 (mm). At 10 (GHz) it is only 0.66 (μm)—a very small distance indeed. Thus a high-frequency electromagnetic wave is attenuated very rapidly as it propagates in a good conductor. The distance δ through which the amplitude of a traveling plane wave decreases by a factor of e^{-1} or 0.368 is called the *skin depth* or the *depth of penetration* of a conductor:

$$\delta = \frac{1}{\alpha} = \frac{1}{\sqrt{\pi f \mu \sigma}} \quad \text{(m)}. \tag{8-57}$$

Since $\alpha = \beta$ for a good conductor, δ can also be written as

$$\delta = \frac{1}{\beta} = \frac{\lambda}{2\pi} \quad \text{(m)}. \tag{8-58}$$

At microwave frequencies the skin depth or depth of penetration of a good conductor is so small that fields and currents can be considered as, for all practical purposes, confined in a very thin layer (that is, in the skin) of the conductor surface.

Table 8–1 lists the skin depths of several types of materials at various frequencies.

━━━ **EXAMPLE 8–4**   The electric field intensity of a linearly polarized uniform plane wave propagating in the $+z$-direction in seawater is $\mathbf{E} = \mathbf{a}_x 100 \cos(10^7 \pi t)$ (V/m) at $z = 0$. The constitutive parameters of seawater are $\epsilon_r = 72$, $\mu_r = 1$, and $\sigma = 4$ (S/m). (a) De-

TABLE 8–1
Skin Depths, δ in (mm), of Various Materials

Material	σ (S/m)	$f = 60$ (Hz)	1 (MHz)	1 (GHz)
Silver	6.17×10^7	8.27 (mm)	0.064 (mm)	0.0020 (mm)
Copper	5.80×10^7	8.53	0.066	0.0021
Gold	4.10×10^7	10.14	0.079	0.0025
Aluminum	3.54×10^7	10.92	0.084	0.0027
Iron ($\mu_r \cong 10^3$)	1.00×10^7	0.65	0.005	0.00016
Seawater	4	32 (m)	0.25 (m)	†

† The ϵ of seawater is approximately $72\epsilon_0$. At $f = 1$ (GHz), $\sigma/\omega\epsilon \cong 1$ (not $\gg 1$). Under these conditions, seawater is not a good conductor, and Eq. (8–57) is no longer applicable.

termine the attenuation constant, phase constant, intrinsic impedance, phase velocity, wavelength, and skin depth. (b) Find the distance at which the amplitude of **E** is 1% of its value at $z = 0$. (c) Write the expressions for $\mathbf{E}(z, t)$ and $\mathbf{H}(z, t)$ at $z = 0.8$ (m) as functions of t.

Solution

$$\omega = 10^7\pi \quad \text{(rad/s)},$$

$$f = \frac{\omega}{2\pi} = 5 \times 10^6 \quad \text{(Hz)},$$

$$\frac{\sigma}{\omega\epsilon} = \frac{\sigma}{\omega\epsilon_0\epsilon_r} = \frac{4}{10^7\pi\left(\dfrac{1}{36\pi} \times 10^{-9}\right)72} = 200 \gg 1.$$

Hence we can use the formulas for good conductors.

a) Attenuation constant:

$$\alpha = \sqrt{\pi f\mu\sigma} = \sqrt{5\pi 10^6 (4\pi 10^{-7})4} = 8.89 \quad \text{(Np/m)}.$$

Phase constant:

$$\beta = \sqrt{\pi f\mu\sigma} = 8.89 \quad \text{(rad/m)}.$$

Intrinsic impedance:

$$\eta_c = (1 + j)\sqrt{\frac{\pi f\mu}{\sigma}}$$

$$= (1 + j)\sqrt{\frac{\pi(5 \times 10^6)(4\pi \times 10^{-7})}{4}} = \pi e^{j\pi/4} \quad (\Omega).$$

Phase velocity:

$$u_p = \frac{\omega}{\beta} = \frac{10^7\pi}{8.89} = 3.53 \times 10^6 \quad \text{(m/s)}.$$

Wavelength:

$$\lambda = \frac{2\pi}{\beta} = \frac{2\pi}{8.89} = 0.707 \quad \text{(m)}.$$

Skin depth:

$$\delta = \frac{1}{\alpha} = \frac{1}{8.89} = 0.112 \quad \text{(m)}.$$

b) Distance z_1 at which the amplitude of wave decreases to 1% of its value at $z = 0$:

$$e^{-\alpha z_1} = 0.01 \qquad \text{or} \qquad e^{\alpha z_1} = \frac{1}{0.01} = 100,$$

$$z_1 = \frac{1}{\alpha} \ln 100 = \frac{4.605}{8.89} = 0.518 \quad \text{(m)}.$$

c) In phasor notation,

$$\mathbf{E}(z) = \mathbf{a}_x 100 e^{-\alpha z} e^{-j\beta z}.$$

The instantaneous expression for $\mathbf{E}$ is

$$\mathbf{E}(z, t) = \mathscr{R}e\left[\mathbf{E}(z)e^{j\omega t}\right]$$
$$= \mathscr{R}e\left[\mathbf{a}_x 100 e^{-\alpha z} e^{j(\omega t - \beta z)}\right] = \mathbf{a}_x 100 e^{-\alpha z} \cos(\omega t - \beta z).$$

At $z = 0.8$ (m) we have

$$\mathbf{E}(0.8, t) = \mathbf{a}_x 100 e^{-0.8\alpha} \cos(10^7 \pi t - 0.8\beta)$$
$$= \mathbf{a}_x 0.082 \cos(10^7 \pi t - 7.11) \quad \text{(V/m)}.$$

We know that a uniform plane wave is a TEM wave with $\mathbf{E} \perp \mathbf{H}$ and that both are normal to the direction of wave propagation $\mathbf{a}_z$. Thus $\mathbf{H} = \mathbf{a}_y H_y$. To find $\mathbf{H}(z, t)$, the instantaneous expression of $\mathbf{H}$ as a function of t, *we must not make the mistake of writing* $H_y(z, t) = E_x(z, t)/\eta_c$ because this would be mixing real time functions $E_x(z, t)$ and $H_z(z, t)$ with a complex quantity η_c. Phasor quantities $E_x(z)$ and $H_y(z)$ must be used. That is,

$$H_y(z) = \frac{E_x(z)}{\eta_c},$$

from which we obtain the relation between instantaneous quantities

$$H_y(z, t) = \mathscr{R}e\left[\frac{E_x(z)}{\eta_c} e^{j\omega t}\right].$$

For the present problem we have, in phasors,

$$H_y(0.8) = \frac{100 e^{-0.8\alpha} e^{-j0.8\beta}}{\pi e^{j\pi/4}} = \frac{0.082 e^{-j7.11}}{\pi e^{j\pi/4}} = 0.026 e^{-j1.61}.$$

Note that *both* angles must be in radians before combining. The instantaneous expression for $\mathbf{H}$ at $z = 0.8$ (m) is then

$$\mathbf{H}(0.8, t) = \mathbf{a}_y 0.026 \cos(10^7 \pi t - 1.61) \quad \text{(A/m)}.$$

We can see that a 5 (MHz) plane wave attenuates very rapidly in seawater and becomes negligibly weak a very short distance from the source. Even at very low frequencies, long-distance radio communication with a submerged submarine is very difficult.   ▬

8–3.3 IONIZED GASES

In the earth's upper atmosphere, roughly from 50 to 500 (km) in altitude, there exist layers of ionized gases called the *ionosphere*. The ionosphere consists of free electrons and positive ions that are produced when the ultraviolet radiation from the sun is absorbed by the atoms and molecules in the upper atmosphere. The charged particles tend to be trapped by the earth's magnetic field. The altitude and character of the ionized layers depend both on the nature of the solar radiation and on the composition of the atmosphere. They change with the sunspot cycle, the season, and the hour of the day in a very complicated way. The electron and ion densities in the individual ionized layers are essentially equal. Ionized gases with equal electron and ion densities are called *plasmas*.

The ionosphere plays an important role in the propagation of electromagnetic waves and affects telecommunication. Because the electrons are much lighter than the positive ions, they are accelerated more by the electric fields of electromagnetic waves passing through the ionosphere. In our analysis we shall ignore the motion of the ions and regard the ionosphere as a free electron gas. Furthermore, we shall neglect the collisions between the electrons and the gas atoms and molecules.[†]

An electron of charge $-e$ and mass m in a time-harmonic electric field $\mathbf{E}$ in the x-direction at an angular frequency ω experiences a force $-eE$, which displaces it from a positive ion by a distance $\mathbf{x}$ such that

$$-e\mathbf{E} = m\frac{d^2\mathbf{x}}{dt^2} = -m\omega^2\mathbf{x} \qquad (8\text{--}59)$$

or

$$\mathbf{x} = \frac{e}{m\omega^2}\mathbf{E}, \qquad (8\text{--}60)$$

where $\mathbf{E}$ and $\mathbf{x}$ are phasors. Such a displacement gives rise to an electric dipole moment:

$$\mathbf{p} = -e\mathbf{x}. \qquad (8\text{--}61)$$

If there are N electrons per unit volume, we have a volume density of electric dipole moment or polarization vector

$$\mathbf{P} = N\mathbf{p} = -\frac{Ne^2}{m\omega^2}\mathbf{E}. \qquad (8\text{--}62)$$

In writing Eq. (8–62) we have implicitly neglected the mutual effect of the induced dipole moments of the electrons on one another. From Eqs. (3–97) and (8–62) we

---

[†] This is not a good assumption in the lowest regions of the ionosphere where the atmospheric pressure is high.

obtain

$$\mathbf{D} = \epsilon_0 \mathbf{E} + \mathbf{P} = \epsilon_0 \left(1 - \frac{Ne^2}{m\omega^2\epsilon_0} \right) \mathbf{E}$$

$$= \epsilon_0 \left(1 - \frac{\omega_p^2}{\omega^2} \right) \mathbf{E},$$

(8–63)

where

$$\omega_p = \sqrt{\frac{Ne^2}{m\epsilon_0}} \qquad \text{(rad/s)}$$

(8–64)

is called the **plasma angular frequency**, a characteristic of the ionized medium. The corresponding **plasma frequency** is

$$f_p = \frac{\omega_p}{2\pi} = \frac{1}{2\pi} \sqrt{\frac{Ne^2}{m\epsilon_0}} \qquad \text{(Hz)}.$$

(8–65)

Thus the equivalent permittivity of the ionosphere or plasma is

$$\boxed{\begin{aligned} \epsilon_p &= \epsilon_0 \left(1 - \frac{\omega_p^2}{\omega^2} \right) \\ &= \epsilon_0 \left(1 - \frac{f_p^2}{f^2} \right) \qquad \text{(F/m)}. \end{aligned}}$$

(8–66)

On the basis of Eq. (8–66) we obtain the propagation constant as

$$\gamma = j\omega\sqrt{\mu\epsilon_0} \sqrt{1 - \left(\frac{f_p}{f} \right)^2},$$

(8–67)

and the intrinsic impedance as

$$\eta_p = \frac{\eta_0}{\sqrt{1 - \left(\dfrac{f_p}{f} \right)^2}},$$

(8–68)

where $\eta_0 = \sqrt{\mu_0/\epsilon_0} = 120\pi$ (Ω).

From Eq. (8–66) we note the peculiar phenomenon of a vanishing ϵ as f approaches f_p. When ϵ becomes zero, electric displacement $\mathbf{D}$ (which depends on free charges only) is zero even when electric field intensity $\mathbf{E}$ (which depends on both free and polarization charges) is not. In that case it would be possible for an oscillating $\mathbf{E}$ to exist in the plasma in the absence of free charges, leading to a so-called **plasma oscillation**.

When $f < f_p$, γ becomes purely real, indicating an attenuation without propagation; at the same time, η_p becomes purely imaginary, indicating a reactive load with no transmission of power. Thus f_p is also referred to as the **cutoff frequency**. We will discuss wave reflection and transmission under various conditions later in this chapter. When $f > f_p$, γ is purely imaginary, and electromagnetic waves propagate unattenuated in the plasma (assuming negligible collision losses).

If the value of e, m, and ϵ_0 are substituted into Eq. (8–65), we find a very simple formula for the plasma (cutoff) frequency:

$$f_p \cong 9\sqrt{N} \qquad \text{(Hz)}. \qquad (8\text{–}69)$$

As we have mentioned before, N at a given altitude is not a constant; it varies with the time of the day, the season, and other factors. The electron density of the ionosphere ranges from about $10^{10}/m^3$ in the lowest layer to $10^{12}/m^3$ in the highest layer. Using these values for N in Eq. (8–69), we find f_p to vary from 0.9 to 9 (MHz). Hence, for communication with a satellite or a space station beyond the ionosphere we must use frequencies much higher than 9 (MHz) to ensure wave penetration through the layer with the largest N at any angle of incidence (see Problem P.8–14). Signals with frequencies lower than 0.9 (MHz) cannot penetrate into even the lowest layer of the ionosphere but may propagate very far around the earth by way of multiple reflections at the ionosphere's boundary and the earth's surface. Signals having frequencies between 0.9 and 9 (MHz) will penetrate partially into the lower ionospheric layers but will eventually be turned back where N is large. We have given here only a very simplified picture of wave propagation in the ionosphere. The actual situation is complicated by the lack of distinct layers of constant electron densities and by the presence of the earth's magnetic field.

EXAMPLE 8–5   When a spacecraft reenters the earth's atmosphere, its speed and temperature ionize the surrounding atoms and molecules and create a plasma. It has been estimated that the electron density is in the neighborhood of 2×10^8 per (cm^3). Discuss the plasma's effect on frequency usage in radio communication between the spacecraft and the mission controllers on earth.

Solution   For

$$N = 2 \times 10^8 \text{ per (cm}^3)$$
$$= 2 \times 10^{14} \text{ per (m}^3),$$

Eq. (8–69) gives $f_p = 9 \times \sqrt{2 \times 10^{14}} = 12.7 \times 10^7$ (Hz), or 127 (MHz). Thus, radio communication cannot be established for frequencies below 127 (MHz).   ■

8–4   Group Velocity

In Section 8–2 we defined the phase velocity, u_p, of a single-frequency plane wave as the velocity of propagation of an equiphase wavefront. The relation between u_p and the phase constant, β, is

$$u_p = \frac{\omega}{\beta} \qquad \text{(m/s)}. \qquad (8\text{–}70)$$

For plane waves in a lossless medium, $\beta = \omega\sqrt{\mu\epsilon}$ is a linear function of ω. As a consequence, the phase velocity $u_p = 1/\sqrt{\mu\epsilon}$ is a constant that is independent of frequency. However, in some cases (such as wave propagation in a lossy dielectric, as discussed previously, or along a transmission line, or in a waveguide to be discussed in later chapters) the phase constant is not a linear function of ω; waves of different frequencies will propagate with different phase velocities. Inasmuch as all information-bearing signals consist of a band of frequencies, waves of the component frequencies travel with different phase velocities, causing a distortion in the signal wave shape. The signal "disperses." The phenomenon of signal distortion caused by a dependence of the phase velocity on frequency is called *dispersion*. In view of Eqs. (8–51) and (7–115), we conclude that a lossy dielectric is obviously a *dispersive medium*.

An information-bearing signal normally has a small spread of frequencies (sidebands) around a high carrier frequency. Such a signal comprises a "group" of frequencies and forms a wave packet. A *group velocity* is the velocity of propagation of the wave-packet envelope (of a group of frequencies).

Consider the simplest case of a wave packet that consists of two traveling waves having equal amplitude and slightly different angular frequencies $\omega_0 + \Delta\omega$ and $\omega_0 - \Delta\omega$ $(\Delta\omega \ll \omega_0)$. The phase constants, being functions of frequency, will also be slightly different. Let the phase constants corresponding to the two frequencies be $\beta_0 + \Delta\beta$ and $\beta_0 - \Delta\beta$. We have

$$
\begin{aligned}
E(z, t) &= E_0 \cos \left[(\omega_0 + \Delta\omega)t - (\beta_0 + \Delta\beta)z \right] \\
&\quad + E_0 \cos \left[(\omega_0 - \Delta\omega)t - (\beta_0 - \Delta\beta)z \right] \\
&= 2E_0 \cos (t\,\Delta\omega - z\,\Delta\beta) \cos (\omega_0 t - \beta_0 z).
\end{aligned}
\tag{8–71}
$$

Since $\Delta\omega \ll \omega_0$, the expression in Eq. (8–71) represents a rapidly oscillating wave having an angular frequency ω_0 and an amplitude that varies slowly with an angular frequency $\Delta\omega$. This is depicted in Fig. 8–6.

The wave inside the envelope propagates with a phase velocity found by setting $\omega_0 t - \beta_0 z = \text{Constant}$:

$$
u_p = \frac{dz}{dt} = \frac{\omega_0}{\beta_0}.
$$

The velocity of the envelope (the *group velocity u_g*) can be determined by setting the argument of the first cosine factor in Eq. (8–71) equal to a constant:

$$
t\,\Delta\omega - z\,\Delta\beta = \text{Constant},
$$

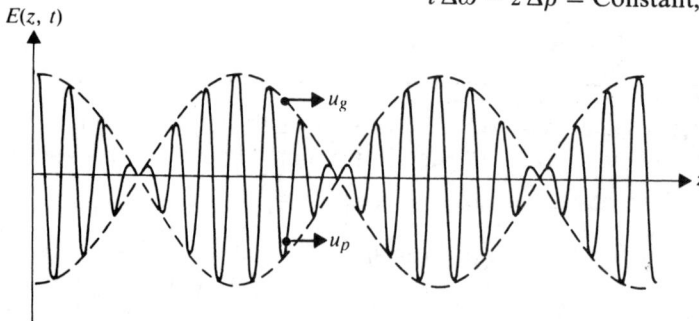

FIGURE 8–6
Sum of two time-harmonic traveling waves of equal amplitude and slightly different frequencies at a given t.

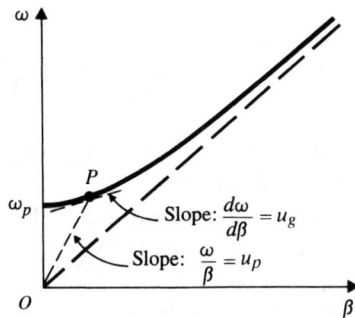

FIGURE 8-7
ω–β graph for ionized gas.

from which we obtain

$$u_g = \frac{dz}{dt} = \frac{\Delta\omega}{\Delta\beta} = \frac{1}{\Delta\beta/\Delta\omega}.$$

In the limit that $\Delta\omega \to 0$, we have the formula for computing the group velocity in a dispersive medium:

$$u_g = \frac{1}{d\beta/d\omega} \qquad \text{(m/s)}. \tag{8-72}$$

This is the velocity of a point on the envelope of the wave packet, as shown in Fig. 8-6, and is identified as the velocity of the narrow-band signal.[†] As we saw in Subsection 8-3.3, β is a function of ω. If ω is plotted versus β, an ω–β graph is obtained. The slope of the straight line drawn from the origin to a point on the graph gives the phase velocity, ω/β, and the local slope of the tangent to the graph at the point is the group velocity, $d\omega/d\beta$.

In Fig. 8-7 an ω–β graph for wave propagation in an ionized medium is plotted, based on Eq. (8-67):

$$\beta = \omega\sqrt{\mu\epsilon_0}\sqrt{1 - \left(\frac{f_p}{f}\right)^2}$$
$$= \frac{\omega}{c}\sqrt{1 - \left(\frac{\omega_p}{\omega}\right)^2}. \tag{8-73}$$

At $\omega = \omega_p$ (the cutoff angular frequency), $\beta = 0$. For $\omega > \omega_p$, wave propagation is possible, and

$$u_p = \frac{\omega}{\beta} = \frac{c}{\sqrt{1 - \left(\frac{\omega_p}{\omega}\right)^2}}. \tag{8-74}$$

[†] The concept of group velocity is not applicable to wide-band signals in a dispersive medium (see Subsection 10-3.2).

Substituting Eq. (8–73) in Eq. (8–72), we have

$$u_g = c \sqrt{1 - \left(\frac{\omega_p}{\omega}\right)^2}. \qquad (8\text{–}75)$$

We note that $u_p \geq c$ and $u_g \leq c$, and for wave propagation in an ionized medium, $u_p u_g = c^2$. A similar situation exists in waveguides (Section 10–2).

A general relation between the group and phase velocities may be obtained by combining Eqs. (8–70) and (8–72). From Eq. (8–70) we have

$$\frac{d\beta}{d\omega} = \frac{d}{d\omega}\left(\frac{\omega}{u_p}\right) = \frac{1}{u_p} - \frac{\omega}{u_p^2} \frac{du_p}{d\omega}.$$

Substitution of the above in Eq. (8–72) yields

$$u_g = \frac{u_p}{1 - \frac{\omega}{u_p} \frac{du_p}{d\omega}}. \qquad (8\text{–}76)$$

From Eq. (8–76) we see three possible cases:

a) No dispersion:

$$\frac{du_p}{d\omega} = 0 \qquad (u_p \text{ independent of } \omega, \ \beta \text{ a linear function of } \omega),$$

$$u_g = u_p.$$

b) Normal dispersion:

$$\frac{du_p}{d\omega} < 0 \qquad (u_p \text{ decreasing with } \omega),$$

$$u_g < u_p.$$

c) Anomalous dispersion:

$$\frac{du_p}{d\omega} > 0 \qquad (u_p \text{ increasing with } \omega),$$

$$u_g > u_p.$$

EXAMPLE 8–6   A narrow-band signal propagates in a lossy dielectric medium which has a loss tangent 0.2 at 550 (kHz), the carrier frequency of the signal. The dielectric constant of the medium is 2.5. (a) Determine α and β. (b) Determine u_p and u_g. Is the medium dispersive?

Solution

a) Since the loss tangent $\epsilon''/\epsilon' = 0.2$ and $\epsilon''^2/8\epsilon'^2 \ll 1$, Eqs. (8–48) and (8–49) can be used to determine α and β respectively. But first we find ϵ'' from the loss tangent:

$$\epsilon'' = 0.2 \, \epsilon' = 0.2 \times 2.5 \, \epsilon_0$$
$$= 4.42 \times 10^{-12} \quad (F/m).$$

Thus,

$$\alpha = \frac{\omega\epsilon''}{2}\sqrt{\frac{\mu}{\epsilon'}} = \pi(550 \times 10^3) \times (4.42 \times 10^{-12}) \times \frac{377}{\sqrt{2.5}} = 1.82 \times 10^{-3} \quad (Np/m);$$

$$\beta = \omega\sqrt{\mu\epsilon'}\left[1 + \frac{1}{8}\left(\frac{\epsilon''}{\epsilon'}\right)^2\right]$$

$$= 2\pi(550 \times 10^3)\frac{\sqrt{2.5}}{3 \times 10^8}\left[1 + \frac{1}{8}(0.2)^2\right]$$

$$= 0.0182 \times 1.005 = 0.0183 \quad (rad/m).$$

b) Phase velocity (from Eq. 8–51):

$$u_p = \frac{\omega}{\beta} = \frac{1}{\sqrt{\mu\epsilon'}\left[1 + \frac{1}{8}\left(\frac{\epsilon''}{\epsilon'}\right)^2\right]} \cong \frac{1}{\sqrt{\mu\epsilon'}}\left[1 - \frac{1}{8}\left(\frac{\epsilon''}{\epsilon'}\right)^2\right]$$

$$= \frac{3 \times 10^8}{\sqrt{2.5}}\left[1 - \frac{1}{8}(0.2)^2\right] = 1.888 \times 10^8 \quad (m/s).$$

c) Group velocity (from Eq. 8–49):

$$\frac{d\beta}{d\omega} = \sqrt{\mu\epsilon'}\left[1 + \frac{1}{8}\left(\frac{\epsilon''}{\epsilon'}\right)^2\right].$$

$$u_g = \frac{1}{(d\beta/d\omega)} \cong \frac{1}{\sqrt{\mu\epsilon'}} \cong u_p.$$

Thus a low-loss dielectric is nearly nondispersive. Here we have assumed ϵ'' to be independent of frequency. For a high-loss dielectric, ϵ'' will be a function of ω and may have a magnitude comparable to ϵ'. The approximation in Eq. (8–49) will no longer hold, and the medium will be dispersive.   ■

8–5   Flow of Electromagnetic Power and the Poynting Vector

Electromagnetic waves carry with them electromagnetic power. Energy is transported through space to distant receiving points by electromagnetic waves. We will now derive a relation between the rate of such energy transfer and the electric and magnetic field intensities associated with a traveling electromagnetic wave.

We begin with the curl equations:

$$\nabla \times \mathbf{E} = -\frac{\partial \mathbf{B}}{\partial t}, \qquad\qquad\qquad (7\text{–}53a) \quad (8\text{–}77)$$

$$\nabla \times \mathbf{H} = \mathbf{J} + \frac{\partial \mathbf{D}}{\partial t}. \qquad\qquad\qquad (7\text{–}53b) \quad (8\text{–}78)$$

The verification of the following identity of vector operations (see Problem P.2–33) is straightforward:

$$\mathbf{V} \cdot (\mathbf{E} \times \mathbf{H}) = \mathbf{H} \cdot (\mathbf{V} \times \mathbf{E}) - \mathbf{E} \cdot (\mathbf{V} \times \mathbf{H}). \qquad (8\text{–}79)$$

Substitution of Eqs. (8–77) and (8–78) in Eq. (8–79) yields

$$\mathbf{V} \cdot (\mathbf{E} \times \mathbf{H}) = -\mathbf{H} \cdot \frac{\partial \mathbf{B}}{\partial t} - \mathbf{E} \cdot \frac{\partial \mathbf{D}}{\partial t} - \mathbf{E} \cdot \mathbf{J}. \qquad (8\text{–}80)$$

In a simple medium, whose constitutive parameters ϵ, μ, and σ do not change with time, we have

$$\mathbf{H} \cdot \frac{\partial \mathbf{B}}{\partial t} = \mathbf{H} \cdot \frac{\partial (\mu \mathbf{H})}{\partial t} = \frac{1}{2} \frac{\partial (\mu \mathbf{H} \cdot \mathbf{H})}{\partial t} = \frac{\partial}{\partial t} \left(\frac{1}{2} \mu H^2 \right),$$

$$\mathbf{E} \cdot \frac{\partial \mathbf{D}}{\partial t} = \mathbf{E} \cdot \frac{\partial (\epsilon \mathbf{E})}{\partial t} = \frac{1}{2} \frac{\partial (\epsilon \mathbf{E} \cdot \mathbf{E})}{\partial t} = \frac{\partial}{\partial t} \left(\frac{1}{2} \epsilon E^2 \right),$$

$$\mathbf{E} \cdot \mathbf{J} = \mathbf{E} \cdot (\sigma \mathbf{E}) = \sigma E^2.$$

Equation (8–80) can then be written as

$$\mathbf{V} \cdot (\mathbf{E} \times \mathbf{H}) = -\frac{\partial}{\partial t} \left(\frac{1}{2} \epsilon E^2 + \frac{1}{2} \mu H^2 \right) - \sigma E^2, \qquad (8\text{–}81)$$

which is a point-function relationship. An integral form of Eq. (8–81) is obtained by integrating both sides over the volume of concern:

$$\oint_S (\mathbf{E} \times \mathbf{H}) \cdot d\mathbf{s} = -\frac{\partial}{\partial t} \int_V \left(\frac{1}{2} \epsilon E^2 + \frac{1}{2} \mu H^2 \right) dv - \int_V \sigma E^2 \, dv, \qquad (8\text{–}82)$$

where the divergence theorem has been applied to convert the volume integral of $\mathbf{V} \cdot (\mathbf{E} \times \mathbf{H})$ to the closed surface integral of $(\mathbf{E} \times \mathbf{H})$.

We recognize that the first and second terms on the right side of Eq. (8–82) represent the time-rate of change of the energy stored in the electric and magnetic fields, respectively. [Compare with Eqs. (3–176b) and (6–172c).] The last term is the ohmic power dissipated in the volume as a result of the flow of conduction current density $\sigma \mathbf{E}$ in the presence of the electric field $\mathbf{E}$. Hence we may interpret the right side of Eq. (8–82) as the *rate of decrease* of the electric and magnetic energies stored, subtracted by the ohmic power dissipated as heat in the volume V. To be consistent with the law of conservation of energy, this must equal the power (rate of energy) *leaving* the volume through its surface. Thus the quantity $(\mathbf{E} \times \mathbf{H})$ is a vector representing the power flow per unit area. Define

$$\boxed{\mathscr{P} = \mathbf{E} \times \mathbf{H} \qquad (\text{W/m}^2).} \qquad (8\text{–}83)$$

Quantity $\mathscr{P}$ is known as the **Poynting vector**, which is a power density vector associated with an electromagnetic field. The assertion that the surface integral of $\mathscr{P}$ over a closed surface, as given by the left side of Eq. (8–82), equals the power leaving the enclosed volume is referred to as ***Poynting's theorem***. This assertion is not limited to plane waves.

Equation (8–82) may be written in another form:

$$-\oint_S \mathscr{P} \cdot d\mathbf{s} = \frac{\partial}{\partial t} \int_V (w_e + w_m)\, dv + \int_V p_\sigma\, dv, \qquad (8\text{–}84)$$

where

$$w_e = \tfrac{1}{2}\epsilon E^2 = \tfrac{1}{2}\epsilon \mathbf{E} \cdot \mathbf{E}^* = \text{Electric energy density,} \qquad (8\text{–}85)$$

$$w_m = \tfrac{1}{2}\mu H^2 = \tfrac{1}{2}\mu \mathbf{H} \cdot \mathbf{H}^* = \text{Magnetic energy density,} \qquad (8\text{–}86)$$

$$p_\sigma = \sigma E^2 = J^2/\sigma = \sigma \mathbf{E} \cdot \mathbf{E}^* = \mathbf{J} \cdot \mathbf{J}^*/\sigma = \text{Ohmic power density.} \qquad (8\text{–}87)$$

In words, Eq. (8–84) states that the total power flowing *into* a closed surface at any instant equals the sum of the rates of increase of the stored electric and magnetic energies and the ohmic power dissipated within the enclosed volume.

Two points concerning the Poynting vector are worthy of note. First, the power relations given in Eqs. (8–82) and (8–84) pertain to the total power flow across a closed surface obtained by the surface integral of $(\mathbf{E} \times \mathbf{H})$. The definition of the Poynting vector in Eq. (8–83) as the power density vector at *every point* on the surface is an *arbitrary*, albeit useful, *concept*. Second, the Poynting vector $\mathscr{P}$ is in a direction normal to both $\mathbf{E}$ and $\mathbf{H}$.

If the region of concern is lossless ($\sigma = 0$), then the last term in Eq. (8–84) vanishes, and the total power flowing into a closed surface is equal to the rate of increase of the stored electric and magnetic energies in the enclosed volume. In a static situation, the first two terms on the right side of Eq. (8–84) vanish, and the total power flowing into a closed surface is equal to the ohmic power dissipated in the enclosed volume.

EXAMPLE 8–7   Find the Poynting vector on the surface of a long, straight conducting wire (of radius b and conductivity σ) that carries a direct current I. Verify Poynting's theorem.

Solution   Since we have a d-c situation, the current in the wire is uniformly distributed over its cross-sectional area. Let us assume that the axis of the wire coincides with the z-axis. Figure 8–8 shows a segment of length ℓ of the long wire. We have

$$\mathbf{J} = \mathbf{a}_z \frac{I}{\pi b^2}$$

and

$$\mathbf{E} = \frac{\mathbf{J}}{\sigma} = \mathbf{a}_z \frac{I}{\sigma \pi b^2}.$$

On the surface of the wire,

$$\mathbf{H} = \mathbf{a}_\phi \frac{I}{2\pi b}.$$

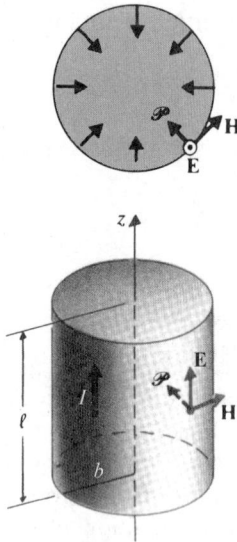

FIGURE 8–8
Illustrating Poynting's theorem (Example 8–7).

Thus the Poynting vector at the surface of the wire is

$$\mathscr{P} = \mathbf{E} \times \mathbf{H} = (\mathbf{a}_z \times \mathbf{a}_\phi) \frac{I^2}{2\sigma\pi^2 b^3}$$

$$= -\mathbf{a}_r \frac{I^2}{2\sigma\pi^2 b^3},$$

which is directed everywhere into the wire surface.

To verify Poynting's theorem, we integrate $\mathscr{P}$ over the wall of the wire segment in Fig. 8–8:

$$-\oint_S \mathscr{P} \cdot d\mathbf{s} = -\oint_S \mathscr{P} \cdot \mathbf{a}_r \, ds = \left(\frac{I^2}{2\sigma\pi^2 b^3} \right) 2\pi b\ell$$

$$= I^2 \left(\frac{\ell}{\sigma\pi b^2} \right) = I^2 R,$$

where the formula for the resistance of a straight wire in Eq. (5–27), $R = \ell/\sigma S$, has been used. The above result affirms that the negative surface integral of the Poynting vector is exactly equal to the $I^2 R$ ohmic power loss in the conducting wire. Hence Poynting's theorem is verified. ▬

8–5.1 INSTANTANEOUS AND AVERAGE POWER DENSITIES

In dealing with time-harmonic electromagnetic waves we have found it convenient to use phasor notation. The instantaneous value of a quantity is then the real part of the product of the phasor quantity and $e^{j\omega t}$ when $\cos \omega t$ is used as the reference. For example, for the phasor

$$\mathbf{E}(z) = \mathbf{a}_x E_x(z) = \mathbf{a}_x E_0 e^{-(\alpha + j\beta)z}, \tag{8–88}$$

the instantaneous expression is

$$\mathbf{E}(z, t) = \mathscr{R}e[\mathbf{E}(z)e^{j\omega t}] = \mathbf{a}_x E_0 e^{-\alpha z} \mathscr{R}e[e^{j(\omega t - \beta z)}]$$
$$= \mathbf{a}_x E_0 e^{-\alpha z} \cos (\omega t - \beta z). \tag{8–89}$$

For a uniform plane wave propagating in a lossy medium in the $+z$-direction, the associated magnetic field intensity phasor is

$$\mathbf{H}(z) = \mathbf{a}_y H_y(z) = \mathbf{a}_y \frac{E_0}{|\eta|} e^{-\alpha z} e^{-j(\beta z + \theta_\eta)}, \tag{8–90}$$

where θ_η is the phase angle of the intrinsic impedance $\eta = |\eta| e^{j\theta_\eta}$ of the medium. The corresponding instantaneous expression for $\mathbf{H}(z)$ is

$$\mathbf{H}(z, t) = \mathscr{R}e[\mathbf{H}(z)e^{j\omega t}] = \mathbf{a}_y \frac{E_0}{|\eta|} e^{-\alpha z} \cos (\omega t - \beta z - \theta_\eta). \tag{8–91}$$

This procedure is permissible as long as the operations and/or the equations involving the quantities with sinusoidal time dependence are *linear*. Erroneous results will be obtained if this procedure is applied to such nonlinear operations as a product of two sinusoidal quantities. (A Poynting vector, being the cross product of $\mathbf{E}$ and $\mathbf{H}$, falls in this category.) The reason is that

$$\mathscr{R}e[\mathbf{E}(z)e^{j\omega t}] \times \mathscr{R}e[\mathbf{H}(z)e^{j\omega t}] \neq \mathscr{R}e[\mathbf{E}(z) \times \mathbf{H}(z)e^{j\omega t}].$$

The instantaneous expression for the Poynting vector or power density vector, from Eqs. (8–88) and (8–90), is

$$\mathscr{P}(z, t) = \mathbf{E}(z, t) \times \mathbf{H}(z, t) = \mathscr{R}e[\mathbf{E}(z)e^{j\omega t}] \times \mathscr{R}e[\mathbf{H}(z)e^{j\omega t}]$$
$$= \mathbf{a}_z \frac{E_0^2}{|\eta|} e^{-2\alpha z} \cos (\omega t - \beta z) \cos (\omega t - \beta z - \theta_\eta) \tag{8–92}$$
$$= \mathbf{a}_z \frac{E_0^2}{2|\eta|} e^{-2\alpha z} [\cos \theta_\eta + \cos (2\omega t - 2\beta z - \theta_\eta)].$$

---

† Consider two general complex vectors $\mathbf{A}$ and $\mathbf{B}$. We know that
$$\mathscr{R}e(\mathbf{A}) = \tfrac{1}{2}(\mathbf{A} + \mathbf{A}^*) \quad \text{and} \quad \mathscr{R}e(\mathbf{B}) = \tfrac{1}{2}(\mathbf{B} + \mathbf{B}^*),$$
where the asterisk denotes "the complex conjugate of." Thus
$$\mathscr{R}e(\mathbf{A}) \times \mathscr{R}e(\mathbf{B}) = \tfrac{1}{2}(\mathbf{A} + \mathbf{A}^*) \times \tfrac{1}{2}(\mathbf{B} + \mathbf{B}^*)$$
$$= \tfrac{1}{4}[(\mathbf{A} \times \mathbf{B}^* + \mathbf{A}^* \times \mathbf{B}) + (\mathbf{A} \times \mathbf{B} + \mathbf{A}^* \times \mathbf{B}^*)]$$
$$= \tfrac{1}{2}\mathscr{R}e(\mathbf{A} \times \mathbf{B}^* + \mathbf{A} \times \mathbf{B}). \tag{8–93}$$
This relation holds also for dot products of vector functions and for products of two complex scalar functions. It is a straightforward exercise to obtain the result in Eq. (8–92) by identifying the vectors $\mathbf{A}$ and $\mathbf{B}$ in Eq. (8–93) with $\mathbf{E}(z)e^{j\omega t}$ and $\mathbf{H}(z)e^{j\omega t}$, respectively.

On the other hand,

$$\mathscr{R}e[\mathbf{E}(z) \times \mathbf{H}(z)e^{j\omega t}] = \mathbf{a}_z \frac{E_0^2}{|\eta|} e^{-2\alpha z} \cos(\omega t - 2\beta z - \theta_\eta),$$

which is obviously not the same as the expression in Eq. (8–92).

As far as the power transmitted by an electromagnetic wave is concerned, its average value is a more significant quantity than its instantaneous value. From Eq. (8–92), we obtain the time-average Poynting vector, $\mathscr{P}_{av}(z)$:

$$\mathscr{P}_{av}(z) = \frac{1}{T} \int_0^T \mathscr{P}(z, t)\, dt = \mathbf{a}_z \frac{E_0^2}{2|\eta|} e^{-2\alpha z} \cos\theta_\eta \qquad (\text{W/m}^2), \qquad (8\text{–}94)^\dagger$$

where $T = 2\pi/\omega$ is the time period of the wave. The second term on the right side of Eq. (8–92) is a cosine function of a double frequency whose average is zero over a fundamental period.

Using Eq. (8–93), we can express the instantaneous Poynting vector in Eq. (8–92) as the real part of the sum of two terms, instead of the product of the real parts of two complex vectors:

$$\begin{aligned}
\mathscr{P}(z, t) &= \mathscr{R}e[\mathbf{E}(z)e^{j\omega t}] \times \mathscr{R}e[\mathbf{H}(z)e^{j\omega t}] \\
&= \tfrac{1}{2}\mathscr{R}e[\mathbf{E}(z) \times \mathbf{H}^*(z) + \mathbf{E}(z) \times \mathbf{H}(z)e^{j2\omega t}].
\end{aligned} \qquad (8\text{–}95)$$

The average power density, $\mathscr{P}_{av}(z)$, can be obtained by integrating $\mathscr{P}(z, t)$ over a fundamental period T. Since the average of the last (second-harmonic) term in Eq. (8–95) vanishes, we have

$$\mathscr{P}_{av}(z) = \tfrac{1}{2}\mathscr{R}e[\mathbf{E}(z) \times \mathbf{H}^*(z)].$$

---

† Equation (8–94) is quite similar to the formula for computing the power dissipated in an impedance $Z = |Z|e^{j\theta_z}$ when a sinusoidal voltage $v(t) = V_0 \cos \omega t$ appears across its terminals. The instantaneous expression for the current $i(t)$ through the impedance is

$$i(t) = \frac{V_0}{|Z|} \cos(\omega t - \theta_z).$$

From the theory of a-c circuits we know that the average power dissipated in Z is

$$P_{av} = \frac{1}{T} \int_0^T v(t)i(t)\, dt = \frac{V_0^2}{2|Z|} \cos\theta_z,$$

where $\cos\theta_z$ is the power factor of the load impedance. The $\cos\theta_\eta$ factor in Eq. (8–94) can be considered as the power factor of the intrinsic impedance of the medium.

In the general case we may not be dealing with a wave propagating in the z-direction. We write

$$\boxed{\mathscr{P}_{av} = \tfrac{1}{2}\mathscr{R}e(\mathbf{E} \times \mathbf{H}^*) \qquad (\text{W/m}^2),}$$

(8–96)[†]

which is a general formula for computing the average power density in a propagating wave.

EXAMPLE 8–8 The far field of a short vertical current element $I\,d\ell$ located at the origin of a spherical coordinate system in free space is

$$\mathbf{E}(R,\theta) = \mathbf{a}_\theta E_\theta(R,\theta) = \mathbf{a}_\theta\!\left(j\,\frac{60\pi I\,d\ell}{\lambda R}\sin\theta\right)e^{-j\beta R} \qquad (\text{V/m})$$

and

$$\mathbf{H}(R,\theta) = \mathbf{a}_\phi\,\frac{E_\theta(R,\theta)}{\eta_0} = \mathbf{a}_\phi\!\left(j\,\frac{I\,d\ell}{2\lambda R}\sin\theta\right)e^{-j\beta R} \qquad (\text{A/m}),$$

where $\lambda = 2\pi/\beta$ is the wavelength.

a) Write the expression for instantaneous Poynting vector.

b) Find the total average power radiated by the current element.

Solution

a) We note that $E_\theta/H_\phi = \eta_0 = 120\,\pi\;(\Omega)$. The instantaneous Poynting vector is

$$\mathscr{P}(R,\theta; t) = \mathscr{R}e\big[\mathbf{E}(R,\theta)e^{j\omega t}\big] \times \mathscr{R}e\big[\mathbf{H}(R,\theta)e^{j\omega t}\big]$$

$$= (\mathbf{a}_\theta \times \mathbf{a}_\phi)30\pi\!\left(\frac{I\,d\ell}{\lambda R}\right)^{2}\sin^2\theta\,\sin^2(\omega t - \beta R)$$

$$= \mathbf{a}_R 15\pi\!\left(\frac{I\,d\ell}{\lambda R}\right)^{2}\sin^2\theta[1 - \cos 2(\omega t - \beta R)] \qquad (\text{W/m}^2).$$

b) The average power density vector is, from Eq. (8–96),

$$\mathscr{P}_{av}(R,\theta) = \mathbf{a}_R 15\pi\!\left(\frac{I\,d\ell}{\lambda R}\right)^{2}\sin^2\theta,$$

which is seen to equal the time-average value of $\mathscr{P}(R,\theta; t)$ given in part (a) of this solution. The total average power radiated is obtained by integrating $\mathscr{P}_{av}(R,\theta)$

[†] We are reminded that in circuit theory if a voltage $v(t) = V_0 \cos(\omega t + \phi) = \mathscr{R}e[V_0 e^{j(\omega t + \phi)}]$ produces a current $i(t) = I_0 \cos(\omega t + \phi - \theta_z) = \mathscr{R}e[I_0 e^{j(\omega t + \phi - \theta_z)}]$ in an impedance, the average power dissipated is $P_{av} = (V_0 I_0/2)\cos\theta_z$. In terms of phasors we have $V = V_0 e^{j\phi}$, $I = I_0 e^{j(\phi - \theta_z)}$, and

$$P_{av} = \tfrac{1}{2}V_0 I_0 \cos\theta_z = \tfrac{1}{2}\mathscr{R}e(VI^*) \qquad (\text{W}),$$

(8–97)

which is analogous to Eq. (8–96).

over the surface of the sphere of radius R:

$$\text{Total } P_{av} = \oint_S \mathscr{P}_{av}(R, \theta) \cdot d\mathbf{s} = \int_0^{2\pi} \int_0^\pi \left[15\pi \left(\frac{I\, d\ell}{\lambda R} \right)^2 \sin^2 \theta \right] R^2 \sin \theta\, d\theta\, d\phi$$

$$= 40\pi^2 \left(\frac{d\ell}{\lambda} \right)^2 I^2 \quad \text{(W)},$$

where I is the amplitude ($\sqrt{2}$ times the effective value) of the sinusoidal current in $d\ell$. ▬

8–6   Normal Incidence at a Plane Conducting Boundary

Up to this point we have discussed the propagation of uniform plane waves in an unbounded homogeneous medium. In practice, waves often propagate in bounded regions where several media with different constitutive parameters are present. When an electromagnetic wave traveling in one medium impinges on another medium with a different intrinsic impedance, it experiences a reflection. In Sections 8–6 and 8–7 we examine the behavior of a plane wave when it is incident upon a plane conducting boundary. Wave behavior at an interface between two dielectric media will be discussed in Sections 8–8, 8–9, and 8–10.

For simplicity we shall assume that the incident wave ($\mathbf{E}_i$, $\mathbf{H}_i$) travels in a lossless medium (medium 1: $\sigma_1 = 0$) and that the boundary is an interface with a perfect conductor (medium 2: $\sigma_2 = \infty$). Two cases will be considered: normal incidence and oblique incidence. In this section we study the field behavior of a uniform plane wave incident normally on a plane conducting boundary.

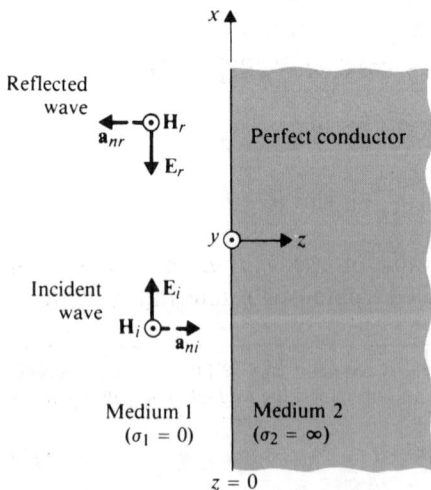

FIGURE 8–9
Plane wave incident normally on a plane conducting boundary.

Consider the situation in Fig. 8–9 where the incident wave travels in the $+z$-direction, and the boundary surface is the plane $z = 0$. The incident electric and magnetic field intensity phasors are

$$\mathbf{E}_i(z) = \mathbf{a}_x E_{i0} e^{-j\beta_1 z}, \tag{8-98}$$

$$\mathbf{H}_i(z) = \mathbf{a}_y \frac{E_{i0}}{\eta_1} e^{-j\beta_1 z}, \tag{8-99}$$

where E_{i0} is the magnitude of $\mathbf{E}_i$ at $z = 0$, and β_1 and η_1 are the phase constant and the intrinsic impedance, respectively, of medium 1. It is noted that the Poynting vector of the incident wave, $\mathscr{P}_i(z) = \mathbf{E}_i(z) \times \mathbf{H}_i(z)$, is in the $\mathbf{a}_z$ direction, which is the direction of energy propagation. The variable z is negative in medium 1.

Inside medium 2 (a perfect conductor), both electric and magnetic fields vanish, $\mathbf{E}_2 = 0$, $\mathbf{H}_2 = 0$; hence no wave is transmitted across the boundary into the $z > 0$ region. The incident wave is reflected, giving rise to a reflected wave $(\mathbf{E}_r, \mathbf{H}_r)$. The reflected electric field intensity can be written as

$$\mathbf{E}_r(z) = \mathbf{a}_x E_{r0} e^{+j\beta_1 z}, \tag{8-100}$$

where the positive sign in the exponent signifies that the reflected wave travels in the $-z$-direction, as discussed in Section 8–2. The total electric field intensity in medium 1 is the sum of $\mathbf{E}_i$ and $\mathbf{E}_r$:

$$\mathbf{E}_1(z) = \mathbf{E}_i(z) + \mathbf{E}_r(z) = \mathbf{a}_x(E_{i0} e^{-j\beta_1 z} + E_{r0} e^{+j\beta_1 z}). \tag{8-101}$$

Continuity of the tangential component of the **E**-field at the boundary $z = 0$ demands that

$$\mathbf{E}_1(0) = \mathbf{a}_x(E_{i0} + E_{r0}) = \mathbf{E}_2(0) = 0,$$

which yields $E_{r0} = -E_{i0}$. Thus, Eq. (8–101) becomes

$$\begin{aligned}
\mathbf{E}_1(z) &= \mathbf{a}_x E_{i0}(e^{-j\beta_1 z} - e^{+j\beta_1 z}) \\
&= -\mathbf{a}_x j 2 E_{i0} \sin \beta_1 z.
\end{aligned} \tag{8-102}$$

The magnetic field intensity $\mathbf{H}_r$ of the reflected wave is related to $\mathbf{E}_r$ by Eq. (8–29):

$$\begin{aligned}
\mathbf{H}_r(z) &= \frac{1}{\eta_1} \mathbf{a}_{nr} \times \mathbf{E}_r(z) = \frac{1}{\eta_1}(-\mathbf{a}_z) \times \mathbf{E}_r(z) \\
&= -\mathbf{a}_y \frac{1}{\eta_1} E_{r0} e^{+j\beta_1 z} = \mathbf{a}_y \frac{E_{i0}}{\eta_1} e^{+j\beta_1 z}.
\end{aligned}$$

Combining $\mathbf{H}_r(z)$ with $\mathbf{H}_i(z)$ in Eq. (8–99), we obtain the total magnetic field intensity in medium 1:

$$\mathbf{H}_1(z) = \mathbf{H}_i(z) + \mathbf{H}_r(z) = \mathbf{a}_y 2 \frac{E_{i0}}{\eta_1} \cos \beta_1 z. \tag{8-103}$$

It is clear from Eqs. (8–102), (8–103), and (8–96) that no average power is associated with the total electromagnetic wave in medium 1, since $\mathbf{E}_1(z)$ and $\mathbf{H}_1(z)$ are in phase quadrature.

In order to examine the space-time behavior of the total field in medium 1, we first write the instantaneous expressions corresponding to the electric and magnetic field intensity phasors obtained in Eqs. (8–102) and (8–103):

$$\mathbf{E}_1(z, t) = \mathscr{R}e[\mathbf{E}_1(z)e^{j\omega t}] = \mathbf{a}_x 2E_{i0} \sin \beta_1 z \sin \omega t, \qquad (8\text{–}104)$$

$$\mathbf{H}_1(z, t) = \mathscr{R}e[\mathbf{H}_1(z)e^{j\omega t}] = \mathbf{a}_y 2\frac{E_{i0}}{\eta_1} \cos \beta_1 z \cos \omega t. \qquad (8\text{–}105)$$

Both $\mathbf{E}_1(z, t)$ and $\mathbf{H}_1(z, t)$ possess zeros and maxima at fixed distances from the conducting boundary for all t, as follows:

$$\left.\begin{array}{l} \text{Zeros of } \mathbf{E}_1(z, t) \\ \text{Maxima of } \mathbf{H}_1(z, t) \end{array}\right\} \text{occur at } \beta_1 z = -n\pi, \quad \text{or } z = -n\frac{\lambda}{2}, \quad n = 0, 1, 2, \ldots,$$

$$\left.\begin{array}{l} \text{Maxima of } \mathbf{E}_1(z, t) \\ \text{Zeros of } \mathbf{H}_1(z, t) \end{array}\right\} \text{occur at } \beta_1 z = -(2n + 1)\frac{\pi}{2}, \quad \text{or } z = -(2n + 1)\frac{\lambda}{4},$$

$$n = 0, 1, 2, \ldots.$$

The total wave in medium 1 is not a traveling wave. It is a **standing wave**, resulting from the superposition of two waves traveling in opposite directions. For a given t, both $\mathbf{E}_1$ and $\mathbf{H}_1$ vary sinusoidally with the distance measured from the boundary plane. The standing waves of $\mathbf{E}_1 = \mathbf{a}_x E_1$ and $\mathbf{H}_1 = \mathbf{a}_y H_1$ are shown in Fig. 8–10 for

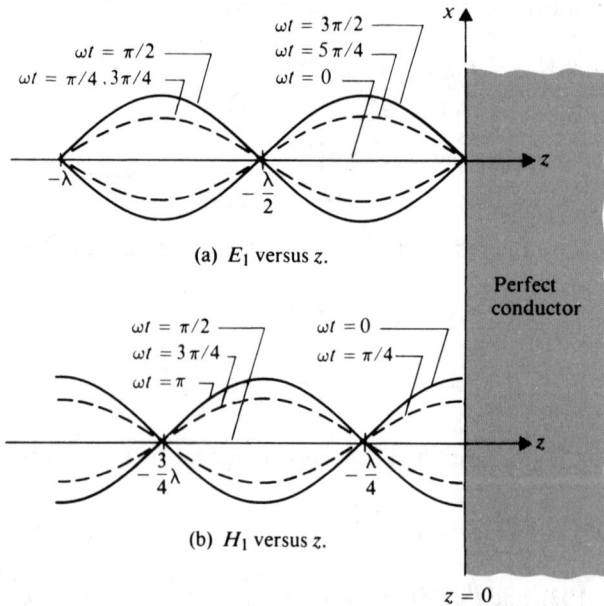

(a) E_1 versus z.

(b) H_1 versus z.

FIGURE 8–10
Standing waves of $\mathbf{E}_1 = \mathbf{a}_x E_1$ and $\mathbf{H}_1 = \mathbf{a}_y H_1$ for several values of ωt.

several values of ωt. Note the following three points: (1) $\mathbf{E}_1$ vanishes on the conducting boundary ($E_{r0} = -E_{i0}$) as well as at points that are multiples of $\lambda/2$ from the boundary; (2) $\mathbf{H}_1$ is a maximum on the conducting boundary ($H_{r0} = H_{i0} = E_{i0}/\eta_1$); (3) the standing waves of $\mathbf{E}_1$ and $\mathbf{H}_1$ are in time quadrature (90° phase difference) and are shifted in space by a quarter wavelength.

EXAMPLE 8–9 A y-polarized uniform plane wave ($\mathbf{E}_i$, $\mathbf{H}_i$) with a frequency 100 (MHz) propagates in air in the $+x$ direction and impinges normally on a perfectly conducting plane at $x = 0$. Assuming the amplitude of $\mathbf{E}_i$ to be 6 (mV/m), write the phasor and instantaneous expressions for (a) $\mathbf{E}_i$ and $\mathbf{H}_i$ of the incident wave; (b) $\mathbf{E}_r$ and $\mathbf{H}_r$ of the reflected wave; and (c) $\mathbf{E}_1$ and $\mathbf{H}_1$ of the total wave in air. (d) Determine the location nearest to the conducting plane where E_1 is zero.

Solution At the given frequency 100 (MHz),

$$\omega = 2\pi f = 2\pi \times 10^8 \quad \text{(rad/s)},$$

$$\beta_1 = k_0 = \frac{\omega}{c} = \frac{2\pi \times 10^8}{3 \times 10^8} = \frac{2\pi}{3} \quad \text{(rad/m)},$$

$$\eta_1 = \eta_0 = \sqrt{\frac{\mu_0}{\epsilon_0}} = 120\pi \quad (\Omega).$$

a) *For the incident wave (a traveling wave):*

i) Phasor expressions:

$$\mathbf{E}_i(x) = \mathbf{a}_y 6 \times 10^{-3} e^{-j2\pi x/3} \quad \text{(V/m)},$$

$$\mathbf{H}_i(x) = \frac{1}{\eta_1} \mathbf{a}_x \times \mathbf{E}_i(x) = \mathbf{a}_z \frac{10^{-4}}{2\pi} e^{-j2\pi x/3} \quad \text{(A/m)}.$$

ii) Instantaneous expressions:

$$\mathbf{E}_i(x, t) = \mathcal{R}e[\mathbf{E}_i(x)e^{j\omega t}]$$

$$= \mathbf{a}_y 6 \times 10^{-3} \cos\left(2\pi \times 10^8 t - \frac{2\pi}{3} x\right) \quad \text{(V/m)},$$

$$\mathbf{H}_i(x, t) = \mathbf{a}_z \frac{10^{-4}}{2\pi} \cos\left(2\pi \times 10^8 t - \frac{2\pi}{3} x\right) \quad \text{(A/m)}.$$

b) *For the reflected wave (a traveling wave):*

i) Phasor expressions:

$$\mathbf{E}_r(x) = -\mathbf{a}_y 6 \times 10^{-3} e^{j2\pi x/3} \quad \text{(V/m)},$$

$$\mathbf{H}_r(x) = \frac{1}{\eta_1}(-\mathbf{a}_x) \times \mathbf{E}_r(x) = \mathbf{a}_z \frac{10^{-4}}{2\pi} e^{j2\pi x/3} \quad \text{(A/m)}.$$

ii) Instantaneous expressions:

$$\mathbf{E}_r(x, t) = \mathscr{R}e[\mathbf{E}_r(x)e^{j\omega t}] = -\mathbf{a}_y 6 \times 10^{-3} \cos\left(2\pi \times 10^8 t + \frac{2\pi}{3} x\right) \quad (\text{V/m}),$$

$$\mathbf{H}_r(x, t) = \mathbf{a}_z \frac{10^{-4}}{2\pi} \cos\left(2\pi \times 10^8 t + \frac{2\pi}{3} x\right) \quad (\text{A/m}).$$

c) *For the total wave (a standing wave):*

  i) Phasor expressions:

$$\mathbf{E}_1(x) = \mathbf{E}_i(x) + \mathbf{E}_r(x) = -\mathbf{a}_y j 12 \times 10^{-3} \sin\left(\frac{2\pi}{3} x\right) \quad (\text{V/m}),$$

$$\mathbf{H}_1(x) = \mathbf{H}_i(x) + \mathbf{H}_r(x) = \mathbf{a}_z \frac{10^{-4}}{\pi} \cos\left(\frac{2\pi}{3} x\right) \quad (\text{A/m}).$$

  ii) Instantaneous expressions:

$$\mathbf{E}_1(x, t) = \mathscr{R}e[\mathbf{E}_1(x)e^{j\omega t}] = \mathbf{a}_y 12 \times 10^{-3} \sin\left(\frac{2\pi}{3} x\right) \sin(2\pi \times 10^8 t) \quad (\text{V/m}),$$

$$\mathbf{H}_1(x, t) = \mathbf{a}_z \frac{10^{-4}}{\pi} \cos\left(\frac{2\pi}{3} x\right) \cos(2\pi \times 10^8 t) \quad (\text{A/m}).$$

d) The electric field vanishes at the surface of the conducting plane at $x = 0$. In medium 1 the first null occurs at

$$x = -\frac{\lambda_1}{2} = -\frac{\pi}{\beta_1} = -\frac{3}{2} \quad (\text{m}). \qquad \blacksquare$$

8–7 Oblique Incidence at a Plane Conducting Boundary

When a uniform plane wave is incident on a plane conducting surface obliquely, the behavior of the reflected wave depends on the polarization of the incident wave. In order to be specific about the direction of $\mathbf{E}_i$ we define a *plane of incidence* as the plane containing the vector indicating the direction of propagation of the incident wave and the normal to the boundary surface. Since an $\mathbf{E}_i$ polarized in an arbitrary direction can always be decomposed into two components—one perpendicular and the other parallel to the plane of incidence—we consider these two cases separately. The general case is obtained by superposing the results of the two component cases.

8–7.1   PERPENDICULAR POLARIZATION[†]

In the case of *perpendicular polarization*, $\mathbf{E}_i$ is perpendicular to the plane of incidence, as illustrated in Fig. 8–11. Noting that

$$\mathbf{a}_{ni} = \mathbf{a}_x \sin \theta_i + \mathbf{a}_z \cos \theta_i, \qquad (8\text{–}106)$$

[†] Also referred to as *horizontal polarization* or *E-polarization*.

where θ_i is the **angle of incidence** measured from the normal to the boundary surface, we obtain, using Eqs. (8–26) and (8–29),

$$\mathbf{E}_i(x, z) = \mathbf{a}_y E_{i0} e^{-j\beta_1 \mathbf{a}_{ni} \cdot \mathbf{R}} = \mathbf{a}_y E_{i0} e^{-j\beta_1(x \sin \theta_i + z \cos \theta_i)}, \tag{8–107}$$

$$\mathbf{H}_i(x, z) = \frac{1}{\eta_1} [\mathbf{a}_{ni} \times \mathbf{E}_i(x, z)]$$

$$= \frac{E_{i0}}{\eta_1} (-\mathbf{a}_x \cos \theta_i + \mathbf{a}_z \sin \theta_i) e^{-j\beta_1(x \sin \theta_i + z \cos \theta_i)}. \tag{8–108}$$

For the reflected wave,

$$\mathbf{a}_{nr} = \mathbf{a}_x \sin \theta_r - \mathbf{a}_z \cos \theta_r, \tag{8–109}$$

where θ_r is the **angle of reflection**, we have

$$\mathbf{E}_r(x, z) = \mathbf{a}_y E_{r0} e^{-j\beta_1(x \sin \theta_r - z \cos \theta_r)}. \tag{8–110}$$

At the boundary surface, $z = 0$, the total electric field intensity must vanish. Thus,

$$\mathbf{E}_1(x, 0) = \mathbf{E}_i(x, 0) + \mathbf{E}_r(x, 0)$$

$$= \mathbf{a}_y(E_{i0} e^{-j\beta_1 x \sin \theta_i} + E_{r0} e^{-j\beta_1 x \sin \theta_r}) = 0.$$

In order for this relation to hold for all values of x, we must have $E_{r0} = -E_{i0}$ and matched phase terms, that is, $\theta_r = \theta_i$. The latter relation, asserting that **the angle of reflection equals the angle of incidence**, is referred to as **Snell's law of reflection**. Thus, Eq. (8–110) becomes

$$\mathbf{E}_r(x, z) = -\mathbf{a}_y E_{i0} e^{-j\beta_1(x \sin \theta_i - z \cos \theta_i)}. \tag{8–111}$$

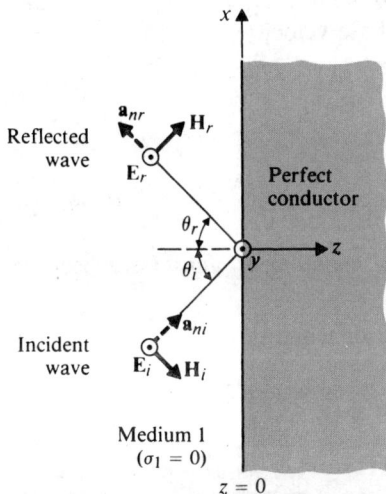

FIGURE 8–11
Plane wave incident obliquely on a plane conducting boundary (perpendicular polarization).

The corresponding $H_r(x, z)$ is

$$
\begin{aligned}
H_r(x, z) &= \frac{1}{\eta_1}[a_{nr} \times E_r(x, z)] \\
&= \frac{E_{i0}}{\eta_1}(-a_x \cos \theta_i - a_z \sin \theta_i)e^{-j\beta_1(x \sin \theta_i - z \cos \theta_i)}.
\end{aligned}
\tag{8-112}
$$

The total field is obtained by adding the incident and reflected fields. From Eqs. (8-107) and (8-111) we have

$$
\begin{aligned}
E_1(x, z) &= E_i(x, z) + E_r(x, z) \\
&= a_y E_{i0}(e^{-j\beta_1 z \cos \theta_i} - e^{j\beta_1 z \cos \theta_i})e^{-j\beta_1 x \sin \theta_i} \\
&= -a_y j2E_{i0} \sin (\beta_1 z \cos \theta_i)e^{-j\beta_1 x \sin \theta_i}.
\end{aligned}
\tag{8-113}
$$

Adding the results in Eqs. (8-108) and (8-112), we get

$$
\begin{aligned}
H_1(x, z) = -2\frac{E_{i0}}{\eta_1}[&a_x \cos \theta_i \cos (\beta_1 z \cos \theta_i)e^{-j\beta_1 x \sin \theta_i} \\
&+ a_z j \sin \theta_i \sin (\beta_1 z \cos \theta_i)e^{-j\beta_1 x \sin \theta_i}].
\end{aligned}
\tag{8-114}
$$

Equations (8-113) and (8-114) are rather complicated expressions, but we can make the following observations about the oblique incidence of a uniform plane wave with perpendicular polarization on a plane conducting boundary:

1. In the direction (z-direction) normal to the boundary, E_{1y} and H_{1x} maintain standing-wave patterns according to $\sin \beta_{1z}z$ and $\cos \beta_{1z}z$, respectively, where $\beta_{1z} = \beta_1 \cos \theta_i$. No average power is propagated in this direction since E_{1y} and H_{1x} are 90° out of time phase.

2. In the direction (x-direction) parallel to the boundary, E_{1y} and H_{1z} are in both time and space phase and propagate with a phase velocity

$$
u_{1x} = \frac{\omega}{\beta_{1x}} = \frac{\omega}{\beta_1 \sin \theta_i} = \frac{u_1}{\sin \theta_i}.
\tag{8-115}
$$

The wavelength in this direction is

$$
\lambda_{1x} = \frac{2\pi}{\beta_{1x}} = \frac{\lambda_1}{\sin \theta_i}.
\tag{8-116}
$$

3. The propagating wave in the x direction is a **nonuniform plane wave** because its amplitude varies with z.

4. Since $E_1 = 0$ for all x when $\sin (\beta_1 z \cos \theta_i) = 0$ or when

$$
\beta_1 z \cos \theta_i = \frac{2\pi}{\lambda_1} z \cos \theta_i = -m\pi, \qquad m = 1, 2, 3, \ldots,
$$

a conducting plate could be inserted at

$$
z = -\frac{m\lambda_1}{2 \cos \theta_i}, \qquad m = 1, 2, 3, \ldots,
\tag{8-117}
$$

without changing the field pattern that exists between the conducting plate and the conducting boundary at $z = 0$. A *transverse electric* (TE) *wave* $(E_{1x} = 0)$ would bounce back and forth between the conducting planes and propagate in the x-direction. We would have, in effect, a parallel-plate waveguide.

An illustration of the bouncing waves and the interference pattern between a conducting plate inserted at $z = -\lambda_1/2 \cos \theta_i$ and the conducting boundary at $z = 0$ is given in Fig. 8–12. The long (thick) dashed lines represent the plane-wave crests with the **E**-vector out of the page, and the short (thin) dashed lines represent wave troughs with the **E**-vector into the page. At the conducting surfaces the reflected **E**-vector has a 180° phase change, cancelling the incident **E**-vector; hence the intersections of the long and short dashed lines (such as points O, A, and A'') are locations of zero electric intensity. The intersections of two long dashed lines (such as B) are locations of maximum electric field intensity directed out of the page, and the intersections of two short dashed lines (such as B') are locations of maximum electric field intensity into the page. The intersections of the two plane waves (incident and reflected) travel in the x-direction with a phase velocity given by Eq. (8–115). From Fig. 8–12 we have

$$\overline{OA'} = \frac{\lambda_1}{2} = \frac{\pi}{\beta_1}, \tag{8–118}$$

$$\overline{OA} = b = \frac{\lambda_1}{2 \cos \theta_i}. \tag{8–119}$$

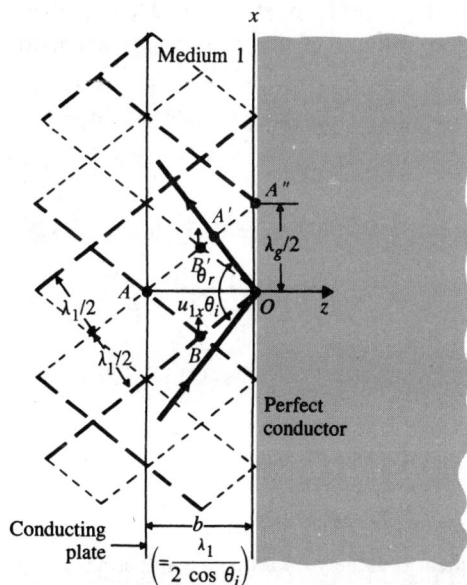

FIGURE 8–12
Illustrating bouncing waves and interference patterns of oblique incidence at a plane conducting boundary (perpendicular polarization).

The traveling wave in the parallel-plate waveguide has a guide wavelength equal to $2\overline{OA}''$, or

$$\lambda_g = 2\overline{OA}'' = 2\frac{\overline{OA}'}{\sin \theta_i}$$

$$= \frac{\lambda_1}{\sin \theta_i} > \lambda_1.$$

(8–120)

At $\theta_i = 0$ there would be no propagating wave in the x-direction. The properties of TE waves between parallel plates will be discussed in Subsection 10–3.2.

EXAMPLE 8–10   A uniform plane wave $(\mathbf{E}_i, \mathbf{H}_i)$ of an angular frequency ω is incident from air on a very large, perfectly conducting wall at an angle of incidence θ_i with perpendicular polarization. Find (a) the current induced on the wall surface, and (b) the time-average Poynting vector in medium 1.

Solution

a) The conditions of this problem are exactly those we have just discussed; hence we could use the formulas directly. Let $z = 0$ be the plane representing the surface of the perfectly conducting wall, and let $\mathbf{E}_i$ be polarized in the y direction, as was shown in Fig. 8–11. At $z = 0$, $\mathbf{E}_1(x, 0) = 0$, and $\mathbf{H}_1(x, 0)$ can be obtained from Eq. (8–114):

$$\mathbf{H}_1(x, 0) = -\frac{E_{i0}}{\eta_0}(\mathbf{a}_x 2 \cos \theta_i)e^{-j\beta_0 x \sin \theta_i}.$$

(8–121)

Inside the perfectly conducting wall, both $\mathbf{E}_2$ and $\mathbf{H}_2$ must vanish. There is then a discontinuity in the magnetic field. The amount of discontinuity is equal to

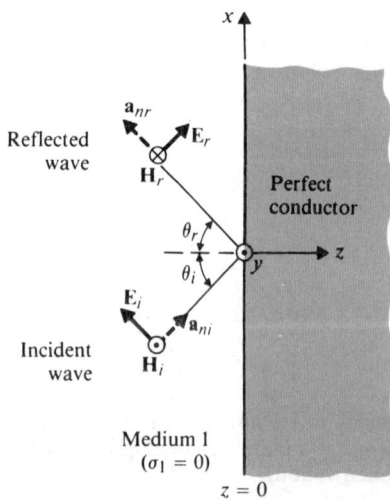

FIGURE 8–13
Plane wave incident obliquely on a plane conducting boundary (parallel polarization).

the surface current. From Eq. (7–68b) we have

$$\mathbf{J}_s(x) = \mathbf{a}_{n2} \times \mathbf{H}_1(x, 0)$$

$$= (-\mathbf{a}_z) \times (-\mathbf{a}_x) \frac{E_{i0}}{\eta_0} (2 \cos \theta_i) e^{-j\beta_0 x \sin \theta_i}$$

$$= \mathbf{a}_y \frac{E_{i0}}{60\pi} (\cos \theta_i) e^{-j(\omega/c)x \sin \theta_i}.$$

The instantaneous expression for the surface current is

$$\mathbf{J}_s(x, t) = \mathbf{a}_y \frac{E_{i0}}{60\pi} \cos \theta_i \cos \omega \left(t - \frac{x}{c} \sin \theta_i \right) \quad \text{(A/m)}. \qquad (8\text{--}122)$$

It is this induced current on the wall surface that gives rise to the reflected wave in medium 1 and cancels the incident wave in the conducting wall.

b) The time-average Poynting vector in medium 1 is found by using Eqs. (8–113) and (8–114) in Eq. (8–96). Since E_{1y} and H_{1x} are in time quadrature, $\mathscr{P}_{av}$ will only have a nonvanishing x-component arising from E_{1y} and H_{1z}:

$$\mathscr{P}_{av_1} = \tfrac{1}{2}\mathscr{R}e[\mathbf{E}_1(x, z) \times \mathbf{H}_1^*(x, z)]$$

$$= \mathbf{a}_x 2 \frac{E_{i0}^2}{\eta_1} \sin \theta_i \sin^2 \beta_{1z} z, \qquad (8\text{--}123)$$

where $\beta_{1z} = \beta_1 \cos \theta_i$. The time-average Poynting vector in medium 2 (a perfect conductor) is, of course, zero.   ▄

8-7.2   PARALLEL POLARIZATION[†]

We now consider the case of $\mathbf{E}_i$ lying in the plane of incidence while a uniform plane wave impinges obliquely on a perfectly conducting plane boundary, as depicted in Fig. 8–13. The unit vectors $\mathbf{a}_{ni}$ and $\mathbf{a}_{nr}$, representing the directions of propagation of the incident and reflected waves, respectively, remain the same as those given in Eqs. (8–106) and (8–109). Both $\mathbf{E}_i$ and $\mathbf{E}_r$ now have components in x- and z-directions, whereas $\mathbf{H}_i$ and $\mathbf{H}_r$ have only a y-component. We have, for the incident wave,

$$\mathbf{E}_i(x, z) = E_{i0}(\mathbf{a}_x \cos \theta_i - \mathbf{a}_z \sin \theta_i)e^{-j\beta_1(x \sin \theta_i + z \cos \theta_i)}, \qquad (8\text{--}124)$$

$$\mathbf{H}_i(x, z) = \mathbf{a}_y \frac{E_{i0}}{\eta_1} e^{-j\beta_1(x \sin \theta_i + z \cos \theta_i)}. \qquad (8\text{--}125)$$

The reflected wave $(\mathbf{E}_r, \mathbf{H}_r)$ have the following phasor expressions:

$$\mathbf{E}_r(x, z) = E_{r0}(\mathbf{a}_x \cos \theta_r + \mathbf{a}_z \sin \theta_r)e^{-j\beta_1(x \sin \theta_r - z \cos \theta_r)}, \qquad (8\text{--}126)$$

$$\mathbf{H}_r(x, z) = -\mathbf{a}_y \frac{E_{r0}}{\eta_1} e^{-j\beta_1(x \sin \theta_r - z \cos \theta_r)}. \qquad (8\text{--}127)$$

[†] Also referred to as *vertical polarization* or *H-polarization*.

At the surface of the perfect conductor, $z = 0$, the tangential component (the x-component) of the total electric field intensity must vanish for all x, or $E_{ix}(x, 0) + E_{rx}(x, 0) = 0$. From Eqs. (8–124) and (8–126) we have

$$(E_{i0} \cos \theta_i)e^{-j\beta_1 x \sin \theta_i} + (E_{r0} \cos \theta_r)e^{-j\beta_1 x \sin \theta_r} = 0,$$

which requires $E_{r0} = -E_{i0}$ and $\theta_r = \theta_i$. The total electric field intensity in medium 1 is the sum of Eqs. (8–124) and (8–126):

$$\mathbf{E}_1(x, z) = \mathbf{E}_i(x, z) + \mathbf{E}_r(x, z)$$

$$= \mathbf{a}_x E_{i0} \cos \theta_i (e^{-j\beta_1 z \cos \theta_i} - e^{j\beta_1 z \cos \theta_i})e^{-j\beta_1 x \sin \theta_i}$$

$$- \mathbf{a}_z E_{i0} \sin \theta_i (e^{-j\beta_1 z \cos \theta_i} + e^{j\beta_1 z \cos \theta_i})e^{-j\beta_1 x \sin \theta_i}$$

or

$$\mathbf{E}_1(x, z) = -2E_{i0}[\mathbf{a}_x j \cos \theta_i \sin (\beta_1 z \cos \theta_i)$$

$$+ \mathbf{a}_z \sin \theta_i \cos (\beta_1 z \cos \theta_i)]e^{-j\beta_1 x \sin \theta_i}. \qquad (8\text{–}128)$$

Adding Eqs. (8–125) and (8–127), we obtain the total magnetic field intensity in medium 1:

$$\mathbf{H}_1(x, z) = \mathbf{H}_i(x, z) + \mathbf{H}_r(x, z)$$

$$= \mathbf{a}_y 2 \frac{E_{i0}}{\eta_1} \cos (\beta_1 z \cos \theta_i)e^{-j\beta_1 x \sin \theta_i}. \qquad (8\text{–}129)$$

The interpretation of Eqs. (8–128) and (8–129) is similar to that of Eqs. (8–113) and (8–114) for the perpendicular-polarization case, except that $\mathbf{E}_1(x, z)$, instead of $\mathbf{H}_1(x, z)$, now has both an x- and a z-component. We conclude, therefore:

1. In the direction (z-direction) normal to the boundary, E_{1x} and H_{1y} maintain standing-wave patterns according to $\sin \beta_{1z}z$ and $\cos \beta_{1z}z$, respectively, where $\beta_{1z} = \beta_1 \cos \theta_i$. No average power is propagated in this direction, since E_{1x} and H_{1y} are 90° out of time phase.

2. In the x-direction parallel to the boundary, E_{1z} and H_{1y} are in both time and space phase and propagate with a phase velocity $u_{1x} = u_1/\sin \theta_i$, which is the same as that in the perpendicular polarization.

3. As in the case of perpendicular polarization, the propagating wave in the x-direction is a nonuniform plane wave.

4. The insertion of a conducting plate at $z = -m\lambda_1/2 \cos \theta_i$ ($m = 1, 2, 3, \ldots$) where $E_{1x} = 0$ for all x would not affect the field pattern that exists between the conducting plate and the conducting boundary at $z = 0$; we would then have a parallel-plate waveguide. A *tranverse magnetic* (TM) *wave* ($H_{1x} = 0$) will propagate in the x-direction. (TM waves between parallel plates will be discussed in Subsection 10–3.1.)

We note here that the $\mathbf{E}_1$ and $\mathbf{H}_1$ expressions for oblique incidence in Eqs. (8–113), (8–114), (8–128), and (8–129) are the sums of the fields of incident and reflected waves. They represent interference patterns. If the incident wave is confined in a narrow beam, the reflected waves will also be a narrow beam propagating in a different direction. There will then be no interference except for a very small region near the conducting surface. Thus, the reflectors on microwave relay towers receive, amplify, and retransmit the original incident wave, not the interference pattern.

8-8 Normal Incidence at a Plane Dielectric Boundary

When an electromagnetic wave is incident on the surface of a dielectric medium that has an intrinsic impedance different from that of the medium in which the wave is originated, part of the incident power is reflected and part is transmitted. We may think of the situation as being like an impedance mismatch in circuits. The case of wave incidence on a perfectly conducting boundary discussed in the two previous sections is like terminating a generator that has a certain internal impedance with a short circuit; no power is transmitted into the conducting region.

As before, we will consider separately the two cases of the normal incidence and the oblique incidence of a uniform plane wave on a plane dielectric medium. Both media are assumed to be dissipationless ($\sigma_1 = \sigma_2 = 0$). We will discuss the wave behavior for normal incidence in this section. The case of oblique incidence will be taken up in Section 8-9.

Consider the situation in Fig. 8-14, where the incident wave travels in the $+z$-direction and the boundary surface is the plane $z = 0$. The incident electric and magnetic field intensity phasors are

$$\mathbf{E}_i(z) = \mathbf{a}_x E_{i0} e^{-j\beta_1 z}, \tag{8-130}$$

$$\mathbf{H}_i(z) = \mathbf{a}_y \frac{E_{i0}}{\eta_1} e^{-j\beta_1 z}. \tag{8-131}$$

These are the same expressions as those given in Eqs. (8-98) and (8-99). Note that z is negative in medium 1.

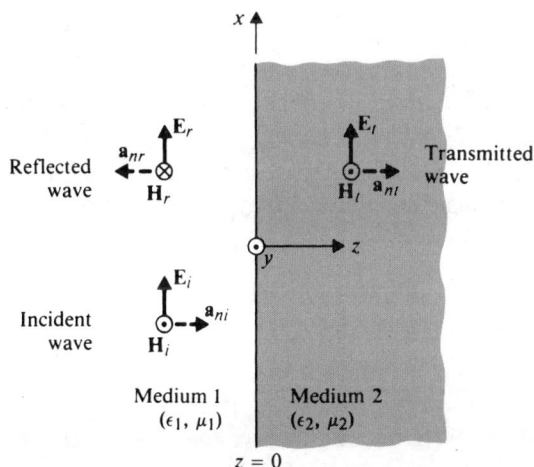

Reflected wave

Incident wave

Transmitted wave

Medium 1 (ϵ_1, μ_1)

Medium 2 (ϵ_2, μ_2)

$z = 0$

FIGURE 8-14
Plane wave incident normally on a plane dielectric boundary.

Because of the medium discontinuity at $z = 0$, the incident wave is partly reflected back into medium 1 and partly transmitted into medium 2. We have

a) *For the reflected wave* $(\mathbf{E}_r, \mathbf{H}_r)$:

$$\mathbf{E}_r(z) = \mathbf{a}_x E_{r0} e^{j\beta_1 z}, \tag{8–132}$$

$$\mathbf{H}_r(z) = (-\mathbf{a}_z) \times \frac{1}{\eta_1} \mathbf{E}_r(z) = -\mathbf{a}_y \frac{E_{r0}}{\eta_1} e^{j\beta_1 z}. \tag{8–133}$$

b) *For the transmitted wave* $(\mathbf{E}_t, \mathbf{H}_t)$:

$$\mathbf{E}_t(z) = \mathbf{a}_x E_{t0} e^{-j\beta_2 z}, \tag{8–134}$$

$$\mathbf{H}_t(z) = \mathbf{a}_z \times \frac{1}{\eta_2} \mathbf{E}_t(z) = \mathbf{a}_y \frac{E_{t0}}{\eta_2} e^{-j\beta_2 z}, \tag{8–135}$$

where E_{t0} is the magnitude of $\mathbf{E}_t$ at $z = 0$, and β_2 and η_2 are the phase constant and the intrinsic impedance, respectively, of medium 2.

Note that the directions of the arrows for $\mathbf{E}_r$ and $\mathbf{E}_t$ in Fig. 8–14 are arbitrarily drawn because E_{r0} and E_{t0} may themselves be positive or negative, depending on the relative magnitudes of the constitutive parameters of the two media.

Two equations are needed for determining the two unknown magnitudes E_{r0} and E_{t0}. These equations are supplied by the boundary conditions that must be satisfied by the electric and magnetic fields. At the dielectric interface $z = 0$ the tangential components (the x-components) of the electric and magnetic field intensities must be continuous. We have

$$\mathbf{E}_i(0) + \mathbf{E}_r(0) = \mathbf{E}_t(0) \qquad \text{or} \qquad E_{i0} + E_{r0} = E_{t0} \tag{8–136}$$

and

$$\mathbf{H}_i(0) + \mathbf{H}_r(0) = \mathbf{H}_t(0) \qquad \text{or} \qquad \frac{1}{\eta_1}(E_{i0} - E_{r0}) = \frac{E_{t0}}{\eta_2}. \tag{8–137}$$

Solving Eqs. (8–136) and (8–137), we obtain

$$E_{r0} = \frac{\eta_2 - \eta_1}{\eta_2 + \eta_1} E_{i0}, \tag{8–138}$$

$$E_{t0} = \frac{2\eta_2}{\eta_2 + \eta_1} E_{i0}. \tag{8–139}$$

The ratios E_{r0}/E_{i0} and E_{t0}/E_{i0} are called **reflection coefficient** and **transmission coefficient**, respectively. In terms of the intrinsic impedances they are

$$\boxed{\Gamma = \frac{E_{r0}}{E_{i0}} = \frac{\eta_2 - \eta_1}{\eta_2 + \eta_1} \qquad \text{(Dimensionless)}} \tag{8–140}$$

and

$$\boxed{\tau = \frac{E_{t0}}{E_{i0}} = \frac{2\eta_2}{\eta_2 + \eta_1} \qquad \text{(Dimensionless)}.} \tag{8–141}$$

Note that the reflection coefficient Γ in Eq. (8–140) can be positive or negative, depending on whether η_2 is greater or less than η_1. The transmission coefficient τ, however, is always positive.

The definitions for Γ and τ in Eqs. (8–140) and (8–141) apply even when the media are dissipative—that is, even when η_1 and/or η_2 are complex. Thus Γ and τ may themselves be complex in the general case. A complex Γ (or τ) simply means that a phase shift is introduced at the interface upon reflection (or transmission). Reflection and transmission coefficients are related by the following equation:

$$\boxed{1 + \Gamma = \tau \qquad \text{(Dimensionless).}} \qquad (8\text{–}142)$$

If medium 2 is a perfect conductor, $\eta_2 = 0$, Eqs. (8–140) and (8–141) yield $\Gamma = -1$ and $\tau = 0$. Consequently, $E_{r0} = -E_{i0}$, and $E_{t0} = 0$. The incident wave will be totally reflected, and a standing wave will be produced in medium 1. The standing wave will have zero and maximum points, as discussed in Section 8–6.

If medium 2 is not a perfect conductor, partial reflection will result. The total electric field in medium 1 can be written as

$$\begin{aligned}
\mathbf{E}_1(z) = \mathbf{E}_i(z) + \mathbf{E}_r(z) &= \mathbf{a}_x E_{i0}(e^{-j\beta_1 z} + \Gamma e^{j\beta_1 z}) \\
&= \mathbf{a}_x E_{i0}[(1 + \Gamma)e^{-j\beta_1 z} + \Gamma(e^{j\beta_1 z} - e^{-j\beta_1 z})] \\
&= \mathbf{a}_x E_{i0}[(1 + \Gamma)e^{-j\beta_1 z} + \Gamma(j2 \sin \beta_1 z)]
\end{aligned}$$

or, in view of Eq. (8–142),

$$\mathbf{E}_1(z) = \mathbf{a}_x E_{i0}[\tau e^{-j\beta_1 z} + \Gamma(j2 \sin \beta_1 z)]. \qquad (8\text{–}143)$$

We see in Eq. (8–143) that $\mathbf{E}_1(z)$ is composed of two parts: a traveling wave with an amplitude τE_{i0} and a standing wave with an amplitude $2\Gamma E_{i0}$. Because of the existence of the traveling wave, $\mathbf{E}_1(z)$ does not go to zero at fixed distances from the interface; it merely has locations of maximum and minimum values.

The locations of maximum and minimum $|\mathbf{E}_1(z)|$ are conveniently found by rewriting $\mathbf{E}_1(z)$ as

$$\mathbf{E}_1(z) = \mathbf{a}_x E_{i0} e^{-j\beta_1 z}(1 + \Gamma e^{j2\beta_1 z}). \qquad (8\text{–}144)$$

For dissipationless media, η_1 and η_2 are real, making both Γ and τ also real. However, Γ can be positive or negative. Consider the following two cases.

1. $\Gamma > 0 \ (\eta_2 > \eta_1)$.

The maximum value of $|\mathbf{E}_1(z)|$ is $E_{i0}(1 + \Gamma)$, which occurs when $2\beta_1 z_{max} = -2n\pi \ (n = 0, 1, 2, \ldots)$, or at

$$z_{max} = -\frac{n\pi}{\beta_1} = -\frac{n\lambda_1}{2}, \qquad n = 0, 1, 2, \ldots \qquad (8\text{–}145)$$

The minimum value of $|\mathbf{E}_1(z)|$ is $E_{i0}(1 - \Gamma)$, which occurs when $2\beta_1 z_{min} = -(2n + 1)\pi$, or at

$$z_{min} = -\frac{(2n + 1)\pi}{2\beta_1} = -\frac{(2n + 1)\lambda_1}{4}, \qquad n = 0, 1, 2, \ldots \qquad (8\text{–}146)$$

2. $\Gamma < 0$ $(\eta_2 < \eta_1)$.

The maximum value of $|E_1(z)|$ is $E_{i0}(1 - \Gamma)$, which occurs at z_{min} given in Eq. (8–146); and the minimum value of $|E_1(z)|$ is $E_{i0}(1 + \Gamma)$, which occurs at z_{max} given in Eq. (8–145). In other words, the locations for $|E_1(z)|_{max}$ and $|E_1(z)|_{min}$ when $\Gamma > 0$ and when $\Gamma < 0$ are interchanged.

The ratio of the maximum value to the minimum value of the electric field intensity of a standing wave is called the *standing-wave ratio* (SWR), S.

$$S = \frac{|E|_{max}}{|E|_{min}} = \frac{1 + |\Gamma|}{1 - |\Gamma|} \qquad \text{(Dimensionless)}. \qquad (8\text{–}147)$$

An inverse relation of Eq. (8–147) is

$$|\Gamma| = \frac{S - 1}{S + 1} \qquad \text{(Dimensionless)}. \qquad (8\text{–}148)$$

While the value of Γ ranges from -1 to $+1$, the value of S ranges from 1 to ∞. It is customary to express S on a logarithmic scale. The standing-wave ratio in decibels is $20 \log_{10} S$. Thus $S = 2$ corresponds to a standing-wave ratio of $20 \log_{10} 2 = 6.02$ dB and $|\Gamma| = (2 - 1)/(2 + 1) = \frac{1}{3}$. A standing-wave ratio of 2 dB is equivalent to $S = 1.26$ and $|\Gamma| = 0.115$.

The magnetic field intensity in medium 1 is obtained by combining $\mathbf{H}_i(z)$ and $\mathbf{H}_r(z)$ in Eqs. (8–131) and (8–133), respectively:

$$\begin{aligned} \mathbf{H}_1(z) &= \mathbf{a}_y \frac{E_{i0}}{\eta_1} (e^{-j\beta_1 z} - \Gamma e^{j\beta_1 z}) \\ &= \mathbf{a}_y \frac{E_{i0}}{\eta_1} e^{-j\beta_1 z}(1 - \Gamma e^{j2\beta_1 z}). \end{aligned} \qquad (8\text{–}149)$$

This should be compared with $\mathbf{E}_1(z)$ in Eq. (8–144). In a dissipationless medium, Γ is real; and $|\mathbf{H}_1(z)|$ will be a minimum at locations where $|\mathbf{E}_1(z)|$ is a maximum, and vice versa.

In medium 2, $(\mathbf{E}_t, \mathbf{H}_t)$ constitute the transmitted wave propagating in $+z$-direction. From Eqs. (8–134) and (8–141) we have

$$\mathbf{E}_t(z) = \mathbf{a}_x \tau E_{i0} e^{-j\beta_2 z}. \qquad (8\text{–}150)$$

And from Eq. (8–135) we obtain

$$\mathbf{H}_t(z) = \mathbf{a}_y \frac{\tau}{\eta_2} E_{i0} e^{-j\beta_2 z}. \qquad (8\text{–}151)$$

━━━ **EXAMPLE 8–11** A uniform plane wave in a lossless medium with intrinsic impedance η_1 is incident normally onto another lossless medium with intrinsic impedance

η_2 through a plane boundary. Obtain the expressions for the time-average power densities in both media.

Solution   Equation (8–96) provides the formula for computing the time-average power density, or time-average Poynting vector:

$$\mathscr{P}_{av} = \tfrac{1}{2}\mathscr{R}e(\mathbf{E} \times \mathbf{H}^*).$$

In medium 1 we use Eqs. (8–144) and (8–149):

$$
\begin{aligned}
(\mathscr{P}_{av})_1 &= \mathbf{a}_z \frac{E_{i0}^2}{2\eta_1} \mathscr{R}e\big[(1 + \Gamma e^{j2\beta_1 z})(1 - \Gamma e^{-j2\beta_1 z})\big] \\
&= \mathbf{a}_z \frac{E_{i0}^2}{2\eta_1} \mathscr{R}e\big[(1 - \Gamma^2) + \Gamma(e^{j2\beta_1 z} - e^{-j2\beta_1 z})\big] \\
&= \mathbf{a}_z \frac{E_{i0}^2}{2\eta_1} \mathscr{R}e\big[(1 - \Gamma^2) + j2\Gamma \sin 2\beta_1 z\big] \\
&= \mathbf{a}_z \frac{E_{i0}^2}{2\eta_1} (1 - \Gamma^2),
\end{aligned}
\tag{8-152}
$$

where Γ is a real number because both media are lossless.

In medium 2 we use Eqs. (8–150) and (8–151) to obtain

$$(\mathscr{P}_{av})_2 = \mathbf{a}_z \frac{E_{i0}^2}{2\eta_2} \tau^2. \tag{8-153}$$

Since we are dealing with lossless media, the power flow in medium 1 must equal that in medium 2; that is,

$$(\mathscr{P}_{av})_1 = (\mathscr{P}_{av})_2, \tag{8-154}$$

or

$$\boxed{1 - \Gamma^2 = \frac{\eta_1}{\eta_2}\, \tau^2.} \tag{8-155}$$

That Eq. (8–155) is true can be readily verified by using Eqs. (8–140) and (8–141). ▬

8-9   Normal Incidence at Multiple Dielectric Interfaces

In certain practical situations a wave may be incident on several layers of dielectric media with different constitutive parameters. One such situation is the use of a dielectric coating on glass to reduce glare from sunlight. Another is a **radome**, which is a dome-shaped enclosure designed not only to protect radar installations from inclement weather but to permit the propagation of electromagnetic waves through

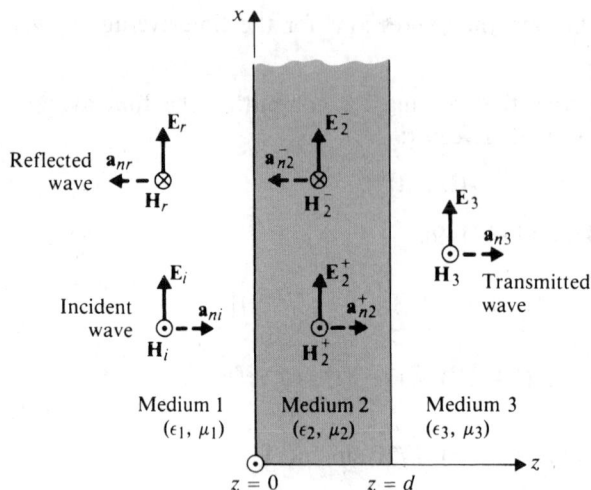

FIGURE 8–15
Normal incidence at multiple dielectric interfaces.

the enclosure with as little reflection as possible. In both situations, determining the proper dielectric material and its thickness is an important design problem.

We now consider the three-region situation depicted in Fig. 8–15. A uniform plane wave traveling in the $+z$-direction in medium 1 (ϵ_1, μ_1) impinges normally at a plane boundary with medium 2 (ϵ_2, μ_2), at $z = 0$. Medium 2 has a finite thickness and interfaces with medium 3 (ϵ_3, μ_3) at $z = d$. Reflection occurs at both $z = 0$ and $z = d$. Assuming an x-polarized incident field, the total electric field intensity in medium 1 can always be written as the sum of the incident component $\mathbf{a}_x E_{i0} e^{-j\beta_1 z}$ and a reflected component $\mathbf{a}_x E_{r0} e^{j\beta_1 z}$:

$$\mathbf{E}_1 = \mathbf{a}_x (E_{i0} e^{-j\beta_1 z} + E_{r0} e^{j\beta_1 z}). \qquad (8-156)$$

However, owing to the existence of a second discontinuity at $z = d$, E_{r0} is no longer related to E_{i0} by Eq. (8–138) or Eq. (8–140). Within medium 2, parts of waves bounce back and forth between the two bounding surfaces, some penetrating into media 1 and 3. The reflected field in medium 1 is the sum of (a) the field reflected from the interface at $z = 0$ as the incident wave impinges on it, (b) the field transmitted back into medium 1 from medium 2 after a first reflection from the interface at $z = d$, (c) the field transmitted back into medium 1 from medium 2 after a second reflection at $z = d$, and so on. The total reflected wave is, in fact, the resultant of the initial reflected component and an infinite sequence of multiply reflected contributions within medium 2 that are transmitted back into medium 1. Since all of the contributions propagate in the $-z$-direction in medium 1 and contain the propagation factor $e^{j\beta_1 z}$, they can be combined into a single term with a coefficient E_{r0}. But how do we determine the relation between E_{r0} and E_{i0} now?

One way to find E_{r0} is to write down the electric and magnetic field intensity vectors in all three regions and apply the boundary conditions. The $\mathbf{H}_1$ in region 1 that corresponds to the $\mathbf{E}_1$ in Eq. (8–156) is, from Eqs. (8–131) and (8–133),

$$\mathbf{H}_1 = \mathbf{a}_y \frac{1}{\eta_1} (E_{i0} e^{-j\beta_1 z} - E_{r0} e^{j\beta_1 z}). \tag{8–157}$$

The electric and magnetic fields in region 2 can also be represented by combinations of forward and backward waves:

$$\mathbf{E}_2 = \mathbf{a}_x (E_2^+ e^{-j\beta_2 z} + E_2^- e^{j\beta_2 z}), \tag{8–158}$$

$$\mathbf{H}_2 = \mathbf{a}_y \frac{1}{\eta_2} (E_2^+ e^{-j\beta_2 z} - E_2^- e^{j\beta_2 z}). \tag{8–159}$$

In region 3, only a forward wave traveling in $+z$-direction exists. Thus,

$$\mathbf{E}_3 = \mathbf{a}_x E_3^+ e^{-j\beta_3 z}, \tag{8–160}$$

$$\mathbf{H}_3 = \mathbf{a}_y \frac{E_3^+}{\eta_3} e^{-j\beta_3 z}. \tag{8–161}$$

On the right side of Eqs. (8–156) through (8–161) there are a total of four unknown amplitudes: E_{r0}, E_2^+, E_2^-, and E_3^+. They can be determined by solving the four boundary-condition equations required by the continuity of the tangential components of the electric and magnetic fields.

At $z = 0$:

$$\mathbf{E}_1(0) = \mathbf{E}_2(0), \tag{8–162}$$

$$\mathbf{H}_1(0) = \mathbf{H}_2(0). \tag{8–163}$$

At $z = d$:

$$\mathbf{E}_2(d) = \mathbf{E}_3(d), \tag{8–164}$$

$$\mathbf{H}_2(d) = \mathbf{H}_3(d). \tag{8–165}$$

The procedure is straightforward and purely algebraic (Problem P.8–29). In the following subsections we introduce the concept of wave impedance and use it in an alternative approach for studying the problem of multiple reflections at normal incidence.

8–9.1 WAVE IMPEDANCE OF THE TOTAL FIELD

We define the *wave impedance of the total field* at any plane parallel to the plane boundary as the ratio of the total electric field intensity to the total magnetic field intensity. With a z-dependent uniform plane wave, as was shown in Fig. 8–15, we write, in general,

$$Z(z) = \frac{\text{Total } E_x(z)}{\text{Total } H_y(z)} \quad (\Omega). \tag{8–166}$$

For a single wave propagating in the $+z$-direction in an unbounded medium, the wave impedance equals the intrinsic impedance, η, of the medium; for a single wave traveling in the $-z$-direction, it is $-\eta$ for all z.

In the case of a uniform plane wave incident from medium 1 normally on a plane boundary with an infinite medium 2, such as that illustrated in Fig. 8–14 and discussed in Section 8–8, the magnitudes of the total electric and magnetic field intensities in medium 1 are, from Eqs. (8–144) and (8–149),

$$E_{1x}(z) = E_{i0}(e^{-j\beta_1 z} + \Gamma e^{j\beta_1 z}), \tag{8-167}$$

$$H_{1y}(z) = \frac{E_{i0}}{\eta_1}(e^{-j\beta_1 z} - \Gamma e^{j\beta_1 z}). \tag{8-168}$$

Their ratio defines the wave impedance of the total field in medium 1 at a distance z from the boundary plane:

$$Z_1(z) = \frac{E_{1x}(z)}{H_{1y}(z)} = \eta_1 \frac{e^{-j\beta_1 z} + \Gamma e^{j\beta_1 z}}{e^{-j\beta_1 z} - \Gamma e^{j\beta_1 z}}, \tag{8-169}$$

which is obviously a function of z.

A distance $z = -\ell$ to the left of the boundary plane,

$$Z_1(-\ell) = \frac{E_{1x}(-\ell)}{H_{1y}(-\ell)} = \eta_1 \frac{e^{j\beta_1 \ell} + \Gamma e^{-j\beta_1 \ell}}{e^{j\beta_1 \ell} - \Gamma e^{-j\beta_1 \ell}}. \tag{8-170}$$

Using the definition of $\Gamma = (\eta_2 - \eta_1)/(\eta_2 + \eta_1)$ in Eq. (8–170), we obtain

$$Z_1(-\ell) = \eta_1 \frac{\eta_2 \cos \beta_1 \ell + j\eta_1 \sin \beta_1 \ell}{\eta_1 \cos \beta_1 \ell + j\eta_2 \sin \beta_1 \ell}, \tag{8-171}$$

which correctly reduces to η_1 when $\eta_2 = \eta_1$. In that case there is no discontinuity at $z = 0$; hence there is no reflected wave and the total-field wave impedance is the same as the intrinsic impedance of the medium.

When we study transmission lines in the next chapter, we will find that Eqs. (8–170) and (8–171) are similar to the formulas for the input impedance of a transmission line of length ℓ that has a characteristic impedance η_1 and terminates in an impedance η_2. There is a close similarity between the behavior of the propagation of uniform plane waves at normal incidence and the behavior of transmission lines.

If the plane boundary is perfectly conducting, $\eta_2 = 0$ and $\Gamma = -1$, and Eq. (8–171) becomes

$$Z_1(-\ell) = j\eta_1 \tan \beta_1 \ell, \tag{8-172}$$

which is the same as the input impedance of a transmission line of length ℓ that has a characteristic impedance η_1 and terminates in a short circuit.

8–9.2 IMPEDANCE TRANSFORMATION WITH MULTIPLE DIELECTRICS

The concept of total-field wave impedance is very useful in solving problems with multiple dielectric interfaces such as the situation shown in Fig. 8–15. The total field in medium 2 is the result of multiple reflections of the two boundary planes at $z = 0$ and $z = d$; but it can be grouped into a wave traveling in the $+z$-direction and another traveling in the $-z$-direction. The wave impedance of the total field in medium

2 at the left-hand interface $z = 0$ can be found from the right side of Eq. (8–171) by replacing η_2 by η_3, η_1 by η_2, β_1 by β_2, and ℓ by d. Thus,

$$Z_2(0) = \eta_2 \frac{\eta_3 \cos \beta_2 d + j\eta_2 \sin \beta_2 d}{\eta_2 \cos \beta_2 d + j\eta_3 \sin \beta_2 d}. \qquad (8\text{–}173)$$

As far as the wave in medium 1 is concerned, it encounters a discontinuity at $z = 0$ and the discontinuity can be characterized by an infinite medium with an intrinsic impedance $Z_2(0)$ as given in Eq. (8–173). The effective reflection coefficient at $z = 0$ for the incident wave in medium 1 is

$$\Gamma_0 = \frac{E_{r0}}{E_{i0}} = -\frac{H_{r0}}{H_{i0}} = \frac{Z_2(0) - \eta_1}{Z_2(0) + \eta_1}. \qquad (8\text{–}174)$$

We note that Γ_0 differs from Γ only in that η_2 has been replaced by $Z_2(0)$. Hence the insertion of a dielectric layer of thickness d and intrinsic impedance η_2 in front of medium 3, which has intrinsic impedance η_3, has the effect of transforming η_3 to $Z_2(0)$. Given η_1 and η_3, Γ_0 can be adjusted by suitable choices of η_2 and d.

Once Γ_0 has been found from Eq. (8–174), E_{r0} of the reflected wave in medium 1 can be calculated: $E_{r0} = \Gamma_0 E_{i0}$. In many applications, Γ_0 and E_{r0} are the only quantities of interest; hence this impedance-transformation approach is conceptually simple and yields the desired answers in a direct manner. If the fields E_2^+, E_2^-, and E_t in media 2 and 3 are also desired, they can be determined from the boundary conditions at $z = 0$ and $z = d$, as indicated in Eqs. (8–162) through (8–165).

EXAMPLE 8-12 A dielectric layer of thickness d and intrinsic impedance η_2 is placed between media 1 and 3 having intrinsic impedances η_1 and η_3, respectively. Determine d and η_2 such that no reflection occurs when a uniform plane wave in medium 1 impinges normally on the interface with medium 2.

Solution With the dielectric layer interposed between media 1 and 3 as shown in Fig. 8–15, the condition of no reflection at interface $z = 0$ requires $\Gamma_0 = 0$, or $Z_2(0) = \eta_1$. From Eq. (8–173) we have

$$\eta_2(\eta_3 \cos \beta_2 d + j\eta_2 \sin \beta_2 d) = \eta_1(\eta_2 \cos \beta_2 d + j\eta_3 \sin \beta_2 d). \qquad (8\text{–}175)$$

Equating the real and imaginary parts separately, we require

$$\eta_3 \cos \beta_2 d = \eta_1 \cos \beta_2 d \qquad (8\text{–}176)$$

and

$$\eta_2^2 \sin \beta_2 d = \eta_1 \eta_3 \sin \beta_2 d. \qquad (8\text{–}177)$$

Equation (8–176) is satisfied if *either*

$$\eta_3 = \eta_1 \qquad (8\text{–}178)$$

or

$$\cos \beta_2 d = 0, \qquad (8\text{–}179)$$

which implies that

$$\beta_2 d = (2n + 1)\frac{\pi}{2}$$

or

$$d = (2n + 1)\frac{\lambda_2}{4}, \qquad n = 0, 1, 2, \ldots \qquad (8\text{–}180)$$

On the one hand, if condition (8–178) holds, Eq. (8–177) can be satisfied when either (a) $\eta_2 = \eta_3 = \eta_1$, which is the trivial case of no discontinuities at all, or (b) $\sin \beta_2 d = 0$, or $d = n\lambda_2/2$.

On the other hand, if relation (8–179) or (8–180) holds, $\sin \beta_2 d$ does not vanish, and Eq. (8–177) can be satisfied when $\eta_2 = \sqrt{\eta_1 \eta_3}$. We have then two possibilities for the condition of no reflection.

1. *When $\eta_3 = \eta_1$, we require*

$$d = n\frac{\lambda_2}{2}, \qquad n = 0, 1, 2, \ldots, \qquad (8\text{–}181)$$

that is, that the thickness of the dielectric layer be a multiple of a half-wavelength in the dielectric at the operating frequency. Such a dielectric layer is referred to as a **half-wave dielectric window**. Since $\lambda_2 = u_{p2}/f = 1/f\sqrt{\mu_2 \epsilon_2}$, where f is the operating frequency, a half-wave dielectric window is a narrow-band device.

2. *When $\eta_3 \neq \eta_1$, we require*

$$\eta_2 = \sqrt{\eta_1 \eta_3} \qquad (8\text{–}182\text{a})$$

and

$$d = (2n + 1)\frac{\lambda_2}{4}, \qquad n = 0, 1, 2, \ldots. \qquad (8\text{–}182\text{b})$$

When media 1 and 3 are different, η_2 should be the geometric mean of η_1 and η_3, and d should be an odd multiple of a quarter wavelength in the dielectric layer at the operating frequency in order to eliminate reflection. Under these conditions the dielectric layer (medium 2) acts like a **quarter-wave impedance transformer**. We will refer to this term again when we study analogous transmission-line problems in Chapter 9.

We see from the above that if a radome is to be constructed around a radar installation ($\eta_1 = \eta_3 = \eta_0$), it should be a half-wave window in order to minimize reflection; that is, it should be a multiple of $\lambda_2/2 \, (= 1/2f_2\sqrt{\mu_2 \epsilon_2})$ thick at the operating radar frequency f_2, where μ_2 and ϵ_2 are the permeability and permittivity, respectively, of the radome material.

8–10 Oblique Incidence at a Plane Dielectric Boundary

We now consider the case of a plane wave that is incident obliquely at an arbitrary angle of incidence θ_i on a plane interface between two dielectric media. The media are assumed to be lossless and to have different constitutive parameters (ϵ_1, μ_1) and

(ϵ_2, μ_2), as indicated in Fig. 8–16. Because of the medium's discontinuity at the interface, a part of the incident wave is reflected and a part is transmitted. Lines AO, $O'A'$, and $O'B$ are the intersections of the wavefronts (surfaces of constant phase) of the incident, reflected, and transmitted waves respectively, with the plane of incidence. Since both the incident and the reflected waves propagate in medium 1 with the same phase velocity u_{p1}, the distances $\overline{O'A'}$ and $\overline{AO'}$ must be equal. Thus,

$$\overline{OO'} \sin \theta_r = \overline{OO'} \sin \theta_i$$

or

$$\boxed{\theta_r = \theta_i.} \tag{8–183}$$

Equation (8–183) assures us that *the angle of reflection is equal to the angle of incidence*, which is *Snell's law of reflection*.

In medium 2 the time it takes for the transmitted wave to travel from O to B equals the time for the incident wave to travel from A to O'. We have

$$\frac{\overline{OB}}{u_{p2}} = \frac{\overline{AO'}}{u_{p1}},$$

$$\frac{\overline{OB}}{\overline{AO'}} = \frac{\overline{OO'} \sin \theta_t}{\overline{OO'} \sin \theta_i} = \frac{u_{p2}}{u_{p1}},$$

from which we obtain

$$\boxed{\frac{\sin \theta_t}{\sin \theta_i} = \frac{u_{p2}}{u_{p1}} = \frac{\beta_1}{\beta_2} = \frac{n_1}{n_2},} \tag{8–184}$$

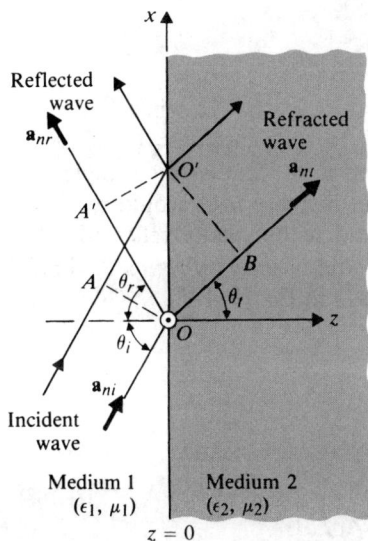

FIGURE 8–16
Uniform plane wave incident obliquely on a plane dielectric boundary.

where n_1 and n_2 are the indices of refraction for media 1 and 2, respectively. The *index of refraction* of a medium is the ratio of the speed of light (electromagnetic wave) in free space to that in the medium; that is, $n = c/u_p$. The relation in Eq. (8–184) is known as *Snell's law of refraction*. It states that *at an interface between two dielectric media, the ratio of the sine of the angle of refraction (transmission) in medium 2 to the sine of the angle of incidence in medium 1 is equal to the inverse ratio of indices of refraction n_1/n_2.*

For *nonmagnetic media*, $\mu_1 = \mu_2 = \mu_0$, Eq. (8–184) becomes

$$\boxed{\frac{\sin \theta_t}{\sin \theta_i} = \sqrt{\frac{\epsilon_1}{\epsilon_2}} = \sqrt{\frac{\epsilon_{r1}}{\epsilon_{r2}}} = \frac{n_1}{n_2} = \frac{\eta_2}{\eta_1},} \qquad (8\text{--}185)$$

where η_1 and η_2 are the intrinsic impedances of the media. Furthermore, if medium 1 is free space such that $\epsilon_{r1} = 1$ and $n_1 = 1$, Eq. (8–185) reduces to

$$\boxed{\frac{\sin \theta_t}{\sin \theta_i} = \frac{1}{\sqrt{\epsilon_{r2}}} = \frac{1}{n_2} = \frac{\eta_2}{120\pi}.} \qquad (8\text{--}186)$$

Since $n_2 \geq 1$, it is clear that a plane wave incident obliquely at an interface with a denser medium will be bent toward the normal.

We have derived here Snell's law of reflection and Snell's law of refraction from a consideration of the ray paths of the incident, reflected, and refracted waves. No mention has been made of the polarization of the waves. Thus Snell's laws are independent of wave polarization. These laws can also be derived by matching the phases of the various propagating waves at the boundary surface $z = 0$, as we shall see when we take up the cases of perpendicular polarization (Subsection 8–10.2) and parallel polarization (Subsection 8–10.3).

8–10.1 TOTAL REFLECTION

Let us now examine Snell's law in Eq. (8–185) for $\epsilon_1 > \epsilon_2$—that is, when the wave in medium 1 is incident on a less dense medium 2. In that case, $\theta_t > \theta_i$. Since θ_t increases with θ_i, an interesting situation airses when $\theta_t = \pi/2$, at which angle the refracted wave will glaze along the interface; a further increase in θ_i would result in no refracted wave, and the incident wave is then said to be totally reflected. The angle of incidence θ_c (which corresponds to the threshold of *total reflection* $\theta_t = \pi/2$) is called the *critical angle*. We have, by setting $\theta_t = \pi/2$ in Eq. (8–185),

$$\sin \theta_c = \sqrt{\frac{\epsilon_2}{\epsilon_1}} \qquad (8\text{--}187)$$

or

$$\boxed{\theta_c = \sin^{-1} \sqrt{\frac{\epsilon_2}{\epsilon_1}} = \sin^{-1} \left(\frac{n_2}{n_1}\right).} \qquad (8\text{--}188)$$

This situation is illustrated in Fig. 8–17, where $\mathbf{a}_{ni}$, $\mathbf{a}_{nr}$, and $\mathbf{a}_{nt}$ are unit vectors denoting the directions of propagation of the incident, reflected, and transmitted waves, respectively.

What happens mathematically if θ_i is larger than the critical angle θ_c ($\sin \theta_i > \sin \theta_c = \sqrt{\epsilon_2/\epsilon_1}$)? From Eq. (8–185) we have

$$\sin \theta_t = \sqrt{\frac{\epsilon_1}{\epsilon_2}} \sin \theta_i > 1, \tag{8–189}$$

which does not yield a real solution for θ_t. Although $\sin \theta_t$ in Eq. (8–189) is still real, $\cos \theta_t$ becomes imaginary when $\sin \theta_t > 1$:

$$\cos \theta_t = \sqrt{1 - \sin^2 \theta_t} = \pm j \sqrt{\frac{\epsilon_1}{\epsilon_2} \sin^2 \theta_i - 1}. \tag{8–190}$$

In medium 2 the unit vector $\mathbf{a}_{nt}$ in the direction of propagation of a typical transmitted (refracted) wave, as shown in Fig. 8–16, is

$$\mathbf{a}_{nt} = \mathbf{a}_x \sin \theta_t + \mathbf{a}_z \cos \theta_t. \tag{8–191}$$

Both $\mathbf{E}_t$ and $\mathbf{H}_t$ vary spatially in accordance with the following factor:

$$e^{-j\beta_2 \mathbf{a}_{nt} \cdot \mathbf{R}} = e^{-j\beta_2(x \sin \theta_t + z \cos \theta_t)},$$

which, when Eqs. (8–189) and (8–190) for $\theta_i > \theta_c$ are used, becomes

$$e^{-\alpha_2 z} e^{-j\beta_{2x} x}, \tag{8–192}$$

where

$$\alpha_2 = \beta_2 \sqrt{(\epsilon_1/\epsilon_2) \sin^2 \theta_i - 1}$$

and

$$\beta_{2x} = \beta_2 \sqrt{\epsilon_1/\epsilon_2} \sin \theta_i.$$

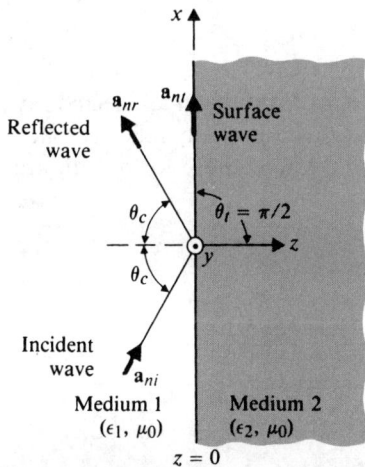

FIGURE 8–17
Plane wave incident at critical angle, $\epsilon_1 > \epsilon_2$.

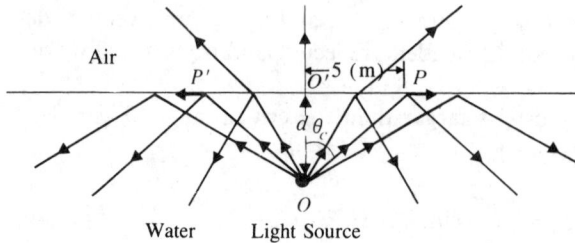

FIGURE 8–18
An underwater light source (Example 8–13).

The upper sign in Eq. (8–190) has been abandoned because it would lead to the impossible result of an increasing field as z increases. We can conclude from (8–192) that for $\theta_i > \theta_c$ an **evanescent wave** exists *along* the interface (in the x-direction), which is attenuated exponentially (rapidly) in medium 2 in the normal direction (z-direction). This wave is tightly bound to the interface and is called a **surface wave**. It is illustrated in Fig. 8–17. Obviously, it is a nonuniform plane wave. No power is transmitted into medium 2 under these conditions. (See Problem P.8–37.)

EXAMPLE 8–13   The permittivity of water at optical frequencies is $1.75\epsilon_0$. It is found that an isotropic light source at a distance d under water yields an illuminated circular area of a radius 5 (m). Determine d.

Solution   The index of refraction of water is $n_w = \sqrt{1.75} = 1.32$. Refer to Fig. 8–18. The radius of illuminated area, $\overline{O'P} = 5$ (m), corresponds to the critical angle

$$\theta_c = \sin^{-1}\left(\frac{1}{n_w}\right) = \sin^{-1}\left(\frac{1}{1.32}\right) = 49.2°.$$

Hence,

$$d = \frac{\overline{O'P}}{\tan \theta_c} = \frac{5}{\tan 49.2°} = 4.32 \quad (\text{m}).$$

As illustrated in Fig. 8–18, an incident ray with $\theta_i = \theta_c$ at P results in a reflected ray and a tangential refracted ray. Incident waves for $\theta_i < \theta_c$ are partially reflected back into the water and partially refracted into the air above, and those for $\theta_i > \theta_c$ are totally reflected (the evanescent surface waves are not shown).

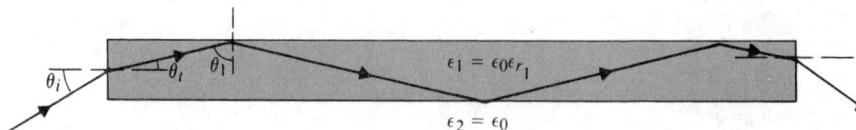

FIGURE 8–19
Dielectric rod or fiber guiding electromagnetic wave by total internal reflection.

EXAMPLE 8–14   A dielectric rod or fiber of a transparent material can be used to guide light or an electromagnetic wave under the conditions of total internal reflection. Determine the minimum dielectric constant of the guiding medium so that a wave incident on one end at any angle will be confined within the rod until it emerges from the other end.

Solution   Refer to Fig. 8–19. For total internal reflection, θ_1 must be greater than or equal to θ_c for the guiding dielectric medium; that is,

$$\sin \theta_1 \geq \sin \theta_c$$

or, since $\theta_1 = \pi/2 - \theta_t$,

$$\cos \theta_t \geq \sin \theta_c. \tag{8–193}$$

From Snell's law of refraction, Eq. (8–186), we have

$$\sin \theta_t = \frac{1}{\sqrt{\epsilon_{r1}}} \sin \theta_i. \tag{8–194}$$

It is important to note here that the dielectric medium has been designated as medium 1 (the denser medium) in order to be consistent with the notation of this subsection. Combining Eqs. (8–193), (8–194), and (8–187), we obtain

$$\sqrt{1 - \frac{1}{\epsilon_{r1}} \sin^2 \theta_i} \geq \sqrt{\frac{\epsilon_0}{\epsilon_1}} = \frac{1}{\sqrt{\epsilon_{r1}}},$$

which requires

$$\epsilon_{r1} \geq 1 + \sin^2 \theta_i. \tag{8–195}$$

Since the largest value of the right side of (8–195) is reached when $\theta_i = \pi/2$, we require the dielectric constant of the guiding medium to be at least 2, which corresponds to an index of refraction $n_1 = \sqrt{2}$. This requirement is satisfied by glass and quartz.   ■

We observe that Snell's law of refraction in Eq. (8–185) and the critical angle for total reflection in Eq. (8–188) are independent of the polarization of the incident electric field. The formulas for the reflection and transmission coefficients, however, are polarization-dependent. In the following two subsections we discuss perpendicular polarization and parallel polarization separately.

8–10.2   PERPENDICULAR POLARIZATION

For oblique incidence with perpendicular polarization we refer to Fig. 8–20. The incident electric and magnetic field intensity phasors in medium 1 are, from Eqs. (8–107) and (8–108),

$$\mathbf{E}_i(x, z) = \mathbf{a}_y E_{i0} e^{-j\beta_1 (x \sin \theta_i + z \cos \theta_i)} \tag{8–196}$$

$$\mathbf{H}_i(x, z) = \frac{E_{i0}}{\eta_1} (-\mathbf{a}_x \cos \theta_i + \mathbf{a}_z \sin \theta_i) e^{-j\beta_1 (x \sin \theta_i + z \cos \theta_i)}. \tag{8–197}$$

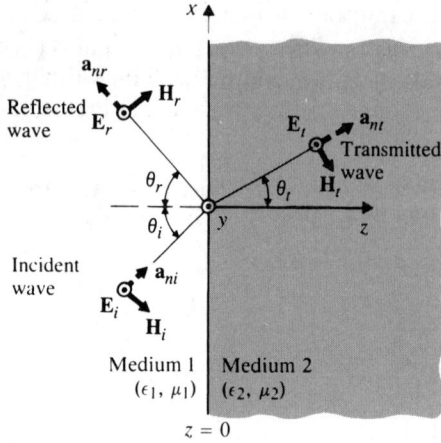

FIGURE 8–20
Plane wave incident on a plane dielectric boundary (perpendicular polarization).

The reflected electric and magnetic fields can be obtained from Eqs. (8–110) and (8–112), but remember that E_{r0} is no longer equal to $-E_{i0}$:

$$\mathbf{E}_r(x, z) = \mathbf{a}_y E_{r0} e^{-j\beta_1(x \sin \theta_r - z \cos \theta_r)} \tag{8–198}$$

$$\mathbf{H}_r(x, z) = \frac{E_{r0}}{\eta_1}(\mathbf{a}_x \cos \theta_r + \mathbf{a}_z \sin \theta_r)e^{-j\beta_1(x \sin \theta_r - z \cos \theta_r)}. \tag{8–199}$$

In medium 2 the transmitted electric and magnetic field intensity phasors can be similarly written as

$$\mathbf{E}_t(x, z) = \mathbf{a}_y E_{t0} e^{-j\beta_2(x \sin \theta_t + z \cos \theta_t)} \tag{8–200}$$

$$\mathbf{H}_t(x, z) = \frac{E_{t0}}{\eta_2}(-\mathbf{a}_x \cos \theta_t + \mathbf{a}_z \sin \theta_t)e^{-j\beta_2(x \sin \theta_t + z \cos \theta_t)}. \tag{8–201}$$

There are four unknown quantities in Eqs. (8–196) through (8–201), namely, E_{r0}, E_{t0}, θ_r, and θ_t. Their determination follows from the requirements that the tangential components of $\mathbf{E}$ and $\mathbf{H}$ be continuous at the boundary $z = 0$. From $E_{iy}(x, 0) + E_{ry}(x, 0) = E_{ty}(x, 0)$ we have

$$E_{i0}e^{-j\beta_1 x \sin \theta_i} + E_{r0}e^{-j\beta_1 x \sin \theta_r} = E_{t0}e^{-j\beta_2 x \sin \theta_t}. \tag{8–202}$$

Similarly, from $H_{ix}(x, 0) + H_{rx}(x, 0) = H_{tx}(x, 0)$ we require

$$\frac{1}{\eta_1}(-E_{i0} \cos \theta_i e^{-j\beta_1 x \sin \theta_i} + E_{r0} \cos \theta_r e^{-j\beta_1 x \sin \theta_r}) = -\frac{E_{t0}}{\eta_2} \cos \theta_t e^{-j\beta_2 x \sin \theta_t}. \tag{8–203}$$

Because Eqs. (8–202) and (8–203) are to be satisfied *for all x*, all three exponential factors that are functions of x must be equal ("phase-matching"). Thus,

$$\beta_1 x \sin \theta_i = \beta_1 x \sin \theta_r = \beta_2 x \sin \theta_t,$$

which leads to Snell's law of reflection ($\theta_r = \theta_i$) and Snell's law of refraction ($\sin \theta_t / \sin \theta_i = \beta_1/\beta_2 = n_1/n_2$). Equations (8–202) and (8–203) can now be written simply as

$$E_{i0} + E_{r0} = E_{t0} \tag{8–204}$$

and

$$\frac{1}{\eta_1}(E_{i0} - E_{r0}) \cos \theta_i = \frac{E_{t0}}{\eta_2} \cos \theta_t, \tag{8–205}$$

from which E_{r0} and E_{t0} can be found in terms of E_{i0}. We have

$$
\Gamma_\perp = \frac{E_{r0}}{E_{i0}} = \frac{\eta_2 \cos \theta_i - \eta_1 \cos \theta_t}{\eta_2 \cos \theta_i + \eta_1 \cos \theta_t}
$$
$$
= \frac{(\eta_2/\cos \theta_t) - (\eta_1/\cos \theta_i)}{(\eta_2/\cos \theta_t) + (\eta_1/\cos \theta_i)} \tag{8–206}†
$$

and

$$
\tau_\perp = \frac{E_{t0}}{E_{i0}} = \frac{2\eta_2 \cos \theta_i}{\eta_2 \cos \theta_i + \eta_1 \cos \theta_t}
$$
$$
= \frac{2(\eta_2/\cos \theta_t)}{(\eta_2/\cos \theta_t) + (\eta_1/\cos \theta_i)}. \tag{8–207}†
$$

Comparing these expressions with the formulas for the reflection and transmission coefficients at normal incidence, Eqs. (8–140) and (8–141), we see that the same formulas apply if η_1 and η_2 are changed to ($\eta_1/\cos \theta_i$) and ($\eta_2/\cos \theta_t$), respectively. When $\theta_i = 0$, making $\theta_r = \theta_t = 0$, these expressions reduce to those for normal incidence, as they should. Furthermore, $\Gamma_\perp$ and $\tau_\perp$ are related in the following way:

$$\boxed{1 + \Gamma_\perp = \tau_\perp,} \tag{8–208}$$

which is similar to Eq. (8–142) for normal incidence.

If medium 2 is a perfect conductor, $\eta_2 = 0$. We have $\Gamma_\perp = -1 \,(E_{r0} = -E_{i0})$ and $\tau_\perp = 0 \,(E_{t0} = 0)$. The tangential **E** field on the surface of the conductor vanishes, and no energy is transmitted across a perfectly conducting boundary, as we have noted in Sections 8–6 and 8–7.

Noting that the numerator for the reflection coefficient in Eq. (8–206) is in the form of a difference of two terms, we inquire whether there is a combination of η_1, η_2, and θ_i, which makes $\Gamma_\perp = 0$ for no reflection. Denoting this particular θ_i by $\theta_{B\perp}$, we require

$$\eta_2 \cos \theta_{B\perp} = \eta_1 \cos \theta_t. \tag{8–209}$$

Using Snell's law of refraction, we have

$$\cos \theta_t = \sqrt{1 - \sin^2 \theta_t} = \sqrt{1 - \frac{n_1^2}{n_2^2} \sin^2 \theta_i} \tag{8–210}$$

---

† These are sometimes referred to as **Fresnel's equations**.

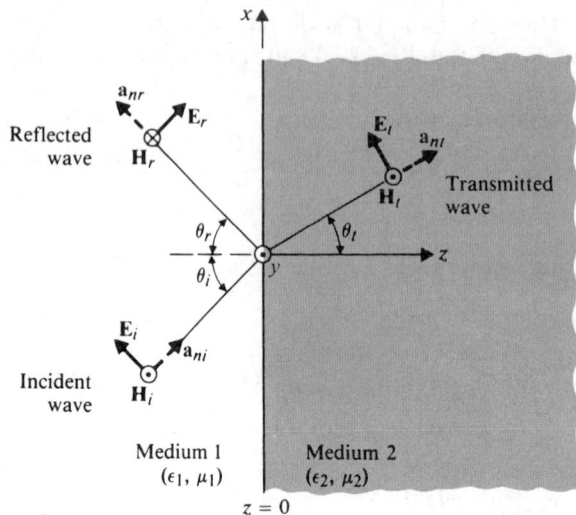

FIGURE 8–21
Plane wave incident obliquely on a plane
dielectric boundary (parallel polarization).

and obtain from Eq. (8–209)

$$\sin^2 \theta_{B\perp} = \frac{1 - \mu_1\epsilon_2/\mu_2\epsilon_1}{1 - (\mu_1/\mu_2)^2}. \tag{8–211}$$

The angle $\theta_{B\perp}$ is called the **Brewster angle** of no reflection for the case of perpendicular polarization. For *nonmagnetic media*, $\mu_1 = \mu_2 = \mu_0$, the right side of Eq. (8–211) becomes infinite, and $\theta_{B\perp}$ *does not exist*. In the case of $\epsilon_1 = \epsilon_2$ and $\mu_1 \neq \mu_2$, Eq. (8–211) reduces to

$$\sin \theta_{B\perp} = \frac{1}{\sqrt{1 + (\mu_1/\mu_2)}}, \tag{8–212}$$

which does have a solution whether μ_1/μ_2 is greater or less than unity. However, it is a very rare situation in electromagnetics that two contiguous media have the same permittivity but different permeabilities.

8–10.3   PARALLEL POLARIZATION

When a uniform plane wave with parallel polarization is incident obliquely on a plane boundary, as illustrated in Fig. 8–21, the incident and reflected electric and magnetic field intensity phasors in medium 1 are, from Eqs. (8–124) through (8–127):

$$\mathbf{E}_i(x, z) = E_{i0}(\mathbf{a}_x \cos \theta_i - \mathbf{a}_z \sin \theta_i)e^{-j\beta_1(x \sin \theta_i + z \cos \theta_i)} \tag{8–213}$$

$$\mathbf{H}_i(x, z) = \mathbf{a}_y \frac{E_{i0}}{\eta_1} e^{-j\beta_1(x \sin \theta_i + z \cos \theta_i)} \tag{8–214}$$

$$\mathbf{E}_r(x, z) = E_{r0}(\mathbf{a}_x \cos \theta_r + \mathbf{a}_z \sin \theta_r)e^{-j\beta_1(x \sin \theta_r - z \cos \theta_r)}, \tag{8-215}$$

$$\mathbf{H}_r(x, z) = -\mathbf{a}_y \frac{E_{r0}}{\eta_1} e^{-j\beta_1(x \sin \theta_r - z \cos \theta_r)}. \tag{8-216}$$

The transmitted electric and magnetic field intensity phasors in medium 2 are

$$\mathbf{E}_t(x, z) = E_{t0}(\mathbf{a}_x \cos \theta_t - \mathbf{a}_z \sin \theta_t)e^{-j\beta_2(x \sin \theta_t + z \cos \theta_t)}, \tag{8-217}$$

$$\mathbf{H}_t(x, z) = \mathbf{a}_y \frac{E_{t0}}{\eta_2} e^{-j\beta_2(x \sin \theta_t + z \cos \theta_t)}. \tag{8-218}$$

Continuity requirements for the tangential components of $\mathbf{E}$ and $\mathbf{H}$ at $z = 0$ lead again to Snell's laws of reflection and refraction, as well as to the following two equations:

$$(E_{i0} + E_{r0}) \cos \theta_i = E_{t0} \cos \theta_t, \tag{8-219}$$

$$\frac{1}{\eta_1}(E_{i0} - E_{r0}) = \frac{1}{\eta_2} E_{t0}. \tag{8-220}$$

Solving for E_{r0} and E_{t0} in terms of E_{i0}, we obtain

$$\Gamma_{\parallel} = \frac{E_{r0}}{E_{i0}} = \frac{\eta_2 \cos \theta_t - \eta_1 \cos \theta_i}{\eta_2 \cos \theta_t + \eta_1 \cos \theta_i} \tag{8-221}†$$

and

$$\tau_{\parallel} = \frac{E_{t0}}{E_{i0}} = \frac{2\eta_2 \cos \theta_i}{\eta_2 \cos \theta_t + \eta_1 \cos \theta_i}. \tag{8-222}†$$

It is easy to verify that

$$\boxed{1 + \Gamma_{\parallel} = \tau_{\parallel}\left(\frac{\cos \theta_t}{\cos \theta_i}\right).} \tag{8-223}$$

Equation (8–223) is seen to be different from Eq. (8–208) for perpendicular polarization except when $\theta_i = \theta_t = 0$, which is the case for normal incidence. At normal incidence, $\Gamma_{\parallel}$ and $\tau_{\parallel}$ reduce to Γ and τ given in Eqs. (8–140) and (8–141), respectively, as did $\Gamma_{\perp}$ and $\tau_{\perp}$.

If medium 2 is a perfect conductor ($\eta_2 = 0$), Eqs. (8–221) and (8–222) simplify to $\Gamma_{\parallel} = -1$ and $\tau_{\parallel} = 0$, respectively, making the tangential component of the total $\mathbf{E}$ field on the surface of the conductor vanish, as expected. We note here that the choices of the reference directions of $\mathbf{E}_r$ and $\mathbf{H}_r$ in Figs. 8–11, 8–13, 8–20, and 8–21 are all arbitrary. The actual directions of $\mathbf{E}_r$ and $\mathbf{H}_r$ in Figs. 8–11 and 8–13 are opposite to those chosen because $E_{r0} = -E_{i0}$. In Figs. 8–20 and 8–21 the actual directions of $\mathbf{E}_r$ and $\mathbf{H}_r$ may or may not be the same as those shown, depending on whether $\Gamma_{\perp}$ in Eq. (8–206) and $\Gamma_{\parallel}$ in Eq. (8–221) is positive or negative, respectively.

If we plot $|\Gamma_{\perp}|^2$ and $|\Gamma_{\parallel}|^2$ versus θ_i, we will find the former always greater than

---

† These are also referred to as *Fresnel's equations*.

the latter except for $\theta_i = 0$. This means that when an unpolarized wave strikes a plane dielectric interface, the reflected wave will contain more power in the component with perpendicular polarization than that with parallel polarization. A popular application of this fact is the design of Polaroid sunglasses to reduce sun glare. Much of the sunlight received by the eye has been reflected from horizontal surfaces on earth. Because $|\Gamma_\perp|^2 > |\Gamma_\|\|^2$, the light reaching the eye is predominantly perpendicular to the plane of reflection (same as the plane of incidence), and hence the electric field is parallel to the earth's surface. Polaroid sunglasses are designed to filter out this component.

From Eq. (8–221) we find that $\Gamma_\|$ goes to zero when the angle of incidence θ_i equals $\theta_{B\|}$, such that

$$\eta_2 \cos \theta_t = \eta_1 \cos \theta_{B\|}, \qquad (8\text{–}224)$$

which, together with Eq. (8–210), requires that

$$\boxed{\sin^2 \theta_{B\|} = \frac{1 - \mu_2\epsilon_1/\mu_1\epsilon_2}{1 - (\epsilon_1/\epsilon_2)^2}.} \quad (\mu_1 = \mu_2) \qquad (8\text{–}225)$$

The angle $\theta_{B\|}$ is known as the **Brewster angle** of no reflection for the case of parallel polarization. A solution for Eq. (8–225) always exists for two contiguous nonmagnetic media. Thus if $\mu_1 = \mu_2$, a reflection-free condition is obtained when the angle of incidence in medium 1 equals the Brewster angle $\theta_{B\|}$, such that

$$\boxed{\sin \theta_{B\|} = \frac{1}{\sqrt{1 + (\epsilon_1/\epsilon_2)}}.} \quad (\mu_1 = \mu_2) \qquad (8\text{–}226)$$

An alternative form for Eq. (8–226) is

$$\boxed{\theta_{B\|} = \tan^{-1} \sqrt{\frac{\epsilon_2}{\epsilon_1}} = \tan^{-1} \left(\frac{n_2}{n_1}\right).} \quad (\mu_1 = \mu_2) \qquad (8\text{–}227)$$

Because of the difference in the formulas for Brewster angles for perpendicular and parallel polarizations, it is possible to separate these two types of polarization in an unpolarized wave. When an unpolarized wave such as random light is incident upon a boundary at the Brewster angle $\theta_{B\|}$ given by Eq. (8–225), only the component with perpendicular polarization will be reflected. Thus a Brewster angle is also referred to as a **polarizing angle**. Based on this principle, quartz windows set at the Brewster angle at the ends of a laser tube are used to control the polarization of an emitted light beam.

━━━ **EXAMPLE 8–15**   The dielectric constant of pure water is 80. (a) Determine the Brewster angle for parallel polarization, $\theta_{B\|}$, and the corresponding angle of transmission. (b) A plane wave with perpendicular polarization is incident from air on water surface at $\theta_i = \theta_{B||}$. Find the reflection and transmission coefficients.

Solution

a) The Brewster angle of no reflection for parallel polarization can be obtained directly from Eq. (8–226):

$$\theta_{B||} = \sin^{-1} \frac{1}{\sqrt{1 + (1/\epsilon_{r2})}}$$

$$= \sin^{-1} \frac{1}{\sqrt{1 + (1/80)}} = 81.0°.$$

The corresponding angle of transmission is, from Eq. (8–186),

$$\theta_t = \sin^{-1} \left(\frac{\sin \theta_{B||}}{\sqrt{\epsilon_{r2}}} \right) = \sin^{-1} \left(\frac{1}{\sqrt{\epsilon_{r2} + 1}} \right)$$

$$= \sin^{-1} \left(\frac{1}{\sqrt{81}} \right) = 6.38°.$$

b) For an incident wave with perpendicular polarization, we use Eqs. (8–206) and (8–207) to find $\Gamma_\perp$ and $\tau_\perp$ at $\theta_i = 81.0°$ and $\theta_t = 6.38°$:

$$\eta_1 = 377 \ (\Omega), \qquad\qquad \eta_1/\cos \theta_i = 2410 \ (\Omega),$$

$$\eta_2 = \frac{377}{\sqrt{\epsilon_{r2}}} = 40.1 \ (\Omega), \qquad \eta_2/\cos \theta_t = 40.4 \ (\Omega).$$

Thus,

$$\Gamma_\perp = \frac{40.4 - 2410}{40.4 + 2410} = -0.967,$$

$$\tau_\perp = \frac{2 \times 40.4}{40.4 + 2410} = 0.033.$$

We note that the relation between $\Gamma_\perp$ and $\tau_\perp$ given in Eq. (8–208) is satisfied.

■

Review Questions

R.8–1 Define *uniform plane wave.*

R.8–2 What is a *wavefront?*

R.8–3 Write the homogeneous vector Helmholtz's equation for **E** in free space.

R.8–4 Define *wavenumber.* How is wavenumber related to wavelength?

R.8–5 Define *phase velocity.*

R.8–6 Define *intrinsic impedance* of a medium. What is the value of the intrinsic impedance of free space?

R.8–7 What is *Doppler effect*?

R.8–8 What is a TEM wave?

R.8–9 Write the phasor expressions for the electric and magnetic field intensity vectors of an x-polarized uniform plane wave propagating in the $+z$-direction.

R.8–10 What is meant by the *polarization* of a wave? When is a wave linearly polarized? Circularly polarized?

R.8–11 Two orthogonal linearly polarized waves are combined. State the conditions under which the resultant will be (a) another linearly polarized wave, (b) a circularly polarized wave, and (c) an elliptically polarized wave.

R.8–12 How is the E-field from AM broadcast stations polarized? From television stations? From FM broadcast stations?

R.8–13 Define (a) *propagation constant*, (b) *attenuation constant*, and (c) *phase constant*.

R.8–14 What is meant by the *skin depth* of a conductor? How is it related to the attenuation constant? How does it depend on σ? On f?

R.8–15 What is the constitution of the ionosphere?

R.8–16 What is a *plasma*?

R.8–17 What is the significance of *plasma frequency*?

R.8–18 When does the equivalent permittivity of the ionosphere become negative? What is the significance of a negative permittivity in terms of wave propagation?

R.8–19 What is meant by the *dispersion* of a signal? Give an example of a dispersive medium.

R.8–20 Define *group velocity*. In what ways is group velocity different from phase velocity?

R.8–21 Define *Poynting vector*. What is the SI unit for this vector?

R.8–22 State Poynting's theorem.

R.8–23 For a time-harmonic electromagnetic field, write the expressions in terms of electric and magnetic field intensity vectors for (a) instantaneous Poynting vector, and (b) time-average Poynting vector.

R.8–24 What is a *standing wave*?

R.8–25 What do we know about the magnitude of the tangential components of **E** and **H** at the interface when a wave impinges normally on a perfectly conducting plane boundary?

R.8–26 Define *plane of incidence*.

R.8–27 What do we mean when we say that an incident wave has (a) perpendicular polarization, and (b) parallel polarization?

R.8–28 Define *reflection coefficient* and *transmission coefficient*. What is the relationship between them?

R.8–29 Under what conditions will reflection and transmission coefficients be real?

R.8–30 What are the values of the reflection and transmission coefficients at an interface with a perfectly conducting boundary?

R.8–31 A plane wave originating in medium 1 (ϵ_1, $\mu_1 = \mu_0$, $\sigma_1 = 0$) is incident normally on a plane interface with medium 2 ($\epsilon_2 \neq \epsilon_1$, $\mu_2 = \mu_0$, $\sigma_2 = 0$). Under what condition will the electric field at the interface be a maximum? A minimum?

R.8–32 Define *standing-wave ratio*. What is its relationship with reflection coefficient?

R.8–33 What is meant by the wave impedance of the total field. When is this impedance equal to the intrinsic impedance of the medium?

R.8–34 A thin dielectric coating is sprayed on optical instruments to reduce glare. What factors determine the thickness of the coating?

R.8–35 How should the thickness of the radome in a radar installation be chosen?

R.8–36 State *Snell's law of reflection.*

R.8–37 State *Snell's law of refraction.*

R.8–38 Define *critical angle.* When does it exist at an interface of two nonmagnetic media?

R.8–39 Define *Brewster angle.* When does it exist at an interface of two nonmagnetic media?

R.8–40 Why is a Brewster angle also called a *polarizing angle?*

R.8–41 Under what conditions will the reflection and transmission coefficients for perpendicular polarization be the same as those for parallel polarization?

Problems

P.8–1 Obtain the wave equations governing the **E** and **H** fields in a source-free conducting medium with constitutive parameters ϵ, μ, and σ.

P.8–2 Prove that the electric field intensity in Eq. (8–22) satisfies the homogeneous Helmholtz's equation provided that the condition in Eq. (8–23) is satisfied.

P.8–3 A Doppler radar is used to determine the speed of a moving vehicle by measuring the frequency shift of the wave reflected from the vehicle.

    a) Assuming that the reflecting surface of the vehicle can be represented by a perfectly conducting plane and that the transmitted signal is a time-harmonic uniform plane wave of a frequency f incident normally on the reflecting surface, find the relation between the frequency shift Δf and the speed u of the vehicle.

    b) Determine u both in (km/hr) and in (miles/hr) if $\Delta f = 2.33$ (kHz) with $f = 10.5$ (GHz).

P.8–4 For a harmonic uniform plane wave propagating in a simple medium, both **E** and **H** vary in accordance with the factor $\exp(-j\mathbf{k} \cdot \mathbf{R})$ as indicated in Eq. (8–26). Show that the four Maxwell's equations for uniform plane wave in a source-free region reduce to the following:

$$\mathbf{k} \times \mathbf{E} = \omega\mu\mathbf{H},$$
$$\mathbf{k} \times \mathbf{H} = -\omega\epsilon\mathbf{E},$$
$$\mathbf{k} \cdot \mathbf{E} = 0,$$
$$\mathbf{k} \cdot \mathbf{H} = 0.$$

P.8–5 The instantaneous expression for the magnetic field intensity of a uniform plane wave propagating in the $+y$ direction in air is given by

$$\mathbf{H} = \mathbf{a}_z 4 \times 10^{-6} \cos\left(10^7 \pi t - k_0 y + \frac{\pi}{4}\right) \quad \text{(A/m)}.$$

    a) Determine k_0 and the location where H_z vanishes at $t = 3$ (ms).

    b) Write the instantaneous expression for **E**.

P.8–6 The E-field of a uniform plane wave propagating in a dielectric medium is given by

$$E(t, z) = \mathbf{a}_x 2 \cos(10^8 t - z/\sqrt{3}) - \mathbf{a}_y \sin(10^8 t - z/\sqrt{3}) \quad \text{(V/m)}.$$

a) Determine the frequency and wavelength of the wave.

b) What is the dielectric constant of the medium?

c) Describe the polarization of the wave.

d) Find the corresponding H-field.

P.8–7 Show that a plane wave with an instantaneous expression for the electric field

$$\mathbf{E}(z, t) = \mathbf{a}_x E_{10} \sin (\omega t - kz) + \mathbf{a}_y E_{20} \sin (\omega t - kz + \psi)$$

is elliptically polarized. Find the polarization ellipse.

P.8–8 Prove the following:

a) An elliptically polarized plane wave can be resolved into right-hand and left-hand circularly polarized waves.

b) A circularly polarized plane wave can be obtained from a superposition of two oppositely directed elliptically polarized waves.

P.8–9 Derive the following general expressions of the attenuation and phase constants for conducting media:

$$\alpha = \omega \sqrt{\frac{\mu\epsilon}{2}} \left[\sqrt{1 + \left(\frac{\sigma}{\omega\epsilon}\right)^2} - 1 \right]^{1/2} \quad \text{(Np/m)}.$$

$$\beta = \omega \sqrt{\frac{\mu\epsilon}{2}} \left[\sqrt{1 + \left(\frac{\sigma}{\omega\epsilon}\right)^2} + 1 \right]^{1/2} \quad \text{(rad/m)}.$$

P.8–10 Determine and compare the intrinsic impedance, attentuation constant (in both Np/m and dB/m), and skin depth of copper $[\sigma_{\text{cu}} = 5.80 \times 10^7 \text{ (S/m)}]$, silver $[\sigma_{\text{ag}} = 6.15 \times 10^7 \text{ (S/m)}]$, and brass $[\sigma_{\text{br}} = 1.59 \times 10^7 \text{ (S/m)}]$ at the following frequencies: (a) 60 (Hz), (b) 1 (MHz), and (c) 1 (GHz).

P.8–11 A 3 (GHz), y-polarized uniform plane wave propagates in the $+x$-direction in a nonmagnetic medium having a dielectric constant 2.5 and a loss tangent 10^{-2}.

a) Determine the distance over which the amplitude of the propagating wave will be cut in half.

b) Determine the intrinsic impedance, the wavelength, the phase velocity, and the group velocity of the wave in the medium.

c) Assuming $\mathbf{E} = \mathbf{a}_y 50 \sin (6\pi 10^9 t + \pi/3)(\text{V/m})$ at $x = 0$, write the instantaneous expression for $\mathbf{H}$ for all t and x.

P.8–12 The magnetic field intensity of a linearly polarized uniform plane wave propagating in the $+y$-direction in seawater $[\epsilon_r = 80, \mu_r = 1, \sigma = 4 \text{ (S/m)}]$ is

$$\mathbf{H} = \mathbf{a}_x 0.1 \sin (10^{10}\pi t - \pi/3) \quad \text{(A/m)}$$

at $y = 0$.

a) Determine the attenuation constant, the phase constant, the intrinsic impedance, the phase velocity, the wavelength, and the skin depth.

b) Find the location at which the amplitude of **H** is 0.01 (A/m).

c) Write the expressions for $\mathbf{E}(y, t)$ and $\mathbf{H}(y, t)$ at $y = 0.5$ (m) as functions of t.

P.8–13 Given that the skin depth for graphite at 100 (MHz) is 0.16 (mm), determine (a) the conductivity of graphite, and (b) the distance that a 1 (GHz) wave travels in graphite such that its field intensity is reduced by 30 (dB).

P.8–14 Assume the ionosphere to be modeled by a plasma region with an electron density that increases with altitude from a low value at the lower boundary toward a value N_{max}

and decreases again as the altitude gets higher. A plane electromagnetic wave impinges on the lower boundary at an angle θ_i with the normal. Determine the highest frequency of the wave that will be turned back toward the earth. (*Hint:* Imagine the ionosphere to be stratified into layers of successively decreasing constant permittivities until the layer containing N_{max}. The frequency to be determined corresponds to that for an emerging angle of $\pi/2$.)

P.8–15 Prove the following relations between group velocity u_g and phase velocity u_p in a dispersive medium:

a) $u_g = u_p + \beta \dfrac{du_p}{d\beta}$ **b)** $u_g = u_p - \lambda \dfrac{du_p}{d\lambda}$.

P.8–16 There is a continuing discussion on radiation hazards to human health. The following calculations will provide a rough comparison.

  a) The U.S. standard for personal safety in a microwave environment is that the power density be less than 10 (mW/cm^2). Calculate the corresponding standard in terms of electric field intensity. In terms of magnetic field intensity.

  b) It is estimated that the earth receives radiant energy from the sun at a rate of about 1.3 (kW/m^2) on a sunny day. Assuming a monochromatic plane wave (which it is not), calculate the equivalent amplitudes of the electric and magnetic field intensity vectors.

P.8–17 Show that the instantaneous Poynting vector of a circularly polarized plane wave propagating in a lossless medium is a constant that is independent of time and distance.

P.8–18 Assuming that the radiation electric field intensity of an antenna system is

$$\mathbf{E} = \mathbf{a}_\theta E_\theta + \mathbf{a}_\phi E_\phi,$$

find the expression for the average outward power flow per unit area.

P.8–19 From the point of view of electromagnetics, the power transmitted by a lossless coaxial cable can be considered in terms of the Poynting vector inside the dielectric medium between the inner conductor and the outer sheath. Assuming that a d-c voltage V_0 applied between the inner conductor (of radius a) and the outer sheath (of inner radius b) causes a current I to flow to a load resistance, verify that the integration of the Poynting vector over the cross-sectional area of the dielectric medium equals the power $V_0 I$ that is transmitted to the load.

P.8–20 A uniform plane electromagnetic wave propagates in the $+z$- (downward) direction and impinges normally at $z = 0$ on an ocean surface. Let the magnetic field at $z = 0$ be $\mathbf{H}(0, t) = \mathbf{a}_y H_0 \cos 10^4 t$ (A/m).

  a) Determine the skin depth. (For the ocean: Conductivity $= \sigma$, permeability $= \mu_0$.)
  b) Find the expressions for $\mathbf{H}(z, t)$ and $\mathbf{E}(z, t)$.
  c) Find the power loss per unit area (in terms of H_0) into the ocean.

P.8–21 A right-hand circularly polarized plane wave represented by the phasor

$$\mathbf{E}(z) = E_0(\mathbf{a}_x - j\mathbf{a}_y)e^{-j\beta z}$$

impinges normally on a perfectly conducting wall at $z = 0$.

  a) Determine the polarization of the reflected wave.
  b) Find the induced current on the conducting wall.
  c) Obtain the instantaneous expression of the total electric intensity based on a cosine time reference.

P.8–22 A uniform sinusoidal plane wave in air with the following phasor expression for electric intensity

$$\mathbf{E}_i(x, z) = \mathbf{a}_y 10 e^{-j(6x + 8z)} \quad \text{(V/m)}$$

is incident on a perfectly conducting plane at $z = 0$.

    a) Find the frequency and wavelength of the wave.

    b) Write the instantaneous expressions for $\mathbf{E}_i(x, z; t)$ and $\mathbf{H}_i(x, z; t)$, using a cosine reference.

    c) Determine the angle of incidence.

    d) Find $\mathbf{E}_r(x, z)$ and $\mathbf{H}_r(x, z)$ of the reflected wave.

    e) Find $\mathbf{E}_1(x, z)$ and $\mathbf{H}_1(x, z)$ of the total field.

P.8–23 Repeat Problem P.8–22 for $\mathbf{E}_i(y, z) = 5(\mathbf{a}_y + \mathbf{a}_z \sqrt{3}) e^{j6(\sqrt{3}y - z)}$ (V/m).

P.8–24 For the case of oblique incidence of a uniform plane wave with perpendicular polarization on a perfectly conducting plane boundary as shown in Fig. 8–11, write (a) the instantaneous expressions

$$\mathbf{E}_1(x, z; t) \qquad \text{and} \qquad \mathbf{H}_1(x, z; t)$$

for the total field in medium 1, using a cosine reference, and (b) the time-average Poynting vector.

P.8–25 For the case of oblique incidence of a uniform plane wave with parallel polarization on a perfectly conducting plane boundary as shown in Fig. 8–13, write (a) the instantaneous expressions

$$\mathbf{E}_1(x, z; t) \qquad \text{and} \qquad \mathbf{H}_1(x, z; t)$$

for the total field in medium 1, using a sine reference, and (b) the time-average Poynting vector.

P.8–26 Determine the condition under which the magnitude of the reflection coefficient equals that of the transmission coefficient for a uniform plane wave at normal incidence on an interface between two lossless dielectric media. What is the standing-wave ratio in dB under this condition?

P.8–27 A uniform plane wave in air with $\mathbf{E}_i(z) = \mathbf{a}_x 10 e^{-j6z}$ (V/m) is incident normally on an interface at $z = 0$ with a lossy medium having a dielectric constant 2.5 and a loss tangent 0.5. Find the following:

    a) The instantaneous expressions for $\mathbf{E}_r(z, t)$, $\mathbf{H}_r(z, t)$, $\mathbf{E}_t(z, t)$, and $\mathbf{H}_t(z, t)$, using a cosine reference.

    b) The expressions for time-average Poynting vectors in air and in the lossy medium.

P.8–28 A uniform plane wave in air with $\mathbf{E}_i(z) = \mathbf{a}_x E_0 \exp(-j\beta_0 z)$ impinges normally onto the surface at $z = 0$ of a highly conducting medium having constitutive parameters ϵ_0, μ, and σ ($\sigma/\omega\epsilon_0 \gg 1$).

    a) Find the reflection coefficient.

    b) Derive the expression for the fraction of the incident power absorbed by the conducting medium.

    c) Obtain the fraction of the power absorbed at 1 (MHz) if the medium is iron.

P.8–29 Consider the situation of normal incidence at a lossless dielectric slab of thickness d in air, as shown in Fig. 8–15 with

$$\epsilon_1 = \epsilon_3 = \epsilon_0 \qquad \text{and} \qquad \mu_1 = \mu_3 = \mu_0.$$

a) Find E_{r0}, E_2^+, E_2^-, and E_{t0} in terms of E_{i0}, d, ϵ_2, and μ_2.

b) Will there be reflection at interface $z = 0$ if $d = \lambda_2/4$? If $d = \lambda_2/2$? Explain.

P.8–30 A transparent dielectric coating is applied to glass ($\epsilon_r = 4$, $\mu_r = 1$) to eliminate the reflection of red light $[\lambda_0 = 0.75\ (\mu m)]$.

a) Determine the required dielectric constant and thickness of the coating.

b) If violet light $[\lambda_0 = 0.42\ (\mu m)]$ is shone normally on the coated glass, what percentage of the incident power will be reflected?

P.8–31 Refer to Fig. 8–15, which depicts three different dielectric media with two parallel interfaces. A uniform plane wave in medium 1 propagates in the $+z$-direction. Let Γ_{12} and Γ_{23} denote the reflection coefficients between media 1 and 2 and between media 2 and 3, respectively. Express the effective reflection coefficient, Γ_0, at $z = 0$ for the incident wave in terms of Γ_{12}, Γ_{23}, and $\beta_2 d$.

P.8–32 A uniform plane wave with

$$\mathbf{E}_i(z, t) = \mathbf{a}_x E_{i0} \cos \omega \left(t - \frac{z}{u_p} \right)$$

in medium 1 (ϵ_1, μ_1) is incident normally onto a lossless dielectric slab (ϵ_2, μ_2) of a thickness d backed by a perfectly conducting plane, as shown in Fig. 8–22. Find

a) $\mathbf{E}_r(z, t)$ b) $\mathbf{E}_1(z, t)$ c) $\mathbf{E}_2(z, t)$ d) $(\mathscr{P}_{av})_1$ e) $(\mathscr{P}_{av})_2$

f) Determine the thickness d that makes $\mathbf{E}_1(z, t)$ the same as if the dielectric slab were absent.

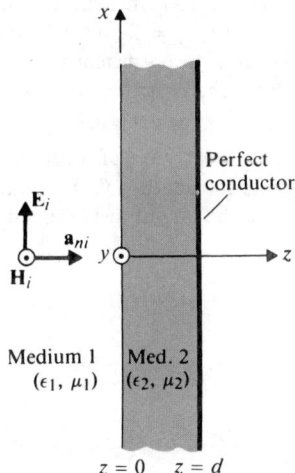

FIGURE 8–22
Plane wave incident normally onto a dielectric slab backed by a perfectly conducting plane (Problem P.8–32).

P.8–33 A uniform plane wave with $\mathbf{E}_i(z) = \mathbf{a}_x E_{i0} e^{-j\beta_0 z}$ in air propagates normally through a thin copper sheet of thickness d, as shown in Fig. 8–23. Neglecting multiple reflections within the copper sheets, find

a) E_2^+, H_2^+ b) E_2^-, H_2^- c) E_{30}, H_{30} d) $(\mathscr{P}_{av})_3/(\mathscr{P}_{av})_i$

Calculate $(\mathscr{P}_{av})_3/(\mathscr{P}_{av})_i$ for a thickness d that equals one skin depth at 10 (MHz). (Note that this pertains to the shielding effectiveness of the thin copper sheet.)

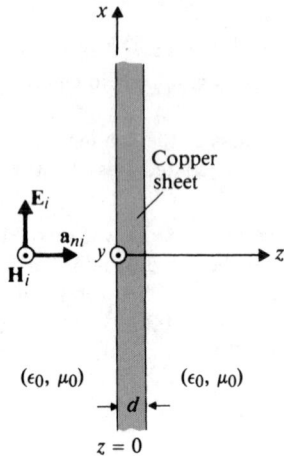

FIGURE 8–23
Plane wave propagating through a thin copper sheet
(Problem P.8–33).

P.8–34 A uniform plane wave is incident on the ionosphere at an angle of incidence
$\theta_i = 60°$. Assuming a constant electron density and a wave frequency equal to one-half of
the plasma frequency of the ionosphere, determine

    a) $\Gamma_\perp$ and $\tau_\perp$,

    b) $\Gamma_\parallel$ and $\tau_\parallel$.

Interpret the significance of these complex quantities.

P.8–35 A 10 (kHz) parallelly polarized electromagnetic wave in air is incident obliquely on
an ocean surface at a near-grazing angle $\theta_i = 88°$. Using $\epsilon_r = 81$, $\mu_r = 1$, and $\sigma = 4$ (S/m)
for seawater, find (a) the angle of refraction θ_t, (b) the transmission coefficient $\tau_\parallel$,
(c) $(\mathscr{P}_{av})_t/(\mathscr{P}_{av})_i$, and (d) the distance below the ocean surface where the field intensity has
been diminished by 30 (dB).

P.8–36 A light ray is incident from air obliquely on a transparent sheet of thickness d with
an index of refraction n, as shown in Fig. 8–24. The angle of incidence is θ_i. Find (a) θ_t,
(b) the distance ℓ_1 at the point of exit, and (c) the amount of the lateral displacement ℓ_2 of
the emerging ray.

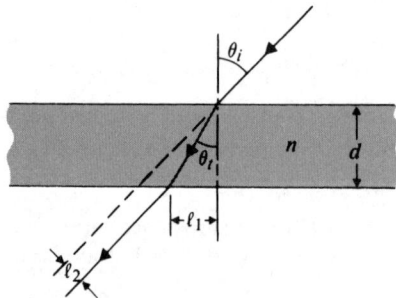

FIGURE 8–24
Light-ray impinging obliquely on a transparent
sheet of refraction index n (Problem P.8–36).

P.8–37 A uniform plane wave with perpendicular polarization represented by Eqs. (8–196)
and (8–197) is incident on a plane interface at $z = 0$, as shown in Fig. 8–16. Assuming

$\epsilon_2 < \epsilon_1$ and $\theta_i > \theta_c$, (a) obtain the phasor expressions for the transmitted field $(\mathbf{E}_t, \mathbf{H}_t)$, and (b) verify that the average power transmitted into medium 2 vanishes.

P.8–38 A uniform plane wave of angular frequency ω in medium 1 having a refractive index n_1 is incident on a plane interface at $z = 0$ with medium 2 having a refractive index n_2 $(<n_1)$ at the critical angle. Let E_{i0} and E_{t0} denote the amplitudes of the incident and refracted electric field intensities, respectively.

   a) Find the ratio E_{t0}/E_{i0} for perpendicular polarization.
   b) Find the ratio E_{t0}/E_{i0} for parallel polarization.
   c) Write the instantaneous expressions of $\mathbf{E}_i(x, z; t)$ and $\mathbf{E}_t(x, z; t)$ for perpendicular polarization in terms of the parameters ω, n_1, n_2, θ_i, and E_{i0}.

P.8–39 An electromagnetic wave from an underwater source with perpendicular polarization is incident on a water–air interface at $\theta_i = 20°$. Using $\epsilon_r = 81$ and $\mu_r = 1$ for fresh water, find (a) critical angle θ_c, (b) reflection coefficient $\Gamma_\perp$, (c) transmission coefficient $\tau_\perp$, and (d) attenuation in dB for each wavelength into the air.

P.8–40 Glass isosceles triangular prisms shown in Fig. 8–25 are used in optical instruments. Assuming $\epsilon_r = 4$ for glass, calculate the percentage of the incident light power reflected back by the prism.

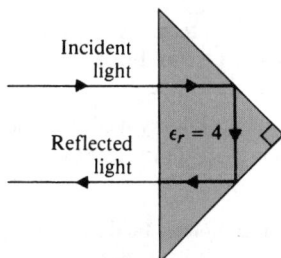

FIGURE 8–25
Light reflection by a right isosceles triangular prism (Problem P.8–40).

P.8–41 For preventing interference of waves in neighboring fibers and for mechanical protection, individual optical fibers are usually cladded by a material of a lower refractive index, as shown in Fig. 8–26, where $n_2 < n_1$.

   a) Express the maximum angle of incidence θ_a in terms of n_0, n_1 and n_2 for meridional rays incident on the core's end face to be trapped inside the core by total internal reflection. (**Meridional rays** are those that pass through the fiber axis. The angle θ_a is called the **acceptance angle**, and $\sin \theta_a$ the **numerical aperture** (N.A.) of the fiber.)
   b) Find θ_a and N.A. if $n_1 = 2$, $n_2 = 1.74$, and $n_0 = 1$.

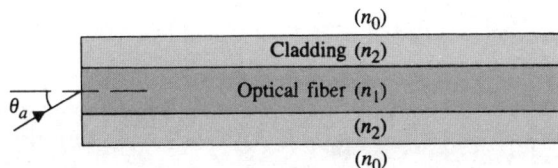

FIGURE 8–26
A cladded-core optical fiber (Problem P.8–41).

P.8–42 An electromagnetic wave in dielectric medium 1 (ϵ_1, μ_0) impinges obliquely on a boundary plane with dielectric medium 2 (ϵ_2, μ_0). Let θ_i and θ_t denote the incident and refraction angles, respectively, and prove the following:

a) For perpendicular polarization:

$$\Gamma_\perp = \frac{\sin(\theta_t - \theta_i)}{\sin(\theta_t + \theta_i)}, \qquad \tau_\perp = \frac{2\sin\theta_t\cos\theta_i}{\sin(\theta_t + \theta_i)}.$$

b) For parallel polarization:

$$\Gamma_\| = \frac{\sin 2\theta_t - \sin 2\theta_i}{\sin 2\theta_t + \sin 2\theta_i}, \qquad \tau_\| = \frac{4\sin\theta_t\cos\theta_i}{\sin 2\theta_t + \sin 2\theta_i}.$$

(These four relations are known as **Fresnel formulas**.)

P.8–43 Prove that, under the condition of no reflection at an interface, the sum of the Brewster angle and the angle of refraction is $\pi/2$ for:

a) perpendicular polarization ($\mu_1 \neq \mu_2$),

b) parallel polarization ($\epsilon_1 \neq \epsilon_2$).

P.8–44 For an incident wave with parallel polarization:

a) Find the relation between the critical angle θ_c and the Brewster angle $\theta_{B\|}$ for nonmagnetic media.

b) Plot θ_c and $\theta_{B\|}$ versus the ratio ϵ_1/ϵ_2.

P.8–45 By using Snell's law of refraction, (a) express Γ and τ in terms of $\epsilon_{r1}, \epsilon_{r2}$, and θ_i; and (b) plot Γ and τ versus θ_i for $\epsilon_{r1}/\epsilon_{r2} = 2.25$ for both perpendicular and parallel polarizations.

P.8–46 A perpendicularly polarized uniform plane wave in air of frequency f is incident obliquely at an angle of incidence θ_i on a plane boundary with a lossy dielectric medium that is characterized by a complex permittivity $\epsilon_2 = \epsilon' - j\epsilon''$. Let the incident electric field be

$$\mathbf{E}_i(x, z) = \mathbf{a}_y E_{i0} e^{-jk_0(x\sin\theta_i - z\cos\theta_i)}.$$

a) Find the expressions of the transmitted electric and magnetic field intensity phasors in terms of the given parameters.

b) Show that the angle of refraction is complex and that $\mathbf{H}_t$ is elliptically polarized.

P.8–47 In some books the reflection and transmission coefficients for parallel polarization are defined as the ratios of the amplitude of the tangential components of the reflected and transmitted **E** fields, respectively, to the amplitude of the tangential component of the incident **E** field. Let the coefficients defined in this manner be designated $\Gamma'_\|$ and $\tau'_\|$, respectively.

a) Find $\Gamma'_\|$ and $\tau'_\|$ in terms of η_1, η_2, θ_i, and θ_t; and compare them with $\Gamma_\|$ and $\tau_\|$ in Eqs. (8–221) and (8–222).

b) Find the relation between $\Gamma'_\|$ and $\tau'_\|$, and compare it with Eq. (8–223).

9

Theory and Applications
of Transmission Lines

9–1 Introduction

We have now developed an electromagnetic model with which we can analyze electromagnetic actions that occur at a distance and are caused by time-varying charges and
currents. These actions are explained in terms of electromagnetic fields and waves.
An isotropic or omnidirectional electromagnetic source radiates waves equally in all
directions. Even when the source radiates through a highly directive antenna, its
energy spreads over a wide area at large distances. This radiated energy is not guided,
and the transmission of power and information from the source to a receiver is in-
efficient. This is especially true at lower frequencies for which directive antennas
would have huge dimensions and therefore would be excessively expensive. For instance, at AM broadcast frequencies a single half-wavelength antenna (which is only
mildly directive[†]) would be over a hundred meters long. At the 60 (Hz) power frequency a wavelength is 5 million meters or 5 (Mm)!

For efficient point-to-point transmission of power and information the source
energy must be directed or guided. In this chapter we study transverse electromagnetic
(TEM) waves guided by transmission lines. The TEM mode of guided waves is one
in which **E** and **H** are perpendicular to each other and both are transverse to the
direction of propagation along the guiding line. We discussed the propagation of
unguided TEM plane waves in the last chapter. We will show in this chapter that
many of the characteristics of TEM waves guided by transmission lines are the same
as those for a uniform plane wave propagating in an unbounded dielectric medium.

The three most common types of guiding structures that support TEM waves are:

a) *Parallel-plate transmission line.* This type of transmission line consists of two
parallel conducting plates separated by a dielectric slab of a uniform thickness.

[†] Principles of antennas and radiating systems will be discussed in Chapter 11.

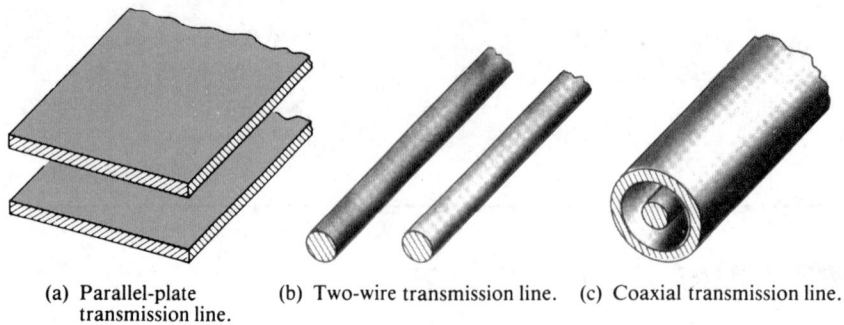

(a) Parallel-plate   (b) Two-wire transmission line.   (c) Coaxial transmission line.
    transmission line.

FIGURE 9–1
Common types of transmission lines.

[See Fig. 9–1(a).] At microwave frequencies, parallel-plate transmission lines can be fabricated inexpensively on a dielectric substrate using printed-circuit technology. They are often called *striplines*.

b) *Two-wire transmission line.* This transmission line consists of a pair of parallel conducting wires separated by a uniform distance. [See Fig. 9–1(b).] Examples are the ubiquitous overhead power and telephone lines seen in rural areas and the flat lead-in lines from a rooftop antenna to a television receiver.

c) *Coaxial transmission line.* This consists of an inner conductor and a coaxial outer conducting sheath separated by a dielectric medium. [See Fig. 9–1(c).] This structure has the important advantage of confining the electric and magnetic fields entirely within the dielectric region. No stray fields are generated by a coaxial transmission line, and little external interference is coupled into the line. Examples are telephone and TV cables and the input cables to high-frequency precision measuring instruments.

We should note that other wave modes more complicated than the TEM mode can propagate on all three of these types of transmission lines when the separation between the conductors is greater than certain fractions of the operating wavelength. These other transmission modes will be considered in the next chapter.

We will show that the TEM wave solution of Maxwell's equations for the parallel-plate guiding structure in Fig. 9–1(a) leads directly to a pair of transmission-line equations. The general transmission-line equations can also be derived from a circuit model in terms of the resistance, inductance, conductance, and capacitance per unit length of a line. The transition from the circuit model to the electromagnetic model is effected from a network with lumped-parameter elements (discrete resistors, inductors, and capacitors) to one with distributed parameters (continuous distributions of R, L, G, and C along the line). From the transmission-line equations, all the characteristics of wave propagation along a given line can be derived and studied.

The study of time-harmonic steady-state properties of transmission lines is greatly facilitated by the use of graphical charts, which avert the necessity of repeated calculations with complex numbers. The best known and most widely used graphical chart is the *Smith chart*. The use of Smith chart for determining wave characteristics on a transmission line and for impedance matching will be discussed.

9–2   Transverse Electromagnetic Wave along a Parallel-Plate Transmission Line

Let us consider a y-polarized TEM wave propagating in the $+z$-direction along a uniform parallel-plate transmission line. Figure 9–2 shows the cross-sectional dimensions of such a line and the chosen coordinate system. For time-harmonic fields the wave equation to be satisfied in the sourceless dielectric region becomes the homogeneous Helmholtz's equation, Eq. (8–46). In the present case the appropriate phasor solution for the wave propagating in the $+z$-direction is

$$\mathbf{E} = \mathbf{a}_y E_y = \mathbf{a}_y E_0 e^{-\gamma z}. \tag{9–1a}$$

The associated $\mathbf{H}$ field is, from Eq. (8–31),

$$\mathbf{H} = \mathbf{a}_x H_x = -\mathbf{a}_x \frac{E_0}{\eta} e^{-\gamma z}, \tag{9–1b}$$

where γ and η are the propagation constant and the intrinsic impedance, respectively, of the dielectric medium. Fringe fields at the edges of the plates are neglected. Assuming perfectly conducting plates and a lossless dielectric, we have, from Chapter 8,

$$\gamma = j\beta = j\omega\sqrt{\mu\epsilon} \tag{9–2}$$

and

$$\eta = \sqrt{\frac{\mu}{\epsilon}}. \tag{9–3}$$

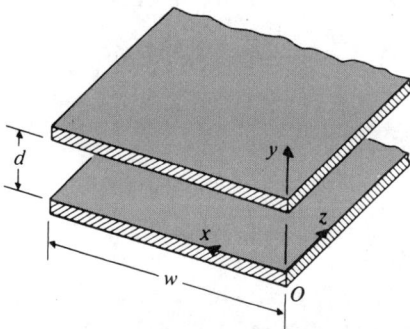

FIGURE 9–2
Parallel-plate transmission line.

The boundary conditions to be satisfied at the interfaces of the dielectric and the perfectly conducting planes are, from Eqs. (7–68a, b, c, and d), as follows:

At both $y = 0$ and $y = d$:

$$E_t = 0 \tag{9–4}$$

and

$$H_n = 0, \tag{9–5}$$

which are obviously satisfied because $E_x = E_z = 0$ and $H_y = 0$.

At $y = 0$ (lower plate), $\mathbf{a}_n = \mathbf{a}_y$:

$$\mathbf{a}_y \cdot \mathbf{D} = \rho_{s\ell} \quad \text{or} \quad \rho_{s\ell} = \epsilon E_y = \epsilon E_0 e^{-j\beta z}; \tag{9–6a}$$

$$\mathbf{a}_y \times \mathbf{H} = \mathbf{J}_{s\ell} \quad \text{or} \quad \mathbf{J}_{s\ell} = -\mathbf{a}_z H_x = \mathbf{a}_z \frac{E_0}{\eta} e^{-j\beta z}. \tag{9–7a}$$

At $y = d$ (upper plate), $\mathbf{a}_n = -\mathbf{a}_y$:

$$-\mathbf{a}_y \cdot \mathbf{D} = \rho_{su} \quad \text{or} \quad \rho_{su} = -\epsilon E_y = -\epsilon E_0 e^{-j\beta z}; \tag{9–6b}$$

$$-\mathbf{a}_y \times \mathbf{H} = \mathbf{J}_{su} \quad \text{or} \quad \mathbf{J}_{su} = \mathbf{a}_z H_x = -\mathbf{a}_z \frac{E_0}{\eta} e^{-j\beta z}. \tag{9–7b}$$

Equations (9–6) and (9–7) indicate that surface charges and surface currents on the conducting planes vary sinusoidally with z, as do E_y and H_x. This is illustrated schematically in Fig. 9–3.

Field phasors $\mathbf{E}$ and $\mathbf{H}$ in Eqs. (9–1a) and (9–1b) satisfy the two Maxwell's curl equations:

$$\nabla \times \mathbf{E} = -j\omega\mu\mathbf{H} \tag{9–8}$$

and

$$\nabla \times \mathbf{H} = j\omega\epsilon\mathbf{E}. \tag{9–9}$$

Since $\mathbf{E} = \mathbf{a}_y E_y$ and $\mathbf{H} = \mathbf{a}_x H_x$, Eqs. (9–8) and (9–9) become

$$\frac{dE_y}{dz} = j\omega\mu H_x \tag{9–10}$$

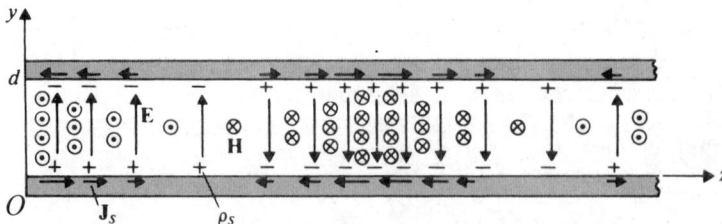

FIGURE 9–3
Field, charge, and current distributions along a parallel-plate transmission line.

and

$$\frac{dH_x}{dz} = j\omega\epsilon E_y. \tag{9–11}$$

Ordinary derivatives appear above because phasors E_y and H_x are functions of z only. Integrating Eq. (9–10) over y from 0 to d, we have

$$\frac{d}{dz} \int_0^d E_y \, dy = j\omega\mu \int_0^d H_x \, dy$$

or

$$-\frac{dV(z)}{dz} = j\omega\mu J_{su}(z)d = j\omega \left(\mu \frac{d}{w} \right) [J_{su}(z)w] \tag{9–12}$$

$$= j\omega L I(z),$$

where

$$V(z) = - \int_0^d E_y \, dy = -E_y(z)d$$

is the potential difference or voltage between the upper and lower plates,

$$I(z) = J_{su}(z)w$$

is the total current flowing in the $+z$ direction in the upper plate (w = plate width), and

$$\boxed{L = \mu \frac{d}{w} \quad \text{(H/m)}} \tag{9–13}$$

is the inductance per unit length of the parallel-plate transmission line. The dependence of phasors $V(z)$ and $I(z)$ on z is noted explicitly in Eq. (9–12) for emphasis.

Similarly, we integrate Eq. (9–11) over x from 0 to w to obtain

$$\frac{d}{dz} \int_0^w H_x \, dx = j\omega\epsilon \int_0^w E_y \, dx$$

or

$$-\frac{dI(z)}{dz} = -j\omega\epsilon E_y(z)w = j\omega \left(\epsilon \frac{w}{d} \right) [-E_y(z)d] \tag{9–14}$$

$$= j\omega C V(z),$$

where

$$\boxed{C = \epsilon \frac{w}{d} \quad \text{(F/m)}} \tag{9–15}$$

is the capacitance per unit length of the parallel-plate transmission line.

Equations (9–12) and (9–14) constitute a pair of *time-harmonic transmission-line equations* for phasors $V(z)$ and $I(z)$. They may be combined to yield second-order differential equations for $V(z)$ and for $I(z)$:

$$\frac{d^2 V(z)}{dz^2} = -\omega^2 LC V(z), \tag{9-16a}$$

$$\frac{d^2 I(z)}{dz^2} = -\omega^2 LC I(z). \tag{9-16b}$$

The solutions of Eqs. (9–16a) and (9–16b) are, for waves propagating in the $+z$-direction,

$$V(z) = V_0 e^{-j\beta z} \tag{9-17a}$$

and

$$I(z) = I_0 e^{-j\beta z}, \tag{9-17b}$$

where the phase constant

$$\boxed{\beta = \omega\sqrt{LC} = \omega\sqrt{\mu\epsilon} \qquad \text{(rad/m)}} \tag{9-18}$$

is the same as that given in Eq. (9–2). The relation between V_0 and I_0 can be found by using either Eq. (9–12) or Eq. (9–14):

$$\boxed{Z_0 = \frac{V(z)}{I(z)} = \frac{V_0}{I_0} = \sqrt{\frac{L}{C}} \qquad (\Omega),} \tag{9-19}$$

which becomes, in view of the results of Eqs. (9–13) and (9–15),

$$\boxed{Z_0 = \frac{d}{w}\sqrt{\frac{\mu}{\epsilon}} = \frac{d}{w}\eta \qquad (\Omega).} \tag{9-20}$$

The quantity Z_0 is the impedance at any location that looks toward an infinitely long (no reflections) transmission line. It is called the *characteristic impedance* of the line. The ratio of $V(z)$ and $I(z)$ at any point on a finite line of any length terminated in Z_0 is Z_0.[†] For a parallel-plate transmission line with perfectly conducting plates of width w and separated by a lossless dielectric slab of thickness d, the characteristic impedance Z_0 is (d/w) times the intrinsic impedance η of the dielectric medium.

The velocity of propagation along the line is

$$\boxed{u_p = \frac{\omega}{\beta} = \frac{1}{\sqrt{LC}} = \frac{1}{\sqrt{\mu\epsilon}} \qquad \text{(m/s)},} \tag{9-21}$$

which is the same as the phase velocity of a TEM plane wave in the dielectric medium.

---

† This statement will be proved in Section 9–4 (see Eq. 9–107).

9–2.1 LOSSY PARALLEL-PLATE TRANSMISSION LINES

We have so far assumed the parallel-plate transmission line to be lossless. In actual situations, loss may arise from two causes. First, the dielectric medium may have a nonvanishing loss tangent; second, the plates may not be perfectly conducting. To characterize these two effects, we define two new parameters: G, the conductance per unit length across the two plates; and R, the resistance per unit length of the *two* plate conductors.

The conductance between two conductors separated by a dielectric medium having a permittivity ϵ and an equivalent conductivity σ can be determined readily by using Eq. (5–81) when the capacitance between the two conductors is known. We have

$$G = \frac{\sigma}{\epsilon} C. \tag{9–22}$$

Use of Eq. (9–15) directly yields

$$\boxed{G = \sigma \frac{w}{d} \quad \text{(S/m)}.} \tag{9–23}$$

If the parallel-plate conductors have a very large but finite conductivity σ_c (which must not be confused with the conductivity σ of the dielectric medium), ohmic power will be dissipated in the plates. This necessitates the presence of a nonvanishing axial electric field $\mathbf{a}_z E_z$ at the plate surfaces, such that the average Poynting vector

$$\mathscr{P}_{av} = \mathbf{a}_y p_\sigma = \tfrac{1}{2}\mathscr{R}e(\mathbf{a}_z E_z \times \mathbf{a}_x H_x^*) \tag{9–24}$$

has a y-component and equals the average power per unit area dissipated in each of the conducting plates. (Obviously the cross product of $\mathbf{a}_y E_y$ and $\mathbf{a}_x H_x$ does not result in a y-component.)

Consider the upper plate where the surface current density is $J_{su} = H_x$. It is convenient to define a ***surface impedance*** of an imperfect conductor, Z_s, as the ratio of the tangential component of the electric field to the surface current density at the conductor surface.

$$\boxed{Z_s = \frac{E_t}{J_s} \quad (\Omega).} \tag{9–25}$$

For the upper plate we have

$$Z_s = \frac{E_z}{J_{su}} = \frac{E_z}{H_x} = \eta_c, \tag{9–26a}$$

where η_c is the intrinsic impedance of the plate conductor. Here we assume that both the conductivity σ_c of the plate conductor and the operating frequency are sufficiently high that the current flows in a very thin surface layer and can be represented by the surface current J_{su}. The intrinsic impedance of a good conductor has been given in Eq. (8–54). We have

$$Z_s = R_s + jX_s = (1 + j)\sqrt{\frac{\pi f \mu_c}{\sigma_c}} \quad (\Omega), \quad (9\text{–}26\text{b})$$

where the subscript c is used to indicate the properties of the conductor.

Substitution of Eq. (9–26a) in Eq. (9–24) gives

$$\begin{aligned} p_\sigma &= \tfrac{1}{2}\mathcal{R}e(|J_{su}|^2 Z_s) \\ &= \tfrac{1}{2}|J_{su}|^2 R_s \quad (\text{W/m}^2). \end{aligned} \quad (9\text{–}27)$$

The ohmic power dissipated in a unit length of the plate having a width w is wp_σ, which can be expressed in terms of the total surface current, $I = wJ_{su}$, as

$$P_\sigma = wp_\sigma = \frac{1}{2}I^2\left(\frac{R_s}{w}\right) \quad (\text{W/m}). \quad (9\text{–}28)$$

Equation (9–28) is the power dissipated when a sinusoidal current of amplitude I flows through a resistance R_s/w. Thus, the effective series resistance per unit length for *both* plates of a parallel-plate transmission line of width w is

$$\boxed{R = 2\left(\frac{R_s}{w}\right) = \frac{2}{w}\sqrt{\frac{\pi f \mu_c}{\sigma_c}} \quad (\Omega/\text{m}).} \quad (9\text{–}29)$$

Table 9–1 lists the expressions for the four *distributed parameters* (R, L, G, and C per unit length) of a parallel-plate transmission line of width w and separation d.

TABLE 9–1
Distributed Parameters of Parallel-Plate Transmission Line (Width = w, Separation = d)

Parameter	Formula	Unit
R	$\dfrac{2}{w}\sqrt{\dfrac{\pi f \mu_c}{\sigma_c}}$	Ω/m
L	$\mu\dfrac{d}{w}$	H/m
G	$\sigma\dfrac{w}{d}$	S/m
C	$\epsilon\dfrac{w}{d}$	F/m

We note from Eq. (9–26b) that surface impedance Z_s has a positive reactance term X_s that is numerically equal to R_s. If the total complex power (instead of its real part, the ohmic power P_σ, only) associated with a unit length of the plate is considered, X_s will lead to an ***internal series inductance*** per unit length $L_i = X_s/\omega = R_s/\omega$. At high frequencies, L_i is negligible in comparison with the external inductance L.

We note in the calculation of the power loss in the plate conductors of a finite conductivity σ_c that a nonvanishing electric field $\mathbf{a}_z E_z$ must exist. The very existence of this axial electric field makes the wave along a lossy transmission line strictly not TEM. However, this axial component is ordinarily very small in comparison to the transverse component E_y. An estimate of their relative magnitudes can be made as follows:

$$\frac{|E_z|}{|E_y|} = \frac{|\eta_c H_x|}{|\eta H_x|} = \sqrt{\frac{\epsilon}{\mu}}\,|\eta_c|$$
$$= \sqrt{\frac{\omega\epsilon\mu_c}{\mu\sigma_c}} = \sqrt{\frac{\omega\epsilon}{\sigma_c}},$$

where Eq. (8–54) has been used. For copper plates $[\sigma_c = 5.80 \times 10^7 \text{ (S/m)}]$ in air $[\epsilon = \epsilon_0 = 10^{-9}/36\pi \text{ (F/m)}]$ at a frequency of 3 (GHz),

$$|E_z| \cong 5.3 \times 10^{-5}|E_y| \ll |E_y|.$$

Hence we retain the designation TEM as well as all its consequences. The introduction of a small E_z in the calculation of p_σ and R is considered a slight perturbation.

9–2.2 MICROSTRIP LINES

The development of solid-state microwave devices and systems has led to the widespread use of a form of parallel-plate transmission lines called microstrip lines or simply ***striplines***. A stripline usually consists of a dielectric substrate sitting on a grounded conducting plane with a thin narrow metal strip on top of the substrate, as shown in Fig. 9–4(a). Since the advent of printed-circuit techniques, striplines can be easily fabricated and integrated with other circuit components. However, because the results that we have derived in this section were based on the assumption of two wide

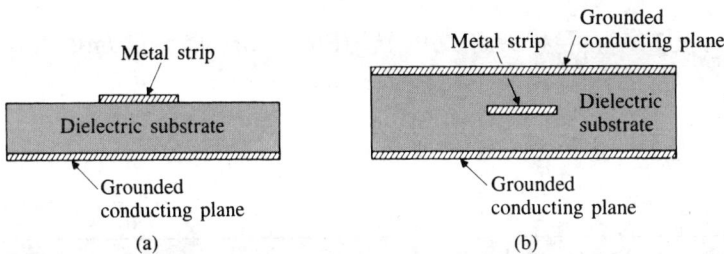

Metal strip · Dielectric substrate · Grounded conducting plane · (a)

Grounded conducting plane · Metal strip · Dielectric substrate · Grounded conducting plane · (b)

FIGURE 9–4
Two types of microstrip lines.

conducting plates (with negligible fringing effect) of equal width, they are not expected to apply here exactly. The approximation is closer if the width of the metal strip is much greater than the substrate thickness.

When the substrate has a high dielectric constant, a TEM approximation is found to be reasonably satisfactory. An exact analytical solution of the stripline in Fig. 9–4(a) satisfying all the boundary conditions is a difficult problem. Not all the fields will be confined in the dielectric substrate; some will stray from the top strip into the region outside of the strip, thus causing interference in the neighboring circuits. Semiempirical modifications to the formulas for the distributed parameters and the characteristic impedance are necessary for more accurate calculations.[†] All of these quantities tend to be frequency-dependent, and striplines are dispersive.

One method for reducing the stray fields of striplines is to have a grounded conducting plane on both sides of the dielectric substrate and to put the thin metal strip in the middle as in Fig. 9–4(b). This arrangement is known as a ***triplate line***. We can appreciate that triplate lines are more difficult and costly to fabricate and that the characteristic impedance of a triplate line is one-half of that of a corresponding stripline.

■■■■ **EXAMPLE 9–1** Neglecting losses and fringe effects and assuming the substrate of a stripline to have a thickness 0.4 (mm) and a dielectric constant 2.25, (a) determine the required width w of the metal strip in order for the stripline to have a characteristic resistance of 50 (Ω); (b) determine L and C of the line; and (c) determine u_p along the line. (d) Repeat parts (a), (b), and (c) for a characteristic resistance of 75 (Ω).

Solution

a) We use Eq. (9–20) directly to find w:

$$w = \frac{d}{Z_0}\sqrt{\frac{\mu}{\epsilon}} = \frac{0.4 \times 10^{-3}}{50}\frac{\eta_0}{\sqrt{\epsilon_r}}$$

$$= \frac{0.4 \times 10^{-3} \times 377}{50\sqrt{2.25}} = 2 \times 10^{-3} \quad (m), \quad \text{or } 2 \quad (mm).$$

b) $L = \mu\dfrac{d}{w} = 4\pi 10^{-7} \times \dfrac{0.4}{2} = 2.51 \times 10^{-7} \quad (H/m), \quad \text{or } 0.251 \quad (\mu H/m).$

$C = \epsilon_0\epsilon_r\dfrac{w}{d} = \dfrac{10^{-9}}{36\pi} \times 2.25 \times \dfrac{2}{0.4} = 99.5 \times 10^{-12} \quad (F/m), \quad \text{or } 99.5 \quad (pF/m).$

c) $u_p = \dfrac{1}{\sqrt{\mu\epsilon}} = \dfrac{c}{\sqrt{\epsilon_r}} = \dfrac{c}{\sqrt{2.25}} = \dfrac{c}{1.5} = 2 \times 10^8 \quad (m/s).$

[†] See, for instance, K. F. Sander and G. A. L. Reed, *Transmission and Propagation of Electromagnetic Waves*, 2nd edition, Sec. 6.5.6, Cambridge University Press, New York, 1986.

d) Since w is inversely proportional to Z_0, we have, for $Z_0' = 75$ (Ω),

$$w' = \left(\frac{Z_0}{Z_0'}\right)w = \frac{50}{75} \times 2 = 1.33 \quad \text{(mm)}.$$

$$L' = \left(\frac{w}{w'}\right)L = \left(\frac{2}{1.33}\right) \times 0.251 = 0.377 \quad (\mu\text{H/m}).$$

$$C' = \left(\frac{w'}{w}\right)C = \left(\frac{1.33}{2}\right) \times 99.5 = 66.2 \quad \text{(pF/m)}.$$

$$u_p' = u_p = 2 \times 10^8 \quad \text{(m/s)}.$$

━

9-3 General Transmission-Line Equations

We will now derive the equations that govern general two-conductor uniform transmission lines that include parallel-plate, two-wire, and coaxial lines. Transmission lines differ from ordinary electric networks in one essential feature. Whereas the physical dimensions of electric networks are very much smaller than the operating wavelength, transmission lines are usually a considerable fraction of a wavelength and may even be many wavelengths long. The circuit elements in an ordinary electric network can be considered discrete and as such may be described by lumped parameters. It is assumed that currents flowing in lumped-circuit elements do not vary spatially over the elements, and that no standing waves exist. A transmission line, on the other hand, is a distributed-parameter network and must be described by circuit parameters that are distributed throughout its length. Except under matched conditions, standing waves exist in a transmission line.

Consider a differential length Δz of a transmission line that is described by the following four parameters:

R, resistance per unit length (both conductors), in Ω/m.

L, inductance per unit length (both conductors), in H/m.

G, conductance per unit length, in S/m.

C, capacitance per unit length, in F/m.

Note that R and L are series elements and G and C are shunt elements. Figure 9-5 shows the equivalent electric circuit of such a line segment. The quantities $v(z, t)$ and $v(z + \Delta z, t)$ denote the instantaneous voltages at z and $z + \Delta z$, respectively. Similarly, $i(z, t)$ and $i(z + \Delta z, t)$ denote the instantaneous currents at z and $z + \Delta z$, respectively. Applying Kirchhoff's voltage law, we obtain

$$v(z, t) - R\,\Delta z\, i(z, t) - L\,\Delta z\, \frac{\partial i(z, t)}{\partial t} - v(z + \Delta z, t) = 0, \qquad (9\text{-}30)$$

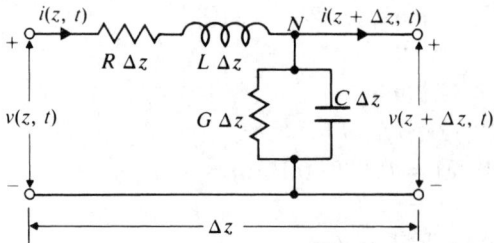

FIGURE 9–5
Equivalent circuit of a differential length Δz of a two-conductor transmission line.

which leads to

$$-\frac{v(z + \Delta z, t) - v(z, t)}{\Delta z} = Ri(z, t) + L\frac{\partial i(z, t)}{\partial t}. \tag{9–30a}$$

In the limit as $\Delta z \to 0$, Eq. (9–30a) becomes

$$-\frac{\partial v(z, t)}{\partial z} = Ri(z, t) + L\frac{\partial i(z, t)}{\partial t}. \tag{9–31}$$

Similarly, applying Kirchhoff's current law to the node N in Fig. 9–5, we have

$$i(z, t) - G\,\Delta z v(z + \Delta z, t) - C\,\Delta z\frac{\partial v(z + \Delta z, t)}{\partial t} - i(z + \Delta z, t) = 0. \tag{9–32}$$

On dividing by Δz and letting Δz approach zero, Eq. (9–32) becomes

$$-\frac{\partial i(z, t)}{\partial z} = Gv(z, t) + C\frac{\partial v(z, t)}{\partial t}. \tag{9–33}$$

Equations (9–31) and (9–33) are a pair of first-order partial differential equations in $v(z, t)$ and $i(z, t)$. They are the *general transmission-line equations*.[†]

For harmonic time dependence the use of phasors simplifies the transmission-line equations to ordinary differential equations. For a cosine reference we write

$$v(z, t) = \mathscr{R}e[V(z)e^{j\omega t}], \tag{9–34a}$$

$$i(z, t) = \mathscr{R}e[I(z)e^{j\omega t}], \tag{9–34b}$$

where $V(z)$ and $I(z)$ are functions of the space coordinate z only and both may be complex. Substitution of Eqs. (9–34a) and (9–34b) in Eqs. (9–31) and (9–33) yields

---

[†] Sometimes referred to as the *telegraphist's equations* or *telegrapher's equations*.

the following ordinary differential equations for phasors $V(z)$ and $I(z)$:

$$-\frac{dV(z)}{dz} = (R + j\omega L)I(z),$$

(9–35a)

$$-\frac{dI(z)}{dz} = (G + j\omega C)V(z).$$

(9–35b)

Equations (9–35a) and (9–35b) are ***time-harmonic transmission-line equations***, which reduce to Eqs. (9–12) and (9–14) under lossless conditions ($R = 0$, $G = 0$).

9–3.1   WAVE CHARACTERISTICS ON AN INFINITE TRANSMISSION LINE

The coupled time-harmonic transmission-line equations, Eqs. (9–35a) and (9–35b), can be combined to solve for $V(z)$ and $I(z)$. We obtain

$$\frac{d^2V(z)}{dz^2} = \gamma^2 V(z)$$

(9–36a)

and

$$\frac{d^2I(z)}{dz^2} = \gamma^2 I(z),$$

(9–36b)

where

$$\gamma = \alpha + j\beta = \sqrt{(R + j\omega L)(G + j\omega C)} \qquad (m^{-1})$$

(9–37)

is the ***propagation constant*** whose real and imaginary parts, α and β, are the ***attenuation constant*** (Np/m) and ***phase constant*** (rad/m) of the line, respectively. The nomenclature here is similar to that for plane-wave propagation in lossy media as defined in Section 8–3. These quantities are not really constants because, in general, they depend on ω in a complicated way.

The solutions of Eqs. (9–36a) and (9–36b) are

$$V(z) = V^+(z) + V^-(z)$$
$$= V_0^+ e^{-\gamma z} + V_0^- e^{\gamma z},$$

(9–38a)

$$I(z) = I^+(z) + I^-(z)$$
$$= I_0^+ e^{-\gamma z} + I_0^- e^{\gamma z},$$

(9–38b)

where the plus and minus superscripts denote waves traveling in the $+z$- and $-z$-directions, respectively. Wave amplitudes (V_0^+, I_0^+) and (V_0^-, I_0^-) are related by Eqs. (9–35a) and (9–35b), and it is easy to verify (Problem P.9–5) that

$$\frac{V_0^+}{I_0^+} = -\frac{V_0^-}{I_0^-} = \frac{R + j\omega L}{\gamma}.$$

(9–39)

For an infinite line (actually a semi-infinite line with the source at the left end) the terms containing the $e^{\gamma z}$ factor must vanish. There are no reflected waves; only the waves traveling in the $+z$-direction exist. We have

$$V(z) = V^+(z) = V_0^+ e^{-\gamma z}, \tag{9-40a}$$

$$I(z) = I^+(z) = I_0^+ e^{-\gamma z}. \tag{9-40b}$$

The ratio of the voltage and the current at any z for an infinitely long line is independent of z and is called the **characteristic impedance** of the line.

$$\boxed{Z_0 = \frac{R + j\omega L}{\gamma} = \frac{\gamma}{G + j\omega C} = \sqrt{\frac{R + j\omega L}{G + j\omega C}} \quad (\Omega).} \tag{9-41}$$

Note that γ and Z_0 are characteristic properties of a transmission line whether or not the line is infinitely long. They depend on R, L, G, C, and ω—not on the length of the line. An infinite line simply implies that there are no reflected waves.

There is a close analogy between the general governing equations and the wave characteristics of a transmission line and those of uniform plane waves in a lossy medium. This analogy will be discussed in the following example.

■■■■ **EXAMPLE 9-2** Demonstrate the analogy between the wave characteristics on a transmission line and uniform plane waves in a lossy medium.

Solution In a lossy medium with a complex permittivity $\epsilon_c = \epsilon' - j\epsilon''$ and a complex permeability $\mu = \mu' - j\mu''$ the Maxwell's curl equations (7–104a) and (7–104b) become

$$\nabla \times \mathbf{E} = -j\omega(\mu' - j\mu'')\mathbf{H}, \tag{9-42a}$$

$$\nabla \times \mathbf{H} = j\omega(\epsilon' - j\epsilon'')\mathbf{E}. \tag{9-42b}$$

If we assume a uniform plane wave characterized by an E_x that varies only with z, Eq. (9–42a) reduces to (see Eq. 8–12b)

$$-\frac{dE_x(z)}{dz} = j\omega(\mu' - j\mu'')H_y$$
$$= (\omega\mu'' + j\omega\mu')H_y. \tag{9-43a}$$

Similarly, we obtain from Eq. (9–42b) the following relation:

$$-\frac{dH_y(z)}{dz} = (\omega\epsilon'' + j\omega\epsilon')E_x. \tag{9-43b}$$

Comparing Eqs. (9–43a) and (9–43b) with Eqs. (9–35a) and (9–35b), respectively, we recognize immediately the analogy of the governing equations for E_x and H_y of a uniform plane wave and those for V and I on a transmission line.

Equations (9–43a) and (9–43b) can be combined to give

$$\frac{d^2 E_x(z)}{dz^2} = \gamma^2 E_x(z) \tag{9-44a}$$

and

$$\frac{d^2 H_y(z)}{dz^2} = \gamma^2 H_y(z), \tag{9-44b}$$

which are entirely similar to Eqs. (9–36a) and (9–36b). The propagation constant of the uniform plane wave is

$$\gamma = \alpha + j\beta = \sqrt{(\omega\mu'' + j\omega\mu')(\omega\epsilon'' + j\omega\epsilon')}, \tag{9-45}$$

which should be compared with Eq. (9–37) for the transmission line. The intrinsic impedance of the lossy medium (the wave impedance of the plane wave traveling in the $+z$-direction) is (see Eq. 8–30)

$$\eta_c = \sqrt{\frac{\mu'' + j\mu'}{\epsilon'' + j\epsilon'}}, \tag{9-46}$$

which is analogous to the expression for the characteristic impedance of a transmission line in Eq. (9–41).

Because of the above analogies, many of the results obtained for normal incidence of uniform plane waves can be adapted to transmission-line problems, and vice versa.

▬

The general expressions for the characteristic impedance in Eq. (9–41) and the propagation constant in Eq. (9–37) are relatively complicated. The following three limiting cases have special significance.

1. *Lossless Line* $(R = 0, G = 0)$.

   a) Propagation constant:

$$\gamma = \alpha + j\beta = j\omega\sqrt{LC}; \tag{9-47}$$

$$\alpha = 0, \tag{9-48}$$

$$\beta = \omega\sqrt{LC} \qquad \text{(a linear function of } \omega\text{).} \tag{9-49}$$

   b) Phase velocity:

$$u_p = \frac{\omega}{\beta} = \frac{1}{\sqrt{LC}} \qquad \text{(constant).} \tag{9-50}$$

   c) Characteristic impedance:

$$Z_0 = R_0 + jX_0 = \sqrt{\frac{L}{C}}; \tag{9-51}$$

$$R_0 = \sqrt{\frac{L}{C}} \qquad \text{(constant),} \tag{9-52}$$

$$X_0 = 0. \tag{9-53}$$

2. *Low-Loss Line* $(R \ll \omega L, G \ll \omega C)$. The low-loss conditions are more easily satisfied at very high frequencies.

a) Propagation constant:

$$\gamma = \alpha + j\beta = j\omega\sqrt{LC}\left(1 + \frac{R}{j\omega L}\right)^{1/2}\left(1 + \frac{G}{j\omega C}\right)^{1/2}$$

$$\cong j\omega\sqrt{LC}\left(1 + \frac{R}{2j\omega L}\right)\left(1 + \frac{G}{2j\omega C}\right) \tag{9-54}$$

$$\cong j\omega\sqrt{LC}\left[1 + \frac{1}{2j\omega}\left(\frac{R}{L} + \frac{G}{C}\right)\right];$$

$$\alpha \cong \frac{1}{2}\left(R\sqrt{\frac{C}{L}} + G\sqrt{\frac{L}{C}}\right), \tag{9-55}$$

$$\beta \cong \omega\sqrt{LC} \quad \text{(approximately a linear function of } \omega). \tag{9-56}$$

b) Phase velocity:

$$u_p = \frac{\omega}{\beta} \cong \frac{1}{\sqrt{LC}} \quad \text{(approximately constant).} \tag{9-57}$$

c) Characteristic impedance:

$$Z_0 = R_0 + jX_0 = \sqrt{\frac{L}{C}}\left(1 + \frac{R}{j\omega L}\right)^{1/2}\left(1 + \frac{G}{j\omega C}\right)^{-1/2}$$

$$\cong \sqrt{\frac{L}{C}}\left[1 + \frac{1}{2j\omega}\left(\frac{R}{L} - \frac{G}{C}\right)\right]; \tag{9-58}$$

$$R_0 \cong \sqrt{\frac{L}{C}}, \tag{9-59}$$

$$X_0 \cong -\sqrt{\frac{L}{C}}\frac{1}{2\omega}\left(\frac{R}{L} - \frac{G}{C}\right) \cong 0. \tag{9-60}$$

3. *Distortionless Line* $(R/L = G/C)$. If the condition

$$\frac{R}{L} = \frac{G}{C} \tag{9-61}$$

is satisfied, the expressions for both γ and Z_0 simplify.

a) Propagation constant:

$$\gamma = \alpha + j\beta = \sqrt{(R + j\omega L)\left(\frac{RC}{L} + j\omega C\right)}$$

$$= \sqrt{\frac{C}{L}}\,(R + j\omega L); \tag{9-62}$$

$$\alpha = R\sqrt{\frac{C}{L}}, \tag{9-63}$$

$$\beta = \omega\sqrt{LC} \quad \text{(a linear function of } \omega). \tag{9-64}$$

b) Phase velocity:

$$u_p = \frac{\omega}{\beta} = \frac{1}{\sqrt{LC}} \quad \text{(constant)}. \tag{9-65}$$

c) Characteristic impedance:

$$Z_0 = R_0 + jX_0 = \sqrt{\frac{R + j\omega L}{(RC/L) + j\omega C}} = \sqrt{\frac{L}{C}}; \tag{9-66}$$

$$R_0 = \sqrt{\frac{L}{C}} \quad \text{(constant)}, \tag{9-67}$$

$$X_0 = 0. \tag{9-68}$$

Thus, except for a nonvanishing attenuation constant, the characteristics of a distortionless line are the same as those of a lossless line—namely, a constant phase velocity $(u_p = 1/\sqrt{LC})$ and a constant real characteristic impedance $(Z_0 = R_0 = \sqrt{L/C})$.

A constant phase velocity is a direct consequence of the linear dependence of the phase constant β on ω. Since a signal usually consists of a band of frequencies, it is essential that the different frequency components travel along a transmission line at the same velocity in order to avoid distortion. This condition is satisfied by a lossless line and is approximated by a line with very low losses. For a lossy line, wave amplitudes will be attentuated, and distortion will result when different frequency components attenuate differently, even when they travel with the same velocity. The condition specified in Eq. (9–61) leads to both a constant α and a constant u_p—thus the name *distortionless line*.

The phase constant of a lossy transmission line is determined by expanding the expression for γ in Eq. (9–37). In general, the phase constant is not a linear function of ω; thus it will lead to a u_p, which depends on frequency. As the different frequency components of a signal propagate along the line with different velocities, the signal suffers *dispersion*. A general, lossy, transmission line is therefore *dispersive*, as is a lossy dielectric.

EXAMPLE 9–3   It is found that the attenuation on a 50 (Ω) distortionless transmission line is 0.01 (dB/m). The line has a capacitance of 0.1 (nF/m).

a) Find the resistance, inductance, and conductance per meter of the line.

b) Find the velocity of wave propagation.

c) Determine the percentage to which the amplitude of a voltage traveling wave decreases in 1 (km) and in 5 (km).

Solution

a) For a distortionless line,

$$\frac{R}{L} = \frac{G}{C}.$$

The given quantities are

$$R_0 = \sqrt{\frac{L}{C}} = 50 \quad (\Omega),$$

$$\alpha = R\sqrt{\frac{C}{L}} = 0.01 \quad (\text{dB/m})$$

$$= \frac{0.01}{8.69} \ (\text{Np/m}) = 1.15 \times 10^{-3} \quad (\text{Np/m}).$$

The three relations above are sufficient to solve for the three unknowns R, L, and G in terms of the given $C = 10^{-10}$ (F/m):

$$R = \alpha R_0 = (1.15 \times 10^{-3}) \times 50 = 0.057 \quad (\Omega/\text{m});$$
$$L = CR_0^2 = 10^{-10} \times 50^2 = 0.25 \quad (\mu\text{H/m});$$
$$G = \frac{RC}{L} = \frac{R}{R_0^2} = \frac{0.057}{50^2} = 22.8 \quad (\mu\text{S/m}).$$

b) The velocity of wave propagation on a distortionless line is the phase velocity given by Eq. (9–65).

$$u_p = \frac{1}{\sqrt{LC}} = \frac{1}{\sqrt{(0.25 \times 10^{-6} \times 10^{-10}}} = 2 \times 10^8 \quad (\text{m/s}).$$

c) The ratio of two voltages a distance z apart along the line is

$$\frac{V_2}{V_1} = e^{-\alpha z}.$$

After 1 (km), $(V_2/V_1) = e^{-1000\alpha} = e^{-1.15} = 0.317$, or 31.7%.
After 5 (km), $(V_2/V_1) = e^{-5000\alpha} = e^{-5.75} = 0.0032$, or 0.32%.   ▬

9–3.2 TRANSMISSION-LINE PARAMETERS

The electrical properties of a transmission line at a given frequency are completely characterized by its four distributed parameters R, L, G, and C. These parameters for a parallel-plate transmission line are listed in Table 9–1. We will now obtain them for two-wire and coaxial transmission lines.

Our basic premise is that the conductivity of the conductors in a transmission line is usually so high that the effect of the series resistance on the computation of the propagation constant is negligible, the implication being that the waves on the line are approximately TEM. We may write, in dropping R from Eq. (9–37),

$$\gamma = j\omega\sqrt{LC}\left(1 + \frac{G}{j\omega C}\right)^{1/2}. \tag{9–69}$$

From Eq. (8–44) we know that the propagation constant for a TEM wave in a medium with constitutive parameters (μ, ϵ, σ) is

$$\gamma = j\omega\sqrt{\mu\epsilon}\left(1 + \frac{\sigma}{j\omega\epsilon}\right)^{1/2}. \tag{9–70}$$

But

$$\frac{G}{C} = \frac{\sigma}{\epsilon} \tag{9–71}$$

in accordance with Eq. (5–81); hence comparison of Eqs. (9–69) and (9–70) yields

$$\boxed{LC = \mu\epsilon.} \tag{9–72}$$

Equation (9–72) is a very useful relation, because if L is known for a line with a given medium, C can be determined, and vice versa. Knowing C, we can find G from Eq. (9–71). Series resistance R is determined by introducing a small axial E_z as a slight perturbation of the TEM wave and by finding the ohmic power dissipated in a unit length of the line, as was done in Subsection 9–2.1.

Equation (9–72), of course, also holds for a lossless line. *The velocity of wave propagation on a lossless transmission line, $u_p = 1/\sqrt{LC}$, therefore, is equal to the velocity of propagation, $1/\sqrt{\mu\epsilon}$, of unguided plane wave in the dielectric of the line.* This fact has been pointed out in connection with Eq. (9–21) for parallel-plate lines.

1. *Two-wire transmission line.* The capacitance per unit length of a two-wire transmission line, whose wires have a radius a and are separated by a distance D, has been found in Eq. (4–47). We have

$$\boxed{C = \frac{\pi\epsilon}{\cosh^{-1}(D/2a)}} \quad \text{(F/m)}. \tag{9–73}$$†

From Eqs. (9–72) and (9–71) we obtain

$$\boxed{L = \frac{\mu}{\pi}\cosh^{-1}\left(\frac{D}{2a}\right)} \quad \text{(H/m)} \tag{9–74}$$†

and

$$\boxed{G = \frac{\pi\sigma}{\cosh^{-1}(D/2a)}} \quad \text{(S/m)}. \tag{9–75}$$†

---

† $\cosh^{-1}(D/2a) \cong \ln(D/a)$ if $(D/2a)^2 \gg 1$.

To determine R, we go back to Eq. (9–28) and express the ohmic power dissipated per unit length of both wires in terms of p_σ. Assuming the current J_s (A/m) to flow in a very thin surface layer, the current in each wire is $I = 2\pi a J_s$, and

$$P_\sigma = 2\pi a p_\sigma = \frac{1}{2} I^2 \left(\frac{R_s}{2\pi a} \right) \quad \text{(W/m)}. \tag{9–76}$$

Hence the series resistance per unit length for both wires is

$$\boxed{R = 2\left(\frac{R_s}{2\pi a} \right) = \frac{1}{\pi a} \sqrt{\frac{\pi f \mu_c}{\sigma_c}} \quad \text{(}\Omega\text{/m)}.} \tag{9–77}$$

In deriving Eqs. (9–76) and (9–77), we have assumed the surface current J_s to be uniform over the circumference of both wires. This is an approximation, inasmuch as the proximity of the two wires tends to make the surface current nonuniform.

2. *Coaxial transmission line.* The external inductance per unit length of a coaxial transmission line with a center conductor of radius a and an outer conductor of inner radius b has been found in Eq. (6–140):

$$\boxed{L = \frac{\mu}{2\pi} \ln \frac{b}{a} \quad \text{(H/m)}.} \tag{9–78}$$

From Eq. (9–72) we obtain

$$\boxed{C = \frac{2\pi\epsilon}{\ln (b/a)} \quad \text{(F/m)},} \tag{9–79}$$

and from Eq. (9–71),

$$\boxed{G = \frac{2\pi\sigma}{\ln (b/a)} \quad \text{(S/m)},} \tag{9–80}$$

where σ is the equivalent conductivity of the lossy dielectric. If one prefers, σ could be replaced by $\omega\epsilon''$ as in Eq. (7–112).

To determine R, we again return to Eq. (9–27), where J_{si} on the surface of the center conductor is different from J_{so} on the inner surface of the outer conductor. We must have

$$I = 2\pi a J_{si} = 2\pi b J_{so}. \tag{9–81}$$

The power dissipated in a unit length of the center and outer conductors are, respectively,

$$P_{\sigma i} = 2\pi a p_{\sigma i} = \frac{1}{2} I^2 \left(\frac{R_s}{2\pi a} \right). \tag{9–82}$$

TABLE 9–2
Distributed Parameters of Two-Wire and Coaxial Transmission Lines

Parameter	Two-Wire Line	Coaxial Line	Unit
R	$\dfrac{R_s}{\pi a}$	$\dfrac{R_s}{2\pi}\left(\dfrac{1}{a}+\dfrac{1}{b}\right)$	Ω/m
L	$\dfrac{\mu}{\pi}\cosh^{-1}\left(\dfrac{D}{2a}\right)$	$\dfrac{\mu}{2\pi}\ln\dfrac{b}{a}$	H/m
G	$\dfrac{\pi\sigma}{\cosh^{-1}(D/2a)}$	$\dfrac{2\pi\sigma}{\ln(b/a)}$	S/m
C	$\dfrac{\pi\epsilon}{\cosh^{-1}(D/2a)}$	$\dfrac{2\pi\epsilon}{\ln(b/a)}$	F/m

Note: $R_s = \sqrt{\pi f\mu_c/\sigma_c}$; $\cosh^{-1}(D/2a) \cong \ln(D/a)$ if $(D/2a)^2 \gg 1$. Internal inductance is not included.

$$P_{\sigma o} = 2\pi b p_{\sigma o} = \frac{1}{2}I^2\left(\frac{R_s}{2\pi b}\right). \qquad (9\text{–}83)$$

From Eqs. (9–82) and (9–83), we obtain the resistance per unit length:

$$R = \frac{R_s}{2\pi}\left(\frac{1}{a}+\frac{1}{b}\right) = \frac{1}{2\pi}\sqrt{\frac{\pi f\mu_c}{\sigma_c}}\left(\frac{1}{a}+\frac{1}{b}\right) \qquad (\Omega/\text{m}). \qquad (9\text{–}84)$$

The R, L, G, C parameters for two-wire and coaxial transmission lines are listed in Table 9–2.

9–3.3 ATTENUATION CONSTANT FROM POWER RELATIONS

The attenuation constant of a traveling wave on a transmission line is the real part of the propagation constant; it can be determined from the basic definition in Eq. (9–37):

$$\alpha = \mathscr{R}e(\gamma) = \mathscr{R}e\left[\sqrt{(R + j\omega L)(G + j\omega C)}\right]. \qquad (9\text{–}85)$$

The attenuation constant can also be found from a power relationship. The phasor voltage and phasor current distributions on an infinitely long transmission line (no reflections) may be written as (Eqs. (9–40a) and (9–40b) with the plus superscript dropped for simplicity):

$$V(z) = V_0 e^{-(\alpha + j\beta)z}, \qquad (9\text{–}86a)$$

$$I(z) = \frac{V_0}{Z_0} e^{-(\alpha + j\beta)z}. \qquad (9\text{–}86b)$$

The time-average power propagated along the line at any z is

$$P(z) = \frac{1}{2}\mathscr{R}e\left[V(z)I^*(z)\right]$$

$$= \frac{V_0^2}{2|Z_0|^2}\,R_0 e^{-2\alpha z}. \tag{9–87}$$

The law of conservation of energy requires that the rate of decrease of $P(z)$ with distance along the line equals the time-average power loss P_L per unit length. Thus,

$$-\frac{\partial P(z)}{\partial z} = P_L(z)$$

$$= 2\alpha P(z),$$

from which we obtain the following formula:

$$\boxed{\alpha = \frac{P_L(z)}{2P(z)} \quad (\text{Np/m}).} \tag{9–88}$$

◼ **EXAMPLE 9–4**

a) Use Eq. (9–88) to find the attenuation constant of a lossy transmission line with distributed parameters R, L, G, and C.

b) Specialize the result in part (a) to obtain the attenuation constants of a low-loss line and of a distortionless line.

Solution

a) For a lossy transmission line the time-average power loss per unit length is

$$P_L(z) = \frac{1}{2}\left[|I(z)|^2 R + |V(z)|^2 G\right]$$

$$= \frac{V_0^2}{2|Z_0|^2}\,(R + G|Z_0|^2)e^{-2\alpha z}. \tag{9–89}$$

Substitution of Eqs. (9–87) and (9–89) in Eq. (9–88) gives

$$\boxed{\alpha = \frac{1}{2R_0}\,(R + G|Z_0|^2) \quad (\text{Np/m}).} \tag{9–90}$$

b) For a low-loss line, $Z_0 \cong R_0 = \sqrt{L/C}$, Eq. (9–90) becomes

$$\alpha \cong \frac{1}{2}\left(\frac{R}{R_0} + GR_0\right)$$

$$= \frac{1}{2}\left(R\sqrt{\frac{C}{L}} + G\sqrt{\frac{L}{C}}\right), \tag{9–91}$$

which checks with Eq. (9–55). For a distortionless line, $Z_0 = R_0 = \sqrt{L/C}$, Eq. (9–91) applies, and

$$\alpha = \frac{1}{2} R \sqrt{\frac{C}{L}} \left(1 + \frac{G}{R}\frac{L}{C}\right),$$

which, in view of the condition in Eq. (9–61), reduces to

$$\alpha = R \sqrt{\frac{C}{L}}. \tag{9–92}$$

Equation (9–92) is the same as Eq. (9–63). ■

9–4 Wave Characteristics on Finite Transmission Lines

In Subsection 9–3.1 we indicated that the general solutions for the time-harmonic one-dimensional Helmholtz equations, Eqs. (9–36a) and (9–36b), for transmission lines are

$$V(z) = V_0^+ e^{-\gamma z} + V_0^- e^{\gamma z} \tag{9–93a}$$

and

$$I(z) = I_0^+ e^{-\gamma z} + I_0^- e^{\gamma z}, \tag{9–93b}$$

where

$$\frac{V_0^+}{I_0^+} = -\frac{V_0^-}{I_0^-} = Z_0. \tag{9–94}$$

For waves launched on an infinitely long line at $z = 0$ there can be only forward waves traveling in the $+z$-direction, and the second terms on the right side of Eqs. (9–93a) and (9–93b), representing reflected waves, vanish. This is also true for finite lines terminated in a characteristic impedance; that is, when the lines are *matched*. From circuit theory we know that *a maximum transfer of power from a given voltage source to a load occurs under "matched conditions" when the load impedance is the complex conjugate of the source impedance* (Problem P.9–11). In transmission line terminology, *a line is matched when the load impedance is equal to the characteristic impedance (not the complex conjugate of the characteristic impedance) of the line.*

Let us now consider the general case of a finite transmission line having a characteristic impedance Z_0 terminated in an arbitrary load impedance Z_L, as depicted in Fig. 9–6. The length of the line is ℓ. A *sinusoidal* voltage source $V_g \underline{/0°}$ with an internal impedance Z_g is connected to the line at $z = 0$. In such a case,

$$\left(\frac{V}{I}\right)_{z=\ell} = \frac{V_L}{I_L} = Z_L, \tag{9–95}$$

which obviously cannot be satisfied without the second terms on the right side of Eqs. (9–93a) and (9–93b) unless $Z_L = Z_0$. Thus reflected waves exist on unmatched lines.

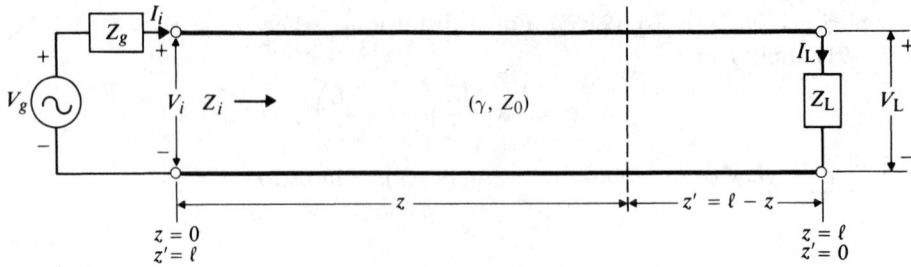

FIGURE 9–6
Finite transmission line terminated with load impedance Z_L.

Given the characteristic γ and Z_0 of the line and its length ℓ, there are four unknowns V_0^+, V_0^-, I_0^+, and I_0^- in Eqs. (9–93a) and (9–93b). These four unknowns are not all independent because they are constrained by the relations at $z = 0$ and at $z = \ell$. Both $V(z)$ and $I(z)$ can be expressed either in terms of V_i and I_i at the input end (Problem P.9–12), or in terms of the conditions at the load end. Consider the latter case.

Let $z = \ell$ in Eqs. (9–93a) and (9–93b). We have

$$V_L = V_0^+ e^{-\gamma\ell} + V_0^- e^{\gamma\ell}, \tag{9–96a}$$

$$I_L = \frac{V_0^+}{Z_0} e^{-\gamma\ell} - \frac{V_0^-}{Z_0} e^{\gamma\ell}. \tag{9–96b}$$

Solving Eqs. (9–96a) and (9–96b) for V_0^+ and V_0^-, we have

$$V_0^+ = \tfrac{1}{2}(V_L + I_L Z_0)e^{\gamma\ell}, \tag{9–97a}$$

$$V_0^- = \tfrac{1}{2}(V_L - I_L Z_0)e^{-\gamma\ell}. \tag{9–97b}$$

Substituting Eq. (9–95) in Eqs. (9–97a) and (9–97b), and using the results in Eqs. (9–93a) and (9–93b), we obtain

$$V(z) = \frac{I_L}{2}\left[(Z_L + Z_0)e^{\gamma(\ell-z)} + (Z_L - Z_0)e^{-\gamma(\ell-z)}\right], \tag{9–98a}$$

$$I(z) = \frac{I_L}{2Z_0}\left[(Z_L + Z_0)e^{\gamma(\ell-z)} - (Z_L - Z_0)e^{-\gamma(\ell-z)}\right]. \tag{9–98b}$$

Since ℓ and z appear together in the combination $(\ell - z)$, it is expedient to introduce a new variable $z' = \ell - z$, which is the distance measured backward from the load. Equations (9–98a) and (9–98b) then become

$$V(z') = \frac{I_L}{2}\left[(Z_L + Z_0)e^{\gamma z'} + (Z_L - Z_0)e^{-\gamma z'}\right], \tag{9–99a}$$

$$I(z') = \frac{I_L}{2Z_0}\left[(Z_L + Z_0)e^{\gamma z'} - (Z_L - Z_0)e^{-\gamma z'}\right]. \tag{9–99b}$$

We note here that although the same symbols V and I are used in Eqs. (9–99a) and (9–99b) as in Eqs. (9–98a) and (9–98b), the dependence of $V(z')$ and $I(z')$ on z' is different from the dependence of $V(z)$ and $I(z)$ on z.

The use of hyperbolic functions simplifies the equations above. Recalling the relations

$$e^{\gamma z'} + e^{-\gamma z'} = 2\cosh \gamma z' \quad \text{and} \quad e^{\gamma z'} - e^{-\gamma z'} = 2\sinh \gamma z',$$

we may write Eqs. (9–99a) and (9–99b) as

$$V(z') = I_L(Z_L \cosh \gamma z' + Z_0 \sinh \gamma z'), \tag{9–100a}$$

$$I(z') = \frac{I_L}{Z_0}(Z_L \sinh \gamma z' + Z_0 \cosh \gamma z'), \tag{9–100b}$$

which can be used to find the voltage and current at any point along a transmission line in terms of I_L, Z_L, γ, and Z_0.

The ratio $V(z')/I(z')$ is the impedance when we look toward the load end of the line at a distance z' from the load.

$$Z(z') = \frac{V(z')}{I(z')} = Z_0 \frac{Z_L \cosh \gamma z' + Z_0 \sinh \gamma z'}{Z_L \sinh \gamma z' + Z_0 \cosh \gamma z'} \tag{9–101}$$

or

$$Z(z') = Z_0 \frac{Z_L + Z_0 \tanh \gamma z'}{Z_0 + Z_L \tanh \gamma z'} \quad (\Omega). \tag{9–102}$$

At the source end of the line, $z' = \ell$, the generator looking into the line sees an **input impedance** Z_i.

$$Z_i = (Z)_{\substack{z=0 \\ z'=\ell}} = Z_0 \frac{Z_L + Z_0 \tanh \gamma \ell}{Z_0 + Z_L \tanh \gamma \ell} \quad (\Omega). \tag{9–103}$$

As far as the conditions at the generator are concerned, the terminated finite transmission line can be replaced by Z_i, as shown in Fig. 9–7. The input voltage V_i and input current I_i in Fig. 9–6 are found easily from the equivalent circuit in Fig. 9–7.

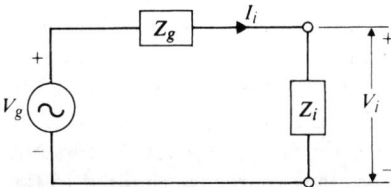

FIGURE 9–7
Equivalent circuit for finite transmission line in Figure 9–6 at generator end.

They are

$$V_i = \frac{Z_i}{Z_g + Z_i} V_g, \tag{9-104a}$$

$$I_i = \frac{V_g}{Z_g + Z_i}. \tag{9-104b}$$

Of course, the voltage and current at any other location on line cannot be determined by using the equivalent circuit in Fig. 9–7.

The average power delivered by the generator to the input terminals of the line is

$$(P_{\text{av}})_i = \tfrac{1}{2}\mathcal{R}e[V_i I_i^*]_{z=0,\, z'=\ell}. \tag{9-105}$$

The average power delivered to the load is

$$(P_{\text{av}})_{\text{L}} = \tfrac{1}{2}\mathcal{R}e[V_{\text{L}} I_{\text{L}}^*]_{z=\ell,\, z'=0}$$
$$= \frac{1}{2}\left|\frac{V_{\text{L}}}{Z_{\text{L}}}\right|^2 R_{\text{L}} = \frac{1}{2}|I_{\text{L}}|^2 R_{\text{L}}. \tag{9-106}$$

For a lossless line, conservation of power requires that $(P_{\text{av}})_i = (P_{\text{av}})_{\text{L}}$.

A particularly important special case is when a line is terminated with its characteristic impedance—that is, when $Z_{\text{L}} = Z_0$. The input impedance, Z_i in Eq. (9–103), is seen to be equal to Z_0. As a matter of fact, the impedance of the line looking toward the load at any distance z' from the load is, from Eq. (9–102),

$$Z(z') = Z_0 \qquad (\text{for } Z_{\text{L}} = Z_0). \tag{9-107}$$

The voltage and current equations in Eqs. (9–98a) and (9–98b) reduce to

$$V(z) = (I_{\text{L}} Z_0 e^{\gamma\ell}) e^{-\gamma z} = V_i e^{-\gamma z}, \tag{9-108a}$$

$$I(z) = (I_{\text{L}} e^{\gamma\ell}) e^{-\gamma z} = I_i e^{-\gamma z}. \tag{9-108b}$$

Equations (9–108a) and (9–108b) correspond to the pair of voltage and current equations—Eqs. (9–40a) and (9–40b)—representing waves traveling in $+z$-direction, and there are no reflected waves. Hence, *when a finite transmission line is terminated with its own characteristic impedance (when a finite transmission line is matched), the voltage and current distributions on the line are exactly the same as though the line has been extended to infinity.*

■■■■■ **EXAMPLE 9–5**   A signal generator having an internal resistance 1 (Ω) and an open-circuit voltage $v_g(t) = 0.3 \cos 2\pi 10^8 t$ (V) is connected to a 50 (Ω) lossless transmission line. The line is 4 (m) long, and the velocity of wave propagation on the line is 2.5×10^8 (m/s). For a matched load, find (a) the instantaneous expressions for the voltage and current at an arbitrary location on the line, (b) the instantaneous expressions for the voltage and current at the load, and (c) the average power transmitted to the load.

Solution

a) In order to find the voltage and current at an arbitrary location on the line, it is first necessary to obtain those at the input end ($z = 0$, $z' = \ell$). The given quantities are as follows:

$$V_g = 0.3\underline{/0^\circ} \quad \text{(V)}, \qquad \text{a phasor with a cosine reference,}$$
$$Z_g = R_g = 1 \quad \text{(Ω)},$$
$$Z_0 = R_0 = 50 \quad \text{(Ω)},$$
$$\omega = 2\pi \times 10^8 \quad \text{(rad/s)},$$
$$u_p = 2.5 \times 10^8 \quad \text{(m/s)},$$
$$\ell = 4 \quad \text{(m)}.$$

Since the line is terminated with a matched load, $Z_i = Z_0 = 50$ (Ω). The voltage and current at the input terminals can be evaluated from the equivalent circuit in Fig. 9–7. From Eqs. (9–104a) and (9–104b) we have

$$V_i = \frac{50}{1 + 50} \times 0.3\underline{/0^\circ} = 0.294\underline{/0^\circ} \quad \text{(V)},$$

$$I_i = \frac{0.3\underline{/0^\circ}}{1 + 50} = 0.0059\underline{/0^\circ} \quad \text{(A)}.$$

Since only forward-traveling waves exist on a matched line, we use Eqs. (9–86a) and (9–86b) for the voltage and current, respectively, at an arbitrary location. For the given line, $\alpha = 0$ and

$$\beta = \frac{\omega}{u_p} = \frac{2\pi \times 10^8}{2.5 \times 10^8} = 0.8\pi \quad \text{(rad/m)}.$$

Thus,

$$V(z) = 0.294e^{-j0.8\pi z} \quad \text{(V)},$$
$$I(z) = 0.0059e^{-j0.8\pi z} \quad \text{(A)}.$$

These are phasors. The corresponding instantaneous expressions are, from Eqs. (9–34a) and (9–34b),

$$v(z, t) = \mathscr{R}e[0.294e^{j(2\pi 10^8 t - 0.8\pi z)}]$$
$$= 0.294 \cos (2\pi 10^8 t - 0.8\pi z) \quad \text{(V)},$$
$$i(z, t) = \mathscr{R}e[0.0059e^{j(2\pi 10^8 t - 0.8\pi z)}]$$
$$= 0.0059 \cos (2\pi 10^8 t - 0.8\pi z) \quad \text{(A)}.$$

b) At the load, $z = \ell = 4$ (m),

$$v(4, t) = 0.294 \cos (2\pi 10^8 t - 3.2\pi) \quad \text{(V)},$$
$$i(4, t) = 0.0059 \cos (2\pi 10^8 t - 3.2\pi) \quad \text{(A)}.$$

c) The average power transmitted to the load on a lossless line is equal to that at the input terminals.

$$(P_{\text{av}})_L = (P_{\text{av}})_i = \tfrac{1}{2}\mathscr{R}e[V(z)I^*(z)]$$
$$= \tfrac{1}{2}(0.294 \times 0.0059) = 8.7 \times 10^{-4} \quad \text{(W)} = 0.87 \quad \text{(mW)}. \qquad \blacksquare$$

9-4.1 TRANSMISSION LINES AS CIRCUIT ELEMENTS

Not only can transmission lines be used as wave-guiding structures for transferring power and information from one point to another, but at ultrahigh frequencies— UHF: frequency from 300 (MHz) to 3 (GHz), wavelength from 1 (m) to 0.1 (m)—they may serve as circuit elements. At these frequencies, ordinary lumped-circuit elements are difficult to make, and stray fields become important. Sections of transmission lines can be designed to give an inductive or capacitive impedance and are used to match an arbitrary load to the internal impedance of a generator for maximum power transfer. The required length of such lines as circuit elements becomes practical in the UHF range. At frequencies much lower than 300 (MHz) the required lines tend to be too long, whereas at frequencies higher than 3 (GHz) the physical dimensions become inconveniently small, and it would be advantageous to use waveguide components.

In most cases, transmission-line segments can be considered lossless: $\gamma = j\beta$, $Z_0 = R_0$, and $\tanh \gamma\ell = \tanh (j\beta\ell) = j \tan \beta\ell$. The formula in Eq. (9–103) for the input impedance Z_i of a lossless line of length ℓ terminated in Z_L becomes

$$Z_i = R_0 \frac{Z_L + jR_0 \tan \beta\ell}{R_0 + jZ_L \tan \beta\ell} \quad (\Omega).$$

(9–109)

(Lossless line)

Comparison of Eq. (9–109) with Eq. (8–171) again confirms the similarity between normal incidence of a uniform plane wave on a plane interface and wave propagation along a terminated transmission line.

We now consider several important special cases.

1. *Open-circuit termination* $(Z_L \to \infty)$. We have, from Eq. (9–109),

$$Z_{io} = jX_{io} = - \frac{jR_0}{\tan \beta\ell} = -jR_0 \cot \beta\ell.$$

(9–110)

Equation (9–110) shows that the input impedance of an open-circuited lossless line is purely reactive. The line can, however, be either capacitive or inductive because the function $\cot \beta\ell$ can be either positive or negative, depending on the value of $\beta\ell$ $(=2\pi\ell/\lambda)$. Figure 9–8 is a plot of $X_{io} = -R_0 \cot \beta\ell$ versus ℓ. We see that X_{io} can assume all values from $-\infty$ to $+\infty$.

When the length of an open-circuited line is very short in comparison with a wavelength, $\beta\ell \ll 1$, we can obtain a very simple formula for its capacitive reactance by noting that $\tan \beta\ell \cong \beta\ell$. From Eq. (9–110) we have

$$Z_{io} = jX_{io} \cong -j \frac{R_0}{\beta\ell} = -j \frac{\sqrt{L/C}}{\omega\sqrt{LC}\ell} = -j \frac{1}{\omega C\ell},$$

(9–111)

which is the impedance of a capacitance of $C\ell$ farads.

In practice, it is not possible to obtain an infinite load impedance at the end of a transmission line, especially at high frequencies, because of coupling to nearby objects and because of radiation from the open end.

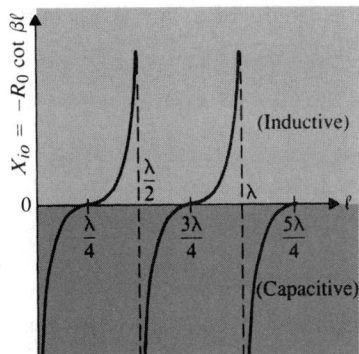

FIGURE 9–8
Input reactance of open-circuited transmission line.

2. *Short-circuit termination* ($Z_L = 0$). In this case, Eq. (9–109) reduces to

$$Z_{is} = jX_{is} = jR_0 \tan \beta\ell. \qquad (9\text{–}112)$$

Since $\tan \beta\ell$ can range from $-\infty$ to $+\infty$, the input impedance of a short-circuited lossless line can also be either purely inductive or purely capacitive, depending on the value of $\beta\ell$. Figure 9–9 is a graph of X_{is} versus ℓ. We note that Eq. (9–112) has exactly the same form as that—Eq. (8–172)—of the wave impedance of the total field at a distance ℓ from a perfectly conducting plane boundary.

Comparing Figs. 9–8 and 9–9, we see that in the range where X_{io} is capacitive X_{is} is inductive, and vice versa. The input reactances of open-circuited and short-circuited lossless transmission lines are the same if their lengths differ by an odd multiple of $\lambda/4$.

When the length of a short-circuited line is very short in comparison with a wavelength, $\beta\ell \ll 1$, Eq. (9–112) becomes approximately

$$Z_{is} = jX_{is} \cong jR_0\beta\ell = j \sqrt{\frac{L}{C}} \, \omega\sqrt{LC}\ell = j\omega L\ell, \qquad (9\text{–}113)$$

which is the impedance of an inductance of $L\ell$ henries.

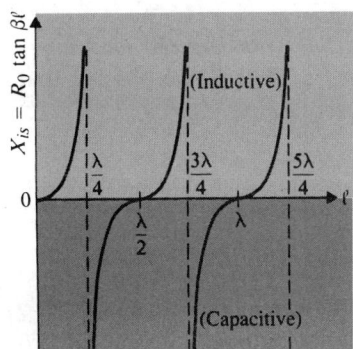

FIGURE 9–9
Input reactance of short-circuited transmission line.

3. *Quarter-wave section* ($\ell = \lambda/4$, $\beta\ell = \pi/2$). When the length of a line is an odd multiple of $\lambda/4$, $\ell = (2n - 1)\lambda/4$, $(n = 1, 2, 3, \ldots)$,

$$\beta\ell = \frac{2\pi}{\lambda}(2n - 1)\frac{\lambda}{4} = (2n - 1)\frac{\pi}{2},$$

$$\tan \beta\ell = \tan\left[(2n - 1)\frac{\pi}{2}\right] \to \pm\infty,$$

and Eq. (9–109) becomes

$$\boxed{Z_i = \frac{R_0^2}{Z_L} \qquad \text{(Quarter-wave line).}} \qquad (9\text{–}114)$$

Hence, *a quarter-wave lossless line transforms the load impedance to the input terminals as its inverse multiplied by the square of the characteristic resistance.* It acts as an impedance inverter and is often referred to as a *quarter-wave transformer.* An open-circuited, quarter-wave line appears as a short circuit at the input terminals, and a short-circuited quarter-wave line appears as an open circuit. Actually, if the series resistance of the line itself is not neglected, the input impedance of a short-circuited, quarter-wave line is an impedance of a very high value similar to that of a parallel resonant circuit. It is interesting to compare Eq. (9–114) with the formula for quarter-wave impedance transformation with multiple dielectrics, Eq. (8–182a).

4. *Half-wave section* ($\ell = \lambda/2$, $\beta\ell = \pi$). When the length of a line is an integral multiple of $\lambda/2$, $\ell = n\lambda/2$ $(n = 1, 2, 3, \ldots)$,

$$\beta\ell = \frac{2\pi}{\lambda}\left(\frac{n\lambda}{2}\right) = n\pi,$$

$$\tan \beta\ell = 0,$$

and Eq. (9–109) reduces to

$$\boxed{Z_i = Z_L \qquad \text{(Half-wave line).}} \qquad (9\text{–}115)$$

Equation (9–115) states that *a half-wave lossless line transfers the load impedance to the input terminals without change.* From Eq. (9–103) we observe that a half-wave line with loss does not have this property unless $Z_L = Z_0$.

By measuring the input impedance of a line section under open- and short-circuit conditions, we can determine the characteristic impedance and the propagation constant of the line. The following expressions follow directly from Eq. (9–103).

$$\text{\textit{Open-circuited line}, } Z_L \to \infty: \qquad Z_{io} = Z_0 \coth \gamma\ell. \qquad (9\text{–}116)$$

$$\text{\textit{Short-circuited line}, } Z_L = 0: \qquad Z_{is} = Z_0 \tanh \gamma\ell. \qquad (9\text{–}117)$$

From Eqs. (9–116) and (9–117) we have

$$\boxed{Z_0 = \sqrt{Z_{io}Z_{is}} \quad (\Omega)}$$ (9–118)

and

$$\boxed{\gamma = \frac{1}{\ell} \tanh^{-1} \sqrt{\frac{Z_{is}}{Z_{io}}} \quad (m^{-1}).}$$ (9–119)

Equations (9–118) and (9–119) apply whether or not the line is lossy.

━━━ **EXAMPLE 9–6** The open-circuit and short-circuit impedances measured at the input terminals of a lossless transmission line of length 1.5 (m), which is less than a quarter wavelength, are $-j54.6$ (Ω) and $j103$ (Ω), respectively. (a) Find Z_0 and γ of the line. (b) Without changing the operating frequency, find the input impedance of a short-circuited line that is twice the given length. (c) How long should the short-circuited line be in order for it to appear as an open circuit at the input terminals?

Solution The given quantities are

$$Z_{io} = -j54.6, \qquad Z_{is} = j103, \qquad \ell = 1.5.$$

a) Using Eqs. (9–118) and (9–119), we find

$$Z_0 = \sqrt{-j54.6(j103)} = 75 \quad (\Omega)$$

$$\gamma = \frac{1}{1.5} \tanh^{-1} \sqrt{\frac{j103}{-j54.6}} = \frac{j}{1.5} \tan^{-1} 1.373 = j0.628 \quad (rad/m).$$

b) For a short-circuited line twice as long, $\ell = 3.0$ (m),

$$\gamma\ell = j0.628 \times 3.0 = j1.884 \quad (rad).$$

The input impedance is, from Eq. (9–117),

$$Z_{is} = 75 \tanh (j1.884) = j75 \tan 108°$$
$$= j75(-3.08) = -j231 \quad (\Omega).$$

Note that Z_{is} for the 3 (m) line is now a capacitive reactance, whereas that for the 1.5 (m) line in part (a) is an inductive reactance. We may conclude from Fig. 9–9 that 1.5 (m) $< \lambda/4 <$ 3.0 (m).

c) In order for a short-circuited line to appear as an open circuit at the input terminals, it should be an odd multiple of a quarter-wavelength long:

$$\lambda = \frac{2\pi}{\beta} = \frac{2\pi}{0.628} = 10 \quad (m).$$

Hence the required line length is

$$\ell = \frac{\lambda}{4} + (n-1)\frac{\lambda}{2}$$

$$= 2.5 + 5(n-1) \quad (m), \qquad n = 1, 2, 3, \ldots. \quad ━━$$

So far in this subsection we have considered only open- and short-circuited loss-less lines as circuit elements. We have seen in Figs. 9–8 and 9–9 that, depending on the length of the line, the input impedance of an open- or short-circuited lossless line can be either purely inductive or purely capacitive. Let us now examine the input impedance of a lossy line with a short-circuit termination. When the line length is a multiple of $\lambda/2$, the input impedance will not vanish as in Fig. 9–9. Instead, we have, from Eq. (9–117),

$$
\begin{aligned}
Z_{is} = Z_0 \tanh \gamma\ell &= Z_0 \frac{\sinh (\alpha + j\beta)\ell}{\cosh (\alpha + j\beta)\ell} \\
&= Z_0 \frac{\sinh \alpha\ell \cos \beta\ell + j \cosh \alpha\ell \sin \beta\ell}{\cosh \alpha\ell \cos \beta\ell + j \sinh \alpha\ell \sin \beta\ell}.
\end{aligned}
\tag{9-120}
$$

For $\ell = n\lambda/2$, $\beta\ell = n\pi$, $\sin \beta\ell = 0$, Eq. (9–120) reduces to

$$
Z_{is} = Z_0 \tanh \alpha\ell \cong Z_0(\alpha\ell),
\tag{9-121}
$$

where we have assumed a low-loss line: $\alpha\ell \ll 1$ and $\tanh \alpha\ell \cong \alpha\ell$. The quantity Z_{is} in Eq. (9–121) is small but not zero. At $\ell = n\lambda/2$ we have the condition of a series-resonant circuit.

When the length of a shorted lossy line is an odd multiple of $\lambda/4$, the input impedance will not go to infinity as indicated in Fig. 9–9. For $\ell = n\lambda/4$, $\beta\ell = n\pi/2$ ($n = \text{odd}$), $\cos \beta\ell = 0$, and Eq. (9–120) becomes

$$
Z_{is} = \frac{Z_0}{\tanh \alpha\ell} \cong \frac{Z_0}{\alpha\ell},
\tag{9-122}
$$

which is large but not infinite. We have the condition of a parallel-resonant circuit. It is a frequency-selective circuit, and we can determine the **quality factor**, or Q, of such a circuit by first finding its **half-power bandwidth**, or simply the **bandwidth**. The bandwidth of a parallel-resonant circuit is the frequency range $\Delta f = f_2 - f_1$ around the resonant frequency f_0, where $f_2 = f_0 + \Delta f/2$ and $f_1 = f_0 - \Delta f/2$ are half-power frequencies at which the voltage across the parallel circuit is $1/\sqrt{2}$ or 70.7% of its maximum value at f_0 (assuming a constant-current source). Hence the associated power, which is proportional to $|Z_{is}|^2$ and is maximum at f_0, is one-half of its value at f_1 and f_2.

Let $f = f_0 + \delta f$, where δf is a small frequency shift from the resonant frequency. We have

$$
\begin{aligned}
\beta\ell = \frac{2\pi f}{u_p} \ell &= \frac{2\pi(f_0 + \delta f)}{u_p} \ell \\
&= \frac{n\pi}{2} + \frac{n\pi}{2}\left(\frac{\delta f}{f_0}\right), \qquad n = \text{odd},
\end{aligned}
\tag{9-123}
$$

$$
\cos \beta\ell = -\sin \left[\frac{n\pi}{2}\left(\frac{\delta f}{f_0}\right)\right] \cong -\frac{n\pi}{2}\left(\frac{\delta f}{f_0}\right),
\tag{9-124}
$$

$$
\sin \beta\ell = \cos \left[\frac{n\pi}{2}\left(\frac{\delta f}{f_0}\right)\right] \cong 1,
\tag{9-125}
$$

where we have assumed $(n\pi/2)(\delta f/f_0) \ll 1$. Substituting Eqs. (9–123), (9–124), and (9–125) in Eq. (9–120), noting that $\alpha\ell \ll 1$, and retaining only small terms of the first order, we obtain

$$Z_{is} = \frac{Z_0}{\alpha\ell + j\dfrac{n\pi}{2}\left(\dfrac{\delta f}{f_0}\right)} \qquad (9\text{--}126)$$

and

$$|Z_{is}|^2 = \frac{|Z_0|^2}{(\alpha\ell)^2 + \left[\dfrac{n\pi}{2}\left(\dfrac{\delta f}{f_0}\right)\right]^2}. \qquad (9\text{--}127)$$

At $f = f_0$, $\delta f = 0$, $|Z_{is}|^2$ is a maximum and equals $|Z_{is}|^2_{max} = |Z_0|^2/(\alpha\ell)^2$. Thus,

$$\frac{|Z_{is}|^2}{|Z_{is}|^2_{max}} = \frac{1}{1 + \left[\dfrac{n\pi}{2\alpha\ell}\left(\dfrac{\delta f}{f_0}\right)\right]^2}. \qquad (9\text{--}128)$$

When $\delta f = \pm\Delta f/2$, we have the half-power frequencies f_2 and f_1, at which the ratio in Eq. (9–128) equals $\frac{1}{2}$, or

$$\frac{n\pi}{2\alpha\ell}\left(\frac{\Delta f}{2f_0}\right) = \frac{\beta}{2\alpha}\left(\frac{\Delta f}{f_0}\right) = 1, \qquad n = \text{odd}. \qquad (9\text{--}129)$$

Therefore, the Q of the parallel-resonant circuit (a shorted lossy line having a length equal to an odd multiple of $\lambda/4$) is

$$\boxed{Q = \frac{f_0}{\Delta f} = \frac{\beta}{2\alpha}.} \qquad (9\text{--}130)$$

Using the expressions of α and β for a low-loss line in Eqs. (9–55) and (9–56), we obtain

$$Q = \frac{\omega L}{R + GL/C} = \frac{1}{[(R/\omega L) + (G/\omega C)]}. \qquad (9\text{--}131)$$

For a well-insulated line, $GL/C \ll R$, and Eq. (9–131) reduces to the familiar expression for the Q of a parallel-resonant circuit:

$$Q = \frac{\omega L}{R}. \qquad (9\text{--}132)$$

In a similar manner an analysis can be made for the resonant behavior of an open-circuited low-loss transmission line whose length is an odd multiple of $\lambda/4$ (series resonance) or a multiple of $\lambda/2$ (parallel resonance). (See Problem P.9–21.)

━━━ EXAMPLE 9–7 The measured attenuation of an air-dielectric coaxial transmission line at 400 (MHz) is 0.01 (dB/m). Determine the Q and the half-power bandwidth of a quarter-wavelength section of the line with a short-circuit termination.

Solution   At $f = 4 \times 10^8$ (Hz),

$$\lambda = \frac{c}{f} = \frac{3 \times 10^8}{4 \times 10^8} = 0.75 \quad \text{(m)},$$

$$\beta = \frac{2\pi}{\lambda} = \frac{2\pi}{0.75} = 8.38 \quad \text{(rad/m)},$$

$$\alpha = 0.01 \text{ (dB/m)} = \frac{0.01}{8.69} \quad \text{(Np/m)}.$$

Therefore,

$$Q = \frac{\beta}{2\alpha} = \frac{8.38 \times 8.69}{2 \times 0.01} = 3641,$$

which is much higher than the Q obtainable from any lumped-element parallel-resonant circuit at 400 (MHz). The half-power bandwidth is

$$\Delta f = \frac{f_0}{Q} = \frac{4 \times 10^8}{3641} = 0.11 \times 10^6 \quad \text{(Hz)}$$

$$= 0.11 \quad \text{(MHz)}, \quad \text{or } 110 \quad \text{(kHz)}. \qquad \blacksquare$$

9–4.2   LINES WITH RESISTIVE TERMINATION

When a transmission line is terminated in a load impedance Z_L different from the characteristic impedance Z_0, both an incident wave (from the generator) and a reflected wave (from the load) exist. Equation (9–99a) gives the phasor expression for the voltage at any distance $z' = \ell - z$ from the load end. Note that in Eq. (9–99a), the term with $e^{\gamma z'}$ represents the incident voltage wave and the term with $e^{-\gamma z'}$ represents the reflected voltage wave. We may write

$$V(z') = \frac{I_L}{2}(Z_L + Z_0)e^{\gamma z'}\left[1 + \frac{Z_L - Z_0}{Z_L + Z_0}e^{-2\gamma z'}\right]$$

$$= \frac{I_L}{2}(Z_L + Z_0)e^{\gamma z'}[1 + \Gamma e^{-2\gamma z'}], \qquad (9\text{--}133\text{a})$$

where

$$\boxed{\Gamma = \frac{Z_L - Z_0}{Z_L + Z_0} = |\Gamma|e^{j\theta_\Gamma} \quad \text{(Dimensionless)}} \qquad (9\text{--}134)$$

is the ratio of the complex amplitudes of the reflected and incident voltage waves at the load ($z' = 0$) and is called the **voltage reflection coefficient** of the load impedance Z_L. It is of the same form as the definition of the reflection coefficient in Eq. (8–140) for a plane wave incident normally on a plane interface between two dielectric media. It is, in general, a complex quantity with a magnitude $|\Gamma| \leq 1$. The current equation corresponding to $V(z')$ in Eq. (9–133a) is, from Eq. (9–99b),

$$I(z') = \frac{I_L}{2Z_0}(Z_L + Z_0)e^{\gamma z'}[1 - \Gamma e^{-2\gamma z'}]. \tag{9–133b}$$

The current reflection coefficient defined as the ratio of the complex amplitudes of the reflected and incident current waves, I_0^-/I_0^+, is different from the voltage reflection coefficient. As a matter of fact, the former is the negative of the latter, inasmuch as $I_0^-/I_0^+ = -V_0^-/V_0^+$, as is evident from Eq. (9–94). In what follows we shall refer only to the voltage reflection coefficient.

For a *lossless* transmission line, $\gamma = j\beta$, Eqs. (9–133a) and (9–133b) become

$$\begin{aligned}
V(z') &= \frac{I_L}{2}(Z_L + R_0)e^{j\beta z'}[1 + \Gamma e^{-j2\beta z'}] \\
&= \frac{I_L}{2}(Z_L + R_0)e^{j\beta z'}[1 + |\Gamma|e^{j(\theta_\Gamma - 2\beta z')}]
\end{aligned} \tag{9–135a}$$

and

$$I(z') = \frac{I_L}{2R_0}(Z_L + R_0)e^{j\beta z'}[1 - |\Gamma|e^{j(\theta_\Gamma - 2\beta z')}]. \tag{9–135b}$$

The voltage and current phasors on a lossless line are more easily visualized from Eqs. (9–100a) and (9–100b) by setting $\gamma = j\beta$ and $V_L = I_L Z_L$. Noting that $\cosh j\theta = \cos \theta$, and $\sinh j\theta = j \sin \theta$, we obtain

$$V(z') = V_L \cos \beta z' + jI_L R_0 \sin \beta z', \tag{9–136a}$$

$$I(z') = I_L \cos \beta z' + j\frac{V_L}{R_0} \sin \beta z'. \tag{9–136b}$$

(Lossless line)

If the terminating impedance is purely resistive, $Z_L = R_L$, $V_L = I_L R_L$, the voltage and current magnitudes are given by

$$|V(z')| = V_L\sqrt{\cos^2 \beta z' + (R_0/R_L)^2 \sin^2 \beta z'}, \tag{9–137a}$$

$$|I(z')| = I_L\sqrt{\cos^2 \beta z' + (R_L/R_0)^2 \sin^2 \beta z'}, \tag{9–137b}$$

where $R_0 = \sqrt{L/C}$. Plots of $|V(z')|$ and $|I(z')|$ as functions of z' are standing waves with their maxima and minima occurring at fixed locations along the line.

Analogously to the plane-wave case in Eq. (8–147), we define the ratio of the maximum to the minimum voltages along a finite, terminated line as the ***standing-wave ratio*** (**SWR**), S:

$$S = \frac{|V_{max}|}{|V_{min}|} = \frac{1 + |\Gamma|}{1 - |\Gamma|} \qquad \text{(Dimensionless).} \tag{9–138}$$

The inverse relation of Eq. (9–138) is

$$\boxed{|\Gamma| = \frac{S-1}{S+1} \quad \text{(Dimensionless).}} \qquad (9\text{–}139)$$

It is clear from Eqs. (9–138) and (9–139) that on a lossless transmission line

$$\Gamma = 0, \qquad S = 1 \qquad \text{when } Z_L = Z_0 \text{ (Matched load)};$$
$$\Gamma = -1, \qquad S \to \infty \qquad \text{when } Z_L = 0 \text{ (Short circuit)};$$
$$\Gamma = +1, \qquad S \to \infty \qquad \text{when } Z_L \to \infty \text{ (Open circuit)}.$$

Because of the wide range of S, it is customary to express it on a logarithmic scale: $20 \log_{10} S$ in (dB). Standing-wave ratio S defined in terms of $|I_{max}|/|I_{min}|$ results in the same expression as that defined in terms of $|V_{max}|/|V_{min}|$ in Eq. (9–138). A high standing-wave ratio on a line is undesirable because it results in a large power loss.

Examination of Eqs. (9–135a) and (9–135b) reveals that $|V_{max}|$ and $|I_{min}|$ occur together when

$$\theta_\Gamma - 2\beta z'_M = -2n\pi, \qquad n = 0, 1, 2, \ldots. \qquad (9\text{–}140)$$

On the other hand, $|V_{min}|$ and $|I_{max}|$ occur together when

$$\theta_\Gamma - 2\beta z'_m = -(2n+1)\pi, \qquad n = 0, 1, 2, \ldots. \qquad (9\text{–}141)$$

For resistive terminations on a lossless line, $Z_L = R_L$, $Z_0 = R_0$, and Eq. (9–134) simplifies to

$$\Gamma = \frac{R_L - R_0}{R_L + R_0} \qquad \text{(Resistive load).} \qquad (9\text{–}142)$$

The voltage reflection coefficient is therefore purely real. Two cases are possible.

1. $R_L > R_0$. In this case, Γ is positive real and $\theta_\Gamma = 0$. At the termination, $z' = 0$, and condition (9–140) is satisfied (for $n = 0$). This means that a voltage maximum (current minimum) will occur at the terminating resistance. Other maxima of the voltage standing wave (minima of the current standing wave) will be located at $2\beta z' = 2n\pi$, or $z' = n\lambda/2$ $(n = 1, 2, 3, \ldots)$ from the load.

2. $R_L < R_0$. Equation (9–142) shows that Γ will be negative real and $\theta_\Gamma = -\pi$. At the termination, $z' = 0$, and condition (9–141) is satisfied (for $n = 0$). A voltage minimum (current maximum) will occur at the terminating resistance. Other minima of the voltage standing wave (maxima of the current standing wave) will be located at $z' = n\lambda/2$ $(n = 1, 2, 3, \ldots)$ from the load. The roles of the voltage and current standing waves are interchanged from those for the case of $R_L > R_0$.

Figure 9–10 illustrates some typical standing waves for a lossless line with resistive termination.

The standing waves on an open-circuited line are similar to those on a resistance-terminated line with $R_L > R_0$, except that the $|V(z')|$ and $|I(z')|$ curves are now magnitudes of sinusoidal functions of the distance z' from the load. This is seen from

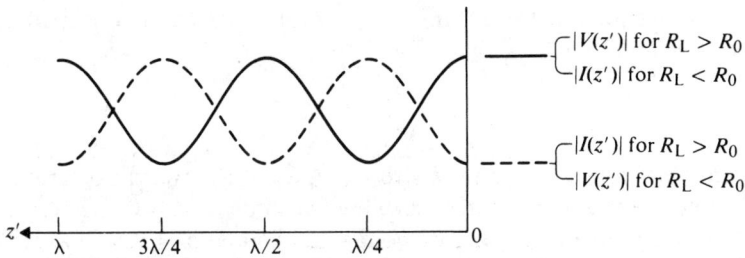

FIGURE 9-10
Voltage and current standing waves on resistance-terminated lossless lines.

Eqs. (9-137a) and (9-137b), by letting $R_L \to \infty$. Of course, $I_L = 0$, but V_L is finite. We have

$$|V(z')| = V_L|\cos \beta z'|, \tag{9-143a}$$

$$|I(z')| = \frac{V_L}{R_0}|\sin \beta z'|. \tag{9-143b}$$

All the minima go to zero. For an open-circuited line, $\Gamma = 1$ and $S \to \infty$.

On the other hand, the standing waves on a short-circuited line are similar to those on a resistance-terminated line with $R_L < R_0$. Here $R_L = 0$, $V_L = 0$, but I_L is finite. Equations (9-137a) and (9-137b) reduce to

$$|V(z')| = I_L R_0|\sin \beta z'|, \tag{9-144a}$$

$$|I(z')| = I_L|\cos \beta z'|. \tag{9-144b}$$

Typical standing waves for open- and short-circuited lossless lines are shown in Fig. 9-11.

EXAMPLE 9-8   The standing-wave ratio S on a transmission line is an easily measurable quantity. (a) Show how the value of a terminating resistance on a lossless line of known characteristic impedance R_0 can be determined by measuring S. (b) What

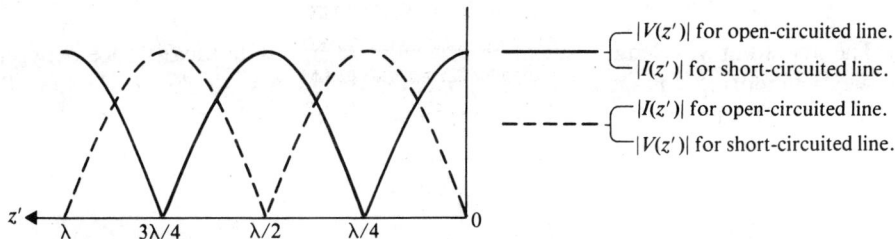

FIGURE 9-11
Voltage and current standing waves on open- and short-circuited lossless lines.

is the impedance of the line looking toward the load at a distance equal to one quarter
of the operating wavelength?

Solution

a) Since the terminating impedance is purely resistive, $Z_L = R_L$, we can determine
whether R_L is greater than R_0 (if there are voltage maxima at $z' = 0, \lambda/2, \lambda$, etc.)
or whether R_L is less than R_0 (if there are voltage minima at $z' = 0, \lambda/2, \lambda$, etc.).
This can be easily ascertained by measurements.

First, if $R_L > R_0$, $\theta_\Gamma = 0$. Both $|V_{max}|$ and $|I_{min}|$ occur at $\beta z' = 0$; and $|V_{min}|$
and $|I_{max}|$ occur at $\beta z' = \pi/2$. We have, from Eqs. (9–136a) and (9–136b),

$$|V_{max}| = V_L, \qquad |V_{min}| = V_L \frac{R_0}{R_L};$$

$$|I_{min}| = I_L, \qquad |I_{max}| = I_L \frac{R_L}{R_0}.$$

Thus,

$$\frac{|V_{max}|}{|V_{min}|} = \frac{|I_{max}|}{|I_{min}|} = S = \frac{R_L}{R_0}$$

or

$$R_L = SR_0. \qquad (9\text{–}145)$$

Second, if $R_L < R_0$, $\theta_\Gamma = -\pi$. Both $|V_{min}|$ and $|I_{max}|$ occur at $\beta z' = 0$; and
$|V_{max}|$ and $|I_{min}|$ occur at $\beta z' = \pi/2$. We have

$$|V_{min}| = V_L, \qquad |V_{max}| = V_L \frac{R_0}{R_L};$$

$$|I_{max}| = I_L, \qquad |I_{min}| = I_L \frac{R_L}{R_0}.$$

Therefore,

$$\frac{|V_{max}|}{|V_{min}|} = \frac{|I_{max}|}{|I_{min}|} = S = \frac{R_0}{R_L}$$

or

$$R_L = \frac{R_0}{S}. \qquad (9\text{–}146)$$

b) The operating wavelength, λ, can be determined from twice the distance between
two neighboring voltage (or current) maxima or minima. At $z' = \lambda/4$, $\beta z' = \pi/2$,
$\cos \beta z' = 0$, and $\sin \beta z' = 1$. Equations (9–136a) and (9–136b) become

$$V(\lambda/4) = jI_L R_0,$$

$$I(\lambda/4) = j\frac{V_L}{R_0}.$$

(*Question:* What is the significance of the j in these equations?) The ratio of $V(\lambda/4)$
to $I(\lambda/4)$ is the input impedance of a quarter-wavelength, resistively terminated,

lossless line.

$$Z_i(z' = \lambda/4) = R_i = \frac{V(\lambda/4)}{I(\lambda/4)}$$

$$= \frac{R_0^2}{R_L}.$$

This result is anticipated because of the impedance-transformation property of a quarter-wave line given in Eq. (9–114).   ■

9–4.3   LINES WITH ARBITRARY TERMINATION

In the preceding subsection we noted that the standing wave on a *resistively terminated* lossless transmission line is such that a voltage maximum (a current minimum) occurs at the termination where $z' = 0$ if $R_L > R_0$, and a voltage minimum (a current maximum) occurs there if $R_L < R_0$. What will happen if the terminating impedance is not a pure resistance? It is intuitively correct to expect that a voltage maximum or minimum will not occur at the termination and that both will be shifted away from the termination. In this subsection we will show that information on the direction and amount of this shift can be used to determine the terminating impedance.

Let the terminating (or load) impedance be $Z_L = R_L + jX_L$, and assume the voltage standing wave on the line to look like that depicted in Fig. 9–12. We note that neither a voltage maximum nor a voltage minimum appears at the load at $z' = 0$. If we let the standing wave continue, say, by an extra distance ℓ_m, it will reach a minimum. The voltage minimum is where it should be if the original terminating impedance Z_L is replaced by a line section of length ℓ_m terminated by a pure resistance

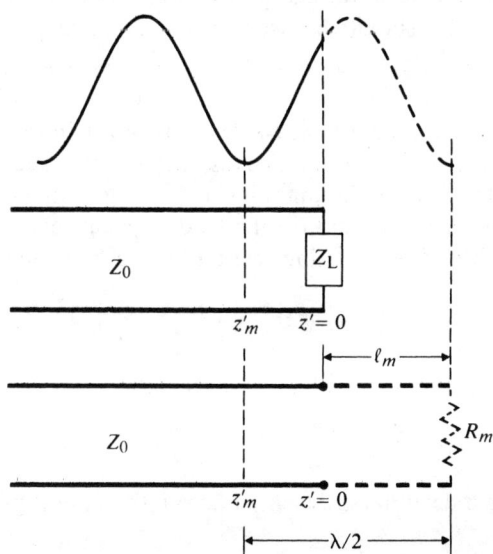

FIGURE 9–12
Voltage standing wave on a line terminated by an arbitrary impedance, and equivalent line section with pure resistive load.

$R_m < R_0$, as shown in the figure. The voltage distribution on the line to the left of the actual termination (where $z' > 0$) is not changed by this replacement.

The fact that any complex impedance can be obtained as the input impedance of a section of lossless line terminated in a resistive load can be seen from Eq. (9–109). Using R_m for Z_L and ℓ_m for ℓ, we have

$$R_i + jX_i = R_0 \frac{R_m + jR_0 \tan \beta \ell_m}{R_0 + jR_m \tan \beta \ell_m}. \tag{9-147}$$

The real and imaginary parts of Eq. (9–147) form two equations, from which the two unknowns, R_m and ℓ_m, can be solved (see Problem P.9–28).

The load impedance Z_L can be determined experimentally by measuring the standing-wave ratio S and the distance z'_m in Fig. 9–12. (Remember that $z'_m + \ell_m = \lambda/2$.) The procedure is as follows:

1. Find $|\Gamma|$ from S. Use $|\Gamma| = \dfrac{S - 1}{S + 1}$ from Eq. (9–139).

2. Find θ_Γ from z'_m. Use $\theta_\Gamma = 2\beta z'_m - \pi$ for $n = 0$ from Eq. (9–141).

3. Find Z_L, which is the ratio of Eqs. (9–135a) and (9–135b) at $z' = 0$:

$$Z_L = R_L + jX_L = R_0 \frac{1 + |\Gamma|e^{j\theta_\Gamma}}{1 - |\Gamma|e^{j\theta_\Gamma}}. \tag{9-148}$$

The value of R_m that, if terminated on a line of length ℓ_m, will yield an input impedance Z_L can be found easily from Eq. (9–147). Since $R_m < R_0$, $R_m = R_0/S$.

The procedure leading to Eq. (9–148) is used to determine Z_L from a measurement of S and of z'_m, the distance from the termination to the first voltage minimum. Of course, the distance from the termination to a voltage maximum, z'_M, could be used instead of z'_m. In that case, Eq. (9–140) should be used to find θ_Γ in Step 2 above.

EXAMPLE 9–9 The standing-wave ratio on a lossless 50 (Ω) transmission line terminated in an unknown load impedance is found to be 3.0. The distance between successive voltage minima is 20 (cm), and the first minimum is located at 5 (cm) from the load. Determine (a) the reflection coefficient Γ, and (b) the load impedance Z_L. In addition, find (c) the equivalent length and terminating resistance of a line such that the input impedance is equal to Z_L.

Solution

a) The distance between successive voltage minima is half a wavelength.

$$\lambda = 2 \times 0.2 = 0.4 \quad \text{(m)}, \qquad \beta = \frac{2\pi}{\lambda} = \frac{2\pi}{0.4} = 5\pi \quad \text{(rad/m)}.$$

Step 1: We find the magnitude of the reflection coefficient, $|\Gamma|$, from the standing-wave ratio $S = 3$.

$$|\Gamma| = \frac{S - 1}{S + 1} = \frac{3 - 1}{3 + 1} = 0.5.$$

Step 2: Find the angle of the reflection coefficient, θ_Γ, from

$$\theta_\Gamma = 2\beta z'_m - \pi = 2 \times 5\pi \times 0.05 - \pi = -0.5\pi \quad \text{(rad)},$$
$$\Gamma = |\Gamma|e^{j\theta_\Gamma} = 0.5e^{-j0.5\pi} = -j0.5.$$

b) The load impedance Z_L is determined from Eq. (9–148):

$$Z_L = 50\left(\frac{1 - j0.5}{1 + j0.5}\right) = 50(0.60 - j0.80) = 30 - j40 \quad (\Omega).$$

c) Now we find R_m and ℓ_m in Fig. 9–12. We may use Eq. (9–147),

$$30 - j40 = 50\left(\frac{R_m + j50 \tan \beta\ell_m}{50 + jR_m \tan \beta\ell_m}\right),$$

and solve the simultaneous equations obtained from the real and imaginary parts for R_m and $\beta\ell_m$. Actually, we know $z'_m + \ell_m = \lambda/2$ and $R_m = R_0/S$. Hence,[†]

$$\ell_m = \frac{\lambda}{2} - z'_m = 0.2 - 0.05 = 0.15 \quad \text{(m)}$$

and

$$R_m = \frac{50}{3} = 16.7 \quad (\Omega).$$

     ■

9–4.4   TRANSMISSION-LINE CIRCUITS

Our discussions on the properties of transmission lines so far have been restricted primarily to the effects of the load on the input impedance and on the characteristics of voltage and current waves. No attention has been paid to the generator at the "other end," which is the source of the waves. Just as the constraint (the boundary condition), $V_L = I_L Z_L$, which the voltage V_L and the current I_L must satisfy at the load end ($z = \ell$, $z' = 0$), a constraint exists at the generator end where $z = 0$ and $z' = \ell$. Let a voltage generator V_g with an internal impedance Z_g represent the source connected to a finite transmission line of length ℓ that is terminated in a load impedance Z_L, as shown in Fig. 9–6. The additional constraint at $z = 0$ will enable the voltage and current anywhere on the line to be expressed in terms of the source characteristics (V_g, Z_g), the line characteristics (γ, Z_0, ℓ), and the load impedance (Z_L).
    The constraint at $z = 0$ is

$$V_i = V_g - I_i Z_g. \tag{9–149}$$

But, from Eqs. (9–133a) and (9–133b),

$$V_i = \frac{I_L}{2}(Z_L + Z_0)e^{\gamma\ell}[1 + \Gamma e^{-2\gamma\ell}] \tag{9–150a}$$

and

$$I_i = \frac{I_L}{2Z_0}(Z_L + Z_0)e^{\gamma\ell}[1 - \Gamma e^{-2\gamma\ell}]. \tag{9–150b}$$

[†] Another set of solutions to part (c) is $\ell'_m = \ell_m - \lambda/4 = 0.05$ (m) and $R'_m = SR_0 = 150$ (Ω). Do you see why?

Substitution of Eqs. (9–150a) and (9–150b) in Eq. (9–149) enables us to find

$$\frac{I_L}{2}(Z_L + Z_0)e^{\gamma\ell} = \frac{Z_0 V_g}{Z_0 + Z_g}\frac{1}{[1 - \Gamma_g\Gamma e^{-2\gamma\ell}]},$$

(9–151)

where

$$\Gamma_g = \frac{Z_g - Z_0}{Z_g + Z_0}$$

(9–152)

is the **voltage reflection coefficient** at the generator end. Using Eq. (9–151) in Eqs. (9–133a) and (9–133b), we obtain

$$\boxed{V(z') = \frac{Z_0 V_g}{Z_0 + Z_g}\, e^{-\gamma z}\left(\frac{1 + \Gamma e^{-2\gamma z'}}{1 - \Gamma_g\Gamma e^{-2\gamma\ell}}\right).}$$

(9–153a)

Similarly,

$$\boxed{I(z') = \frac{V_g}{Z_0 + Z_g}\, e^{-\gamma z}\left(\frac{1 - \Gamma e^{-2\gamma z'}}{1 - \Gamma_g\Gamma e^{-2\gamma\ell}}\right).}$$

(9–153b)

Equations (9–153a) and (9–153b) are analytical phasor expressions for the voltage and current at any point on a finite line fed by a sinusoidal voltage source V_g. These are rather complicated expressions, but their significance can be interpreted in the following way. Let us concentrate our attention on the voltage equation (9–153a); obviously, the interpretation of the current equation (9–153b) is quite similar. We expand Eq. (9–153a) as follows:

$$\begin{aligned}V(z') &= \frac{Z_0 V_g}{Z_0 + Z_g}\, e^{-\gamma z}(1 + \Gamma e^{-2\gamma z'})(1 - \Gamma_g\Gamma e^{-2\gamma\ell})^{-1}\\[2mm]
&= \frac{Z_0 V_g}{Z_0 + Z_g}\, e^{-\gamma z}(1 + \Gamma e^{-2\gamma z'})(1 + \Gamma_g\Gamma e^{-2\gamma\ell} + \Gamma_g^2\Gamma^2 e^{-4\gamma\ell} + \cdots)\\[2mm]
&= \frac{Z_0 V_g}{Z_0 + Z_g}\, [e^{-\gamma z} + (\Gamma e^{-\gamma\ell})e^{-\gamma z'} + \Gamma_g(\Gamma e^{-2\gamma\ell})e^{-\gamma z} + \cdots]\\[2mm]
&= V_1^+ + V_1^- + V_2^+ + V_2^- + \cdots,\end{aligned}$$

(9–154)

where

$$V_1^+ = \frac{V_g Z_0}{Z_0 + Z_g}\, e^{-\gamma z} = V_M e^{-\gamma z},$$

(9–154a)

$$V_1^- = \Gamma(V_M e^{-\gamma\ell})e^{-\gamma z'},$$

(9–154b)

$$V_2^+ = \Gamma_g(\Gamma V_M e^{-2\gamma\ell})e^{-\gamma z}.$$

(9–154c)

$$\vdots$$

The quantity

$$V_M = \frac{Z_0 V_g}{Z_0 + Z_g}$$

(9–155)

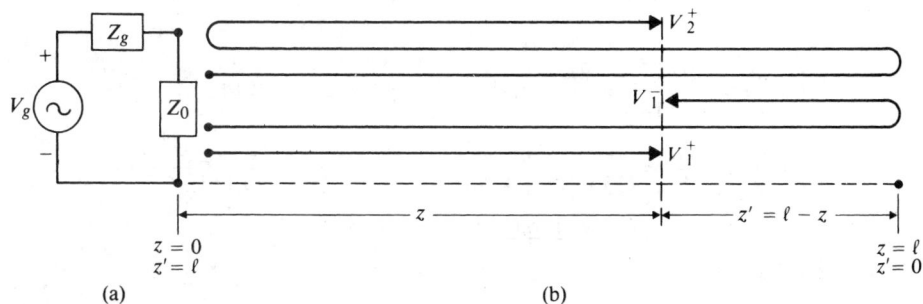

FIGURE 9–13
A transmission-line circuit and traveling waves.

is the complex amplitude of the voltage wave initially sent down the transmission line from the generator. It is obtained directly from the simple circuit shown in Fig. 9–13(a). The phasor V_1^+ in Eq. (9–154a) represents the initial wave traveling in the $+z$-direction. Before this wave reaches the load impedance Z_L, it sees Z_0 of the line as if the line were infinitely long.

When the first wave $V_1^+ = V_M e^{-\gamma z}$ reaches Z_L at $z = \ell$, it is reflected because of mismatch, resulting in a wave V_1^- with a complex amplitude $\Gamma(V_M e^{-\gamma \ell})$ traveling in the $-z$-direction. As the wave V_1^- returns to the generator at $z = 0$, it is again reflected for $Z_g \neq Z_0$, giving rise to a second wave V_2^+ with a complex amplitude $\Gamma_g(\Gamma V_M e^{-2\gamma \ell})$ traveling in $+z$-direction. This process continues indefinitely with reflections at both ends, and the resulting standing wave $V(z')$ is the sum of all the waves traveling in both directions. This is illustrated schematically in Fig. 9–13(b). In practice, $\gamma = \alpha + j\beta$ has a real part, and the attenuation effect of $e^{-\alpha \ell}$ diminishes the amplitude of a reflected wave each time the wave transverses the length of the line.

When the line is terminated with a matched load, $Z_L = Z_0$, $\Gamma = 0$, only V_1^+ exists, and it stops at the matched load with no reflections. If $Z_L \neq Z_0$ but $Z_g = Z_0$ (if the internal impedance of the generator is matched to the line), then $\Gamma \neq 0$ and $\Gamma_g = 0$. As a consequence, both V_1^+ and V_1^- exist, and V_2^+, V_2^- and all higher-order reflections vanish.

EXAMPLE 9–10   A 100 (MHz) generator with $V_g = 10\underline{/0°}$ (V) and internal resistance 50 (Ω) is connected to a lossless 50 (Ω) air line that is 3.6 (m) long and terminated in a 25 + j25 (Ω) load. Find (a) $V(z)$ at a location z from the generator, (b) V_i at the input terminals and V_L at the load, (c) the voltage standing-wave ratio on the line, and (d) the average power delivered to the load.

Solution   Referring to Fig. 9–6, the given quantities are

$$V_g = 10\underline{/0°} \ \text{(V)}, \qquad Z_g = 50 \ \text{(Ω)}, \qquad f = 10^8 \ \text{(Hz)},$$
$$R_0 = 50 \ \text{(Ω)}, \qquad Z_L = 25 + j25 = 35.36\underline{/45°} \ \text{(Ω)}, \qquad \ell = 3.6 \ \text{(m)}.$$

Thus,

$$\beta = \frac{\omega}{c} = \frac{2\pi 10^8}{3 \times 10^8} = \frac{2\pi}{3} \quad \text{(rad/m)}, \qquad \beta\ell = 2.4\pi \quad \text{(rad)},$$

$$\Gamma = \frac{Z_L - Z_0}{Z_L + Z_0} = \frac{(25 + j25) - 50}{(25 + j25) + 50} = \frac{-25 + j25}{75 + j25} = \frac{35.36\underline{/135°}}{79.1\underline{/18.4°}}$$

$$= 0.447\underline{/116.6°} = 0.447\underline{/0.648\pi},$$

$$\Gamma_g = 0.$$

a) From Eq. (9–153a) we have

$$V(z) = \frac{Z_0 V_g}{Z_0 + Z_g} e^{-j\beta z}[1 + \Gamma e^{-j2\beta(\ell - z)}]$$

$$= \frac{50(10)}{100} e^{-j2\pi z/3}[1 + 0.447 e^{j(0.648 - 4.8)\pi} e^{j4\pi z/3}]$$

$$= 5[e^{-j2\pi z/3} + 0.447 e^{j(2z/3 - 0.152)\pi}] \quad \text{(V)}.$$

We see that, because $\Gamma_g = 0$, $V(z)$ is the superposition of only two traveling waves, V_1^+ and V_1^-, as defined in Eq. (9–154).

b) At the input terminals,

$$V_i = V(0) = 5(1 + 0.447 e^{-j0.152\pi})$$

$$= 5(1.396 - j0.207)$$

$$= 7.06\underline{/-8.43°} \quad \text{(V)}.$$

At the load,

$$V_L = V(3.6) = 5[e^{-j0.4\pi} + 0.447 e^{j0.248\pi}]$$

$$= 5(0.627 - j0.637) = 4.47\underline{/-45.5°} \quad \text{(V)}.$$

c) The voltage standing-wave ratio (VSWR) is

$$S = \frac{1 + |\Gamma|}{1 - |\Gamma|} = \frac{1 + 0.447}{1 - 0.447} = 2.62.$$

d) The average power delivered to the load is

$$P_{av} = \frac{1}{2}\left|\frac{V_L}{Z_L}\right|^2 R_L = \frac{1}{2}\left(\frac{4.47}{35.36}\right)^2 \times 25 = 0.200 \quad \text{(W)}.$$

It is interesting to compare this result with the case of a matched load when $Z_L = Z_0 = 50 + j0$ (Ω). In that case, $\Gamma = 0$,

$$|V_L| = |V_i| = \frac{V_g}{2} = 5 \quad \text{(V)},$$

and a maximum average power is delivered to the load:

$$\text{Maximum } P_{av} = \frac{V_L^2}{2R_L} = \frac{5^2}{2 \times 50} = 0.25 \quad \text{(W)},$$

which is larger than the P_{av} calculated for the unmatched load in part (d) by an amount equal to the power reflected, $|\Gamma|^2 \times 0.25 = 0.05$ (W). ▬

9–5 Transients on Transmission Lines

The discussion of the wave characteristics on transmission lines in the previous section was based on steady-state, single-frequency, time-harmonic sources and signals. We worked with voltage and current phasors. Quantities such as reactances (X), wavelength (λ), wavenumber (k), and phase constant (β) would lose their meaning under transient conditions. However, there are important practical situations in which the sources and signals are not time-harmonic and the conditions are not steady-state. Examples are digital (pulse) signals in computer networks and sudden surges in power and telephone lines. In this section we will consider the transient behavior of lossless transmission lines. For such lines ($R = 0$, $G = 0$), characteristic impedance becomes characteristic resistance $R_0 = 1/\sqrt{LC}$, and voltage and current waves propagate along the line with a velocity $u = 1/\sqrt{LC}$.

The simplest case is shown in Fig. 9–14(a), where a d-c voltage source V_0 is applied through a series (internal) resistance R_g at $t = 0$ to the input terminals of a lossless line terminated in a characteristic resistance R_0. Since the impedance looking into the terminated line is R_0, a voltage wave of magnitude

$$V_1^+ = \frac{R_0}{R_0 + R_g} V_0 \tag{9–156}$$

travels down the line in the $+z$-direction with a velocity $u = 1/\sqrt{LC}$. The corresponding magnitude, I_1^+, of the current wave is

$$I_1^+ = \frac{V_1^+}{R_0} = \frac{V_0}{R_0 + R_g}. \tag{9–157}$$

If we plot the voltage across the line at $z = z_1$ as a function of time, we obtain a delayed step function at $t = z_1/u$ as in Fig. 9–14(b). The current in the line at $z = z_1$

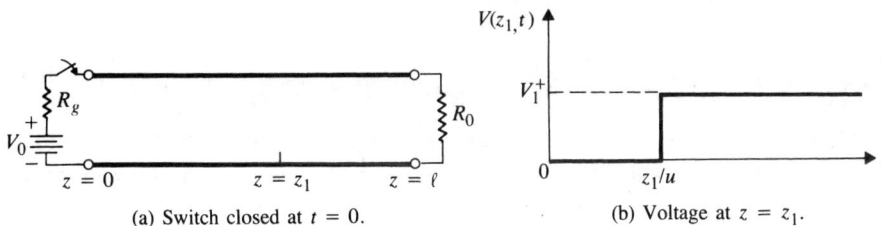

(a) Switch closed at $t = 0$.

(b) Voltage at $z = z_1$.

FIGURE 9–14
A d-c source applied to a line terminated in characteristic resistance R_0 through a series resistance R_g.

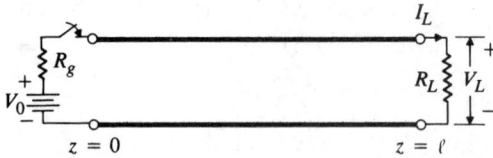

FIGURE 9–15
A d-c source applied to a terminated lossless line
at $t = 0$ (general case).

has the same shape with a magnitude I_1^+ given in Eq. (9–157). When the voltage
and current waves reach the termination at $z = \ell$, there are no reflected waves because $\Gamma = 0$. A steady state is established, and the entire line is charged to a voltage
equal to V_1^+.

If both the series resistance R_g and the load resistance R_L are not equal to R_0,
as in Fig. 9–15, the situation is more complicated. When the switch is closed at $t = 0$,
the d-c source sends a voltage wave of magnitude

$$V_1^+ = \frac{R_0}{R_0 + R_g} V_0 \qquad (9–158)$$

in the $+z$-direction with a velocity $u = 1/\sqrt{LC}$ as before because the V_1^+ wave has
no knowledge of the length of the line or the nature of the load at the other end; it
proceeds as if the line were infinitely long. At $t = T = \ell/u$ this wave reaches the load
end $z = \ell$. Since $R_L \neq R_0$, a reflected wave will travel in the $-z$-direction with a
magnitude

$$V_1^- = \Gamma_L V_1^+, \qquad (9–159)$$

where

$$\Gamma_L = \frac{R_L - R_0}{R_L + R_0} \qquad (9–160)$$

is the reflection coefficient of the load resistance R_L. This reflected wave arrives at
the input end at $t = 2T$, where it is reflected by $R_g \neq R_0$. A new voltage wave having
a magnitude V_2^+ then travels down the line, where

$$V_2^+ = \Gamma_g V_1^- = \Gamma_g \Gamma_L V_1^+. \qquad (9–161)$$

In Eq. (9–161),

$$\Gamma_g = \frac{R_g - R_0}{R_g + R_0} \qquad (9–162)$$

is the reflection coefficient of the series resistance R_g. This process will go on indefinitely with waves traveling back and forth, being reflected at each end at $t = nT$
$(n = 1, 2, 3, \ldots)$.

Two points are worth noting here. First, some of the reflected waves traveling
in either direction may have a negative amplitude, since Γ_L or Γ_g (or both) may be
negative. Second, except for an open circuit or a short circuit, Γ_L and Γ_g are less
than unity. Thus the magnitude of the successive reflected waves becomes smaller

and smaller, leading to a convergent process. The progression of the transient voltage waves on the lossless line in Fig. 9–15 for $R_L = 3R_0$ ($\Gamma_L = \frac{1}{2}$) and $R_g = 2R_0$ ($\Gamma_g = \frac{1}{3}$) is illustrated in Figs. 9–16(a), 9–16(b), and 9–16(c) for three different time intervals. The corresponding current waves are given in Figs. 9–16(d), 9–16(e), and 9–16(f), which are self-explanatory. The voltage and current at any particular location on the line in any particular time interval are just the *algebraic sums* $(V_1^+ + V_1^- + V_2^+ + V_2^- + \ldots)$ and $(I_1^+ + I_1^- + I_2^+ + I_2^- + \ldots)$, respectively.

It is interesting to check the ultimate value of the voltage across the load, $V_L = V(\ell)$, as t increases indefinitely. We have

$$
\begin{aligned}
V_L &= V_1^+ + V_1^- + V_2^+ + V_2^- + V_3^+ + V_3^- + \cdots \\
&= V_1^+ (1 + \Gamma_L + \Gamma_g \Gamma_L + \Gamma_g \Gamma_L^2 + \Gamma_g^2 \Gamma_L^2 + \Gamma_g^2 \Gamma_L^3 + \cdots) \\
&= V_1^+ [(1 + \Gamma_g \Gamma_L + \Gamma_g^2 \Gamma_L^2 + \cdots) + \Gamma_L(1 + \Gamma_g \Gamma_L + \Gamma_g^2 \Gamma_L^2 + \cdots)] \\
&= V_1^+ \left[\left(\frac{1}{1 - \Gamma_g \Gamma_L} \right) + \left(\frac{\Gamma_L}{1 - \Gamma_g \Gamma_L} \right) \right] \\
&= V_1^+ \left(\frac{1 + \Gamma_L}{1 - \Gamma_g \Gamma_L} \right).
\end{aligned}
\tag{9–163}
$$

For the present case, $V_1^+ = V_0/3$, $\Gamma_L = \frac{1}{2}$, and $\Gamma_g = 1/3$, Eq. (9–163) gives

$$
V_L = \tfrac{9}{5} V_1^+ = \tfrac{3}{5} V_0
\tag{9–163a}
$$

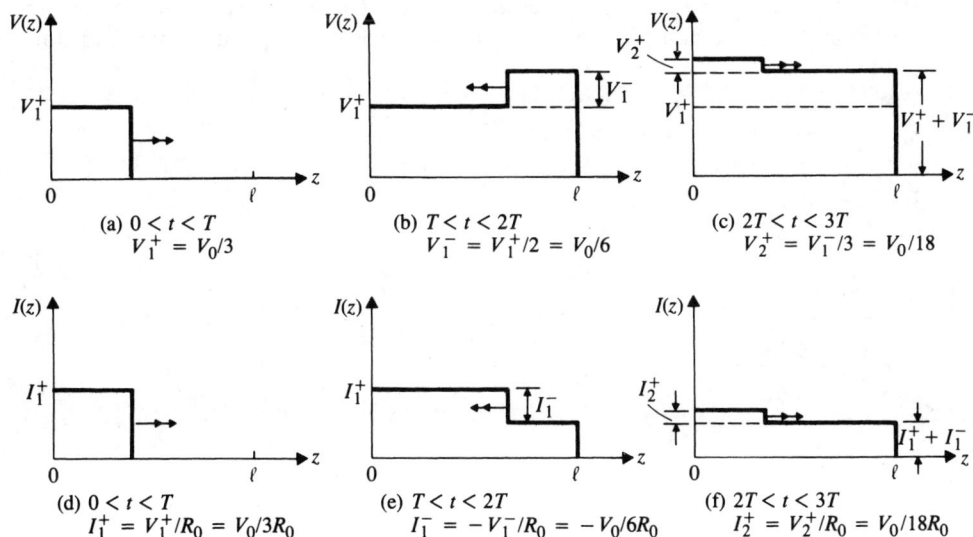

(a) $0 < t < T$
$V_1^+ = V_0/3$

(b) $T < t < 2T$
$V_1^- = V_1^+/2 = V_0/6$

(c) $2T < t < 3T$
$V_2^+ = V_1^-/3 = V_0/18$

(d) $0 < t < T$
$I_1^+ = V_1^+/R_0 = V_0/3R_0$

(e) $T < t < 2T$
$I_1^- = -V_1^-/R_0 = -V_0/6R_0$

(f) $2T < t < 3T$
$I_2^+ = V_2^+/R_0 = V_0/18R_0$

FIGURE 9–16
Transient voltage and current waves on transmission line in Fig. 9–15 for $R_L = 3R_0$ and $R_g = 2R_0$.

as $t \to \infty$. This result is obviously correct because, in the steady state, V_0 is divided between R_L and R_g in a ratio of 3 to 2. Similarly, we find

$$I_L = \left(\frac{1 - \Gamma_L}{1 - \Gamma_g \Gamma_L}\right) \frac{V_1^+}{R_0},$$

which yields

$$I_L = \frac{3}{5}\left(\frac{V_1^+}{R_0}\right) = \frac{V_0}{5R_0}, \qquad (9\text{--}164)$$

as expected.

9–5.1   REFLECTION DIAGRAMS

The preceding step-by-step construction and calculation procedure of the voltage and current at a particular time and location on a transmission line with arbitrary resistive terminations tends to be tedious and difficult to visualize when it is necessary to consider many reflected waves. In such cases the graphical construction of a reflection diagram is very helpful. Let us first construct a *voltage reflection diagram*. A *reflection diagram* plots the time elapsed after the change of circuit conditions versus the distance z from the source end. The voltage reflection diagram for the transmission-line circuit in Fig. 9–15 is given in Fig. 9–17. It starts with a wave V_1^+ at $t = 0$ traveling from the source end ($z = 0$) in the $+z$-direction with a velocity $u = 1/\sqrt{LC}$. This wave is represented by the directed straight line marked V_1^+ from the origin. This line has a positive slope equal to $1/u$. When the V_1^+ wave reaches the load at $z = \ell$, a reflected wave $V_1^- = \Gamma_L V_1^+$ is created if $R_L \neq R_0$. The V_1^- wave travels in the $-z$-direction and is represented by the directed line marked $\Gamma_L V_1^+$ with a negative slope equal to $-1/u$.

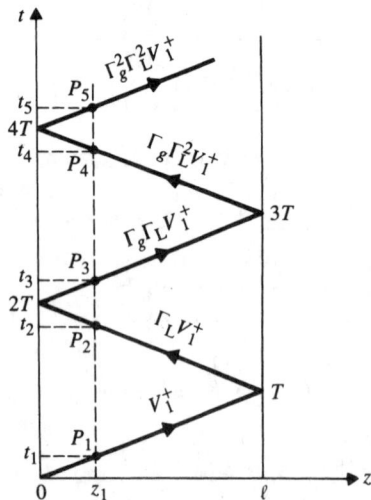

FIGURE 9–17
Voltage reflection diagram for transmission-line circuit in Fig. 9–15.

The V_1^- wave returns to the source end at $t = 2T$ and gives rise to a reflected wave $V_2^+ = \Gamma_g V_1^- = \Gamma_g \Gamma_L V_1^+$, which is represented by a second directed line with a positive slope. This process continues back and forth indefinitely. The voltage reflection diagram can be used conveniently to determine the voltage distribution along the transmission line at a given time as well as the variation of the voltage as a function of time at an arbitrary point on the line.

Suppose we wish to know the voltage distribution along the line at $t = t_4$ ($3T < t_4 < 4T$). We proceed as follows:

1. Mark t_4 on the vertical t-axis of the voltage reflection diagram.

2. Draw a horizontal line from t_4, intersecting the directed line marked $\Gamma_g \Gamma_L^2 V_1^+$ at P_4. (All directed lines above P_4 are irrelevant to our problem because they pertain to $t > t_4$.)

3. Draw a vertical line through P_4, intersecting the horizontal z-axis at z_1. The significance of z_1 is that in the range $0 < z < z_1$ (to the left of the vertical line) the voltage has a value equal to $V_1^+ + V_1^- + V_2^+ = V_1^+(1 + \Gamma_L + \Gamma_g \Gamma_L)$; and in the range $z_1 < z < \ell$ (to the right of the vertical line) the voltage is $V_1^+ + V_1^- + V_2^+ + V_2^- = V_1^+(1 + \Gamma_L + \Gamma_g \Gamma_L + \Gamma_g \Gamma_L^2)$. There is a voltage discontinuity equal to $\Gamma_g \Gamma_L^2 V_1^+$ at $z = z_1$.

4. The voltage distribution along the line at $t = t_4$, $V(z, t_4)$, is then as shown in Fig. 9–18(a), plotted for $R_L = 3R_0$ ($\Gamma_L = \frac{1}{2}$) and $R_g = 2R_0$ ($\Gamma_g = \frac{1}{3}$).

Next let us find the variation of the voltage as a function of time at the point $z = z_1$. We use the following procedure:

1. Draw a vertical line at z_1, intersecting the directed lines at points P_1, P_2, P_3, P_4, P_5, and so on. (There would be an infinite number of such intersection points if $R_L \neq R_0$ and $R_g \neq R_0$, as there would be an infinite number of directed lines if $\Gamma_L \neq 0$ and $\Gamma_g \neq 0$.)

2. From these intersection points, draw horizontal lines intersecting the vertical t-axis at t_1, t_2, t_3, t_4, t_5, and so on. These are the instants at which a new voltage wave arrives and abruptly changes the voltage at $z = z_1$.

3. The voltage at $z = z_1$ as a function of t can be read from the voltage reflection diagram as follows:

Time Range	Voltage	Voltage Discontinuity
$0 \leq t < t_1$ ($t_1 = z_1/u$)	0	0
$t_1 \leq t < t_2$ ($t_2 = 2T - t_1$)	V_1^+	V_1^+ at t_1
$t_2 \leq t < t_3$ ($t_3 = 2T + t_1$)	$V_1^+(1 + \Gamma_L)$	$\Gamma_L V_1^+$ at t_2
$t_3 \leq t < t_4$ ($t_4 = 4T - t_1$)	$V_1^+(1 + \Gamma_L + \Gamma_g \Gamma_L)$	$\Gamma_g \Gamma_L V_1^+$ at t_3
$t_4 \leq t < t_5$ ($t_5 = 4T + t_1$)	$V_1^+(1 + \Gamma_L + \Gamma_g \Gamma_L + \Gamma_g \Gamma_L^2)$	$\Gamma_g \Gamma_L^2 V_1^+$ at t_4
$\vdots$	$\vdots$	$\vdots$

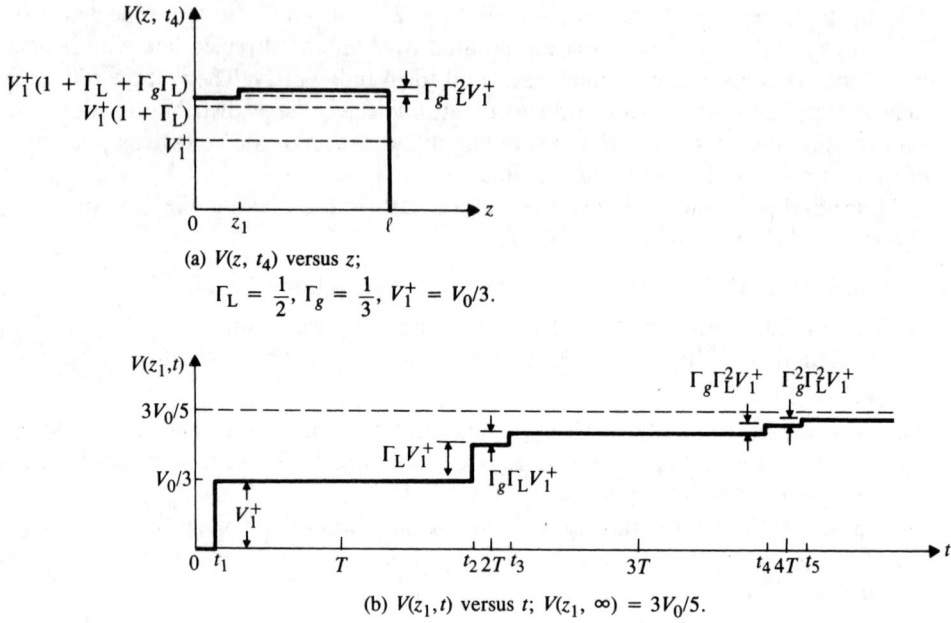

(a) $V(z, t_4)$ versus z;

$$\Gamma_L = \frac{1}{2}, \ \Gamma_g = \frac{1}{3}, \ V_1^+ = V_0/3.$$

(b) $V(z_1, t)$ versus t; $V(z_1, \infty) = 3V_0/5$.

FIGURE 9–18
Transient voltage on lossless transmission line for $R_L = 3R_0$ and $R_g = 2R_0$.

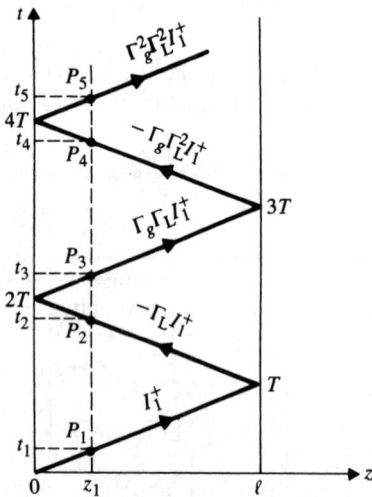

FIGURE 9–19
Current reflection diagram for transmission-line circuit in Fig. 9–15.

4. The graph of $V(z_1, t)$ is plotted in Fig. 9–18(b) for $\Gamma_L = \frac{1}{2}$ and $\Gamma_g = \frac{1}{3}$. When t increases indefinitely, the voltage at z_1 (and at all other points along the lossless line) will assume the value $3V_0/5$, as given in Eq. (9–163a).

Similar to the voltage reflection diagram in Fig. 9–17, a *current reflection diagram* for the transmission-line circuit of Fig. 9–15 can be constructed. This is shown in Fig. 9–19. Here we draw directed lines representing current waves. The essential difference between the voltage and current reflection diagrams is in the negative sign associated with the current waves traveling in the $-z$-direction on account of Eq. (9–94). The current reflection diagram can be used to determine the current distribution along the transmission line at a given time as well as the variation of the current as a function of time at a particular point on the line, following the same procedures outlined previously for voltage. For example, we can determine the current at $z = z_1$ by drawing a vertical line through z_1 in Fig. 9–19, intersecting the directed lines at points P_1, P_2, P_3, P_4, P_5, and so on, and by finding the corresponding times t_1, t_2, t_3, t_4, t_5, and so on, as before. Figure 9–20 is a plot of $I(z_1, t)$ versus t, which accompanies the $V(z_1, t)$ graph in Fig. 9–18(b). We see that they are quite dissimilar. The current along the line oscillates around the steady-state value of $V_0/5R_0$ (see Eq. 9–164) with successively smaller discontinuous jumps at t_1, t_2, t_3, t_4, t_5, etc.

We note two special cases here.

1. When $R_L = R_0$ (matched load, $\Gamma_L = 0$), the voltage and current reflection diagrams will each have only a single directed line, existing in the interval $0 < t < T$, irrespective of what R_g is.

2. When $R_g = R_0$ (matched source, $\Gamma_g = 0$) and $R_L \neq R_0$, the voltage and current reflection diagrams will each have only two directed lines, existing in the intervals $0 < t < T$ and $T < t < 2T$.

In both cases the determination of the transient behavior on the transmission line is much simplified.

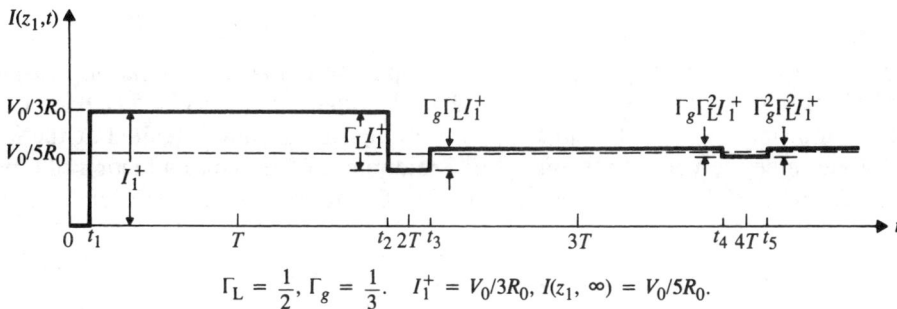

$$\Gamma_L = \frac{1}{2}, \ \Gamma_g = \frac{1}{3}. \quad I_1^+ = V_0/3R_0, \ I(z_1, \infty) = V_0/5R_0.$$

FIGURE 9–20
Transient current on lossless transmission line for $R_L = 3R_0$ and $R_g = 2R_0$.

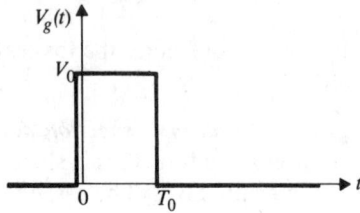

FIGURE 9–21
A rectangular pulse.

9–5.2 PULSE EXCITATION

So far, we have discussed the transient behavior of lossless transmission lines when the source is a sudden voltage surge in the form of a step function; that is,

$$v_g(t) = V_0 U(t), \tag{9–165}$$

where $U(t)$ denotes the unit step function

$$U(t) = \begin{cases} 0, & t < 0, \\ 1, & t > 0. \end{cases} \tag{9–166}$$

In many instances, such as in computer networks and pulse-modulation systems, the excitation may be in the form of pulses. The analysis of the transient behavior of a line with pulse excitation, however, does not present special difficulties because a rectangular pulse can be decomposed into two step functions. For example, the pulse of an amplitude V_0 lasting from $t = 0$ to $t = T_0$ shown in Fig. 9–21 can be written as

$$v_g(t) = V_0[U(t) - U(t - T_0)]. \tag{9–166a}$$

If $v_g(t)$ in Eq. (9–166a) is applied to a transmission line, the transient response is simply the superposition of the result obtained from a d-c voltage V_0 applied at $t = 0$ and that obtained from another d-c voltage $-V_0$ applied at $t = T_0$. We will illustrate this process by an example.

EXAMPLE 9–11 A rectangular pulse of an amplitude 15 (V) and a duration 1 (μs) is applied through a series resistance of 25 (Ω) to the input terminals of a 50 (Ω) lossless coaxial transmission line. The line is 400 (m) long and is short-circuited at the far end. Determine the voltage response at the midpoint of the line as a fuction of time up to 8 (μs). The dielectric constant of the insulating material in the cable is 2.25.

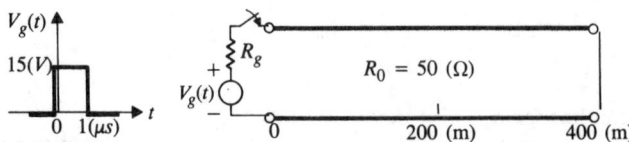

FIGURE 9–22
A pulse applied to a short-circuited line.

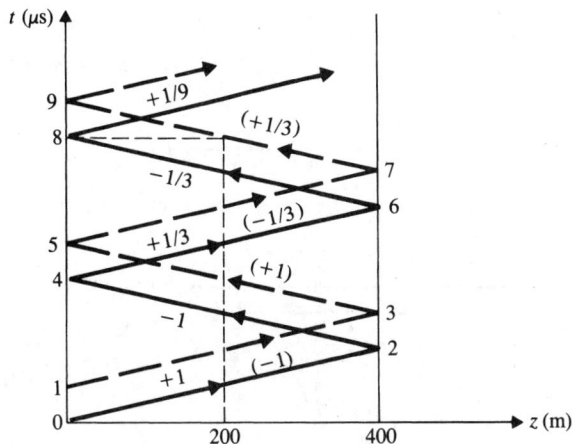

FIGURE 9–23
Voltage reflection diagram for Example
9–11.

Solution We have a situation as given in Fig. 9–22, where $R_g = 25$ (Ω) and $R_L = 0$. Also,

$$\Gamma_L = -1, \qquad \Gamma_g = \frac{25 - 50}{25 + 50} = -\frac{1}{3},$$

$$v_g(t) = 10[U(t) - U(t - 10^{-6})],$$

$$u = \frac{c}{\sqrt{\epsilon_r}} = \frac{3 \times 10^8}{\sqrt{2.25}} = 2 \times 10^8 \quad (\text{m/s}),$$

$$T = \frac{\ell}{u} = \frac{400}{2 \times 10^8} = 2 \times 10^{-6} \text{ (s)} = 2 \quad (\mu\text{s}),$$

$$V_1^+ = \frac{15R_0}{R_0 + R_g} = \frac{15 \times 50}{50 + 25} = 10 \quad (\text{V}).$$

A voltage reflection diagram is constructed in Fig. 9–23 for this problem. There are two sets of directed lines: The solid lines are for $+15$ (V) applied at $t = 0$, and the dashed lines are for -15 (V) applied at $t = 1$ (μs). Along each directed line is marked the amplitude of the wave (with the appropriate sign) normalized with respect to $V_1^+ = 10$ (V). The markings for the applied voltage $-15U(t - 10^{-6})$ are enclosed in brackets for easy reference. To obtain the voltage variation at the line's midpoint for the interval $0 < t \leq 8$ (μs), we draw a vertical line at $z = 200$ (m) and a horizontal line at $t = 8$ (μs). The voltage function due to $15U(t)$ can be read from the intersections of the vertical line with the solid directed lines. This is sketched as v_a in Fig. 9–24(a). Similarly, the voltage function due to $-15U(t - 10^{-6})$ is read from the intersections of the vertical line with the dashed directed lines; it is sketched as v_b in Fig. 9–24(b). The required response $v(200, t)$ for $0 < t \leq 8$ (μs) is then the sum of the responses, $v_a + v_b$, and is given in Fig. 9–24(c). ▬

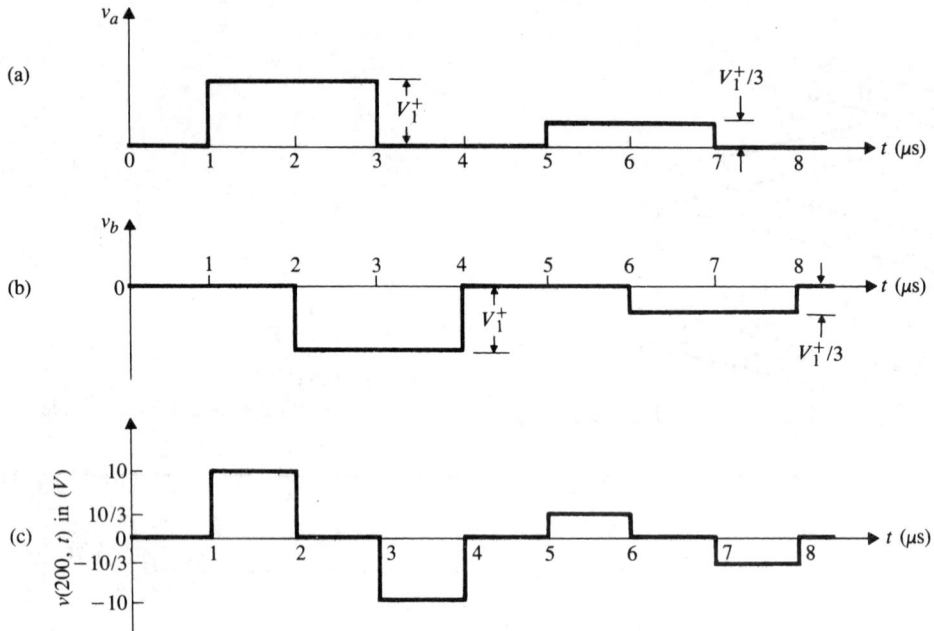

FIGURE 9–24
Voltage responses at the midpoint of the short-circuited line in Fig. 9–22
(Example 9–11).

9–5.3 INITIALLY CHARGED LINE

In our discussion of transients on transmission lines we have assumed that the lines themselves have no initial voltages or currents when an external source is applied. Actually, any disturbance or change in a transmission-line circuit will start transients along the line even without an external source if initial voltages and/or currents exist. We examine in this subsection a situation involving an initially charged line and develop a method of analysis.

Consider the following example.

EXAMPLE 9–12   A lossless, air-dielectric, open-circuited transmission line of characteristic resistance R_0 and length ℓ is initially charged to a voltage V_0. At $t = 0$ the line is connected to a resistance R. Determine the voltage across and the current in R as functions of time. Assume that $R = R_0$.

Solution   This problem, as depicted in Fig. 9–25(a), can be analyzed by examining the circuits in Figs. 9–25(b), 9–25(c), and 9–25(d). The circuit in Fig. 9–25(b) is equivalent to that in Fig. 9–25(a). After the switch is closed, the conditions in the circuit in Fig. 9–25(b) are the same as the superposition of those shown in Figs. 9–25(c) and 9–25(d). But the circuit in Fig. 9–25(c) does not give rise to transients because of the opposing voltages; hence we use the circuit in Fig. 9–25(d) to study the transient behavior of the original circuit in Fig. 9–25(a). The line in the circuit in Fig. 9–25(d) is uncharged, and our problem has then been reduced to one with which we are already familiar.

When the switch is closed, a voltage wave of amplitude V_1^+ will be sent down the line in the $+z$-direction, where

$$V_1^+ = -\frac{R_0}{R + R_0} V_0 = -\frac{V_0}{2}.$$

At $t = \ell/c$, the V_1^+ wave reaches the open end, having reduced the voltage along the whole line from V_0 to $V_0/2$. At the open end, $\Gamma = 1$, and a reflected V_1^- wave is sent back in the $-z$-direction with $V_1^- = V_1^+ = -V_0/2$. This reflected wave returns to the sending end at $t = 2\ell/c$, reducing the voltage on the line to zero.

From Fig. 9–25(d),

where
$$I_R = -I_1,$$

$$I_1 = I_1^+ = \frac{V_1^+}{R_0} = -\frac{V_0}{2R_0} \qquad \text{for} \quad 0 \le t < 2\ell/c.$$

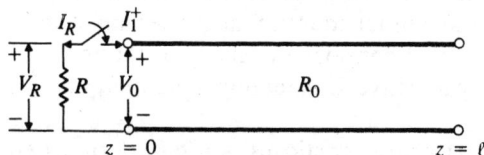

(a)

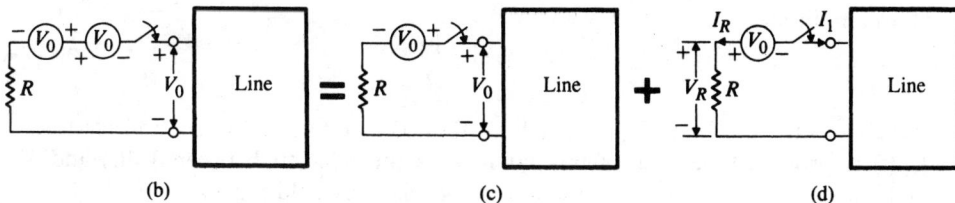

(b)   (c)   (d)

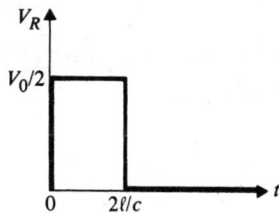

(e)

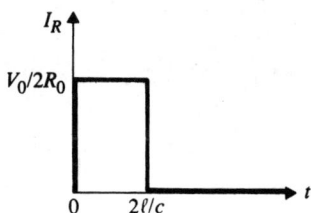

(f)

FIGURE 9–25
Transient problem of an open-circuited, initially charged line, $R = R_0$ (Example 9–12).

At $t = \ell/c$, I_1^+ reaches the open end, and the reflected I_1^- must make the total current there zero. Hence,

$$I_1^- = -I_1^+ = \frac{V_0}{2R_0},$$

which reaches the sending end at $t = 2\ell/c$ and reduces both I_1 and I_R to zero. Since $R = R_0$, there is no further reflection, and the transient state ends. As shown in Figs. 9–25(e) and 9–25(f), both V_R and I_R are a pulse of duration $2\ell/c$. We then have a way of generating a pulse by discharging a charged open-circuited transmission line, the width of the pulse being adjustable by changing ℓ. ■

9–5.4   LINE WITH REACTIVE LOAD

When the termination on a transmission line is a resistance different from the characteristic resistance, an incident voltage or current wave will produce a reflected wave of the same time dependence. The ratio of the amplitudes of the reflected and incident waves is a constant, which is defined as the reflection coefficient. If, however, the termination is a reactive element such as an inductance or a capacitance, the reflected wave will no longer have the same time dependence (no longer be of the same shape) as the incident wave. The use of a constant reflection coefficient is not feasible in such cases, and it is necessary to solve a differential equation at the termination in order to study the transient behavior. We shall consider the effect on the reflected wave of an inductive termination and a capacitive termination separately in this subsection.

Figure 9–26(a) shows a lossless line with a characteristic resistance R_0, terminated at $z = \ell$ with an inductance L_L. A d-c voltage V_0 is applied to the line at $z = 0$ through a series resistance R_0. When the switch is closed at $t = 0$, a voltage wave of an amplitude

$$V_1^+ = \frac{V_0}{2} \tag{9–167}$$

travels toward the load. Upon reaching the load at $t = \ell/u = T$, a reflected wave $V_1^-(t)$ is produced because of mismatch. It is the relation between $V_1^-(t)$ and V_1^+ that we wish to find. At $z = \ell$, the following relations hold for all $t \geq T$:

$$v_L(t) = V_1^+ + V_1^-(t), \tag{9–168}$$

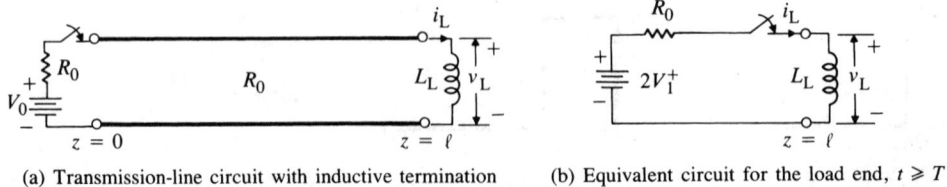

(a) Transmission-line circuit with inductive termination   (b) Equivalent circuit for the load end, $t \geqslant T$

FIGURE 9–26
Transient calculations for a lossless line with an inductive termination.

$$i_L(t) = \frac{1}{R_0}\left[V_1^+ - V_1^-(t)\right], \tag{9-169}$$

$$v_L(t) = L_L \frac{di_L(t)}{dt}. \tag{9-170}$$

Eliminating $V_1^-(t)$ from Eqs. (9–168) and (9–169), we obtain

$$v_L(t) = 2V_1^+ - R_0 i_L(t). \tag{9-171}$$

It is seen that Eq. (9–171) describes the application of Kirchhoff's voltage law to the circuit in Fig. 9–26(b), which is then the equivalent circuit at the load end for $t \geq T$. In view of Eq. (9–170), Eq, (9–171) leads to a first-order differential equation with constant coefficients:

$$L_L \frac{di_L(t)}{dt} + R_0 i_L(t) = 2V_1^+, \qquad t \geq T. \tag{9-172}$$

The solution of Eq. (9–172) is

$$i_L(t) = \frac{2V_1^+}{R_0}\left[1 - e^{-(t-T)R_0/L_L}\right], \qquad t \geq T, \tag{9-173}$$

which correctly gives $i_L(T) = 0$ and $i_L(\infty) = 2V_1^+/R_0$. The voltage across the inductive load is

$$v_L(t) = L_L \frac{di_L(t)}{dt} = 2V_1^+ e^{-(t-T)R_0/L_L}, \qquad t \geq T. \tag{9-174}$$

The amplitude of the reflected wave, $V_1^-(t)$, can be found from Eq. (9–168):

$$\begin{aligned}V_1^-(t) &= v_L(t) - V_1^+ \\ &= 2V_1^+\left[e^{-(t-T)R_0/L_L} - \tfrac{1}{2}\right], \qquad t > T.\end{aligned} \tag{9-175}$$

This reflected wave travels in the $-z$-direction. The voltage at any point $z = z_1$ along the line is V_1^+ before the reflected wave from the load end reaches that point, $(t - T) < (\ell - z_1)/u$, and equals $V_1^+ + V_1^-(t - T)$ after that.

In Figs. 9–27(a), 9–27(b), and 9–27(c) are plotted $i_L(t)$, $v_L(t)$, and $V_1^-(t)$ at $z = \ell$ using Eqs. (9–173), (9–174), and (9–175). The voltage distribution along the line for $T < t_1 < 2T$ is shown in Fig. 9–27(d). Obviously, the transient behavior on a transmission line with a reactive termination is more complicated than that with a resistive termination.

We follow a similar procedure in examining the transient behavior of a lossless line with a capacitive termination, shown in Fig. 9–28(a). The same Eqs. (9–167), (9–168), (9–169), and (9–171) apply at $z = \ell$, but Eq. (9–170) relating the load current $i_L(t)$ and load voltage $v_L(t)$ must now be changed to

$$i_L(t) = C_L \frac{dv_L(t)}{dt}. \tag{9-176}^\dagger$$

---

† The subscript roman L denotes load; it has nothing to do with inductance.

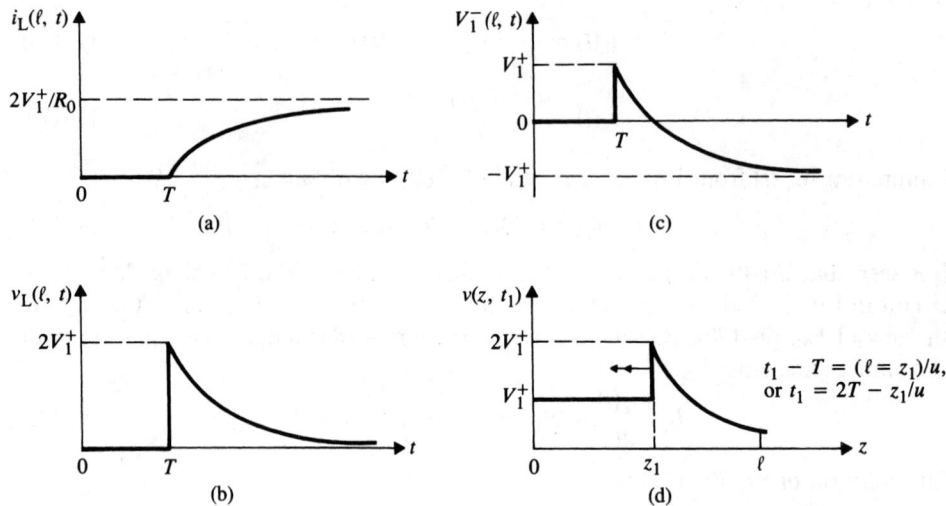

(a)

(b)

(c)

(d)

FIGURE 9–27
Transient responses of a lossless line with an inductive termination.

The differential equation to be solved at the load end is, by substituting Eq. (9–176) in Eq. (9–171),

$$C_L \frac{dv_L(t)}{dt} + \frac{1}{R_0} v_L(t) = \frac{2}{R_0} V_1^+, \qquad t \geq T, \tag{9–177}$$

where $V_1^+ = V_0/2$, as given in Eq. (9–167). The solution of Eq. (9–177) is

$$v_L(t) = 2V_1^+ \left[1 - e^{-(t-T)/R_0 C_L} \right], \qquad t \geq T. \tag{9–178}$$

The current in the load capacitance is obtained from Eq. (9–176):

$$i_L(t) = \frac{2V_1}{R_0} e^{-(t-T)/R_0 C_L}, \qquad t \geq T. \tag{9–179}$$

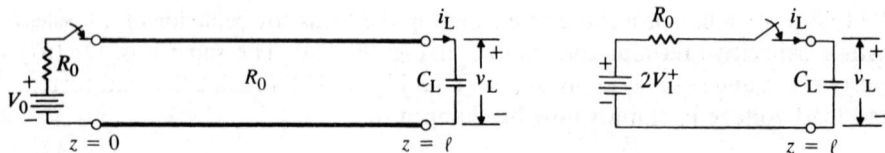

(a) Transmission-line circuit with capacitive termination.   (b) Equivalent circuit for the load end, $t \geq T$.

FIGURE 9–28
Transient calculations for a lossless line with a capacitive termination.

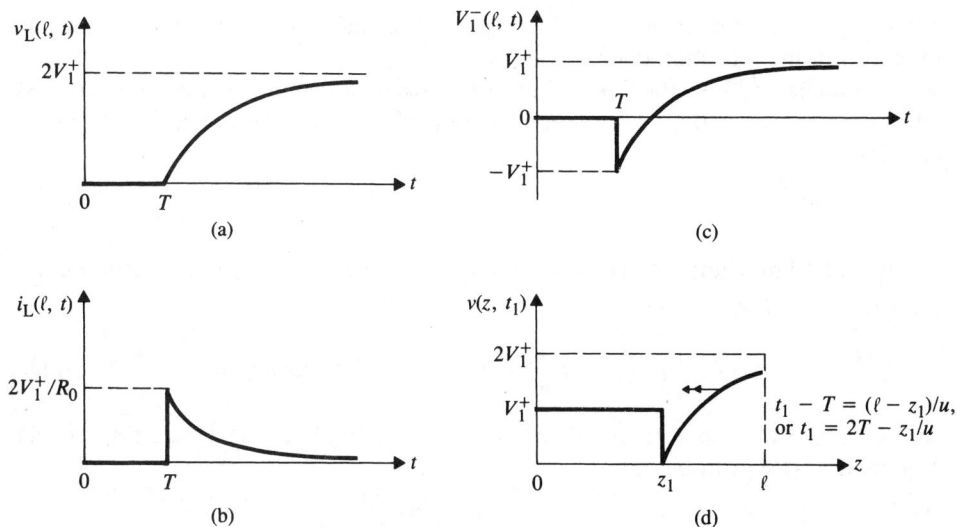

FIGURE 9–29
Transient responses of a lossless line with a capacitive termination.

Using Eq. (9–178) in Eq. (9–168), we find the amplitude of the reflected wave as a function of t:

$$V_1^-(t) = 2V_1^+[\tfrac{1}{2} - e^{-(t-T)/R_0 C_L}], \qquad t \geq T. \qquad (9\text{–}180)$$

The graphs of $v_L(t)$, $i_L(t)$, and $V_1^-(t)$ at $z = \ell$ are plotted in Figs. 9–29(a), 9–29(b), and 9–29(c) using Eqs. (9–178), (9–179), and (9–180), respectively. The voltage distribution along the line for $T < t_1 < 2T$ is shown in Fig. 9–29(d).

In this section we have discussed the transient behavior of only lossless transmission lines. For lossy lines, both the voltage and the current waves traveling in either direction will be attenuated as they proceed. This situation introduces additional complication in numerical computation, but the basic concept remains the same.

9–6 The Smith Chart

Transmission-line calculations—such as the determination of input impedance by Eq. (9–109), reflection coefficient by Eq. (9–134), and load impedance by Eq. (9–148)—often involve tedious manipulations of complex numbers. This tedium can be alleviated by using a graphical method of solution. The best known and most widely used graphical chart is the *Smith chart* devised by P. H. Smith.[†] Stated

---

[†] P H. Smith, "Transmission-line calculator," *Electronics*, vol. 12, p. 29, January 1939; and "An improved transmission-line calculator," *Electronics*, vol. 17, p. 130, January 1944.

succinctly, a Smith chart is a graphical plot of normalized resistance and reactance functions in the reflection-coefficient plane.

To understand how the Smith chart for a *lossless* transmission line is constructed, let us examine the voltage reflection coefficient of the load impedance defined in Eq. (9–134):

$$\Gamma = \frac{Z_L - R_0}{Z_L + R_0} = |\Gamma|e^{j\theta_\Gamma}. \tag{9–181}$$

Let the load impedance Z_L be normalized with respect to the characteristic impedance $R_0 = \sqrt{L/C}$ of the line.

$$z_L = \frac{Z_L}{R_0} = \frac{R_L}{R_0} + j\frac{X_L}{R_0} = r + jx \qquad \text{(Dimensionless)}, \tag{9–182}$$

where r and x are the normalized resistance and normalized reactance, respectively. Equation (9–181) can be rewritten as

$$\Gamma = \Gamma_r + j\Gamma_i = \frac{z_L - 1}{z_L + 1}, \tag{9–183}$$

where Γ_r and Γ_i are the real and imaginary parts, respectively, of the voltage reflection coefficient Γ. The inverse relation of Eq. (9–183) is

$$z_L = \frac{1 + \Gamma}{1 - \Gamma} = \frac{1 + |\Gamma|e^{j\theta_\Gamma}}{1 - |\Gamma|e^{j\theta_\Gamma}} \tag{9–184}$$

or

$$r + jx = \frac{(1 + \Gamma_r) + j\Gamma_i}{(1 - \Gamma_r) - j\Gamma_i}. \tag{9–185}$$

Multiplying both the numerator and the denominator of Eq. (9–185) by the complex conjugate of the denominator and separating the real and imaginary parts, we obtain

$$r = \frac{1 - \Gamma_r^2 - \Gamma_i^2}{(1 - \Gamma_r)^2 + \Gamma_i^2} \tag{9–186}$$

and

$$x = \frac{2\Gamma_i}{(1 - \Gamma_r)^2 + \Gamma_i^2}. \tag{9–187}$$

If Eq. (9–186) is plotted in the $\Gamma_r - \Gamma_i$ plane for a given value of r, the resulting graph is the locus for this r. The locus can be recognized when the equation is rearranged as

$$\left(\Gamma_r - \frac{r}{1 + r}\right)^2 + \Gamma_i^2 = \left(\frac{1}{1 + r}\right)^2. \tag{9–188}$$

It is the equation for a circle having a radius $1/(1 + r)$ and centered at $\Gamma_r = r/(1 + r)$ and $\Gamma_i = 0$. Different values of r yield circles of different radii with centers at different positions on the Γ_r-axis. A family of r-circles are shown in solid lines in Fig. 9–30. Since $|\Gamma| \leq 1$ for a lossless line, only that part of the graph lying within the unit circle on the $\Gamma_r - \Gamma_i$ plane is meaningful; everything outside can be disregarded.

Several salient properties of the *r*-circles are noted as follows:

1. The centers of all *r*-circles lie on the Γ_r-axis.
2. The $r = 0$ circle, having a unity radius and centered at the origin, is the largest.
3. The *r*-circles become progressively smaller as *r* increases from 0 toward ∞, ending at the $(\Gamma_r = 1, \Gamma_i = 0)$ point for open-circuit.
4. All *r*-circles pass through the $(\Gamma_r = 1, \Gamma_i = 0)$ point.

Similarly, Eq. (9–187) may be rearranged as

$$(\Gamma_r - 1)^2 + \left(\Gamma_i - \frac{1}{x}\right)^2 = \left(\frac{1}{x}\right)^2. \tag{9–189}$$

This is the equation for a circle having a radius $1/|x|$ and centered at $\Gamma_r = 1$ and $\Gamma_i = 1/x$. Different values of *x* yield circles of different radii with centers at different positions on the $\Gamma_r = 1$ line. A family of the portions of *x*-circles lying inside the $|\Gamma| = 1$ boundary are shown in dashed lines in Fig. 9–30. The following is a list of several salient properties of the *x*-circles.

1. The centers of all *x*-circles lie on the $\Gamma_r = 1$ line; those for $x > 0$ (inductive reactance) lie above the Γ_r-axis, and those for $x < 0$ (capacitive reactance) lie below the Γ_r-axis.
2. The $x = 0$ circle becomes the Γ_r-axis.

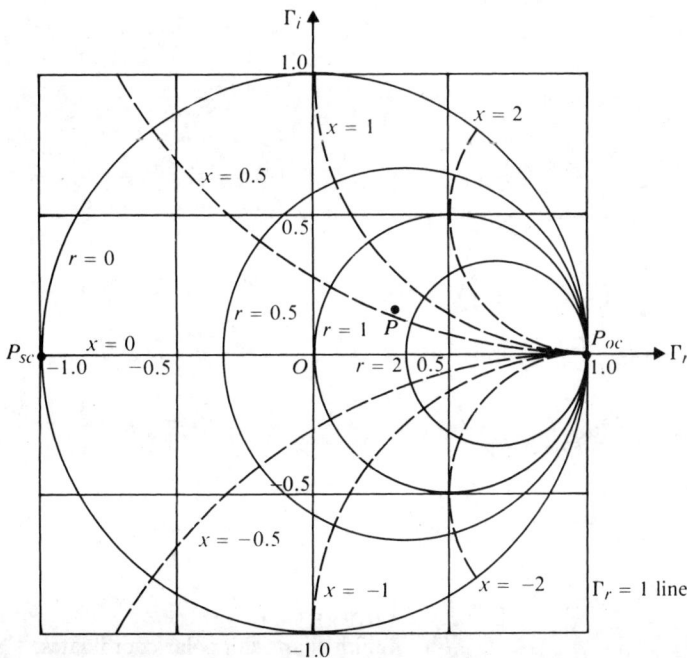

FIGURE 9–30
Smith chart with rectangular coordinates.

3. The x-circle becomes progressively smaller as $|x|$ increases from 0 toward ∞, ending at the $(\Gamma_r = 1, \Gamma_i = 0)$ point for open-circuit.

4. All x-circles pass through the $(\Gamma_r = 1, \Gamma_i = 0)$ point.

A Smith chart is a chart of r- and x-circles in the $\Gamma_r - \Gamma_i$ plane for $|\Gamma| \le 1$. It can be proved that the r- and x-circles are everywhere orthogonal to one another. The intersection of an r-circle and an x-circle defines a point that represents a normalized load impedance $z_L = r + jx$. The actual load impedance is $Z_L = R_0(r + jx)$. Since a Simith chart plots the normalized impedance, it can be used for calculations concerning a lossless transmission line with an arbitrary characteristic impedance (resistance).

As an illustration, point P in Fig. 9–30 is the intersection of the $r = 1.7$ circle and the $x = 0.6$ circle. Hence it represents $z_L = 1.7 + j0.6$. The point P_{sc} at $(\Gamma_r = -1, \Gamma_i = 0)$ corresponds to $r = 0$ and $x = 0$ and therefore represents a short-circuit. The point P_{oc} at $(\Gamma_r = 1, \Gamma_i = 0)$ corresponds to an infinite impedance and represents an open-circuit.

The Smith chart in Fig. 9–30 is marked with Γ_r and Γ_i rectangular coordinates. The same chart can be marked with polar coordinates, such that every point in the Γ-plane is specified by a magnitude $|\Gamma|$ and a phase angle θ_Γ. This is illustrated in Fig. 9–31, where several $|\Gamma|$-circles are shown in dashed lines and some θ_Γ-angles are marked around the $|\Gamma| = 1$ circle. The $|\Gamma|$-circles are normally not shown on commercially available Smith charts; but once the point representing a certain $z_L = r + jx$ is located, it is a simple matter to draw a circle centered at the origin through the

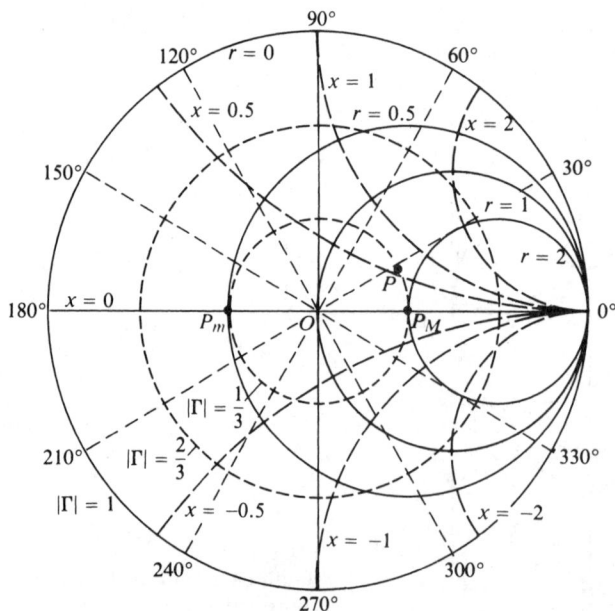

FIGURE 9–31
Smith chart with polar coordinates.

point. The fractional distance from the center to the point (compared with the unity radius to the edge of the chart) is equal to the magnitude $|\Gamma|$ of the load reflection coefficient; and the angle that the line to the point makes with the real axis is θ_Γ. This graphical determination circumvents the need for computing Γ by Eq. (9–183).

Each $|\Gamma|$-circle intersects the real axis at two points. In Fig. 9–31 we designate the point on the positive-real axis (OP_{oc}) as P_M and the point on the negative-real axis (OP_{sc}) as P_m. Since $x = 0$ along the real axis, P_M and P_m both represent situations with a purely resistive load, $Z_L = R_L$. Obviously, $R_L > R_0$ at P_M, where $r > 1$; and $R_L < R_0$ at P_m, where $r < 1$. In Eq. (9–145) we found that $S = R_L/R_0 = r$ for $R_L > R_0$. This relation enables us to say immediately, without using Eq. (9–138), that *the value of the r-circle passing through the point P_M is numerically equal to the standing-wave ratio*. Similarly, we conclude from Eq. (9–146) that *the value of the r-circle passing through the point P_m on the negative-real axis is numerically equal to $1/S$*. For the $z_L = 1.7 + j0.6$ point, marked P in Fig. 9–31, we find $|\Gamma| = \frac{1}{3}$ and $\theta_\Gamma = 28°$. At P_M, $r = S = 2.0$. These results can be verified analytically.

In summary, we note the following:

1. All $|\Gamma|$-circles are centered at the origin, and their radii vary uniformly from 0 to 1.

2. The angle, measured from the positive real axis, of the line drawn from the origin through the point representing z_L equals θ_Γ.

3. The value of the r-circle passing through the intersection of the $|\Gamma|$-circle and the positive-real axis equals the standing-wave ratio S.

So far we have based the construction of the Smith chart on the definition of the voltage reflection coefficient of the load impedance, as given in Eq. (9–134). The input impedance looking toward the load at a distance z' from the load is the ratio of $V(z')$ and $I(z')$. From Eqs. (9–133a) and (9–133b) we have, by writing $j\beta$ for γ for a lossless line,

$$Z_i(z') = \frac{V(z')}{I(z')} = Z_0 \left[\frac{1 + \Gamma e^{-j2\beta z'}}{1 - \Gamma e^{-j2\beta z'}} \right]. \tag{9–190}$$

The normalized input impedance is

$$z_i = \frac{Z_i}{Z_0} = \frac{1 + \Gamma e^{-j2\beta z'}}{1 - \Gamma e^{-j2\beta z'}}$$

$$= \frac{1 + |\Gamma| e^{j\phi}}{1 - |\Gamma| e^{j\phi}}, \tag{9–191}$$

where

$$\phi = \theta_\Gamma - 2\beta z'. \tag{9–192}$$

We note that Eq. (9–191) relating z_i and $\Gamma e^{-j2\beta z'} = |\Gamma| e^{j\phi}$ is of exactly the same form as Eq. (9–184) relating z_L and $\Gamma = |\Gamma| e^{j\theta_\Gamma}$. In fact, the latter is a special case of the former for $z' = 0$ ($\phi = \theta_\Gamma$). The magnitude, $|\Gamma|$, of the reflection coefficient and therefore the standing-wave ratio S, are not changed by the additional line length z'. Thus, just as we can use the Smith chart to find $|\Gamma|$ and θ_Γ for a given z_L at the load, we can keep $|\Gamma|$ constant and *subtract* (rotate in the *clockwise* direction) from θ_Γ an angle equal to $2\beta z' = 4\pi z'/\lambda$. This will locate the point for $|\Gamma|e^{j\phi}$, which determines z_i, the normalized input impedance looking into a lossless line of characteristic impedance R_0, length z', and a normalized load impedance z_L. Two additional scales in $\Delta z'/\lambda$ are usually provided along the perimeter of the $|\Gamma| = 1$ circle for easy reading of the

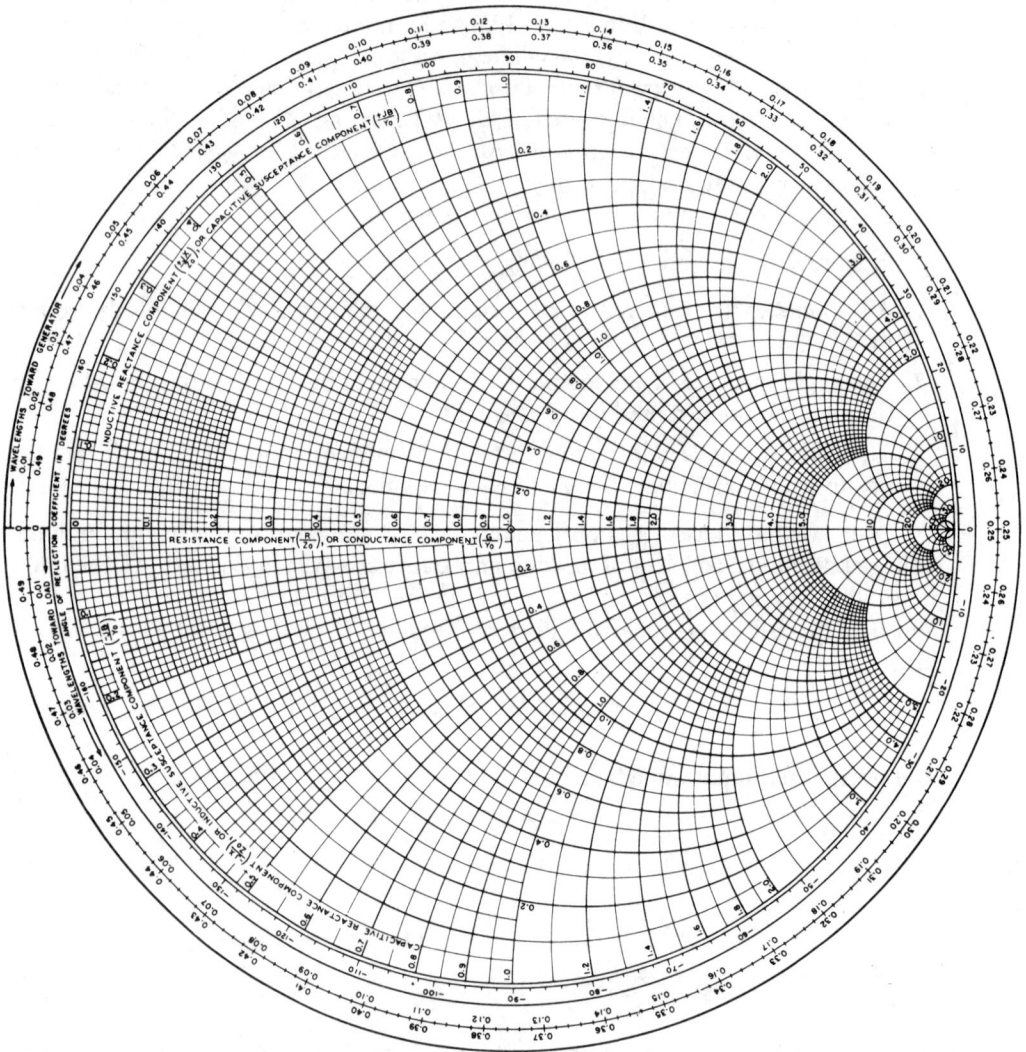

FIGURE 9–32
The Smith chart.

phase change $2\beta(\Delta z')$ due to a change in line length $\Delta z'$: The outer scale is marked "wavelengths toward generator" in the clockwise direction (increasing z'); and the inner scale is marked "wavelengths toward load" in the counterclockwise direction (decreasing z'). Figure 9–32 is a typical Smith chart, which is commercially available.[†] It has a complicated appearance, but it actually consists merely of constant-r and constant-x circles. We note that a change of half a wavelength in line length ($\Delta z' = \lambda/2$) corresponds to a $2\beta(\Delta z') = 2\pi$ change in ϕ. A complete revolution around a $|\Gamma|$-circle returns to the same point and results in no change in impedance, as was asserted in Eq. (9–115).

We shall illustrate the use of the Smith chart for solving some typical transmission-line problems by several examples.

EXAMPLE 9–13   Use the Smith chart to find the input impedance of a section of a 50 (Ω) lossless transmission line that is 0.1 wavelength long and is terminated in a short-circuit.

Solution   Given

$$z_L = 0,$$
$$R_0 = 50 \quad (\Omega),$$
$$z' = 0.1\lambda.$$

1. Enter the Smith chart at the intersection of $r = 0$ and $x = 0$ (point P_{sc} on the extreme left of chart; see Fig. 9–33).
2. Move along the perimeter of the chart ($|\Gamma| = 1$) by 0.1 "wavelengths toward generator" in a clockwise direction to P_1.
3. At P_1, read $r = 0$ and $x \cong 0.725$, or $z_i = j0.725$. Thus, $Z_i = R_0 z_i = 50(j0.725) = j36.3$ (Ω). (The input impedance is purely inductive.)

This result can be checked readily by using Eq. (9–112):

$$Z_i = jR_0 \tan \beta\ell = j50 \tan\left(\frac{2\pi}{\lambda}\right)0.1\lambda$$
$$= j50 \tan 36° = j36.3 \quad (\Omega).$$

EXAMPLE 9–14   A lossless transmission line of length 0.434λ and characteristic impedance 100 (Ω) is terminated in an impedance $260 + j180$ (Ω). Find (a) the voltage reflection coefficient, (b) the standing-wave ratio, (c) the input impedance, and (d) the location of a voltage maximum on the line.

[†] All of the Smith charts used in this book are reprinted with permission of Emeloid Industries, Inc., New Jersey.

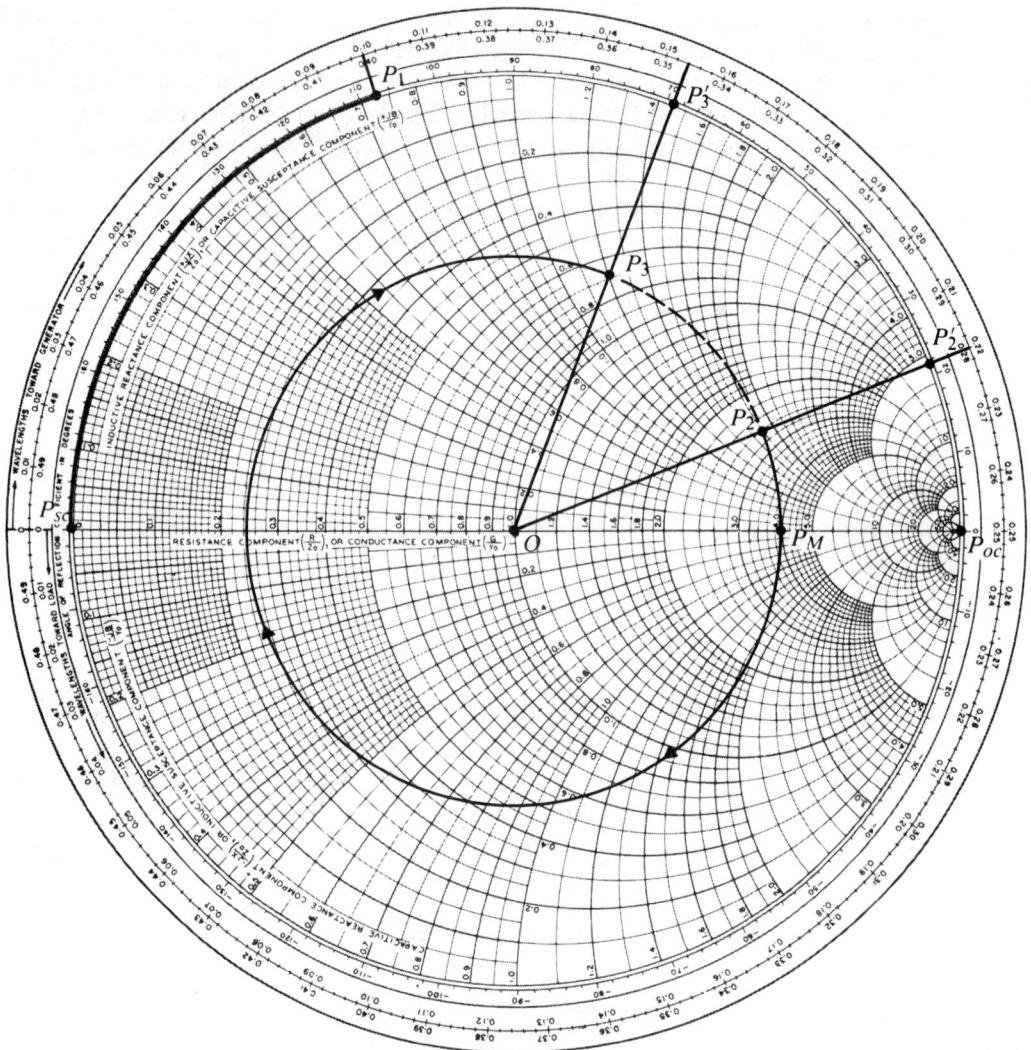

FIGURE 9–33
Smith-chart calculations for Examples 9–13 and 9–14.

Solution Given

$$z' = 0.434\lambda,$$

$$R_0 = 100 \quad (\Omega),$$

$$Z_L = 260 + j180 \quad (\Omega).$$

a) We find the voltage reflection coefficient in several steps:

  1. Enter the Smith chart at $z_L = Z_L/R_0 = 2.6 + j1.8$ (point P_2 in Fig. 9–33).

2. With the center at the origin, draw a circle of radius $\overline{OP}_2 = |\Gamma| = 0.60$. (The radius of the chart $\overline{OP}_{sc}$ equals unity.)

3. Draw the straight line OP_2 and extend it to P'_2 on the periphery. Read 0.220 on "wavelengths toward generator" scale. The phase angle θ_Γ of the reflection coefficient is $(0.250 - 0.220) \times 4\pi = 0.12\pi$ (rad) or $21°$. (We multiply the change in wavelengths by 4π because angles on the Smith chart are measured in $2\beta z'$ or $4\pi z'/\lambda$. A half-wavelength change in line length corresponds to a complete revolution on the Smith chart.) The answer to part (a) is then

$$\Gamma = |\Gamma|e^{j\theta_\Gamma} = 0.60\underline{/21°}.$$

b) The $|\Gamma| = 0.60$ circle intersects with the positive-real axis OP_{oc} at $r = S = 4$. Thus the voltage standing-wave ratio is 4.

c) To find the input impedance, we proceed as follows:

1. Move P'_2 at 0.220 by a total of 0.434 "wavelengths toward generator," first to 0.500 (same as 0.000) and then further to $0.154 [(0.500 - 0.220) + 0.154 = 0.434]$ to P'_3.

2. Join O and P'_3 by a straight line which intersects the $|\Gamma| = 0.60$ circle at P_3.

3. Read $r = 0.69$ and $x = 1.2$ at P_3. Hence,

$$Z_i = R_0 z_i = 100(0.69 + j1.2) = 69 + j120 \quad (\Omega).$$

d) In going from P_2 to P_3, the $|\Gamma| = 0.60$ circle intersects the positive-real axis OP_{oc} at P_M where the voltage is a maximum. Thus a voltage maximum appears at $(0.250 - 0.220)\lambda$ or 0.030λ from the load. ▄▄

▄▄▄▄▄ **EXAMPLE 9–15**   Solve Example 9–9 by using the Smith chart. Given

$$R_0 = 50 \quad (\Omega),$$
$$S = 3.0,$$
$$\lambda = 2 \times 0.2 = 0.4 \quad (m),$$

First voltage minimum at $z'_m = 0.05$ (m),

find (a) Γ, (b) Z_L, (c) ℓ_m, and R_m (Fig. 9–12).

Solution

a) On the positive-real axis OP_{oc}, locate the point P_M at which $r = S = 3.0$ (see Fig. 9–34). Then $\overline{OP}_M = |\Gamma| = 0.5$ ($\overline{OP}_{oc} = 1.0$). We cannot find θ_Γ until we have located the point that represents the normalized load impedance.

b) We use the following procedure to find the load impedance on the Smith chart:

1. Draw a circle centered at the origin with radius $\overline{OP}_M$, which intersects with the negative-real axis OP_{sc} at P_m where there will be a voltage minimum.

2. Since $z'_m/\lambda = 0.05/0.4 = 0.125$, move from P_{sc} 0.125 "wavelengths toward load" in the counterclockwise direction to P'_L.

3. Join O and P'_L by a straight line, intersecting the $|\Gamma| = 0.5$ circle at P_L. This is the point representing the normalized load impedance.

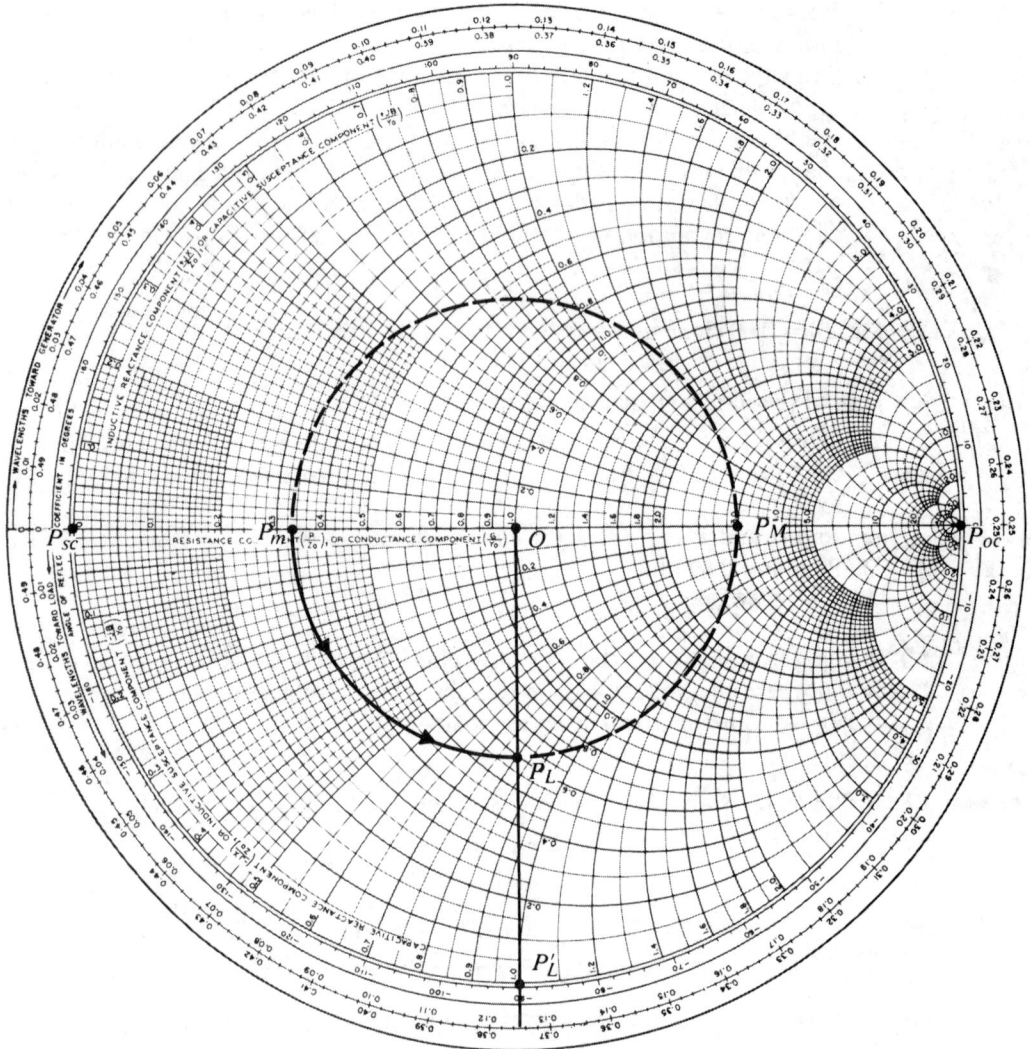

FIGURE 9–34
Smith-chart calculations for Example 9–15.

4. Read the angle $\angle P_{oc}OP'_L = 90° = \pi/2$ (rad). There is no need to use a pro-
tractor because $\angle P_{oc}OP'_L = 4\pi(0.250 - 0.125) = \pi/2$. Hence $\theta_\Gamma = -\pi/2$ (rad),
or $\Gamma = 0.5\underline{/-90°} = -j0.5$.

5. Read at P_L, $z_L = 0.60 - j0.80$, which gives

$$Z_L = 50(0.60 - j0.80) = 30 - j40 \quad (\Omega).$$

c) The equivalent line length and the terminating resistance can be found easily:

$$\ell_m = \frac{\lambda}{2} - z'_m = 0.2 - 0.05 = 0.15 \quad (\text{m}),$$

$$R_m = \frac{R_0}{S} = \frac{50}{3} = 16.7 \quad (\Omega).$$

All the above results are the same as those obtained in Example 9–9, but no calculations with complex numbers are needed in using the Smith chart. ∎

9-6.1   SMITH-CHART CALCULATIONS FOR LOSSY LINES

In discussing the use of the Smith chart for transmission-line calculations we have assumed the line to be lossless. This is normally a satisfactory approximation, since we generally deal with relatively short sections of low-loss lines. The lossless assumption enables us to say, following Eq. (9–191), that the magnitude of the $\Gamma e^{-j2\beta z'}$ term does not change with line length z' and that we can find z_i from z_L, and vice versa, by moving along the $|\Gamma|$-circle by an angle equal to $2\beta z'$.

For a lossy line of a sufficient length ℓ, such that $2\alpha\ell$ is not negligible in comparison to unity, Eq. (9–191) must be amended to read

$$
\begin{aligned}
z_i &= \frac{1 + \Gamma e^{-2\alpha z'}e^{-j2\beta z'}}{1 - \Gamma e^{-2\alpha z'}e^{-j2\beta z'}} \\
&= \frac{1 + |\Gamma|e^{-2\alpha z'}e^{j\phi}}{1 - |\Gamma|e^{-2\alpha z'}e^{j\phi}}, \qquad \phi = \theta_\Gamma - 2\beta z'.
\end{aligned}
\tag{9-193}
$$

Hence, to find z_i from z_L, we cannot simply move along the $|\Gamma|$-circle; auxiliary calculations are necessary to account for the $e^{-2\alpha z'}$ factor. The following example illustrates what has to be done.

EXAMPLE 9-16   The input impedance of a short-circuited lossy transmission line of length 2 (m) and characteristic impedance 75 (Ω) (approximately real) is $45 + j225$ (Ω). (a) Find α and β of the line. (b) Determine the input impedance if the short-circuit is replaced by a load impedance $Z_L = 67.5 - j45$ (Ω).

Solution

a) The short-circuit load is represented by the point P_{sc} on the extreme left of the Smith impedance chart.

   1. Enter $z_{i1} = (45 + j225)/75 = 0.60 + j3.0$ in the chart as P_1 (Fig. 9–35).
   2. Draw a straight line from the origin O through P_1 to P'_1.
   3. Measure $\overline{OP}_1/\overline{OP'}_1 = 0.89 = e^{-2\alpha\ell}$. It follows that

$$\alpha = \frac{1}{2\ell}\ln\left(\frac{1}{0.89}\right) = \frac{1}{4}\ln 1.124 = 0.029 \quad (\text{Np/m}).$$

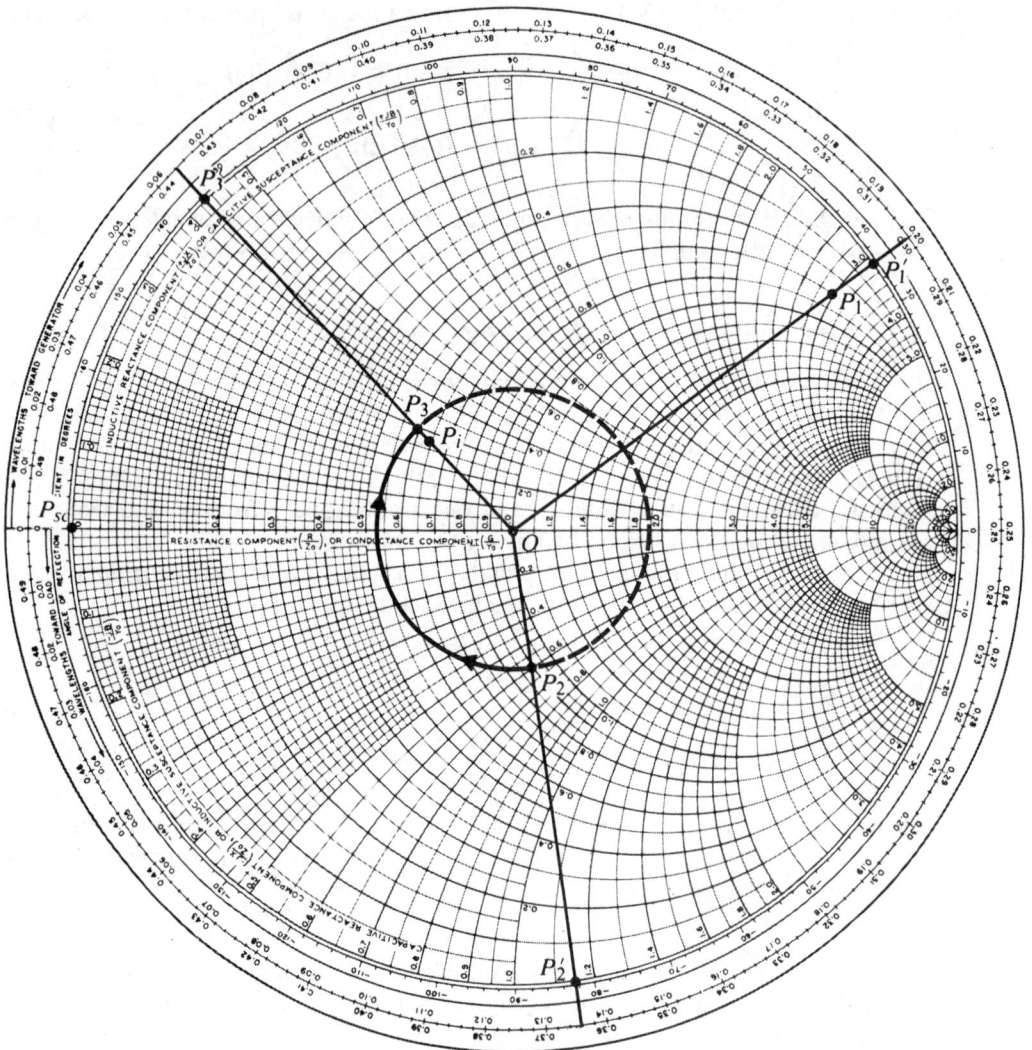

FIGURE 9–35
Smith-chart calculations for a lossy transmission line (Example 9–16).

---

4. Record that the arc $P_{sc}P'_1$ is 0.20 "wavelengths toward generator." We have $\ell/\lambda = 0.20$ and $2\beta\ell = 4\pi\ell/\lambda = 0.8\pi$. Thus,

$$\beta = \frac{0.8\pi}{2\ell} = \frac{0.8\pi}{4} = 0.2\pi \quad (\text{rad/m}).$$

b) To find the input impedance for $Z_L = 67.5 - j45$ (Ω):

1. Enter $z_L = Z_L/Z_0 = (67.5 - j45)/75 = 0.9 - j0.6$ on the Smith chart as P_2.

2. Draw a straight line from O through P_2 to P'_2 where the "wavelengths toward generator" reading is 0.364.
3. Draw a $|\Gamma|$-circle centered at O with radius $\overline{OP_2}$.
4. Move P'_2 along the perimeter by 0.20 "wavelengths toward generator" to P'_3 at $0.364 + 0.20 = 0.564$ or 0.064.
5. Join P'_3 and O by a straight line, intersecting the $|\Gamma|$-circle at P_3.
6. Mark on line OP_3 a point P_i such that $\overline{OP_i}/\overline{OP_3} = e^{-2\alpha\ell} = 0.89$.
7. At P_i, read $z_i = 0.64 + j0.27$. Hence,

$$Z_i = 75(0.64 + j0.27) = 48.0 + j20.3 \quad (\Omega).$$ ▬

9–7 Transmission-Line Impedance Matching

Transmission lines are used for the transmission of power and information. For radio-frequency power transmission it is highly desirable that as much power as possible is transmitted from the generator to the load and as little power as possible is lost on the line itself. This will require that the load be matched to the characteristic impedance of the line so that the standing-wave ratio on the line is as close to unity as possible. For information transmission it is essential that the lines be matched because reflections from mismatched loads and junctions will result in echoes and will distort the information-carrying signal. In this section we discuss several methods for impedance-matching on lossless transmission lines. We note parenthetically that the methods we develop will be of little consequence to power transmission by 60 (Hz) lines inasmuch as these lines are generally very short in comparison to the 5 (Mm) wavelength and the line losses are appreciable. Sixty-hertz power-line circuits are usually analyzed in terms of equivalent lumped electrical networks.

9–7.1 IMPEDANCE MATCHING BY QUARTER-WAVE TRANSFORMER

A simple method for matching a resistive load R_L to a lossless transmission line of a characteristic impedance R_0 is to insert a quarter-wave transformer with a characteristic impedance R'_0 such that

$$R'_0 = \sqrt{R_0 R_L}. \tag{9–194}$$

Since the length of the quarter-wave line depends on wavelength, this matching method is frequency-sensitive, as are all the other methods to be discussed.

▬▬▬ **EXAMPLE 9–17** A signal generator is to feed equal power through a lossless air transmission line with a characteristic impedance 50 (Ω) to two separate resistive loads, 64 (Ω) and 25 (Ω). Quarter-wave transformers are used to match the loads to the 50 (Ω) line, as shown in Fig. 9–36. (a) Determine the required characteristic impedances of the quarter-wave lines. (b) Find the standing-wave ratios on the matching line sections.

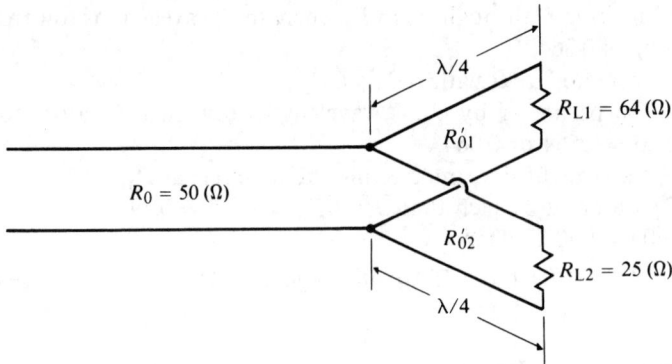

FIGURE 9–36
Impedance matching by quarter-wave lines (Example 9–17).

Solution

a) To feed equal power to the two loads, the input resistance at the junction with the main line looking toward each load must be equal to $2R_0$. $R_{i1} = R_{i2} = 2R_0 = 100$ (Ω):

$$R'_{01} = \sqrt{R_{i1}R_{L1}} = \sqrt{100 \times 64} = 80 \quad (\Omega),$$
$$R'_{02} = \sqrt{R_{i2}R_{L2}} = \sqrt{100 \times 25} = 50 \quad (\Omega).$$

b) Under matched conditions there are no standing waves on the main transmission line ($S = 1$). The standing-wave ratios on the two matching line sections are as follows.

Matching section No. 1:

$$\Gamma_1 = \frac{R_{L1} - R'_{01}}{R_{L1} + R'_{01}} = \frac{64 - 80}{64 + 80} = -0.11,$$
$$S_1 = \frac{1 + |\Gamma_1|}{1 - |\Gamma_1|} = \frac{1 + 0.11}{1 - 0.11} = 1.25.$$

Matching section No. 2:

$$\Gamma_2 = \frac{R_{L2} - R'_{02}}{R_{L2} + R'_{02}} = \frac{25 - 50}{25 + 50} = -0.33,$$
$$S_2 = \frac{1 + |\Gamma_2|}{1 - |\Gamma_2|} = \frac{1 + 0.33}{1 - 0.33} = 1.99. \qquad\blacksquare$$

Ordinarily, the main transmission line and the matching line sections are essentially lossless. In that case, both R_0 and R'_0 are purely real, and Eq. (9–194) will have no solution if R_L is replaced by a complex Z_L. Hence quarter-wave transformers are not useful for matching a complex load impedance to a low-loss line.

In the following subsection we will discuss a method for matching an arbitrary load impedance to a line by using a single open- or short-circuited line section (a single stub) in parallel with the main line and at an appropriate distance from the load. Since it is more convenient to use admittances instead of impedances for parallel connections, we first examine how the Smith chart can be used to make admittance calculations.

Let $Y_L = 1/Z_L$ denote the load admittance. The normalized load impedance is

$$z_L = \frac{Z_L}{R_0} = \frac{1}{R_0 Y_L} = \frac{1}{y_L},\qquad (9\text{--}195)$$

where

$$\begin{aligned} y_L &= Y_L/Y_0 = Y_L/G_0 \\ &= R_0 Y_L = g + jb \qquad \text{(Dimensionless)} \end{aligned} \qquad (9\text{--}196)$$

is the normalized load admittance having normalized conductance g and normalized susceptance b as its real and imaginary parts, respectively. Equation (9–195) suggests that a quarter-wave line with a unity normalized characteristic impedance will transform z_L to y_L, and vice versa. On the Smith chart we need only move the point representing z_L along the $|\Gamma|$-circle by a quarter-wavelength to locate the point representing y_L. Since a $\lambda/4$-change in line length ($\Delta z'/\lambda = \frac{1}{4}$) corresponds to a change of π radians ($2\beta\,\Delta z' = \pi$) on the Smith chart, *the points representing z_L and y_L are* then *diametrically opposite to each other on the $|\Gamma|$-circle.* This observation enables us to find y_L from z_L, and z_L from y_L, on the Smith chart in a very simple manner.

EXAMPLE 9–18 Given $Z_L = 95 + j20$ (Ω), find Y_L.

Solution This problem has nothing to do with any transmission line. In order to use the Smith chart we can choose an arbitrary normalizing constant; for instance, $R_0 = 50$ (Ω). Thus,

$$z_L = \tfrac{1}{50}(95 + j20) - 1.9 + j0.4.$$

Enter z_L as point P_1 on the Smith chart in Fig. 9–37. The point P_2 on the other side of the line joining P_1 and O represents y_L: $\overline{OP}_2 = \overline{OP}_1$.

$$Y_L = \frac{1}{R_0}\, y_L = \frac{1}{50}\,(0.5 - j0.1) = 10 - j2 \quad (\text{mS}).$$

EXAMPLE 9–19 Find the input admittance of an open-circuited line of characteristic impedance 300 (Ω) and length 0.04λ.

Solution

1. For an open-circuited line we start from the point P_{oc} on the extreme right of the impedance Smith chart, at 0.25 in Fig. 9–38.
2. Move along the perimeter of the chart by 0.04 "wavelengths toward generator" to P_3 (at 0.29).
3. Draw a straight line from P_3 through O, intersecting at P_3' on the opposite side.

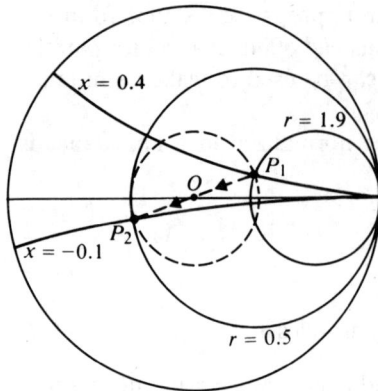

FIGURE 9–37
Finding admittance from impedance (Example 9–18).

4. Read at P'_3

$$y_i = 0 + j0.26.$$

Thus,

$$Y_i = \frac{1}{300}(0 + j0.26) = j0.87 \quad \text{(mS)}.$$

■

In the preceding two examples we have made admittance calculations by using the Smith chart as an impedance chart. The Smith chart can also be used as an admittance chart, in which case the r- and x-circles would be g- and b-circles. The points representing an open- and a short-circuit termination would be the points on the extreme left and the extreme right, respectively, on an admittance chart. For Example 9–19, we could then start from extreme left point on the chart, at 0.00 in Fig. 9–38, and move 0.04 "wavelengths toward generator" to P'_3 directly.

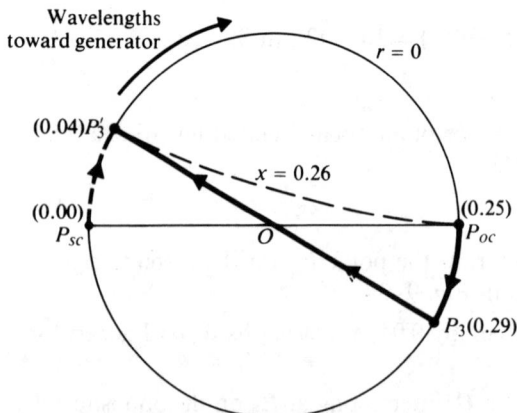

FIGURE 9–38
Finding input admittance of open-circuited line (Example 9–19).

9–7.2 SINGLE-STUB MATCHING

We now tackle the problem of matching a load impedance Z_L to a lossless line that has a characteristic impedance R_0 by placing a single short-circuited stub in parallel with the line, as shown in Fig. 9–39. This is the ***single-stub method*** for impedance matching. We need to determine the length of the stub, ℓ, and the distance from the load, d, such that the impedance of the parallel combination to the right of points B–B' equals R_0. Short-circuited stubs are usually used in preference to open-circuited stubs because an infinite terminating impedance is more difficult to realize than a zero terminating impedance for reasons of radiation from an open end and coupling effects with neighboring objects. Moreover, a short-circuited stub of an adjustable length and a constant characteristic resistance is much easier to construct than an open-circuited one. Of course, the difference in the required length for an open-circuited stub and that for a short-circuited stub is an odd multiple of a quarter-wavelength.

The parallel combination of a line terminated in Z_L and a stub at points B–B' in Fig. 9–39 suggest that it is advantageous to analyze the matching requirements in terms of admittances. The basic requirement is

$$Y_i = Y_B + Y_s$$
$$= Y_0 = \frac{1}{R_0}. \tag{9–197}$$

In terms of normalized admittances, Eq. (9–197) becomes

$$1 = y_B + y_s, \tag{9–198}$$

where $y_B = R_0 Y_B$ is for the load section and $y_s = R_0 Y_s$ is for the short-circuited stub. However, since the input admittance of a short-circuited stub is purely susceptive, y_s is purely imaginary. As a consequence, Eq. (9–198) can be satisfied only if

$$y_B = 1 + jb_B \tag{9–199}$$

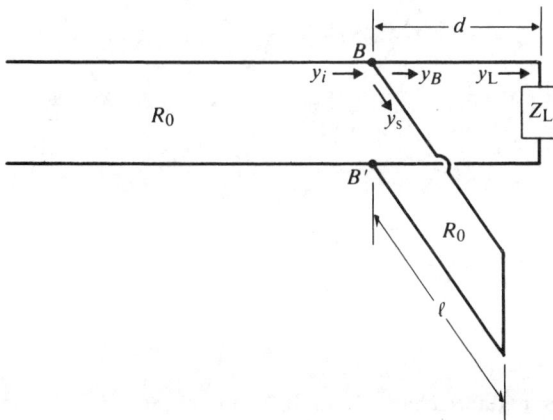

FIGURE 9–39
Impedance matching by single-stub method.

and

$$y_s = -jb_B, \qquad\qquad (9\text{-}200)$$

where b_B can be either positive or negative. Our objectives, then, are to find the length d such that the admittance, y_B, of the load section looking to the right of terminals B–B' has a *unity real part* and to find the length ℓ_B of the stub required to *cancel the imaginary part*.

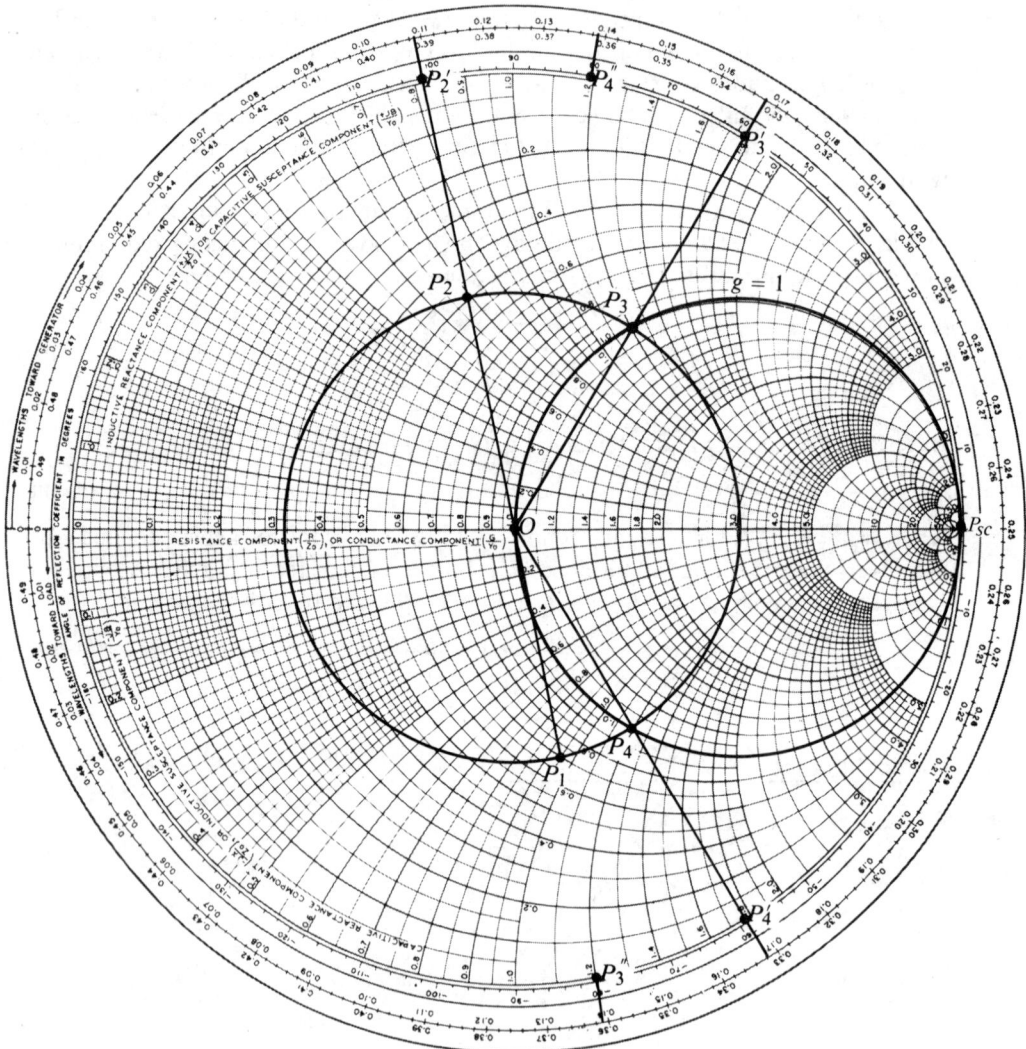

FIGURE 9–40
Construction for single-stub matching on Smith admittance chart (Example 9–20).

Using the Smith chart as an admittance chart, we proceed as follows for single-stub matching:

1. Enter the point representing the normalized load admittance y_L.
2. Draw the $|\Gamma|$-circle for y_L, which will intersect the $g = 1$ circle at two points. At these points, $y_{B1} = 1 + jb_{B1}$ and $y_{B2} = 1 + jb_{B2}$. Both are possible solutions.
3. Determine load-section lengths d_1 and d_2 from the angles between the point representing y_L and the points representing y_{B1} and y_{B2}.
4. Determine stub lengths ℓ_{B1} and ℓ_{B2} from the angles between the short-circuit point on the extreme right of the chart to the points representing $-jb_{B1}$ and $-jb_{B2}$, respectively.

The following example will illustrate the necessary steps.

━━━ **EXAMPLE 9–20**   A 50 (Ω) transmission line is connected to a load impedance $Z_L = 35 - j47.5$ (Ω). Find the position and length of a short-circuited stub required to match the line.

Solution   Given

$$R_0 = 50 \ (\Omega)$$
$$Z_L = 35 - j47.5 \ (\Omega)$$
$$z_L = Z_L/R_0 = 0.70 - j0.95.$$

1. Enter z_L on the Smith chart as P_1 (Fig. 9–40).
2. Draw a $|\Gamma|$-circle centered at O with radius $\overline{OP}_1$.
3. Draw a straight line from P_1 through O to point P'_2 on the perimeter, intersecting the $|\Gamma|$-circle at P_2, which represents y_L. Note 0.109 at P'_2 on the "wavelengths toward generator" scale.
4. Note the two points of intersection of the $|\Gamma|$-circle with the $g = 1$ circle.

$$\text{At } P_3: \quad y_{B1} = 1 + j1.2 = 1 + jb_{B1};$$
$$\text{At } P_4: \quad y_{B2} = 1 - j1.2 = 1 + jb_{B2}.$$

5. Solutions for the position of the stub:

$$\text{For } P_3 \text{ (from } P'_2 \text{ to } P'_3\text{):} \quad d_1 = (0.168 - 0.109)\lambda = 0.059\lambda;$$
$$\text{For } P_4 \text{ (from } P'_2 \text{ to } P'_4\text{):} \quad d_2 = (0.332 - 0.109)\lambda = 0.223\lambda.$$

6. Solutions for the length of short-circuited stub to provide $y_s = -jb_B$:

For P_3 (from P_{sc} on the extreme right of chart to P''_3, which represents $-jb_{B1} = -j1.2$):

$$\ell_{B1} = (0.361 - 0.250)\lambda = 0.111\lambda;$$

For P_4 (from P_{sc} to P''_4, which represents $-jb_{B2} = j1.2$):

$$\ell_{B2} = (0.139 + 0.250)\lambda = 0.389\lambda.$$

In general, the solution with the shorter lengths is preferred unless there are other practical constraints. The exact length, ℓ_B, of the short-circuited stub may require fine adjustments in the actual matching procedure; hence the shorted matching sections are sometimes called **stub tuners**. ▬

The use of Smith chart in solving impedance-matching problems avoids the manipulation of complex numbers and the computation of tangent and arc-tangent functions; but graphical constructions are needed, and graphical methods have limited accuracy. Actually, the analytical solutions of impedance-matching problems are relatively simple, and easy access to a computer may diminish the reliance on the Smith chart and, at the same time, yield more accurate results.

For the single-stub matching problem illustrated in Fig. 9–39 we have, from Eq. (9–109).

$$z_B = \frac{(r_L + jx_L) + jt}{1 + j(r_L + jx_L)t},$$

(9–201)

where

$$t = \tan \beta d.$$

(9–202)

The normalized input admittance to the right of points B–B' is

$$y_B = \frac{1}{z_B} = g_B + jb_B,$$

(9–203)

where

$$g_B = \frac{r_L(1 - x_L t) + r_L t(x_L + t)}{r_L^2 + (x_L + t)^2}$$

(9–204)

and

$$b_B = \frac{r_L^2 t - (1 - x_L t)(x_L + t)}{r_L + (x_L + t)^2}.$$

(9–205)

A perfect match requires the simultaneous satisfaction of Eqs. (9–199) and (9–200). Equating g_B in Eq. (9–204) to unity, we have

$$(r_L - 1)t^2 - 2x_L t + (r_L - r_L^2 - x_L^2) = 0.$$

(9–206)

Solving Eq. (9–206), we obtain

$$t = \begin{cases} \dfrac{1}{r_L - 1} \{x_L \pm \sqrt{r_L[(1 - r_L)^2 + x_L^2]}\}, & r_L \neq 1, \quad \text{(9–207a)} \\[4mm] -\dfrac{x_L}{2}, & r_L = 1. \quad \text{(9–207b)} \end{cases}$$

The required length d can be found from Eqs. (9–202), (9–207a), and (9–207b):

$$\frac{d}{\lambda} = \begin{cases} \dfrac{1}{2\pi} \arctan^{-1} t, & t \geq 0, \quad \text{(9–208a)} \\[4mm] \dfrac{1}{2\pi} (\pi + \arctan^{-1} t), & t < 0. \quad \text{(9–208b)} \end{cases}$$

Similarly, from Eqs. (9–200) and (9–205), we obtain

$$\frac{\ell}{\lambda} = \begin{cases} \dfrac{1}{2\pi} \arctan^{-1}\left(\dfrac{1}{b_B}\right), & b_B \geq 0, \quad\quad (9\text{–}209\text{a}) \\[2em] \dfrac{1}{2\pi}\left[\pi + \arctan^{-1}\left(\dfrac{1}{b_B}\right)\right], & b_B < 0. \quad\quad (9\text{–}209\text{b}) \end{cases}$$

For a given load impedance, both d/λ and ℓ/λ can be determined easily on a scientific calculator. It is also a simple matter to write a general computer program for the single-stub matching problem. More accurate answers to the problem in Example 9–20 ($r_L = 0.70$ and $x_L = -0.95$) are

$$d_1 = 0.05894469\lambda, \quad \ell_{B1} = 0.11117792\lambda,$$
$$d_2 = 0.22347730\lambda, \quad \ell_{B2} = 0.38882208\lambda.$$

Of course, such accuracies are seldom needed in an actual problem; but these answers have been obtained easily without a Smith chart.

9–7.3 DOUBLE-STUB MATCHING

The method of impedance matching by means of a single stub described in the preceding subsection can be used to match any arbitrary, nonzero, finite load impedance to the characteristic resistance of a line. However, the single-stub method requires that the stub be attached to the main line at a specific point, which varies as the load impedance or the operating frequency is changed. This requirement often presents practical difficulties because the specified junction point may occur at an undesirable location from a mechanical viewpoint. Furthermore, it is very difficult to build a variable-length coaxial line with a constant characteristic impedance. In such cases an alternative method for impedance-matching is to use two short-circuited stubs attached to the main line at fixed positions, as shown in Fig. 9–41. Here, the distance d_o is fixed and arbitrarily chosen (such as $\lambda/16$, $\lambda/8$, $3\lambda/16$, $3\lambda/8$, etc.), and the lengths of the two stub tuners are adjusted to match a given load impedance Z_L to the main line. This scheme is the ***double-stub method*** for impedance matching.

In the arrangement in Fig. 9–41 a stub of length ℓ_A is connected directly in parallel with the load impedance Z_L at terminals A–A', and a second stub of length ℓ_B is attached at terminals B–B' at a fixed distance d_o away. For impedance matching with a main line that has a characteristic resistance R_0, we demand the total input admittance at terminals B–B', looking toward the load, to equal the characteristic conductance of the line; that is,

$$Y_i = Y_B + Y_{sB}$$
$$= Y_0 = \frac{1}{R_0}. \quad\quad (9\text{–}210)$$

In terms of normalized admittances, Eq. (9–210) becomes

$$1 = y_B + y_{sB}. \quad\quad (9\text{–}211)$$

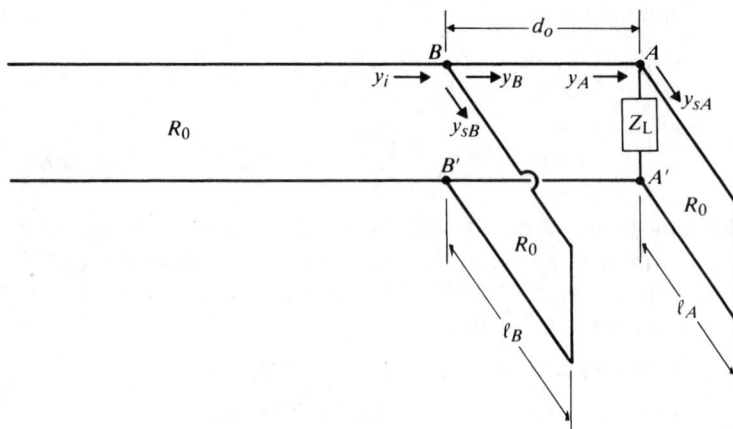

FIGURE 9–41
Impedance matching by double-stub method.

Now, since the input admittance y_{sB} of a short-circuited stub is purely imaginary, Eq. (9–211) can be satisfied only if

$$y_B = 1 + jb_B \tag{9–212}$$

and

$$y_{sB} = -jb_B. \tag{9–213}$$

Note that these requirements are exactly the same as those for single-stub matching.

On the Smith admittance chart the point representing y_B must lie on the $g = 1$ circle. This requirement must be translated by a distance d_o/λ "wavelengths toward load"; that is, y_A at terminals A–A' must lie on the $g = 1$ circle rotated by an angle $4\pi d_o/\lambda$ in the counterclockwise direction. Again, since the input admittance y_{sA} of the short-circuited stub is purely imaginary, the real part of y_A must be solely contributed by the real part of the normalized load admittance, g_L. The solution (or solutions) of the double-stub matching problem is then determined by the intersection (or intersections) of the g_L-circle with the rotated $g = 1$ circle. The procedure for solving a double-stub matching problem on the Smith admittance chart is as follows.

1. Draw the $g = 1$ circle. This is where the point representing y_B should be located.

2. Draw this circle rotated in the counterclockwise direction by d_o/λ "wavelengths toward load." This is where the point representing y_A should be located.

3. Enter the $y_L = g_L + jb_L$ point.

4. Draw the $g = g_L$ circle, intersecting the rotated $g = 1$ circle at one or two points where $y_A = g_L + jb_A$.

5. Mark the corresponding y_B-points on the $g = 1$ circle: $y_B = 1 + jb_B$.

6. Determine stub length ℓ_A from the angle between the point representing y_A and the point representing y_L.

7. Determine stub length ℓ_B from the angle between the point representing $-jb_B$ and P_{sc} on the extreme right.

▰▰▰▰ **EXAMPLE 9-21**   A 50 (Ω) transmission line is connected to a load impedance $Z_L = 60 + j80$ (Ω). A double-stub tuner spaced an eighth of a wavelength apart is used to match the load to the line, as shown in Fig. 9-41. Find the required lengths of the short-circuited stubs.

Solution   Given $R_0 = 50$ (Ω) and $Z_L = 60 + j80$ (Ω), it is easy to calculate

$$y_L = \frac{1}{z_L} = \frac{R_0}{Z_L} = \frac{50}{60 + j80} = 0.30 - j0.40.$$

(We could find y_L on the Smith chart by locating the point diametrically opposite to $z_L = (60 + j80)/50 = 1.20 + j1.60$, but this would clutter up the chart too much.) We follow the procedure outlined above, using a Smith admittance chart.

1. Draw the $g = 1$ circle (Fig. 9-42).
2. Rotate this $g = 1$ circle by $\frac{1}{8}$ "wavelengths toward load" in the counterclockwise direction. The angle of rotation is $4\pi/8$ (rad) or $90°$.
3. Enter $y_L = 0.30 - j0.40$ as P_L.
4. Mark the two points of intersection, P_{A1} and P_{A2}, of the $g_L = 0.30$ circle with the rotated $g = 1$ circle.

$$\text{At } P_{A1}, \text{read} \quad y_{A1} = 0.30 + j0.29;$$
$$\text{At } P_{A2}, \text{read} \quad y_{A2} = 0.30 + j1.75.$$

5. Use a compass centered at the origin O to mark the points P_{B1} and P_{B2} on the $g = 1$ circle corresponding to the points P_{A1} and P_{A2}, respectively.

$$\text{At } P_{B1}, \text{read} \quad y_{B1} = 1 + j1.38;$$
$$\text{At } P_{B2}, \text{read} \quad y_{B2} = 1 - j3.5.$$

6. Determine the required stub lengths ℓ_{A1} and ℓ_{A2} from

$$(y_{sA})_1 = y_{A1} - y_L = j0.69, \quad \ell_{A1} = (0.096 + 0.250)\lambda = 0.346\lambda \text{ (Point } A_1),$$
$$(y_{sA})_2 = y_{A2} - y_L = j2.15, \quad \ell_{A2} = (0.181 + 0.250)\lambda = 0.431\lambda \text{ (Point } A_2).$$

7. Determine the required stub lengths ℓ_{B1} and ℓ_{B2} from

$$(y_{sB})_1 = -j1.38, \quad \ell_{B1} = (0.350 - 0.250)\lambda = 0.100\lambda \text{ (Point } B_1),$$
$$(y_{sB})_2 = j3.5, \quad \ell_{B2} = (0.206 + 0.250)\lambda = 0.456\lambda \text{ (Point } B_2). \quad ▰▰$$

Examination of the construction in Fig. 9-42 reveals that if the point P_L, representing the normalized load admittance $y_L = g_L + jb_L$, lies within the $g = 2$ circle (if $g_L > 2$), then the $g = g_L$ circle does not intersect with the rotated $g = 1$ circle, and no solution exists for double-stub matching with $d_o = \lambda/8$. This region for no solution varies with the chosen distance d_o between the stubs (Problem P.9-52). In such cases,

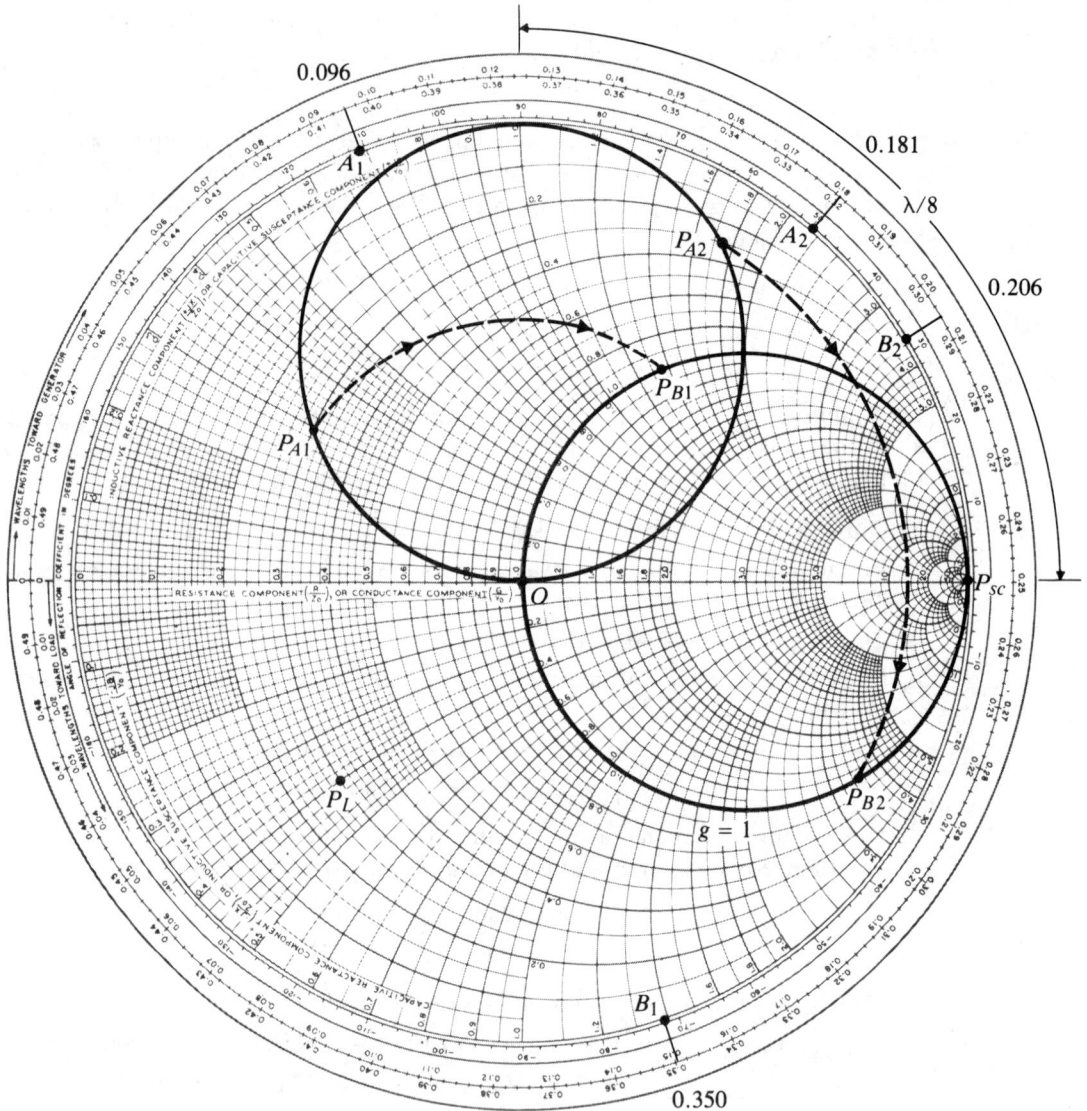

FIGURE 9–42
Construction for double-stub matching on Smith admittance chart.

impedance matching by the double-stub method can be achieved by adding an appropriate line section between Z_L and terminals A–A', as illustrated in Fig. 9–43 (Problem P.9–51).

An analytical solution of the double-stub impedance matching problem is, of course, also possible, albeit more involved than that of the single-stub problem devel-

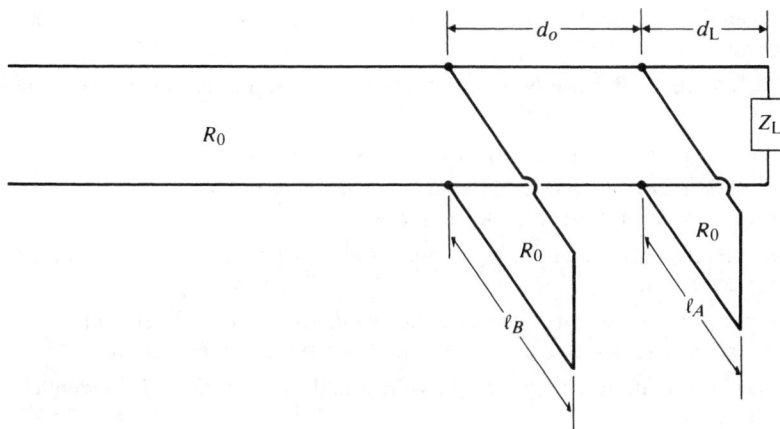

FIGURE 9–43
Double-stub impedance matching with added load-line section.

oped in the preceding subsection. The more ambitious reader may wish to obtain such an analytical solution and write a computer program for determining $d_L/\lambda, \ell_A/\lambda,$ and ℓ_B/λ in terms of z_L and $d_o/\lambda.$[†]

Review Questions

R.9–1 Discuss the similarities and dissimilarities of uniform plane waves in an unbounded media and TEM waves along transmission lines.

R.9–2 What are the three most common types of guiding structures that support TEM waves?

R.9–3 Compare the advantages and disadvantages of coaxial cables and two-wire transmission lines.

R.9–4 Write the transmission-line equations for a lossless parallel-plate line supporting TEM waves.

R.9–5 What are *striplines*?

R.9–6 Describe how the characteristic impedance of a parallel-plate transmission line depends on plate width and dielectric thickness.

R.9–7 Compare the velocity of TEM-wave propagation along a parallel-plate transmission line with that in an unbounded medium.

R.9–8 Define *surface impedance*. How is surface impedance related to the power dissipated in a plate conductor?

[†] D. K. Cheng and C. H. Liang, "Computer solution of double-stub impedance-matching problems," *IEEE Transactions on Education*, November 1982, vol. E-25, pp. 120–123.

R.9–9 State the difference between the surface resistance and the resistance per unit length of a parallel-plate transmission line.

R.9–10 What is the essential difference between a transmission line and an ordinary electric network?

R.9–11 Explain why waves along a lossy transmission line cannot be purely TEM.

R.9–12 What is a *triplate line*? How does the characteristic impedance of a triplate line compare with that of a corresponding stripline? Explain.

R.9–13 Write the general transmission-line equations for arbitrary time dependence and for time-harmonic time dependence.

R.9–14 Define *propagation constant* and *characteristic impedance* of a transmission line. Write their general expressions in terms of R, L, G, and C for sinusoidal excitation.

R.9–15 What is the phase relationship between the voltage and current waves on an infinitely long transmission line?

R.9–16 What is meant by a "distortionless line"? What relation must the distributed parameters of a line satisfy in order for the line to be distortionless?

R.9–17 Is a distortionless line lossless? Is a lossy transmission line dispersive? Explain.

R.9–18 Outline a procedure for determining the distributed parameters of a transmission line.

R.9–19 Show how the attenuation constant of a transmission line is determined from the propagated power and the power lost in the line per unit length.

R.9–20 What does "matched transmission line" mean?

R.9–21 On what factors does the input impedance of a transmission line depend?

R.9–22 What is the input impedance of an open-circuited lossless transmission line if the length of the line is (a) $\lambda/4$, (b) $\lambda/2$, and (c) $3\lambda/4$?

R.9–23 What is the input impedance of a short-circuited lossless transmission line if the length of the line is (a) $\lambda/4$, (b) $\lambda/2$, and (c) $3\lambda/4$?

R.9–24 Is the input reactance of a transmission line $\lambda/8$ long inductive or capacitive if it is (a) open-circuited, and (b) short-circuited?

R.9–25 On a line of length ℓ, what is the relation between the line's characteristic impedance and propagation constant and its open- and short-circuit input impedances?

R.9–26 What is a "quarter-wave transformer"? Why is it not useful for matching a complex load impedance to a low-loss line?

R.9–27 What is the input impedance of a lossless transmission line of length ℓ that is terminated in a load impedance Z_L if (a) $\ell = \lambda/2$, and (b) $\ell = \lambda$?

R.9–28 Discuss how a section of an open-circuited or short-circuited low-loss transmission line can be used to provide a parallel-resonant circuit.

R.9–29 Define the *bandwidth* and the *quality factor*, Q, of a parallel resonant circuit.

R.9–30 Define *voltage reflection coefficient*. Is it the same as "current reflection coefficient"? Explain.

R.9–31 Define *standing-wave ratio*. How is it related to voltage and current reflection coefficients?

R.9–32 What are Γ and S for a line with an open-circuit termination? A short-circuit termination?

R.9–33 Where do the minima of the voltage standing wave on a lossless line with a resistive termination occur (a) if $R_L > R_0$ and (b) if $R_L < R_0$?

R.9–34 Explain how the value of a terminating resistance can be determined by measuring the standing-wave ratio on a lossless transmission line.

R.9–35 Explain how the value of an arbitrary terminating impedance on a lossless transmission line can be determined by standing-wave measurements on the line.

R.9–36 A voltage generator having an internal impedance Z_g is connected at $t = 0$ to the input terminals of a lossless transmission line of length ℓ. The line has a characteristic impedance Z_0 and is terminated with a load impedance Z_L. At what time will a steady state on the line be reached if (a) $Z_g = Z_0$ and $Z_L = Z_0$, (b) $Z_L = Z_0$ but $Z_g \neq Z_0$, (c) $Z_g = Z_0$ but $Z_L \neq Z_0$, and (d) $Z_g \neq Z_0$ and $Z_L \neq Z_0$?

R.9–37 A battery of voltage V_0 is applied through a series resistance R_g to the input terminals of a lossless transmission line having a characteristic resistance R_0 and a load resistance R_L at the far end. What is the amplitude of the first transient voltage wave traveling from the battery to the load? What is the amplitude of the first reflected voltage wave from the load to the battery?

R.9–38 In Question R.9–37, what are the amplitudes of the first current wave traveling from the battery to the load and the first reflected current wave from the load to the battery?

R.9–39 What are *reflection diagrams* of transmission lines? For what purposes are they useful?

R.9–40 How do the voltage and current reflection diagrams of a terminated line differ?

R.9–41 A d-c voltage is applied to a lossless transmission line. Under what conditions will the transient voltage and current distributions along the line have different shapes? Under what conditions will they have the same shape?

R.9–42 Why is the concept of reflection coefficients not useful in analyzing the transient behavior of a transmission line terminated in a reactive load?

R.9–43 What is a Smith chart and why is it useful in making transmission-line calculations?

R.9–44 Where is the point representing a matched load on a Smith chart?

R.9–45 For a given load impedance Z_L on a lossless line of characteristic impedance Z_0, how do we use a Smith chart to determine (a) the reflection coefficient, and (b) the standing-wave ratio?

R.9–46 Why does a change of half a wavelength in line length correspond to a complete revolution on a Smith chart?

R.9–47 Given an impedance $Z = R + jX$, what procedure do we follow to find the admittance $Y = 1/Z$ on a Smith chart?

R.9–48 Given an admittance $Y = G + jB$, how do we use a Smith chart to find the impedance $Z = 1/Y$?

R.9–49 Where is the point representing a short-circuit on a Smith admittance chart?

R.9–50 Is the standing-wave ratio constant on a transmission line even when the line is lossy? Explain.

R.9–51 Can a Smith chart be used for impedance calculations on a lossy transmission line? Explain.

R.9–52 Why is it more convenient to use a Smith chart as an admittance chart for solving impedance-matching problems than to use it as an impedance chart?

R.9–53 Why is it desirable to achieve an impedance match in a transmission line?

R.9–54 Explain the single-stub method for impedance matching on a transmission line.

R.9–55 Explain the double-stub method for impedance matching on a transmission line.

R.9–56 Compare the relative advantages and disadvantages of the single-stub and the double-stub methods of impedance matching.

R.9–57 Why are the stubs used in impedance matching usually of the short-circuited type instead of the open-circuited type?

Problems

P.9–1 Neglecting fringe fields, prove analytically that a y-polarized TEM wave that propagates along a parallel-plate transmission line in $+z$-direction has the following properties: $\partial E_y/\partial x = 0$ and $\partial H_x/\partial y = 0$.

P.9–2 The electric and magnetic fields of a general TEM wave traveling in the $+z$-direction along a transmission line may have both x- and y-components, and both components may be functions of the transverse dimensions.
   a) Find the relations among $E_x(x, y)$, $E_y(x, y)$, $H_x(x, y)$, and $H_y(x, y)$.
   b) Verify that all the four field components in part (a) satisfy the two-dimensional Laplace's equation for static fields.

P.9–3 Consider lossless stripline designs for a given characteristic impedance.
   a) How should the dielectric thickness, d, be changed for a given plate width, w, if the dielectric constant, ϵ_r, is doubled?
   b) How should w be changed for a given d if ϵ_r is doubled?
   c) How should w be changed for a given ϵ_r if d is doubled?
   d) Will the velocity of propagation remain the same as that for the original line after the changes specified in parts (a), (b), and (c)? Explain.

P.9–4 Consider a transmission line made of two parallel brass strips—$\sigma_c = 1.6 \times 10^7$ (S/m)—of width 20 (mm) and separated by a lossy dielectric slab—$\mu = \mu_0$, $\epsilon_r = 3$, $\sigma = 10^{-3}$ (S/m)—of thickness 2.5 (mm). The operating frequency is 500 MHz.
   a) Calculate the R, L, G, and C per unit length.
   b) Compare the magnitudes of the axial and transverse components of the electric field.
   c) Find γ and Z_0.

P.9–5 Verify Eq. (9–39).

P.9–6 Show that the attenuation and phase constants for a transmission line with perfect conductors separated by a lossy dielectric that has a complex permittivity $\epsilon = \epsilon' - j\epsilon''$ are, respectively,

$$\alpha = \omega \sqrt{\frac{\mu\epsilon'}{2}} \left[\sqrt{1 + \left(\frac{\epsilon''}{\epsilon'}\right)^2} - 1 \right]^{1/2} \quad \text{(Np/m)}, \tag{9–214}$$

$$\beta = \omega \sqrt{\frac{\mu\epsilon'}{2}} \left[\sqrt{1 + \left(\frac{\epsilon''}{\epsilon'}\right)^2} + 1 \right]^{1/2} \quad \text{(rad/m)}. \tag{9–215}$$

P.9–7 In the derivation of the approximate formulas of γ and Z_0 for low-loss lines in Subsection 9–3.1, all terms containing the second and higher powers of $(R/\omega L)$ and $(G/\omega C)$

were neglected in comparison with unity. At lower frequencies, better approximations than those given in Eqs. (9–54) and (9–58) may be required. Find new formulas for γ and Z_0 for low-loss lines that retain terms containing $(R/\omega L)^2$ and $(G/\omega C)^2$. Obtain the corresponding expression for phase velocity.

P.9–8 Obtain approximate expressions for γ and Z_0 for a lossy transmission line at very low frequencies such that $\omega L \ll R$ and $\omega C \ll G$.

P.9–9 The following characteristics have been measured on a lossy transmission line at 100 MHz:

$$Z_0 = 50 + j0 \quad (\Omega),$$

$$\alpha = 0.01 \quad (\text{dB/m}),$$

$$\beta = 0.8\pi \quad (\text{rad/m}).$$

Determine R, L, G, and C for the line.

P.9–10 It is desired to construct uniform transmission lines using polyethylene ($\epsilon_r = 2.25$) as the dielectric medium. Assuming negligible losses, (a) find the distance of separation for a 300 (Ω) two-wire line, where the radius of the conducting wires is 0.6 (mm); and (b) find the inner radius of the outer conductor for a 75 (Ω) coaxial line, where the radius of the center conductor is 0.6 (mm).

P.9–11 Prove that a maximum power is transferred from a voltage source with an internal impedance Z_g to a load impedance Z_L over a lossless transmission line when $Z_i = Z_g^*$, where Z_i is the impedance looking into the loaded line. What is the maximum power-transfer efficiency?

P.9–12 Express $V(z)$ and $I(z)$ in terms of the voltage V_i and current I_i at the input end and γ and Z_0 of a transmission line (a) in exponential form, and (b) in hyperbolic form.

P.9–13 Consider a section of a uniform transmission line of length ℓ, characteristic impedance Z_0, and propagation constant γ between terminal pairs 1–1' and 2–2' shown in Fig. 9–44(a). Let (V_1, I_1) and (V_2, I_2) be the phasor voltages and phasor currents at terminals 1–1' and 2–2', respectively.

a) Use Eqs. (9–100a) and (9–100b) to write the equations relating (V_1, I_1) and (V_2, I_2) in the form

$$\begin{bmatrix} V_1 \\ I_1 \end{bmatrix} = \begin{bmatrix} A & B \\ C & D \end{bmatrix} \begin{bmatrix} V_2 \\ I_2 \end{bmatrix}. \qquad (9\text{–}216)$$

Determine A, B, C, and D, and note the following relations:

$$A = D \qquad (9\text{–}217)$$

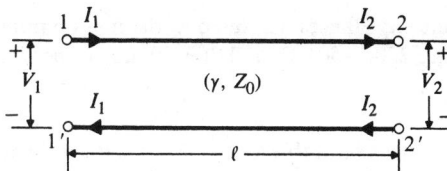

(a) A line section of length ℓ.   (b) An equivalent two-port symmetrical T-network.

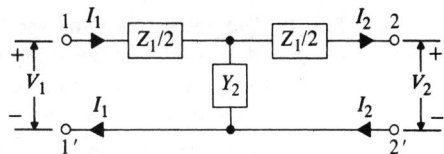

FIGURE 9–44
Equivalence of a line section and a symmetrical two-port network.

and

$$AD - BC = 1. \tag{9-218}$$

b) Because of Eqs. (9–216), (9–217), and (9–218), the line section in Fig. 9–44(a) can be replaced by an equivalent two-port symmetrical T-network shown in Fig. 9–44(b). Prove that

$$Z_1 = \frac{2}{C}(A - 1) = 2Z_0 \tanh \frac{\gamma \ell}{2} \tag{9-219}$$

and

$$Y_2 = C = \frac{1}{Z_0} \sinh \gamma \ell. \tag{9-220}$$

P.9–14 A d-c generator of voltage V_g and internal resistance R_g is connected to a lossy transmission line characterized by a resistance per unit length R and a conductance per unit length G.

    a) Write the governing voltage and current transmission-line equations.

    b) Find the general solutions for $V(z)$ and $I(z)$.

    c) Specialize the solutions in part (b) to those for an infinite line.

    d) Specialize the solutions in part (b) to those for a finite line of length ℓ that is terminated in a load resistance R_L.

P.9–15 A generator with an open-circuit voltage $v_g(t) = 10 \sin 8000\pi t$ (V) and internal impedance $Z_g = 40 + j30$ (Ω) is connected to a 50 (Ω) distortionless line. The line has a resistance of 0.5 (Ω/m), and its lossy dielectric medium has a loss tangent of 0.18%. The line is 50 (m) long and is terminated in a matched load. Find (a) the instantaneous expressions for the voltage and current at an arbitrary location on the line, (b) the instantaneous expressions for the voltage and current at the load, and (c) the average power transmitted to the load.

P.9–16 The input impedance of an open- or short-circuited *lossy* transmission line has both a resistive and a reactive component. Prove that the input impedance of a very short section ℓ of a slightly lossy line ($\alpha\ell \ll 1$ and $\beta\ell \ll 1$) is approximately

    a) $Z_{in} = (R + j\omega L)\ell$ with a short-circuit termination.

    b) $Z_{in} = (G - j\omega C)/[G^2 + (\omega C)^2]\ell$ with an open-circuit termination.

P.9–17 Find the input impedance of a low-loss quarter-wavelength line ($\alpha\lambda \ll 1$)

    a) terminated in a short circuit.

    b) terminated in an open circuit.

P.9–18 A 2 (m) lossless air-spaced transmission line having a characteristic impedance 50 (Ω) is terminated with an impedance $40 + j30$ (Ω) at an operating frequency of 200 (MHz). Find the input impedance.

P.9–19 The open-circuit and short-circuit impedances measured at the input terminals of an air-spaced transmission line 4 (m) long are $250\underline{/-50^\circ}$ (Ω) and $360\underline{/20^\circ}$ (Ω), respectively.

    a) Determine Z_0, α, and β of the line.

    b) Determine R, L, G, and C.

P.9–20 Measurements on a 0.6 (m) lossless coaxial cable at 100 (kHz) show a capacitance of 54 (pF) when the cable is open-circuited and an inductance of 0.30 (μH) when it is short-circuited.

    a) Determine Z_0 and the dielectric constant of its insulating medium.

    b) Calculate the X_{io} and X_{is} at 10 (MHz).

P.9–21 Starting from the input impedance of an open-circuited lossy transmission line in Eq. (9–116), find the expressions for the half-power bandwidth and the Q of a low-loss line with $\ell = n\lambda/2$.

P.9–22 A lossless quarter-wave line section of characteristic impedance R_0 is terminated with an inductive load impedance $Z_L = R_L + jX_L$.
   a) Prove that the input impedance is effectively a resistance R_i in parallel with a capacitive reactance X_i. Determine R_i and X_i in terms of R_0, R_L, and X_L.
   b) Find the ratio of the magnitude of the voltage at the input to that at the load (*voltage transformation ratio*, $|V_{in}|/|V_L|$) in terms of R_0 and Z_L.

P.9–23 A 75 (Ω) lossless line is terminated in a load impedance $Z_L = R_L + jX_L$.
   a) What must be the relation between R_L and X_L in order that the standing-wave ratio on the line be 3?
   b) Find X_L, if $R_L = 150$ (Ω).
   c) Where does the voltage minimum nearest to the load occur on the line for part (b)?

P.9–24 Consider a lossless transmission line.
   a) Determine the line's characteristic resistance so that it will have a minimum possible standing-wave ratio for a load impedance $40 + j30$ (Ω).
   b) Find this minimum standing-wave ratio and the corresponding voltage reflection coefficient.
   c) Find the location of the voltage minimum nearest to the load.

P.9–25 A lossy transmission line with characteristic impedance Z_0 is terminated in an arbitrary load impedance Z_L.
   a) Express the standing-wave ratio S on the line in terms of Z_0 and Z_L.
   b) Find in terms of S and Z_0 the impedance looking toward the load at the location of a voltage maximum.
   c) Find the impedance looking toward the load at a location of a voltage minimum.

P.9–26 A transmission line of characteristic impedance $R_0 = 50$ (Ω) is to be matched to a load impedance $Z_L = 40 + j10$ (Ω) through a length ℓ' of another transmission line of characteristic impedance R_0'. Find the required ℓ' and R_0' for matching.

P.9–27 The standing-wave ratio on a lossless 300 (Ω) transmission line terminated in an unknown load impedance is 2.0, and the nearest voltage minimum is at a distance 0.3λ from the load. Determine (a) the reflection coefficient Γ of the load, (b) the unknown load impedance Z_L, and (c) the equivalent length and terminating resistance of a line, such that the input impedance is equal to Z_L.

P.9–28 Obtain from Eq. (9–147) the formulas for finding the length ℓ_m and the terminating resistance R_m of a lossless line having a characteristic impedance R_0 such that the input impedance equals $Z_i = R_i + jX_i$.

P.9–29 Obtain an analytical expression for the load impedance Z_L connected to a line of characteristic impedance Z_0 in terms of standing-wave ratio S and the distance, z_m'/λ, of the voltage minimum closest to the load.

P.9–30 A sinusoidal voltage generator with $V_g = 0.1\underline{/0^\circ}$ (V) and internal impedance $Z_g = R_0$ is connected to a lossless transmission line having a characteristic impedance $R_0 = 50$ (Ω). The line is ℓ meters long and is terminated in a load resistance $R_L = 25$ (Ω). Find (a) V_i, I_i, V_L, and I_L; (b) the standing-wave ratio on the line; and (c) the average power delivered to the load. Compare the result in part (c) with the case where $R_L = 50$ (Ω).

P.9–31 Consider a lossless transmission line of a characteristic impedance R_0. A time-harmonic voltage source of an amplitude V_g and an internal impedance $R_g = R_0$ is connected to the input terminals of the line, which is terminated with a load impedance $Z_L = R_L + jX_L$. Let P_{inc} be the average incident power associated with the wave traveling in the $+z$-direction.

    a) Find the expression for P_{inc} in terms of V_g and R_0.

    b) Find the expression for the average power P_L delivered to the load in terms of V_g and the reflection coefficient Γ.

    c) Express the ratio P_L/P_{inc} in terms of the standing-wave ratio S.

    d) For $V_g = 100$ (V), $R_g = R_0 = 50$ (Ω), $Z_L = 50 - j25$ (Ω) determine P_{inc}, Γ, S, P_L, $|V_L|$, and $|I_L|$.

P.9–32 A sinusoidal voltage generator $v_g = 110 \sin \omega t$ (V) and internal impedance $Z_g = 50$ (Ω) is connected to a quarter-wave lossless line having a characteristic impedance $R_0 = 50$ (Ω) that is terminated in a purely reactive load $Z_L = j50$ (Ω).

    a) Obtain the voltage and current phasor expressions $V(z')$ and $I(z')$.

    b) Write the instantaneous voltage and current expressions $v(z', t)$ and $i(z', t)$.

    c) Obtain the instantaneous power and the average power delivered to the load.

P.9–33 A d-c voltage V_0 is applied at $t = 0$ directly to the input terminals of an open-circuited lossless transmission line of length ℓ as in Fig. 9–45. Sketch the voltage and current waves on the line (in the manner of Fig. 9–16) for the following time intervals:

    a) $0 < t < T \,(=\ell/u)$

    b) $T < t < 2T$

    c) $2T < t < 3T$

    d) $3T < t < 4T$

What happens after $t = 4T$?

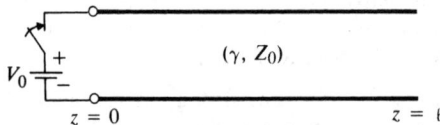

FIGURE 9–45
A d-c voltage applied to an open-circuited line (Problem P.9–33).

P.9–34 A 100 (V) d-c voltage is applied at $t = 0$ to the input terminals of a lossless coaxial cable ($R_{01} = 50$ (Ω), dielectric constant of insulation $\epsilon_{r1} = 2.25$) through an internal resistance $R_g = R_{01}$. The cable is 200 (m) long and is connected to a lossless two-wire line ($R_{02} = 200$ (Ω), $\epsilon_{r2} = 1$), which is 400 (m) long and is terminated in its characteristic resistance.

    a) Describe the transient behavior of the system and find the amplitudes of all reflected and transmitted voltage and current waves.

    b) Sketch the voltage and current as functions of t at the midpoint of the coaxial cable.

    c) Repeat part (b) at the midpoint of the two-wire line.

P.9–35 A d-c voltage V_0 is applied at $t = 0$ to the input terminals of an open-circuited air-dielectric line of a length ℓ through a series resistance equal to $R_0/2$, where R_0 is the characteristic resistance of the line.

a) Draw the voltage and current reflection diagrams.
b) Sketch $V(0, t)$ and and $I(0, t)$.
c) Sketch $V(\ell/2, t)$ and $I(\ell/2, t)$.

P.9–36 A d-c voltage V_0 is applied at $t = 0$ directly to the input terminals of a lossless air-dielectric transmission line of a length ℓ. The line has a characteristic resistance R_0 and is terminated in a load resistance $R_L = 2R_0$.
a) Draw the voltage and currrent reflection diagrams.
b) Sketch $V(\ell, t)$ and $I(\ell, t)$.
c) Sketch $V(z, 2.5T)$ and $I(z, 2.5T)$, where $T = \ell/u$.

P.9–37 For the problem in Example 9–11, determine and sketch $i(200, t)$.

P.9–38 A lossless, air-dielectric, open-circuited transmission line of characteristic resistance R_0 and length ℓ is initially charged to a voltage V_0. At $t = 0$ the line is connected to a resistance R, as shown in Fig. 9–46. Determine $V_R(t)$ and $I_R(t)$ for $0 < t < 5\ell/c$:
a) if $R = 2R_0$,
b) if $R = R_0/2$.

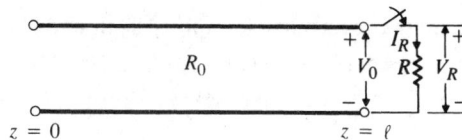

FIGURE 9–46
An initially charged line connected to a resistance (Problem 9–38).

P.9–39 Refer to Fig. 9–26(a) but change the load from a pure inductance to a series combination of $R_L = 10\ (\Omega)$ and $L_L = 48\ (\mu H)$. Assume that $V_0 = 100\ (V)$, $R_0 = 50\ (\Omega)$, $\ell = 900\ (m)$, and $u = c$.
a) Find the expressions for the current in and the voltage across the load as functions of t.
b) Sketch the current and voltage distributions along the transmission line at $t_1 = 4\ (\mu s)$.

P.9–40 Refer to Fig. 9–28(a) but change the load from a pure capacitance to a parallel combination of $C_L = 14\ (nF)$ and $R_L = 1000\ (\Omega)$. Assume that $V_0 = 100\ (V)$, $R_0 = 50\ (\Omega)$, $\ell = 900\ (m)$, and $u = c$.
a) Find the expressions for the current in and the voltage across the load as functions of t.
b) Sketch the current and voltage distributions along the transmission line at $t_1 = 4\ (\mu s)$.

P.9–41 The Smith chart, constructed on the basis of Eqs. (9–188) and (9–189) for a lossless transmission line, is restricted to a unit circle because $|\Gamma| \le 1$. In the case of a lossy line, Z_0 is a complex quantity, and so, in general, is the normalized load impedance $z_L = Z_L/Z_0$.
a) Show that the phase angle of z_L, θ_L, lies between $\pm 3\pi/4$.
b) Show that $|\Gamma|$ may be greater than unity.
c) Prove that max. $|\Gamma| = 2.414$.

P.9–42 The characteristic impedance of a given lossless transmission line is 75 (Ω). Use a Smith chart to find the input impedance at 200 (MHz) of such a line that is (a) 1 (m) long and open-circuited, and (b) 0.8 (m) long and short-circuited. Then (c) determine the corresponding input admittances for the lines in parts (a) and (b).

P.9–43 A load impedance $30 + j10$ (Ω) is connected to a lossless transmission line of length 0.101λ and characteristic impedance 50 (Ω). Use a Smith chart to find (a) the standing-wave ratio, (b) the voltage reflection coefficient, (c) the input impedance, (d) the input admittance, and (e) the location of the voltage minimum on the line.

P.9–44 Repeat Problem P.9–43 for a load impedance $30 - j10$ (Ω).

P.9–45 In a laboratory experiment conducted on a 50 (Ω) lossless transmission line terminated in an unknown load impedance, it is found that the standing-wave ratio is 2.0. The successive voltage minima are 25 (cm) apart, and the first minimum occurs at 5 (cm) from the load. Find (a) the load impedance, and (b) the reflection coefficient of the load. (c) Where would the first voltage minimum be located if the load were replaced by a short-circuit?

P.9–46 The input impedance of a short-circuited lossy transmission line of length 1.5 (m) ($< \lambda/2$) and characteristic impedance 100 (Ω) (approximately real) is $40 - j280$ (Ω).

   a) Find α and β of the line.
   b) Determine the input impedance if the short-circuit is replaced by a load impedance $Z_L = 50 + j50$ (Ω).
   c) Find the input impedance of the short-circuited line for a line length 0.15λ.

P.9–47 A dipole antenna having an input impedance of 73 (Ω) is fed by a 200 (MHz) source through a 300 (Ω) two-wire transmission line. Design a quarter-wave two-wire air line with a 2 (cm) spacing to match the antenna to the 300 (Ω) line.

P.9–48 The single-stub method is used to match a load impedance $25 + j25$ (Ω) to a 50 (Ω) transmission line.

   a) Find the required length and position of a short-circuited stub made of a section of the same 50 (Ω) line.
   b) Repeat part (a) assuming that the short-circuited stub is made of a section of a line that has a characteristic impedance of 75 (Ω).

P.9–49 A load impedance can be matched to a transmission line also by using a single stub placed in series with the load at an appropriate location, as shown in Fig. 9–47. Assuming that $Z_L = 25 + j25$ (Ω), $R_0 = 50$ (Ω), and $R_0' = 35$ (Ω), find d and ℓ required for matching.

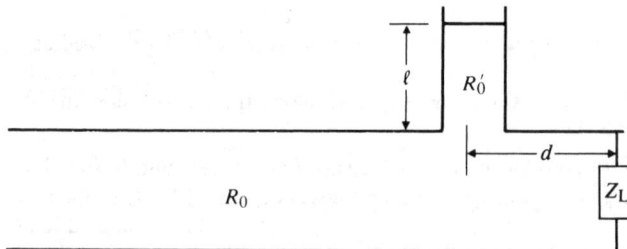

FIGURE 9–47
Impedance matching by a series stub (Problem P.9–49).

P.9–50 The double-stub method is used to match a load impedance $100 + j100$ (Ω) to a lossless transmission line of characteristic impedance 300 (Ω). The spacing between the stubs is $3\lambda/8$, with one stub connected directly in parallel with the load. Determine the lengths of the stub tuners (a) if they are both short-circuited, and (b) if they are both open-circuited.

P.9–51 If the load impedance in Problem P.9–50 is changed to $100 + j50 \, (\Omega)$, one discovers that a perfect match using the double-stub method with $d_0 = 3\lambda/8$ and one stub connected directly across the load is not possible. However, the modified arrangement shown in Fig. 9–43 can be used to match this load with the line.

    a) Find the minimum required additional line length d_L.

    b) Find the required lengths of the short-circuited stub tuners, using the minimum d_L found in part (a).

P.9–52 The double-stub method shown in Fig. 9–41 cannot be used to match certain loads to a line with a given characteristic impedance. Determine the regions of load admittances on a Smith admittance chart for which the double-stub arrangement in Fig. 9–41 cannot lead to a match for $d_o = \lambda/16$, $\lambda/4$, $3\lambda/8$, and $7\lambda/16$.

10

Waveguides and Cavity Resonators

10–1 Introduction

In the preceding chapter we studied the characteristic properties of transverse electromagnetic (TEM) waves guided by transmission lines. The TEM mode of guided waves is one in which the electric and magnetic fields are perpendicular to each other and both are transverse to the direction of propagation along the guiding line. One of the salient properties of TEM waves guided by conducting lines of negligible resistance is that the velocity of propagation of a wave of any frequency is the same as that in an unbounded dielectric medium. This was pointed out in connection with Eq. (9–21) and was reinforced by Eq. (9–72).

TEM waves, however, are not the only mode of guided waves that can propagate on transmission lines; nor are the three types of transmission lines (parallel-plate, two-wire, and coaxial) mentioned in Section 9–1 the only possible wave-guiding structures. As a matter of fact, we see from Eqs. (9–55) and (9–63) that the attenuation constant resulting from the finite conductivity of the lines increases with R, the resistance per unit line length, which, in turn, is proportional to $\sqrt{f}$ in accordance with Tables 9–1 and 9–2. Hence the attenuation of TEM waves tends to increase monotonically with frequency and would be prohibitively high in the microwave range.

In this chapter we first present a general analysis of the characteristics of the waves propagating along uniform guiding structures. Waveguiding structures are called *waveguides*, of which the three types of transmission lines are special cases. The basic governing equations will be examined. We will see that, in addition to *transverse electromagnetic (TEM) waves*, which have no field components in the direction of propagation, both *transverse magnetic (TM) waves* with a longitudinal electric-field component and *transverse electric (TE) waves* with a longitudinal magnetic-field component can also exist. Both TM and TE modes have characteristic *cutoff frequencies*. Waves of frequencies below the cutoff frequency of a particular mode cannot propagate, and power and signal transmission at that mode is possible only for frequencies higher than the cutoff frequency. Thus waveguides operating in TM and TE modes are like high-pass filters.

Also in this chapter we will reexamine the field and wave characteristics of parallel-plate waveguides with emphasis on TM and TE modes and show that all transverse field components can be expressed in terms of E_z (z being the direction of propagation) for TM waves and in terms of H_z for TE waves. The attenuation constants resulting from imperfectly conducting walls will be determined for TM and TE waves, and we will find that the attenuation constant depends, in a complicated way, on the mode of the propagating wave, as well as on frequency. For some modes the attenuation may decrease as the frequency increases; for other modes the attenuation may reach a minimum as the frequency exceeds the cutoff frequency by a certain amount.

Electromagnetic waves can propagate through hollow metal pipes of an arbitrary cross section. Without electromagnetic theory it would not be possible to explain the properties of hollow waveguides. We will see that single-conductor waveguides cannot support TEM waves. We will examine in detail the fields, the current and charge distributions, and the propagation and attenuation characteristics of rectangular and circular cylindrical waveguides. Both TM and TE modes will be discussed.

Electromagnetic waves can also be guided by an open dielectric-slab waveguide. The fields are essentially confined within the dielectric region and decay rapidly away from the slab surface in the transverse plane. For this reason the waves supported by a dielectric-slab waveguide are called *surface waves*. Both TM and TE modes are possible. We will examine the field characteristics and cutoff frequencies of those surface waves. Cylindrical optical fibers will also be discussed.

At microwave frequencies, ordinary lumped-parameter elements (such as inductances and capacitances) connected by wires are no longer practical as circuit elements or as resonant circuits because the dimensions of the elements would have to be extremely small, because the resistance of the wire circuits becomes very high as a result of the skin effect, and because of radiation. We will briefly discuss irises and posts as waveguide reactive elements. A hollow conducting box with proper dimensions can be used as a resonant device. The box walls provide large areas for current flow, and losses are extremely small. Consequently, an enclosed conducting box can be a resonator of a very high Q. Such a box, which is essentially a segment of a waveguide with closed end faces, is called a *cavity resonator*. We will discuss the different mode patterns of the fields inside rectangular as well as circular cylindrical cavity resonators.

10–2 General Wave Behaviors along Uniform Guiding Structures

In this section we examine some general characteristics for waves propagating along straight guiding structures with a uniform cross section. We will assume that the waves propagate in the $+z$-direction with a propagation constant $\gamma = \alpha + j\beta$ that is yet to be determined. For harmonic time dependence with an angular frequency ω, the dependence on z and t for all field components can be described by the exponential

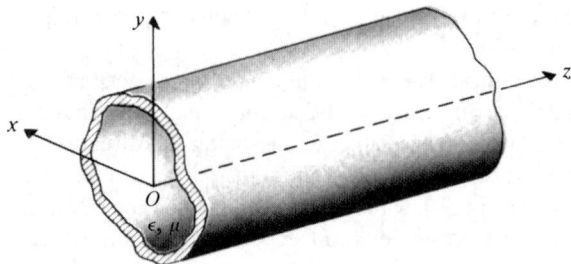

FIGURE 10–1
A uniform waveguide with an arbitrary cross section.

factor

$$e^{-\gamma z}e^{j\omega t} = e^{(j\omega t - \gamma z)} = e^{-\alpha z}e^{j(\omega t - \beta z)}. \tag{10–1}$$

As an example, for a cosine reference we may write the instantaneous expression for the **E** field in Cartesian coordinates as

$$\mathbf{E}(x, y, z; t) = \mathscr{R}e\big[\mathbf{E}^0(x, y)e^{(j\omega t - \gamma z)}\big], \tag{10–2}$$

where $\mathbf{E}^0(x, y)$ is a two-dimensional vector phasor that depends only on the cross-sectional coordinates. The instantaneous expression for the **H** field can be written in a similar way. Hence, in using a phasor representation in equations relating field quantities we may replace partial derivatives with respect to t and z simply by products with $(j\omega)$ and $(-\gamma)$, respectively; the common factor $e^{(j\omega t - \gamma z)}$ can be dropped.

We consider a straight waveguide in the form of a dielectric-filled metal tube having an arbitrary cross section and lying along the z-axis, as shown in Fig. 10–1. According to Eqs. (7–105) and (7–106), the electric and magnetic field intensities in the charge-free dielectric region inside satisfy the following homogeneous vector Helmholtz's equations:

$$\nabla^2\mathbf{E} + k^2\mathbf{E} = 0 \tag{10–3}$$

and

$$\nabla^2\mathbf{H} + k^2\mathbf{H} = 0, \tag{10–4}$$

where **E** and **H** are three-dimensional vector phasors, and k is the wavenumber:

$$k = \omega\sqrt{\mu\epsilon}. \tag{10–5}$$

The three-dimensional Laplacian operator ∇^2 may be broken into two parts: $\nabla^2_{u_1 u_2}$ for the cross-sectional coordinates and ∇^2_z for the longitudinal coordinate. For waveguides with a rectangular cross section we use Cartesian coordinates:

$$\nabla^2\mathbf{E} = (\nabla^2_{xy} + \nabla^2_z)\mathbf{E} = \left(\nabla^2_{xy} + \frac{\partial^2}{\partial z^2}\right)\mathbf{E}$$
$$= \nabla^2_{xy}\mathbf{E} + \gamma^2\mathbf{E}. \tag{10–6}$$

Combination of Eqs. (10–3) and (10–6) gives

$$\nabla^2_{xy}\mathbf{E} + (\gamma^2 + k^2)\mathbf{E} = 0. \tag{10–7}$$

Similarly, from Eq. (10–4) we have

$$\nabla_{xy}^2 \mathbf{H} + (\gamma^2 + k^2)\mathbf{H} = 0. \tag{10-8}$$

We note that each of Eqs. (10–7) and (10–8) is really three second-order partial differential equations, one for each component of $\mathbf{E}$ and $\mathbf{H}$. The exact solution of these component equations depends on the cross-sectional geometry and the boundary conditions that a particular field component must satisfy at conductor-dielectric interfaces. We note further that by writing $\nabla_{r\phi}^2$ for the transversal operator ∇_{xy}^2, Eqs. (10–7) and (10–8) become the governing equations for waveguides with a circular cross section.

Of course, the various components of $\mathbf{E}$ and $\mathbf{H}$ are not all independent, and it is not necessary to solve all six second-order partial differential equations for the six components of $\mathbf{E}$ and $\mathbf{H}$. Let us examine the interrelationships among the six components in Cartesian coordinates by expanding the two source-free curl equations, Eqs. (7–104a) and (7–104b):

From $\nabla \times \mathbf{E} = -j\omega\mu\mathbf{H}$:		From $\nabla \times \mathbf{H} = j\omega\epsilon\mathbf{E}$:	
$\dfrac{\partial E_z^0}{\partial y} + \gamma E_y^0 = -j\omega\mu H_x^0$	(10-9a)	$\dfrac{\partial H_z^0}{\partial y} + \gamma H_y^0 = j\omega\epsilon E_x^0$	(10-10a)
$-\gamma E_x^0 - \dfrac{\partial E_z^0}{\partial x} = -j\omega\mu H_y^0$	(10-9b)	$-\gamma H_x^0 - \dfrac{\partial H_z^0}{\partial x} = j\omega\epsilon E_y^0$	(10-10b)
$\dfrac{\partial E_y^0}{\partial x} - \dfrac{\partial E_x^0}{\partial y} = -j\omega\mu H_z^0$	(10-9c)	$\dfrac{\partial H_y^0}{\partial x} - \dfrac{\partial H_x^0}{\partial y} = j\omega\epsilon E_z^0$	(10-10c)

Note that partial derivatives with respect to z have been replaced by multiplications by $(-\gamma)$. All the component field quantities in the equations above are phasors that depend only on x and y, the common $e^{-\gamma z}$ factor for z-dependence having been omitted. By manipulating these equations we can express the transverse field components H_x^0, H_y^0, and E_x^0, and E_y^0 in terms of the two longitudinal components E_z^0 and H_z^0. For instance, Eqs. (10–9a) and (10–10b) can be combined to eliminate E_y^0 and obtain H_x^0 in terms of E_z^0 and H_z^0. We have

$$H_x^0 = -\frac{1}{h^2}\left(\gamma\frac{\partial H_z^0}{\partial x} - j\omega\epsilon\frac{\partial E_z^0}{\partial y}\right), \tag{10-11}$$

$$H_y^0 = -\frac{1}{h^2}\left(\gamma\frac{\partial H_z^0}{\partial y} + j\omega\epsilon\frac{\partial E_z^0}{\partial x}\right), \tag{10-12}$$

$$E_x^0 = -\frac{1}{h^2}\left(\gamma\frac{\partial E_z^0}{\partial x} + j\omega\mu\frac{\partial H_z^0}{\partial y}\right), \tag{10-13}$$

$$E_y^0 = -\frac{1}{h^2}\left(\gamma\frac{\partial E_z^0}{\partial y} - j\omega\mu\frac{\partial H_z^0}{\partial x}\right), \tag{10-14}$$

where

$$h^2 = \gamma^2 + k^2. \qquad (10-15)$$

The wave behavior in a waveguide can be analyzed by solving Eqs. (10–7) and (10–8) for the longitudinal components, E_z^0 and H_z^0, respectively, subject to the required boundary conditions, and then by using Eqs. (10–11) through (10–14) to determine the other components.

It is convenient to classify the propagating waves in a uniform waveguide into three types according to whether E_z or H_z exists.

1. *Transverse electromagnetic (TEM) waves.* These are waves that contain neither E_z nor H_z. We encountered TEM waves in Chapter 8 when we discussed plane waves and in Chapter 9 on waves along transmission lines.

2. *Transverse magnetic (TM) waves.* These are waves that contain a nonzero E_z but $H_z = 0$.

3. *Transverse electric (TE) waves.* These are waves that contain a nonzero H_z but $E_z = 0$.

The propagation characteristics of the various types of waves are different; they will be discussed in subsequent subsections.

10–2.1 TRANSVERSE ELECTROMAGNETIC WAVES

Since $E_z = 0$ and $H_z = 0$ for TEM waves within a guide, we see that Eqs. (10–11) through (10–14) constitute a set of trivial solutions (all field components vanish) unless the denominator h^2 also equals zero. In other words, TEM waves exist only when

$$\gamma_{\text{TEM}}^2 + k^2 = 0 \qquad (10-16)$$

or

$$\gamma_{\text{TEM}} = jk = j\omega\sqrt{\mu\epsilon}, \qquad (10-17)$$

which is exactly the same expression for the propagation constant of a uniform plane wave in an unbounded medium characterized by constitutive parameters ϵ and μ. We recall that Eq. (10–17) also holds for a TEM wave on a lossless transmission line. It follows that the velocity of propagation (phase velocity) for TEM waves is

$$\boxed{u_{p(\text{TEM})} = \frac{\omega}{k} = \frac{1}{\sqrt{\mu\epsilon}} \qquad (\text{m/s}).} \qquad (10-18)$$

We can obtain the ratio between E_x^0 and H_y^0 from Eqs. (10–9b) and (10–10a) by setting E_z and H_z to zero. This ratio is called the *wave impedance*. We have

$$Z_{\text{TEM}} = \frac{E_x^0}{H_y^0} = \frac{j\omega\mu}{\gamma_{\text{TEM}}} = \frac{\gamma_{\text{TEM}}}{j\omega\epsilon}, \qquad (10-19)$$

which becomes, in view of Eq. (10–17),

$$Z_{TEM} = \sqrt{\frac{\mu}{\epsilon}} = \eta \qquad (\Omega).$$

(10–20)

We note that Z_{TEM} is the same as the intrinsic impedance of the dielectric medium, as given in Eq. (8–30). Equations (10–18) and (10–20) assert that *the phase velocity and the wave impedance for TEM waves are independent of the frequency of the waves.*

Letting $E_z^0 = 0$ in Eq. (10–9a) and $H_z^0 = 0$ in Eq. (10–10b), we obtain

$$\frac{E_y^0}{H_x^0} = -Z_{TEM} = -\sqrt{\frac{\mu}{\epsilon}}.$$

(10–21)

Equations (10–19) and (10–21) can be combined to obtain the following formula for a TEM wave propagating in the $+z$-direction:

$$\mathbf{H} = \frac{1}{Z_{TEM}} \mathbf{a}_z \times \mathbf{E} \qquad (A/m),$$

(10–22)

which again reminds us of a similar relation for a uniform plane wave in an unbounded medium—see Eq. (8–29).

Single-conductor waveguides cannot support TEM waves. In Section 6–2 we pointed out that magnetic flux lines always close upon themselves. Hence if a TEM wave were to exist in a waveguide, the field lines of **B** and **H** would form closed loops in a transverse plane. However, the generalized Ampère's circuital law, Eq. (7–54b), requires that the line integral of the magnetic field (the magnetomotive force) around any closed loop in a transverse plane must equal the sum of the longitudinal conduction and displacement currents through the loop. Without an inner conductor there is no longitudinal conduction current inside the waveguide. By definition, a TEM wave does not have an E_z-component; consequently, there is no longitudinal displacement current. The total absence of a longitudinal current inside a waveguide leads to the conclusion that there can be no closed loops of magnetic field lines in any transverse plane. Therefore, we conclude that *TEM waves cannot exist in a single-conductor hollow (or dielectric-filled) waveguide of any shape.* On the other hand, *assuming perfect conductors,* a coaxial transmission line having an inner conductor *can* support TEM waves; so can a two-conductor stripline and a two-wire transmission line. When the conductors have losses, waves along transmission lines are strictly no longer TEM, as noted in Section 9–2.

10–2.2   TRANSVERSE MAGNETIC WAVES

Transverse magnetic (TM) waves do not have a component of the magnetic field in the direction of propagation, $H_z = 0$. The behavior of TM waves can be analyzed by solving Eq. (10–7) for E_z subject to the boundary conditions of the guide and using Eqs. (10–11) through (10–14) to determine the other components. Writing Eq. (10–7) for E_z, we have

$$\nabla_{xy}^2 E_z^0 + (\gamma^2 + k^2)E_z^0 = 0 \tag{10–23}$$

or

$$\boxed{\nabla_{xy}^2 E_z^0 + h^2 E_z^0 = 0.} \tag{10–24}$$

Equation (10–24) is a second-order partial differential equation, which can be solved for E_z^0. In this section we wish to discuss only the general properties of the various wave types. The actual solution of Eq. (10–24) will wait until subsequent sections when we examine particular waveguides.

For TM waves we set $H_z = 0$ in Eqs. (10–11) through (10–14) to obtain

$$H_x^0 = \frac{j\omega\epsilon}{h^2}\frac{\partial E_z^0}{\partial y}, \tag{10–25}$$

$$H_y^0 = -\frac{j\omega\epsilon}{h^2}\frac{\partial E_z^0}{\partial x}, \tag{10–26}$$

$$E_x^0 = -\frac{\gamma}{h^2}\frac{\partial E_z^0}{\partial x}, \tag{10–27}$$

$$E_y^0 = -\frac{\gamma}{h^2}\frac{\partial E_z^0}{\partial y}. \tag{10–28}$$

It is convenient to combine Eqs. (10–27) and (10–28) and write

$$\boxed{(\mathbf{E}_T^0)_{\text{TM}} = \mathbf{a}_x E_x^0 + \mathbf{a}_y E_y^0 = -\frac{\gamma}{h^2}\nabla_T E_z^0 \qquad (\text{V/m}),} \tag{10–29}$$

where

$$\nabla_T E_z^0 = \left(\mathbf{a}_x\frac{\partial}{\partial x} + \mathbf{a}_y\frac{\partial}{\partial y}\right)E_z^0 \tag{10–30}$$

denotes the gradient of E_z^0 in the transverse plane. Equation (10–29) is a concise formula for finding E_x^0 and E_y^0 from E_z^0.

The transverse components of magnetic field intensity, H_x^0 and H_y^0, can be determined simply from E_x^0 and E_y^0 on the introduction of the wave impedance for the TM mode. We have, from Eqs. (10–25) through (10–28),

$$\boxed{Z_{\text{TM}} = \frac{E_x^0}{H_y^0} = -\frac{E_y^0}{H_x^0} = \frac{\gamma}{j\omega\epsilon} \qquad (\Omega).} \tag{10–31}$$

It is important to note that Z_{TM} is *not* equal to $j\omega\mu/\gamma$, because γ for TM waves, unlike γ_{TEM}, is *not* equal to $j\omega\sqrt{\mu\epsilon}$. The following relation between the electric and magnetic field intensities holds for TM waves:

$$\boxed{\mathbf{H} = \frac{1}{Z_{TM}} (\mathbf{a}_z \times \mathbf{E}) \qquad (\text{A/m}).}$$

(10–32)

Equation (10–32) is seen to be of the same form as Eq. (10–22) for TEM waves.

When we undertake to solve the two-dimensional homogeneous Helmholtz equation, Eq. (10–24), subject to the boundary conditions of a given waveguide, we will discover that solutions are possible only for *discrete values of h*. There may be an infinity of these discrete values, but solutions are not possible for all values of h. The values of h for which a solution of Eq. (10–24) exists are called the **characteristic values** or **eigenvalues** of the boundary-value problem. Each of the eigenvalues determines the characteristic properties of a particular TM mode of the given waveguide.

In the following sections we will also discover that the eigenvalues of the various waveguide problems are real numbers. From Eq. (10–15) we have

$$\begin{aligned} \gamma &= \sqrt{h^2 - k^2} \\ &= \sqrt{h^2 - \omega^2 \mu\epsilon}. \end{aligned}$$

(10–33)

Two distinct ranges of the values for the propagation constant are noted, the dividing point being $\gamma = 0$, where

$$\omega_c^2 \mu\epsilon = h^2$$

(10–34)

or

$$\boxed{f_c = \frac{h}{2\pi\sqrt{\mu\epsilon}} \qquad (\text{Hz}).}$$

(10–35)

The frequency, f_c, at which $\gamma = 0$ is called a **cutoff frequency**. *The value of f_c for a particular mode in a waveguide depends on the eigenvalue of this mode.* Using Eq. (10–35), we can write Eq. (10–33) as

$$\gamma = h\sqrt{1 - \left(\frac{f}{f_c}\right)^2}.$$

(10–36)

The two distinct ranges of γ can be defined in terms of the ratio $(f/f_c)^2$ as compared to unity.

a) $\left(\dfrac{f}{f_c}\right)^2 > 1$, or $f > f_c$. In this range, $\omega^2\mu\epsilon > h^2$ and γ is imaginary. We have, from Eq. (10–33),

$$\gamma = j\beta = jk\sqrt{1 - \left(\frac{h}{k}\right)^2} = jk\sqrt{1 - \left(\frac{f_c}{f}\right)^2}.$$

(10–37)

It is a propagating mode with a phase constant β:

$$\beta = k \sqrt{1 - \left(\frac{f_c}{f}\right)^2} \qquad \text{(rad/m)}. \tag{10-38}$$

The corresponding wavelength in the guide is

$$\lambda_g = \frac{2\pi}{\beta} = \frac{\lambda}{\sqrt{1 - (f_c/f)^2}} > \lambda, \tag{10-39}$$

where

$$\lambda = \frac{2\pi}{k} = \frac{1}{f\sqrt{\mu\epsilon}} = \frac{u}{f} \tag{10-40}$$

is the wavelength of a plane wave with a frequency f in an unbounded dielectric medium characterized by μ and ϵ, and $u = 1/\sqrt{\mu\epsilon}$ is the velocity of light in the medium. Equation (10–39) can be rearranged to give a simple relation connecting λ, the guide wavelength λ_g, and the cutoff wavelength $\lambda_c = u/f_c$:

$$\frac{1}{\lambda^2} = \frac{1}{\lambda_g^2} + \frac{1}{\lambda_c^2}. \tag{10-41}$$

The phase velocity of the propagating wave in the guide is

$$u_p = \frac{\omega}{\beta} = \frac{u}{\sqrt{1 - (f_c/f)^2}} = \frac{\lambda_g}{\lambda} u > u. \tag{10-42}$$

We see from Eq. (10–42) that the phase velocity within a waveguide is always higher than that in an unbounded medium and is frequency-dependent. Hence *single-conductor waveguides are dispersive transmission systems*, although an unbounded lossless dielectric medium is nondispersive. The group velocity for a propagating wave in a waveguide can be determined by using Eq. (8–72):

$$u_g = \frac{1}{d\beta/d\omega} = u \sqrt{1 - \left(\frac{f_c}{f}\right)^2} = \frac{\lambda}{\lambda_g} u < u. \tag{10-43}$$

Thus,

$$u_g u_p = u^2. \tag{10-44}$$

For air dielectric, $u = c$, Eq. (10–44) becomes $u_g u_p = c^2$. In a lossless waveguide the velocity of signal propagation (the *velocity of energy transport*) is equal to the group velocity. An illustration of this statement can be found later, in Subsection 10–3.3.

Substitution of Eq. (10–37) in Eq. (10–31) yields

$$Z_{\text{TM}} = \eta \sqrt{1 - \left(\frac{f_c}{f}\right)^2} \qquad (\Omega). \tag{10-45}$$

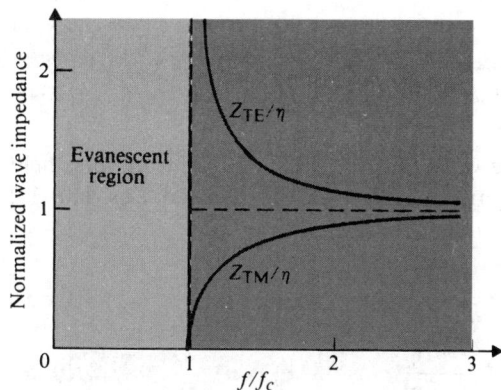

FIGURE 10-2
Normalized wave impedances for propagating TM and TE waves.

The wave impedance of propagating TM modes in a waveguide with a lossless dielectric is purely resistive and is always less than the intrinsic impedance of the dielectric medium. The variation of Z_{TM} versus f/f_c for $f > f_c$ is sketched in Fig. 10-2.

b) $\left(\dfrac{f}{f_c}\right)^2 < 1$, or $f < f_c$. When the operating frequency is lower than the cutoff frequency, γ is real and Eq. (10-36) can be written as

$$\gamma = \alpha = h\sqrt{1 - \left(\frac{f}{f_c}\right)^2}, \qquad f < f_c, \tag{10-46}$$

which is, in fact, an attenuation constant. Since all field components contain the propagation factor $e^{-\gamma z} = e^{-\alpha z}$, the wave diminishes rapidly with z and is said to be *evanescent*. Therefore, *a waveguide exhibits the property of a high-pass filter. For a given mode, only waves with a frequency higher than the cutoff frequency of the mode can propagate in the guide.*

Substitution of Eq. (10-46) in Eq. (10-31) gives the wave impedance of TM modes for $f < f_c$:

$$Z_{TM} = -j\frac{h}{\omega\epsilon}\sqrt{1 - \left(\frac{f}{f_c}\right)^2}, \qquad f < f_c. \tag{10-47}$$

Thus, the wave impedance of evanescent TM modes at frequencies below cutoff is purely reactive, indicating that there is no power flow associated with evanescent waves.

10-2.3 TRANSVERSE ELECTRIC WAVES

Transverse electric (TE) waves do not have a component of the electric field in the direction of propagation, $E_z = 0$. The behavior of TE waves can be analyzed by first solving Eq. (10–8) for H_z:

$$\nabla^2_{xy}H_z + h^2 H_z = 0. \qquad (10\text{–}48)$$

Proper boundary conditions at the guide walls must be satisfied. The transverse field components can then be found by substituting H_z into the reduced Eqs. (10–11) through (10–14) with E_z set to zero. We have

$$H^0_x = -\frac{\gamma}{h^2}\frac{\partial H^0_z}{\partial x}, \qquad (10\text{–}49)$$

$$H^0_y = -\frac{\gamma}{h^2}\frac{\partial H^0_z}{\partial y}, \qquad (10\text{–}50)$$

$$E^0_x = -\frac{j\omega\mu}{h^2}\frac{\partial H^0_z}{\partial y}, \qquad (10\text{–}51)$$

$$E^0_y = \frac{j\omega\mu}{h^2}\frac{\partial H^0_z}{\partial x}. \qquad (10\text{–}52)$$

Combining Eqs. (10–49) and (10–50), we obtain

$$(\mathbf{H}^0_T)_{TE} = \mathbf{a}_x H^0_x + \mathbf{a}_y H^0_y = -\frac{\gamma}{h^2}\nabla_T H^0_z \qquad (A/m). \qquad (10\text{–}53)$$

We note that Eq. (10–53) is entirely similar to Eq. (10–29) for TM modes.

The transverse components of electric field intensity, E^0_x and E^0_y, are related to those of magnetic field intensity through the wave impedance. We have, from Eqs. (10–49) through (10–52),

$$Z_{TE} = \frac{E^0_x}{H^0_y} = -\frac{E^0_y}{H^0_x} = \frac{j\omega\mu}{\gamma} \qquad (\Omega). \qquad (10\text{–}54)$$

Note that Z_{TE} in Eq. (10–54) is quite different from Z_{TM} in Eq. (10–31) because γ for TE waves, unlike γ_{TEM}, is *not* equal to $j\omega\sqrt{\mu\epsilon}$. Equations (10–51), (10–52), and (10–54) can now be combined to give the following vector formula:

$$\mathbf{E} = -Z_{TE}(\mathbf{a}_z \times \mathbf{H}) \qquad (V/m). \qquad (10\text{–}55)$$

Inasmuch as we have not changed the relation between γ and h, Eqs. (10–33) through (10–44) pertaining to TM waves also apply to TE waves. There are also two distinct ranges of γ, depending on whether the operating frequency is higher or lower than the cutoff frequency, f_c, given in Eq. (10–35).

a) $\left(\dfrac{f}{f_c}\right)^2 > 1$, or $f > f_c$. In this range, γ is imaginary, and we have a propagating mode. The expression for γ is the same as that given in Eq. (10–37):

$$\gamma = j\beta = jk\sqrt{1 - \left(\frac{f_c}{f}\right)^2}. \tag{10–56}$$

Consequently, the formulas for β, λ_g, u_p, and u_g in Eqs. (10–38), (10–39), (10–42), and (10–43), respectively, also hold for TE waves. Using Eq. (10–56) in Eq. (10–54), we obtain

$$\boxed{Z_{\text{TE}} = \frac{\eta}{\sqrt{1 - (f_c/f)^2}} \quad (\Omega),} \tag{10–57}$$

which is obviously different from the expression for Z_{TM} in Eq. (10–45). Equation (10–57) indicates that *the wave impedance of propagating TE modes in a waveguide with a lossless dielectric is purely resistive and is always larger than the intrinsic impedance of the dielectric medium*. The variation of Z_{TE} versus f/f_c for $f > f_c$ is also sketched in Fig. 10–2.

b) $\left(\dfrac{f}{f_c}\right)^2 < 1$, or $f < f_c$. In this case, γ is real and we have an evanescent or non-propagating mode:

$$\gamma = \alpha = h\sqrt{1 - \left(\frac{f}{f_c}\right)^2}, \qquad f < f_c. \tag{10–58}$$

Substitution of Eq. (10–58) in Eq. (10–54) gives the wave impedance of TE modes for $f < f_c$:

$$Z_{\text{TE}} = j\frac{\omega\mu}{h\sqrt{1 - (f/f_c)^2}}, \qquad f < f_c, \tag{10–59}$$

which is purely reactive, indicating again that there is no power flow for evanescent waves at $f < f_c$.

EXAMPLE 10–1   (a) Determine the wave impedance and guide wavelength at a frequency equal to twice the cutoff frequency in a waveguide for TM and TE modes. (b) Repeat part (a) for a frequency equal to one-half of the cutoff frequency. (c) What are the wave impedance and guide wavelength for the TEM mode?

Solution

a) At $f = 2f_c$, which is above the cutoff frequency, we have propagating modes. The appropriate formulas are Eqs. (10–45), (10–57), and (10–39).

For $f = 2f_c$, $(f_c/f)^2 = \frac{1}{4}$, $\sqrt{1 - (f_c/f)^2} = \sqrt{3}/2 = 0.866$. Thus,

$$Z_{\text{TM}} = 0.866\eta < \eta, \qquad \lambda_{\text{TM}} = 1.155\lambda > \lambda,$$
$$Z_{\text{TE}} = 1.155\eta > \eta, \qquad \lambda_{\text{TE}} = 1.155\lambda > \lambda,$$

TABLE 10-1
Wave Impedances and Guide Wavelengths for $f > f_c$

Mode	Wave Impedance, Z	Guide Wavelength, λ_g
TEM	$\eta = \sqrt{\dfrac{\mu}{\epsilon}}$	$\lambda = \dfrac{1}{f\sqrt{\mu\epsilon}}$
TM	$\eta\sqrt{1 - \left(\dfrac{f_c}{f}\right)^2}$	$\dfrac{\lambda}{\sqrt{1 - (f_c/f)^2}}$
TE	$\dfrac{\eta}{\sqrt{1 - (f_c/f)^2}}$	$\dfrac{\lambda}{\sqrt{1 - (f_c/f)^2}}$

where η is the intrinsic impedance of the guide medium. These results are summarized in Table 10-1.

b) At $f = f_c/2 < f_c$, the waveguide modes are evanescent, and guide wavelength has no significance. We now have

$$Z_{TM} = -j\frac{h}{\omega\epsilon}\sqrt{1 - \left(\frac{f}{f_c}\right)^2} = -j0.276h/f_c\epsilon,$$

$$Z_{TE} = j\frac{\omega\mu}{h\sqrt{1 - (f/f_c)^2}} = j3.63f_c\mu/h.$$

We note that both Z_{TM} and Z_{TE} become imaginary (reactive) for evanescent modes at $f < f_c$; their values depend on the eigenvalue h, which is a characteristic of the particular TM or TE mode.

c) The TEM mode does not exhibit a cutoff property and $h = 0$. The wave impedance and guide wavelength are independent of frequency. From Eqs. (10-20) and (10-18) we have

$$Z_{TEM} = \eta$$

and

$$\lambda_{TEM} = \lambda.$$ ∎

For propagating modes, $\gamma = j\beta$ and the variation of β versus frequency determines the characteristics of a wave along a guide. It is therefore useful to plot and examine an $\omega-\beta$ diagram.[†] Figure 10-3 is such a diagram in which the dashed line through the origin represents the $\omega-\beta$ relationship for TEM mode. The constant slope of this straight line is $\omega/\beta = u = 1/\sqrt{\mu\epsilon}$, which is the same as the velocity of light in an unbounded dielectric medium with constitutive parameters μ and ϵ.

---

† Also referred to as a **Brillouin diagram**.

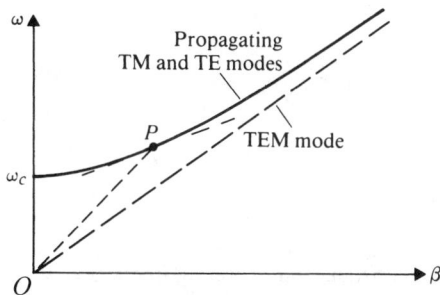

FIGURE 10–3
ω–β diagram for waveguide.

The solid curve above the dashed line depicts a typical ω–β relation for either a TM or a TE propagating mode, given by Eq. (10–38). We can write

$$\omega = \frac{\beta u}{\sqrt{1 - (\omega_c/\omega)^2}}. \tag{10–60}$$

The ω–β curve intersects the ω-axis ($\beta = 0$) at $\omega = \omega_c$. The slope of the line joining the origin and any point, such as P, on the curve is equal to the phase velocity, u_p, for a particular mode having a cutoff frequency f_c and operating at a particular frequency. The local slope of the ω–β curve at P is the group velocity, u_g. We note that, for propagating TM and TE waves in a waveguide, $u_p > u$, $u_g < u$, and Eq. (10–44) holds. As the operating frequency increases much above the cutoff frequency, both u_p and u_g approach u asymptotically. The exact value of ω_c depends on the eigenvalue h in Eq. (10–35)—that is, on the particular TM or TE mode in a waveguide of a given cross section. Methods for determining h will be discussed when we examine different types of waveguides. We recall that the ω–β graph for wave propagation in an ionized medium (Fig. 8–7) was quite similar to the ω–β diagram for a waveguide shown in Fig. 10–3.

EXAMPLE 10–2 Obtain a graph showing the relation between the attenuation constant α and the operating frequency f for evanescent modes in a waveguide.

Solution For evanescent TM or TE modes, $f < f_c$ and Eq. (10–46) or (10–58) applies. We have

$$\left(\frac{f_c}{h}\alpha\right)^2 + f^2 = f_c^2. \tag{10–61}$$

Hence the graph of $(f_c\alpha/h)$ plotted versus f is a circle centered at the origin and having a radius f_c. This is shown in Fig. 10–4. The value of α for any $f < f_c$ can be found from this quarter of a circle.

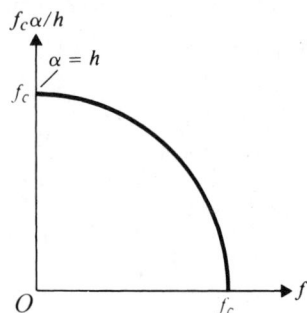

FIGURE 10–4
Relation between attenuation constant and operating frequency for evanescent modes (Example 10–2).

10–3 Parallel-Plate Waveguide

In Section 9–2 we discussed the characteristics of TEM waves propagating along a parallel-plate transmission line. It was then pointed out, and again emphasized in Subsection 10–2.1, that the field behavior for TEM modes bears a very close resemblance to that for uniform plane waves in an unbounded dielectric medium. However, TEM modes are not the only type of waves that can propagate along perfectly conducting parallel-plates separated by a dielectric. A parallel-plate waveguide can also support TM and TE waves. The characteristics of these waves are examined separately in following subsections.

10–3.1   TM WAVES BETWEEN PARALLEL PLATES

Consider the parallel-plate waveguide of two perfectly conducting plates separated by a dielectric medium with constitutive parameters ϵ and μ, as shown in Fig. 10–5. The plates are assumed to be infinite in extent in the x-direction. This is tantamount to assuming that the fields do not vary in the x-direction and that edge effects are negligible. Let us suppose that TM waves ($H_z = 0$) propagate in the $+z$-direction. For harmonic time dependence it is expedient to work with equations relating field

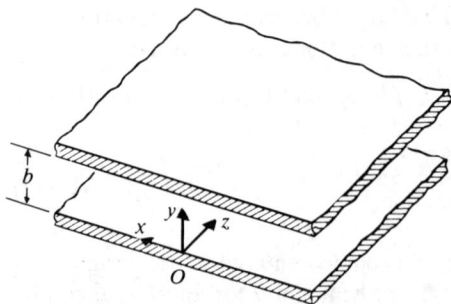

FIGURE 10–5
An infinite parallel-plate waveguide.

quantities with the common factor $e^{(j\omega t - \gamma z)}$ omitted. We write the phasor $E_z(y, z)$ as $E_z^0(y)e^{-\gamma z}$. Equation (10–24) then becomes

$$\frac{d^2 E_z^0(y)}{dy^2} + h^2 E_z^0(y) = 0. \tag{10–62}$$

The solution of Eq. (10–62) must satisfy the boundary conditions

$$E_z^0(y) = 0 \quad \text{at } y = 0 \quad \text{and} \quad y = b.$$

From Section 4–5 we conclude that $E_z^0(y)$ must be of the following form $(h = n\pi/b)$:

$$E_z^0(y) = A_n \sin\left(\frac{n\pi y}{b}\right), \tag{10–63}$$

where the amplitude A_n depends on the strength of excitation of the particular TM wave. The only other nonzero field components are obtained from Eqs. (10–25) and (10–28). Keeping in mind that $\partial E_z/\partial x = 0$ and omitting the $e^{-\gamma z}$ factor, we have

$$H_x^0(y) = \frac{j\omega\epsilon}{h} A_n \cos\left(\frac{n\pi y}{b}\right), \tag{10–64}$$

$$E_y^0(y) = -\frac{\gamma}{h} A_n \cos\left(\frac{n\pi y}{b}\right). \tag{10–65}$$

The γ in Eq. (10–65) is the propagation constant that can be determined from Eq. (10–33):

$$\gamma = \sqrt{\left(\frac{n\pi}{b}\right)^2 - \omega^2 \mu\epsilon}. \tag{10–66}$$

The cutoff frequency is the frequency that makes $\gamma = 0$. We have

$$\boxed{f_c = \frac{n}{2b\sqrt{\mu\epsilon}} \quad \text{(Hz),}} \tag{10–67}$$

which, of course, checks with Eq. (10–35). Waves with $f > f_c$ propagate with a phase constant β, given in Eq. (10–38); and waves with $f \leq f_c$ are evanescent.

Depending on the values of n, there are different possible propagating TM modes (eigenmodes) corresponding to the different eigenvalues h. Thus there are the TM_1 mode $(n = 1)$ with cutoff frequency $(f_c)_1 = 1/2b\sqrt{\mu\epsilon}$, the TM_2 mode $(n = 2)$ with $(f_c)_2 = 1/b\sqrt{\mu\epsilon}$, and so on. Each mode has its own characteristic phase constant, guide wavelength, phase velocity, group velocity, and wave impedance; they can be determined from Eqs. (10–38), (10–39), (10–42), (10–43), and (10–45), respectively. When $n = 0$, $E_z = 0$ and only the transverse components H_x and E_y exist. Hence TM_0 mode is the TEM mode, for which $f_c = 0$. The mode having the lowest cutoff *frequency is called the **dominant mode** of the waveguide. **For parallel-plate waveguides the dominant mode is the TEM mode.**

EXAMPLE 10-3 (a) Write the instantaneous field expressions for TM$_1$ mode in a parallel-plate waveguide. (b) Sketch the electric and magnetic field lines in the yz-plane.

Solution

a) The instantaneous field expressions for the TM$_1$ mode are obtained by multiplying the phasor expressions in Eqs. (10–63), (10–64), and (10–65) with $e^{j(\omega t - \beta z)}$ and taking the real part of the product. We have, for $n = 1$,

$$E_z(y, z; t) = A_1 \sin\left(\frac{\pi y}{b}\right) \cos(\omega t - \beta z), \tag{10–68}$$

$$E_y(y, z; t) = \frac{\beta b}{\pi} A_1 \cos\left(\frac{\pi y}{b}\right) \sin(\omega t - \beta z), \tag{10–69}$$

$$H_x(y, z; t) = -\frac{\omega \epsilon b}{\pi} A_1 \cos\left(\frac{\pi y}{b}\right) \sin(\omega t - \beta z), \tag{10–70}$$

where

$$\beta = \sqrt{\omega^2 \mu \epsilon - \left(\frac{\pi}{b}\right)^2}. \tag{10–71}$$

b) In the yz-plane, **E** has both a y- and a z-component, and the equation of the electric field lines at a given t can be found from the relation

$$\frac{dy}{E_y} = \frac{dz}{E_z}. \tag{10–72}$$

For example, at $t = 0$, Eq. (10–72) can be written as

$$\frac{dy}{dz} = \frac{E_y(y, z; 0)}{E_z(y, x; 0)} = -\frac{\beta b}{\pi} \cot\left(\frac{\pi y}{b}\right) \tan \beta z, \tag{10–73}$$

which gives the slope of the electric field lines. Equation (10–73) can be integrated to give

$$\cos\left(\frac{\pi y}{b}\right) \cos \beta z = \text{Constant}, \qquad 0 \le y \le b. \tag{10–74}^\dagger$$

---

† Equation (10–73) can be rearranged as

$$\frac{dy}{dz} = -\left(\frac{\beta b}{\pi}\right) \frac{\cos(\pi y/b) \sin \beta z}{\sin(\pi y/b) \cos \beta_z}$$

or

$$\frac{(\pi/b) \sin(\pi y/b)\, dy}{\cos(\pi y/b)} = \frac{-\beta \sin \beta z\, dz}{\cos \beta z}$$

or

$$-\frac{d[\cos(\pi y/b)]}{\cos(\pi y/b)} = \frac{d(\cos \beta z)}{\cos \beta z}.$$

Integration gives

$$-\ln[\cos(\pi y/b)] = \ln(\cos \beta z) + c_1$$

or

$$\cos(\pi y/b) \cos \beta z = c_2,$$

which is Eq. (10–74). c_1 and c_2 are constants.

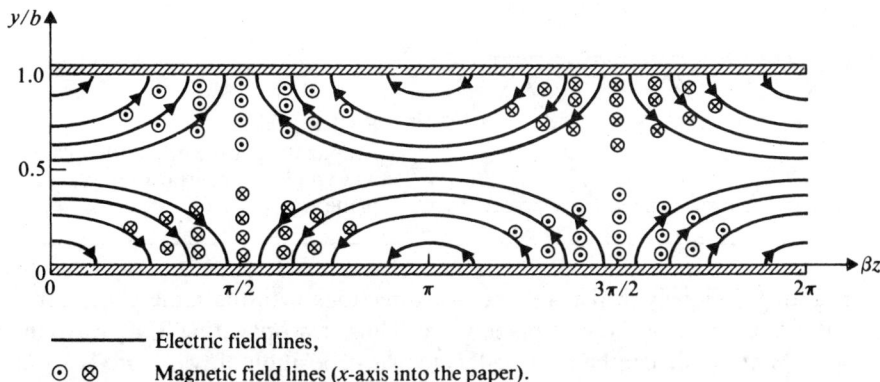

--------- Electric field lines,

⊙ ⊗   Magnetic field lines (x-axis into the paper).

FIGURE 10-6
Field lines for TM_1 mode in parallel-plate waveguide.

Several such electric field lines are drawn in Fig. 10-6. The field lines repeat themselves for every change of 2π (rad) in βz and reverse their directions for every change of π (rad).

Since **H** has only an x-component, the magnetic field lines are everywhere perpendicular to the yz-plane. For the TM_1 mode at $t = 0$, Eq. (10-70) becomes

$$H_x(y, z; 0) = \frac{\omega \epsilon b}{\pi} A_1 \cos\left(\frac{\pi y}{b}\right) \sin \beta z. \qquad (10-75)$$

The density of H_x lines varies as $\cos(\pi y/b)$ in the y-direction and as $\sin \beta z$ in the z-direction. This is also sketched in Fig. 10-6. At the conducting plates ($y = 0$ and $y = b$) there are surface currents because of a discontinuity in the tangential magnetic field, and surface charges because of the presence of a normal electric field. (Problem P.10-4).
■

EXAMPLE 10-4 Show that the field solution of a propagating TM_1 wave in a parallel-plate waveguide can be interpreted as the superposition of two plane waves bouncing back and forth obliquely between the two conducting plates.

Solution This can be seen readily by writing the phasor expression of $E_z^0(y)$ from Eq. (10-63) for $n = 1$ and with the factor $e^{-j\beta z}$ restored. We have

$$E_z(y, z) = A_1 \sin\left(\frac{\pi y}{b}\right) e^{-j\beta z} = \frac{A_1}{2j} (e^{j\pi y/b} - e^{-j\pi y/b}) e^{-j\beta z}$$

$$= \frac{A_1}{2j} \left[e^{-j(\beta z - \pi y/b)} - e^{-j(\beta z + \pi y/b)} \right]. \qquad (10-76)$$

From Chapter 8 we recognize that the first term on the right side of Eq. (10-76) represents a plane wave propagating obliquely in the $+z$ and $-y$ directions with phase constants β and π/b, respectively. Similarly, the second term represents a plane

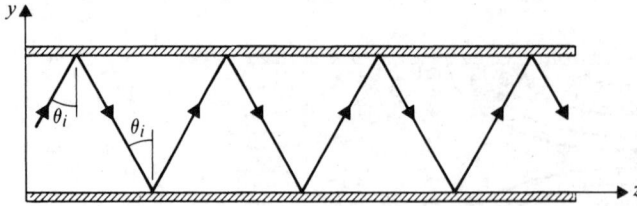

FIGURE 10-7
Propagating wave in parallel-plate
waveguide as superposition of two
plane waves.

wave propagating obliquely in the $+z$ and $+y$ directions with the same phase constants β and π/b as those of the first plane wave. Thus, a propagating TM_1 wave in
a parallel-plate waveguide can be regarded as the superposition of two plane waves,
as depicted in Fig. 10-7.   ▬

In Subsection 8-7.2 on reflection of a parallelly polarized (TM) plane wave incident obliquely at a conducting boundary plane, we obtained an expression for the
longitudinal component of the total $\mathbf{E}_1$ field that is the sum of the longitudinal
components of the incident $\mathbf{E}_i$ and the reflected $\mathbf{E}_r$. To adapt the coordinate designations of Fig. 8-13 to those of Fig. 10-5, x and z must be changed to z and $-y$,
respectively. We rewrite E_x of Eq. (8-128) as

$$E_z(y, z) = E_{i0} \cos \theta_i (e^{j\beta_1 y \cos \theta_i} - e^{-j\beta_1 y \cos \theta_i})e^{-j\beta_1 z \sin \theta_i}.$$

Comparing the exponents of the terms in this equation with those in Eq. (10-76),
we obtain two equations:

$$\beta_1 \sin \theta_i = \beta, \tag{10-77}$$

$$\beta_1 \cos \theta_i = \frac{\pi}{b}. \tag{10-78}$$

(The field amplitudes involved in these equations are of no importance in the present
consideration.) Solution of Eqs. (10-77) and (10-78) gives

$$\beta = \sqrt{\beta_1^2 - \left(\frac{\pi}{b}\right)^2} = \sqrt{\omega^2 \mu\epsilon - \left(\frac{\pi}{b}\right)^2},$$

which is the same as Eq. (10-71), and

$$\cos \theta_i = \frac{\pi}{\beta_1 b} = \frac{\lambda}{2b}, \tag{10-79}$$

where $\lambda = 2\pi/\beta_1$ is the wavelength in the unbounded dielectric medium.

We observe that a solution of Eq. (10-79) for θ_i exists only when $\lambda/2b \le 1$. At
$\lambda/2b = 1$, or $f = u/\lambda = 1/2b\sqrt{\mu\epsilon}$, which is the cutoff frequency in Eq. (10-67) for
$n = 1$, $\cos \theta_i = 1$, and $\theta_i = 0$. This corresponds to the case in which the waves bounce
back and forth in the y-direction, normal to the parallel plates, and there is no propagation in the z-direction ($\beta = \beta_1 \sin \theta_i = 0$). Propagation of TM_1 mode is possible
only when $\lambda < \lambda_c = 2b$ or $f > f_c$. Both $\cos \theta_i$ and $\sin \theta_i$ can be expressed in terms of

cutoff frequency f_c. From Eqs. (10–79) and (10–77) we have

$$\cos \theta_i = \frac{\lambda}{\lambda_c} = \frac{f_c}{f} \tag{10–80}$$

and

$$\sin \theta_i = \frac{\lambda}{\lambda_g} = \frac{u}{u_p} = \sqrt{1 - \left(\frac{f_c}{f}\right)^2}. \tag{10–81}$$

Equation (10–81) is in agreement with Eqs. (10–39) and (10–42).

We studied traveling waves in a parallel-plate waveguide in terms of bouncing plane waves in Section 8–7 with the aid of Fig. 8–12. We note here that Eqs. (10–79) and (10–81) are consistent respectively with Eqs. (8–119) and (8–120), which hold for both perpendicular and parallel polarizations.

10-3.2   TE WAVES BETWEEN PARALLEL PLATES

For transverse electric waves, $E_z = 0$, we solve the following equation for $H_z^0(y)$, which is a simplified version of Eq. (10–48) with no x-dependence:

$$\frac{d^2 H_z^0(y)}{dy^2} + h^2 H_z^0(y) = 0. \tag{10–82}$$

We note that $H_z(y, z) = H_z^0(y)e^{-\gamma z}$. The boundary conditions to be satisfied by $H_z^0(y)$ are obtained from Eq. (10–51). Since E_x must vanish at the surfaces of the conducting plates, we require

$$\frac{dH_z^0(y)}{dy} = 0 \quad \text{at } y = 0 \quad \text{and} \quad y = b.$$

Therefore the proper solution of Eq. (10–82) is of the form

$$H_z^0(y) = B_n \cos \left(\frac{n\pi y}{b}\right), \tag{10–83}$$

where the amplitude B_n depends on the strength of excitation of the particular TE wave. We obtain the only other nonzero field components from Eqs. (10–50) and (10–51), keeping in mind that $\partial H_z/\partial x = 0$:

$$H_y^0(y) = \frac{\gamma}{h} B_n \sin \left(\frac{n\pi y}{b}\right), \tag{10–84}$$

$$E_x^0(y) = \frac{j\omega\mu}{h} B_n \sin \left(\frac{n\pi y}{b}\right). \tag{10–85}$$

The propagation constant γ in Eq. (10–84) is the same as that for TM waves given in Eq. (10–66). Inasmuch as cutoff frequency is the frequency that makes $\gamma = 0$, *the cutoff frequency for the TE$_n$ mode in a parallel-plate waveguide is exactly the same as that for the TM$_n$ mode given in Eq. (10–67)*. For $n = 0$, both H_y and E_x vanish; hence the TE$_0$ mode does not exist in a parallel-plate waveguide.

EXAMPLE 10–5   (a) Write the instantaneous field expressions for the TE$_1$ mode in a parallel-plate waveguide. (b) Sketch the electric and magnetic field lines in the yz-plane.

Solution

a) The instantaneous field expressions for the TE$_1$ mode are obtained by taking the real part of the products of the phasor expressions in Eqs. (10–83), (10–84), and (10–85) with $e^{j(\omega t - \beta z)}$. We have, for $n = 1$,

$$H_z(y, z; t) = B_1 \cos\left(\frac{\pi y}{b}\right) \cos(\omega t - \beta z), \tag{10–86}$$

$$H_y(y, z; t) = -\frac{\beta b}{\pi} B_1 \sin\left(\frac{\pi y}{b}\right) \sin(\omega t - \beta z), \tag{10–87}$$

$$E_x(y, z; t) = -\frac{\omega \mu b}{\pi} B_1 \sin\left(\frac{\pi y}{b}\right) \sin(\omega t - \beta z), \tag{10–88}$$

where the phase constant β is given by Eq. (10–71), same as that for the TM$_1$ mode.

b) The electric field has only an x-component. At $t = 0$, Eq. (10–88) becomes

$$E_x(y, z; 0) = \frac{\omega \mu b}{\pi} B_1 \sin\left(\frac{\pi y}{b}\right) \sin \beta z. \tag{10–89}$$

Thus the density of E_x lines varies as $\sin(\pi y/b)$ in the y direction and as $\sin \beta z$ in the z direction; E_x lines are sketched as dots and crosses in Fig. 10–8.

The magnetic field has both a y- and a z-component. The equation of the magnetic field lines at $t = 0$ can be found from the following relation:

$$\frac{dy}{dz} = \frac{H_y(y, z; 0)}{H_z(y, z; 0)} = \frac{\beta b}{\pi} \tan\left(\frac{\pi y}{b}\right) \tan \beta z. \tag{10–90}$$

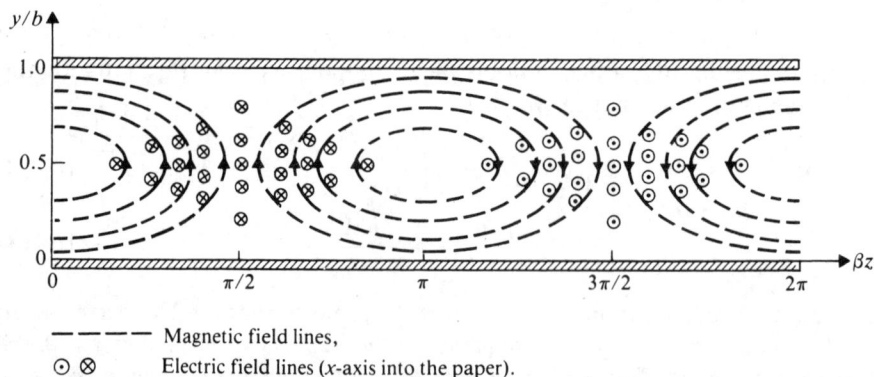

——— Magnetic field lines,

⊙ ⊗   Electric field lines (x-axis into the paper).

FIGURE 10–8
Field lines for TE$_1$ mode in parallel-plate waveguide.

Using the procedure illustrated in Example 10–3, we can integrate Eq. (10–90) to obtain

$$\sin\left(\frac{\pi y}{b}\right)\cos\beta z = \text{Constant}, \qquad 0 \le y \le b, \tag{10–91}$$

which is the equation for magnetic field lines in the yz-plane at $t = 0$. The constant in Eq. (10–91) lies between -1 and $+1$. According to Eq. (10–86), the density of the H_z line varies as $|\cos(\pi y/b)|$. Several magnetic field lines are drawn in Fig. 10–8. The lines repeat themselves for every change of 2π (rad) in βz. ▬

10–3.3 ENERGY-TRANSPORT VELOCITY

In Subsections 10–2.2 and 10–2.3 we noted that signals having a frequency higher than the cutoff frequency will propagate in a waveguide with a phase velocity u_p given by Eq. (10–42) and a group velocity u_g given by Eq. (10–43). When the concept of group velocity was introduced in Section 8–4, it was defined as the velocity of the envelope of a narrow-band signal. For signals with a broad frequency spectrum, such as pulses of short durations, group velocity loses its significance because the low-frequency components may be below cutoff (therefore cannot propagate) and the high-frequency components will travel with widely different velocities. These wide-band signals will then be badly distorted, and no single group velocity can represent the signal-propagation velocity. In such cases we examine the velocity at which energy propagates along a waveguide, or *energy-transport velocity*.

For signal transmission in a lossless waveguide we define energy-transport velocity, u_{en}, as the ratio of the time-average propagated power to the time-average stored energy per unit guide length:

$$\boxed{u_{en} = \frac{(P_z)_{av}}{W'_{av}} \qquad \text{(m/s)},} \tag{10–92}$$

where the time-average power $(P_z)_{av}$ is equal to the time-average Poynting vector $\mathscr{P}_{av}$ integrated over the guide cross section:

$$(P_z)_{av} = \int_S \mathscr{P}_{av} \cdot d\mathbf{s}, \tag{10–93}$$

and the time-average stored energy per unit length W'_{av} is the sum of the time-average stored electric energy density $(w_e)_{av}$ and the time-average stored magnetic energy density $(w_m)_{av}$ integrated over the guide cross section:

$$W'_{av} = \int_S \left[(w_e)_{av} + (w_m)_{av}\right] ds. \tag{10–94}$$

For a particular mode of propagation in a waveguide we calculate $(P_z)_{av}$ and W'_{av} from Eqs. (10–93) and (10–94), respectively, and substitute into Eq. (10–92) to find energy-transport velocity.

EXAMPLE 10-6   Determine the energy-transport velocity of the TM_n mode in a lossless parallel-plate waveguide.

Solution   We first obtain the time-average Poynting vector by using Eqs. (8–96), (10–63), (10–64), and (10–65):

$$\mathscr{P}_{av} = \tfrac{1}{2}\mathscr{R}e(\mathbf{E} \times \mathbf{H}^*)$$
$$= \tfrac{1}{2}\mathscr{R}e(-\mathbf{a}_z E_y^0 H_x^{0*} + \mathbf{a}_y E_z^0 H_x^{0*}). \tag{10-95}$$

Thus,

$$\mathscr{P}_{av} \cdot \mathbf{a}_z = -\tfrac{1}{2}\mathscr{R}e(E_y^0 H_x^{0*})$$
$$= \frac{\omega\epsilon\beta}{2h^2} A_n^2 \cos^2\left(\frac{n\pi y}{b}\right), \tag{10-96}$$

where we have replaced γ by $j\beta$. For a unit width of the parallel-plate waveguide, substitution of Eq. (10–96) in Eq. (10–93) yields

$$(P_z)_{av} = \int_0^b \mathscr{P}_{av} \cdot \mathbf{a}_z \, dy$$
$$= \frac{\omega\epsilon\beta b}{4h^2} A_n^2. \tag{10-97}$$

Following the procedure leading to Eq. (8–96) from Eq. (8–83), we can readily prove from Eqs. (8–85) and (8–86) that

$$(w_e)_{av} = \frac{\epsilon}{4} \mathscr{R}e(\mathbf{E} \cdot \mathbf{E}^*), \tag{10-98}$$

and

$$(w_m)_{av} = \frac{\mu}{4} \mathscr{R}e(\mathbf{H} \cdot \mathbf{H}^*). \tag{10-99}$$

Substituting Eqs. (10–63) and (10–65) in Eq. (10–98), we have

$$(w_e)_{av} = \frac{\epsilon}{4} A_n^2 \left[\sin^2\left(\frac{n\pi y}{b}\right) + \frac{\beta^2}{h^2} \cos^2\left(\frac{n\pi y}{b}\right) \right] \tag{10-100}$$

and

$$\int_0^b (w_e)_{av} \, dy = \frac{\epsilon b}{8} A_n^2 \left[1 + \frac{\beta^2}{h^2} \right]$$
$$= \frac{\epsilon b}{8h^2} k^2 A_n^2, \tag{10-101}$$

where Eq. (10–15) has been used to replace $\beta^2 + h^2$ by k^2. Similarly, using Eq. (10–64) in Eq. (10–99), we obtain

$$(w_m)_{av} = \frac{\mu}{4}\left(\frac{\omega^2\epsilon^2}{h^2}\right) A_n^2 \cos^2\left(\frac{n\pi y}{b}\right) \tag{10-102}$$

and

$$\int_0^b (w_m)_{av} \, dy = \frac{\mu b}{8h^2} (\omega^2\epsilon^2) A_n^2 = \frac{\epsilon b}{8h^2} k^2 A_n^2, \tag{10-103}$$

which is seen to be equal to the time-average stored electric energy per unit guide width obtained in Eq. (10–101).

We are now ready to find u_{en} from Eq. (10–92) by dividing $(P_z)_{av}$ in Eq. (10–97) by the sum of the stored energies in Eqs. (10–101) and (10–103):

$$
\begin{aligned}
u_{en} &= \frac{\omega\beta}{k^2} = \frac{\omega}{k}\left(\frac{\beta}{k}\right) \\
&= u\sqrt{1 - \left(\frac{f_c}{f}\right)^2},
\end{aligned}
\tag{10–104}
$$

where we have made use of Eqs. (10–5) and (10–38). We recognize that the energy-transport velocity in Eq. (10–104) is equal to the group velocity given in Eq. (10–43).

10–3.4 ATTENUATION IN PARALLEL-PLATE WAVEGUIDES

Attenuation in any waveguide (not just the parallel-plate waveguide) arises from two sources: lossy dielectric and imperfectly conducting walls. Losses modify the electric and magnetic fields within the guide, making exact solutions difficult to obtain. However, in practical waveguides the losses are usually very small, and we will assume that the transverse field patterns of the propagating modes are not affected by them. A real part of the propagation constant now appears as the attenuation constant, which accounts for power losses. The attenuation constant consists of two parts:

$$
\alpha = \alpha_d + \alpha_c,
\tag{10–105}
$$

where α_d is the attenuation constant due to losses in the dielectric and α_c is that due to ohmic power loss in the imperfectly conducting walls.

We will now consider the attenuation constants for TEM, TM, and TE modes separately.

TEM Modes The attenuation constant for TEM modes on a parallel-plate transmission line has been discussed in Subsection 9–3.4. From Eq. (9–90) and Table 9–1 we have approximately

$$
\alpha_d = \frac{G}{2}R_0 = \frac{\sigma}{2}\sqrt{\frac{\mu}{\epsilon}} = \frac{\sigma}{2}\eta \qquad \text{(Np/m)},
\tag{10–106}
$$

where ϵ, μ, and σ are the permittivity, permeability, and conductivity, respectively, of the dielectric medium. In Eq. (10–106), $\eta = \sqrt{\mu/\epsilon}$ is the intrinsic impedance of the dielectric if the dielectric is lossless. If the losses in the dielectric are represented by the imaginary part, $-\epsilon''$, of a complex permittivity as in Eq. (7–111), we may replace σ by $\omega\epsilon''$ and write Eq. (10–106) alternatively as

$$
\alpha_d \cong \frac{\omega\epsilon''}{2}\eta \qquad \text{(Np/m)}.
\tag{10–107}
$$

Also from Eq. (9–90) and Table 9–1 we have

$$\alpha_c = \frac{R}{2R_0} = \frac{1}{b}\sqrt{\frac{\pi f \epsilon}{\sigma_c}} \qquad \text{(Np/m)}, \qquad (10\text{--}108)$$

where σ_c is the conductivity of the metal plates. We note that, for TEM modes, α_d is independent of frequency, and α_c is proportional to $\sqrt{f}$. We note further that $\alpha_d \to 0$ as $\sigma \to 0$ and that $\alpha_c \to 0$ as $\sigma_c \to \infty$, as expected.

TM Modes   The attenuation constant due to losses in the dielectric at frequencies above f_c can be found from Eq. (10–66) by substituting $\epsilon_d = \epsilon + (\sigma/j\omega)$ for ϵ. We have

$$\begin{aligned}
\gamma &= j\left[\omega^2\mu\epsilon\left(1 - \frac{j\sigma}{\omega\epsilon}\right) - \left(\frac{n\pi}{b}\right)^2\right]^{1/2} \\
&= j\sqrt{\omega^2\mu\epsilon - \left(\frac{n\pi}{b}\right)^2}\left\{1 - j\omega\mu\sigma\left[\omega^2\mu\epsilon - \left(\frac{n\pi}{b}\right)^2\right]^{-1}\right\}^{1/2} \qquad (10\text{--}109) \\
&\cong j\sqrt{\omega^2\mu\epsilon - \left(\frac{n\pi}{b}\right)^2}\left\{1 - \frac{j\omega\mu\sigma}{2}\left[\omega^2\mu\epsilon - \left(\frac{n\pi}{b}\right)^2\right]^{-1}\right\}.
\end{aligned}$$

Only the first two terms in the binomial expansion for the second line in Eq. (10–109) are retained in the third line under the assumption that

$$\omega\mu\sigma \ll \omega^2\mu\epsilon - \left(\frac{n\pi}{b}\right)^2.$$

From Eq. (10–67) we see that

$$\frac{n\pi}{b} = 2\pi f_c\sqrt{\mu\epsilon},$$

which enables us to write

$$\begin{aligned}
\sqrt{\omega^2\mu\epsilon - \left(\frac{n\pi}{b}\right)^2} &= \omega\sqrt{\mu\epsilon}\sqrt{1 - (\omega_c/\omega)^2} \\
&= \omega\sqrt{\mu\epsilon}\sqrt{1 - (f_c/f)^2}.
\end{aligned}$$

With the above relation, Eq. (10–109) becomes

$$\gamma = \alpha_d + j\beta = \frac{\sigma}{2}\sqrt{\frac{\mu}{\epsilon}}\frac{1}{\sqrt{1 - (f_c/f)^2}} + j\omega\sqrt{\mu\epsilon}\sqrt{1 - (f_c/f)^2},$$

from which we obtain

$$\alpha_d = \frac{\sigma\eta}{2\sqrt{1 - (f_c/f)^2}} \qquad \text{(Np/m)} \qquad (10\text{--}110)$$

and

$$\beta = \omega\sqrt{\mu\epsilon}\sqrt{1 - (f_c/f)^2} \qquad \text{(rad/m)}. \qquad (10\text{--}111)$$

Thus, α_d for TM modes decreases when frequency increases.

To find the attenuation constant due to losses in the imperfectly conducting plates, we use Eq. (9–88), which was derived from the law of conservation of energy. Thus,

$$\alpha_c = \frac{P_L(z)}{2P(z)}, \tag{10-112}$$

where $P(z)$ is the time-average power flowing through a cross section (say, of width w) of the waveguide, and $P_L(z)$ is the time-average power lost in the two plates per unit length. For TM modes we use Eqs. (10–64) and (10–65):

$$
\begin{aligned}
P(z) &= w \int_0^b -\tfrac{1}{2}(E_y^0)(H_x^0)^* \, dy \\
&= \frac{w\omega\epsilon\beta}{2} \left(\frac{bA_n}{n\pi}\right)^2 \int_0^b \cos^2\left(\frac{n\pi y}{b}\right) dy \\
&= w\omega\epsilon\beta b \left(\frac{bA_n}{2n\pi}\right)^2.
\end{aligned}
\tag{10-113}
$$

The surface current densities on the upper and lower plates have the same magnitude. On the lower plate where $y = 0$ we have

$$|J_{sz}^0| = |H_x^0(y = 0)| = \frac{\omega\epsilon bA_n}{n\pi}.$$

The total power loss per unit length in two plates of width w is

$$P_L(z) = 2w\left(\frac{1}{2}|J_{sz}^0|^2 R_s\right) = w\left(\frac{\omega\epsilon bA_n}{n\pi}\right)^2 R_s. \tag{10-114}$$

Substitution of Eqs. (10–113) and (10–114) in Eq. (10–112) yields

$$\alpha_c = \frac{2\omega\epsilon R_s}{\beta b} = \frac{2R_s}{\eta b \sqrt{1 - (f_c/f)^2}} \qquad \text{(Np/m),} \tag{10-115}$$

where, from Eq. (9–26b),

$$R_s = \sqrt{\frac{\pi f \mu_c}{\sigma_c}} \qquad (\Omega). \tag{10-116}$$

The use of Eq. (10–116) in Eq. (10–115) gives the explicit dependence of α_c on f for TM modes:

$$\alpha_c = \frac{2}{\eta b} \sqrt{\frac{\pi \mu_c f_c}{\sigma_c}} \frac{1}{\sqrt{(f_c/f)[1 - (f_c/f)^2]}}. \tag{10-117}$$

A sketch of the normalized α_c is shown in Fig. 10–9, which reveals the existence of a minimum.

TE Modes   In Subsection 10–3.2 we noted that the expression for the propagation constant for TE waves between parallel plates is the same as that for TM waves. It follows that the formula for α_d in Eq. (10–110) holds for TE modes as well.

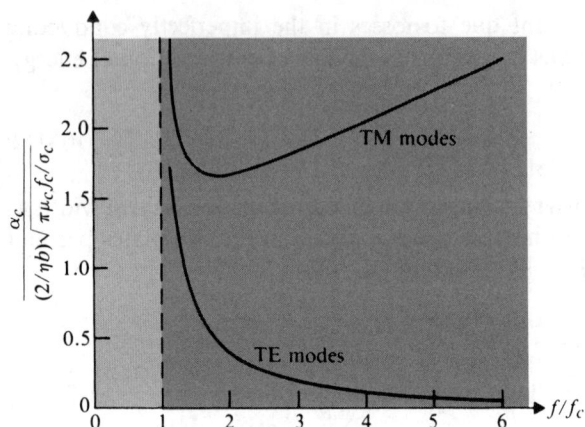

FIGURE 10–9
Normalized attenuation constant due to finite conductivity of the plates in parallel-plate waveguide.

In order to determine the attenuation constant α_c due to losses in the imperfectly conducting plates, we again apply Eq. (10–112). Of course, the field expressions in Eqs. (10–83), (10–84), and (10–85) for TE modes must now be used. We have

$$
\begin{aligned}
P(z) &= w \int_0^b \tfrac{1}{2}(E_x^0)(H_y^0)^* \, dy \\
&= \frac{w\omega\mu\beta}{2}\left(\frac{bB_n}{n\pi}\right)^2 \int_0^b \sin^2\left(\frac{n\pi y}{b}\right) dy \\
&= w\omega\mu\beta b\left(\frac{bB_n}{2n\pi}\right)^2
\end{aligned}
\tag{10–118}
$$

and

$$
\begin{aligned}
P_L(z) &= 2w(\tfrac{1}{2}|J_{sx}^0|^2 R_s) \\
&= w|H_z^0(y=0)|^2 R_z = wB_n^2 R_s.
\end{aligned}
\tag{10–119}
$$

Consequently,

$$
\begin{aligned}
\alpha_c &= \frac{P_L(z)}{2P(z)} = \frac{2R_s}{\omega\mu\beta b}\left(\frac{n\pi}{b}\right)^2 \\
&= \frac{2R_s f_c^2}{\eta b f^2 \sqrt{1-(f_c/f)^2}}.
\end{aligned}
\tag{10–120}
$$

A normalized α_c curve based on Eq. (10–120) is also sketched in Fig. 10–9. Unlike α_c for TM modes, α_c for TE modes does not have a minimum but decreases monotonically as f increases.

10-4 Rectangular Waveguides

The analysis of parallel-plate waveguides in Section 10-3 assumed the plates to be of an infinite extent in the transverse x direction; that is, the fields do not vary with x. In practice, these plates are always finite in width, with fringing fields at the edges. Electromagnetic energy will leak through the sides of the guide and create undesirable stray couplings to other circuits and systems. Thus practical waveguides are usually uniform structures of a cross section of the enclosed variety. The simplest of such cross sections, in terms of ease both in analysis and in manufacture, are rectangular and circular. In this section we will analyze the wave behavior in hollow rectangular waveguides. Circular waveguides will be treated in the next section. Rectangular waveguides are more commonly used in practice than circular waveguides.

In the following discussion we draw on the material in Section 10-2 concerning general wave behaviors along uniform guiding structures. Propagation of time-harmonic waves in the $+z$ direction with a propagation constant γ is considered. TM and TE modes will be discussed separately. As we have noted previously, TEM waves cannot exist in a single-conductor hollow or dielectric-filled waveguide.

10-4.1   TM WAVES IN RECTANGULAR WAVEGUIDES

Consider the waveguide sketched in Fig. 10-10, with its rectangular cross section of sides a and b. The enclosed dielectric medium is assumed to have constitutive parameters ϵ and μ. For TM waves, $H_z = 0$ and E_z is to be solved from Eq. (10-24). Writing $E_z(x, y, z)$ as

$$E_z(x, y, z) = E_z^0(x, y)e^{-\gamma z}, \tag{10-121}$$

we solve the following second-order partial differential equation:

$$\left(\frac{\partial^2}{\partial x^2} + \frac{\partial^2}{\partial y^2} + h^2\right)E_z^0(x, y) = 0. \tag{10-122}$$

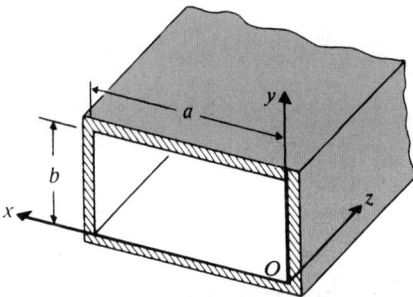

FIGURE 10-10
A rectangular waveguide.

Here we use the method of separation of variables discussed in Section 4–5 by letting

$$E_z^0(x, y) = X(x)Y(y). \tag{10-123}$$

Substituting Eq. (10–123) in Eq. (10–122) and dividing the resulting equation by $X(x)Y(y)$, we have

$$-\frac{1}{X(x)} \frac{d^2 X(x)}{dx^2} = \frac{1}{Y(y)} \frac{d^2 Y(y)}{dy^2} + h^2. \tag{10-124}$$

Now we argue that, since the left side of Eq. (10–124) is a function of x only and the right side is a function of y only, both sides must equal a constant in order for the equation to hold for all values of x and y. Calling this constant (separation constant) k_x^2, we obtain two separate ordinary differential equations:

$$\frac{d^2 X(x)}{dx^2} + k_x^2 X(x) = 0, \tag{10-125}$$

$$\frac{d^2 Y(y)}{dy^2} + k_y^2 Y(y) = 0, \tag{10-126}$$

where

$$k_y^2 = h^2 - k_x^2. \tag{10-127}$$

The possible solutions of Eqs. (10–125) and (10–126) are listed in Table 4–1, Section 4–5. The appropriate forms to be chosen must satisfy the following boundary conditions.

1. In the x-direction:

$$E_z^0(0, y) = 0, \tag{10-128}$$

$$E_z^0(a, y) = 0. \tag{10-129}$$

2. In the y-direction:

$$E_z^0(x, 0) = 0, \tag{10-130}$$

$$E_z^0(x, b) = 0. \tag{10-131}$$

Obviously, then, we must choose:

$X(x)$ in the form of $\sin k_x x$,

$$k_x = \frac{m\pi}{a}, \qquad m = 1, 2, 3, \ldots$$

$Y(y)$ in the form of $\sin k_y y$,

$$k_y = \frac{n\pi}{b}, \qquad n = 1, 2, 3, \ldots,$$

and the proper solution for $E_z^0(x, y)$ is

$$\boxed{E_z^0(x, y) = E_0 \sin\left(\frac{m\pi}{a} x\right) \sin\left(\frac{n\pi}{b} y\right) \quad \text{(V/m)}.} \tag{10-132}$$

From Eq. (10–127) we have

$$h^2 = \left(\frac{m\pi}{a}\right)^2 + \left(\frac{n\pi}{b}\right)^2. \tag{10–133}$$

The other field components are obtained from Eqs. (10–25) through (10–28):

$$E_x^0(x, y) = -\frac{\gamma}{h^2}\left(\frac{m\pi}{a}\right)E_0 \cos\left(\frac{m\pi}{a}x\right)\sin\left(\frac{n\pi}{b}y\right), \tag{10–134}$$

$$E_y^0(x, y) = -\frac{\gamma}{h^2}\left(\frac{n\pi}{b}\right)E_0 \sin\left(\frac{m\pi}{a}x\right)\cos\left(\frac{n\pi}{b}y\right), \tag{10–135}$$

$$H_x^0(x, y) = \frac{j\omega\epsilon}{h^2}\left(\frac{n\pi}{b}\right)E_0 \sin\left(\frac{m\pi}{a}x\right)\cos\left(\frac{n\pi}{b}y\right), \tag{10–136}$$

$$H_y^0(x, y) = -\frac{j\omega\epsilon}{h^2}\left(\frac{m\pi}{a}\right)E_0 \cos\left(\frac{m\pi}{a}x\right)\sin\left(\frac{n\pi}{b}y\right), \tag{10–137}$$

where

$$\gamma = j\beta = j\sqrt{\omega^2\mu\epsilon - \left(\frac{m\pi}{a}\right)^2 - \left(\frac{n\pi}{b}\right)^2}. \tag{10–138}$$

Every combination of the integers m and n defines a possible mode that may be designated as the TM_{mn} mode; thus there are a double infinite number of TM modes. The first subscript denotes the number of half-cycle variations of the fields in the x-direction, and the second subscript denotes the number of half-cycle variations of the fields in the y-direction. The cutoff of a particular mode is the condition that makes γ vanish. For the TM_{mn} mode the cutoff frequency is

$$(f_c)_{mn} = \frac{1}{2\sqrt{\mu\epsilon}}\sqrt{\left(\frac{m}{a}\right)^2 + \left(\frac{n}{b}\right)^2} \quad \text{(Hz)}, \tag{10–139}$$

which checks with Eq. (10–35). Alternatively, we may write

$$(\lambda_c)_{mn} = \frac{2}{\sqrt{\left(\frac{m}{a}\right)^2 + \left(\frac{n}{b}\right)^2}} \quad \text{(m)}, \tag{10–140}$$

where λ_c is the *cutoff wavelength*.

For TM modes in rectangular waveguides, neither m nor n can be zero. (Do you know why?) Hence, the TM_{11} mode has the lowest cutoff frequency of all TM modes in a rectangular waveguide. The expressions for the phase constant β and the wave impedance Z_{TM} for propagating modes in Eqs. (10–38) and (10–45), respectively, apply here directly.

■■■■ **EXAMPLE 10–7** (a) Write the instantaneous field expressions for the TM_{11} mode in a rectangular waveguide of sides a and b. (b) Sketch the electric and magnetic field lines in a typical xy-plane and in a typical yz-plane.

Solution

a) The instantaneous field expressions for the TM_{11} mode are obtained by multiplying the phasor expressions in Eqs. (10–132) and (10–134) through (10–137) with $e^{j(\omega t - \beta z)}$ and then taking the real part of the product. We have, for $m = n = 1$,

$$E_x(x, y, z; t) = \frac{\beta}{h^2}\left(\frac{\pi}{a}\right)E_0 \cos\left(\frac{\pi}{a}x\right)\sin\left(\frac{\pi}{b}y\right)\sin(\omega t - \beta z), \qquad (10\text{–}141)$$

$$E_y(x, y, z; t) = \frac{\beta}{h^2}\left(\frac{\pi}{b}\right)E_0 \sin\left(\frac{\pi}{a}x\right)\cos\left(\frac{\pi}{b}y\right)\sin(\omega t - \beta z), \qquad (10\text{–}142)$$

$$E_z(x, y, z; t) = E_0 \sin\left(\frac{\pi}{a}x\right)\sin\left(\frac{\pi}{b}y\right)\cos(\omega t - \beta z), \qquad (10\text{–}143)$$

$$H_x(x, y, z; t) = -\frac{\omega\epsilon}{h^2}\left(\frac{\pi}{b}\right)E_0 \sin\left(\frac{\pi}{a}x\right)\cos\left(\frac{\pi}{b}y\right)\sin(\omega t - \beta z), \qquad (10\text{–}144)$$

$$H_y(x, y, z; t) = \frac{\omega\epsilon}{h^2}\left(\frac{\pi}{a}\right)E_0 \cos\left(\frac{\pi}{a}x\right)\sin\left(\frac{\pi}{b}y\right)\sin(\omega t - \beta z), \qquad (10\text{–}145)$$

$$H_z(x, y, z; t) = 0, \qquad (10\text{–}146)$$

where

$$\beta = \sqrt{k^2 - h^2} = \sqrt{\omega^2\mu\epsilon - \left(\frac{\pi}{a}\right)^2 - \left(\frac{\pi}{b}\right)^2}. \qquad (10\text{–}147)$$

b) In a typical xy-plane, the slopes of the electric field and magnetic field lines are

$$\left(\frac{dy}{dx}\right)_E = \frac{a}{b}\tan\left(\frac{\pi}{a}x\right)\cot\left(\frac{\pi}{b}y\right), \qquad (10\text{–}148)$$

$$\left(\frac{dy}{dx}\right)_H = -\frac{b}{a}\cot\left(\frac{\pi}{a}x\right)\tan\left(\frac{\pi}{b}y\right). \qquad (10\text{–}149)$$

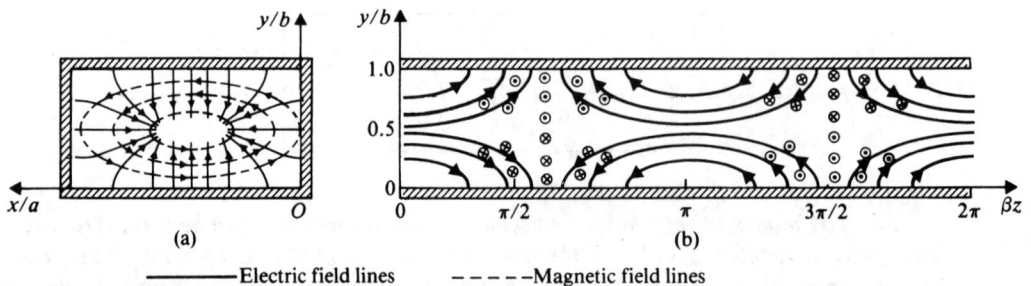

FIGURE 10–11
Field lines for TM_{11} mode in rectangular waveguide.

These equations are quite similar to Eq. (10–73) and can be used to sketch the **E** and **H** lines shown in Fig. 10–11(a). Note that from Eqs. (10–148) and (10–149),

$$\left(\frac{dy}{dx}\right)_E \left(\frac{dy}{dx}\right)_H = -1, \tag{10–150}$$

indicating that **E** and **H** lines are everywhere perpendicular to one another. Note also that **E** lines are normal and that **H** lines are parallel to conducting guide walls.

Similarly, in a typical yz-plane, say, for $x = a/2$ or $\sin(\pi x/a) = 1$ and $\cos(\pi x/a) = 0$, we have

$$\left(\frac{dy}{dz}\right)_E = \frac{\beta}{h^2}\left(\frac{\pi}{b}\right)\cot\left(\frac{\pi}{b}y\right)\tan(\omega t - \beta z), \tag{10–151}$$

and **H** has only an x-component. Some typical **E** and **H** lines are drawn in Fig. 10–11(b) for $t = 0$.   ▬

10–4.2 TE WAVES IN RECTANGULAR WAVEGUIDES

For transverse electric waves, $E_z = 0$, we solve Eq. (10–48) for H_z. We write

$$H_z(x, y, z) = H_z^0(x, y)e^{-\gamma z}, \tag{10–152}$$

where $H_z^0(x, z)$ satisfies the following second-order partial differential equation:

$$\left(\frac{\partial^2}{\partial x^2} + \frac{\partial^2}{\partial y^2} + h^2\right)H_z^0(x, y) = 0. \tag{10–153}$$

Equation (10–153) is seen to be of exactly the same form as Eq. (10–122). The solution for $H_z^0(x, y)$ must satisfy the following boundary conditions.

1. In the x-direction:

$$\frac{\partial H_z^0}{\partial x} = 0 \;(E_y = 0) \quad \text{at} \quad x = 0, \tag{10–154}$$

$$\frac{\partial H_z^0}{\partial x} = 0 \;(E_y = 0) \quad \text{at} \quad x = a. \tag{10–155}$$

2. In the y-direction:

$$\frac{\partial H_z^0}{\partial y} = 0 \;(E_x = 0) \quad \text{at} \quad y = 0, \tag{10–156}$$

$$\frac{\partial H_z^0}{\partial y} = 0 \;(E_x = 0) \quad \text{at} \quad y = b. \tag{10–157}$$

It is readily verified that the appropriate solution for $H_z^0(x, y)$ is

$$\boxed{H_z^0(x, y) = H_0 \cos\left(\frac{m\pi}{a}x\right)\cos\left(\frac{n\pi}{b}y\right) \quad \text{(A/m)}.} \tag{10–158}$$

The relation between the eigenvalue h and $(m\pi/a)$ and $(n\pi/b)$ is the same as that given in Eq. (10–133) for TM modes.

The other field components are obtained from Eqs. (10–49) through (10–52):

$$E_x^0(x, y) = \frac{j\omega\mu}{h^2}\left(\frac{n\pi}{b}\right)H_0 \cos\left(\frac{m\pi}{a}x\right)\sin\left(\frac{n\pi}{b}y\right), \tag{10–159}$$

$$E_y^0(x, y) = -\frac{j\omega\mu}{h^2}\left(\frac{m\pi}{a}\right)H_0 \sin\left(\frac{m\pi}{a}x\right)\cos\left(\frac{n\pi}{b}y\right), \tag{10–160}$$

$$H_x^0(x, y) = \frac{\gamma}{h^2}\left(\frac{m\pi}{a}\right)H_0 \sin\left(\frac{m\pi}{a}x\right)\cos\left(\frac{n\pi}{b}y\right), \tag{10–161}$$

$$H_y^0(x, y) = \frac{\gamma}{h^2}\left(\frac{n\pi}{b}\right)H_0 \cos\left(\frac{m\pi}{a}x\right)\sin\left(\frac{n\pi}{b}y\right), \tag{10–162}$$

where γ has the same expression as that given in Eq. (10–138) for TM modes.

Equation (10–139) for cutoff frequency also applies here. For TE modes, either m or n (but not both) can be zero. If $a > b$, the cutoff frequency is the *lowest* when $m = 1$ and $n = 0$:

$$\boxed{(f_c)_{TE_{10}} = \frac{1}{2a\sqrt{\mu\epsilon}} = \frac{u}{2a} \quad \text{(Hz)}.} \tag{10–163}$$

The corresponding cutoff wavelength is

$$\boxed{(\lambda_c)_{TE_{10}} = 2a \quad \text{(m)}.} \tag{10–164}$$

Hence *the TE$_{10}$ mode is the dominant mode of a rectangular waveguide with a > b*. Because the TE$_{10}$ mode has the lowest attenuation of all modes in a rectangular waveguide and its electric field is definitely polarized in one direction everywhere, it is of particular practical importance (see Subsection 10–4.3).

EXAMPLE 10–8   (a) Write the instantaneous field expressions for the TE$_{10}$ mode in a rectangular waveguide having sides a and b. (b) Sketch the electric and magnetic field lines in typical xy-, yz-, and xz-planes. (c) Sketch the surface currents on the guide walls.

Solution

a) The instantaneous field expressions for the dominant TE$_{10}$ mode are obtained by multiplying the phasor expressions in Eqs. (10–158) through (10–162) with $e^{j(\omega t - \beta z)}$ and then taking the real part of the product. We have, for $m = 1$ and $n = 0$,

$$E_x(x, y, z; t) = 0, \tag{10–165}$$

$$E_y(x, y, z; t) = \frac{\omega\mu}{h^2}\left(\frac{\pi}{a}\right)H_0 \sin\left(\frac{\pi}{a}x\right)\sin(\omega t - \beta z), \qquad (10\text{–}166)$$

$$E_z(x, y, z; t) = 0, \qquad (10\text{–}167)$$

$$H_x(x, y, z; t) = -\frac{\beta}{h^2}\left(\frac{\pi}{a}\right)H_0 \sin\left(\frac{\pi}{a}x\right)\sin(\omega t - \beta z), \qquad (10\text{–}168)$$

$$H_y(x, y, z; t) = 0, \qquad (10\text{–}169)$$

$$H_z(x, y, z; t) = H_0 \cos\left(\frac{\pi}{a}x\right)\cos(\omega t - \beta z), \qquad (10\text{–}170)$$

where

$$\beta = \sqrt{k^2 - h^2} = \sqrt{\omega^2\mu\epsilon - \left(\frac{\pi}{a}\right)^2}. \qquad (10\text{–}171)$$

b) We see from Eqs. (10–165) through (10–170) that the TE_{10} mode has only three nonzero field components—namely, E_y, H_x, and H_z. In a typical xy-plane, say, when $\sin(\omega t - \beta z) = 1$, both E_y and H_x vary as $\sin(\pi x/a)$ and are independent of y, as shown in Fig. 10–12(a).

In a typical yz-plane, for example at $x = a/2$ or $\sin(\pi x/a) = 1$ and $\cos(\pi x/a) = 0$, we only have E_y and H_x, both of which vary sinusoidally with βz. A sketch of E_y and H_x at $t = 0$ is given in Fig. 10–12(b).

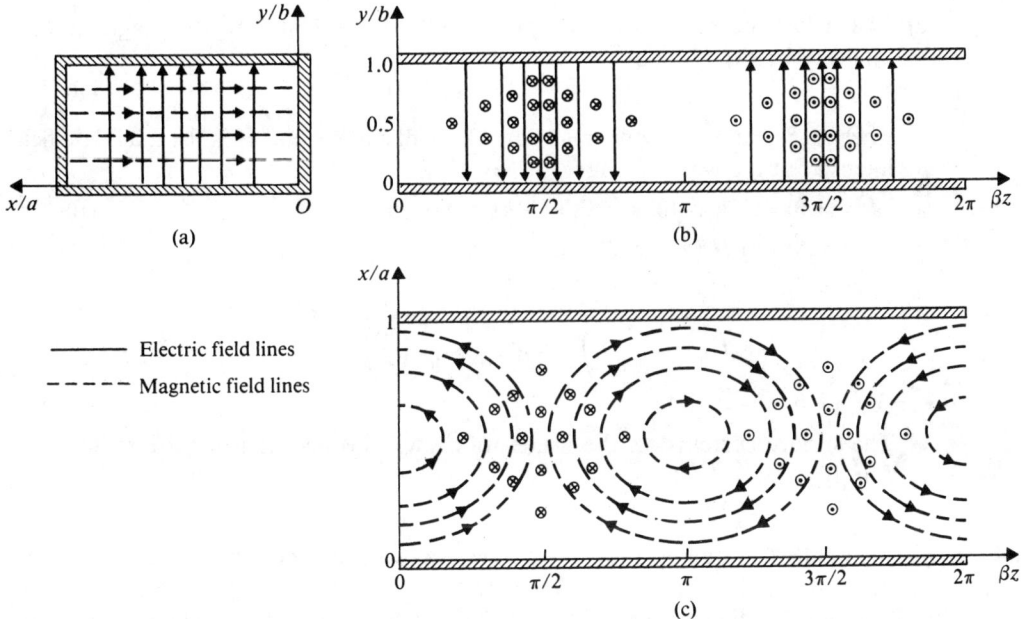

Electric field lines
Magnetic field lines

FIGURE 10–12
Field lines for TE_{10} mode in rectangular waveguide.

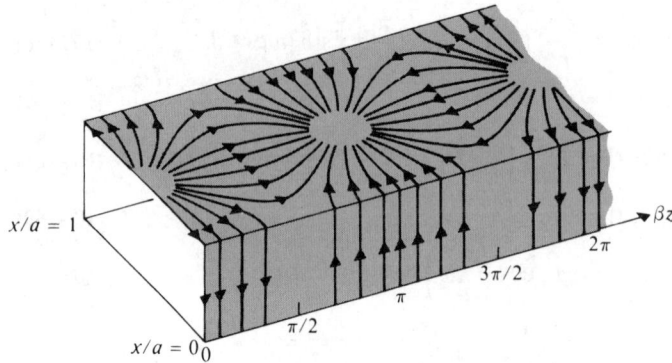

$x/a = 1$

$x/a = 0$

FIGURE 10–13
Surface currents on guide walls
for TE_{10} mode in rectangular
waveguide.

The sketch in an xz-plane will show all three nonzero field components—
E_y, H_x, and H_z. The slope of the **H** lines at $t = 0$ is governed by the following
equation:

$$\left(\frac{dx}{dz}\right)_H = \frac{\beta}{h^2}\left(\frac{\pi}{a}\right)\tan\left(\frac{\pi}{a}x\right)\tan\beta z, \qquad (10\text{–}172)$$

which can be used to draw the **H** lines in Fig. 10–12(c). These lines are independent of y.

c) The surface current density on guide walls, $\mathbf{J}_s$, is related to the magnetic field
intensity by Eq. (7–66b):

$$\mathbf{J}_s = \mathbf{a}_n \times \mathbf{H}, \qquad (10\text{–}173)$$

where $\mathbf{a}_n$ is the *outward* normal from the wall surface and **H** is the magnetic field
intensity at the wall. We have, at $t = 0$,

$$\mathbf{J}_s(x = 0) = -\mathbf{a}_y H_z(0, y, z; 0) = -\mathbf{a}_y H_0 \cos\beta z, \qquad (10\text{–}174)$$

$$\mathbf{J}_s(x = a) = \mathbf{a}_y H_z(a, y, z; 0) = \mathbf{J}_s(x = 0), \qquad (10\text{–}175)$$

$$\mathbf{J}_s(y = 0) = \mathbf{a}_x H_z(x, 0, z; 0) - \mathbf{a}_z H_x(x, 0, z; 0)$$

$$= \mathbf{a}_x H_0 \cos\left(\frac{\pi}{a}x\right)\cos\beta z - \mathbf{a}_z \frac{\beta}{h^2}\left(\frac{\pi}{a}\right)H_0 \sin\left(\frac{\pi}{a}x\right)\sin\beta z, \qquad (10\text{–}176)$$

$$\mathbf{J}_s(y = b) = -\mathbf{J}_s(y = 0). \qquad (10\text{–}177)$$

The surface currents on the inside walls at $x = 0$ and at $y = b$ are sketched in
Fig. 10–13. ■

■ **EXAMPLE 10–9** Standard air-filled waveguides have been designed for the radar
bands listed in Subsection 7–7.4. One type, designated WG-16, is suitable for X-band
applications. Its dimensions are: $a = 2.29$ cm (0.90 in.) and $b = 1.02$ cm (0.40 in.). If
it is desired that a WG-16 waveguide operate only in the dominant TE_{10} mode and

that the operating frequency be at least 25% above the cutoff frequency of the TE_{10} mode but no higher than 95% of the next higher cutoff frequency, what is the allowable operating-frequency range?

Solution For $a = 2.29 \times 10^{-2}$ (m) and $b = 1.02 \times 10^{-2}$ (m), the two modes having the lowest cutoff frequencies are TE_{10} and TE_{20}. Using Eq. (10–139), we find

$$(f_c)_{10} = \frac{c}{2a} = \frac{3 \times 10^8}{2 \times 2.29 \times 10^{-2}} = 6.55 \times 10^9 \quad (\text{Hz}),$$

$$(f_c)_{20} = \frac{c}{a} = 13.10 \times 10^9 \quad (\text{Hz}).^{\dagger}$$

Thus the allowable operating-frequency range under the specified conditions is

$$1.25(f_c)_{\text{TE}_{10}} \leq f \leq 0.95(f_c)_{\text{TE}_{20}}$$

or

$$8.19 \quad (\text{GHz}) \leq f \leq 12.45 \quad (\text{GHz}).$$

10–4.3 ATTENUATION IN RECTANGULAR WAVEGUIDES

Attenuation for propagating modes results when there are losses in the dielectric and in the imperfectly conducting guide walls. Because these losses are usually very small, we will assume, as in the case of parallel-plate waveguides, that the transverse field patterns are not appreciably affected by the losses. The attenuation constant due to losses in the dielectric can be obtained by substituting $\epsilon_d = \epsilon + (\sigma/j\omega)$ for ϵ in Eq. (10–138). The result is exactly the same as that given in Eq. (10–110), which is repeated below:

$$\alpha_d = \frac{\sigma\eta}{2\sqrt{1 - (f_c/f)^2}}, \tag{10–178}$$

where σ and η are the equivalent conductivity (see Eq. 7–112) and intrinsic impedance of the dielectric medium, respectively, and f_c is given by Eq. (10–139). It is easy to see from Eq. (10–178) that the attenuation constant of propagating waves due to losses in the dielectric decreases monotonically from an infinitely large value toward the value $\sigma\eta/2$ as the frequency increases from the cutoff frequency.

To determine the attenuation constant due to wall losses, we make use of Eq. (10–112). The derivations of α_c for the general TM_{mn} and TE_{mn} modes tend to be tedious. Below we obtain the formula for the dominant TE_{10} mode, which is the most important of all propagating modes in a rectangular waveguide.

For the TE_{10} mode the only nonzero field components are E_y, H_x, and H_z. Letting $m = 1$, $n = 0$, and $h = (\pi/a)$ in Eqs. (10–160) and (10–161), we calculate the

---

† Note that $(f_c)_{01} = (c/2b) > (f_c)_{20}$ and $(f_c)_{11} = (c/2a)\sqrt{1 + (a/b)^2} > (f_c)_{20}$.

time-average power flowing through a cross section of the waveguide:

$$
\begin{aligned}
P(z) &= \int_0^b \int_0^a -\tfrac{1}{2}(E_y^0)(H_x^0)^* \, dx \, dy \\
&= \frac{1}{2}\,\omega\mu\beta\left(\frac{a}{\pi}\right)^2 H_0^2 \int_0^b \int_0^a \sin^2\left(\frac{\pi}{a}x\right) dx \, dy \qquad (10\text{–}179) \\
&= \omega\mu\beta ab\left(\frac{aH_0}{2\pi}\right)^2.
\end{aligned}
$$

In order to calculate the time-average power lost in the conducting walls per unit length, we must consider all four walls. From Eqs. (10–173), (10–158), and (10–161) we see that

$$
\mathbf{J}_s^0(x = 0) = \mathbf{J}_s^0(x = a) = -\mathbf{a}_y H_z^0(x = 0) = -\mathbf{a}_y H_0 \qquad (10\text{–}180)
$$

and

$$
\begin{aligned}
\mathbf{J}_s^0(y = 0) &= -\mathbf{J}_s^0(y = b) = \mathbf{a}_x H_z^0(y = 0) - \mathbf{a}_z H_x^0(y = 0) \\
&= \mathbf{a}_x H_0 \cos\left(\frac{\pi}{a}x\right) - \mathbf{a}_z \frac{\beta a}{\pi} H_0 \sin\left(\frac{\pi}{a}x\right). \qquad (10\text{–}181)
\end{aligned}
$$

The total power loss is then double the sum off the losses in the walls at $x = 0$ and at $y = 0$. We have

$$
P_L(z) = 2[P_L(z)]_{x=0} + 2[P_L(z)]_{y=0}, \qquad (10\text{–}182)
$$

where

$$
[P_L(z)]_{x=0} = \int_0^b \frac{1}{2}|J_s^0(x=0)|^2 R_s \, dy = \frac{b}{2} H_0^2 R_s \qquad (10\text{–}183)
$$

and

$$
\begin{aligned}
[P_L(z)]_{y=0} &= \int_0^a \frac{1}{2}\left[|J_{sx}^0(y=0)|^2 + |J_{sz}^0(y=0)|^2\right] R_s \, dx \\
&= \frac{a}{4}\left[1 + \left(\frac{\beta a}{\pi}\right)^2\right] H_0^2 R_s.
\end{aligned}
\qquad (10\text{–}184)
$$

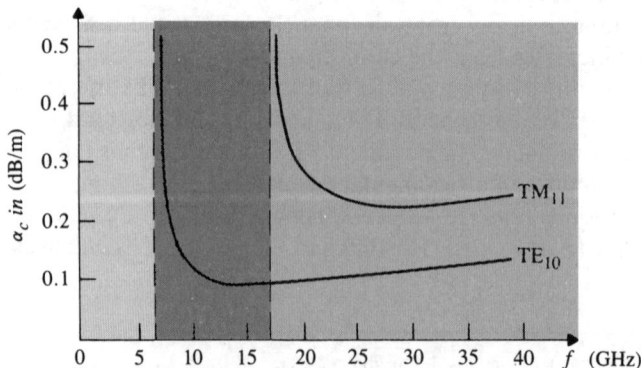

FIGURE 10–14
Attenuation due to wall losses in rectangular copper waveguide for TE_{10} and TM_{11} modes. $a = 2.29$ (cm), $b = 1.02$ (cm).

Substitution of Eqs. (10–183) and (10–184) in Eq. (10–182) yields

$$P_L(z) = \left\{ b + \frac{a}{2}\left[1 + \left(\frac{\beta a}{\pi}\right)^2\right]\right\} H_0^2 R_s$$

$$= \left[b + \frac{a}{2}\left(\frac{f}{f_c}\right)^2\right] H_0^2 R_s. \tag{10–185}$$

The last expression is the result of recognizing that

$$\beta = \sqrt{\omega^2\mu\epsilon - \left(\frac{\pi}{a}\right)^2} = \omega\sqrt{\mu\epsilon}\,\sqrt{1 - \left(\frac{f_c}{f}\right)^2}. \tag{10–186}$$

Inserting Eqs. (10–179) and (10–185) in Eq. (10–112), we obtain

$$\boxed{\begin{aligned}(\alpha_c)_{\text{TE}_{10}} &= \frac{R_s[1 + (2b/a)(f_c/f)^2]}{\eta b\sqrt{1 - (f_c/f)^2}} \\ &= \frac{1}{\eta b}\sqrt{\frac{\pi f \mu_c}{\sigma_c[1 - (f_c/f)^2]}}\left[1 + \frac{2b}{a}\left(\frac{f_c}{f}\right)^2\right] \quad (\text{Np/m}).\end{aligned}} \tag{10–187}$$

Equation (10–187) reveals a rather complicated dependence of $(\alpha_c)_{\text{TE}_{10}}$ on the ratio (f_c/f). It tends to infinity when f is close to the cutoff frequency, decreases toward a minimum as f increases, and increases again steadily for further increases in f.

For a given guide width a, the attenuation decreases as b increases. However, increasing b also decreases the cutoff frequency of the next higher-order mode TE_{11} (or TM_{11}), with the consequence that the available bandwidth for the dominant TE_{10} mode (the range of frequencies over which TE_{10} is the only possible propagating mode) is reduced. The usual compromise is to choose the ratio b/a in the neighborhood of $\frac{1}{2}$.

If we follow a similar procedure that led to Eq. (10–187), the attenuation constant due to wall losses for TM modes can be derived. For the TM_{11} mode we obtain

$$(\alpha_c)_{\text{TM}_{11}} = \frac{2R_s(b/a^2 + a/b^2)}{\eta ab\sqrt{1 - (f_c/f)^2}(1/a^2 + 1/b^2)}. \tag{10–188}$$

In Fig. 10–14 are plotted the graphs of $(\alpha_c)_{\text{TE}_{10}}$ and $(\alpha_c)_{\text{TM}_{11}}$ for a standard air-filled WR-16 rectangular copper waveguide with $a = 2.29$ (cm) and $b = 1.02$ (cm). From Eq. (10–139) we find $(f_c)_{10} = 6.55$ (GHz) and $(f_c)_{11} = 16.10$ (GHz). The curves show that the attenuation constant increases rapidly toward infinity as the operating frequency approaches the cutoff frequency. In the operating range $(f > f_c)$, both curves possess a broad minimum. The attenuation constant of the TE_{10} mode is everywhere lower than that of the TM_{11} mode. These facts have direct relevance in the choice of operating modes and frequencies.

EXAMPLE 10–10   A TE_{10} wave at 10 (GHz) propagates in a brass—$\sigma_c = 1.57 \times 10^7$ (S/m)—rectangular waveguide with inner dimensions $a = 1.5$ (cm) and $b = 0.6$ (cm), which is filled with polyethylene—$\epsilon_r = 2.25$, $\mu_r = 1$, loss tangent $= 4 \times 10^{-4}$. Determine (a) the phase constant, (b) the guide wavelength, (c) the phase velocity, (d) the wave impedance, (e) the attenuation constant due to loss in the dielectric, and (f) the attenuation constant due to loss in the guide walls.

Solution   At $f = 10^{10}$ (Hz) the wavelength in *unbounded* polyethylene is

$$\lambda = \frac{u}{f} = \frac{3 \times 10^8}{\sqrt{2.25} \times 10^{10}} = \frac{2 \times 10^8}{10^{10}} = 0.02 \quad \text{(m)}.$$

The cutoff frequency for the TE_{10} mode is, from Eq. (10–163),

$$f_c = \frac{u}{2a} = \frac{2 \times 10^8}{2 \times (1.5 \times 10^{-2})} = 0.667 \times 10^{10} \quad \text{(Hz)}.$$

a) The phase contant is, from Eq. (10–186),

$$\beta = \frac{\omega}{u} \sqrt{1 - \left(\frac{f_c}{f}\right)^2} = \frac{2\pi 10^{10}}{2 \times 10^8} \sqrt{1 - 0.667^2}$$

$$= 74.5\pi = 234 \quad \text{(rad/m)}.$$

b) The guide wavelength is, from Eq. (10–39),

$$\lambda_g = \frac{\lambda}{\sqrt{1 - (f_c/f)^2}} = \frac{0.02}{0.745} = 0.0268 \quad \text{(m)}.$$

c) The phase velocity is, from Eq. (10–42),

$$u_p = \frac{u}{\sqrt{1 - (f_c/f)^2}} = \frac{2 \times 10^8}{0.745} = 2.68 \times 10^8 \quad \text{(m/s)}.$$

d) The wave impedance is, from Eq. (10–57),

$$(Z_{TE})_{10} = \frac{\sqrt{\mu/\epsilon}}{\sqrt{1 - (f_c/f)^2}} = \frac{377/\sqrt{2.25}}{0.745} = 337.4 \quad \text{(\Omega)}.$$

e) The attenuation constant due to loss in dielectric is obtained from Eq. (10–178). The effective conductivity for polyethylene at 10 (GHz) can be determined from the given loss tangent by using Eq. (7–115):

$$\sigma = 4 \times 10^{-4}\omega\epsilon = 4 \times 10^{-4} \times (2\pi \times 10^{10}) \times \left(\frac{2.25}{36\pi} \times 10^{-9}\right)$$

$$= 5 \times 10^{-4} \quad \text{(S/m)}.$$

Thus,

$$\alpha_d = \frac{\sigma}{2} Z_{TE} = \frac{5 \times 10^{-4}}{2} \times 337.4 = 0.084 \quad \text{(Np/m)}$$

$$= 0.73 \quad \text{(dB/m)}.$$

f) The attenuation constant due to loss in the guide walls is found from Eq. (10–187). We have, from Eq. (9–26b),

$$R_s = \sqrt{\frac{\pi f \mu_c}{\sigma_c}} = \sqrt{\frac{\pi 10^{10}(4\pi 10^{-7})}{1.57 \times 10^7}} = 0.0501 \quad (\Omega),$$

$$\alpha_c = \frac{R_s[1 + (2b/a)(f_c/f)^2]}{\eta b \sqrt{1 - (f_c/f)^2}} = \frac{0.0501[1 + (1.2/1.5)(0.667)^2]}{251 \times 0.006 \times 0.745} = 0.0605 \quad (\text{Np/m})$$

$$= 0.526 \quad (\text{dB/m}). \quad \blacksquare$$

10–4.4 DISCONTINUITIES IN RECTANGULAR WAVEGUIDES

Just as in the case of transmission lines, it is desirable to have impedance match for wave propagation in waveguides in order to achieve maximum power transfer and to reduce local power loss due to a high standing-wave ratio. There is a need to introduce shunt susceptances at appropriate points along a waveguide. These shunt susceptances often take the form of a thin metal diaphragm with an iris such as those shown in Figs. 10–15(a) and 10–15(b). When a diaphragm with an iris is in place, the electric and magnetic fields must satisfy the additional boundary conditions on the metal surface. If the waveguide operates in the dominant TE_{10} mode, the additional boundary conditions require the presence of all higher-order modes, and the situation is vastly more complicated. However, the waveguide is usually designed so that only the dominant mode can propagate. The higher-order modes are then all cutoff modes; they are evanescent and are localized near the iris. An analytical determination of the effective shunt susceptance of an iris would necessitate the solution

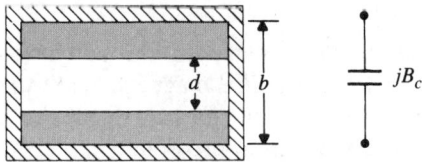

(a) Capacitive iris and equivalent susceptance

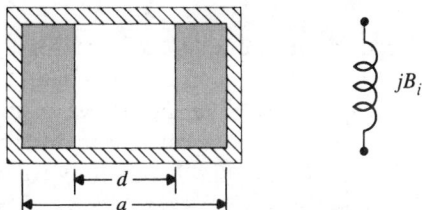

(b) Inductive iris and equivalent susceptance.

FIGURE 10–15
Irises in waveguide as susceptances.

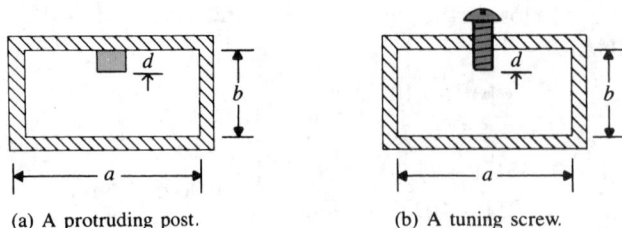

(a) A protruding post.   (b) A tuning screw.

FIGURE 10–16
Post or screw in a waveguide.

of a difficult electromagnetics problem. We will offer only a qualitative discussion here and give the approximate formulas[†] for the irises in Figs. 10–15(a) and 10–15(b).

The iris in Fig. 10–15(a) is made of thin conducting diaphragms extending from one narrow wall to the other. As seen in Fig. 10–12(a), the electric field lines of the dominant TE_{10} mode in a cross section are in the y-direction, going across the narrow dimension. Reducing this dimension from b to d may be expected to have the effect of increasing this field as well as the stored electric energy locally. Consequently, the equivalent shunt susceptance is expected to be capacitive. An approximate expression for the normalized capacitive susceptance is

$$b_c = \frac{B_c}{Y_{10}} = \frac{4b}{\lambda_g} \ln \left[\csc \left(\frac{\pi d}{2b} \right) \right], \qquad (10\text{--}189)$$

where Y_{10} is the reciprocal of $Z_{TE_{10}}$ from Eq. (10–57) and λ_g is the guide wavelength given in Eq. (10–39). As we have indicated before, the actual situation is much more complicated, owing to the presence of the evanescent higher-order modes near the iris. A more accurate analysis will show that b_c is not strictly proportional to (b/λ_g). The approximate formula in Eq. (10–189) is accurate to within 5% in the normal range of operating frequencies.

The iris in Fig. 10–15(b) provides additional current paths through the conducting diaphragms in the y-direction, causing new longitudinal magnetic field to exist in the iris opening and increasing the stored mangetic energy locally. Hence the equivalent shunt susceptance is expected to be inductive. An approximate expression for the normalized inductive susceptance of the iris is

$$b_i = \frac{B_i}{Y_{10}} = -\frac{\lambda_g}{a} \cot^2 \left(\frac{\pi d}{2a} \right). \qquad (10\text{--}190)$$

Another type of discontinuity that provides a shunt susceptance is a conducting post protruding into the waveguide on a broad face, as in Fig. 10–16(a). If the post length d is small, the shunt susceptance is capacitive. When d becomes an appreciable fraction of b, considerable current can flow along the post, causing an inductive effect. A resonance occurs when d is in the neighborhood of $(3/4)b$. Still longer d will result

---

† For more details, see R. E. Collin, *Field Theory of Guided Waves*, McGraw-Hill, New York, 1960, Chapter 8; C. C. Johnson, *Field and Wave Electrodynamics*, McGraw-Hill, New York, 1965, Chapter 5.

in an inductive susceptance. In practical usage the post usually takes the form of a metal screw, as shown in Fig. 10–16(b). The screw could be inserted in a slit cut axially in the center of the broad face. The center slit does not appreciably disturb the field pattern in the waveguide, and the sliding screw with a variable d can be used for tuning and matching a given load to the waveguide. This is a technique similar to the single-stub matching scheme discussed in Subsection 9–7.2.

EXAMPLE 10–11   Measurements on a WG-10 S-band waveguide ($a = 7.21$ cm, $b = 3.40$ cm) feeding a horn antenna show a standing-wave ratio (SWR) of 2.00 at the 3 (GHz) operating frequency, and the existence of a maximum electric field at 12 (cm) from the neck of the horn. Find the location and the dimensions of a symmetrical inductive iris necessary to achieve a perfect match. Assume the waveguide to be lossless.

Solution   With $a = 7.21 \times 10^{-2}$ (m) and $b = 3.40 \times 10^{-2}$ (m), the cutoff frequency for the dominant TE_{10} mode is

$$f_c = \frac{c}{2a}$$

$$= \frac{3 \times 10^8}{2 \times 7.21 \times 10^{-2}} = 2.08 \times 10^9 \quad (Hz).$$

The guide wavelength is, from Eq. (10–39),

$$\lambda_g = \frac{\lambda}{\sqrt{1 - (f_c/f)^2}} = \frac{c}{\sqrt{f^2 - f_c^2}}$$

$$= \frac{3 \times 10^8}{10^9 \sqrt{3^2 - 2.08^2}} = 0.139 \ (m) = 13.9 \quad (cm).$$

Thus, the measured maximum of the electric field is at a distance $12/13.9 = 0.863\lambda_g$ from the neck of the horn. At that location the normalized effective load resistance is (see Eq. 9–145)

$$r_L = \frac{R_L}{R_0} = S.$$

The corresponding normalized conductance is

$$g_L = \frac{Y_L}{Y_0} = \frac{1}{S}$$

$$= \frac{1}{2.00} = 0.50.$$

The rest of the problem is that of single-stub matching discussed in Subsection 9–7.2. We use the Smith admittance chart and proceed as follows (see Fig. 10–17):

1. Enter $g_L = 0.50$ on an Smith admittance chart as P_M (point of maximum electric field).

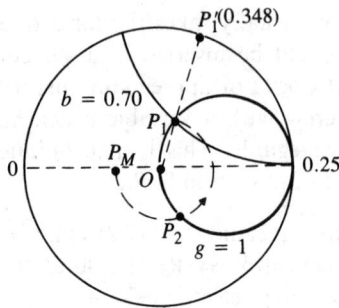

FIGURE 10–17
Construction on Smith admittance chart (Example 10–11).

2. Draw a $|\Gamma|$-circle centered at O with radius $\overline{OP_M}$, intersecting the $g = 1$ circle at two points, P_1 and P_2. Read

$$\text{at } P_1: \quad y_1 = 1 + j0.70,$$
$$\text{at } P_2: \quad y_2 = 1 - j0.70.$$

Point P_2 is not useful to us because it requires a capacitive (positive) susceptance to achieve matching.

3. Draw a straight line from O through P_1 to point P_1' on the perimeter. Read 0.348 on the "wavelength toward load" scale at P_1'. This is $(0.863 - 0.348)\lambda_g = 7.16$ (cm) from the neck of the horn and is where an inductive iris of a normalized susceptance -0.70 should be placed.

4. Using Eq. (10–190), we determine the distance d of the required inductive iris shown in Fig. 10–15(b):

$$-0.70 = -\frac{13.9}{7.21} \cot^2\left(\frac{\pi d}{2 \times 7.21}\right),$$

from which we find $d = 4.72$ (cm). ∎

10–5 Circular Waveguides

Electromagnetic waves can also propagate inside round metal pipes. In this section we will study wave behaviors in circular waveguides—metal pipes having a uniform circular cross section and filled with a dielectric medium.

The basic equations to be satisfied by time-harmonic electric and magnetic field intensities in the charge-free dielectric region inside a waveguide are Eqs. (10–3) and (10–4), which are repeated below:

$$\mathbf{V}^2\mathbf{E} + k^2\mathbf{E} = 0 \tag{10–191}$$

and

$$\mathbf{V}^2\mathbf{H} + k^2\mathbf{H} = 0. \tag{10–192}$$

For a straight waveguide with a uniform circular cross section and having its axis in the z-direction, it is expedient to decompose the three-dimensional Laplacian operator $\mathbf{V}^2$ into two parts: $\mathbf{V}_{r\phi}^2$ for the transverse coordinates, and $\mathbf{V}_z^2$ for the longitudinal z-component. Similarly, both $\mathbf{E}$ and $\mathbf{H}$ vectors can be written as the sum of a transverse component and an axial component:

$$\mathbf{E} = \mathbf{E}_T + \mathbf{a}_z E_z \qquad (10\text{–}193)$$

and

$$\mathbf{H} = \mathbf{H}_T + \mathbf{a}_z H_z, \qquad (10\text{–}194)$$

where the subscript T denotes the two-dimensional transverse component. We already know from Subsection 10–2.1 that TEM waves cannot exist in such a waveguide without an inner conductor. The propagating waves can be classified into two groups, as in rectangular waveguides: transverse magnetic (TM) and transverse electric (TE). For TM waves, $H_z = 0$, $E_z \neq 0$, and all field components can be expressed in terms of $E_z = E_z^0 e^{-\gamma z}$, where E_z^0 satisfies the homogeneous Helmholtz's equation

$$\mathbf{V}_{r\phi}^2 E_z^0 + (\gamma^2 + k^2)E_z^0 = 0 \qquad (10\text{–}195)$$

or

$$\mathbf{V}_{r\phi}^2 E_z^0 + h^2 E_z^0 = 0. \qquad (10\text{–}196)$$

For TE waves, $E_z = 0$, $H_z \neq 0$, and all field components can be expressed in terms of $H_z = H_z^0 e^{-\gamma z}$, where H_z^0 satisfies exactly the same homogeneous Helmholtz's equation required of E_z^0 above.

   Although Eq. (10–196) is similar in form to Eq. (10–24), their solutions are quite different. We will consider the solution of Eq. (10–196) in the following subsection.

10–5.1   BESSEL'S DIFFERENTIAL EQUATION AND BESSEL FUNCTIONS

In cylindrical coordinates the expansion of Eq. (10–196) gives (see Eq. (4–8))

$$\frac{1}{r}\frac{\partial}{\partial r}\left(r\frac{\partial E_z^0}{\partial r}\right) + \frac{1}{r^2}\frac{\partial^2 E_z^0}{\partial \phi^2} + h^2 E_z^0 = 0. \qquad (10\text{–}197)$$

To solve Eq. (10–197), we apply the method of separation of variables by assuming a product solution.

$$E_z^0(r, \phi) = R(r)\Phi(\phi), \qquad (10\text{–}198)$$

where $R(r)$ and $\Phi(\phi)$ are functions only of r and ϕ, respectively. Substituting solution (10–198) in Eq. (10–197) and dividing by the product $R(r)\Phi(\phi)$, we obtain

$$\frac{r}{R(r)}\frac{d}{dr}\left[r\frac{dR(r)}{dr}\right] + h^2 r^2 = -\frac{1}{\Phi(\phi)}\frac{d^2\Phi(\phi)}{d\phi^2}. \qquad (10\text{–}199)$$

Now the left side of Eq. (10–199) is a function of r only, and the right side is a function of ϕ only. For Eq. (10–199) to hold for all values of r and ϕ, both sides must be equal to the same constant. Let this constant (separation constant) be n^2. We can separate Eq. (10–199) into two ordinary differential equations:

$$\frac{d^2\Phi(\phi)}{d\phi^2} + n^2\Phi(\phi) = 0 \qquad (10\text{–}200)$$

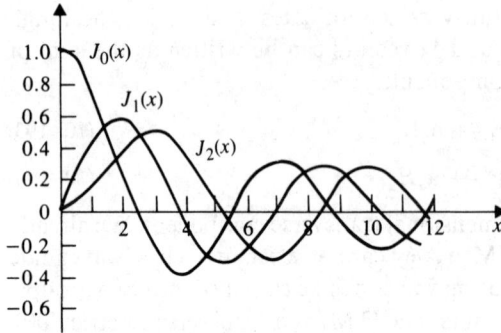

FIGURE 10–18
Bessel functions of the first kind.

and

$$\frac{r}{R(r)} \frac{d}{dr} \left[r \frac{dR(r)}{dr} \right] + h^2 r^2 = n^2$$

or

$$\frac{d^2 R(r)}{dr^2} + \frac{1}{r} \frac{dR(r)}{dr} + \left(h^2 - \frac{n^2}{r^2} \right) R(r) = 0. \qquad (10\text{–}201)$$

Equation (10–201) is known as **Bessel's differential equation.**

A solution of Eq. (10–201) can be obtained by assuming $R(r)$ to be a power series in r with unknown coefficients,

$$R(r) = \sum_{p=0}^{\infty} C_p (hr)^p, \qquad (10\text{–}202)$$

substituting it into the equation, and equating the sum of the coefficients of each power of r to zero. The actual work is tedious.[†] The result is

$$R(r) = C_n J_n(hr), \qquad (10\text{–}203)$$

where C_n is an arbitrary constant, and

$$J_n(hr) = \sum_{m=0}^{\infty} \frac{(-1)^m (hr)^{n+2m}}{m!(n+m)!2^{n+2m}} \qquad (10\text{–}204)$$

is called the **Bessel function of the first kind** of nth order with an argument hr. Equation (10–204) holds only for *integral values of n*, which is true for cases of our interest, as we shall see later. $J_n(x)$ versus x curves of the first few orders have been plotted in Fig. 10–18. Several things are worth noting. First, $J_n(0) = 0$ for all n except when $n = 0$; for the zeroth order, $J_0(0) = 1$. Second, $J_n(x)$ are alternating functions

---

[†] N. W. MaLachlan, *Bessel Functions for Engineers*, 2d ed, Oxford University Press, New York, 1946.

TABLE 10–2
Zeros of $J_n(x)$, x_{np}

p ╲ n	$n = 0$	$n = 1$	$n = 2$
1	2.405	3.832	5.136
2	5.520	7.016	8.417

of decreasing amplitudes that cross the zero level at progressively shorter intervals. As x becomes very large, all $J_n(x)$ approach a sinusoidal form. Table 10–2 lists the values of the first several x_{np}, which denotes the pth zero of $J_n(x)$: $J_n(x_{np}) = 0$. In the next subsection we will find that the values of x_{np} determine the eigenvalues of TM modes in a circular waveguide. The eigenvalues of TE modes, on the other hand, depend on the zeros of the derivative of Bessel functions of the first kind—that is, on the values of x'_{np}, which make $J'_n(x'_{np}) = 0$ (see Subsection 10–5.3). The values of the first several x'_{np} are tabulated in Table 10–3.

So far, we have obtained only one solution—Bessel function of the first kind, $J_n(hr)$—for the Bessel's differential equation (10–201). But Bessel's equation is a second-order equation; there should be two linearly independent solutions for each value of n. In other words, there should be another solution that is not linearly dependent on $J_n(hr)$. Such a solution exists. It is called **Bessel function of the second kind** or **Neumann function** and is usually denoted by $N_n(hr)$:

$$N_n(hr) = \frac{(\cos n\pi)J_n(hr) - J_{-n}(hr)}{\sin n\pi}. \tag{10–205}$$

The general solution of Eq. (10–201) can then be written as

$$R(r) = C_n J_n(hr) + D_n N_n(hr), \tag{10–206}$$

where C_n and D_n are arbitrary constants to be determined from boundary conditions.

A distinctive property of Bessel function of the second kind of all orders is that they become infinite when the argument is zero. When we study wave propagation in a circular waveguide, our region of interest includes the axis where $r = 0$. Since an infinite field is a physical impossibility, the solution $R(r)$ in Eq. (10–206) cannot contain a $N_n(hr)$ term. This means that the coefficient D_n must be zero for all n. Thus,

TABLE 10–3
Zeros of $J'_n(x)$, x'_{np}

p ╲ n	$n = 0$	$n = 1$	$n = 2$
1	3.832	1.841	3.054
2	7.016	5.331	6.706

for wave-mode problems inside a circular waveguide there is no need to be concerned with the $N_n(hr)$ term.

In the study of circular waveguides that follows, the preceding short summary of Bessel's differential equation and Bessel functions should suffice. The rest of this subsection discusses some additional aspects for completeness. It may be skipped if the material on dielectric-rod waveguides in Subsection 10–6.3 is to be omitted.

In case the region of interest of a problem in cylindrical coordinates does not include the axis where $r = 0$ (such as the problem of a coaxial waveguide with an inner conductor), the radial solution $R(r)$ in Eq. (10–206) must consist of both $J_n(hr)$ and $N_n(hr)$ terms, and the coefficients C_n and D_n are to be determined from boundary conditions. Furthermore, if a problem does not involve the entire 2π range of ϕ (such as the problem of a wedge-shaped waveguide), the constant n in Eq. (10–200) will not be an integer. Let it be denoted by v. We write the solution of the Bessel's differential equation as

$$R(r) = CJ_v(hr) + DN_v(hr). \tag{10–207}^\dagger$$

In some wave problems it is convenient to define linear combinations of the Bessel functions:

$$H_v^{(1)}(hr) = J_v(hr) + jN_v(hr), \tag{10–208}$$

$$H_v^{(2)}(hr) = J_v(hr) - jN_v(hr), \tag{10–209}$$

where $H_v^{(1)}$ and $H_v^{(2)}$ are called **Hankel functions** of the first and second kind, respectively. When the argument hr is very large, the asymptotic expressions for $H_v^{(1)}$ and $H_v^{(2)}$ are

$$H_v^{(1)}(hr) \rightarrow \sqrt{\frac{2}{\pi hr}}\, e^{j(hr - \pi/4 - v\pi/2)}, \tag{10–210}$$

$$H_v^{(2)}(hr) \rightarrow \sqrt{\frac{2}{\pi hr}}\, e^{-j(hr - \pi/4 - v\pi/2)}. \tag{10–211}$$

These expressions with imaginary exponential coefficients and decreasing amplitudes place in evidence the wave character of the Hankel functions. They are useful in problems of radiation.

When h^2 is negative ($h = j\zeta$), two other functions $I_v(\zeta)$ and $K_v(\zeta)$, related to J_v and $H_v^{(1)}$, respectively, are defined:

$$I_v(\zeta r) = j^{-v}J_v(j\zeta r), \tag{10–212}$$

$$K_v(\zeta r) = \frac{\pi}{2}j^{v+1}H_v^{(1)}(j\zeta r). \tag{10–213}$$

---

† The expression for $J_v(hr)$ for a nonintegral v is that given in Eq. (10–204) with $(n + m)!$ replaced by the gamma function $\Gamma(v + m + 1)$.

I_v and K_v are called **modified Bessel functions** of the first and second kind, respectively. For large arguments the following asymptotic expressions are obtained:

$$I_v(\zeta r) \rightarrow \sqrt{\frac{1}{2\pi\zeta r}}\, e^{\zeta r}, \tag{10-214}$$

$$K_v(\zeta r) \rightarrow \sqrt{\frac{\pi}{2\zeta r}}\, e^{-\zeta r}. \tag{10-215}$$

It is seen that at large r, $K_v(\zeta r)$ shows an exponential decay with distance, characteristic of an evenescent wave. It is useful in surface-wave problems such as dielectric-rod waveguides and optical fibers. The choice of the appropriate form as a solution for the Bessel's differential equation depends on the type of the problem and on convenience.

10-5.2 TM WAVES IN CIRCULAR WAVEGUIDES

Figure 10–19 shows a circular waveguide of radius a. It consists of a metal pipe centered around the z-axis. The enclosed dielectric medium is assumed to have constitutive parameters ϵ and μ. For TM waves, $H_z = 0$. We write

$$E_z(r, \phi, z) = E_z^0(r, \phi)e^{-\gamma z}, \tag{10-216}$$

where $E_z^0(r, \phi)$ satisfies Eq. (10–196). The solution is written in the form of Eq. (10–198), in which

$$R(r) = C_n J_n(hr), \tag{10-217}$$

and $\Phi(\phi)$ is the solution of Eq. (10–200). Since all field components are periodic with respect to ϕ (period $= 2\pi$), the only admissible solution for Eq. (10–200) is $\sin n\phi$ or $\cos n\phi$, or a linear combination of the two (see Table 4–1). It is because of this requirement of periodicity that we demand n to be an integer, as indicated previously. Whether $\sin n\phi$ or $\cos n\phi$ is chosen is immaterial; it changes only the location of the

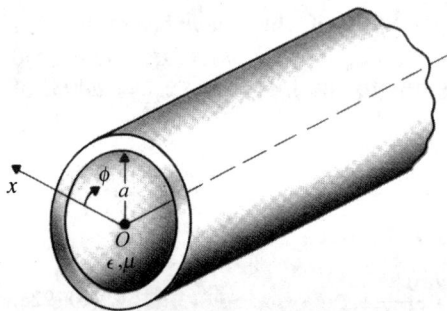

FIGURE 10–19
A circular waveguide.

reference $\phi = 0$ angle. Customarily, we write $E_z^0(r, \phi)$ for TM modes as

$$\boxed{E_z^0 = C_n J_n(hr) \cos n\phi. \qquad \text{(TM modes)}} \qquad (10\text{--}218)$$

The transverse components E_r^0 and E_ϕ^0 can be found from an adaptation of Eq. (10–29) to the cross-sectional polar coordinates (Problem P.10–26):

$$(\mathbf{E}_T^0)_{\text{TM}} = \mathbf{a}_r E_r^0 + \mathbf{a}_\phi E_\phi^0 = -\frac{\gamma}{h^2}\, \mathbf{\nabla}_T E_z^0, \qquad (10\text{--}219)$$

where

$$\mathbf{\nabla}_T E_z^0 = \left(\mathbf{a}_r \frac{\partial}{\partial r} + \mathbf{a}_\phi \frac{\partial}{r\partial\phi} \right) E_z^0. \qquad (10\text{--}220)$$

The magnetic field components can then be obtained by using Eq. (10–32).

We have for TM modes, in addition to E_z^0 in Eq. (10–218),

$$E_r^0 = -\frac{j\beta}{h}\, C_n J_n'(hr) \cos n\phi, \qquad (10\text{--}221)$$

$$E_\phi^0 = \frac{j\beta n}{h^2 r}\, C_n J_n(hr) \sin n\phi, \qquad (10\text{--}222)$$

$$H_r^0 = -\frac{j\omega\epsilon n}{h^2 r}\, C_n J_n(hr) \sin n\phi, \qquad (10\text{--}223)$$

$$H_\phi^0 = -\frac{j\omega\epsilon}{h}\, C_n J_n'(hr) \cos n\phi, \qquad (10\text{--}224)$$

$$H_z^0 = 0, \qquad (10\text{--}225)$$

where γ has been replaced by $j\beta$, J_n' is the derivative of J_n with respect to its argument (hr), and the coefficient C_n depends on the field strength of the excitation.

The eigenvalues of TM modes (the admissible values of h) are determined from the boundary condition that E_z^0 must vanish at $r = a$; that is,

$$\boxed{J_n(ha) = 0. \qquad \text{(TM modes)}} \qquad (10\text{--}226)$$

There are infinitely many zeros of $J_n(x)$, the first several of which have been tabulated in Table 10–2. The cutoff frequency is given by Eq. (10–35) as before. Hence the eigenvalue for the TM_{01} mode that corresponds to the first zero ($x_{01} = 2.405$) of $J_0(x)$ is

$$(h)_{\text{TM}_{01}} = \frac{2.405}{a}, \qquad (10\text{--}227)$$

which yields the lowest cutoff frequency for a TM mode:

$$(f_c)_{\text{TM}_{01}} = \frac{(h)_{\text{TM}_{01}}}{2\pi\sqrt{\mu\epsilon}} = \frac{0.383}{a\sqrt{\mu\epsilon}}. \qquad (10\text{--}228)$$

The phase constant β and the guide wavelength λ_g can be found from Eqs. (10–38) and (10–39), respectively.

For the TM_{01} mode $(n = 0)$, E_z^0, E_r^0, and H_ϕ^0 are the only nonzero field components. A sketch of the electric and magnetic field lines in a typical transverse plane is given in Fig. 10–20. According to Eq. (10–224), H_ϕ^0 varies with r as $J_0'(hr)$, which equals $-J_1(hr)$. Thus the density of the magnetic field lines increases from $r = 0$ to $r = a$.

Note that in rectangular waveguides the first and second numbers of the mode index denote the number of half-wave field variations in the x- and y-directions, respectively, in a transverse xy-plane. By convention *the first number of the mode index for circular waveguides always represents the number of half-wave field variations in the ϕ-direction*, and the second number represents the number of half-wave field variations in the r-direction. Hence the transverse field pattern of the TM_{01} mode in a circular waveguide is analogous to the TM_{11} mode (instead of the TM_{01} mode, which does not exist) in a rectangular waveguide.

10–5.3   TE WAVES IN CIRCULAR WAVEGUIDES

For TE modes, $E_z = 0$, and

$$H_z(r, \phi, z) = H_z^0(r, \phi)e^{-\gamma z}, \tag{10–229}$$

where H_z^0 satisfies the homogeneous Helmholtz's equation

$$\nabla_{r\phi}^2 H_z^0 + h^2 H_z^0 = 0. \tag{10–230}$$

Analogously to the TM case, we write the solution as

$$\boxed{H_z^0 = C_n' J_n(hr) \cos n\phi. \qquad \text{(TE modes)}} \tag{10–231}$$

From H_z^0 we find the transverse magnetic field components H_r^0 and H_ϕ^0 by using Eq. (10–53), and we find the electric field components E_r^0 and E_ϕ^0 by applying Eq. (10–55)— similar to Eq. (10–219).

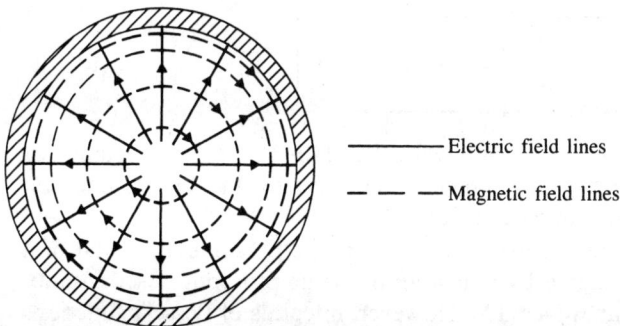

——————— Electric field lines

— — — — Magnetic field lines

FIGURE 10–20
Field lines for TM_{01} mode in a transverse plane of circular waveguide.

We have for TE modes, in addition to H_z^0 in Eq. (10–229),

$$H_r^0 = -\frac{j\beta}{h} C_n' J_n'(hr) \cos n\phi, \tag{10–232}$$

$$H_\phi^0 = \frac{j\beta n}{h^2 r} C_n' J_n(hr) \sin n\phi, \tag{10–233}$$

$$E_r^0 = \frac{j\omega\mu n}{h^2 r} C_n' J_n(hr) \sin n\phi, \tag{10–234}$$

$$E_\phi^0 = \frac{j\omega\mu}{h} C_n' J_n'(hr) \cos n\phi, \tag{10–235}$$

$$E_z^0 = 0. \tag{10–236}$$

The required boundary condition for TE waves is that the normal derivative of H_z^0 must vanish at $r = a$; that is,

$$\boxed{J_n'(ha) = 0. \qquad \text{(TE modes)}} \tag{10–237}$$

The first several zeros of $J_n'(x)$ are listed in Table 10–3, from which we see that the smallest x_{np}' is $x_{11}' = 1.841$. This corresponds to the *smallest eigenvalue*

$$(h)_{TE_{11}} = \frac{1.841}{a}, \tag{10–238}$$

and the *lowest cutoff frequency*

$$\boxed{(f_c)_{TE_{11}} = \frac{h_{TE_{11}}}{2\pi\sqrt{\mu\epsilon}} = \frac{0.293}{a\sqrt{\mu\epsilon}} \qquad \text{(Hz)},} \tag{10–239}$$

which is lower than $(f_c)_{TM_{01}}$ given in Eq. (10–228). Hence **the TE$_{11}$ mode is the dominant mode in a circular waveguide**. In an air-filled circular waveguide of radius a, the cutoff wavelength for the dominant mode is

$$\boxed{(\lambda_c)_{TE_{11}} = \frac{a}{0.293} = 3.41a \qquad \text{(m).}} \tag{10–240}$$

It is interesting to compare Eq. (10–240) with Eq. (10–164) for a rectangular waveguide. A sketch of the electric and magnetic field lines for the TE$_{11}$ mode in a typical transverse plane is shown in Fig. 10–21.

The attenuation constant due to losses in the imperfectly conducting wall of a circular waveguide can be calculated by following the same procedure used in Subsection 10–4.3 for a rectangular waveguide. However, integrals of Bessel's functions would be involved, and we shall not pursue this aspect further in this book. Suffice

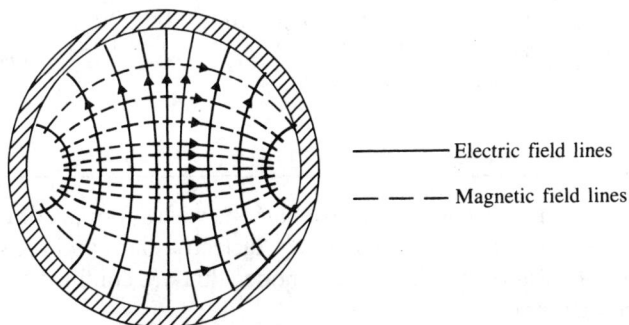

———— Electric field lines

— — — Magnetic field lines

FIGURE 10–21
Field lines for TE_{11} mode in a transverse plane of a circular waveguide.

it to say that the attenuation constants of the dominant-mode propagating waves in circular and rectangular waveguides having comparable dimensions are of the same order of magnitude. A point of special interest for circular waveguides is that the attenuation constant of TE_{0p} waves decreases monotonically with frequency—the absence of a minimum point in $\alpha_c \sim f$ curves. No other waves in circular or rectangular waveguides have this property.

EXAMPLE 10–12 (a) A 10 (GHz) signal is to be transmitted inside a hollow circular conducting pipe. Determine the inside diameter of the pipe such that its lowest cutoff frequency is 20% below this signal frequency. (b) If the pipe is to operate at 15 (GHz), what waveguide modes can propagate in the pipe?

Solution

a) The cutoff frequency of the dominant mode in a circular waveguide of radius a is, from Eq. (10–239),

$$(f_c)_{TE_{11}} = \frac{0.293c}{a} = \frac{0.879}{a} \times 10^8 \quad \text{(Hz)}$$

$$= \frac{0.0879}{a} \quad \text{(GHz)}.$$

This is to be equated to $0.80 \times 10 = 8$ (GHz). Hence the required inside diameter of the pipe is $2a = 2 \times (0.0879/8) = 0.022$ (m), or 2.2 (cm).

b) Cutoff frequencies for waveguide modes in a hollow circular pipe of inner radius $a = 0.011$ (m) that are lower than 15 (GHz) are, from Tables 10–1 and 10–2,

$$(f_c)_{TE_{11}} = 8 \quad \text{(GHz)},$$

$$(f_c)_{TM_{01}} = 8 \times \left(\frac{x_{01}}{x'_{11}}\right) = 8 \times \left(\frac{2.405}{1.841}\right) = 10.45 \quad \text{(GHz)},$$

$$(f_c)_{TE_{21}} = 8 \times \left(\frac{x'_{21}}{x'_{11}}\right) = 8 \times \left(\frac{3.054}{1.841}\right) = 13.27 \quad \text{(GHz)}.$$

The f_c of all other modes are higher than 15 (GHz); hence only $\mathrm{TE_{11}}$, $\mathrm{TM_{01}}$, and $\mathrm{TE_{21}}$ modes can propagate in the pipe.   ∎

10–6 Dielectric Waveguides

In previous sections we discussed the behavior of electromagnetic waves propagating along waveguides with conducting walls. We now show that dielectric slabs and rods without conducting walls can also support guided-wave modes that are confined essentially within the dielectric medium.

Figure 10–22 shows a longitudinal cross section of a dielectric-slab waveguide of thickness d. For simplicity we consider this a problem with no dependence on the x-coordinate. Let ϵ_d and μ_d be the permittivity and permeability, respectively, of the dielectric slab, which is situated in free space (ϵ_0, μ_0). We assume that the dielectric is lossless and that waves propagate in the $+z$-direction. The behavior of TM and TE modes will now be analyzed separately.

10–6.1   TM WAVES ALONG A DIELECTRIC SLAB

For transverse magnetic waves, $H_z = 0$. Since there is no x-dependence, Eq. (10–62) applies. We have

$$\frac{d^2 E_z^0(y)}{dy^2} + h^2 E_z^0(y) = 0, \tag{10–241}$$

where

$$h^2 = \gamma^2 + \omega^2 \mu \epsilon. \tag{10–242}$$

Solutions of Eq. (10–241) must be considered in both the slab and the free-space regions, and they must be matched at the boundaries.

In the slab region we assume that the waves propagate in the $+z$-direction without attenuation (lossless dielectric); that is, we assume

$$\gamma = j\beta. \tag{10–243}$$

The solution of Eq. (10–241) in the dielectric slab may contain both a sine term and a cosine term, which are an odd and an even function, respectively, of y:

$$E_z^0(y) = E_o \sin k_y y + E_e \cos k_y y, \qquad |y| \le \frac{d}{2}, \tag{10–244}$$

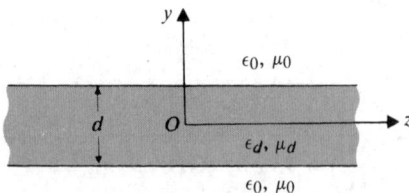

FIGURE 10–22
A longitudinal cross-section of a dielectric-slab waveguide.

where

$$k_y^2 = \omega^2 \mu_d \epsilon_d - \beta^2 = h_d^2. \qquad (10\text{-}245)$$

In the free-space regions ($y > d/2$ and $y < -d/2$) the waves must decay exponentially so that they are guided along the slab and do not radiate away from it. We have

$$E_z^0(y) = \begin{cases} C_u e^{-\alpha(y-d/2)}, & y \geq \dfrac{d}{2}, & (10\text{-}246a) \\[3mm] C_l e^{\alpha(y+d/2)}, & y \leq -\dfrac{d}{2}, & (10\text{-}246b) \end{cases}$$

where

$$\alpha^2 = \beta^2 - \omega^2 \mu_0 \epsilon_0 = -h_0^2. \qquad (10\text{-}247)$$

Equations (10–245) and (10–247) are called **dispersion relations** because they show the nonlinear dependence of the phase constant β on ω.

At this stage we have not yet determined the values of k_y and α; nor have we found the relationships among the amplitudes E_o, E_e, C_u, and C_l. In the following, we will consider the odd and even TM modes separately.

a) *Odd TM modes.* For odd TM modes, $E_z^0(y)$ is described by a sine function that is antisymmetric with respect to the $y = 0$ plane. The only other field components, $E_y^0(y)$ and $H_x^0(y)$, are obtained from Eqs. (10–28) and (10–25), respectively.

i) In the dielectric region, $|y| \leq d/2$:

$$E_z^0(y) = E_o \sin k_y y, \qquad (10\text{-}248)$$

$$E_y^0(y) = -\frac{j\beta}{k_y} E_o \cos k_y y, \qquad (10\text{-}249)$$

$$H_x^0(y) = \frac{j\omega\epsilon_d}{k_y} E_o \cos k_y y. \qquad (10\text{-}250)$$

ii) In the upper free-space region, $y \geq d/2$:

$$E_z^0(y) = \left(E_o \sin \frac{k_y d}{2} \right) e^{-\alpha(y-d/2)}, \qquad (10\text{-}251)$$

$$E_y^0(y) = -\frac{j\beta}{\alpha} \left(E_o \sin \frac{k_y d}{2} \right) e^{-\alpha(y-d/2)}, \qquad (10\text{-}252)$$

$$H_x^0(y) = \frac{j\omega\epsilon_0}{\alpha} \left(E_o \sin \frac{k_y d}{2} \right) e^{-\alpha(y-d/2)}, \qquad (10\text{-}253)$$

where C_u in Eq. (10–246a) has been set to equal $E_0 \sin (k_y d/2)$, which is the value of $E_z^0(y)$ in Eq. (10–248) at the upper interface, $y = d/2$.

iii) In the lower free-space region, $y \leq -d/2$:

$$E_z^0(y) = -\left(E_o \sin \frac{k_y d}{2} \right) e^{\alpha(y+d/2)}, \qquad (10\text{-}254)$$

$$E_y^0(y) = -\frac{j\beta}{\alpha}\left(E_o \sin \frac{k_y d}{2}\right)e^{\alpha(y+d/2)}, \tag{10–255}$$

$$H_x^0(y) = \frac{j\omega\epsilon_0}{\alpha}\left(E_o \sin \frac{k_y d}{2}\right)e^{\alpha(y+d/2)}, \tag{10–256}$$

where C_l in Eq. (10–246b) has been set to equal $-E_o \sin (k_y d/2)$, which is the value of $E_z^0(y)$ in Eq. (10–248) at the lower interface $y = -d/2$.

Now we must determine k_y and α for a given angular frequency of excitation ω. The continuity of H_x at the dielectric surface requires that $H_x^0(d/2)$ computed from Eqs. (10–250) and (10–253) be the same. We have

$$\boxed{\frac{\alpha}{k_y} = \frac{\epsilon_0}{\epsilon_d} \tan \frac{k_y d}{2}} \qquad \text{(Odd TM modes).} \tag{10–257}$$

By adding dispersion relations Eqs. (10–245) and (10–247) we find

$$\alpha^2 + k_y^2 = \omega^2(\mu_d\epsilon_d - \mu_0\epsilon_0) \tag{10–258}$$

or

$$\boxed{\alpha = [\omega^2(\mu_d\epsilon_d - \mu_0\epsilon_0) - k_y^2]^{1/2}.} \tag{10–259}$$

Equations (10–257) and (10–259) can be combined to give an expression in which k_y is the only unknown:

$$[\omega^2(\mu_d\epsilon_d - \mu_0\epsilon_0) - k_y^2]^{1/2} = \frac{\epsilon_0}{\epsilon_d} k_y \tan \frac{k_y d}{2}. \tag{10–260}$$

Unfortunately, the transcendental equation, Eq. (10–260), cannot be solved analytically. But for a given ω and given values of ϵ_d, μ_d, and d of the dielectric slab, both the left and the right sides of Eq. (10–260) can be plotted versus k_y. The intersections of the two curves give the values of k_y for odd TM modes, of which there are only a finite number, indicating that there are only a finite number of possible modes. This is in contrast with the infinite number of modes possible in waveguides with conducting walls.

We note from Eq. (10–248) that $E_z^0 = 0$ for $y = 0$. Hence a perfectly conducting plane may be introduced to coincide with the $y = 0$ plane without affecting the existing fields. It follows that the characteristics of odd TM waves propagating along a dielectric-slab waveguide of thickness d are the same as those of the corresponding TM modes supported by a dielectric slab of a thickness $d/2$ that is backed by a perfectly conducting plane.

The *surface impedance* looking down from above on the surface of dielectric slab is

$$Z_s = -\frac{E_z^0}{H_x^0} = j\frac{\alpha}{\omega\epsilon_0} \qquad \text{(TM modes),} \tag{10–261}$$

which is an inductive reactance. Thus *a TM surface wave can be supported by an inductive surface.*

b) *Even TM modes.* For even TM modes, $E_z^0(y)$ is described by a cosine function that is symmetric with respect to the $y = 0$ plane:

$$E_z^0(y) = E_e \cos k_y y, \qquad |y| \le \frac{d}{2}. \tag{10–262}$$

The other nonzero field components, E_y^0 and H_x^0, both inside and outside the dielectric slab can be obtained in exactly the same manner as in the case of odd TM modes (see Problem P.10–33). Instead of Eq. (10–257), the characteristic relation between k_y and α now becomes

$$\boxed{\frac{\alpha}{k_y} = -\frac{\epsilon_0}{\epsilon_d} \cot \frac{k_y d}{2} \qquad \text{(Even TM modes)}, } \tag{10–263}$$

which can be used in conjunction with Eq. (10–259) to determine the transverse wavenumber k_y and the transverse attenuation constant α. The several solutions correspond to the several even TM modes that can exist in the dielectric slab waveguide of thickness d. Of course, in this case a conducting plane *cannot* be placed at $y = 0$ without disturbing the whole field structure.

From Eqs. (10–245) and (10–247), it is easy to see that the phase constant, β, of propagating TM waves lies between the intrinsic phase constant of the free space, $k_0 = \omega \sqrt{\mu_0 \epsilon_0}$, and that of the dielectric, $k_d = \omega \sqrt{\mu_d \epsilon_d}$; that is,

$$\omega \sqrt{\mu_0 \epsilon_0} < \beta < \omega \sqrt{\mu_d \epsilon_d}.$$

As β approaches the value of $\omega \sqrt{\mu_0 \epsilon_0}$, Eq. (10–247) indicates that α approaches zero. An absence of attenuation means that the waves are no longer bound to the slab. The limiting frequencies under this condition are called the ***cutoff frequencies*** of the dielectric waveguide. From Eq. (10–245) we have $k_y = \omega_c \sqrt{\mu_d \epsilon_d - \mu_0 \epsilon_0}$ at cutoff. Substitution into Eqs. (10–257) and (10–263) with α set to zero yields the following relations for TM modes. At cutoff:

Odd TM Modes	Even TM Modes
$\tan\left(\dfrac{\omega_{co} d}{2} \sqrt{\mu_d \epsilon_d - \mu_0 \epsilon_0}\right) = 0$	$\cot\left(\dfrac{\omega_{ce} d}{2} \sqrt{\mu_d \epsilon_d - \mu_0 \epsilon_0}\right) = 0$
$\pi f_{co} d \sqrt{\mu_d \epsilon_d - \mu_0 \epsilon_0} = (n-1)\pi,$ $n = 1, 2, 3, \ldots$	$\pi f_{ce} d \sqrt{\mu_d \epsilon_d - \mu_0 \epsilon_0} = (n - \tfrac{1}{2})\pi,$ $n = 1, 2, 3, \ldots$
$\boxed{f_{co} = \dfrac{(n-1)}{d\sqrt{\mu_d \epsilon_d - \mu_0 \epsilon_0}}}$ (10–264)	$\boxed{f_{ce} = \dfrac{(n - \tfrac{1}{2})}{d\sqrt{\mu_d \epsilon_d - \mu_0 \epsilon_0}}}$ (10–265)

It is seen that $f_{co} = 0$ for $n = 1$. This means that *the lowest-order odd TM mode can propagate along a dielectric-slab waveguide regardless of the thickness of the slab.*

As the frequency of a given TM wave increases beyond the corresponding cutoff frequency, α increases and the wave clings more tightly to the slab.

10–6.2   TE WAVES ALONG A DIELECTRIC SLAB

For transverse electric waves, $E_z = 0$, and Eq. (10–82) applies

$$\frac{d^2 H_z^0(y)}{dy^2} + h^2 H_z^0(y) = 0, \tag{10–266}$$

where h^2 is the same as that given in Eq. (10–242). The solution for $H_z^0(y)$ may also contain both a sine term and a cosine term:

$$H_y^0(y) = H_o \sin k_y y + H_e \cos k_y y, \qquad |y| \le \frac{d}{2}, \tag{10–267}$$

where k_y has been defined in Eq. (10–245). In the free-space regions ($y > d/2$ and $y < -d/2$) the waves must decay exponentially. We write

$$H_z^0(y) = \begin{cases} C_u' e^{-\alpha(y - d/2)}, & y \ge \dfrac{d}{2}, & (10\text{–}268a) \\[2mm] C_l' e^{\alpha(y + d/2)}, & y \le -\dfrac{d}{2}, & (10\text{–}268b) \end{cases}$$

where α is defined in Eq. (10–247). Following the same procedure as used for TM waves, we consider the odd and even TE modes separately. Besides $H_z^0(y)$, the only other field components are $H_y^0(y)$ and $E_x^0(y)$, which can be obtained from Eqs. (10–50) and (10–51).

a) *Odd TE modes.*

  i) In the dielectric region, $|y| \le d/2$:

$$H_z^0(y) = H_o \sin k_y y, \tag{10–269}$$

$$H_y^0(y) = -\frac{j\beta}{k_y} H_o \cos k_y y, \tag{10–270}$$

$$E_x^0(y) = -\frac{j\omega\mu_d}{k_y} H_o \cos k_y y. \tag{10–271}$$

  ii) In the upper free-space region, $y \ge d/2$:

$$H_z^0(y) = \left(H_o \sin \frac{k_y d}{2} \right) e^{-\alpha(y - d/2)}, \tag{10–272}$$

$$H_y^0(y) = -\frac{j\beta}{\alpha} \left(H_o \sin \frac{k_y d}{2} \right) e^{-\alpha(y - d/2)}, \tag{10–273}$$

$$E_x^0(y) = -\frac{j\omega\mu_0}{\alpha} \left(H_o \sin \frac{k_y d}{2} \right) e^{-\alpha(y - d/2)}. \tag{10–274}$$

iii) In the lower free-space region, $y \le -d/2$:

$$H_z^0(y) = -\left(H_o \sin \frac{k_y d}{2}\right) e^{\alpha(y + d/2)}, \tag{10-275}$$

$$H_y^0(y) = -\frac{j\beta}{\alpha}\left(H_o \sin \frac{k_y d}{2}\right) e^{\alpha(y + d/2)}, \tag{10-276}$$

$$E_x^0(y) = -\frac{j\omega\mu_0}{\alpha}\left(H_o \sin \frac{k_y d}{2}\right) e^{\alpha(y + d/2)}. \tag{10-277}$$

A relation between k_y and α can be obtained by equating $E_x^0(y)$, given in Eqs. (10-271) and (10-274), at $y = d/2$. Thus,

$$\boxed{\frac{\alpha}{k_y} = \frac{\mu_0}{\mu_d} \tan \frac{k_y d}{2}} \quad \text{(Odd TE modes)}, \tag{10-278}$$

which is seen to be closely analogous to the characteristic equation, Eq. (10-257), for odd TM modes. Equations (10-259) and (10-278) can be combined in the manner of Eq. (10-260) to find k_y graphically. From k_y, α can be found from Eq. (10-259).

From a position of looking down from above, the surface impedance of the dielectric slab is

$$Z_s = \frac{E_x^0}{H_z^0} = -j\frac{\omega\mu_0}{\alpha} \quad \text{(TE modes)}, \tag{10-279}$$

which is a capacitive reactance. Hence *a TE surface wave can be supported by a capacitive surface.*

b) *Even TE modes.* For even TE modes, $H_z^0(y)$ is described by a cosine function that is symmetric with respect to the $y = 0$ plane.

$$H_z^0(y) = H_e \cos k_y y, \qquad |y| \le d/2. \tag{10-280}$$

The other nonzero field components, H_y^0 and E_x^0, both inside and outside the dielectric slab can be obtained in the same manner as for odd TE modes (see Problem P.10-35). The characteristic relation between k_y and α is closely analogous to that for even TM modes as given in Eq. (10-263):

$$\boxed{\frac{\alpha}{k_y} = -\frac{\mu_0}{\mu_d} \cot \frac{k_y d}{2}} \quad \text{(Even TE modes)}. \tag{10-281}$$

It is easy to see that the expressions for the cutoff frequencies given in Eqs. (10-264) and (10-265) apply also to TE modes. Like the lowest-order ($n = 1$) TM mode, the lowest-order odd TE mode has no cutoff frequency. The characteristic relations for all the propagating modes along a dielectric-slab waveguide of a thickness d are listed in Table 10-4.

TABLE 10–4
Characteristic Relations for Dielectric-Slab Waveguide [†]

Mode		Characteristic Relation	Cutoff Frequency
TM	Odd	$(\alpha/k_y) = (\epsilon_0/\epsilon_d) \tan (k_y d/2)$	$f_{co} = (n - 1)/d\sqrt{\mu_d\epsilon_d - \mu_0\epsilon_0}$
	Even	$(\alpha/k_y) = -(\epsilon_0/\epsilon_d) \cot (k_y d/2)$	$f_{ce} = (n - \frac{1}{2})/d\sqrt{\mu_d\epsilon_d - \mu_0\epsilon_0}$
TE	Odd	$(\alpha/k_y) = (\mu_0/\mu_d) \tan (k_y d/2)$	$f_{co} = (n - 1)/d\sqrt{\mu_d\epsilon_d - \mu_0\epsilon_0}$
	Even	$(\alpha/k_y) = -(\mu_0/\mu_d) \cot (k_y d/2)$	$f_{ce} = (n - \frac{1}{2})/d\sqrt{\mu_d\epsilon_d - \mu_0\epsilon_0}$

[†] $\alpha = [\omega^2(\mu_d\epsilon_d - \mu_0\epsilon_0) - k_y^2]^{1/2}$.

EXAMPLE 10–13   A dielectric-slab waveguide with constitutive parameters $\mu_d = \mu_0$ and $\epsilon_d = 2.50\epsilon_0$ is situated in free space. Determine the minimum thickness of the slab so that a TM or TE wave of the even type at a frequency 20 GHz may propagate along the guide.

Solution   The lowest TM and TE waves of the even type have the same cutoff frequency along a dielectric-slab waveguide:

$$f_c = \frac{n - \frac{1}{2}}{d\sqrt{\mu_d\epsilon_d - \mu_0\epsilon_0}}.$$

Letting $n = 1$, we have

$$f_c = \frac{c}{2d\sqrt{\dfrac{\mu_d\epsilon_d}{\mu_0\epsilon_0} - 1}}.$$

Therefore,

$$d_{min} = \frac{c}{2f_c\sqrt{\dfrac{\mu_d\epsilon_d}{\mu_0\epsilon_0} - 1}}$$

$$= \frac{3 \times 10^8}{2 \times 20 \times 10^9\sqrt{2.5 - 1}} = 6.12 \times 10^{-3} \text{ (m) or 6.12   (mm)}.$$

EXAMPLE 10–14   (a) Obtain an approximate expression for the decaying rate of the dominant TM surface wave outside of a very thin dielectric-slab waveguide. (b) Find the time-average power per unit slab width transmitted along the guide. (c) What is the time-average power transmitted in the transverse direction?

Solution

a) The dominant TM wave is the odd mode having a zero cutoff frequency—$f_{co} = 0$ for $n = 1$, independent of the slab thickness (see Table 10–4). With a slab that is very thin in comparison to the operating wavelength, $k_y d/2 \ll 1$, $\tan(k_y d/2) \cong k_y d/2$, and Eq. (10–257) becomes

$$\alpha \cong \frac{\epsilon_0}{2\epsilon_d} k_y^2 d. \qquad (10\text{–}282)$$

Using Eq. (10–258), Eq. (10–282) can be written approximately as

$$\alpha \cong \frac{\epsilon_0}{2\epsilon_d} \omega^2 (\mu_d \epsilon_d - \mu_0 \epsilon_0) d \qquad (\text{Np/m}). \qquad (10\text{–}283)$$

In Eq. (10–283) it has been assumed that $\alpha d/2 \ll \epsilon_d/\epsilon_0$.

b) The time-average Poynting vector in the $+z$-direction in the dielectric slab is

$$\mathscr{P}_{av} = \tfrac{1}{2}\mathscr{R}e(-\mathbf{a}_y E_y \times \mathbf{a}_x H_x).$$

Using Eqs. (10–249) and (10–250), we have $\mathbf{P}_{av} = \mathbf{a}_z P_{av}$ and

$$P_{av} = 2\int_0^{d/2} \mathscr{P}_{av}\, dy = \frac{\omega\epsilon_d\beta}{k_y^2} E_o^2 \int_0^{d/2} \cos^2(k_y y)\, dy$$

$$= \frac{\omega\epsilon_d\beta}{4k_y^2} E_o^2 \left[d + \frac{1}{k_y}\sin(k_y d) \right] \qquad (\text{W/m}), \qquad (10\text{–}284)$$

where

$$k_y \cong \omega\sqrt{\mu_d\epsilon_d - \mu_0\epsilon_0} \qquad (10\text{–}284a)$$

and

$$\beta \cong \omega\sqrt{\mu_0\epsilon_0}. \qquad (10\text{–}284b)$$

c) The time-average Poynting vector in the transverse direction is calculated from

$$\mathscr{P}_{av} = \tfrac{1}{2}\mathscr{R}e(\mathbf{a}_z E_z \times \mathbf{a}_x H_x).$$

From Subsection 10–6.1 we see that the expressions of E_z^0 and H_x^0 are 90° out of time phase. Their product has no real part, yielding a zero $\mathscr{P}_{av}$. Hence no average power is transmitted in the transverse direction normal to the reactive surface. ■

10–6.3 ADDITIONAL COMMENTS ON DIELECTRIC WAVEGUIDES

In the preceding subsection we studied the characteristics of electromagnetic waves guided by dielectric slabs with an analysis based on Maxwell's equations and the associated boundary conditions. We can gain some physical insight from the concept of total reflection in plane-wave theory that we discussed in Section 8–10.

Consider the dielectric slab in Fig. 10–23. From Section 8–10 we know that if a plane wave in the slab with a permittivity $\epsilon_d > \epsilon_0$ is incident obliquely on the lower

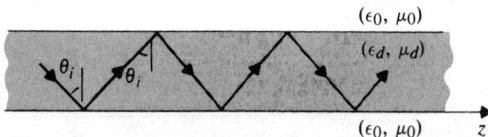

FIGURE 10–23
Bouncing-wave interpretation of propagating waves along a dielectric waveguide.

boundary at an angle of incidence θ_i greater than the critical angle (see Eq. 8–188)

$$\theta_c = \sin^{-1} \sqrt{\frac{\epsilon_0}{\epsilon_d}}, \qquad (10\text{--}285)$$

it will be totally reflected toward the upper boundary. Moreover, an evanescent wave exists along the interface (in z-direction) that is attenuated exponentially in the transverse direction outside of the boundary. The reflected wave from the lower boundary will be incident on the upper boundary at the same angle of incidence $\theta_i > \theta_c$ and will be similarly totally reflected. This process will continue so that there will be two sets of multiply reflected waves: one set going from the upper boundary toward the lower boundary, and the other set from the lower boundary toward the upper boundary. Under the condition that the points on the same wavefront have the same phase, each set of reflected waves forms a single uniform plane wave. We then have two interfering uniform plane waves, giving rise to an interference pattern, which is the mode pattern of the propagating wave. It is clear that the phase requirements at both reflecting boundaries depend on the angle of incidence θ_i, since θ_i determines the phase shifts caused by total internal reflections. Analysis shows that the required phase conditions correspond precisely to the dispersion and characteristic relations obtained in the preceding section.[†] Thus the results based on Maxwell's equations and boundary conditions can be interpreted by bouncing waves due to total internal reflections.

So far our attention has been directed toward the wave behavior in dielectric-slab waveguides. Similar analyses apply to round dielectric-rod waveguides. In particular, they can be used to study the transmission of light waves along quartz or glass fibers that form optical waveguides. Optical fiber waveguides are of great importance as transmission media for communication or control systems because of their low-attenuation and large-bandwidth properties. They also are extremely compact and flexible. A study of circular dielectric waveguides necessitates the use of cylindrical coordinates that lead to Bessel's differential equation and Bessel functions. The study is complicated by the fact that pure TM or TE modes are possible only if the fields are circularly symmetrical; that is, if the fields are independent of the angle coordinate ϕ. When the fields are dependent on ϕ, separation into TM and TE modes is no longer possible, and it is necessary to assume the existence of both E_z and H_z components simultaneously and study the so-called **hybrid modes**.

As a simple example, consider the circularly symmetrical TM modes for a round dielectric rod of radius a and permittivity ϵ_d, situated in air. The transverse distribution of the axial component of electric field intensity, E_z^0, in the dielectric rod ($r \leq a$) is, from Eq. (10–218) by setting $n = 0$,

$$E_{zi}^0 = C_0 J_0(hr), \qquad r \leq a, \qquad (10\text{--}286)$$

where

$$h^2 = \gamma^2 + k_d^2 = \omega^2 \mu_0 \epsilon_d - \beta^2. \qquad (10\text{--}287)$$

[†] S. R. Seshadri, *Fundamentals of Transmission Lines and Electromagnetic Fields*, Addison-Wesley, Reading, Mass., 1971, Chapter 8.

The corresponding $H^0_{\phi i}$ is, from Eq. (10–224),

$$H^0_{\phi i} = -\frac{j\omega\epsilon_d}{h} C_0 J'_0(hr), \qquad r \le a. \tag{10–288}$$

Outside the dielectric rod, the fields are required to be evanescent and must decrease exponentially with distance. An appropriate choice for E^0_{zo} is $K_0(\zeta r)$, the modified Bessel function of the second kind of order zero, whose asymptotic expansion for large arguments is given in Eq. (10–215). We write

$$E^0_{zo} = D_0 K_0(\zeta r), \qquad r \ge a, \tag{10–289}$$

where

$$\zeta^2 = \beta^2 - k_0^2 = \beta^2 - \omega^2 \mu_0 \epsilon_0, \tag{10–290}$$

and D_0 is a constant. The corresponding $H^0_{\phi o}$ is

$$H^0_{\phi o} = \frac{j\omega\epsilon_0}{\zeta} D_0 K'_0(\zeta r), \qquad r \ge a. \tag{10–291}$$

The field components E^0_z and H^0_ϕ must be continuous at $r = a$, which requires

$$C_0 J_0(ha) = D_0 K_0(\zeta a) \tag{10–291}$$

and

$$\frac{\epsilon_d}{h} C_0 J'_0(ha) = -\frac{\epsilon_0}{\zeta} D_0 K'_0(\zeta a). \tag{10–292}$$

Combination of Eqs. (10–291) and (10–292) gives the following characteristic equation for circularly symmetrical TM modes:

$$\frac{J_0(ha)}{J'_0(ha)} = -\frac{\epsilon_d \zeta}{\epsilon_0 h} \frac{K_0(\zeta a)}{K'_0(\zeta a)}, \tag{10–293}$$

where ζ and h are related through Eqs. (10–287) and (10–290):

$$h^2 + \zeta^2 = \omega^2 \mu_0(\epsilon_d - \epsilon_0). \tag{10–294}$$

Equations (10–293) and (10–294) can be solved for h and ξ either graphically or on a computer. Once the eigenvalues have been found, the cutoff frequencies and other properties of the corresponding circularly symmetrical TM modes can be determined.

In the above example we discussed only the analysis procedure for circularly symmetrical TM modes in an unclad homogeneous optical fiber. In practice, commercially available optical fibers are mainly of two types: step-index fibers that consist of a central homogeneous dielectric core and an outer sheath of a material having a lower refractive index and graded-index fibers whose center core has a nonhomogeneous refractive-index profile. Detailed studies of these types do not fall into the scope of this book.[†]

[†] See, for instance, D. Marcuse, *Theory of Dielectric Waveguides*, Academic Press, New York, 1974; A. W. Snyder and J. D. Love, *Optical Waveguide Theory*, Methuen Inc., New York, 1984.

10–7 Cavity Resonators

We have previously pointed out that at UHF (300 MHz to 3 GHz) and higher frequencies, ordinary lumped-circuit elements such as R, L, and C are difficult to make, and stray fields become important. Circuits with dimensions comparable to the operating wavelength become efficient radiators and will interfere with other circuits and systems. Furthermore, conventional wire circuits tend to have a high effective resistance both because of energy loss through radiation and as a result of skin effect. To provide a resonant circuit at UHF and higher frequencies, we look to an enclosure (a cavity) completely surrounded by conducting walls. Such a shielded enclosure confines electromagnetic fields inside and furnishes large areas for current flow, thus eliminating radiation and high-resistance effects. These enclosures have natural resonant frequencies and a very high Q (quality factor), and are called *cavity resonators*. In this section we will study the properties of rectangular and circular cylindrical cavity resonators.

10–7.1   RECTANGULAR CAVITY RESONATORS

Consider a rectangular waveguide with both ends closed by a conducting wall. The interior dimensions of the cavity are a, b, and d, as shown in Fig. 10–24. Let us disregard for the moment the probe-excitation part of the figure. Since both TM and TE modes can exist in a rectangular guide, we expect TM and TE modes in a rectangular resonator too. However, the designation of TM and TE modes in a resonator is *not unique* because we are free to choose x or y or z as the "direction of propagation"; that is, there is no unique "longitudinal direction." For example, a TE mode with respect to the z-axis could be a TM mode with respect to the y-axis.

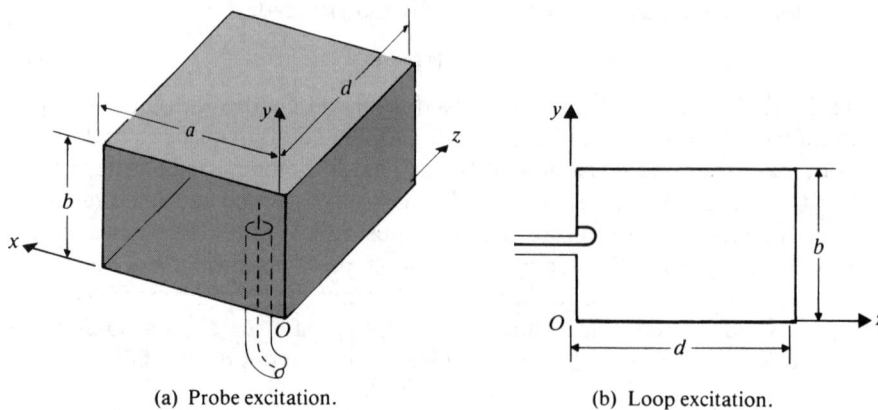

(a) Probe excitation.   (b) Loop excitation.

FIGURE 10–24
Excitation of cavity modes by a coaxial line.

For our purposes *we choose the z-axis as the reference* "direction of propagation." In actuality the existence of conducting end walls at $z = 0$ and $z = d$ gives rise to multiple reflections and sets up standing waves; no wave propagates in an enclosed cavity. A three-symbol (*mnp*) subscript is needed to designate a TM or TE standing wave pattern in a cavity resonator.

TM$_{mnp}$ Modes   The expressions for the transverse variations of the field components for TM$_{mn}$ modes in a waveguide have been given in Eqs. (10–132) and (10–134) through (10–137). Note that the longitudinal variation for a wave traveling in the $+z$-direction is described by the factor $e^{-\gamma z}$ or $e^{-j\beta z}$, as indicated in Eq. (10–121). This wave will be reflected by the end wall at $z = d$; and the reflected wave, going in the $-z$-direction, is described by a factor $e^{j\beta z}$. The superposition of a term with $e^{-j\beta z}$ and another of the same amplitude[†] with $e^{j\beta z}$ results in a standing wave of the sin βz or cos βz type. Which should it be? The answer to this question depends on the particular field component.

Consider the transverse component $E_y(x, y, z)$. Boundary conditions at the conducting surfaces require that it be zero at $z = 0$ and $z = d$. This means that (1) its z-dependence must be of the sin βz type, and that (2) $\beta = p\pi/d$. The same argument applies to the other transverse electric field component $E_x(x, y, z)$.

Recalling that the appearance of the factor $(-\gamma)$ in Eqs. (10–134) and (10–135) is the result of a differentiation with respect to z, we conclude that the other components $E_z(x, y, z)$, $H_x(x, y, z)$, and $H_y(x, y, z)$, which do not contain the factor $(-\gamma)$, must vary according to cos βz. We have then, from Eqs. (10–132) and (10–134) through (10–137), the following *phasors* of the field components for TM$_{mnp}$ modes in a rectangular cavity resonator:

$$E_z(x, y, z) = E_0 \sin\left(\frac{m\pi}{a} x\right) \sin\left(\frac{n\pi}{b} y\right) \cos\left(\frac{p\pi}{d} z\right), \tag{10-295}$$

$$E_x(x, y, z) = -\frac{1}{h^2}\left(\frac{m\pi}{a}\right)\left(\frac{p\pi}{d}\right) E_0 \cos\left(\frac{m\pi}{a} x\right) \sin\left(\frac{n\pi}{b} y\right) \sin\left(\frac{p\pi}{d} z\right), \tag{10-296}$$

$$E_y(x, y, z) = -\frac{1}{h^2}\left(\frac{n\pi}{b}\right)\left(\frac{p\pi}{d}\right) E_0 \sin\left(\frac{m\pi}{a} x\right) \cos\left(\frac{n\pi}{b} y\right) \sin\left(\frac{p\pi}{d} z\right), \tag{10-297}$$

$$H_x(x, y, z) = \frac{j\omega\epsilon}{h^2}\left(\frac{n\pi}{b}\right) E_0 \sin\left(\frac{m\pi}{a} x\right) \cos\left(\frac{n\pi}{b} y\right) \cos\left(\frac{p\pi}{d} z\right), \tag{10-298}$$

$$H_y(x, y, z) = -\frac{j\omega\epsilon}{h^2}\left(\frac{m\pi}{a}\right) E_0 \cos\left(\frac{m\pi}{a} x\right) \sin\left(\frac{n\pi}{b} y\right) \cos\left(\frac{p\pi}{d} z\right), \tag{10-299}$$

where

$$h^2 = \left(\frac{m\pi}{a}\right)^2 + \left(\frac{n\pi}{b}\right)^2. \tag{10-300}$$

[†] The reflection coefficient at a perfect conductor is -1.

It is clear that the integers m, n, and p denote the number of half-wave variations in the x-, y-, and z-direction, respectively.

From Eq. (10–138) we obtain the following expression for the resonant frequency:

$$\omega_{mnp} = \frac{1}{\sqrt{\mu\epsilon}} \sqrt{\left(\frac{m\pi}{a}\right)^2 + \left(\frac{n\pi}{b}\right)^2 + \left(\frac{p\pi}{d}\right)^2}$$

or

$$f_{mnp} = \frac{u}{2} \sqrt{\left(\frac{m}{a}\right)^2 + \left(\frac{n}{b}\right)^2 + \left(\frac{p}{d}\right)^2} \quad \text{(Hz)}. \qquad (10\text{--}301)$$

Equation (10–301) states the obvious fact that the resonant frequency increases as the order of a mode becomes higher.

TE_{mnp} Modes   For TE_{mnp} modes ($E_z = 0$) the phasor expressions for the standing-wave field components can be written from Eqs. (10–158) and (10–159) through (10–162). We follow the same rules as those we used for TM_{mnp} modes; namely, (1) the transverse (tangential) electric field components must vanish at $z = 0$ and $z = d$, and (2) the factor γ indicates a negative partial differentiation with respect to z. The first rule requires a $\sin (p\pi z/d)$ factor in $E_x(x, y, z)$ and $E_y(x, y, z)$, as well as in $H_z(x, y, z)$; and the second rule indicates a $\cos (p\pi z/d)$ factor in $H_x(x, y, z)$ and $H_y(x, y, z)$, and the replacement of γ by $-(p\pi/d)$. Thus,

$$H_z(x, y, z) = H_0 \cos\left(\frac{m\pi}{a} x\right) \cos\left(\frac{n\pi}{b} y\right) \sin\left(\frac{p\pi}{d} z\right), \qquad (10\text{--}302)$$

$$E_x(x, y, z) = \frac{j\omega\mu}{h^2}\left(\frac{n\pi}{b}\right) H_0 \cos\left(\frac{m\pi}{a} x\right) \sin\left(\frac{n\pi}{b} y\right) \sin\left(\frac{p\pi}{d} z\right), \qquad (10\text{--}303)$$

$$E_y(x, y, z) = -\frac{j\omega\mu}{h^2}\left(\frac{m\pi}{a}\right) H_0 \sin\left(\frac{m\pi}{a} x\right) \cos\left(\frac{n\pi}{b} y\right) \sin\left(\frac{p\pi}{d} z\right), \qquad (10\text{--}304)$$

$$H_x(x, y, z) = -\frac{1}{h^2}\left(\frac{m\pi}{a}\right)\left(\frac{p\pi}{d}\right) H_0 \sin\left(\frac{m\pi}{a} x\right) \cos\left(\frac{n\pi}{b} y\right) \cos\left(\frac{p\pi}{d} z\right), \qquad (10\text{--}305)$$

$$H_y(x, y, z) = -\frac{1}{h^2}\left(\frac{n\pi}{b}\right)\left(\frac{p\pi}{d}\right) H_0 \cos\left(\frac{m\pi}{a} x\right) \sin\left(\frac{n\pi}{b} y\right) \cos\left(\frac{p\pi}{d} z\right), \qquad (10\text{--}306)$$

The value of h^2 has been given in Eq. (10–300). The expression for resonant frequency, f_{mnp}, remains the same as that obtained for TM_{mnp} modes in Eq. (10–301). Different modes having the same resonant frequency are called **degenerate modes**. Thus TM_{mnp} and TE_{mnp} modes are always degenerate if none of the mode indices is zero. The mode with the lowest resonant frequency for a given cavity size is referred to as the **dominant mode** (see Example 10–15).

Examination of the field expressions, Eqs. (10–295) through (10–299), for TM modes in a cavity reveals that the longitudinal and transverse electric field components are in time phase with one another and in time quadrature with the magnetic field components. Hence the time-average Poynting vector and time-average power transmitted in any direction are zero, as they should be in a lossless cavity. This is in contrast to the field expressions Eqs. (10–132) and (10–134) through (10–137) for TM modes in a waveguide, where the transverse electric field components are in time phase with the transverse magnetic field components, resulting in a time-average power flow in the direction of wave propagation. The same contrasting phase relationships between electric and magnetic field components for TE modes in a cavity resonator (Eqs. (10–302) through (10–306)) and those in a waveguide (Eqs. 10–158 through (10–162)) are also in evidence.

A particular mode in a cavity resonator (or a waveguide) may be excited from a coaxial line by means of a small probe or loop antenna. In Fig. 10–24(a) a probe is shown that is the tip of the inner conductor of a coaxial cable and protrudes into a cavity at a location where the electric field is a maximum for the desired mode. The probe is, in fact, an antenna that couples electromagnetic energy into the resonator. Alternatively, a cavity resonator may be excited through the introduction of a small loop at a place where the magnetic flux of the desired mode linking the loop is a maximum. Figure 10–24(b) illustrates such an arrangement. Of course, the source frequency from the coaxial line must be the same as the resonant frequency of the desired mode in the cavity.

As an example, for the TE_{101} mode in an $a \times b \times d$ rectangular cavity, there are only three nonzero field components:

$$E_y = -\frac{j\omega\mu a}{\pi} H_0 \sin\left(\frac{\pi}{a}x\right)\sin\left(\frac{\pi}{d}z\right), \tag{10–307}$$

$$H_x = -\frac{a}{d} H_0 \sin\left(\frac{\pi}{a}x\right)\cos\left(\frac{\pi}{d}z\right), \tag{10–308}$$

$$H_z = H_0 \cos\left(\frac{\pi}{a}x\right)\sin\left(\frac{\pi}{d}z\right). \tag{10–309}$$

This mode may be excited by a probe inserted in the center region of the top or bottom face where E_y is maximum, as shown in Fig. 10–24(a), or by a loop to couple a maximum H_x placed inside the front or back face, as shown in Fig. 10–24(b). The best location of a probe or a loop is affected by the impedance-matching requirements of the microwave circuit of which the resonator is a part.

A commonly used method for coupling energy from a waveguide to a cavity resonator is the introduction of a hole or iris at an appropriate location in the cavity wall. The field in the waveguide at the hole must have a component that is favorable in exciting the desired mode in the resonator.

■■■■ **EXAMPLE 10–15** Determine the dominant modes and their frequencies in an air-filled rectangular cavity resonator for (a) $a > b > d$, (b) $a > d > b$, and (c) $a = b = d$, where a, b, and d are the dimensions in the x-, y-, and z-directions, respectively.

Solution   With the z-axis chosen as the reference "direction of propagation": First, for TM$_{mnp}$ modes, Eqs. (10–295) through (10–299) show that neither m nor n can be zero, but that p can be zero; second, for TE$_{mnp}$ modes, Eqs. (10–302) through (10–306) show that either m or n (but not both m and n) can be zero, but that p cannot be zero. Thus the modes of the lowest orders are

$$\text{TM}_{110}, \qquad \text{TE}_{011}, \qquad \text{and TE}_{101}.$$

The resonant frequency for both TM and TE modes is given by Eq. (10–301).

a) For $a > b > d$: The lowest resonant frequency is

$$f_{110} = \frac{c}{2}\sqrt{\frac{1}{a^2} + \frac{1}{b^2}}, \tag{10–310}$$

where c is the velocity of light in free space. Therefore TM$_{110}$ is the dominant mode.

b) For $a > d > b$: The lowest resonant frequency is

$$f_{101} = \frac{c}{2}\sqrt{\frac{1}{a^2} + \frac{1}{d^2}}, \tag{10–311}$$

and TE$_{101}$ is the dominant mode.

c) For $a = b = d$, all three of the lowest-order modes (namely, TM$_{110}$, TE$_{011}$, and TE$_{101}$) have the same field patterns. The resonant frequency of these degenerate modes is

$$f_{110} = \frac{c}{\sqrt{2}\,a}. \tag{10–312}$$

▬

10–7.2   QUALITY FACTOR OF CAVITY RESONATOR

A cavity resonator stores energy in the electric and magnetic fields for any particular mode pattern. In any practical cavity the walls have a finite conductivity; that is, a nonzero surface resistance, and the resulting power loss causes a decay of the stored energy. The **quality factor**, or Q, of a resonator, like that of any resonant circuit, is a measure of the bandwidth of the resonator and is defined as

$$\boxed{Q = 2\pi\, \frac{\text{Time-average energy stored at a resonant frequency}}{\text{Energy dissipated in one period of this frequency}}.} \tag{10–313}$$

(Dimensionless)

Let W be the total time-average energy in a cavity resonator. We write

$$W = W_e + W_m, \tag{10–314}$$

where W_e and W_m denote the energies stored in the electric and magnetic fields, respectively. If P_L is the time-average power dissipated in the cavity, then the energy dissipated in one period is P_L divided by frequency, and Eq. (10–313) can be written as

$$Q = \frac{\omega W}{P_L} \quad \text{(Dimensionless).} \tag{10–315}$$

In determining the Q of a cavity at a resonant frequency, it is customary to assume that the loss is small enough to allow the use of the field patterns without loss.

We will now find the Q of an $a \times b \times d$ cavity for the TE_{101} mode that has three nonzero field components given in Eqs. (10–307), (10–308), and (10–309). The time-average stored electric energy is

$$W_e = \frac{\epsilon_0}{4} \int |E_y|^2 \, dv$$

$$= \frac{\epsilon_0 \omega^2 \mu_0^2 \pi^2}{4h^4 a^2} H_0^2 \int_0^d \int_0^b \int_0^a \sin^2 \left(\frac{\pi}{a} x \right) \sin^2 \left(\frac{\pi}{d} z \right) dx \, dy \, dz \tag{10–316}$$

$$= \frac{\epsilon_0 \omega_{101}^2 \mu_0^2 a^2}{4\pi^2} H_0^2 \left(\frac{a}{2} \right) b \left(\frac{d}{2} \right) = \frac{1}{4} \epsilon_0 \mu_0^2 a^3 b d f_{101}^2 H_0^2,$$

where we have used $h^2 = (\pi/a)^2$ from Eq. (10–300). The total time-average stored magnetic energy is

$$W_m = \frac{\mu_0}{4} \int \{|H_x|^2 + |H_z|^2\} \, dv$$

$$= \frac{\mu_0}{4} H_0^2 \int_0^d \int_0^b \int_0^a \left\{ \frac{\pi^4}{h^4 a^2 d^2} \sin^2 \left(\frac{\pi}{a} x \right) \cos^2 \left(\frac{\pi}{d} z \right) \right.$$

$$\left. + \cos^2 \left(\frac{\pi}{a} x \right) \sin^2 \left(\frac{\pi}{d} z \right) \right\} dx \, dy \, dz \tag{10–317}$$

$$= \frac{\mu_0}{4} H_0^2 \left\{ \frac{a^2}{d^2} \left(\frac{a}{2} \right) b \left(\frac{d}{2} \right) + \left(\frac{a}{2} \right) b \left(\frac{d}{2} \right) \right\} = \frac{\mu_0}{16} abd \left(\frac{a^2}{d^2} + 1 \right) H_0^2.$$

From Eq. (10–311) the resonant frequency for the TE_{101} mode is

$$f_{101} = \frac{1}{2\sqrt{\mu_0 \epsilon_0}} \sqrt{\frac{1}{a^2} + \frac{1}{d^2}}. \tag{10–318}$$

Substitution of f_{101} from Eq. (10–318) in Eq. (10–316) proves that, *at the resonant frequency*, $W_e = W_m$. Thus,

$$W = 2W_e = 2W_m = \frac{\mu_0 H_0^2}{8} abd \left(\frac{a^2}{d^2} + 1 \right). \tag{10–319}$$

To find P_L, we note that the power loss per unit area is

$$\mathscr{P}_{av} = \tfrac{1}{2} |J_s|^2 R_s = \tfrac{1}{2} |H_t|^2 R_s, \tag{10–320}$$

where $|H_t|$ denotes the magnitude of the tangential component of the magnetic field at the cavity walls. The power loss in the $z = d$ (back) wall is the same as that in the $z = 0$ (front) wall. Similarly, the power loss in the $x = a$ (left) wall is the same as that in the $x = 0$ (right) wall; and the power loss in the $y = b$ (upper) wall is the same as that in the $y = 0$ (lower) wall. We have

$$P_L = \oint \mathscr{P}_{av}\,ds = R_s \left\{ \int_0^b \int_0^a |H_x(z = 0)|^2\,dx\,dy + \int_0^d \int_0^b |H_z(x = 0)|^2\,dy\,dz \right.$$

$$\left. + \int_0^d \int_0^a |H_x|^2\,dx\,dz + \int_0^d \int_0^a |H_z|^2\,dx\,dz \right\} \tag{10-321}$$

$$= \frac{R_s H_0^2 a}{2} \left\{ \frac{a^2}{d} \left(\frac{b}{d} + \frac{1}{2} \right) + d\left(\frac{b}{a} + \frac{1}{2} \right) \right\}.$$

Using Eqs. (10–319) and (10–321) in Eq. (10–315), we obtain

$$Q_{101} = \frac{\pi f_{101}\mu_0 abd(a^2 + d^2)}{R_s[2b(a^3 + d^3) + ad(a^2 + d^2)]} \qquad (\text{TE}_{101} \text{ mode}), \tag{10-322}$$

where f_{101} has been given in Eq. (10–318).

EXAMPLE 10–16   (a) What should be the size of a hollow cubic cavity made of copper in order for it to have a dominant resonant frequency of 10 (GHz)? (b) Find the Q at that frequency.

Solution

a) For a cubic cavity, $a = b = d$: From Example 10–15 we know that TM_{110}, TE_{011}, and TE_{101} are degenerate dominant modes having the same field patterns and that

$$f_{101} = \frac{3 \times 10^8}{\sqrt{2}a} = 10^{10} \quad (\text{Hz}).$$

Therefore,

$$a = \frac{3 \times 10^8}{\sqrt{2} \times 10^{10}} = 2.12 \times 10^{-2} \quad (\text{m})$$

$$= 21.2 \quad (\text{mm}).$$

b) The expression of Q in Eq. (10–322) for a cubic cavity reduces to

$$Q_{101} = \frac{\pi f_{101}\mu_0 a}{3R_s} = \frac{a}{3}\sqrt{\pi f_{101}\mu_0 \sigma}. \tag{10-323}$$

For copper, $\sigma = 5.80 \times 10^7$ (S/m), we have

$$Q_{101} = \left(\frac{2.12}{3} \times 10^{-2} \right)\sqrt{\pi 10^{10}(4\pi 10^{-7})(5.80 \times 10^7)} = 10{,}700.$$

The Q of a cavity resonator is thus extremely high in comparison with that obtainable from lumped L–C resonant circuits. In practice, the preceding value is somewhat lower owing to losses through feed connections and surface irregularities.

10–7.3   CIRCULAR CAVITY RESONATOR

In a manner similar to the construction of a rectangular cavity resonator from a rectangular waveguide, a circular cylindrical resonator can be formed by placing conducting walls at both ends of a cylindrical waveguide. For simplicity, let us consider the TM_{01} mode in a circular waveguide of radius a at cutoff so that there is no variation in the z-direction. The ends of the waveguide are shorted by conducting plates at a distance d ($<2a$) apart, forming a circular cylindrical cavity. The field components inside the cavity are, from Eqs. (10–218) and (10–224) by setting $n = 0$ and recalling Eq. (10–227),

$$E_z = C_0 J_0(hr) = C_0 J_0\left(\frac{2.405}{a}r\right), \qquad (10\text{–}324)$$

$$H_\phi = -\frac{jC_0}{\eta_0}J_0'(hr) = \frac{jC_0}{\eta_0}J_1\left(\frac{2.405}{a}r\right), \qquad (10\text{–}325)$$

where the relation $J_0'(hr) = -J_1(hr)$ has been used. The electric and magnetic field patterns for the TM_{010} mode in the circular cavity in both transverse and longitudinal sections are sketched in Fig. 10–25. Note from Eqs. (10–324) and (10–325) again that

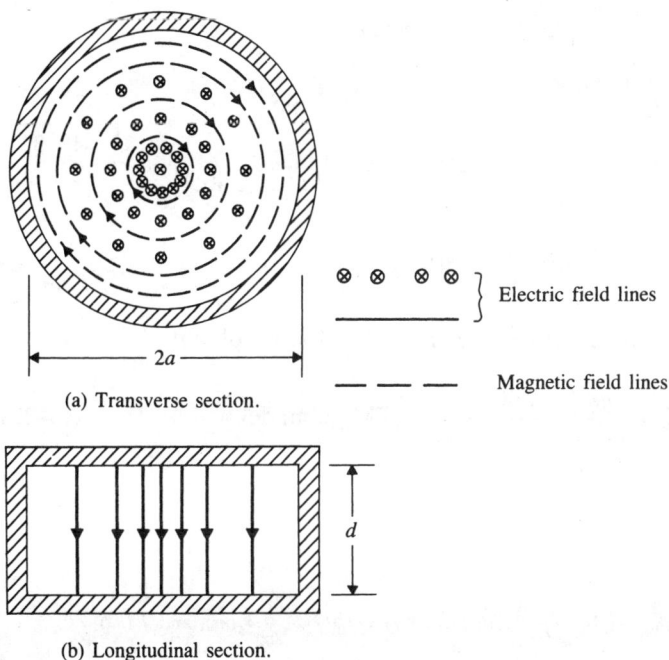

(a) Transverse section.

⊗ ⊗   ⊗ ⊗ } Electric field lines
_____

— — —   Magnetic field lines

(b) Longitudinal section.

FIGURE 10–25
TM_{010} field patterns in a circular cylindrical cavity resonator.

the electric and magnetic fields are in time quadrature, resulting in no power loss in the cavity walls.

In actuality the cavity walls do have a finite conductivity and a nonzero surface resistance. There will be power loss in the walls, and the cavity Q is not infinite. To calculate the cavity Q, we apply Eq. (10–315) and follow the same procedure as that used in the preceding subsection for the rectangular resonator. We will assume that the field intensities inside a low-loss cavity remain approximately the same as those for a lossless cavity.

Let us find the Q of a circular cylindrical cavity of radius a and length d for the TM_{010} mode. The field components have been given in Eqs. (10–324) and (10–325). The time-average stored energy is

$$
\begin{aligned}
W = 2W_e &= \frac{\epsilon_0}{2} \int_V |E_z|^2 \, dv \\
&= \frac{\epsilon_0 C_0^2}{2} (2\pi d) \int_0^a J_0^2\left(\frac{2.405}{a} r\right) r \, dr \qquad (10\text{–}326) \\
&= (\pi\epsilon_0 d) C_0^2 \left[\frac{a^2}{2} J_1^2(2.405)\right].
\end{aligned}
$$

The average power loss per unit area is given by Eq. (10–320). Here $H_t = H_\phi$, and there are radial surface currents $\mathbf{J}_r$ on the flat end faces and uniform longitudinal surface currents $\mathbf{J}_z$ on the inside of the cylindrical walls. We have

$$
\begin{aligned}
P_L &= \frac{R_s}{2} \left\{ 2 \int_0^a |J_r|^2 2\pi r \, dr + (2\pi a d)|J_z|^2 \right\} \\
&= \pi R_s \left\{ 2 \int_0^a |H_\phi|^2 r \, dr + (ad)|H_\phi(r=a)|^2 \right\} \\
&= \frac{\pi R_s C_0^2}{\eta_0^2} \left\{ 2 \int_0^a r J_1^2\left(\frac{2.405}{a} r\right) dr + (ad)J_1^2(2.405) \right\} \qquad (10\text{–}327)^\dagger \\
&= \frac{\pi a R_s C_0^2}{\eta_0^2} (a + d)J_1^2(2.405).
\end{aligned}
$$

Substituting Eqs. (10–326) and (10–327) in Eq. (10–315), we obtain

$$
Q = \left(\frac{\eta_0}{R_s}\right) \frac{2.405}{2(1 + a/d)} \qquad (TM_{010} \text{ mode}), \qquad (10\text{–}328)
$$

---

† The following relations have been used:

$$
\int J_n^2(hr) r \, dr = \frac{r^2}{2}\left[J_n'^2(hr) + \left(1 - \frac{n^2}{h^2 r^2}\right) J_n^2(hr)\right], \quad J_1'(hr) = J_0(hr) - \frac{1}{hr} J_1(hr), \text{ and } J_0(ha) = 0.
$$

where $R_s = \sqrt{\pi f \mu_0 / \sigma}$ is to be calculated at the resonant frequency for the TM_{010} mode, which is, from Eqs. (10–227) and (10–228),

$$(f)_{TM_{010}} = \frac{2.405}{2\pi a \sqrt{\mu_0 \epsilon_0}} = \frac{0.115}{a} \quad \text{(GHz)}. \tag{10-329}$$

EXAMPLE 10–17 A hollow circular cylindrical cavity resonator is to be constructed of copper such that its length d equals its diameter $2a$. (a) Determine a and d for a resonant frequency of 10 (GHz) at the TM_{010} mode. (b) Find the Q of the cavity at resonance.

Solution

a) From Eq. (10–329) we have

$$\frac{0.115}{a} = 10,$$

or

$$a = 1.15 \times 10^{-2} \text{ (m)} = 1.15 \quad \text{(cm)}.$$

Thus,

$$d = 2a = 2.30 \quad \text{(cm)}.$$

b)

$$R_s = \sqrt{\frac{\pi f \mu_0}{\sigma}}$$

$$= \sqrt{\frac{\pi \times 10^{10} \times (4\pi 10^{-7})}{5.80 \times 10^7}} = 2.61 \times 10^{-2} \quad (\Omega).$$

From Eq. (10–328) we obtain

$$Q = \left(\frac{377}{2.61 \times 10^{-2}}\right) \frac{2.405}{2(1 + 1/2)} = 11,580.$$

It is interesting to compare the results of this example with those obtained in Example 10–16 for a rectangular cavity resonator of a comparable size that resonates at the same frequency.

	Circular Cavity	Rectangular Cavity
Resonant mode at frequency	TM_{010} 10 (GHz)	TE_{101} 10 (GHz)
Dimensions	Diameter $2a = 2.30$ (cm) Length $d = 2.30$ (cm)	$a = b = d = 2.12$ (cm)
Volume	$\pi a^2 d = 9.56$ (cm³)	$a \times b \times d = 9.53$ (cm³)
Total area	$2(\pi a^2) + (2\pi a d) = 24.93$ (cm²)	$6a^2 = 26.97$ (cm²)
Q	11,580	10,700

We see that these two cavities have approximately the same volume, but the total surface area of the rectangular cavity is about 8.2% larger. The larger surface area leads to a higher power loss and a lower Q. The Q of the circular cavity is approximately 8.2% higher.   ■

Review Questions

R.10–1 Why are the common types of transmission lines not useful for the long-distance signal transmission at microwave frequencies in the TEM mode?

R.10–2 What is meant by a *cutoff frequency* of a waveguide?

R.10–3 Why are lumped-parameter elements connected by wires not useful as resonant circuits at microwave frequencies?

R.10–4 What is the governing equation for electric and magnetic field intensity phasors in the dielectric region of a straight waveguide with a uniform cross section?

R.10–5 What are the three basic types of propagating waves in a uniform waveguide?

R.10–6 Define *wave impedance*.

R.10–7 Explain why single-conductor hollow or dielectric-filled waveguides cannot support TEM waves.

R.10–8 Discuss the analytical procedure for studying the characteristics of TM waves in a waveguide.

R.10–9 Discuss the analytical procedure for studying the characteristics of TE waves in a waveguide.

R.10–10 What are *eigenvalues* of a boundary-value problem?

R.10–11 Can a waveguide have more than one cutoff frequency? On what factors does the cutoff frequency of a waveguide depend.

R.10–12 What is an *evanescent mode*?

R.10–13 Is the guide wavelength of a propagating wave in a waveguide longer or shorter than the wavelength in the corresponding unbounded dielectric medium?

R.10–14 In what way does the wave impedance in a waveguide depend on frequency:
  a) For a propagating TEM wave?
  b) For a propagating TM wave?
  c) For a propagating TE wave?

R.10–15 What is the significance of a purely reactive wave impedance?

R.10–16 Can one tell from an ω–β diagram whether a certain propagating mode in a waveguide is dispersive? Explain.

R.10–17 Explain how one determines the phase velocity and the group velocity of a propagating mode from its ω–β diagram.

R.10–18 What is meant by an *eigenmode*?

R.10–19 On what factors does the cutoff frequency of a parallel-plate waveguide depend?

R.10–20 What is meant by the *dominant mode* of a waveguide? What is the dominant mode of a parallel-plate waveguide?

R.10–21 Can a TM or TE wave with a wavelength 3 (cm) propagate in a parallel-plate waveguide whose plate separation is 1 (cm)? 2 (cm)? Explain.

R.10–22 Compare the cutoff frequencies of TM_0, TM_n, TM_m $(m > n)$, and TE_n modes in a parallel-plate waveguide.

R.10–23 Define *energy-transport velocity*.

R.10–24 Does the attenuation constant due to dielectric losses increase or decrease with frequency for TM and TE modes in a parallel-plate waveguide?

R.10–25 Discuss the essential differences in the frequency behavior of the attenuation caused by finite plate conductivity in a parallel-plate waveguide for TEM, TM, and TE modes.

R.10–26 State the boundary conditions to be satisfied by E_z for TM waves in a rectangular waveguide.

R.10–27 Which TM mode has the lowest cutoff frequency of all the TM modes in a rectangular waveguide?

R.10–28 State the boundary conditions to be satisfied by H_z for TE waves in a rectangular waveguide.

R.10–29 Which mode is the dominant mode in a rectangular waveguide if (a) $a > b$, (b) $a < b$, and (c) $a = b$?

R.10–30 What is the cutoff wavelength of the TE_{10} mode in a rectangular waveguide?

R.10–31 Which are the nonzero field components for the TE_{10} mode in a rectangular waveguide?

R.10–32 Discuss the frequency-dependence of the attenuation constant caused by losses in the dielectric medium in a waveguide.

R.10–33 Discuss the general attenuation behavior caused by wall losses as a function of frequency for the TE_{10} mode in a rectangular waveguide.

R.10–34 Discuss the general attenuation behavior caused by wall losses as a function of frequency for the TM_{11} mode in a rectangular waveguide.

R.10–35 Discuss the factors that affect the choice of the linear dimensions a and b for the cross section of a rectangular waveguide.

R.10–36 What type of conducting diaphragm with an iris in a waveguide can provide a shunt capacitive susceptance? A shunt inductive susceptance? Explain.

R.10–37 Under what circumstances does a Bessel's differential equation arise?

R.10–38 Describe some general properties of Bessel functions of the first kind.

R.10–39 Why are Bessel functions of the second kind not useful in the analysis of wave propagation in a hollow circular waveguide?

R.10–40 Which mode is the dominant mode in a circular waveguide?

R.10–41 It is claimed that the TE_{11} wave of a given frequency will propagate in a circular cylindrical pipe having a diameter only 76.5% of that required to support a TM_{01} wave of the same frequency. Explain.

R.10–42 What is the distinctive characteristic of the attenuation constant of TE_{0n} modes in a circular waveguide?

R.10–43 Why is it necessary that the permittivity of the dielectric slab in a dielectric waveguide be larger than that of the surrounding medium?

R.10–44 What are dispersion relations?

R.10–45 Can a dielectric-slab waveguide support an infinite number of discrete TM and TE modes? Explain.

R.10–46 What kind of surface can support a TM surface wave? A TE surface wave?

R.10–47 What is the dominant mode in a dielectric-slab waveguide? What is its cutoff frequency?

R.10–48 Does the attenuation of the waves outside a dielectric slab waveguide increase or decrease with slab thickness?

R.10–49 How does the time-average power transmitted in the transverse direction of a dielectric waveguide depend on the propagating mode in the guide?

R.10–50 What kinds of Bessel functions are appropriate in the analysis of wave behavior in and around optical fibers? Explain.

R.10–51 What are cavity resonators? What are their most desirable properties?

R.10–52 Are the field patterns in a cavity resonator traveling waves or standing waves? How do they differ from those in a waveguide?

R.10–53 In terms of field patterns, what does the TM_{110} mode signify? The TE_{123} mode?

R.10–54 What is the expression for the resonant frequency of TM_{mnp} modes in a rectangular cavity resonator of dimensions $a \times b \times d$? Of TE_{mnp} modes?

R.10–55 What is meant by *degenerate modes*?

R.10–56 What are the modes of the lowest orders in a rectangular cavity resonator?

R.10–57 Define the quality factor, Q, of a resonator.

R.10–58 What fundamental assumption is made in the derivation of the formulas for the Q of cavity resonators?

R.10–59 What field components exist in a circular cylindrical cavity operating in the TM_{010} mode?

R.10–60 Will the Q of a circular cylindrical cavity resonator be higher or lower by increasing its length? Explain by physical reasoning.

R.10–61 Explain why the measured Q of a cavity resonator is lower than the calculated value.

Problems

P.10–1 In studying the wave behavior in a straight waveguide having a uniform but arbitrary cross section it is expedient to find general formulas expressing the transverse field components in terms of their longitudinal components. We write

$$\mathbf{E} = \mathbf{E}_T + \mathbf{a}_z E_z,$$

$$\mathbf{H} = \mathbf{H}_T + \mathbf{a}_z H_z;$$

$$\mathbf{V} = \mathbf{V}_T + \mathbf{a}_z \frac{\partial}{\partial z},$$

where the subscript T denotes "transverse." Prove the following relations for time-harmonic excitation:

a) $\mathbf{E}_T = -\dfrac{1}{h^2}(\gamma \nabla_T E_z - \mathbf{a}_z j\omega\mu \times \nabla_T H_z)$ (10–330)

b) $\mathbf{H}_T = -\dfrac{1}{h^2}(\gamma \nabla_T H_z + \mathbf{a}_z j\omega\epsilon \times \nabla_T E_z),$ (10–331)

where h^2 is that given in Eq. (10–15).

P.10–2 For rectangular waveguides, use appropriate relations in Section 10–2 to:
  a) plot the universal circle diagrams relating u_g/u and β/k versus f_c/f,
  b) plot the universal graphs of u/u_p, β/k, and λ_g/λ versus f/f_c,
  c) find u_p/u, u_g/u, β/k, and λ_g/λ at $f = 1.25f_c$.

P.10–3 Sketch the ω–β diagrams of a parallel-plate waveguide separated by a dielectric slab of thickness b and constitutive parameters (ϵ, μ) for TM_1, TM_2, and TM_3 modes. Discuss
  a) how b and the constitutive parameters affect the diagrams,
  b) whether the same curves apply to TE modes.

P.10–4 Obtain the expressions for the surface charge density and the surface current density for TM_n modes on the conducting plates of a parallel-plate waveguide. Do the currents on the two plates flow in the same direction or in opposite directions?

P.10–5 Obtain the expressions for the surface current density for TE_n modes on the conducting plates of a parallel-plate waveguide. Do the currents on the two plates flow in the same direction or in opposite directions?

P.10–6 Sketch the electric and magnetic field lines for (a) the TM_2 mode and (b) the TE_2 mode in a parallel-plate waveguide.

P.10–7 Determine the energy-transport velocity of the TE_n mode in a lossless parallel-plate waveguide in terms of its cutoff frequency.

P.10–8 A waveguide is formed by two parallel copper sheets—$\sigma_c = 5.80 \times 10^7$ (S/m)—separated by a 5 (cm) thick lossy dielectric—$\epsilon_r = 2.25$, $\mu_r = 1$, $\sigma = 10^{-10}$ (S/m). For an operating frequency of 10 (GHz), find β, α_d, α_c, u_p, u_g, and λ_g for (a) the TEM mode, (b) the TM_1 mode, and (c) the TM_2 mode.

P.10–9 Repeat Problem P.10–8 for (a) the TE_1 mode and (b) the TE_2 mode.

P.10–10 For a parallel-plate waveguide,
  a) find the frequency (in terms of the cutoff frequency f_c) at which the attenuation constant due to conductor losses for the TM_n mode is a minimum,
  b) obtain the formula for this minimum attenuation constant,
  c) calculate this minimum α_c for the TM_1 mode if the parallel plates are made of copper and spaced 5 (cm) apart in air.

P.10–11 A parallel-plate waveguide made of two perfectly conducting infinite planes spaced 3 (cm) apart in air operates at a frequency 10 (GHz). Find the maximum time-average power that can be propagated per unit width of the guide without a voltage breakdown for
  a) the TEM mode,   **b)** the TM_1 mode,   **c)** the TE_1 mode.

P.10–12 Without deriving any new equations, roughly sketch the electric and magnetic field lines in a typical xy-plane of a rectangular waveguide for
  a) TM_{21} mode by an extension of Fig. 10–11(a).
  b) TE_{11} mode by an extension of Fig. 10–12(a).
The densities of the field lines should show the proper sine or cosine variations.

P.10–13 For an $a \times b$ rectangular waveguide operating at the TM_{11} mode,
  a) derive the expressions for the surface current densities on the conducting walls,
  b) sketch the surface currents on the walls at $x = 0$ and at $y = b$.

P.10–14 A standard air-filled S-band rectangular waveguide has dimensions $a = 7.21$ (cm) and $b = 3.40$ (cm). What mode types can be used to transmit electromagnetic waves having the following wavelengths?
  a) $\lambda = 10$ (cm),   **b)** $\lambda = 5$ (cm).

P.10–15 Determine the energy-transport velocity of the TE_{10} mode in a lossless $a \times b$ rectangular waveguide in terms of its cutoff frequency.

P.10–16 Calculate and list in ascending order the cutoff frequencies (in terms of the cutoff frequency of the dominant mode) of an $a \times b$ rectangular waveguide for the following modes: TE_{01}, TE_{10}, TE_{11}, TE_{02}, TE_{20}, TM_{11}, TM_{12}, and TM_{22} (a) if $a = 2b$ and (b) if $a = b$.

P.10–17 An air-filled $a \times b$ ($b < a < 2b$) rectangular waveguide is to be constructed to operate at 3 (GHz) in the dominant mode. We desire the operating frequency to be at least 20% higher than the cutoff frequency of the dominant mode and also at least 20% below the cutoff frequency of the next higher-order mode.
  a) Give a typical design for the dimensions a and b.
  b) Calculate for your design β, u_p, λ_g, and the wave impedance at the operating
       frequency.

P.10–18 Calculate and compare the values of β, u_p, u_g, λ_g, and $Z_{\text{TE}_{10}}$ for a 2.5 (cm) $\times$ 1.5 (cm) rectangular waveguide operating at 7.5 (GHz)
  a) if the waveguide is hollow,
  b) if the waveguide is filled with a dielectric medium characterized by $\epsilon_r = 2$, $\mu_r = 1$
       and $\sigma = 0$.

P.10–19 An air-filled rectangular waveguide made of copper and having transverse dimensions $a = 7.20$ (cm) and $b = 3.40$ (cm) operates at a frequency 3 (GHz) in the dominant mode. Find (a) f_c, (b) λ_g, (c) α_c, and (d) the distance over which the field intensities of the propagating wave will be attenuated by 50%.

P.10–20 An average power of 1 (kW) at 10 (GHz) is to be *delivered to* an antenna at the TE_{10} mode by an air-filled rectangular copper waveguide 1 (m) long and having sides $a = 2.25$ (cm) and $b = 1.00$ (cm). Find
  a) the attenuation constant due to conductor losses,
  b) the maximum values of the electric and magnetic field intensities within the
       waveguide,
  c) the maximum value of the surface current density on the conducting walls,
  d) the total amount of average power dissipated in the waveguide.

P.10–21 Find the maximum amount of 10 (GHz) average power that can be transmitted through an air-filled rectangular waveguide—$a = 2.25$ (cm), $b = 1.00$ (cm)—at the TE_{10} mode without a breakdown.

P.10–22 Determine the value of (f/f_c) at which the attenuation constant due to conductor losses in an $a \times b$ rectangular waveguide for the TE_{10} mode is a minimum. What is the minimum obtainable α_c in a 2 (cm) $\times$ 1 (cm) guide? At what frequency?

P.10–23 Derive Eq. (10–188), the formula for the attenuation constant due to conductor losses in an $a \times b$ rectangular waveguide for the TM_{11} mode. Determine the value of (f/f_c) at which this attenuation constant is a minimum.

P.10–24 Measurements at 10 (GHz) on an X-band air-filled rectangular waveguide ($a = 2.29$ cm, $b = 1.02$ cm) connected to an unknown load indicate a minimum electric field at 6 (cm) from the load and a standing-wave ratio (SWR) of 1.80. Find the location and the dimensions of a symmetrical capacitive iris required to bring the SWR to unity.

P.10–25 A solution of the Bessel's differential equation

$$\frac{d^2 R(r)}{dr^2} + \frac{1}{r}\frac{dR(r)}{dr} + R(r) = 0 \qquad (10–332)$$

can be obtained by assuming $R(r)$ to be a power series in r as in Eq. (10–202), substituting it in the equation, and equating the sum of the coefficients of each power of r to zero. Find the solution and verify that it is consistent with $J_0(r)$ given in Eq. (10–204).

P.10–26 Starting from Maxwell's curl equations in simple media, verify Eq. (10–219) for TM modes in a circular waveguide.

P.10–27 Without deriving any new equations, roughly sketch the electric and magnetic field lines in a typical transverse plane of a circular waveguide
   a) for TM_{11} mode by an extension of Fig. 10–20, and
   b) for TE_{01} mode.
   c) Determine the cutoff frequencies for TM_{11} and TE_{01} modes in an air-filled circular waveguide of radius a.

P.10–28 Sketch the ω–β diagrams for TE_{11} and TM_{01} modes in a hollow circular waveguide of radius a. Discuss how the diagrams will be affected
   a) if a is doubled,
   b) if the waveguide is filled with a nonmagnetic medium having a dielectric constant ϵ_r.

P.10–29 For a straight waveguide with a semicircular cross section shown in Fig. 10–26,
   a) write the appropriate expression of E_z^0 for TM modes,
   b) write the appropriate expression of H_z^0 for TE modes.
   c) Explain how the eigenvalues of the respective modes can be determined.

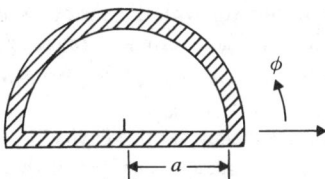

FIGURE 10–26
Cross section of a semicircular waveguide
(Problem P.10–29).

P.10–30 Show that electromagnetic waves propagate along a dielectric waveguide with a velocity between that of plane-wave propagation in the dielectric medium and that in the medium outside.

P.10–31 Find the solutions of Eq. (10–260) for k_y by plotting Eqs. (10–257) and (10–258) with $\alpha d/2$ versus $k_y d/2$ for $d = 1$ (cm) and $\epsilon_r = 3.25$ if (a) $f = 200$ (MHz), and (b) $f = 500$ (MHz). Determine β and α for the lowest-order odd TM modes at the two frequencies.

P.10–32 Repeat problem P.10–31 using Eq. (10–263). What can you conclude about the even TM modes?

P.10–33 For an infinite dielectric-slab waveguide of thickness d situated in air, obtain the instantaneous expressions of all the nonzero field components for even TM modes in the slab, as well as in the upper and lower free-space regions.

P.10–34 When the slab thickness of a dielectric-slab waveguide is very small in terms of the operating wavelength, the field intensities decay very slowly away from the slab surface, and the propagation constant is nearly equal to that of the surrounding medium.
   a) Show that if $k_y d \ll 1$, the following relations hold approximately for the dominant TE mode:

$$\beta \cong k_0,$$

$$\alpha \cong \frac{\mu_0 d}{2\mu_d} (k_d^2 - k_0^2),$$

   where $k_d = \omega\sqrt{\mu_d \epsilon_d}$ and $k_0 = \omega\sqrt{\mu_0 \epsilon_0}$.
   b) For a slab of thickness 5 (mm) and dielectric constant 3, estimate the distance from the slab surface at which the field intensities have decayed to 36.8% of their values at the surface for an operating frequency of 300 (MHz).

P.10–35 For an infinite dielectric-slab waveguide of thickness d situated in free space, obtain the instantaneous expressions of all the nonzero field components for even TE modes in the slab, as well as in the upper and lower free-space regions. Derive Eq. (10–281).

P.10–36 A waveguide consists of an infinite dielectric slab (ϵ_d, μ_d) of thickness d that is sitting on a perfect conductor.
   a) What are the propagating modes and what are their cutoff frequencies?
   b) Obtain the phasor expressions for the surface current and surface charge densities on the conducting base for the propagating modes.

P.10–37 A round dielectric-rod waveguide of radius a, permittivity ϵ_1, and permeability μ_1 is enveloped in a homogeneous medium characterized by permittivity ϵ_2 and permeability μ_2.
   a) Write the expressions of all the field amplitudes for circularly symmetrical TE modes.
   b) Obtain the characteristic equation for these modes.

P.10–38 Given an air-filled lossless rectangular cavity resonator with dimensions 8 (cm) $\times$ 6 (cm) $\times$ 5 (cm), find the first twelve lowest-order modes and their resonant frequencies.

P.10–39 An air-filled rectangular cavity with brass walls—ϵ_0, μ_0, $\sigma = 1.57 \times 10^7$ (S/m)— has the following dimensions: $a = 4$ (cm), $b = 3$ (cm), and $d = 5$ (cm).
   a) Determine the dominant mode and its resonant frequency for this cavity.
   b) Find the Q and the time-average stored electric and magnetic energies at the resonant frequency, assuming H_0 to be 0.1 (A/m).

P.10–40 If the rectangular cavity in Problem P.10–39 is filled with a lossless dielectric material having a dielectric constant 2.5, find
   a) the resonant frequency of the dominant mode,
   b) the Q,
   c) the time-average stored electric and magnetic energies at the resonant frequency, assuming H_0 to be 0.1 (A/m).

P.10–41 A rectangular cavity resonator of length d is constructed from an $a \times b$ rectangular waveguide. It is to be operated at the TE_{101} mode.
   a) For a fixed b, determine the relative magnitudes of a and d such that the cavity Q is maximized.
   b) Obtain an expression for Q as a function of a/b under the above conditions.

P.10–42 For an air-filled rectangular copper cavity resonator,

    a) calculate its Q for the TE_{101} mode if its dimensions are $a = d = 1.8b = 3.6$ (cm),

    b) determine how much b should be increased in order to make Q 20% higher.

P.10–43 Derive an expression for the Q of an air-filled $a \times b \times d$ rectangular resonator for the TM_{110} mode.

P.10–44 For an air-filled circular cylindrical cavity resonator of radius a and length d:

    a) Write the general expressions for the resonant frequencies and the corresponding wavelengths for TM_{mnp} and TE_{mnp} modes.

    b) For $d = a$, list the first seven modes that have the lowest resonant frequencies.

P.10–45 In some microwave applications, ring-shaped cavity resonators with a very narrow center part are used. A cross section of such a resonator is shown in Fig. 10–27, in which d is very small in comparison with the resonant wavelength. Assuming that this resonator can be represented approximately by a parallel combination of the capacitance of the narrow center part and the inductance of the rest of the structure, find

    a) the approximate resonant frequency,

    b) the approximate resonant wavelength.

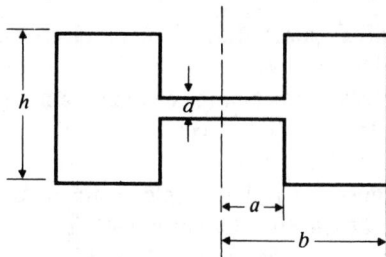

FIGURE 10–27
A ring-shaped resonator with a narrow center part (Problem P.10–45).

11

Antennas and
Radiating Systems

11-1 Introduction

In Chapter 8 we studied the propagation characteristics of plane electromagnetic waves in source-free media without considering how the waves were generated. Of course, the waves must originate from sources, which in electromagnetic terms are time-varying charges and currents. In order to radiate electromagnetic energy efficiently in prescribed directions, the charges and currents must be distributed in specific ways. *Antennas* are structures designed for radiating electromagnetic energy effectively in a prescribed manner. Without an efficient antenna, electromagnetic energy would be localized, and wireless transmission of information over long distances would be impossible.

An antenna may be a single straight wire or a conducting loop excited by a voltage source, an aperture at the end of a waveguide, or a complex array of these properly arranged radiating elements. Reflectors and lenses may be used to accentuate certain radiation characteristics. Among radiation characteristics of importance are field pattern, directivity, impedance, and bandwidth. These parameters will be examined when particular antenna types are studied in this chapter.

To study electromagnetic radiation, we must call upon our knowledge of Maxwell's equations and relate electric and magnetic fields to time-varying charge and current distributions. A primary difficulty of this task is that the charge and current distributions on antenna structures resulting from given excitations are generally unknown and very difficult to determine. In fact, the geometrically simple case of a straight conducting wire (linear antenna) excited by a voltage source in the middle[†] has been a subject of extensive research for many years, and the exact charge and current distributions on a wire of a finite radius are extremely complicated even when the wire is assumed to be perfectly conducting. Fortunately, the radiation field

---

[†] This arrangement is called a *dipole antenna*.

of such an antenna is relatively insensitive to slight deviations in the current distribution, and a physically plausible approximate current on the wire yields useful results for nearly all practical purposes. We will examine the radiation properties of linear antennas with assumed currents.

By combining Maxwell's equations we can derive nonhomogeneous wave equations in **E** and in **H** (see Problem P.11–1). However, these equations tend to involve the charge and current densities in a complicated way. It is generally simpler to solve for the auxiliary potential functions **A** and V first. Using **A** and V in Eqs. (7–55) and (7–57), we can determine **H** and **E**. For harmonic time variation in a simple medium we have

$$\mathbf{H} = \frac{1}{\mu} \mathbf{\nabla} \times \mathbf{A} \tag{11-1}$$

and

$$\mathbf{E} = -\mathbf{\nabla}V - j\omega\mathbf{A}. \tag{11-2}$$

The potential functions **A** and V are themselves solutions of nonhomogeneous wave equations, Eqs. (7–63) and (7–65), and the solutions are given in Eqs. (7–78) and (7–77), respectively. For harmonic time dependence the *phasor retarded potentials* are, from Eqs. (7–100) and (7–99),

$$\mathbf{A} = \frac{\mu}{4\pi} \int_{V'} \frac{\mathbf{J}e^{-jkR}}{R}\, dv', \tag{11-3}$$

$$V = \frac{1}{4\pi\epsilon} \int_{V'} \frac{\rho e^{-jkR}}{R}\, dv', \tag{11-4}$$

where $k = \omega\sqrt{\mu\epsilon} = 2\pi/\lambda$ is the wavenumber. Of course, **A** and V are related by the Lorentz condition, Eq. (7–98), for potentials, just as **J** and ρ are related by the equation of continuity Eq. (7–48), or

$$\mathbf{\nabla} \cdot \mathbf{J} = -j\omega\rho. \tag{11-5}$$

Hence there is no need for evaluating the integrals in both Eqs. (11–3) and (11–4). As a matter of fact, since **E** and **H** are related by Eq. (7–104b),

$$\mathbf{E} = \frac{1}{j\omega\epsilon} \mathbf{\nabla} \times \mathbf{H}. \tag{11-6}$$

We follow three steps in the determination of electromagnetic fields from a current distribution: (1) determine **A** from **J** using Eq. (11–3); (2) find **H** from **A** using Eq. (11–1); and (3) find **E** from **H** using Eq. (11–6). Note that only Step 1 requires an integration and that Steps 2 and 3 involve only straightforward differentiation. This is the procedure we will use in finding the radiation pattern of antennas.

We will first study the radiation fields and characteristic properties of an elemental electric dipole and of a small current loop (or magnetic dipole). We then consider finite-length thin linear antennas, of which the half-wavelength dipole is an important special case. The radiation characteristics of a linear antenna are largely determined by its length and the manner in which it is excited. To obtain more directivity and other desirable properties, a number of such antennas may be arranged together to form an **antenna array**. The geometrical configuration, the spacings between the array elements, as well as the relative amplitudes and phases of the excitations in the elements all affect the field pattern of the array. Some basic properties of simple arrays will be considered.

When an antenna is used as a receiving device, its function is to collect energy from an incoming electromagnetic wave and deliver it to a receiver. Any antenna that is useful for radiation is also useful for reception. We will use the reciprocity theorem to show that the pattern, directivity, input impedance, effective height, and effective aperture of an antenna are the same for transmitting as for receiving. We will define backscatter cross section and study the radar equation and the effect of wave propagation near the earth's surface. Finally, we will discuss such antenna types as traveling-wave antennas, Yagi-Uda antennas, helical antennas, broadband antennas and arrays, and aperture antennas.

11–2 Radiation Fields of Elemental Dipoles

In this section we study the radiation fields of the simplest types of all radiating systems—namely, elemental oscillating electric and magnetic dipoles. We will find that the field solutions for electric and magnetic dipoles are duals of each other. As a consequence, the radiation properties of one can be deduced from those of the other without recalculation.

11–2.1 THE ELEMENTAL ELECTRIC DIPOLE

Consider the elemental oscillating electric dipole (in free space), as shown in Fig. 11–1, which consists of a short conducting wire of length $d\ell$ terminated in two small conductive spheres or disks (capacitive loading). We assume the current in the wire to be uniform and to vary sinusoidally with time:

$$i(t) = I \cos \omega t = \mathscr{R}e[Ie^{j\omega t}]. \tag{11–7}$$

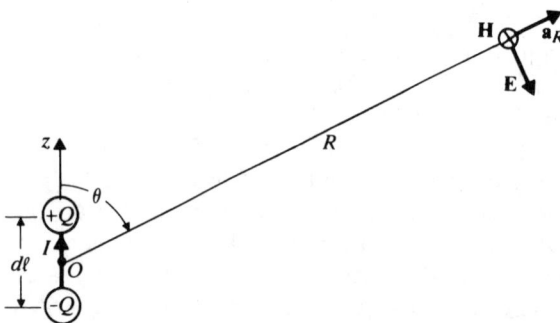

FIGURE 11–1
A Hertzian dipole.

Since the current vanishes at the ends of the wire, charge must be deposited there. The relation between the charge and the current is

$$i(t) = \pm \frac{dq(t)}{dt}.$$ (11–8)

In phasor notation, $q(t) = \mathscr{R}e[Qe^{j\omega t}]$, we have

$$I = \pm j\omega Q$$ (11–9)

or

$$Q = \pm \frac{I}{j\omega},$$ (11–10)

where, for the indicated current direction in Fig. 11–1, the positive sign is for the charge on the upper end and the negative sign for the charge on the lower end. The pair of equal and opposite charges separated by a short distance effectively form an electric dipole with a vector phasor electric moment

$$\mathbf{p} = \mathbf{a}_z Q \, d\ell \qquad (\text{C·m}).$$ (11–11)

Such an oscillating dipole is called a **Hertzian dipole**.

To determine the electromagnetic field of a Hertzian dipole, we follow the three steps outlined in Section 11–1. The phasor representation of the retarded vector potential is, from Eq. (11–3),

$$\mathbf{A} = \mathbf{a}_z \frac{\mu_0 I \, d\ell}{4\pi} \left(\frac{e^{-j\beta R}}{R} \right),$$ (11–12)

where $\beta = k_0 = \omega/c = 2\pi/\lambda$. Since

$$\mathbf{a}_z = \mathbf{a}_R \cos\theta - \mathbf{a}_\theta \sin\theta,$$ (11–13)

the spherical components of $\mathbf{A} = \mathbf{a}_R A_R + \mathbf{a}_\theta A_\theta + \mathbf{a}_\phi A_\phi$ are

$$A_R = A_z \cos\theta = \frac{\mu_0 I \, d\ell}{4\pi} \left(\frac{e^{-j\beta R}}{R} \right) \cos\theta,$$ (11–14a)

$$A_\theta = -A_z \sin\theta = -\frac{\mu_0 I \, d\ell}{4\pi} \left(\frac{e^{-j\beta R}}{R} \right) \sin\theta,$$ (11–14b)

$$A_\phi = 0.$$ (11–14c)

From the geometry of Fig. 11–1 we expect no variation with respect to the coordinate ϕ. We have, from Eq. (2–139)

$$\mathbf{H} = \frac{1}{\mu_0} \nabla \times \mathbf{A} = \mathbf{a}_\phi \frac{1}{\mu_0 R} \left[\frac{\partial}{\partial R}(R A_\theta) - \frac{\partial A_R}{\partial \theta} \right]$$

$$= -\mathbf{a}_\phi \frac{I \, d\ell}{4\pi} \beta^2 \sin\theta \left[\frac{1}{j\beta R} + \frac{1}{(j\beta R)^2} \right] e^{-j\beta R}.$$ (11–15)

The electric field intensity can be obtained from Eq. (11–6):

$$\mathbf{E} = \frac{1}{j\omega\epsilon_0} \nabla \times \mathbf{H}$$

$$= \frac{1}{j\omega\epsilon_0} \left[\mathbf{a}_R \frac{1}{R\sin\theta} \frac{\partial}{\partial\theta} (H_\phi \sin\theta) - \mathbf{a}_\theta \frac{1}{R} \frac{\partial}{\partial R} (RH_\phi) \right], \qquad (11\text{–}16)$$

which gives

$$E_R = -\frac{I\,d\ell}{4\pi} \eta_0 \beta^2 2 \cos\theta \left[\frac{1}{(j\beta R)^2} + \frac{1}{(j\beta R)^3} \right] e^{-j\beta R}, \qquad (11\text{–}16\text{a})$$

$$E_\theta = -\frac{I\,d\ell}{4\pi} \eta_0 \beta^2 \sin\theta \left[\frac{1}{j\beta R} + \frac{1}{(j\beta R)^2} + \frac{1}{(j\beta R)^3} \right] e^{-j\beta R}, \qquad (11\text{–}16\text{b})$$

$$E_\phi = 0, \qquad (11\text{–}16\text{c})$$

where $\eta_0 = \sqrt{\mu_0/\epsilon_0} \cong 120\pi \ (\Omega)$.

Equations (11–15) and (11–16) constitute the electromagnetic field of a Hertzian dipole. Note that in deriving these expressions we used only the current in the dipole to find the vector potential $\mathbf{A}$; the charges at the ends of the dipole did not enter into the calculations. We could, however, take an alternative approach by finding both $\mathbf{A}$ from $I\,d\ell$, as in Eq. (11–12), and the scalar potential V from the pair of equal and opposite charges using Eq. (11–4). The electric field intensity could then be determined from Eq. (11–2), instead of from Eq. (11–6). The result would be exactly the same as that obtained above (see Problem P.11–2).

The complete field expressions in Eqs. (10–15) and (10–16) are fairly complicated. It is advantageous to examine their behavior in regions near to and far from the dipole separately.

Near Field   In the region near to the Hertzian dipole (in the *near zone*), $\beta R = 2\pi R/\lambda \ll 1$, the leading term in Eq. (11–15) is

$$H_\phi = \frac{I\,d\ell}{4\pi R^2} \sin\theta, \qquad (11\text{–}17)$$

where we have approximated the factor $e^{-j\beta R} = 1 - j\beta R - (\beta R)^2/2 + \cdots$ by unity. Equation (11–17) is exactly what would be obtained for the magnetic field intensity due to a current element $I\,d\ell$ by applying the Biot-Savart law in magnetostatics as given in Eq. (6–33b).

The leading near-zone terms for the electric field intensity are, from Eqs. (11–16a) and (11–16b),

$$E_R = \frac{p}{4\pi\epsilon_0 R^3} 2 \cos\theta \qquad (11\text{–}18\text{a})$$

and

$$E_\theta = \frac{p}{4\pi\epsilon_0 R^3} \sin\theta, \qquad (11\text{–}18\text{b})$$

where the phasor relations (11–10) and (11–11) have been used. These expressions are identical to those of the electric field intensity due to an elemental electric dipole of a moment p in the z-direction, as given in Eq. (3–31), obtained by an application of the laws of electrostatics. The *near-zone fields* of an oscillating time-varying dipole are then *quasi-static fields*.

Far Field The region where $\beta R = 2\pi R/\lambda \gg 1$ is the *far zone*. The far-zone leading terms in Eqs. (11–15) and (11–16) are

$$H_\phi = j\,\frac{I\,d\ell}{4\pi}\left(\frac{e^{-j\beta R}}{R}\right)\beta \sin\theta \qquad \text{(A/m)}, \qquad (11\text{–}19\text{a})$$

$$E_\theta = j\,\frac{I\,d\ell}{4\pi}\left(\frac{e^{-j\beta R}}{R}\right)\eta_0 \beta \sin\theta \qquad \text{(V/m)}. \qquad (11\text{–}19\text{b})$$

Several important observations can be made on these *far-zone fields*. First, E_θ and H_ϕ are in space quadrature and in time phase. Second, their ratio $E_\theta/H_\phi = \eta_0$ is a constant equal to the intrinsic impedance of the medium (which is, in the present case, free space). The far-zone fields, then, have the same properties as those of a plane wave. This is not unexpected, since at very large distances from the dipole a spherical wavefront closely resembles a plane wavefront.

A third observation from Eqs. (11–19a, b) is that the magnitude of the far-zone fields varies inversely with the distance from the source. The phase of both E_θ and H_ϕ is a periodic function of R with a period that is the wavelength:

$$\lambda = \frac{2\pi}{\beta} = \frac{c}{f}. \qquad (11\text{–}20)$$

Note that the far-zone condition $\beta R \gg 1$ translates into $R \gg \lambda/2\pi$; hence one has to be farther away from the dipole at lower frequencies in order to be in the far zone. (Other characteristics of far-zone fields will be discussed in Section 11–3.)

11–2.2 THE ELEMENTAL MAGNETIC DIPOLE

Let us now consider a small filamentary loop of radius b carrying a uniform time-harmonic current $i(t) = I \cos \omega t$ around its circumference, as shown in Fig. 11–2. This is an elemental magnetic dipole with a vector phasor magnetic moment

$$\mathbf{m} = \mathbf{a}_z I\pi b^2 = \mathbf{a}_z m \qquad \text{(A·m}^2). \qquad (11\text{–}21)$$

To determine the electromagnetic field, we first find the vector potential. The procedure is the same as that used in Section 6–5, except for the time-dependent nature of the current. Instead of starting from Eq. (6–39), we have

$$\mathbf{A} = \frac{\mu_0 I}{4\pi} \oint \frac{e^{-j\beta R_1}}{R_1}\, d\ell'. \qquad (11\text{–}22)$$

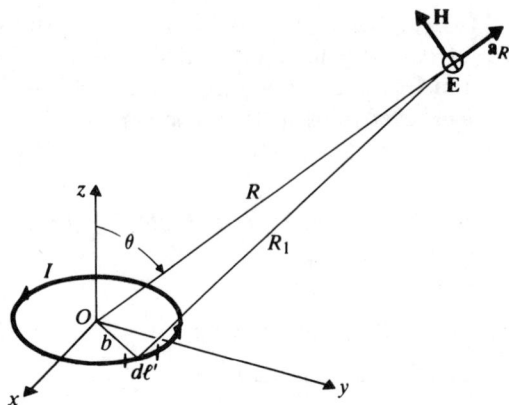

FIGURE 11-2
A magnetic dipole.

The integral in Eq. (11–22) is relatively difficult to carry out exactly because R_1 changes with the location of $d\boldsymbol{\ell}'$ on the loop. For a small loop the exponential factor in the numerator can be written as

$$e^{-j\beta R_1} = e^{-j\beta R}e^{-j\beta(R_1 - R)}$$
$$\cong e^{-j\beta R}[1 - j\beta(R_1 - R)]. \tag{11-23}$$

Substitution of Eq. (11–23) in Eq. (11–22) yields approximately

$$\mathbf{A} = \frac{\mu_0 I}{4\pi} e^{-j\beta R}\left[(1 + j\beta R) \oint \frac{d\boldsymbol{\ell}'}{R_1} - j\beta \oint d\boldsymbol{\ell}' \right]. \tag{11-24}$$

The second integral in Eq. (11–24) obviously vanishes. The first integral is the same as that in Eq. (6–39), except for the multiplying factor $(1 + j\beta R)e^{-j\beta R}$. In view of the result in Eq. (6–43) we have

$$\mathbf{A} = \mathbf{a}_\phi \frac{\mu_0 m}{4\pi R^2} (1 + j\beta R)e^{-j\beta R} \sin\theta. \tag{11-25}$$

The electric and magnetic field intensities can be determined by straightforward differentiation using Eqs. (11–6) and (11–1), respectively:

$$E_\phi = \frac{j\omega\mu_0 m}{4\pi} \beta^2 \sin\theta \left[\frac{1}{j\beta R} + \frac{1}{(j\beta R)^2} \right] e^{-j\beta R}, \tag{11-26a}$$

$$H_R = -\frac{j\omega\mu_0 m}{4\pi\eta_0} \beta^2 2\cos\theta \left[\frac{1}{(j\beta R)^2} + \frac{1}{(j\beta R)^3} \right] e^{-j\beta R}, \tag{11-26b}$$

$$H_\theta = -\frac{j\omega\mu_0 m}{4\pi\eta_0} \beta^2 \sin\theta \left[\frac{1}{j\beta R} + \frac{1}{(j\beta R)^2} + \frac{1}{(j\beta R)^3} \right] e^{-j\beta R}. \tag{11-26c}$$

Comparison of Eqs. (11–26a, b, c) with Eqs. (11–15) and (11–16a, b) reveals immediately the dual nature of the electromagnetic fields of electric and magnetic dipoles.

Let $(\mathbf{E}_e, \mathbf{H}_e)$ denote the electric and magnetic fields of the electric dipole and $(\mathbf{E}_m, \mathbf{H}_m)$ the electric and magnetic fields of the magnetic dipole. We have

$$\mathbf{E}_e = \eta_0 \mathbf{H}_m \tag{11-27}$$

and

$$\mathbf{H}_e = -\frac{\mathbf{E}_m}{\eta_0} \tag{11-28}$$

if the electric and magnetic dipole moments are related as follows:

$$I\,d\ell = j\beta m, \tag{11-29}$$

where $\beta = \omega\mu_0/\eta_0 = \omega\sqrt{\mu_0\epsilon_0}$. Equations (11–27) and (11–28) are results expected from the principle of duality, which was introduced in connection with Example 7–7. Thus Hertzian electric dipole and elemental magnetic dipole are dual devices, and their electromagnetic fields are dual solutions of source-free Maxwell's equations. As a consequence of this duality, the discussions about the nature of the near and far fields of an electric dipole apply to the dual quantities of a magnetic dipole. In particular, the far-zone ($\beta R \gg 1$) fields of a magnetic dipole are

$$\boxed{E_\phi = \frac{\omega\mu_0 m}{4\pi}\left(\frac{e^{-j\beta R}}{R}\right)\beta \sin\theta \qquad \text{(V/m)},} \tag{11-30a}$$

$$\boxed{H_\theta = -\frac{\omega\mu_0 m}{4\pi\eta_0}\left(\frac{e^{-j\beta R}}{R}\right)\beta \sin\theta \qquad \text{(A/m)}.} \tag{11-30b}$$

We can see that the far-field intensities vary inversely as R and their ratio E_ϕ/H_θ equals the intrinsic impedance η_0 of free space.

Examination of the far-field E_θ in Eq. (11–19b) of the electric dipole and E_ϕ in Eq. (11–30a) of the magnetic dipole reveals that they have the same pattern function $|\sin\theta|$ and are in both space and time quadrature. Thus it is possible to combine electric and magnetic dipoles to form an antenna that produces circular polarization (see Problem P.11–4).

11–3 Antenna Patterns and Antenna Parameters

In antenna problems we are primarily interested in the far-zone fields. These are also called **radiation fields**. No physical antennas radiate uniformly in all directions in space. The graph that describes the relative far-zone field strength versus direction at a fixed distance from an antenna is called the **radiation pattern** of the antenna, or

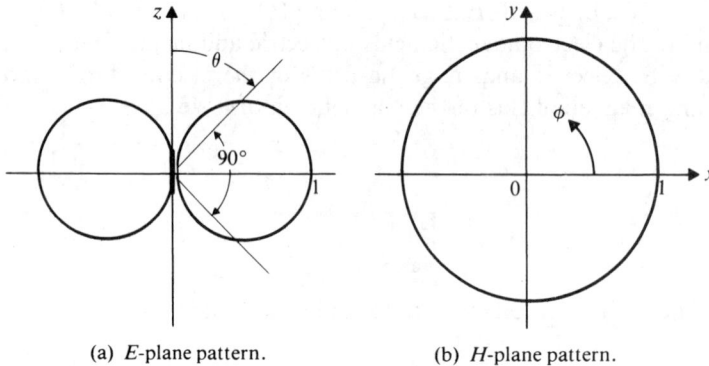

(a) *E*-plane pattern.   (b) *H*-plane pattern.

FIGURE 11–3
Radiation patterns of a Hertzian dipole.

simply the ***antenna pattern***. In general, an antenna pattern is three-dimensional, varying with both θ and ϕ in a spherical coordinate system. The difficulties of making three-dimensional plots can be avoided—as is the usual practice—by plotting separately the magnitude of the normalized field strength (with respect to the peak value) versus θ for a constant ϕ (an ***E-plane pattern***) and the magnitude of the normalized field strength versus ϕ for $\theta = \pi/2$ (the ***H-plane pattern***).

EXAMPLE 11–1   Plot the *E*-plane and *H*-plane radiation patterns of a Hertzian dipole.

Solution   Since E_θ and H_ϕ in the far zone are proportional to each other, we need only consider the normalized magnitude of E_θ.

a) *E-plane pattern*. At a given R, E_θ is independent of ϕ; and from Eq. (11–19b) the normalized magnitude of E_θ is

$$\text{Normalized } |E_\theta| = |\sin \theta|. \tag{11–31}$$

This is the *E*-plane ***pattern function*** of a Hertzian dipole. For any given ϕ, Eq. (11–31) represents a pair of circles, as shown in Fig. 11–3(a).

b) *H-plane pattern*. At a given R and for $\theta = \pi/2$ the normalized magnitude of E_θ is $|\sin \theta| = 1$. The *H*-plane pattern is then simply a circle of unity radius centered at the z-directed dipole, as shown in Fig. 11–3(b).

The radiation pattern of practical antennas are usually more complicated than those shown in Fig. 11–3. A typical *H*-plane pattern might look like the one illustrated in Fig. 11–4(a), which is plotted in polar coordinates with normalized $|E_\theta|$ versus ϕ. It generally has a major maximum and several minor maxima. The region of maximum radiation between the first null points around it is the ***main beam***, and the regions of minor maxima are ***sidelobes***.

Sometimes it is convenient to plot antenna patterns in rectangular coordinates. The polar plot in Fig. 11–4(a) will appear as Fig. 11–4(b) in rectangular coordinates. Since the field intensities in the main-beam and sidelobe directions may differ by many orders of magnitude, antenna patterns are frequently plotted in a logarithmic scale measured in decibels down from the main-beam level. The pattern in Fig. 11–4(b) converted to a decibel scale will have the shape shown in Fig. 11–4(c).

In the comparison of various antenna patterns the following characteristic parameters are of importance: (1) width of main beam, (2) sidelobe levels, and (3) directivity.

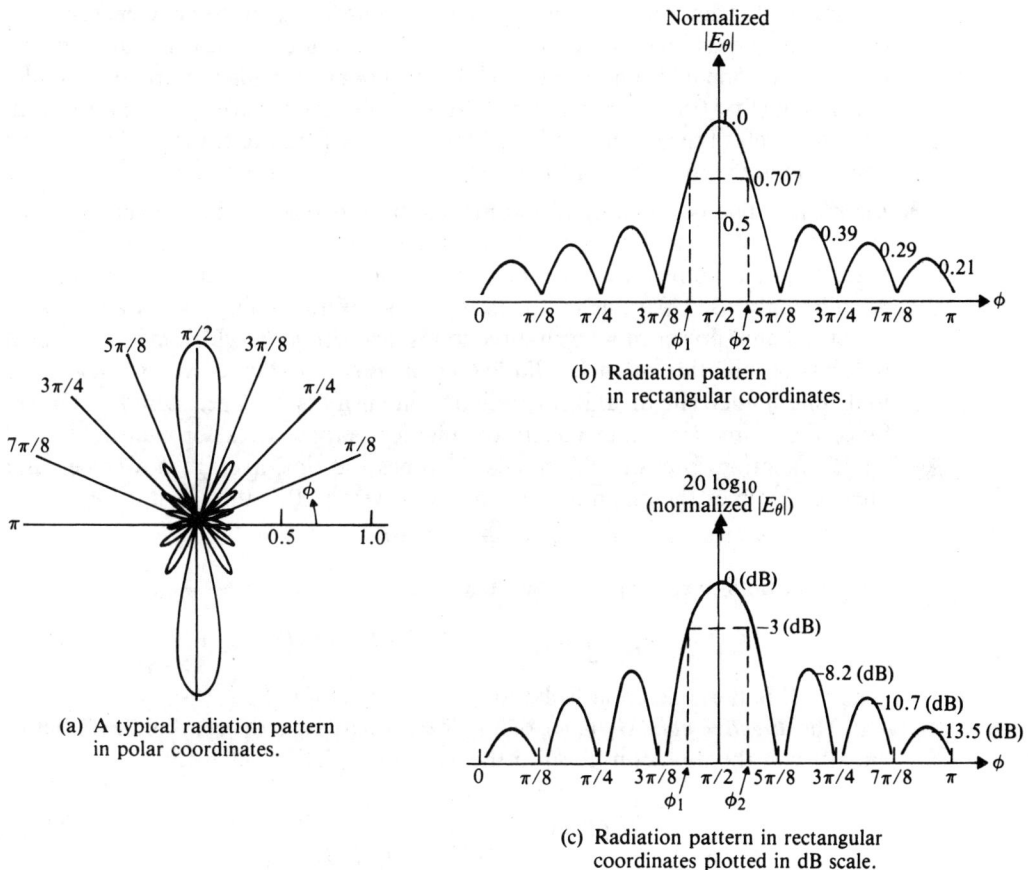

(b) Radiation pattern
in rectangular coordinates.

(a) A typical radiation pattern
in polar coordinates.

(c) Radiation pattern in rectangular
coordinates plotted in dB scale.

FIGURE 11–4

Typical *H*-plane radiation patterns.

The significance of each of these parameters is explained below.

1. *Width of main beam* (or simply **beamwidth**). The main-beam beamwidth describes the sharpness of the main radiation region. It is generally taken to be the angular width of a pattern between the half-power, or -3 (dB), points. In electric-intensity plots it is the angular width between points that are $1/\sqrt{2}$ or 0.707 times the maximum intensity. Thus, the H-plane pattern in Fig. 11–4 has a 3 (dB) beamwidth equal to $(\phi_2 - \phi_1)$, and the E-plane pattern of the Hertzian dipole in Fig. 11–3(a) has a 3 (dB) beamwidth of 90°. Occasionally the angular width of the main beam between -10 (dB) points or between the first nulls is also of interest. Of course, the main beam must point in the direction where the antenna is designed to have its maximum radiation.

2. *Sidelobe levels.* Sidelobes of a directive (nonisotropic) pattern represent regions of unwanted radiation; they should have levels as low as possible. Generally, the levels of distant sidelobes are lower than the levels of those near the main beam. Hence, when one talks about the sidelobe level of an antenna pattern, one usually refers to the first (the nearest and highest) sidelobe. In modern radar applications, sidelobe levels of the order of minus 40 or more decibels are required. In practical applications the locations of the sidelobes are also of importance.

3. *Directivity.* The beamwidth of an antenna pattern specifies the sharpness of the main beam, but it does not provide us with any information about the rest of the pattern. For example, the sidelobes may be very high—an undesirable feature. A commonly used parameter to measure the overall ability of an antenna to direct radiated power in a given direction is **directive gain**, which may be defined in terms of radiation intensity. **Radiation intensity** is the time-average power per unit solid angle. The SI unit for radiation intensity is watt per steradian (W/sr). Since there are R^2 square meters of spherical surface area for each unit solid angle, radiation intensity, U, equals R^2 times the time-average power per unit area or R^2 times the magnitude of the time-average Poynting vector, $\mathscr{P}_{av}$:

$$U = R^2 \mathscr{P}_{av} \quad \text{(W/sr).} \qquad (11\text{–}32)$$

The total time-average power radiated is

$$P_r = \oint \mathscr{P}_{av} \cdot d\mathbf{s} = \oint U \, d\Omega \quad \text{(W),} \qquad (11\text{–}33)$$

where $d\Omega$ is the differential solid angle, $d\Omega = \sin\theta \, d\theta \, d\phi$.

The **directive gain**, $G_D(\theta, \phi)$, of an antenna pattern is the ratio of the radiation intensity in the direction (θ, ϕ) to the average radiation intensity:

$$G_D(\theta, \phi) = \frac{U(\theta, \phi)}{P_r/4\pi} = \frac{4\pi U(\theta, \phi)}{\oint U \, d\Omega}. \qquad (11\text{–}34)$$

Obviously, the directive gain of an isotropic or omnidirectional antenna (an antenna that radiates uniformly in all directions) is unity. However, an isotropic antenna does not exist in practice.

The maximum directive gain of an antenna is called the *directivity* of the antenna. It is the ratio of the maximum radiation intensity to the average radiation intensity and is usually denoted by D:

$$D = \frac{U_{max}}{U_{av}} = \frac{4\pi U_{max}}{P_r} \quad \text{(Dimensionless).}$$

(11–35)

In terms of electric field intensity, D can be expressed as

$$D = \frac{4\pi |E_{max}|^2}{\int_0^{2\pi} \int_0^{\pi} |E(\theta, \phi)|^2 \sin \theta \, d\theta \, d\phi} \quad \text{(Dimensionless).}$$

(11–36)

Directivity is frequently expressed in decibels, referring to unity.

━━━ **EXAMPLE 11–2**   Find the directive gain and the directivity of a Hertzian dipole.

Solution   For a Hertzian dipole the magnitude of the time-average Poynting vector is

$$\mathscr{P}_{av} = \tfrac{1}{2}\mathscr{R}e|\mathbf{E} \times \mathbf{H}^*| = \tfrac{1}{2}|E_\theta| \, |H_\phi|.$$

(11–37)

Hence from Eqs. (11–19a, b) and (11–32),

$$U = \frac{(I \, d\ell)^2}{32\pi^2} \eta_0 \beta^2 \sin^2 \theta.$$

(11–38)

The directive gain can be obtained from Eq. (11–34):

$$G_D(\theta, \phi) = \frac{4\pi \sin^2 \theta}{\int_0^{2\pi} \int_0^{\pi} (\sin^2 \theta) \sin \theta \, d\theta \, d\phi}$$

$$= \tfrac{3}{2} \sin^2 \theta.$$

The directivity is the maximum value of $G_D(\theta, \phi)$:

$$D = G_D\left(\frac{\pi}{2}, \phi\right) = 1.5,$$

which corresponds to $10 \log_{10} 1.5$ or 1.76 (dB).   ━━━

We note that beamwidth, sidelobe levels, and directive gain are parameters of an antenna pattern; they do not convey information about the efficiency or the input impedance of the antenna. A measure of antenna efficiency is the power gain. The *power gain*, or simply the *gain*, G_P, of an antenna referred to an isotropic source is the ratio of its maximum radiation intensity to the radiation intensity of a lossless isotropic source with the same power input. The directive gain as defined in Eq. (11–34) is based on radiated power P_r. Because of ohmic power loss, P_ℓ, in the antenna itself as well as in nearby lossy structures including the ground, P_r is less than the total input power P_i. We have

$$P_i = P_r + P_\ell. \tag{11-39}$$

The power gain of an antenna is then

$$\boxed{G_P = \frac{4\pi U_{max}}{P_i} \quad \text{(Dimensionless).}} \tag{11-40}$$

The ratio of the gain to the directivity of an antenna is the *radiation efficiency*, η_r:

$$\boxed{\eta_r = \frac{G_P}{D} = \frac{P_r}{P_i} \quad \text{(Dimensionless).}} \tag{11-41}$$

Normally, the efficiency of well-constructed antennas is very close to 100%.

A useful measure of the amount of power radiated by an antenna is radiation resistance. The *radiation resistance* of an antenna is the value of a hypothetical resistance that would dissipate an amount of power equal to the radiated power P_r when the current in the resistance is equal to the maximum current along the antenna. Naturally, a high radiation resistance is a desirable property for an antenna.

EXAMPLE 11-3   Find the radiation resistance of a Hertzian dipole.

Solution   If we assume no ohmic losses, the time-average power radiated by a Hertzian dipole for an input time-harmonic current with an amplitude I is

$$P_r = \tfrac{1}{2} \int_0^{2\pi} \int_0^{\pi} E_\theta H_\phi^* R^2 \sin\theta \, d\theta \, d\phi. \tag{11-42}$$

Using the far-zone fields in Eqs. (11-19a, b), we find

$$P_r = \frac{I^2 (d\ell)^2}{32\pi^2} \eta_0 \beta^2 \int_0^{2\pi} \int_0^{\pi} \sin^3\theta \, d\theta \, d\phi$$
$$= \frac{I^2 (d\ell)^2}{12\pi} \eta_0 \beta^2 = \frac{I^2}{2} \left[80\pi^2 \left(\frac{d\ell}{\lambda} \right)^2 \right]. \tag{11-43}$$

In this last expression we have used 120π for the intrinsic impedance of free space, η_0, and substituted $2\pi/\lambda$ for β.

Since the current along the short Hertzian dipole is uniform, we refer the power dissipated in the radiation resistance R_r to I. Equating $I^2 R_r / 2$ to P_r, we obtain

$$\boxed{R_r = 80\pi^2 \left(\frac{d\ell}{\lambda} \right)^2 \quad (\Omega).} \tag{11-44}$$

As an example, if $d\ell = 0.01\lambda$, R_r is only about 0.08 (Ω), an extremely small value. Hence a short dipole antenna is a poor radiator of electromagnetic power. However, it is erroneous to say without qualification that the radiation resistance of a dipole antenna increases as the square of its length because Eq. (11–44) holds only if $d\ell \ll \lambda$.

Radiation resistance may be quite different from the real part of the input impedance because the latter includes ohmic losses in the antenna structure itself as well as losses in the ground. The input impedance of a short dipole antenna has a large capacitive reactance, which makes it difficult to match and therefore difficult to feed power to the antenna efficiently.

■■■■■ **EXAMPLE 11–4**   Find the radiation efficiency of an isolated Hertzian dipole made of a metal wire of radius a, length d, and conductivity σ.

Solution   Let I be the amplitude of the current in the wire dipole having a loss resistance R_ℓ. Then the ohmic power loss is

$$P_\ell = \tfrac{1}{2}I^2 R_\ell. \tag{11–45}$$

In terms of radiation resistance R_r the radiated power is

$$P_r = \tfrac{1}{2}I^2 R_r. \tag{11–46}$$

From Eqs. (11–39) and (11–41) we have

$$
\begin{aligned}
\eta_r &= \frac{P_r}{P_r + P_\ell} = \frac{R_r}{R_r + R_\ell} \\
&= \frac{1}{1 + (R_\ell/R_r)},
\end{aligned}
\tag{11–47}
$$

where R_r has been found in Eq. (11–44). The loss resistance R_ℓ of the metal wire can be expressed in terms of the surface resistance R_s:

$$R_\ell = R_s\left(\frac{d\ell}{2\pi a}\right), \tag{11–48}$$

where

$$R_s = \sqrt{\frac{\pi f \mu_0}{\sigma}} \tag{11–49}$$

as given in Eq. (9–26b). Using Eqs. (11–44) and (11–48) in Eq. (11–47), we obtain the radiation efficiency of an isolated Hertzian dipole:

$$\eta_r = \frac{1}{1 + \dfrac{R_s}{160\pi^3}\left(\dfrac{\lambda}{a}\right)\left(\dfrac{\lambda}{d\ell}\right)}. \tag{11–50}$$

Assume that $a = 1.8$ (mm), $d\ell = 2$ (m), operating frequency $f = 1.5$ (MHz), and σ (for copper) $= 5.80 \times 10^7$ (S/m). We find that

$$\lambda = \frac{c}{f} = \frac{3 \times 10^8}{1.5 \times 10^6} = 200 \quad (\text{m}),$$

$$R_s = \sqrt{\frac{\pi \times (1.50 \times 10^6) \times (4\pi 10^{-7})}{5.80 \times 10^7}} = 3.20 \times 10^{-4} \quad (\Omega),$$

$$R_\ell = 3.20 \times 10^{-4} \times \left(\frac{2}{2\pi 1.8 \times 10^{-3}}\right) = 0.057 \quad (\Omega),$$

$$R_r = 80\pi^2 \left(\frac{2}{200}\right)^2 = 0.079 \quad (\Omega),$$

and

$$\eta_r = \frac{0.079}{0.079 + 0.057} = 58\%,$$

which is very low. Equation (11–50) shows that smaller values of (a/λ) and $(d\ell/\lambda)$ lower the radiation efficiency. ∎

11–4 Thin Linear Antennas

We have just indicated that a short dipole antenna is not a good radiator of electromagnetic power because of its low radiation resistance and low radiation efficiency. We now examine the radiation characteristics of a center-fed thin straight antenna having a length comparable to a wavelength, as shown in Fig. 11–5. Such an antenna is a *linear dipole antenna*. If the current distribution along the antenna is known, we can find its radiation field by integrating over the entire length of the antenna the radiation field due to an elemental dipole. The determination of the exact current distribution on such a seemingly simple geometrical configuration (a straight wire

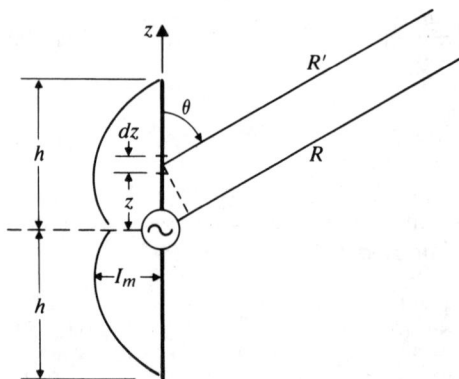

FIGURE 11–5
A center-fed linear dipole with sinusoidal current distribution.

of a finite radius) is, however, a very difficult boundary-value problem even if the wire is assumed to be perfectly conducting. The current must be zero at the ends of the wire where charges are deposited, and the tangential electric field due to all currents and charges must vanish at every point on the wire surface. An analytical formulation of the problem leads to an integral equation in which the current distribution along the antenna is the unknown function under the integral. Unfortunately, an exact solution of the integral equation does not exist. Various approximate solutions have been attempted. With the advent of high-speed digital computers, numerical solutions for current distributions and input impedances can be obtained for linear antennas of specific lengths and thicknesses. The ratio of the voltage and the current at the feed points is the input impedance. Both the solution procedure and the numerical results are quite involved, and we shall not delve into them in this book. For our purposes the knowledge of the exact current distribution on the linear antenna is not of prime importance; a good estimate will give us considerable useful information on the radiation characteristics of the antenna. We assume a sinusoidal current distribution on a very thin, straight dipole. Such a current distribution constitutes a kind of standing wave over the dipole and represents a good approximation.

Since the dipole is center-driven, the currents on the two halves of the dipole are symmetrical and go to zero at the ends. We write the current phasor as

$$
\begin{aligned}
I(z) &= I_m \sin \beta(h - |z|), \\
&= \begin{cases} I_m \sin \beta(h - z), & z > 0, \\ I_m \sin \beta(h + z), & z < 0. \end{cases}
\end{aligned}
\tag{11–51}
$$

We are interested only in the far-zone fields. The far-field contribution from the differential current element $I\,dz$ is, from Eqs. (11–19a, b),

$$
dE_\theta = \eta_0\, dH_\phi = j\frac{I\,dz}{4\pi}\left(\frac{e^{-j\beta R'}}{R'}\right)\eta_0\beta \sin \theta.
\tag{11–52}
$$

Now R' in Eq. (11–52) is slightly different from R measured to the origin of the spherical coordinates, which coincides with the center of the dipole. In the far zone, $R \gg h$,

$$
R' = (R^2 + z^2 - 2Rz \cos \theta)^{1/2} \cong R - z \cos \theta.
\tag{11–53}
$$

The magnitude difference between $1/R'$ and $1/R$ is insignificant, but the approximate relation in Eq. (11–53) must be retained in the phase term. Using Eqs. (11–51) and (11–53) in Eq. (11–52) and integrating, we have

$$
\begin{aligned}
E_\theta &= \eta_0 H_\phi \\
&= j\frac{I_m\eta_0\beta \sin \theta}{4\pi R}\, e^{-j\beta R}\int_{-h}^{h} \sin \beta(h - |z|)e^{j\beta z \cos \theta}\,dz.
\end{aligned}
\tag{11–54}
$$

The integrand in Eq. (11–54) is the product of an even function of z, $\sin \beta(h - |z|)$, and

$$
e^{j\beta z \cos \theta} = \cos (\beta z \cos \theta) + j \sin (\beta z \cos \theta),
$$

where $\sin(\beta z \cos\theta)$ is an odd function of z. Integrating between symmetrical limits $-h$ and h, we find that only the part of the integrand containing the product $\sin\beta(h-|z|)\cos(\beta z \cos\theta)$ does not vanish. Equation (11–54) then reduces to

$$E_\theta = \eta_0 H_\phi = j\frac{I_m\eta_0\beta\sin\theta}{2\pi R}e^{-j\beta R}\int_0^h \sin\beta(h-z)\cos(\beta z \cos\theta)\,dz$$

$$= \frac{j60I_m}{R}e^{-j\beta R}F(\theta),$$

(11–55)

where

$$F(\theta) = \frac{\cos(\beta h \cos\theta) - \cos\beta h}{\sin\theta}.$$

(11–56)

The factor $|F(\theta)|$ is the E-plane **pattern function** of a linear dipole antenna. It describes the radiation pattern or the variation of the normalized far field, $|E_\theta|$, versus the angle θ. The exact shape of the radiation pattern represented by $|F(\theta)|$ in Eq. (11–56) depends on the value of $\beta h = 2\pi h/\lambda$ and can be quite different for different antenna lengths. The radiation pattern, however, is always symmetrical with respect to the $\theta = \pi/2$ plane. Figure 11–6 shows the E-plane patterns for four different dipole lengths measured in terms of wavelength: $2h/\lambda = \frac{1}{2}, 1, \frac{3}{2}$ and 2. The H-plane patterns are circles inasmuch as $F(\theta)$ is independent of ϕ. From the patterns in Fig. 11–6 we see that the direction of maximum radiation tends to shift away from the $\theta = 90°$ plane when the dipole length approaches $3\lambda/2$. For $2h = 2\lambda$ there is no radiation in the $\theta = 90°$ plane.

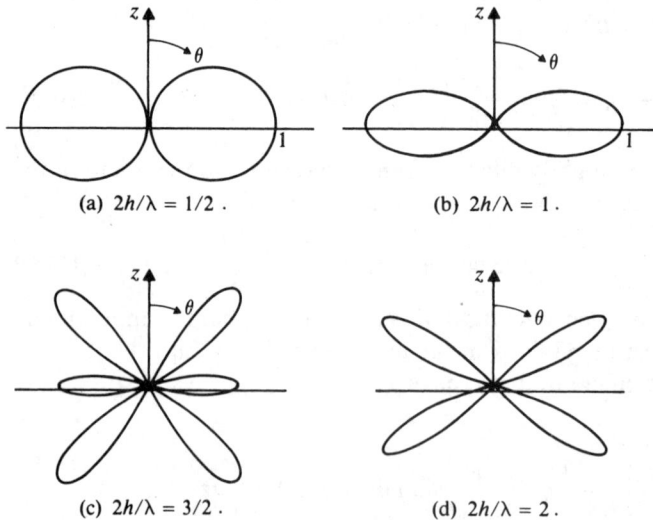

(a) $2h/\lambda = 1/2$.

(b) $2h/\lambda = 1$.

(c) $2h/\lambda = 3/2$.

(d) $2h/\lambda = 2$.

FIGURE 11–6
E-plane radiation patterns for center-fed dipole antennas.

11-4.1   THE HALF-WAVE DIPOLE

The half-wave dipole having a length $2h = \lambda/2$ is of particular practical importance because of its desirable pattern and impedance characteristics. We shall now examine its properties in more detail.

For a half-wave dipole, $\beta h = 2\pi h/\lambda = \pi/2$, the pattern function in Eq. (11–56) becomes

$$F(\theta) = \frac{\cos\left[(\pi/2)\cos\theta\right]}{\sin\theta}. \tag{11-57}$$

This function has a maximum equal to unity at $\theta = 90°$ and has nulls at $\theta = 0°$ and $180°$. The corresponding E-plane radiation pattern is sketched in Fig. 11–6(a). The far-zone field phasors are, from Eq. (11–55),

$$E_\theta = \frac{j60I_m}{R}\, e^{-j\beta R}\left\{\frac{\cos\left[(\pi/2)\cos\theta\right]}{\sin\theta}\right\} \tag{11-58}$$

and

$$H_\phi = \frac{jI_m}{2\pi R}\, e^{-j\beta R}\left\{\frac{\cos\left[(\pi/2)\cos\theta\right]}{\sin\theta}\right\}. \tag{11-59}$$

The magnitude of the time-average Poynting vector is

$$\mathscr{P}_{av} = \frac{1}{2}\, E_\theta H_\phi^* = \frac{15I_m^2}{\pi R^2}\left\{\frac{\cos\left[(\pi/2)\cos\theta\right]}{\sin\theta}\right\}^2. \tag{11-60}$$

The total power radiated by a half-wave dipole is obtained by integrating $\mathscr{P}_{av}$ over the surface of a great sphere:

$$\begin{aligned}
P_r &= \int_0^{2\pi}\int_0^\pi \mathscr{P}_{av}R^2 \sin\theta\, d\theta\, d\phi \\
&= 30I_m^2 \int_0^\pi \frac{\cos^2\left[(\pi/2)\cos\theta\right]}{\sin\theta}\, d\theta.
\end{aligned} \tag{11-61}$$

The integral in Eq. (11–61) can be evaluated numerically to give a value 1.218. Hence

$$P_r = 36.54I_m^2 \quad \text{(W)}, \tag{11-62}$$

from which we obtain the radiation resistance of a free-standing half-wave dipole:

$$R_r = \frac{2P_r}{I_m^2} = 73.1 \quad (\Omega). \tag{11-63}$$

Neglecting losses, we find that the input resistance of a thin half-wave dipole equals 73.1 (Ω) and that the input reactance is a small positive number that can be made to vanish when the dipole length is adjusted to be slightly shorter than $\lambda/2$. (As we have indicated before, the actual calculation of the input impedance is tedious and is beyond the scope of this book.)

The directivity of a half-wave dipole can be found by using Eq. (11–35). We have, from Eqs. (11–32) and (11–60),

$$U_{max} = R^2 \mathscr{P}_{av}(90°) = \frac{15}{\pi} I_m^2 \qquad (11\text{–}64)$$

and

$$D = \frac{4\pi U_{max}}{P_r} = \frac{60}{36.54} = 1.64, \qquad (11\text{–}65)$$

which corresponds to $10 \log_{10} 1.64$ or 2.15 (dB) referring to an omnidirectional radiator.

The half-power beamwidth of the radiation pattern is the angle between the two solutions of the equation

$$\frac{\cos\left[(\pi/2)\cos\theta\right]}{\sin\theta} = \frac{1}{\sqrt{2}}, \qquad 0 < \theta < \pi,$$

which can be solved either numerically or graphically to give a beamwidth of 78°. Thus a half-wave dipole is only slightly more directive than a short Hertzian dipole that has a directivity of 1.76 (dB) and a beamwidth of 90°.

EXAMPLE 11–5   A thin quarter-wavelength vertical antenna over a perfectly conducting ground is excited by a sinusoidal source at its base. Find its radiation pattern,
radiation resistance, and directivity.

Solution   Since current is charge in motion, we can use the method of images discussed in Section 4–4 and replace the conducting ground by the image of the vertical
antenna. A little thought will convince us that the image of a vertical antenna carrying
a current I is another vertical antenna. The image antenna has the same length, is
equidistant from the ground, and carries the same current in the *same direction* as

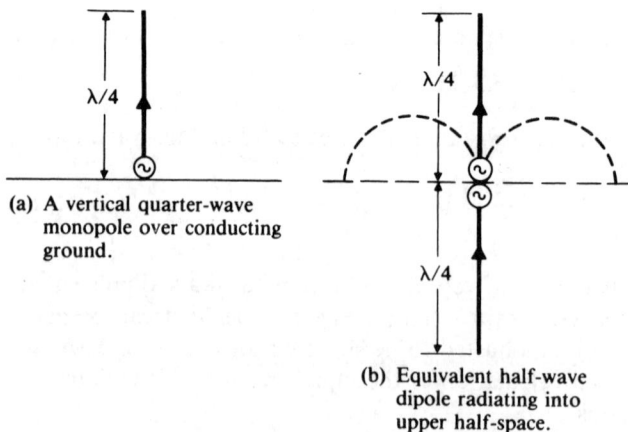

λ/4

(a) A vertical quarter-wave
monopole over conducting
ground.

λ/4

λ/4

(b) Equivalent half-wave
dipole radiating into
upper half-space.

FIGURE 11–7
Quarter-wave monopole over a
conducting ground and its equivalent half-wave dipole.

the original antenna. The electromagnetic field *in the upper half-space* due to the quarter-wave vertical antenna in Fig. 11–7(a) is, then, the same as that of the half-wave antenna in Fig. 11–7(b). The pattern function in Eq. (11–57) applies here for $0 \le \theta \le \pi/2$, and the radiation pattern drawn in dashed lines in Fig. 11–7(b) is the upper half of that in Fig. 11–6(a).

The magnitude of the time-average Poynting vector, $\mathscr{P}_{av}$, in Eq. (11–60), holds for $0 \le \theta \le \pi/2$. Inasmuch as the quarter-wave antenna (a **monopole**) radiates only into the upper half-space, its total radiated power is only one-half that given in Eq. (11–62):

$$P_r = 18.27 I_m^2 \quad \text{(W)}.$$

Consequently, the radiation resistance is

$$R_r = \frac{2P_r}{I_m^2} = 36.54 \quad (\Omega), \tag{11–66}$$

which is one-half of the radiation resistance of a half-wave antenna in free-space.

To calculate directivity, we note that although the maximum radiation intensity $U_{\max}$ remains the same as that given in Eq. (11–64), the average radiation intensity is now $P_r/2\pi$. Thus,

$$D = \frac{U_{\max}}{U_{av}} = \frac{U_{\max}}{P_r/2\pi} = 1.64, \tag{11–67}$$

which is the same as the directivity of a half-wave antenna.   ▬

11–4.2   EFFECTIVE ANTENNA LENGTH

For thin linear antennas with a given current distribution it is sometimes convenient to define a quantity called the **effective length**, to which the far-zone field is proportional. Let us refer to the dipole antenna in Fig. 11–5 and assume a general phasor current distribution $I(z)$. The far-zone field is then, from Eq. (11–54),

$$E_\theta = \eta_0 H_\phi = \frac{j30}{R} \beta e^{-j\beta R} \left\{ \sin \theta \int_{-h}^{h} I(z) e^{j\beta z \cos \theta} \, dz \right\}. \tag{11–68}$$

Let $I(0)$ be the input current at the feed point of the antenna. We write Eq. (11–68) as

$$E_\theta = \eta_0 H_\phi = \frac{j30 I(0)}{R} \beta e^{-j\beta R} \ell_e(\theta), \tag{11–69}$$

where

$$\ell_e(\theta) = \frac{\sin \theta}{I(0)} \int_{-h}^{h} I(z) e^{j\beta z \cos \theta} \, dz \tag{11–70}$$

is the **effective length** of the transmitting antenna. (We will discuss the effective length of a receiving antenna presently.) As we see from Eq. (11–69), ℓ_e measures the effectiveness of the antenna as a radiator, and for a given current distribution the far-zone field is proportional to ℓ_e, which contains all the information about the directional properties of the antenna. In most practical situations the important value of the effective length is that at $\theta = \pi/2$, where

$$\ell_e = \frac{1}{I(0)} \int_{-h}^{h} I(z)\, dz \qquad \text{(m)}. \tag{11-71}$$

Equation (11-71) indicates that ℓ_e is the length of an equivalent linear antenna with a uniform current $I(0)$ such that it radiates the same far-zone field in the $\theta = \pi/2$ plane.

EXAMPLE 11-6   Assume a sinusoidal current distribution on a center-fed, thin, straight half-wave dipole. Find its effective length. What is its maximum value?

Solution   For the assumed sinusoidal current distribution we use Eq. (11-51) for $I(z)$ and substitute it in Eq. (11-70), where $I(0) = I_m$ and $h = \lambda/4$. We have

$$\ell_e(\theta) = \sin\theta \int_{-\lambda/4}^{\lambda/4} \sin\beta\left(\frac{\lambda}{4} - |z|\right) e^{j\beta z \cos\theta}\, dz. \tag{11-72}$$

The above integral has been evaluated in Eq. (11-56). Thus,

$$\ell_e(\theta) = \frac{2}{\beta} \left[\frac{\cos\left(\dfrac{\pi}{2}\cos\theta\right)}{\sin\theta} \right]. \tag{11-73}$$

The maximum value of $\ell_e(\theta)$ is at $\theta = \pi/2$, where the effective length is

$$\ell_e\left(\frac{\pi}{2}\right) = \frac{2}{\beta} = \frac{\lambda}{\pi}. \tag{11-74}$$

We note from Eq. (11-74) that the maximum effective length of a half-wave dipole is less than its physical length, $\lambda/2$.

A careful examination of Eq. (11-71) reveals a potential anomaly in the appearance of $I(0)$ in the denominator. When the half-length of a dipole is greater than $\lambda/4$ and approaches $\lambda/2$, $I(0)$ would be progressively less than I_m, which would not occur at $z = 0$. This could make ℓ_e much greater than $2h$. Thus the definition of effective length as given in Eqs. (11-70) and (11-71) is meaningful only for relatively short antennas that have a current maximum at the feed point.

The effective length of a receiving linear antenna is defined as the ratio of the open-circuit voltage V_{oc} induced at the antenna terminals and the electric field intensity $E_i = |\mathbf{E}_i|$ at the antenna that induces it:

$$\ell_e(\theta) = -\frac{V_{oc}}{E_i}, \tag{11-75}$$

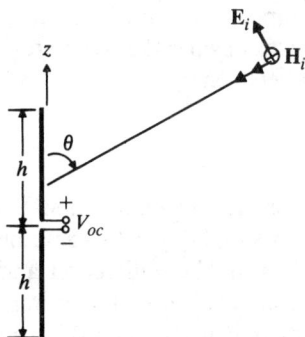

FIGURE 11-8
A linear antenna in the receiving mode.

where the negative sign is to conform with the convention that the electric potential increases in a direction opposite to that of the electric field. The situation is illustrated in Fig. 11–8. We will assume that $\mathbf{E}_i$ lies in the plane of incidence, since the component of $\mathbf{E}_i$ normal to the antenna does not induce a voltage across the antenna terminals. Obviously, the open-circuit voltage V_{oc} depends on E_i, θ, and βh in a complicated way. It is possible to use a reciprocity theorem to prove formally that *the effective length of an antenna for receiving is the same as that for transmitting* [14].[†] In Section 11–6 we shall prove that both the impedance and the directional pattern of an isolated antenna in the receiving mode are the same as those of the antenna in the transmitting mode. We may also conclude the equality of the effective lengths operating under these two modes.

If the incoming electric field $\mathbf{E}_i$ is not parallel to the dipole, there is a polarization mismatch, and the magnitude of the open-circuit voltage will be

$$|V_{oc}| = |\boldsymbol{\ell}_e \cdot \mathbf{E}_i|, \qquad (11\text{–}76)$$

where $\boldsymbol{\ell}_e$ denotes the vector effective length. Obviously, $|V_{oc}|$ will be maximum when $\mathbf{E}_i$ is parallel to the dipole and will be zero if $\mathbf{E}_i$ is perpendicular to the dipole.

11–5 Antenna Arrays

Antenna arrays are groups of similar antennas arranged in various configurations (straight lines, circles, triangles, and so on) with proper amplitude and phase relations to give certain desired radiation characteristics. Frequently, the radiation characteristics of importance are the direction and width of the main beam, sidelobe levels, and/or directivity. In this section we examine the basic theories and characteristics of linear antenna arrays (radiating elements arranged along a straight line). The electromagnetic field of an array is the vector superposition of the fields produced by

---

[†] Bracketed numbers refer to the literature listed in the reference section at the end of this chapter.

the individual antenna elements. We first consider the simplest case of two-element arrays. After some experience has been gained with them, we consider the basic properties of uniform linear arrays made up of many identical elements.

11–5.1 TWO-ELEMENT ARRAYS

The simplest array is one consisting of two identical radiating elements (antennas) spaced a distance apart. This is illustrated in Fig. 11–9. For simplicity, let us assume that the far-zone electric field of the individual antennas be in the θ-direction and that the antennas are lined along the x-axis. The antennas are excited with a current of the same magnitude, but the phase in antenna 1 leads that in antenna 0 by an angle ξ. We have

$$E_0 = E_m F(\theta, \phi) \frac{e^{-j\beta R_0}}{R_0}, \tag{11–77}$$

$$E_1 = E_m F(\theta, \phi) \frac{e^{j\xi} e^{-j\beta R_1}}{R_1}, \tag{11–78}$$

where $F(\theta, \phi)$ is the pattern function of the individual antennas, and E_m is an amplitude function. The electric field of the two-element array is the sum of E_0 and E_1. Hence,

$$E = E_0 + E_1 = E_m F(\theta, \phi) \left[\frac{e^{-j\beta R_0}}{R_0} + \frac{e^{j\xi} e^{-j\beta R_1}}{R_1} \right]. \tag{11–79}$$

In the far zone, $R_0 \gg d/2$, and the factor $1/R_1$ in the magnitude may be replaced approximately by $1/R_0$. However, a small difference between R_0 and R_1 in the exponents may lead to a significant phase difference, and a better approximation must be used. Because the lines joining the field point P and the two antennas are nearly parallel, we may write

$$R_1 \cong R_0 - d \sin \theta \cos \phi. \tag{11–80}$$

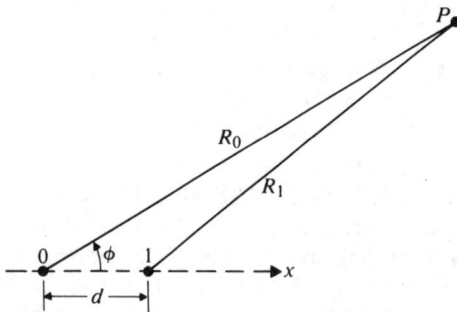

FIGURE 11–9
A two-element array.

Substitution of Eq. (11–80) in Eq. (11–79) yields

$$E = E_m \frac{F(\theta, \phi)}{R_0} e^{-j\beta R_0}\left[1 + e^{j\beta d \sin \theta \cos \phi}e^{j\xi}\right]$$

$$= E_m \frac{F(\theta, \phi)}{R_0} e^{-j\beta R_0}e^{j\psi/2}\left(2 \cos \frac{\psi}{2}\right),$$

(11–81)

where

$$\psi = \beta d \sin \theta \cos \phi + \xi.$$

(11–82)

The magnitude of the electric field of the array is

$$|E| = \frac{2E_m}{R_0} |F(\theta, \phi)|\left|\cos \frac{\psi}{2}\right|,$$

(11–83)

where $|F(\theta, \phi)|$ may be called the **element factor**, and $|\cos (\psi/2)|$ the normalized **array factor**. The element factor is the magnitude of the pattern function of the individual radiating elements, and the array factor depends on array geometry as well as on the relative amplitudes and phases of the excitations in the elements. (In this particular case the excitation amplitudes are equal.) The array factor is that of an array of *isotropic* elements, the directional property of the elements having been accounted for by the element factor. From Eq. (11–83) we may conclude that *the pattern function of an array of identical elements is described by the product of the element factor and the array factor*. This property is called the **principle of pattern multiplication**.

For an array of two parallel z-directed half-wave dipoles the magnitude of the total electric field is, from Eqs. (11–57) and (11–83),

$$|E| = \frac{2E_m}{R_0} \left|\frac{\cos \left[(\pi/2) \cos \theta\right]}{\sin \theta}\right|\left|\cos \frac{\psi}{2}\right|.$$

(11–84)

Since ψ is also a function of θ, we see that the pattern in an E-plane is not the same as that of a single dipole, except when $\phi = \pm\pi/2$. In the H-plane, $\theta = \pi/2$, and the pattern is determined entirely by the array factor $|\cos (\psi/2)|$.

EXAMPLE 11–7 Plot the H-plane radiation patterns of two parallel dipoles for the following two cases: (a) $d = \lambda/2$, $\xi = 0$; (b) $d = \lambda/4$, $\xi = -\pi/2$.

Solution Let the dipoles be z-directed and placed along the x-axis, as shown in Fig. 11–9. In the H-plane ($\theta = \pi/2$), each dipole is omnidirectional, and the normalized pattern function is equal to the normalized array factor $|A(\phi)|$. Thus

$$|A(\phi)| = \left|\cos \frac{\psi}{2}\right| = \left|\cos \frac{1}{2}(\beta d \cos \phi + \xi)\right|.$$

a) $d = \lambda/2$ $(\beta d = \pi)$, $\xi = 0$:

$$|A(\phi)| = \left|\cos \left(\frac{\pi}{2} \cos \phi\right)\right|.$$

(11–85a)

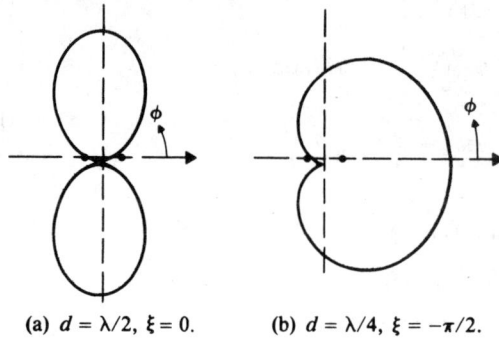

FIGURE 11–10
H-plane radiation patterns of two-element
parallel dipole array.

(a) $d = \lambda/2$, $\xi = 0$.   (b) $d = \lambda/4$, $\xi = -\pi/2$.

The pattern has its maximum at $\phi_0 = \pm\pi/2$—that is, in the broadside direction. This is a type of **broadside array**. Figure 11–10(a) shows this broadside pattern. Since the excitations in the two dipoles are in phase, their electric fields add in the broadside directions, $\phi = \pm\pi/2$. At $\phi = 0$ and π the electric fields cancel each other because the $\lambda/2$ separation leads to a phase difference of 180°.

b) $d = \lambda/4$ $(\beta d = \pi/2)$, $\xi = -\pi/2$:

$$|A(\phi)| = \left| \cos \frac{\pi}{4} (\cos \phi - 1) \right|, \tag{11–85b}$$

which has a maximum at $\phi_0 = 0$ and vanishes at $\phi = \pi$. The pattern maximum is now in a direction *along* the line of the array, and the two dipoles constitute an **endfire array**. Figure 11–10(b) shows this endfire pattern. In this case the phase in the right-hand dipole *lags* by $\pi/2$, which exactly compensates for the fact that its electric field arrives in the $\phi = 0$ direction a quarter of a cycle *earlier* than the electric field of the left-hand dipole. As a consequence, the electric fields add in the $\phi = 0$ direction. In the $\phi = \pi$ direction, the $\pi/2$ phase lag in the right-hand dipole plus the quarter-cycle delay results in a complete cancellation of the fields.

━━

EXAMPLE 11–8   Discuss the radiation pattern of a linear array of the three isotropic sources spaced $\lambda/2$ apart. The excitations in the sources are in-phase and have amplitude ratios 1:2:1.

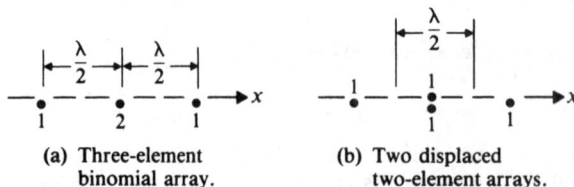

(a) Three-element
binomial array.

(b) Two displaced
two-element arrays.

FIGURE 11–11
A three-element array and its equivalent
pair of displaced two-element arrays.

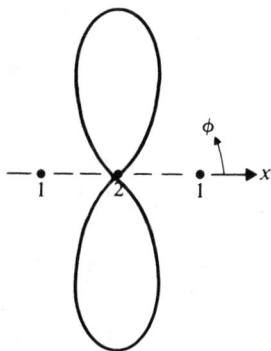

FIGURE 11–12
Radiation pattern of three-element broadside binomial array.

Solution This three-source array is equivalent to two two-element arrays displaced $\lambda/2$ from each other as depicted in Fig. 11–11. Each two-element array can be considered as as a radiating source with an element factor as given by Eq. (11–85a) and an array factor, which is also given by the same equation. By the principle of pattern multiplication we obtain

$$|E| = \frac{4E_m}{R_0}\left|\cos\left(\frac{\pi}{2}\cos\phi\right)\right|^2. \tag{11–86}$$

The radiation pattern represented by the pattern function $|\cos[(\pi/2)\cos\phi]|^2$ is sketched in Fig. 11–12. Compared to the pattern of the uniform two-element array in Fig. 11–10(a), this three-element broadside pattern is sharper (more directive). Both patterns have no sidelobes. ■

The three-element broadside array is a special case of a class of *sidelobeless* arrays called **binomial arrays**. In a binomial array of N elements the excitation amplitudes vary according to the coefficients of a binomial expansion $\binom{N-1}{n}$, $n = 0, 1, 2, \ldots,$ $N-1$. For $N = 3$ the relative excitation amplitudes are $\binom{2}{0} = 1$, $\binom{2}{1} = 2$ and $\binom{2}{2} = 1$, as in Example 11–8. To obtain a directive pattern without sidelobes, d in a binomial array is normally restricted to be $\lambda/2$. The feature of no sidelobes in the array pattern of a binomial array is accompanied by a wider beamwidth and a lower directivity compared to those of a uniform array with the same number of elements.

11–5.2 GENERAL UNIFORM LINEAR ARRAYS

We now consider an array of identical antennas equally spaced along a straight line. The antennas are fed with currents of equal magnitude and have a uniform progressive phase shift along the line. Such an array is called a **uniform linear array.** An example is shown in Fig. 11–13, where N antenna elements are aligned along the x-axis. Since the antenna elements are identical, the array pattern function is the product of the element factor and the array factor. Our attention here will be concentrated on the

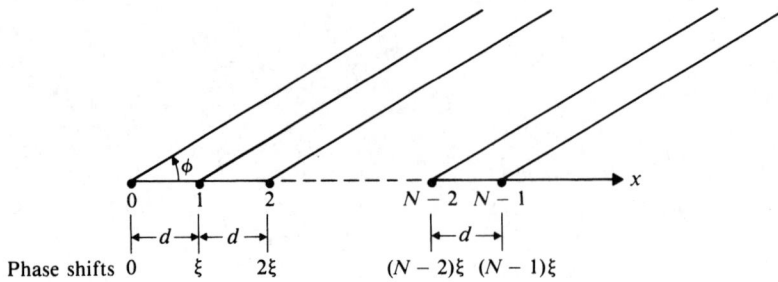

FIGURE 11–13
A general uniform linear array.

manner in which the array factor depends on the parameter βd $(=2\pi d/\lambda)$ and the progressive phase shift ξ between neighboring elements. The normalized array factor in the xy-plane is

$$|A(\psi)| = \frac{1}{N} \left|1 + e^{j\psi} + e^{j2\psi} + \cdots + e^{j(N-1)\psi}\right|, \tag{11–87}$$

where

$$\psi = \beta d \cos \phi + \xi. \tag{11–88}$$

The polynomial on the right side of Eq. (11–87) is a geometric progression and can be summed up in a closed form:

$$|A(\psi)| = \frac{1}{N} \left|\frac{1 - e^{jN\psi}}{1 - e^{j\psi}}\right|$$

or

$$\boxed{|A(\psi)| = \frac{1}{N} \left|\frac{\sin (N\psi/2)}{\sin (\psi/2)}\right| \qquad \text{(Dimensionless).}} \tag{11–89}$$

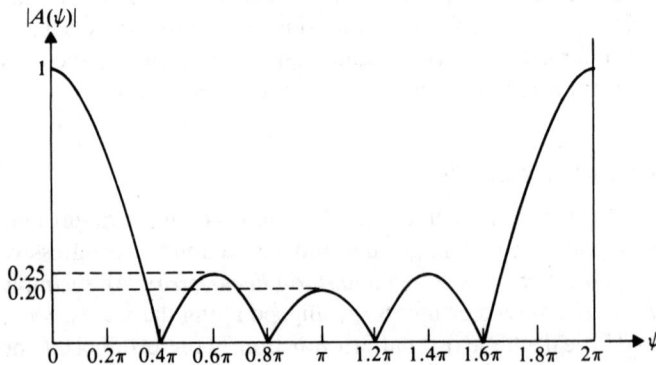

FIGURE 11–14
Normalized array factor of a five-element uniform linear array.

This is the general expression of the normalized array factor for a uniform linear array. Figure 11–14 is a sketch of the normalized array factor for a five-element array. The actual radiation pattern as a function of ϕ depends on the values of βd and ξ (see Problem P.11–17). As ϕ varies from 0 to 2π, the value of ψ changes from $\beta d + \xi$ to $-\beta d + \xi$, covering a range of $2\beta d$ or $4\pi d/\lambda$. This defines the **visible range** of the radiation pattern.

We may derive several significant properties from $|A(\psi)|$ as given in Eq. (11–89).

1. *Main-beam direction.* The maximum value occurs when $\psi = 0$ or when

$$\beta d \cos \phi_0 + \xi = 0,$$

which leads to

$$\cos \phi_0 = -\frac{\xi}{\beta d}. \tag{11–90}$$

Two special cases are of particular importance.

a) *Broadside array.* For a broadside array, maximum radiation occurs at a direction perpendicular to the line of the array—that is, at $\phi_0 = \pm \pi/2$. This requires $\xi = 0$, which means that all the elements in a linear broadside array should be excited in phase, as was the case in Example 11–7(a).

b) *Endfire array.* For an endfire array, maximum radiation occurs at $\phi_0 = 0$. Equation (11–90) gives

$$\xi = -\beta d \cos \phi_0 = -\beta d.$$

We note that this condition is satisfied by the two-element array in Example 11–7(b).

2. *Null locations.* The array pattern has nulls when $|A(\phi)| = 0$ or when

$$\frac{N\psi}{2} = \pm k\pi, \qquad k = 1, 2, 3, \dots . \tag{11 91}$$

It is obvious that the corresponding null locations in ϕ are different for broadside and endfire arrays because of the different values of ξ implicit in ψ.

3. *Width of main beam.* The angular width of the main beam between the first nulls can be determined approximately for large N. Let ψ_{01} denote the values of ψ at the first nulls:

$$\frac{N\psi_{01}}{2} = \pm \pi \qquad \text{or} \qquad \psi_{01} = \pm \frac{2\pi}{N}.$$

In order to see how ψ_{01} converts to an angle between the first nulls in ϕ, we need to know the direction of the main beam.

a) *Broadside array* ($\xi = 0$, $\phi_0 = \pi/2$). For a broadside array, $\psi = \beta d \cos \phi$. If the first null occurs at ϕ_{01}, then the width of the main beam between the first nulls is $2 \Delta\phi = 2(\phi_{01} - \phi_0)$. At ϕ_{01} we have

$$\cos \phi_{01} = \cos (\phi_0 + \Delta\phi) = \frac{\psi_{01}}{\beta d},$$

which, for $\phi_0 = \pi/2$, gives

$$\cos\left(\frac{\pi}{2} + \Delta\phi\right) = -\sin\Delta\phi = -\frac{2\pi}{N\beta d}$$

or

$$\Delta\phi = \sin^{-1}\left(\frac{\lambda}{Nd}\right) \cong \frac{\lambda}{Nd}. \qquad (11\text{-}92)$$

The last approximation is obtained when $Nd \gg \lambda$. Equation (11–92) leads to a useful rule of thumb that the width of the main beam (in radians) of a long uniform broadside array is approximately *twice* the reciprocal of the array length in wavelengths.

b) *Endfire array* $(\xi = -\beta d, \phi_0 = 0)$. For an endfire array, $\psi = \beta d(\cos\phi - 1)$, and

$$\cos\phi_{01} - 1 = \frac{\psi_{01}}{\beta d} = -\frac{2\pi}{N\beta d} = -\frac{\lambda}{Nd}.$$

But $\cos\phi_{01} = \cos\Delta\phi \cong 1 - (\Delta\phi)^2/2$ for small $\Delta\phi$. Thus,

$$\frac{(\Delta\phi)^2}{2} \cong \frac{\lambda}{Nd}$$

or

$$\Delta\phi \cong \sqrt{\frac{2\lambda}{Nd}}. \qquad (11\text{-}93)$$

Comparing Eq. (11–93) with Eq. (11–92), we may conclude that the width of the main beam of a uniform endfire array is greater than that of a uniform broadside array of the same length (because $Nd > \lambda/2$).

4. *Sidelobe locations.* Sidelobes are minor maxima that occur approximately when the numerator on the right side of Eq. (11–89) is a maximum—that is, when $|\sin(N\psi/2)| = 1$ or when

$$\frac{N\psi}{2} = \pm(2m + 1)\frac{\pi}{2}, \qquad m = 1, 2, 3, \ldots. \qquad (11\text{-}94)$$

The first sidelobes occur when

$$\frac{N\psi}{2} = \pm\frac{3}{2}\pi, \qquad (m = 1). \qquad (11\text{-}95)$$

Note that $N\psi/2 = \pm\pi/2\ (m = 0)$ does not represent locations of sidelobes because they are still within the main-lobe region.

5. *First sidelobe level.* An important characteristic of the radiation pattern of an array is the level of the first sidelobes compared to that of the main beam, since the former is usually the highest of all sidelobes. All sidelobes should be kept as low as possible in order that most of the radiated power be concentrated in the main-beam direction and not be diverted to sidelobe regions. Substituting Eq.

(11–95) in Eq. (11–89), we find the amplitude of the first sidelobes to be

$$\frac{1}{N}\left|\frac{1}{\sin{(3\pi/2N)}}\right| \cong \frac{1}{N}\left|\frac{1}{3\pi/2N}\right| = \frac{2}{3\pi} = 0.212$$

for large N. In logarithmic terms the first sidelobes of a uniform linear antenna array of many elements are $20 \log_{10} (1/0.212)$ or 13.5 (dB) down from the principal maximum. This number is almost independent of N as long as N is large.

One way to reduce the sidelobe level in the radiation pattern of a linear array is to taper the current distribution in the array elements—that is, to make the excitation amplitudes in the elements in the center portion of an array higher than those in the end elements. This method is illustrated in the following example.

■■■■■ **EXAMPLE 11–9** Find the array factor and plot the normalized radiation pattern of a broadside array of five isotropic elements spaced $\lambda/2$ apart and having excitation amplitude ratios $1:2:3:2:1$. Compare the first sidelobe level with that of a five-element uniform array.

Solution The normalized array factor of the five-element tapered array is

$$|A(\psi)| = \tfrac{1}{9}|1 + 2e^{j\psi} + 3e^{j2\psi} + 2e^{j3\psi} + e^{j4\psi}|$$
$$= \tfrac{1}{9}|e^{j2\psi}[3 + 2(e^{j\psi} + e^{-j\psi}) + (e^{j2\psi} + e^{-j2\psi})]| \qquad (11\text{–}96)$$
$$= \tfrac{1}{9}|3 + 4\cos\psi + 2\cos 2\psi|.$$

The graph of $|A(\psi)|$ versus ψ is shown in Fig. 11–15(a). Note that this figure holds for a general $\psi = \beta d \cos\phi + \xi$; the values of βd and ξ have not yet been specified.

In order to plot the desired radiation pattern we use the following additional information:

$$\text{Broadside radiation, } \zeta = 0: \quad \psi = \beta d \cos\phi;$$

$$\text{Element spacing, } d = \frac{\lambda}{2}: \quad \psi = \pi \cos\phi.$$

The normalized radiation pattern can be plotted from

$$|A(\phi)| = \tfrac{1}{9}|3 + 4\cos{(\pi \cos\phi)} + 2\cos{(2\pi \cos\phi)}|.$$

However, having calculated and plotted $|A(\psi)|$, we do not need to recalculate the array factor as a function of ϕ. This conversion can be effected graphically as follows (see Fig. 11–15):

1. Extend the vertical axis of the array-factor graph downward, and let it intersect with a horizontal line (which represents the line for $\phi = 0$ and $\phi = \pi$). The point of intersection is the point for $\xi = 0$.

2. Locate the point, P_0, on the horizontal line that is ξ radians to the right or left of the point of intersection, depending on whether ξ is positive or negative. (In the present case, $\xi = 0$ and P_0 is at the point of intersection.)

3. Using P_0 as the center, draw a circle with βd as the radius.

4. For any angle ϕ_1, draw the radius vector P_0P_1. (The projection P_0P_1' is equal to $\psi_1 = \beta d \cos \phi_1$.)

5. At ψ_1, measure the magnitude of $|A(\psi_1)|$, which is marked as P_2 on the radius vector P_0P_1. (P_2 is a point on the normalized radiation pattern.)

Repeat this process until the entire radiation pattern is obtained.

Figure 11–15(b) shows the normalized radiation pattern of this five-element broadside array with tapered excitation. The first sidelobe level is found to be 0.11

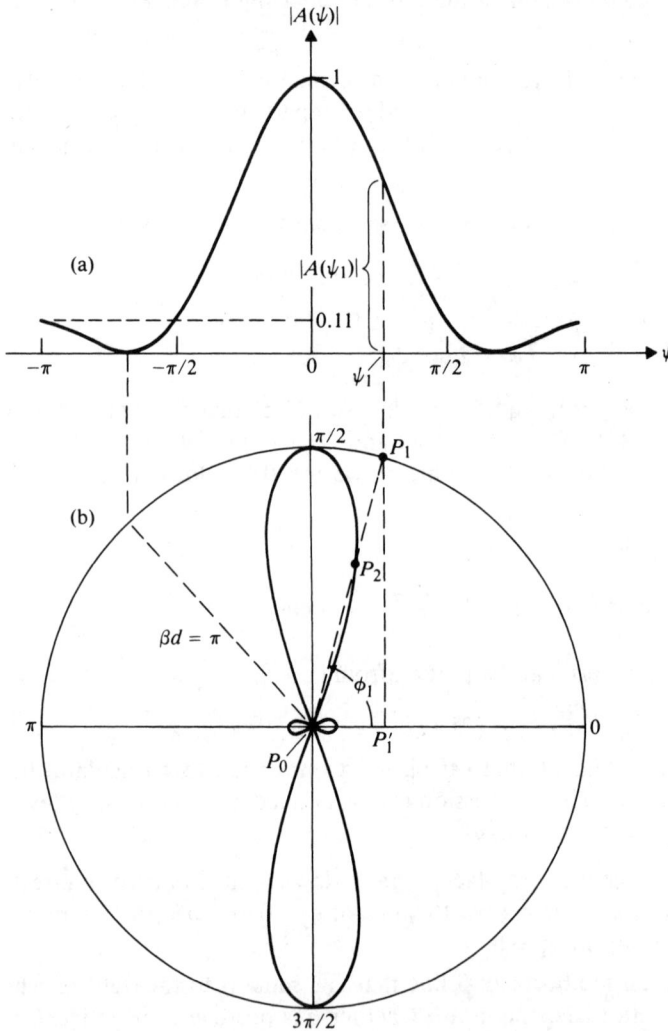

FIGURE 11–15
(a) Graph for normalized array factor as a function of ψ, and (b) normalized polar radiation pattern of a five-element broadside array with $d = \lambda/2$ and tapered excitation amplitude ratios 1:2:3:2:1 (Example 11–9).

or 20 $\log_{10}$ (1/0.11) = 19.2 (dB) down from the main-beam radiation. This compares with 0.25 or 12 (dB) down for the five-element uniform broadside array shown in Fig. 11–14. ▬

   In the discussion of uniform linear arrays we started out with the assumptions of equal spacing, equal excitation amplitude, and constant progressive phase shifts. The main reason for making these assumptions is mathematical simplicity in analyzing radiation characteristics. The preceding example shows that a tapered nonuniform amplitude distribution in the array elements produces the desirable result of a reduction in the sidelobe level. In a similar manner the spacings between neighboring elements may be made unequal [1]–[4], and the phase shifts do not have to be constant [5]. In two-dimensional arrays the elements need not be arranged in a rectangular lattice [6], [7]. We have, then, many additional parameters that can be adjusted to achieve desirable results. Adjustments in these parameters, however, destroy the simplicity of the analysis. There are techniques for synthesizing an antenna array to approximate a specified radiation pattern closely. It is not possible to examine all the various possible array designs in this book, but they do exist and present themselves as interesting problems [8]–[12].
   Our discussions on linear arrays can be extended to two-dimensional rectangular arrays. A rectangular array can be studied as an array of linear arrays, to which the principle of pattern multiplication applies. From Eq. (11–90) we note that the direction of the main beam of a uniform linear array can be changed by simply changing the amount of progressive phase shift ξ. In fact, the radiation pattern can be changed from broadside ($\xi = 0$) to endfire ($\xi = -\beta d$) or to somewhere in between. We see here a possibility of *scanning* the main beam by simply varying ξ. This can be achieved in practice by using electronically controlled phase shifters. Antenna arrays equipped with phase shifters to steer the main beam electronically are called ***phased arrays***. The main beam of a two-dimensional array can be made to scan in both θ (elevation) and ϕ (azimuth) directions. Scanning phased arrays are of great practical importance in radar and radioastronomy work, in which the antenna system may be arrays of many thousands of elements that are not amenable to rapid mechanical motion for beam steering. Time-delay circuits may also be used to furnish the required phase shifts to the various array elements. By changing the frequency the time-delays are translated into varying phase shifts. This scheme is called ***frequency scanning***.

11–6 Receiving Antennas

In the discussion of antennas and antenna arrays so far we have implied that they operate in a transmitting mode. In the transmitting mode a voltage source is applied to the input terminals of an antenna, setting up currents and charges on the antenna structure. The time-varying currents and charges, in turn, radiate electromagnetic waves, which carry energy and/or information. A transmitting antenna can then be regarded as a device that transforms energy from a source (a generator) to energy

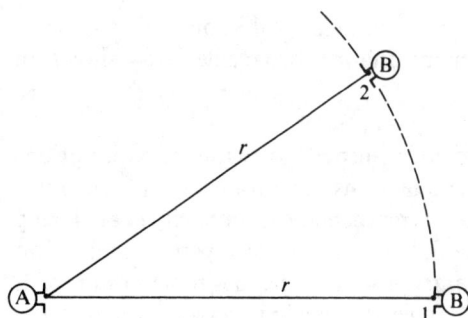

FIGURE 11–16
Two coupled antennas.

associated with an electromagnetic wave. A receiving antenna, on the other hand, extracts energy from an incident electromagnetic wave and delivers it to a load. In the receiving mode the external electromagnetic field that causes currents and charges to flow is incident on the entire antenna structure, not just at the input terminals. Moreover, the induced currents and charges, which depend on the direction of arrival of the incident electromagnetic wave, will produce reradiation or scattering of electromagnetic energy, making the situation very complicated. We may reasonably expect that the current and charge distributions on an antenna in the receiving mode are different from those in the transmitting mode. Nevertheless, despite these differences, reciprocity relations enable us to conclude that (1) the equivalent generator impedance of an antenna in the receiving mode is equal to the input impedance of the antenna in the transmitting mode, and (2) the directional pattern of an antenna for reception is identical with that for transmission. We will justify these two important conclusions by using equivalent network representations. Also in this section we will discuss the concepts of effective area and backscatter cross section.

11–6.1   INTERNAL IMPEDANCE AND DIRECTIONAL PATTERN

Let us assume that a transmitter with antenna A radiates electromagnetic energy, which is absorbed by a distant receiver with antenna B. Antenna B moves about antenna A at a constant distance $r^\dagger$ and is always oriented in such a way as to receive maximum power, as illustrated in Fig. 11–16. The two coupled antennas and the space between can be represented as a two-port T-network shown in Fig. 11–17. The terminal characteristics, (V_1, I_1) and (V_2, I_2) of antennas A and B, respectively, are linearly related by the following equations:

$$V_1 = Z_{11}I_1 + Z_{12}I_2, \tag{11–97}$$

$$V_2 = Z_{21}I_1 + Z_{22}I_2, \tag{11–98}$$

where Z_{11}, Z_{12}, Z_{21}, and Z_{22} are open-circuit impedance coefficients.

---

† The symbol r, instead of R, is used here to denote distance in order to avoid possible confusion of the latter with the symbol for resistance used later in this chapter.

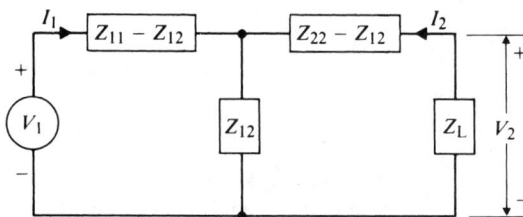

FIGURE 11–17
Equivalent two-port network of coupled transmitting and receiving antennas.

When the medium in the transmission path between antennas A and B is bilateral such that reciprocity relations hold, the transfer or coupling impedances Z_{12} and Z_{21} are equal.[†] Under normal circumstances, transmitting and receiving antennas are separated by very large distances, and the coupling impedances are negligibly small as far as *the reaction on the transmitting antenna* owing to scattering by the receiving antenna is concerned. In the limit $r \rightarrow \infty$,

$$\lim_{r \to \infty} Z_{12} = 0. \tag{11-99}$$

The parallel arm of the T-network in Fig. 11–17 is almost a short-circuit, and the impedance coefficients Z_{11} and Z_{22} are nearly equal to the input impedances Z_A and Z_B, respectively, of isolated antennas A and B in the transmitting mode. Equation (11–97) can be written approximately as

$$V_1 \cong Z_{11}I_1 \cong Z_A I_1. \tag{11-100}$$

An equivalent circuit representing Eq. (11–100) is drawn in Fig. 11–18(a).

The coupling *from* the transmitting antenna *to* the receiving antenna, however, cannot be neglected inasmuch as it is through this coupling that the latter extracts energy from the electromagnetic wave originated from the former. Thévenin's theorem can be applied to the left of the load impedance Z_L in the network in Fig. 11–17 to determine an open-circuit voltage V_{oc} and an internal impedance Z_g. An equivalent circuit at the receiving end is shown in Fig. 11–18(b). We have

$$V_{oc} = \frac{Z_{12}}{Z_{11}} V_1, \tag{11-101}$$

$$Z_g = (Z_{22} - Z_{12}) + \frac{Z_{12}}{Z_{11}}(Z_{11} - Z_{12}) = Z_{22} - \frac{Z_{12}^2}{Z_{11}}. \tag{11-102}$$

Because of the weak coupling, we conclude that *the equivalent generator internal impedance Z_g for antenna B in the receiving mode is approximately equal to its input impedance when it is transmitting*; that is,

$$Z_g \cong Z_{22} \cong Z_B. \tag{11-103}$$

[†] An example of a nonbilateral medium for which $Z_{12} \neq Z_{21}$ is the ionosphere under the influence of the earth's magnetic field.

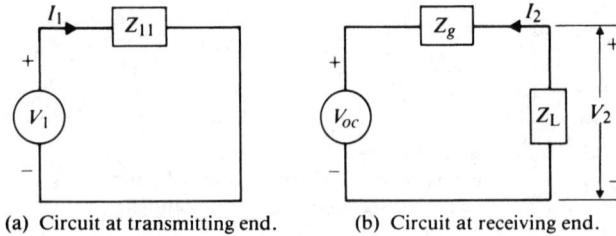

(a) Circuit at transmitting end.   (b) Circuit at receiving end.

FIGURE 11–18
Approximate equivalent circuits for
weakly coupled antennas.

When antenna B is receiving, $V_2 = -I_2 Z_L$, and Eq. (11–98) becomes

$$I_2 = -\frac{Z_{21}}{Z_{22} + Z_L} I_1. \qquad (11\text{–}104)$$

The time-average power absorbed in Z_L is

$$P_L = \frac{1}{2} \mathcal{R}e[-V_2 I_2^*] = \frac{|I_1|^2}{2} \left| \frac{Z_{21}}{Z_{22} + Z_L} \right|^2 \mathcal{R}e(Z_L). \qquad (11\text{–}105)$$

For two successive positions of antenna B as indicated in Fig. 11–16 the ratio of the absorbed powers in Z_L is

$$\frac{P_L(\theta_1, \phi_1)}{P_L(\theta_2, \phi_2)} = \left| \frac{Z_{21}(\theta_1, \phi_1)}{Z_{21}(\theta_2, \phi_2)} \right|^2. \qquad (11\text{–}106)$$

Thus the absorbed power is proportional to the square of the transfer impedance coefficient.

If we consider the situation in which antenna B is transmitting and antenna A is receiving, then the ratio of the absorbed powers in Z_L connected to antenna A for the two successive locations of antenna B would be the same as that given in Eq. (11–106), except that Z_{21} would be replaced by Z_{12}. Because of the reciprocity relation $Z_{12} = Z_{21}$, we conclude that *the directional pattern of an antenna for reception is identical with that for transmission.*

11–6.2 EFFECTIVE AREA

In discussing receiving antennas it is convenient to define a quantity called the *effective area.*[†] The effective area, A_e, of a receiving antenna is the ratio of the average power delivered to a *matched load* to the time-average power density (time-average Poynting vector) of the incident electromagnetic wave at the antenna. We write

$$P_L = A_e \mathcal{P}_{av}, \qquad (11\text{–}107)$$

where P_L is the maximum average power transferred to the load (under matched conditions) with the receiving antenna properly oriented with respect to the polariza-

---

[†] Also called *effective aperture* or *receiving cross section.*

tion of the incident wave. We will now show that the effective area bears a definite relationship with the directive gain of an antenna.

When the load impedance is matched to the internal impedance,

$$Z_L = Z_g^* \cong R_B - jX_B, \tag{11–108}$$

the maximum power delivered to the load is, from Eq. (11–105),

$$P_L = \frac{|I_1 Z_{21}|^2}{8R_B}. \tag{11–109}$$

Let R_A be the input resistance of transmitting antenna A. The transmitted power is then

$$P_t = \tfrac{1}{2}|I_1|^2 R_A. \tag{11–110}$$

Combining Eqs. (11–109) and (11–110), we have

$$\frac{P_L}{P_t} = \frac{|Z_{21}|^2}{4R_A R_B}. \tag{11–111}$$

When antenna B is receiving, the time-average power density at B depends on the directive gain of transmitting antenna A in that direction:

$$\mathscr{P}_{av} = \frac{P_t}{4\pi r^2} G_{DA}. \tag{11–112}$$

Using Eq. (11–112) in Eq. (11–107), we obtain

$$\frac{P_L}{P_t} = \frac{A_{eB} G_{DA}}{4\pi r^2}. \tag{11–113}$$

Comparison of Eqs. (11–111) and (11–113) yields

$$|Z_{21}|^2 = \frac{R_A R_B A_{eB} G_{DA}}{\pi r^2}. \tag{11–114}$$

If antenna B is transmitting and antenna A is receiving, a similar derivation leads to

$$|Z_{12}|^2 = \frac{R_B R_A A_{eA} G_{DB}}{\pi r^2}. \tag{11–115}$$

Since $Z_{12} = Z_{21}$, Eqs. (11–114) and (11–115) lead to the following important relation:

$$\frac{G_{DA}}{A_{eA}} = \frac{G_{DB}}{A_{eB}}. \tag{11–116}$$

Inasmuch as we have not specified the types of transmitting and receiving antennas in obtaining Eq. (11–116), we conclude that *the ratio of the directive gain and the effective area of an antenna is a universal constant*. This constant can be found by determining the directive gain and effective area of any antenna—for instance, those of an elemental dipole as illustrated in the following example.

EXAMPLE 11–10   Determine the effective area, $A_e(\theta)$, of an elemental electric dipole of a length $d\ell$ ($\ll \lambda$) used to receive an incident plane electromagnetic wave of wavelength λ that is polarized in a direction shown in Fig. 11–8.

Solution   Let E_i be the amplitude of the electric field intensity at an elemental dipole of length $d\ell$. Then the time-average power density is

$$\mathscr{P}_{av} = \frac{E_i^2}{2\eta_0}. \tag{11–117}$$

The average power delivered to a matched load ($Z_L = Z_g^*$) is

$$P_L = \frac{1}{2}\left|\frac{E_i\, d\ell \sin\theta}{Z_g + Z_g^*}\right|^2 R_r = \frac{(E_i\, d\ell)^2 \sin^2\theta}{8R_r}, \tag{11–118}$$

where $R_r = 80(\pi\, d\ell/\lambda)^2$ has been given in Eq. (11–44). The ratio $P_L/\mathscr{P}_{av}$ gives the effective area of the elemental dipole:

$$\begin{aligned} A_e(\theta) &= \frac{P_L}{\mathscr{P}_{av}} = \frac{\eta_0}{4R_r}(d\ell)^2 \sin^2\theta \\ &= \frac{3}{8\pi}(\lambda \sin\theta)^2 \qquad (\text{m}^2). \end{aligned} \tag{11–119}$$

It is interesting to note that the effective area of an elemental electric dipole is independent of its length.   ■

From Example 11–2 we have $G_D(\theta) = \frac{3}{2}\sin^2\theta$ for an elemental electric dipole. Thus,

$$\begin{aligned} G_D(\theta) &= \frac{3}{2}\sin^2\theta = \frac{4\pi}{\lambda^2}\frac{3}{8\pi}(\lambda \sin\theta)^2 \\ &= \frac{4\pi}{\lambda^2} A_e(\theta), \end{aligned} \tag{11–120}$$

which indicates that the universal constant for Eq. (11–116) is $4\pi/\lambda^2$, and we may write the following relation for an antenna under matched impedance conditions:

$$\boxed{G_D(\theta, \phi) = \frac{4\pi}{\lambda^2} A_e(\theta, \phi) \qquad (\text{Dimensionless}).} \tag{11–121}$$

In the case of thin linear antennas the concept of effective area may seem arbitrary. Nevertheless, its definition is useful in measuring the power available to a particular antenna. Of course, we expect the effective area $A_e(\theta)$ to be related to the effective length $\ell_e(\theta)$. The available power to the antenna load under matched conditions is

$$P_L = \frac{V_{oc}^2}{8R_r} = \frac{(-\ell_e E_i)^2}{8R_r}, \tag{11–122}$$

where the relation in Eq. (11–75) has been used. Substitution of Eqs. (11–117) and (11–122) into Eq. (11–107) yields

$$A_e(\theta) = \frac{30\pi}{R_r} \ell_e^2(\theta).$$

(11–123)

11-6.3 BACKSCATTER CROSS SECTION

As we saw in the preceding subsection, the concept of effective area pertains to the power available to the matched load of a receiving antenna for a given incident power density. In cases in which the incident wave impinges on a passive object whose purpose is not to extract energy from the incident wave but whose presence creates a scattered field, it is appropriate to define a quantity called the **backscatter cross section**, or **radar cross section**. The backscatter cross section of an object is the equivalent area that would intercept that amount of incident power in order to produce the same scattered power density at the receiver site if the object scattered uniformly (isotropically) in all directions. Let

$\mathscr{P}_i$ = Time-average incident power density at the object   (W/m²),

$\mathscr{P}_s$ = Time-average scattered power density at the receiver site   (W/m²),

σ_{bs} = Backscatter cross section   (m²),

r = Distance between scatterer and receiver   (m).

Then,

$$\frac{\sigma_{bs}\mathscr{P}_i}{4\pi r^2} = \mathscr{P}_s$$

or

$$\sigma_{bs} = 4\pi r^2 \frac{\mathscr{P}_s}{\mathscr{P}_i} \quad (\text{m}).$$

(11–124)

Note that $\mathscr{P}_s$ is inversely proportional to r^2 for large r and that σ_{bs} does not change with r.

The backscatter cross section is a measure of the detectability of the object (target) by **radar** (**ra**dio **d**etection **a**nd **r**anging); hence the term radar cross section. It is a composite measure, depending on the geometry, orientation, and constitutive parameters of the object, and on the frequency and polarization of the incident wave in a complicated way.

■■■■ **EXAMPLE 11–11**   A uniform plane wave with electric field intensity $\mathbf{E}_i = \mathbf{a}_z E_i$ impinges on a small dielectric sphere of radius b ($\ll \lambda$) and dielectric constant ϵ_r, as shown in Fig. 11–19. Assume the polarization produced in the sphere to be the same

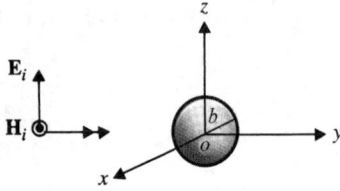

FIGURE 11–19
Plane wave incident on a small dielectric sphere.

as that produced in a uniform static electric field $\mathbf{E}_i$ and to be given by (see Problem P.4–29)

$$\mathbf{P} = \epsilon_0(\epsilon_r - 1)\mathbf{E}$$

$$= \mathbf{a}_z 3\epsilon_0 \left(\frac{\epsilon_r - 1}{\epsilon_r + 2}\right) E_i \qquad (\text{C/m}^2). \qquad (11\text{--}125)$$

a) Find the backscatter cross section σ_{bs}.

b) Determine σ_{bs} for a spherical raindrop of diameter 3 (mm) at 15 (GHz), assuming the dielectric constant of water to be 55 at that frequency.

Solution

a) Since the induced polarization vector (the volume density of electric dipole moment) $\mathbf{P}$ is constant within the dielectric sphere, the total electric dipole moment induced in the sphere of radius $b\,(\ll \lambda)$ is

$$\mathbf{p} = \tfrac{4}{3}\pi b^3 \mathbf{P}$$

$$= \mathbf{a}_z 4\pi b^3 \epsilon_0 \left(\frac{\epsilon_r - 1}{\epsilon_r + 2}\right) E_i \qquad (\text{C·m}). \qquad (11\text{--}126)$$

Thus the dielectric sphere acts electromagnetically like an elemental electric dipole of moment $\mathbf{p}$ given in Eq. (11–126). The scattered electric field intensity in the far-zone is then, from Eq. (11–19b) and using Eqs. (11–10) and (11–11),

$$\mathbf{E}_s = \mathbf{a}_\theta E_s = -\mathbf{a}_\theta \frac{\omega p}{4\pi}\left(\frac{e^{-j\beta r}}{r}\right)\eta_0 \beta \sin\theta$$

$$= -\mathbf{a}_\theta \beta^2 b^3 \left(\frac{e^{-j\beta r}}{r}\right)\left(\frac{\epsilon_r - 1}{\epsilon_r + 2}\right) E_i \sin\theta \qquad (\text{V/m}). \qquad (11\text{--}127)$$

The time-average backscattered power density is

$$\mathscr{P}_s = \frac{1}{2\eta_0}|E_s|^2_{\theta=\pi/2} = \frac{b^2(\beta b)^4}{2\eta_0 r^2}\left(\frac{\epsilon_r - 1}{\epsilon_r + 2}\right)^2 E_i^2 \qquad (\text{W/m}^2). \qquad (11\text{--}128)$$

The time-average incident power density is

$$\mathscr{P}_i = \frac{1}{2\eta_0} E_i^2 \qquad (\text{W/m}^2). \qquad (11\text{--}129)$$

Substitution of Eqs. (11–128) and (11–129) in Eq. (11–124) yields the backscatter cross section:

$$\sigma_{bs} = 4\pi b^2 (\beta b)^4 \left(\frac{\epsilon_r - 1}{\epsilon_r + 2}\right)^2 \quad (\text{m}^2). \qquad (11\text{–}130)$$

b) For $f = 15$ (GHz), $\lambda = 20$ (mm), the radius of the raindrop $b = \frac{3}{2}$ (mm) $\ll \lambda$. We obtain

$$\sigma_{bs} = 1.25 \times 10^{-6} \quad (\text{m}^2)$$
$$= 1.25 \quad (\text{mm}^2),$$

which is a fraction of the geometrical cross section πb^2 of the sphere:

$$\frac{\sigma_{bs}}{\pi b^2} = \frac{1.25}{1.5^2 \pi} = 0.177.$$

$\blacksquare$

Of course, raindrops do not exist singly; nor is their shape strictly spherical. Meaningful calculations of backscatter from rain require a knowledge of the rainfall rate and the distribution of the drop size, which are mutually dependent. The assumption of an equivalent spherical drop for nonspherical droplets has been found to be acceptable as long as their sizes are much smaller than the wavelength. Of equal importance to the calculation of backscatter from rainfall is the estimation of the attenuation suffered by an electromagnetic wave propagating through rain due to an imaginary part of the permittivity of raindrops. Interested readers should refer to the literature for details. [13]

11–7 Transmit-Receive Systems

In the preceding section we discussed the concepts of effective area for receiving antennas and backscatter cross section for scattering objects. We shall now examine the power transmission relation between transmitting and receiving antennas. When the same antenna is used for transmitting short pulses of radiation and for receiving them after they have been reflected (scattered) back by a target, the transmit-receive system is a *radar*; it is a special case. Measurement of the time elapsed Δt between the transmitted pulse and the received pulse determines the distance r of the target to the antenna site through the relation $\Delta t = 2r/c$, where c is the velocity of light.

If the transmission path between the transmitting and receiving antennas is near the earth's surface, the effect of the conducting earth must be considered. We shall also discuss the transmit-receive arrangement over a flat earth in this section.

11–7.1   FRIIS TRANSMISSION FORMULA AND RADAR EQUATION

Consider a communication circuit between stations 1 and 2 with antennas having effective areas A_{e1} and A_{e2}, respectively. The antennas are separated by a distance r. We wish to find a relation between the transmitted and the received powers.

Let P_L and P_t be the received and transmitted powers, respectively. Combining Eqs. (11–113) and (11–121), we obtain

$$\frac{P_L}{P_t} = \left(\frac{A_{e2}}{4\pi r^2}\right) G_{D1} = \left(\frac{A_{e2}}{4\pi r^2}\right)\left(\frac{4\pi A_{e1}}{\lambda^2}\right)$$

or

$$\boxed{\frac{P_L}{P_t} = \frac{A_{e1}A_{e2}}{r^2\lambda^2}.}$$ (11–131)

The relation in Eq. (11–131) is referred to as the **Friis transmission formula**. For a given transmitted power the received power is directly proportional to the product of the effective areas of the transmitting and receiving antennas and is inversely proportional to the square of the product of the distance of separation and wavelength.

Noting Eq. (11–121), we may write the Friis transmission formula in the following alternative form:

$$\boxed{\frac{P_L}{P_t} = \frac{G_{D1}G_{D2}\lambda^2}{(4\pi r)^2}.}$$ (11–132)

The received power P_L in Eqs. (11–131) and (11–132) assumes a matched condition and disregards the power dissipated in the antenna itself. From Eq. (11–131) we see that for a given transmitted power the received power increases as the square of the operating frequency (decreases as the inverse square of wavelength). But, at progressively increasing frequencies, P_t is limited by available technology, and the minimum detectable power over electromagnetic noise also increases. It is incorrect to conclude from Eq. (11–132) that P_L increases as the square of the wavelength because the directive gains usually decrease as the wavelength increases.

Now consider a radar system that uses the same antenna for transmitting short pulses of time-harmonic radiation and for receiving the energy scattered back from a target. For a transmitted power P_t the power density at a target at a distance r away is (see Eq. 11–112)

$$\mathcal{P}_T = \frac{P_t}{4\pi r^2} G_D(\theta, \phi),$$ (11–133)

where $G_D(\theta, \phi)$ is the directive gain of the antenna in the direction of the target. If σ_{bs} denotes the backscatter or radar cross section of the target, then the equivalent power that is scattered isotropically is $\sigma_{bs}\mathcal{P}_T$, which results in a power density at the antenna $\sigma_{bs}\mathcal{P}_T/4\pi r^2$. Let A_e be the effective area of the antenna. We have the following expression for the received power:

$$P_L = A_e\sigma_{bs}\frac{\mathcal{P}_T}{4\pi r^2}$$

$$= A_e\sigma_{bs}\frac{P_t}{(4\pi r^2)^2} G_D(\theta, \phi).$$ (11–134)

By using Eq. (11–121), Eq. (11–134) becomes

$$\frac{P_L}{P_t} = \frac{\sigma_{bs}\lambda^2}{(4\pi)^3 r^4} G_D^2(\theta, \phi),$$

(11–135)

which is called the **radar equation**. In terms of the antenna effective area A_e instead of the directive gain $G_D(\theta, \phi)$, the radar equation can be written as

$$\frac{P_L}{P_t} = \frac{\sigma_{bs}}{4\pi}\left(\frac{A_e}{\lambda r^2}\right)^2.$$

(11–136)

Because radar signals have to make round trips from the antenna to the target and then back to the antenna, the received power is inversely proportional to the fourth power of the distance r of the target from the antenna.

EXAMPLE 11–12   Assume that 50 (kW) is fed into the antenna of a radar system operating at 3 (GHz). The antenna has an effective area of 4 (m²) and a radiation efficiency of 90%. The minimum detectable signal power (over noise inherent in the receiving system and from the environment) is 1.5 (pW), and the power reflection coefficient for the antenna on receiving is 0.05. Determine the maximum usable range of the radar for detecting a target with a backscatter cross section of 1 (m²).

Solution   At $f = 3 \times 10^9$ (Hz), $\lambda = 0.1$ (m):

$$A_e = 4 \quad (\text{m}^2),$$
$$P_t = 0.90 \times 5 \times 10^4 = 4.5 \times 10^4 \quad (\text{W}),$$
$$P_L = 1.5 \times 10^{-12}\left(\frac{1}{1 - 0.05}\right) = 1.58 \times 10^{-12} \quad (\text{W}),$$
$$\sigma_{bs} = 1 \quad (\text{m}^2).$$

From Eq. (11–136),

$$r^4 = \frac{\sigma_{bs} A_e^2}{4\pi\lambda^2}\left(\frac{P_t}{P_L}\right),$$

and

$$r = 4.36 \times 10^4 \text{ (m)} = 43.6 \quad (\text{km}).$$

A satellite communication system makes use of satellites traveling in orbits in the earth's equatorial plane. The speed of the satellites and the radius of their orbits are such that the period of rotation of the satellites around the earth is the same as that of the earth. Thus the satellites appear to be stationary with respect to the earth's surface, and they are said to be geostationary. The radius of the geosynchronous orbit is 42,300 (km). With an earth radius of 6,380 (km) the satellites are about 36,000 (km) from the earth's surface.

Signals are transmitted from a high-gain antenna at an earth station toward a satellite, which receives the signals, amplifies them, and retransmits them back toward the earth station at a different frequency. Three satellites equally spaced around the geosynchronous orbit would cover almost the entire earth's surface except the polar regions (see Problem P.11–27). A quantitative analysis of the power and antenna gain relations for a satellite communication circuit requires the application of the Friis transmission formula twice, once for the uplink (earth station to satellite) and once for the downlink (satellite to earth station).

11–7.2   WAVE PROPAGATION NEAR EARTH'S SURFACE

Consider a transmitting antenna A at a height h_1 and a receiving antenna B at a height h_2 above the flat earth's surface with a distance of separation d, as shown in Fig. 11–20. If antenna A is an elemental electric dipole, then the electric field intensity at B is the sum of the direct contribution $\mathbf{E}_{\theta 1}$ from A and the indirect contribution $\mathbf{E}_{\theta 2}$ after reflection at point C. We write

$$\mathbf{E} = \mathbf{E}_{\theta 1} + \mathbf{E}_{\theta 2}, \tag{11–137}$$

where the magnitudes of $\mathbf{E}_{\theta 1}$ and $\mathbf{E}_{\theta 2}$ are

$$E_{\theta 1} = K\left(\frac{e^{-j\beta R}}{R}\right)\sin\theta, \tag{11–137a}$$

$$E_{\theta 2} = K\left(\frac{e^{-j\beta R'}}{R'}\right)\sin\theta'. \tag{11–137b}$$

The constant K equals $jI\,d\ell\eta_0\beta/4\pi$ (see Eq. 11–19b), and the distance $R' = \overline{AC} + \overline{CB} = \overline{A'B}$. The effect of the perfectly conducting (assumed) earth's surface is replaced by an image antenna at A'. In the general case, $\mathbf{E}_{\theta 1}$ and $\mathbf{E}_{\theta 2}$ are not parallel; but if $d \gg h_1, h_2$, then $\theta \cong \theta'$, and Eqs. (11–137a) and (11–137b) may be combined to give

$$E_\theta \cong \mathbf{a}_\theta K\left(\frac{e^{-j\beta R}}{R}\right)(\sin\theta)F, \tag{11–138}$$

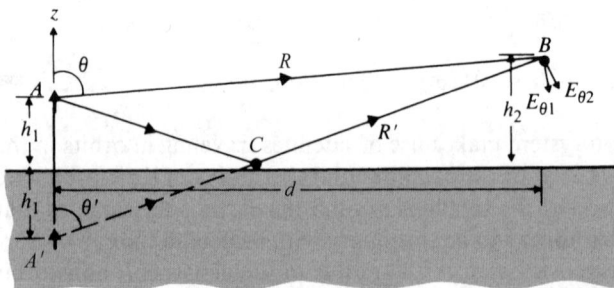

FIGURE 11–20
Transmit-receive system near the earth's surface.

where

$$F = 1 + e^{-j\beta(R' - R)}. \qquad (11\text{–}139)$$

The distances

$$R = [d^2 + (h_2 - h_1)^2]^{1/2}$$

$$= d\left[1 + \frac{(h_2 - h_1)^2}{d^2}\right]^{1/2} \cong d + \frac{(h_2 - h_1)^2}{2d} \qquad (11\text{–}140\text{a})$$

and

$$R' = [d^2 + (h_2 + h_1)^2]^{1/2} \cong d + \frac{(h_2 + h_1)^2}{2d} \qquad (11\text{–}140\text{b})$$

yield approximately

$$R' - R \cong \frac{(h_2 + h_1)^2}{2d} - \frac{(h_2 - h_1)^2}{2d}$$

$$= \frac{2h_1 h_2}{d}. \qquad (11\text{–}141)$$

Substituting Eq. (11–141) in Eq. (11–139), we obtain

$$|F| = \left|1 + e^{-j\beta 2h_1 h_2/d}\right|, \qquad (11\text{–}142)$$

which is like the array factor of a two-element array.

Equation (11–142) may be written as

$$|F| = \left|e^{-j\beta h_1 h_2/d}\left(e^{j\beta h_1 h_2/d} + e^{-j\beta h_1 h_2/d}\right)\right| = 2\left|\cos\left(\frac{2\pi h_1 h_2}{\lambda d}\right)\right|. \qquad (11\text{–}143)$$

Equation (11–143) shows that for fixed values of h_1 and λ the electric field intensity E_θ at the receiving site B will have nulls and maximum values as the ratio h_2/d is changed. The quantity $|F|$ varies from 0 to 2 and is called the ***path-gain factor***. Calculation of the path-gain factor for a spherical earth is a much more involved task.

11–8 Some Other Antenna Types

Practical antennas take many different shapes and sizes, each designed to fulfill certain desired performance characteristics. Our attention so far has been focused on the radiation properties of linear antennas having a current distribution in the form of a standing wave. In this section we shall discuss several other types of antennas of practical importance.

11–8.1   TRAVELING-WAVE ANTENNAS

In the analysis of thin linear antennas in Section 11–4 we assumed that the center-driven dipole antennas were not terminated and that the currents from the excitation

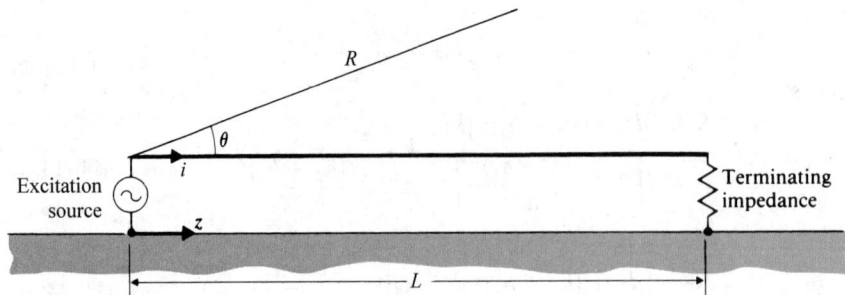

FIGURE 11–21
A traveling-wave antenna with termination.

sources were reflected at the ends, resulting in a standing-wave distribution as given in Eq. (11–51). If an antenna is several wavelengths long and properly terminated, as illustrated in Fig. 11–21, we have a situation similar to that of a transmission line terminated in its characteristic impedance. No reflection results, and the current distribution along the antenna is in the form of a traveling wave:

$$I(z) = I_0 e^{-j\beta z}. \tag{11-144}$$

Disregarding for the moment the effect of the nearby ground, we can find the far-zone electric field of an isolated antenna by using Eq. (11–144) in Eq. (11–19b) and integrating:

$$
\begin{aligned}
E_\theta &= \frac{j\eta_0 \beta \sin\theta}{4\pi r} e^{-j\beta R} \int_0^L I(z) e^{j\beta z \cos\theta} \, dz \\
&= \frac{j\eta_0 \beta I_0 \sin\theta}{4\pi R} e^{-j\beta R} \int_0^L e^{-j\beta z(1 - \cos\theta)} \, dz \\
&= \frac{j60 I_0}{R} e^{-j\beta [R + (L/2)(1 - \cos\theta)]} F(\theta),
\end{aligned}
\tag{11-145}
$$

where

$$F(\theta) = \frac{\sin\theta \sin\left[\beta L(1 - \cos\theta)/2\right]}{1 - \cos\theta} \tag{11-146}$$

is the pattern function of an isolated traveling-wave antenna of length L. Comparison of Eqs. (11–146) and (11–56) shows that the pattern functions of traveling-wave and standing-wave antennas have quite different characteristics. One especially notable difference is that the pattern represented by Eq. (11–146) is no longer symmetrical with respect to the $\theta = \pi/2$ plane.

A typical radiation pattern of an isolated traveling-wave antenna that is several wavelengths long may look like that shown in Fig. 11–22. In general, the main beams tilt toward the direction of the traveling current wave: The longer the antenna, the more the tilt. The sidelobes are generally only a few decibels down from the main beam level.

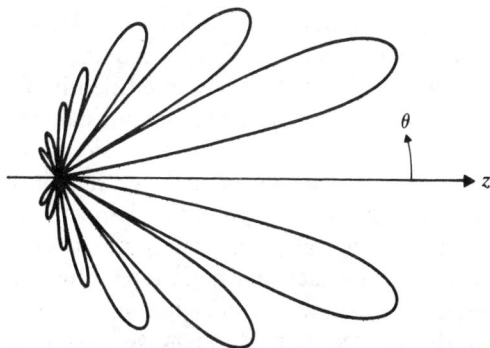

FIGURE 11–22
A typical radiation pattern of a traveling-wave antenna.

We can examine the effect of the ground by applying the method of images. The flat, perfectly conducting ground may be replaced by an image antenna carrying a current $I(z)$ given by Eq. (11–144) in the *opposite* direction, as shown in Fig. 11–23. We have, in effect, an array of two traveling-wave antennas separated by a distance $2h$ and carrying equal and opposite currents. By the principle of pattern multiplication the resultant pattern function is then the product of $F(\theta)$ in Eq. (11–146) and the array factor $|\cos(\psi/2)|$ of two elements given in Eq. (11–83). In this case, $d = 2h$ and $\zeta = \pi$. We have

$$\left|\cos\frac{\psi}{2}\right| = \left|\cos\left(\beta h \sin\theta \cos\phi + \frac{\pi}{2}\right)\right|$$
$$= |\sin(\beta h \sin\theta \cos\phi)|. \qquad (11\text{–}147)$$

The long linear antenna excited by a progressive current wave is only one of many traveling-wave antennas.

11–8.2 HELICAL ANTENNAS

We have seen that the far-zone electromagnetic field produced by linear antennas and small loop antennas are linearly polarized; that is, the electric field at a given

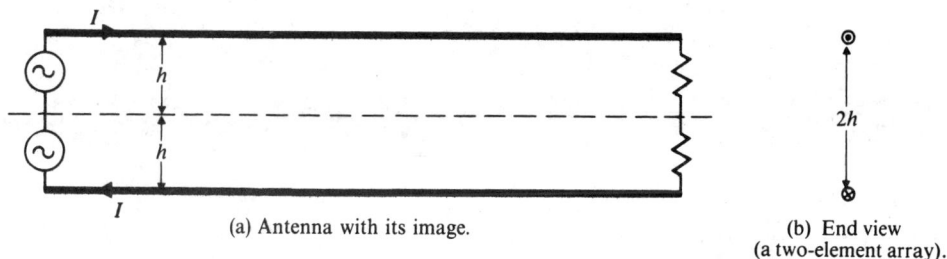

(a) Antenna with its image.

(b) End view
(a two-element array).

FIGURE 11–23
Method of images applied to traveling-wave antenna above perfectly conducting ground.

646 11 Antennas and Radiating Systems

location has a fixed direction that does not change with time. For instance, the electric field in the far zone of a vertical dipole is $\mathbf{E} = \mathbf{a}_\theta E_\theta$, and that of a horizontal loop is $\mathbf{E} = \mathbf{a}_\phi E_\phi$. The signals radiated from these antennas can be received efficiently by a linear antenna oriented in the direction of the electric field. There are circumstances, however, in which the direction of polarization of the incoming radiation is unknown or the orientation of the receiving antenna changes (such as the antennas on satellites or space vehicles). No reception will occur when the receiving antenna happens to be perpendicular to the direction of polarization of the signal. Earth communication with satellites and space vehicles is complicated by the fact that radiations from the earth must propagate through the ionosphere, which, under the influence of the earth magnetic field, becomes anisotropic and causes the electric field to change direction, the extent of this change being dependent on the electron density of the ionosphere, the strength of the earth magnetic field, and the propagation path. In these circumstances the use of antennas with circular polarization is advantageous because they are able to intercept waves polarized in any direction not normal to the plane of circular polarization.

Since the superposition of two equal-magnitude linearly polarized waves in both space and time quadrature yields a circularly polarized wave (see Subsection 8–2.3), we can obviously construct an antenna radiating (and receiving, by virtue of reciprocity) circularly polarized waves by feeding two perpendicularly oriented dipoles with currents that are equal in magnitude but 90° out of phase. This type of composite crossed-dipole arrangement is called a *turnstile antenna*. If the currents are of different magnitudes, an elliptically polarized wave is radiated. A circularly or elliptically polarized wave can also be generated by combining electric and magnetic dipoles (see Problem P.11–4).

A *helical antenna* is a wire antenna wound in the form of a helix. It is usually installed over a flat grounded conducting plane and fed by a coaxial transmission line, as shown in Fig. 11–24. Depending on the dimensions of the helix, the helical antenna has two quite different modes of operation. When the dimensions are very small in comparison to the operating wavelength, its radiation pattern is like that of an elemental electric dipole, given in Fig. 11–3. Maximum radiation occurs in the

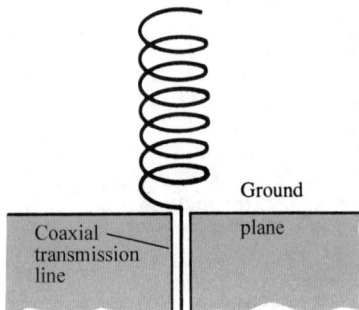

Coaxial
transmission
line

Ground
plane

FIGURE 11–24
A helical antenna.

plane perpendicular to the helix axis. The helical antenna is said to be in the ***normal mode.*** An approximate analysis of the helical antenna radiating in the normal mode can be made with two assumptions. First, it is assumed that the helix can be replaced by a combination of elemental electric and magnetic dipoles, as illustrated in Fig. 11–25(a). Second, the current along the helix is assumed to be uniform both in amplitude and in phase. (Some kind of top loading would then be necessary.) The far-zone electric field for an N-turn helical antenna is then a combination of Eqs. (11–19b) and (11–30a):

$$
\begin{aligned}
\mathbf{E} &= \mathbf{a}_\theta E_\theta + \mathbf{a}_\phi E_\phi \\
&= \frac{N\omega\mu_0 I}{4\pi} \left(\frac{e^{-j\beta R}}{R} \right) [\mathbf{a}_\theta js + \mathbf{a}_\phi \beta\pi b^2] \sin\theta.
\end{aligned}
\tag{11–148}
$$

Thus, the θ- and ϕ-components are in both space and time quadrature, resulting in elliptical polarization—circular polarization if

$$
s = \beta\pi b^2
\tag{11–148a}
$$

or

$$
b = \frac{1}{\pi} \sqrt{\frac{s\lambda}{2}}.
\tag{11–148b}
$$

The maximum radiation is in the broadside direction, and the radiation pattern has the shape of a doughnut with zero inner diameter. Fig. 11–25(b) shows a section in the E-plane. Normal-mode helical antennas are seldomly used in practice on account of their low radiation efficiency and low directive gain.

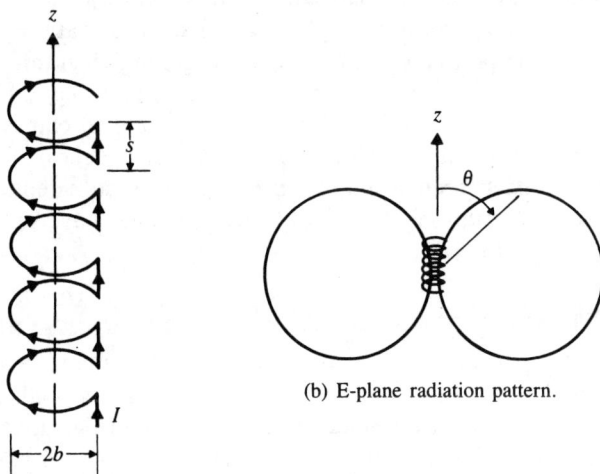

(b) E-plane radiation pattern.

(a) Combination of elemental
electric and magnetic dipoles.

FIGURE 11–25
Analysis of a normal-mode
helical antenna.

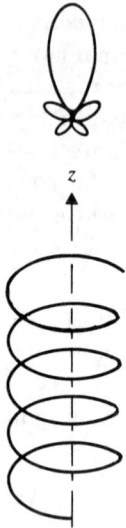

FIGURE 11–26
Axial-mode helical antenna and radiation pattern.

When both the turn circumference and the spacing between turns of a helix are comparable to a wavelength, the antenna behaves in an entirely different manner. The main beam of radiation will be in the end-fire direction, operating in the *axial mode*. A theoretical analysis of an axial-mode helical antenna is very difficult because of its geometry. The boundary-value problem of determining the current distribution along the helix can only be solved numerically in a limited way. The usual approach is to use some experimentally observed results for an assumed traveling-wave current and find the radiation pattern [14]. We will simply assert that the radiation pattern of an axial-mode helical antenna takes the form of a main beam in the end-fire direction together with some sidelobes, as indicated in Fig. 11–26. The radiation in the main beam is elliptically polarized, the axial ratio of the polarization ellipse being dependent on the frequency and the various dimensions of the helix. Helical antennas operating in the axial mode have been installed on communication satellites and space vehicles as well as on earth stations. Arrays of helical antennas have been used at radio telescope sites.

11–8.3 YAGI-UDA ANTENNAS

One type of antenna that is of particular practical importance is the ubiquitous Yagi-Uda antenna [15], [16], which one sees on the rooftops of many households for reception of television signals. A *Yagi-Uda antenna* in the transmitting mode is an array of parallel linear antennas, of which only one is driven by an excitation source and the rest are parasitic (not directly connected to the source). In the receiving mode an electromagnetic wave impinges on all elements of the array, but the received signal is collected from one "active element." The simplicity in the feed structure is a major advantage of Yagi-Uda antennas over other linear arrays.

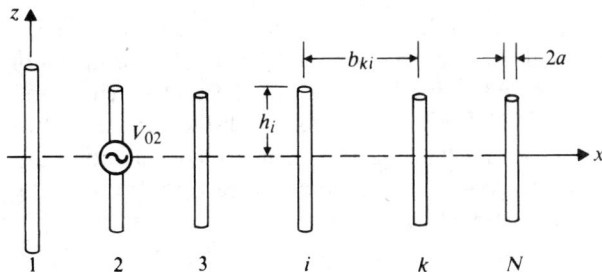

FIGURE 11–27
A typical Yagi-Uda array.

Figure 11–27 is a sketch of a typical Yagi-Uda antenna, which is actually an array. Element 2 is the dirven or active element and is normally a dipole of approximately (slightly less than) a half-wavelength long tuned to resonance. All other elements are parasitic. Element 1 is a *reflector* that is usually slightly longer than the driven element, whereas elements 3 to N are *directors* and are shorter than the driven element. Because all the elements are coupled, the current distribution in each element depends on the length and spacing of all the other elements. Consequently, a Yagi-Uda antenna of many elements presents a formidable analytical problem.

Experience has shown that little advantage is gained by using more than one reflector, but directivity can be improved by increasing the number of directors. A Yagi-Uda antenna is an endfire array with its main beam pointing away from the reflector. A good Yagi-Uda antenna should have a high directivity, a narrow beamwidth, low sidelobes, and a high front-to-back ratio. With the dipole radius assumed to be $a = 0.003369\lambda$ (ln $\lambda/2a = 5$) a typical well-designed six-element Yagi-Uda array with four uniformly spaced directors of equal length may have the following data:

Antenna Dimensions

Element lengths	$2h_1$ 0.510λ	$2h_2$ 0.490λ	$2h_3 = 2h_4 = 2h_5 = 2h_6$ 0.430λ
Element spacings		b_{12} 0.250λ	$b_{23} = b_{34} = b_{45} = b_{56}$ 0.310λ

Pattern Characteristics

Directivity (Referring to $\lambda/2$ Dipole)	Half-power Beamwidth	First Sidelobes	Front-to-back Ratio
7.54 (8.77 dB)	45°	−7.2 (dB)	9.52 (dB)

The directivity of a half-wave dipole is 1.64 or 2.15 (dB)—see Eq. (11–65).

It has been found that uniformly spaced directors of equal lengths do not make an optimum array. Analytical methods have been developed for the maximization of the directivity of a Yagi-Uda array by adjusting both the interelement spacings and the lengths of all the array elements [17], [18]. The effects of a finite element radius and the mutual coupling between the array elements are taken into consideration. An application of these methods to the foregoing six-element array leads to the following optimized array (for dipole radius $a = 0.003369\lambda$):

Antenna Dimensions

Element lengths	$2h_1$ 0.476λ	$2h_2$ 0.452λ	$2h_3$ 0.436λ	$2h_4$ 0.430λ	$2h_5$ 0.434λ	$2h_6$ 0.430λ
Element spacings	b_{12} 0.250λ	b_{23} 0.289λ	b_{34} 0.406λ	b_{45} 0.323λ	b_{56} 0.422λ	

Pattern Characteristics

Directivity (Referring to $\lambda/2$-Dipole)	Half-power Beamwidth	First Sidelobes	Front-to-back Ratio
13.36 (12.58 dB)	37°	−10.9 (dB)	10.04 (dB)

The pattern characteristics of the optimized array are better in all aspects than those of the array with uniformly spaced directors of equal length. The optimized array sustains a predominantly traveling wave in the sense that the amplitudes of the currents in the driven and director elements decrease smoothly and the phases change progressively.

11–8.4 BROADBAND ANTENNAS

To provide flexibility and versatility, antennas are often required to operate over a wide frequency range with satisfactory pattern, impedance, and polarization characteristics. It is difficult to define the useful bandwidth of an antenna in general terms because what is useful depends critically on applications. Normally, one refers to the characteristics of the radiation pattern—namely, the directivity, the main-lobe beamwidth, and/or the sidelobe levels. However, for antennas of relatively small dimensions in terms of wavelength, the impedance characteristics become very important. For antennas with circular polarization the polarization characteristics may be the limiting factor on useful bandwidth. The bandwidth of linear dipoles is very narrow. Increasing the thickness of the dipoles improves the bandwidth slightly, but the latter can seldom be more than a few percent of the designed center frequency. In this subsection we discuss briefly two types of broadband antennas that have come to be known as *frequency-independent antennas* and *log-periodic antennas*. The design concepts of log-periodic dipole arrays will also be introduced.

The concept of frequency-independent antennas evolved from the observation that the pattern and impedance characteristics of an antenna depend critically on its dimensions measured in wavelengths. An examination of the pattern function in Eq. (11–56) and the radiation patterns in Fig. 11–6 will confirm this observation.[†] Antennas having similar geometric structures will then retain the same radiation characteristics if a frequency change does not change the ratio of antenna dimensions to wavelength—that is, if the dimensions of the antenna structures are scaled in wavelengths. This observation led to the suggestion that, if an antenna structure could be described entirely by *angles* without specifying any characteristic length dimensions, its pattern and impedance characteristics would be frequency-independent [19]. The *equiangular spiral* defined by the equation

$$r = r_0 e^{a(\phi - \delta)} \tag{11–149a}$$

is such a structure. In Eq. (11–149a), r and ϕ are the usual polar coordinates (cylindrical coordinates for a constant z); and r_0, a, and δ are design constants. The spiral is equiangular in the sense that it makes a constant angle with the radius vector at all points on the curve.

The structure defined by Eq. (11–149a) is also called a *logarithmic spiral* because the angle change is proportional to $\ln (r/r_0)$:

$$\phi - \delta = \frac{1}{a} \ln \left(\frac{r}{r_0} \right). \tag{11–149b}$$

The three constants r_0, a, and δ in Eqs. (11–149a) and (11–149b) determine the size of the terminal region, the reciprocal of the rate of spiral, and the arm width, respectively. Figure 11–28 shows a planar equiangular spiral antenna consisting of two symmetrical arms. The four edges of the two spirals are defined by the relations $r = r_0 \exp (a\phi)$, $r = r_0 \exp [a(\phi - \delta)]$, $r = r_0 \exp [a(\phi - \pi)]$, and $r = r_0 \exp [a(\phi - \pi - \delta)]$.

The antenna is excited at the terminals by a voltage source that causes currents to flow. One viewpoint is that the currents flow outward along the spiral arms until they reach a region where most of the radiation occurs. Another viewpoint is that the electric field vector between the spiral arms travels outward until the space between the arms is approximately a half-wavelength at the operating frequency. In this region, resonance occurs and strong radiation takes place. Beyond this region, currents and fields diminish rapidly, and the truncation of the infinite spiral at a finite boundary is of little consequence. An increase or a decrease in the operating frequency simply moves the radiating region inward or outward along the spiral, but the effective radiating aperture in terms of wavelength does not change. As a result, an automatic scaling process takes place, and the pattern and impedance characteristics

---

[†] We have not studied the method for calculating the input impedance of antennas. The impedance characteristics depend on the current distribution on an antenna, which, in turn, depends critically on antenna dimensions measured in wavelengths.

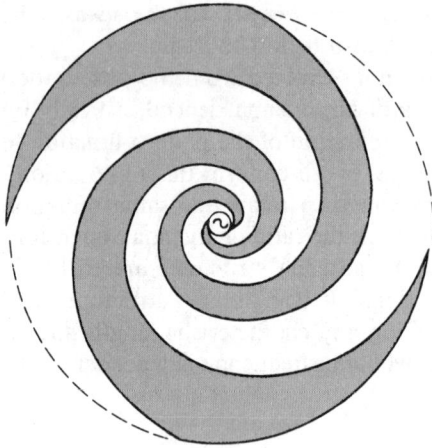

FIGURE 11–28
A planar two-arm equiangular spiral antenna.

remain almost indepedent of frequency [20]. The two-arm equiangular spiral antenna is circularly polarized. As the frequency is changed, the radiation pattern rotates about the axis normal to the spirals. Strictly speaking, the spirals must extend to infinity in order to be truly frequency-independent.

The planar equiangular spiral antenna of Fig. 11–28 is bidirectional in that it radiates a broadside main beam on both sides of the plane. This is sometimes undesirable. By wrapping a balanced two-arm equiangular spiral on the surface of a cone of revolution we can obtain a unidirectional radiation pattern with a single main beam in the direction of the cone apex. The pattern is still broadband and substantially circularly polarized. Both planar and conical equiangular spiral antennas can be designed to cover a 30-to-1 frequency range or more.

Can a broadband antenna be designed with linear (instead of circular) polarization? The search for an answer to this question led to a distinctive class of broadband linearly polarized antennas called *log-periodic antennas* [21], [22]. The basic log-periodic antenna has a toothed design cut out of sheet metal as shown in Fig. 11–29. The teeth are discontinuities that tend to localize the region of maximum radiation and cause the current to diminish rapidly beyond this region. The lengths of the teeth (the distances between the tips and the triangular support section) are determined by the angles between the lines from the origin. The region of strongest radiation is where the teeth are approximately a quarter-wavelength long.

The spacings between the successive edges of the teeth follow the rule that governs the distance between neighboring conductors in an equiangular spiral. From Eq. (11–149a),

$$\frac{r_{n+1}}{r_n} = \frac{r_0 e^{a(\phi - \delta)}}{r_0 e^{a(\phi + 2\pi - \delta)}} = e^{-2\pi a} \tag{11–150}$$

$$= \tau \quad \text{(a constant).}$$

This constant ratio is used as a design parameter for log-periodic antennas. Changing the operating frequency changes the tooth, whose length corresponds to a definite fraction of a wavelength. In terms of wavelength a scaling scheme is in effect that does not rely on a specification of length dimensions. This is the basis of the broadband nature of these antennas.

For an infinite structure, antenna characteristics would be identical at a discrete number of frequencies that are related by the parameter τ:

$$f_n = \tau f_{n+1} \tag{11–151a}$$

or

$$\ln (f_{n+1}) = \ln (f_n) + \ln \left(\frac{1}{\tau}\right). \tag{11–151b}$$

Antenna characteristics will vary somewhat between the discrete frequencies f_n and f_{n+1}; but, plotted versus the logarithm of frequency, they are periodic with a period equal to $\ln (1/\tau)$. Thus the name **_log-periodic antennas_**.

It has been found that, instead of the basic sheet metal structure, a log-periodic antenna can be made with wires or tubes that outline the cut-out design. Theoretically, the sheet thickness and the wire diameter should increase linearly with the distance from the feed point in accordance with Eq. (11–150). This consideration becomes important when the demand on bandwidth is severe. Bandwidths covering a 30-to-1 frequency range or more are achievable.

The planar log-periodic antenna of Fig. 11–29 is bidirectional as was the planar equiangular spiral antenna. It can be made unidirectional if the two halves of the antenna are folded to form a wedge-like structure. The main beam will point off the

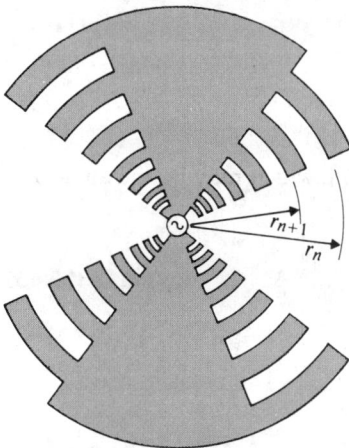

FIGURE 11–29
A planar log-periodic antenna.

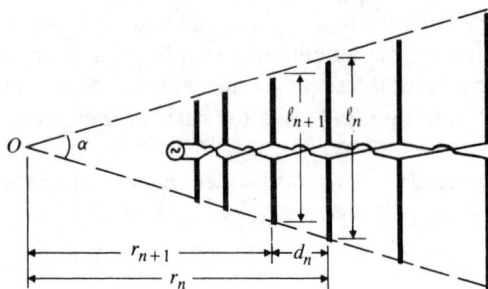

FIGURE 11–30
A log-periodic dipole array.

direction of the apex. It is interesting to note here that the planar log-periodic antenna in Fig. 11–29 without the teeth becomes a bow-tie antenna, which can be described entirely by the apex angle and is therefore expected to have broadband properties. This is true to a limited degree and explains the appearance of the bow-tie antenna as a form of commercially available UHF television antennas; but its finite length and lack of distinct resonance regions limit the bandwidth.

Broadband characteristics can also be achieved by a class of linear dipole arrays called *log-periodic dipole arrays* [23], [24], an example of which is shown in Fig. 11–30. The dipoles are of unequal lengths and are nonuniformly spaced according to the following relations:

$$\frac{\ell_{n+1}}{\ell_n} = \frac{r_{n+1}}{r_n} = \tau, \tag{11-152}$$

where τ is a design parameter, as in Eq. (11–150). Since the element spacings are related to the distances to the imaginary apex point O,

$$d_n = r_n - r_{n+1} = r_n(1 - \tau), \tag{11-153}$$

we also have

$$\frac{d_{n+1}}{d_n} = \tau. \tag{11-154}$$

Besides τ, only one other design parameter is required, which may be the angle α or the spacing factor κ:

$$\kappa = \frac{d_n}{2\ell_n}. \tag{11-155}$$

The relation among τ, α, and κ is as follows:

$$\tan\frac{\alpha}{2} = \frac{\ell_n}{2r_n} = \frac{\ell_n(1 - \tau)}{2d_n}$$

$$= \frac{1 - \tau}{4\kappa}. \tag{11-156}$$

Hence only two of the three parameters are independent. Because of the scaling relations in Eqs. (11–152) and (11–154), a change in the operating frequency changes only the particular dipole whose length is a certain fraction of a wavelength.[†] The remarks pertaining to Eqs. (11–150), (11–151a), and (11–151b) apply, and we have a log-periodic dipole array.

The array is usually fed by a source connected to a transmission line. An important discovery is that neighboring elements must be fed at opposite phases. This is accomplished by transposing the wires of the transmission line leading to alternate dipoles, as illustrated in Fig. 11–30. At the operating frequency the active region of the array consists mainly of the several dipoles whose lengths are approximately a half-wavelength and where the dipole currents are large. The currents in the dipoles outside this region are relatively very small. The array operates in an end-fire fashion with its main beam of radiation in the direction of short dipoles.

11–9 Aperture Radiators

Our analysis of the radiation characteristics of antennas has generally proceeded from a current distribution on the antenna structure. From the current distribution the retarded vector potential is determined by using Eq. (11–3). The magnetic and electric field intensities can then be found from Eqs. (11–1) and (11–6), respectively. In many cases, electromagnetic radiation may be viewed as emanating from an opening or an aperture in a conducting enclosure. To be sure, the source of radiation can always be traced to some time-varying currents somewhere; but the current distributions are often unknown and difficult to determine or approximate. Such radiating systems are quite unlike dipole antennas and must be analyzed in a different way. They are aperture radiators or aperture antennas. Examples are slots, horns, reflectors, and lenses, some of which are illustrated in Fig. 11–31.

In our analysis we will use an approximate aperture-field method, assuming that electric and magnetic fields exist only in the aperture area and that the field elsewhere in an infinite screen containing the aperture is zero. In the case of the slot radiator shown in Fig. 11–31(a), the field for dominant TE_{10}-mode excitation is usually assumed to be a half-sine having a maximum at the center of the slot and tapering to zero at the edges. For the horn in Fig. 11–31(b) the aperture field is derived from the waveguide mode propagating into a horn of infinite extent. The aperture fields of the reflector in Fig. 11–31(c) and the lens in Fig. 11–31(d) are found by methods of geometrical optics from the reflection and refraction of rays emanating from the primary feed.

For TE_{10}-mode excitation the field in a plane aperture is approximately linearly polarized, and deviations from the results obtained by geometrical optics are small.

[†] Strictly speaking, it is also necessary to scale the radius a_n of the dipoles according to $a_{n+1}/a_n = \tau$.

(a) Slot.

(b) Horn.

Aperture

(c) Reflector.

Aperture

(d) Lens.

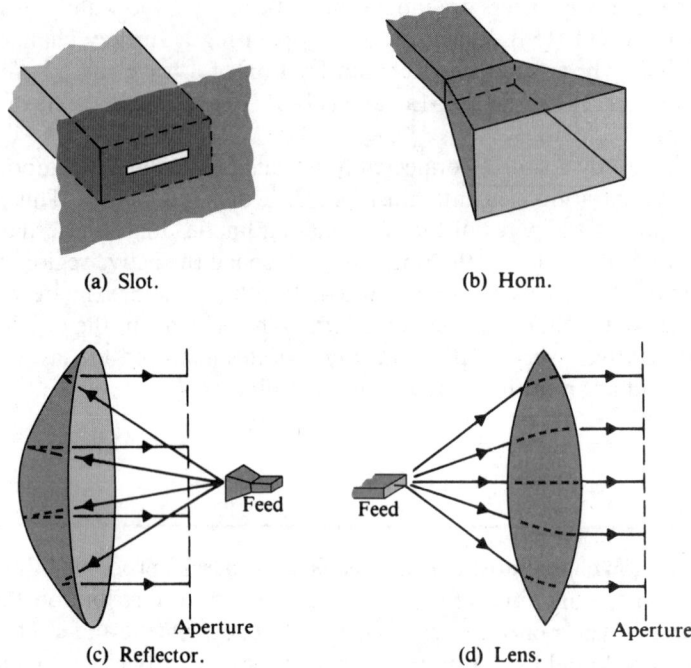

FIGURE 11–31
Aperture antennas.

With a nearly uniform phase over the aperture, the far-zone field is a two-dimensional Fourier integral of the field distribution in the aperture. Let the electric field distribution in the aperture outlined in Fig. 11–32 be linearly polarized, say in the x-direction, with no phase variation:

$$\mathbf{E}_a = \mathbf{a}_x E_a. \tag{11–157}$$

If the aperture dimensions are large in comparison to the operating wavelength, then almost all the energy of the radiated field will be contained in a small angular region around the z-axis, and the far-zone electric field at a distant point $P(R_0, \theta, \phi)$ can be written as $\mathbf{E}_P = \mathbf{a}_x E_P$, where [13], [25]

$$E_P = \frac{j}{\lambda R_0} \iint\limits_{\text{aper.}} E_a(x', y') e^{-j\beta R} \, dx' \, dy'. \tag{11–158}$$

For $\beta R \gg 1$ we have

$$R \cong R_0 - (\mathbf{a}_x x' + \mathbf{a}_y y') \cdot (\mathbf{a}_x \sin\theta \cos\phi + \mathbf{a}_y \sin\theta \sin\phi)$$
$$= R_0 - (x' \sin\theta \cos\phi + y' \sin\theta \sin\phi). \tag{11–159}$$

Substitution of Eq. (11–159) in Eq. (11–158) yields

$$E_P = \frac{j}{\lambda R_0} e^{-j\beta R_0} F(\theta, \phi), \tag{11–160}$$

where

$$F(\theta, \phi) = \iint\limits_{\text{aper.}} E_a(x', y')e^{j\beta \sin\theta(x'\cos\phi + y'\sin\phi)}\,dx'\,dy' \qquad (11\text{–}161)$$

is the pattern function of the aperture antenna. Equation (11–161) expresses the rather simple relation between the aperture distribution and the pattern function; namely, they are the Fourier transform of each other. The inverse relation, expressing $E_a(x', y')$ in terms of $F(\theta, \phi)$ enables us to determine the aperture field required for a specified pattern function. This is a synthesis problem.

For a rectangular aperture with dimensions $a \times b$ and separable field distributions:

$$E_a(x', y') = f_1(x')f_2(y'), \qquad (11\text{–}162)$$

the pattern function in Eq. (11–161) is also separable:

$$F(\theta, \phi) = \int_{-a/2}^{a/2} f_1(x')e^{j\beta x' \sin\theta\cos\phi}\,dx' \int_{-b/2}^{b/2} f_2(y')e^{j\beta y' \sin\theta\sin\phi}\,dy'. \quad (11\text{–}163)$$

If we are interested only in the patterns in the principal planes, Eq. (11–163) can be further simplified.

1. *In the xz-plane,* $\phi = 0$:

$$
\begin{aligned}
F_{xz}(\theta) &= \left[\int_{-b/2}^{b/2} f_2(y')\,dy'\right]\int_{-a/2}^{a/2} f_1(x')e^{j\beta x' \sin\theta}\,dx' \\
&= C_1 \int_{-a/2}^{a/2} f_1(x')e^{j\beta x' \sin\theta}\,dx',
\end{aligned}
\qquad (11\text{–}164)
$$

where C_1 is a constant. We see that the radiation pattern in the xz-plane depends only on the aperture field distribution in the x'-direction.

2. *In the yz-plane,* $\phi = \pi/2$:

$$
\begin{aligned}
F_{yz}(\theta) &= \left[\int_{-a/2}^{a/2} f_1(x')\,dx'\right]\int_{-b/2}^{b/2} f_2(y')e^{j\beta y' \sin\theta}\,dy' \\
&= C_2 \int_{-b/2}^{b/2} f_2(y')e^{j\beta y' \sin\theta}\,dy',
\end{aligned}
\qquad (11\text{–}165)
$$

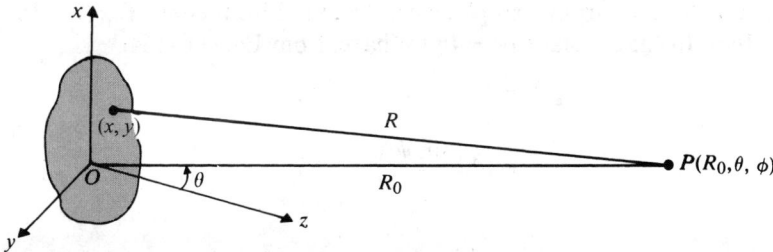

FIGURE 11–32
Pattern calculation from aperture-field distribution.

where C_2 is a constant. The radiation pattern in the yz-plane depends only on the aperture-field distribution in the y'-direction.

The directivity of an aperture radiator is obtained by using Eq. (11–35), which for convenience, is repeated below:

$$D = \frac{4\pi U_{max}}{P_r}, \tag{11–166}$$

where

$$U_{max} = \frac{1}{2\eta_0} R_0^2 |E_p|_{max}^2$$

$$= \frac{1}{2\eta_0 \lambda^2} \left| \iint_{aper.} E_a(x', y')\, dx'\, dy' \right|^2 \tag{11–167}$$

and

$$P_r = \text{Total power radiated}$$

$$= \frac{1}{2\eta_0} \iint_{aper.} |E_a(x', y')|^2\, dx'\, dy'. \tag{11–168}$$

Combining Eqs. (11–166), (11–167), and (11–168), we have

$$D = \frac{4\pi}{\lambda^2} \frac{\left| \iint_{aper.} E_a(x', y')\, dx'\, dy' \right|^2}{\iint_{aper.} |E_a(x', y')|^2\, dx'\, dy'} \quad \text{(Dimensionless).} \tag{11–169}$$

It is interesting to note that, when $E_a(x', y') =$ a constant (uniform aperture-field distribution), D is a maximum and equals $4\pi/\lambda^2$ times the area of the aperture. This is in agreement with Eq. (11–121).

EXAMPLE 11–13   For an $a \times b$ rectangular aperture with a uniform field distribution, find (a) the pattern function in a principal plane, (b) the half-power beamwidth, (c) the location of the first nulls, and (d) the level of the first sidelobes.

Solution   For simplicity we set $E_a(x', y') = 1$.

a) The pattern function in a principal plane can be found from either Eq. (11–164) or Eq. (11–165). In the xz-plane ($\phi = 0$) we have, from Eq. (11–164),

$$F_{xz}(\theta) = b \int_{-a/2}^{a/2} e^{j\beta x' \sin\theta}\, dx'$$

$$= ab\left(\frac{\sin\psi}{\psi}\right), \tag{11–170}$$

where

$$\psi = \frac{\pi a}{\lambda} \sin\theta. \tag{11–171}$$

Exactly the same pattern function is obtained for $F_{yz}(\theta)$ in the other principal plane ($\phi = \pi/2$) except that b will replace a in Eq. (11–171). Note that the pattern function in Eq. (11–170) is similar to the array factor of a uniform linear array given in Eq. (11–89) when ψ is small.

b) The half-power points are determined by setting

$$\frac{\sin \psi_{1/2}}{\psi_{1/2}} = \frac{1}{\sqrt{2}},$$

from which we find

$$\psi_{1/2} = \frac{\pi a}{\lambda} \sin \theta_{1/2} = 1.39$$

or

$$\sin \theta_{1/2} = 0.442 \frac{\lambda}{a}. \tag{11–172}$$

For sufficiently large apertures, $\sin \theta_{1/2}$ is nearly equal to $\theta_{1/2}$,[†] and the half-power beamwidth is approximately

$$2\theta_{1/2} \cong 0.88 \frac{\lambda}{a} \qquad \text{(rad)}$$

$$\cong 50 \frac{\lambda}{a} \qquad \text{(deg)}.$$

c) The first null occurs at

$$\psi_{n1} = \frac{\pi a}{\lambda} \sin \theta_{n1} = \pi$$

or

$$\theta_{n1} \cong \sin \theta_{n1} = \frac{\lambda}{a} \qquad \text{(rad)}. \tag{11–173}$$

d) The location of the first sidelobes is found by setting

$$\frac{\partial}{\partial \psi} \left(\frac{\sin \psi}{\psi} \right) = 0,$$

which requires $\tan \psi_1 = \psi_1$ or $\psi_1 = \pm 1.43\pi$. Thus,

$$\left| \frac{\sin \psi_1}{\psi_1} \right| = \left| \frac{\sin 1.43\pi}{1.43\pi} \right| = 0.217.$$

Referring to unity at $\psi = 0$, we find that the first sidelobes are $20 \log_{10} (1/0.217) = 13.3$ (dB) down from the level of maximum radiation. ∎

[†] For example, when $a = 5\lambda$, $\sin \theta_{1/2} = 0.442/5 = 0.0884$ and $\theta_{1/2} = \sin^{-1} (0.0884) = 0.0885$, an error of only 0.11%. The narrow beamwidth of the main lobe confirms our previous statement that almost all of the radiated energy is confined in a small angular region around the z-axis.

EXAMPLE 11–14   A linearly polarized uniform electric field $\mathbf{E}_a = \mathbf{a}_x E_0$ exists in a circular aperture of radius b in a conducting plane at $z = 0$. Assuming b to be large in comparison to wavelength, (a) find an expression for the far-zone electric field, and (b) determine the width of the main beam between first nulls.

Solution

a) For a circular aperture we use polar coordinates $x' = \rho' \cos \phi'$, $y' = \rho' \sin \phi'$, and $x' \cos \phi + y' \sin \phi = \rho'(\cos \phi \cos \phi' + \sin \phi \sin \phi') = \rho'(\cos (\phi - \phi')$. The integrand in Eq. (11–161) is to be integrated over the circular aperture. We have

$$
\begin{aligned}
F(\theta, \phi) &= E_0 \int_0^b \int_0^{2\pi} e^{j\beta\rho' \sin \theta \cos (\phi - \phi')} \rho' \, d\phi' \, d\rho' \\
&= E_0 \int_0^b 2\pi J_0(\beta\rho' \sin \theta) \rho' \, d\rho' \\
&= E_0 2\pi b^2 \left[\frac{J_1(\beta b \sin \theta)}{\beta b \sin \theta} \right],
\end{aligned}
$$

(11–174)[†]

where $J_1(u)$ is the Bessel function of the first kind of the first order. The far-zone electric field is then, from Eq. (11–160),

$$
\mathbf{E}_P = \mathbf{a}_x j E_0 \frac{2\pi b^2}{\lambda R_0} e^{-j\beta R_0} \left[\frac{J_1(u)}{u} \right],
$$

(11–175)

where

$$
u = \beta b \sin \theta = \frac{2\pi b}{\lambda} \sin \theta.
$$

(11–176)

b) The first null of the radiation pattern occurs at the first zero, u_{11}, of $J_1(u)$. From Table 10–2 we find $u_{11} = 3.832$, which corresponds to an angle

$$
\begin{aligned}
\theta_1 &= \sin^{-1} \left(\frac{3.832\lambda}{2\pi b} \right) \cong \frac{3.832\lambda}{2\pi b} \\
&= 1.22 \frac{\lambda}{D} \quad \text{(rad)},
\end{aligned}
$$

(11–177)

where $D = 2b$ is the diameter of the circular aperture. Hence the width of the main beam between the first nulls is $2\theta_1 = 2.44\lambda/D$ (rad). Comparing θ_1 in Eq. (11–177) with θ_{n1} in Eq. (11–173) for a rectangular aperture with width a equaling the diameter D of the circular aperture, we find that the main-lobe beamwidth for the circular aperture is wider. On the other hand, the first sidelobe level for the circular aperture is found to be 0.13, which is $20 \log_{10} (1/0.13) = 17.7$ (dB) down from the maximum radiation. This is lower than the 13.3 (dB) first sidelobes for the rectangular aperture with $a = D$.   ∎

[†] We have made use of the following two integral relations:

$$
\int_0^{2\pi} e^{jw \cos \phi'} \, d\phi' = 2\pi J_0(w) \quad \text{and} \quad \int wJ_0(w) \, dw = wJ_1(w).
$$

In this section we have considered the radiation properties of only relatively simple cases of rectangular and circular apertures in conducting planes. The analysis of other aperture-type antennas such as horns, reflectors, and lenses is more difficult and requires the use of more advanced concepts. Slots cut in the walls of a waveguide that interrupt current flow will radiate. Suitably arranged, they will form antenna arrays in a manner analogous to dipole arrays. These and other radiation problems are topics for more specialized books on antennas [9], [11]–[13].

References

[1] H. Unz, "Linear arrays with arbitrarily distributed elements," *IRE Transactions on Antennas and Propagation*, vol. AP-8, pp. 222–223, March 1960.

[2] R. F. Harrington, "Sidelobe reduction by nonuniform element spacing," *IRE Transactions on Antennas and Propagation*, vol. AP-9, pp. 187–192, March 1961.

[3] A. Ishimaru and Y. S. Chen, "Thinning and broadbanding antenna arrays by unequal spacings," *IEEE Transactions on Antennas and Propagation*, vol. AP-13, pp. 34–42, January 1965.

[4] F. I. Tseng and D. K. Cheng, "Spacing perturbation techniques for array optimization," *Radio Science*, vol. 3 (New Series), pp. 451–457, May 1968.

[5] D. K. Cheng and P. D. Raymond, Jr., "Optimization of array directivity by phase adjustments," *Electronics Letters*, vol. 7, pp. 552–553, September 9, 1971.

[6] E. D. Sharp, "A triangular arrangement of planar array elements that reduces the number needed," *IRE Transactions on Antennas and Propagation*, vol. AP-9, pp. 126–129, March 1961.

[7] N. Goto, "Pattern synthesis of hexagonal planar array," *IEEE Transactions on Antennas and Propagation*, vol. AP-20, pp. 104–106, January 1972.

[8] D. K. Cheng, "Optimization techniques for antenna arrays," *Proceedings of the IEEE*, vol. 59, pp. 1664–1674, December 1971.

[9] E. C. Jordan and K. G. Balmain, *Electromagnetic Waves and Radiating Systems*, Prentice-Hall, Englewood Cliffs, N.J., 1968.

[10] M. T. Ma, *Theory and Application of Antenna Arrays*, Wiley, New York, 1974.

[11] R. S. Elliott, *Antenna Theory and Design*, Prentice-Hall, Englewood Cliffs, N.J., 1981.

[12] W. L. Stutzman and G. A. Thiele, *Antenna Theory and Design*, Wiley, New York, 1981.

[13] R. E. Collin, *Antennas and Radiowave Propagation*, McGraw-Hill, New York, 1985.

[14] K. F. Lee, *Principles of Antenna Theory*, Wiley, New York, 1984.

[15] H. Yagi, "Beam transmission of ultra short waves," *Proceedings of the IEEE*, vol. 16, pp. 715–741, June 1928.

[16] S. Uda and Y. Mushiaki, *Yagi-Uda Antenna*, Maruzan, Tokyo, 1954.

[17] D. K. Cheng and C. A. Chen, "Optimum element spacings for Yagi-Uda arrays," *IEEE Transactions on Antennas and Propagation*, vol. AP-21, pp. 615–623, September 1973.

[18] C. A. Chen and D. K. Cheng, "Optimum element lengths for Yagi-Uda arrays," *IEEE Transactions on Antennas and Propagation*, vol. AP-23, pp. 8–15, January 1975.

[19] V. H. Rumsey, *Frequency-Independent Antennas*, Academic Press, New York, 1966.

[20] J. D. Dyson, "The equiangular spiral," *IRE Transactions on Antennas and Propagation*, vol. AP-7, pp. 181–187, April 1959.

[21] R. H. DuHamel and D. E. Isbell, "Broadband logarithmically periodic antenna structures," *IRE National Convention Record,* Part I, pp. 119–128, 1957.

[22] R. H. DuHamel and F. R. Ore, "Logarithmically periodic antenna design," *IRE National Convention Record*, Part I, pp. 139–151, 1958.

[23] R. Carrel, "The design of log-periodic dipole antennas," *IRE International Convention Record*, Part I, pp. 61–75, 1961.

[24] E. C. Jordan et al., "Developments in broadband antennas," *IEEE Spectrum*, vol. 1, pp. 58–71, April 1964.

[25] S. Silver (ed.), *Microwave Antenna Theory and Design,* M.I.T. Radiation Laboratory Series, vol. 12, Chapters 5 and 6, McGraw-Hill, New York, 1949.

Review Questions

R.11–1 Give a general definition for *antenna*.

R.11–2 Why are antennas important for wireless communication over long distances?

R.11–3 State the procedure for finding the electromagnetic field due to an assumed time-harmonic current distribution on an antenna structure.

R.11–4 What is a *Hertzian dipole*?

R.11–5 What constitutes an elemental magnetic dipole?

R.11–6 Define the *near zone* and the *far zone* of an antenna.

R.11–7 Why are the near-zone fields called quasi-static fields?

R.11–8. Explain how the magnitude of far fields varies with distance.

R.11–9 In what ways does the electromagnetic field of a radiating magnetic dipole differ from that of a Hertzian dipole?

R.11–10 What are *radiation fields*?

R.11–11 Define *antenna pattern*.

R.11–12 Describe the *E*-plane and *H*-plane patterns of a Hertzian dipole.

R.11–13 Define *beamwidth* of an antenna pattern.

R.11–14 Define *sidelobe level* of an antenna pattern.

R.11–15 Define *radiation intensity*.

R.11–16 Define *directive gain* and *directivity* of an antenna.

R.11–17 Define *power gain* and *radiation efficiency* of an antenna.

R.11–18 Define *radiation resistance* of an antenna.

R.11–19 Discuss how the ratios (a/λ) and $(d\ell/\lambda)$ of a Hertzian dipole affect its radiation resistance and radiation efficiency.

R.11–20 Describe the radiation pattern of a half-wave dipole antenna.

R.11–21 What are the radiation resistance and directivity of a half-wave dipole antenna?

R.11–22 What is the image of a horizontal dipole over a conducting ground?

R.11–23 What are the radiation resistance and directivity of a vertical quarter-wave monopole over a conducting ground?

R.11–24 Define the *effective length* of a linear antenna for transmitting. Upon what factors does it depend?

R.11–25 Define the *effective length* of a linear antenna for receiving.

R.11–26 What is meant by the *normalized array factor* of an antenna array? How is it different from the pattern function of the individual antennas?

R.11–27 State the *principle of pattern multiplication*.

R.11–28 State the difference between a *broadside array* and an *endfire array*.

R.11–29 What is a *binomial array*? What are the relative excitation amplitudes of a six-element binomial array?

R.11–30 Is the radiation pattern of all linear binomial arrrays sidelobeless? Explain.

R.11–31 In the radiation pattern of a uniform linear array of many elements, how many decibels down from the principal maximum are the first sidelobes?

R.11–32 How can the sidelobes of an equally spaced linear array be made lower than those of a uniform linear array?

R.11–33 What is a *phased array*?

R.11–34 What is a *frequency-scanning array*?

R.11–35 What are the important consequences of reciprocity relations concerning antennas that operate in the transmitting and receiving modes?

R.11–36 Define *effective area* of an antenna.

R.11–37 What is the universal constant that is the ratio of the directive gain and the effective area of an antenna?

R.11–38 Define *backscatter cross section* of an object.

R.11–39 Explain the principle of *radar*.

R.11–40 What does the *Friis transmission formula* say?

R.11–41 Define *path gain factor* concerning wave propagation near the earth's surface.

R.11–42 In what essential ways does the radiation pattern of a long traveling-wave antenna differ from that of an unterminated dipole antenna?

R.11–43 What is the essential difference between the radiation characteristics of a helical antenna and a dipole antenna?

R.11–44 What are the two different operating modes of a helical antenna? Explain.

R.11–45 What is a *Yagi-Uda antenna*?

R.11–46 How should the lengths of the reflector and director elements in a Yagi-Uda array compare with the length of the driven element?

R.11–47 What is the principle of *frequency-independent antennas*?

R.11–48 What is an *equiangular spiral*? Why does it have broadband properties?

R.11–49 What is a *log-periodic antenna*?

R.11–50 Explain the principle of operation of log-periodic dipole arrays.

R.11–51 Give three examples of aperture radiators.

R.11–52 For a linearly polarized aperture field with uniform phase, what is the relation between the aperture's field distribution and the pattern function?

R.11–53 What is the directivity of an aperture having an area A and a linearly polarized uniform field distribution at frequency f?

R.11–54 Describe the manner in which the beamwidth in a principal plane of a rectangular aperture with a uniform field distribution depends on its dimensions.

R.11–55 Assume that a linearly polarized constant excitation field exists in a rectangular aperture with width b and a circular aperture with diameter $D = b$. Compare the main-lobe beamwidths and the first sidelobe levels of their radiation patterns.

Problems

P.11–1 Starting from Maxwell's equations, derive the nonhomogeneous wave equations (a) for **E**, and (b) for **H** in a simple medium.

P.11–2 Obtain the electric field intensity of a Hertzian dipole by finding both **A** and V and using Eq. (11–2). Check your result with Eqs. (11–16a, b, c).

P.11–3 A small filamentary rectangular loop of dimensions L_x and L_y lies in the xy-plane with its center at the origin and sides parallel to the x- and y-axes. The loop carries a current $i(t) = I_0 \cos \omega t$. Assuming L_x and L_y to be much less than the wavelength, find the instantaneous expressions for the following quantities at a point in the far zone:

  a) vector magnetic potential **A**,
  b) electric field intensity **E**,
  c) magnetic field intensity **H**.

Compare the results in parts (b) and (c) with Eqs. (11–30a) and (11–30b), respectively.

P.11–4 A composite antenna consists of an elemental Hertzian electric dipole of length L along the z-axis and an elemental magnetic dipole of area S lying in the xy-plane. Equal time-harmonic currents of amplitude I_0 and angular frequency ω flow in the dipoles.

  a) Verify that the far field of the composite antenna is elliptically polarized.
  b) Determine the condition for circular polarization.

P.11–5 (a) Assume the spatial distribution of the current on a very thin center-fed half-wave dipole lying along the z-axis to be $I_0 \cos \beta z$, where $\beta = \omega/c = 2\pi/\lambda$. Find the charge distribution on the dipole. (b) Repeat part (a), assuming the current distribution along the dipole to be a triangular function described by

$$I(z) = I_0 \left(1 - \frac{4}{\lambda} |z| \right).$$

P.11–6 A 1 (MHz) uniform current flows in a vertical antenna of length 15 (m). The antenna is a center-fed copper rod having a radius of 2 (cm). Find:

  a) the radiation resistance,
  b) the radiation efficiency,
  c) the maximum electric field intensity at a distance of 20 (km) if the radiated power of the antenna is 1.6 (kW).

P.11–7 The amplitude of the time-harmonic current distribution on a center-fed short dipole antenna of length $2h(h \ll \lambda)$ can be approximated by a triangular function

$$I(z) = I_0 \left(1 - \frac{|z|}{h} \right).$$

Find (a) the far-zone electric and magnetic field intensities, (b) the radiation resistance, and (c) the directivity.

P.11–8 The transmitting antenna of a radio navigation system is a vertical metal mast 40 (m) in height insulated from the earth. A 180 (kHz) source sends a current having an amplitude of 100 (A) into the base of the mast. Assuming the current amplitude in the antenna to decrease linearly toward zero at the top of the mast and the earth to be a perfectly conducting plane, determine:

   a) the effective length of the antenna,
   b) the maximum field intensity at a distance 160 (km) from the antenna,
   c) the time-average radiated power,
   d) the radiation resistance.

P.11–9 A time-harmonic uniform current $I_0 \cos \omega t$ flows in a small circular loop of radius $b \, (\ll \lambda)$ lying in the xy-plane.

   a) Find the radiation resistance R_r of the magnetic dipole.
   b) Obtain an expression for its radiation efficiency η_r if the loop is made of copper wire of radius a,
   c) Calculate R_r and η_r for $f = 1$ (MHz), $b = 50$ (cm), and $a = 3$ (mm).
   d) Rework part (c) if the loop has ten closely wound insulated turns.

P.11–10 Repeat parts (a) and (b) of Problem P.11–9 for a small rectangular loop of sides L_x and L_y. Repeat part (c) for $f = 1$ (MHz), $L_x = L_y = 2b = 1$ (m), a = 3 (mm), and compare results.

P.11–11 Use the total field expressions in Eqs. (11–15) and (11–16) to find the time-average power radiated by a Hertzian dipole, and compare it with the result in Eq. (11–43) using only the far-zone fields.

P.11–12 Sketch the polar radiation pattern versus θ for a thin dipole antenna of total length $2h = 1.25\lambda$. Determine the width of the main beam between the first nulls.

P.11–13 Assuming a triangular current distribution on a center-fed $\lambda/6$ dipole ($h = \lambda/12$), find an expression for its effective length. What is its maximum value?

P.11–14 A 1.5 (MHz) uniform plane wave having a peak electric field intensity E_0 is incident on a half-wave dipole at an angle θ.

   a) Find the expression for the open-circuit voltage V_{oc} at the terminals of the dipole.
   b) If the dipole is connected to a matched load, what is the maximum power P_L delivered to the load?
   c) Calculate V_{oc} and P_L for $E_0 = 50$ (mV/m) and for $\theta = \pi/2$ and $\pi/4$.

P.11–15 Two elemental dipole antennas, each of length $2h \ll \lambda$, are aligned colinearly along the z-axis with their centers spaced a distance d ($d > 2h$) apart. The excitations in the two antennas are of equal amplitude and equal phase.

   a) Write the general expression for the far-zone electric field of this two-element colinear array.
   b) Plot the normalized E-plane pattern for $d = \lambda/2$.
   c) Repeat part (b) for $d = \lambda$.

P.11–16 A horizontal elemental electric dipole of length $d\ell$ and carrying a time-harmonic current of amplitude I_0 in the $+y$-direction is situated at a distance d above a perfectly conducting ground. Find its pattern functions (a) in the xy-plane, (b) in the xz-plane, and (c) in the yz-plane. (d) Sketch the patterns for parts (a), (b), and (c) for $d = \lambda/4$.

P.11–17 Plot the *H*-plane polar radiation pattern of two parallel dipoles for
   a) $d = \lambda/4,\ \xi = \pi/2$;   **b)** $d = 3\lambda/4,\ \xi = \pi/2$.

P.11–18 For a five-element broadside binomial array:
   a) Determine the relative excitation amplitudes in the array elements.
   b) Plot the array factor for $d = \lambda/2$.
   c) Determine the half-power beamwidth and compare it with that of a five-element uniform array having the same element spacings.

P.11–19 For a uniform linear array of 12 elements spaced $\lambda/2$ apart:
   a) Sketch the normalized array pattern $|A(\psi)|$ in Eq. (11–89) versus ψ.
   b) Find the widths of the main beam at half-power points and between the first nulls when the array is operated in the broadside mode.
   c) Repeat part (b) for an endfire operation.

P.11–20 For a uniform linear array with a large number of elements the denominator $\sin(\psi/2)$ in Eq. (11–89) remains small over a large portion of the normalized array pattern near the main beam and can be approximated by $(\psi/2)$. Use this approximation to determine the directivity of the array of a large uniform array with many elements.

P.11–21 Using the graph in Fig. 11–15(a) for the normalized array factor of a five-element broadside linear array with $d = \lambda/2$ and amplitude ratios 1:2:3:2:1, plot the polar radiation pattern for $d = \lambda/4$ and $\xi = -\pi/2$.

P.11–22 Letting $\zeta = \exp(j\psi)$, we can write the array factor of an equally spaced array as a polynomial, $A(\psi)$, in ψ, and many characteristics of the array pattern can be estimated by examining the distribution of the zeros of the array polynomial on a unit circle. In general, an N-element linear array has $N - 1$ zeros, ψ_{0m} ($m = 1, 2, \ldots, N - 1$), distributed around the unit circle. Find $A(\psi)$ and locate all ψ_{0m} on a unit circle for the following linear arrays:
   a) a two-element array,
   b) a three-element binomial array,
   c) a five-element uniform array,
   d) a five-element array having amplitude ratios 1:2:3:2:1 (as in Example 11–9).
   e) Based on the locations of ψ_{0m} for the two arrays in parts (c) and (d), explain why the pattern for the array in part (d) has lower sidelobes but a wider beamwidth.

P.11–23 Obtain the pattern function of a uniformly excited rectangular array of $N_1 \times N_2$ parallel half-wave dipoles. Assume that the dipoles are parallel to the z-axis and their centers are spaced d_1 and d_2 apart in the x- and y-directions, respectively.

P.11–24 Assume that a linearly polarized plane electromagnetic wave is incident on a half-wave dipole, as in Fig. 11–8.
   a) Obtain an expression for the effective area, $A_e(\theta)$.
   b) Calculate the maximum value of A_e for 100 (MHz).

P.11–25 A uniform plane wave with electric field intensity $\mathbf{E}_i = \mathbf{a}_z E_i$ impinges on a small dielectric sphere of radius $b\ (\ll \lambda)$ and dielectric constant ϵ_r.
   a) Find the total time-average power scattered by the sphere.
   b) Obtain the expression for the *total scattering cross section* σ_s, which is the ratio of the total scattered power to the incident power density. Compare σ_s with the backscatter cross section σ_{bs}.

P.11–26 Communication is to be established between two stations 1.5 (km) apart that operate at 300 (MHz). Each is equipped with a half-wave dipole.

    a) If 100 (W) is transmitted from one station, how much power is received by a matched load at the other station?

    b) Repeat part (a) assuming that both antennas are Hertzian dipoles.

P.11–27 (a) Show that three satellites equally spaced around the geosynchronous orbit in the equatorial plane would cover almost the entire earth's surface. Explain why the polar regions are not covered. (b) Assuming the main beam of the radiation pattern of the satellite antenna to have the shape of a circular cone that just covers the earth with no spillover, find a relation between the main-lobe beamwidth and the directive gain of the antenna.

P.11–28 The antenna at the earth station of a satellite communication link having a gain of 55 (dB) at 14 (GHz) is aimed at a geostationary satellite 36,500 (km) away. Assume that the antenna on the satellite has a gain of 35 (dB) in transmitting the signal back toward the earth station at 12 (GHz). The minimum usable signal is 8 (pW).

    a) Neglecting antenna ohmic and mismatch losses, find the minimum satellite transmitting power required.

    b) Find the peak transmitting pulse power needed at the earth station in order to detect the satellite as a passive object, assuming the backscatter cross section of the satellite including its solar panels as 25 (m^2) and the minimum detectable return pulse power to be 0.5 (pW).

P.11–29 A transmitting vertical half-wave dipole 60 (m) above the ground radiates 400 (W) at 100 (MHz). Assume the ground to be perfectly conducting.

    a) Calculate the power available at a vertical half-wave receiving antenna 50 (km) away at a height 30 (m) above the ground.

    b) At a distance 50 (km) from the transmitting antenna, where (at what altitudes) would there be a null field?

P.11–30 The current along an isolated and terminated traveling-wave antenna of length L is given as

$$I(z) = I_0 e^{-j\beta z}.$$

    a) Find the far-zone vector potential, $\mathbf{A}(R, \theta)$.

    b) Determine $\mathbf{H}(R, \theta)$ and $\mathbf{E}(R, \theta)$ from $\mathbf{A}(R, \theta)$.

    c) Sketch the radiation pattern for $L = \lambda/2$.

P.11–31 A turnstile antenna consists of two perpendicular half-wave dipoles, one (antenna A) lying along the x-axis and the other (antenna B) along the y-axis. The output of antenna B, after a 90° phase retardation, is combined with that of antenna A. A right-hand elliptically polarized plane wave $\mathbf{E}_i = E_0(\mathbf{a}_x + \mathbf{a}_y jp) \exp (jkz)$ is incident on the antennas.

    a) Determine the open-circuit voltage at the output terminals of the turnstile antenna. What is its value if $p = 1$?

    b) Repeat part (a) for a left-hand elliptically polarized incident wave $\mathbf{E}_i = E_0(\mathbf{a}_x - \mathbf{a}_y jp) \exp (jkz)$.

    c) Repeat part (a) for a linearly polarized incident wave $\mathbf{E}_i = \mathbf{a}_x E_0 \exp (jkz)$.

(*Hint:* Find the complex effective length of the turnstile antenna and use Eq. 11–76.)

P.11–32 A helical antenna operating in the normal mode has N turns with diameter $2b$ and interturn spacing s. Both $2b$ and s are very small in comparison to λ/N and are adjusted to radiate circularly polarized waves. Find:

    a) its directive gain and directivity,

    b) its radiation resistance.

P.11–33 For the problem in Example 11–13, write the expression for the far-zone electric field $\mathbf{E}_P(\theta, \phi)$ at a point $P(\theta, \phi)$ located near the z-axis ($\cos\theta \cong 1$) but not in either of the principal planes.

P.11–34 Assume that the field in an $a \times b$ rectangular aperture in an xy-plane is linearly polarized in the y-direction, and that the aperture excitation has a uniform phase and a triangular amplitude distribution

$$f(x) = 1 - \left|\frac{2}{a}x\right|, \qquad |x| \leq \frac{a}{2}.$$

Find (a) the pattern function in the xz-plane, (b) the half-power beamwidth, (c) the location of the first nulls, and (d) the level of the first sidelobes. Compare the results with those obtained in Example 11–13 for uniform field distribution.

P.11–35 Do Problem P.11–34 for a uniform-phased cosinusoidal amplitude distribution

$$f(x) = \cos\left(\frac{\pi x}{a}\right), \qquad |x| \leq \frac{a}{2},$$

and compare your results with those obtained in Example 11–13 for a uniform field distribution.

Appendixes

A

Symbols and Units

A–1 Fundamental SI (Rationalized MKSA) Units[†]

Quantity	Symbol	Unit	Abbreviation
Length	ℓ	meter	m
Mass	m	kilogram	kg
Time	t	second	s
Current	I, i	ampere	A

[†] Besides the MKSA system for the units of length, mass, time, and current, the SI adopted by the International Committee on Weights and Measures consists of two other fundamental units. They are Kelvin degree (K) for thermodynamic temperature and candela (cd) for luminous intensity.

A–2 Derived Quantities

Quantity	Symbol	Unit	Abbreviation
Admittance	Y	siemens	S
Angular frequency	ω	radián/second	rad/s
Attenuation constant	α	neper/meter	Np/m
Capacitance	C	farad	F
Charge	Q, q	coulomb	C
Charge density (linear)	ρ_ℓ	coulomb/meter	C/m
Charge density (surface)	ρ_s	coulomb/meter^2	C/m^2
Charge density (volume)	ρ	coulomb/meter^3	C/m^3

(continued)

A–2 **Derived Quantities** (continued)

Quantity	Symbol	Unit	Abbreviation
Conductance	G	siemens	S
Conductivity	σ	siemens/meter	S/m
Current density (surface)	$\mathbf{J}_s$	ampere/meter	A/m
Current density (volume)	$\mathbf{J}$	ampere/meter^2	A/m^2
Dielectric constant (relative permittivity)	ϵ_r	(dimensionless)	—
Directivity	D	(dimensionless)	—
Electric dipole moment	$\mathbf{p}$	coulomb-meter	C·m
Electric displacement (Electric flux density)	$\mathbf{D}$	coulomb/meter^2	C/m^2
Electric field intensity	$\mathbf{E}$	volt/meter	V/m
Electric potential	V	volt	V
Electric susceptibility	χ_e	(dimensionless)	—
Electromotive force	$\mathscr{V}$	volt	V
Energy (work)	W	joule	J
Energy density	w	joule/meter^3	J/m^3
Force	$\mathbf{F}$	newton	N
Frequency	f	hertz	Hz
Impedance	Z, η	ohm	Ω
Inductance	L	henry	H
Magnetic dipole moment	$\mathbf{m}$	ampere-meter^2	A·m^2
Magnetic field intensity	$\mathbf{H}$	ampere/meter	A/m
Magnetic flux	Φ	weber	Wb
Magnetic flux density	$\mathbf{B}$	tesla	T
Magnetic potential (vector)	$\mathbf{A}$	weber/meter	Wb/m
Magnetic susceptibility	χ_m	(dimensionless)	—
Magnetization	$\mathbf{M}$	ampere/meter	A/m
Magnetomotive force	$\mathscr{V}_m$	ampere	A
Permeability	μ, μ_0	henry/meter	H/m
Permittivity	ϵ, ϵ_0	farad/meter	F/m
Phase	ϕ	radian	rad
Phase constant	β	radian/meter	rad/m
Polarization vector	$\mathbf{P}$	coulomb/meter^2	C/m^2
Power	P	watt	W
Poynting vector (power density)	$\mathscr{P}$	watt/meter^2	W/m^2
Propagation constant	γ	meter^{-1}	m^{-1}

(continued)

A–2 **Derived Quantities** (continued)

Quantity	Symbol	Unit	Abbreviation
Radiation intensity	U	watt/steradian	W/sr
Reactance	X	ohm	Ω
Relative permeability	μ_r	(dimensionless)	—
Relative permittivity (dielectric constant)	ϵ_r	(dimensionless)	—
Reluctance	$\mathscr{R}$	henry^{-1}	H^{-1}
Resistance	R	ohm	Ω
Susceptance	B	siemens	S
Torque	T	newton-meter	N·m
Velocity	u	meter/second	m/s
Voltage	V	volt	V
Wavelength	λ	meter	m
Wavenumber	k	radian/meter	rad/m
Work (energy)	W	joule	J

A–3 **Multiples and Submultiples of Units**

Factor by Which Unit Is Multiplied	Prefix	Symbol
$1\ 000\ 000\ 000\ 000\ 000\ 000 = 10^{18}$	exa	E
$1\ 000\ 000\ 000\ 000\ 000 = 10^{15}$	peta	P
$1\ 000\ 000\ 000\ 000 = 10^{12}$	tera	T
$1\ 000\ 000\ 000 = 10^{9}$	giga	G
$1\ 000\ 000 = 10^{6}$	mega	M
$1\ 000 = 10^{3}$	kilo	k
$100 = 10^{2}$	hecto†	h
$10 = 10^{1}$	deka†	da
$0.1 = 10^{-1}$	deci†	d
$0.01 = 10^{-2}$	centi†	c
$0.001 = 10^{-3}$	milli	m
$0.000\ 001 = 10^{-6}$	micro	μ
$0.000\ 000\ 001 = 10^{-9}$	nano	n
$0.000\ 000\ 000\ 001 = 10^{-12}$	pico	p
$0.000\ 000\ 000\ 000\ 001 = 10^{-15}$	femto	f
$0.000\ 000\ 000\ 000\ 000\ 001 = 10^{-18}$	atto	a

† These prefixes are generally not used except for measurements of length, area, and volume.

B

Some Useful Material Constants

B–1 Constants of Free Space

Constant	Symbol	Value
Velocity of light	c	$\sim 3 \times 10^8$ (m/s)
Permittivity	ϵ_0	$\sim \dfrac{1}{36\pi} \times 10^{-9}$ (F/m)
Permeability	μ_0	$4\pi \times 10^{-7}$ (H/m)
Intrinsic impedance	η_0	$\sim 120\pi$ or 377 (Ω)

B–2 Physical Constants of Electron and Proton

Constant	Symbol	Value
Rest mass of electron	m_e	9.107×10^{-31} (kg)
Charge of electron	$-e$	-1.602×10^{-19} (C)
Charge-to-mass ratio of electron	$-e/m_e$	-1.759×10^{11} (C/kg)
Radius of electron	R_e	2.81×10^{-15} (m)
Rest mass of proton	m_p	1.673×10^{-27} (kg)

B–3   Relative Permittivities (Dielectric Constants)[†]

Material	Relative Permittivity, ϵ_r
Air	1.0
Bakelite	5.0
Glass	4~10
Mica	6.0
Oil	2.3
Paper	2~4
Parafin wax	2.2
Plexiglass	3.4
Polyethylene	2.3
Polystyrene	2.6
Porcelain	5.7
Rubber	2.3~4.0
Soil (dry)	3~4
Teflon	2.1
Water (distilled)	80
Seawater	72

B–4   Conductivities[†]

Material	Conductivity, $\sigma(S/m)$	Material	Conductivity, $\sigma(S/m)$
Silver	6.17×10^7	Fresh water	10^{-3}
Copper	5.80×10^7	Distilled water	2×10^{-4}
Gold	4.10×10^7	Dry soil	10^{-5}
Aluminum	3.54×10^7	Transformer oil	10^{-11}
Brass	1.57×10^7	Glass	10^{-12}
Bronze	10^7	Porcelain	2×10^{-13}
Iron	10^7	Rubber	10^{-15}
Seawater	4	Fused quartz	10^{-17}

[†] Note that the constitutive parameters of some of the materials are frequency and temperature dependent. The listed constants are average low-frequency values at room temperature.

B–5   Relative Permeabilities[†]

Material	Relative Permeability, μ_r
Ferromagnetic (nonlinear)	
Nickel	250
Cobalt	600
Iron (pure)	4,000
Mumetal	100,000
Paramagnetic	
Aluminum	1.000021
Magnesium	1.000012
Palladium	1.00082
Titanium	1.00018
Diamagnetic	
Bismuth	0.99983
Gold	0.99996
Silver	0.99998
Copper	0.99999

[†] Note that the constitutive parameters of some of the materials are frequency and temperature dependent. The listed constants are average low-frequency values at room temperature.

C

Index of Tables

General Bibliography

In addition to the references included in the footnotes throughout the book and at the end of Chapter 11, the following books on electromagnetic fields and waves at a comparable level have been found to be useful.

Bewley, L. V., *Two Dimensional Fields in Electrical Engineering*, Dover Publications, New York, 1963.

Collin, R. E., *Antennas and Radiowave Propagation*, McGraw-Hill, New York, 1985.

Crowley, J. M., *Fundamentals of Applied Electrostatics*, Wiley, New York, 1986.

Feynman, R. P.; Leighton, R. O.; and Sands, M., *Lectures on Physics*, vol. 2, Addison-Wesley, Reading, Mass., 1964.

Javid, M., and Brown, P. M., *Field Analysis and Electromagnetics*, McGraw-Hill, New York, 1963.

Jordan, E. C., and Balmain, K. G., *Electromagnetic Waves and Radiating Systems*, 2nd ed., Prentice-Hall, Englewood Cliffs, N.J., 1968.

Kraus, J. D., *Electromagnetics*, 3rd ed., McGraw-Hill, New York, 1984.

Lorrain, P., and Corson, D., *Electromagnetic Fields and Waves*, 2nd ed., Freeman, San Franscisco, Calif., 1970.

Paris, D. T., and Hurd, F. K., *Basic Electromagnetic Theory*, McGraw-Hill, New York, 1969.

Parton, J. E.; Owen, S. J. T.; and Raven, M. S., *Applied Electromagnetics*, 2nd ed., Macmillan, London, 1986.

Plonsey, R., and Collin, R. E., *Principles and Applications of Electromagnetic Fields*, 2nd ed., McGraw-Hill, New York, 1982.

Popović B. D. *Introductory Engineering Electromagnetics*, Addison-Wesley, Reading, Mass., 1971.

Ramo, S.; Whinnery, J. R.; and Van Duzer, T., *Fields and Waves in Communication Electronics*, 2nd ed., Wiley, New York, 1984.

Sander K. F., and Reed, G. A. L., *Transmission and Propagation of Electromagnetic Waves*, 2nd ed., Cambridge University Press, Cambridge, England, 1986.

Seshadri, S. R., *Fundamentals of Transmission Lines and Electromagnetic Fields*, Addison-Wesley, Reading, Mass., 1971.

Shen, L. C., and Kong, J. A., *Applied Electromagnetism*, 2nd ed., PWS Engineering, Boston, Mass., 1987.

Zahn, M., *Electromagnetic Field Theory*, Wiley, New York, 1979.

Answers to Selected Problems

Chapter 2

P.2–1 a) $(\mathbf{a}_x + \mathbf{a}_y 2 - \mathbf{a}_z 3)/\sqrt{14}$.　　**b)** $\sqrt{53}$.　　**c)** -11.　　**d)** $135.5°$.

e) $11/\sqrt{29}$.　　**f)** $-(\mathbf{a}_x 4 + \mathbf{a}_y 13 + \mathbf{a}_z 10)$.　　**g)** -42.

h) $\mathbf{a}_x 2 - \mathbf{a}_y 40 + \mathbf{a}_z 5$ and $\mathbf{a}_x 55 - \mathbf{a}_y 44 - \mathbf{a}_z 11$.

P.2–5 $\mathbf{X} = (p\mathbf{A} + \mathbf{B} \times \mathbf{A})/A^2$.

P.2–9 a) $\cos(\alpha - \beta) = \cos\alpha\cos\beta + \sin\alpha\sin\beta$.

P.2–15 1.12.

P.2–17 a) $|\mathbf{E}| = 1/2, E_x = -0.212$.　　**b)** $\theta = 154°$.

P.2–21 a) 14.　　**b)** 14.

P.2–23 a) $(\nabla V)_P = -(\mathbf{a}_y 0.026 + \mathbf{a}_z 0.043)$.　　**b)** 0.0485.

P.2–25 $\ell = 0, m = p = 1/\sqrt{2}$; $\int_S \mathbf{F} \cdot d\mathbf{s} = 20$.

P.2–29 $\oint_S \mathbf{A} \cdot d\mathbf{s} = \int_V \nabla \cdot \mathbf{A}\, dv = 1{,}200\pi$.

P.2–31 See Eq. (2–114).

P.2–35 $\dfrac{1}{R\sin\theta}\left[\dfrac{\partial}{\partial\theta}(A_\phi \sin\theta) - \dfrac{\partial A_\theta}{\partial\phi}\right]$.

P.2–39 a) $c_1 = 1, c_2 = 0, c_3 = -3$.　　**b)** $c_4 = -1$.　　**c)** $V = -\dfrac{x^2}{2} - xz + 3yz + \dfrac{z^2}{2}$.

Chapter 3

P.3–1 a) $\alpha = \tan^{-1}(mu_0^2/ewE_d)$.　　**b)** $L/w = 10.5$.

P.3–5 a) $Q_1/Q_2 = -3/4\sqrt{2}$.　　**b)** $Q_1/Q_2 = 1/2\sqrt{2}$.

P.3–7 $|\mathbf{F}| = Q\rho_\ell bh/2\epsilon(b^2 + h^2)^{3/2}$.

P.3–9 $\mathbf{a}_y 3\rho_{\ell 1}/4\pi\epsilon_0 L$.

P.3–11 (1) $0 \le R \le b: E_{R1} = \dfrac{\rho_0 R}{\epsilon_0}\left(\dfrac{1}{3} - \dfrac{R^2}{5b^2}\right).$ (2) $b \le R < R_i: E_{R2} = \dfrac{2\rho_0 b^3}{15\epsilon_0 R^2}.$

    (3) $R_i < R < R_o: E_{R3} = 0.$ (4) $R > R_o: E_{R4} = \dfrac{2\rho_0 b^3}{15\epsilon_0 R^2}.$

P.3–13 a) $28(\mu J).$ b) $28(\mu J).$

P.3–15 a) $V = \dfrac{qd^2}{16\pi\epsilon_0 R^3}(3\cos^2\theta - 1), \mathbf{E} = \dfrac{3qd^2}{16\pi\epsilon_0 R^4}[\mathbf{a}_R(3\cos^2\theta - 1) + \mathbf{a}_\theta \sin 2\theta].$

    b) $R^3 = C_V(3\cos^2\theta - 1).$

    c) $R^2 = C_E \sin^2\theta \cos\theta.$

P.3–17 $V_P = \dfrac{\rho_\ell}{4\pi\epsilon}\left[\sinh^{-1}\left(\dfrac{L-x}{b}\right) + \sinh^{-1}\left(\dfrac{x}{b}\right)\right].$

P.3–19 If origin is chosen at the center of the base of the circular tube:

    a) $z \ge h, V_o = \dfrac{b\rho_s}{2\epsilon_0}\ln\dfrac{z + \sqrt{b^2 + z^2}}{(z - h) + \sqrt{b^2 + (z - h)^2}}.$

    b) $z \le h, V_i = \dfrac{b\rho_s}{2\epsilon_0}\ln\dfrac{1}{b^2}(z + \sqrt{b^2 + z^2})[(h - z) + \sqrt{b^2 + (h - z)^2}],$

    where $\rho_s = Q/2\pi bh.$

P.3–21 $r_o = \dfrac{4\pi\epsilon_0 b^3}{Ne}E_o.$

P.3–23 $P/3\epsilon_0.$

P.3–25 $\mathbf{E}_2(z = 0) = \mathbf{a}_x 2y - \mathbf{a}_y 3x + \mathbf{a}_z(10/3).$

P.3–29 a) $19.3\,(kV).$ b) $1.82\,(kV).$

P.3–31 a) $\mathbf{E}(a) = \mathbf{a}_r \dfrac{V_0}{a\ln(b/a)}.$ b) $b/a = e = 2.718.$ c) $\min E(a) = eV_0/b.$

    d) $C = 2\pi\epsilon\,(F/m).$

P.3–33 $C = \dfrac{\pi\epsilon_0(\epsilon_{r1} + \epsilon_{r2})L}{\ln(r_o/r_i)}.$

P.3–35 a) $0.708\,(mF).$ b) $1.35 \times 10^{10}\,(C).$

P.3–37 a) $\mathbf{D} = \begin{cases} \mathbf{a}_R \dfrac{\epsilon_0\epsilon_r V}{R^2\left(\dfrac{1}{R_i} - \dfrac{1}{2b} - \dfrac{1}{2R_o}\right)}, & \text{for } R_i < R < R_o; \\ 0, & \text{for } R < R_i \text{ and } R > R_o. \end{cases}$

    b) $C = \dfrac{4\pi\epsilon_0\epsilon_r}{\dfrac{1}{R_i} - \dfrac{1}{2b} - \dfrac{1}{2R_o}}.$

P.3–39 Designate the wires as conductors 0, 1, and 2 with wire 1 in the center.

    $C_{10} = C_{12} = 3.36\,(pF/m), C_{20} = 2.35\,(pF/m).$

P.3–41 $1.69 \times 10^{-15}\,(m).$

P.3–47 $F_\ell = \dfrac{\pi\epsilon_0 V_0^2}{2D\,[\ln(D/b)]^2}.$

Chapter 4

P.4–1 a) $V_d = \dfrac{5yV_0}{(4 + \epsilon_r)d}$, $\quad \mathbf{E}_d = -\mathbf{a}_y \dfrac{5V_0}{(4 + \epsilon_r)d}$.

b) $V_a = \dfrac{5\epsilon_r y - 4(\epsilon_r - 1)d}{(4 + \epsilon_r)d} V_0$, $\quad \mathbf{E}_a = -\mathbf{a}_y \dfrac{5\epsilon_r V_0}{(4 + \epsilon_r)d}$.

c) $(\rho_s)_{y=d} = \dfrac{5\epsilon_0\epsilon_r V_0}{(4 + \epsilon_r)d} = -(\rho_s)_{y=0}$.

P.4–7 a) $\rho_s = -\dfrac{Qd}{2\pi(d^2 + r^2)^{3/2}}$.

b) $-Q$.

P.4–11 $C = \dfrac{\pi\epsilon_0}{\ln\{d/[a\sqrt{1 + (d/2h)^2}]\}}$.

P.4–13 $C = \dfrac{2\pi\epsilon_0}{\ln\left[\dfrac{1}{2}\left(\dfrac{D^2}{a_1 a_2} - \dfrac{a_1}{a_2} - \dfrac{a_2}{a_1}\right) - \sqrt{\dfrac{1}{4}\left(\dfrac{D^2}{a_1 a_2} - \dfrac{a_1}{a_2} - \dfrac{a_2}{a_1}\right)^2 - 1}\right]}$

$\qquad = \dfrac{2\pi\epsilon_0}{\cosh^{-1}\left[\dfrac{1}{2}\left(\dfrac{D^2}{a_1 a_2} - \dfrac{a_1}{a_2} - \dfrac{a_2}{a_1}\right)\right]}$.

P.4–15 b) $\rho_s = -\dfrac{Q(b^2 - d^2)}{4\pi b(b^2 + d^2 - 2bd\cos\theta)^{3/2}}$.

P.4–17 $Q_1 = Q_2 = \dfrac{\epsilon_2 - \epsilon_1}{\epsilon_2 + \epsilon_1} Q$.

P.4–19 $V_n(x, y) = C_n \cosh\dfrac{n\pi}{b}(x - a)\cos\dfrac{n\pi}{b}y$.

P.4–21 $V_n(x, y) = \sin\dfrac{n\pi}{a}x\left[A_n \sinh\dfrac{n\pi}{a}y + B_n \cosh\dfrac{n\pi}{a}y\right]$.

P.4–23 a) $V(\phi) = \dfrac{V_0}{\alpha}\phi$. $\qquad$ **b)** $V(\phi) = \dfrac{V_0}{2\pi - \alpha}(2\pi - \phi)$.

P.4–25 $V(r, \phi) = -E_0 r\left(1 - \dfrac{b^2}{r^2}\right)\cos\phi$.

$\qquad \mathbf{E}(r, \phi) = \mathbf{a}_r E_0\left(1 + \dfrac{b^2}{r^2}\right)\cos\phi - \mathbf{a}_\phi E_0\left(1 - \dfrac{b^2}{r^2}\right)\sin\phi$.

P.4–27 a) $V(\theta) = V_0\dfrac{\ln\left(\tan\dfrac{\theta}{2}\right)}{\ln\left(\tan\dfrac{\alpha}{2}\right)}$.

b) $\mathbf{E}(\theta) = -\mathbf{a}_\theta \dfrac{V_0}{R\ln[\tan(\alpha/2)]\sin\theta}$.

P.4–29 $V_i(R, \theta) = -\dfrac{3E_0}{\epsilon_r + 2} R \cos \theta, \quad V_o(R, \theta) = -\left[R - \dfrac{(\epsilon_r - 1) b^3}{(\epsilon_r + 2) R^2} \right] E_0 \cos \theta.$

$\mathbf{E}_i(R, \theta) = (\mathbf{a}_R \cos \theta - \mathbf{a}_\theta \sin \theta) \dfrac{3 E_0}{\epsilon_r + 2} = \mathbf{a}_z \dfrac{3 E_0}{\epsilon_r + 2}.$

$\mathbf{E}_o(R, \theta) = \mathbf{a}_R \left[1 + \dfrac{2(\epsilon_r - 1) b^3}{(\epsilon_r + 2) R^3} \right] E_0 \cos \theta - \mathbf{a}_\theta \left[1 - \dfrac{(\epsilon_r - 1) b^3}{(\epsilon_r + 2) R^3} \right] E_0 \sin \theta.$

Chapter 5

P.5–1 a) $V(y) = V_0 (y/d)^{4/3}, \quad E(y) = -(4V_0/3d)(y/d)^{1/3}.$

b) $Q = -(4V_0/3d) \epsilon_0 S.$

c) Charge on cathode = 0; charge on anode = $-Q$.

d) 3.58 (ns).

P.5–3 a) 2.32 a.

b) $E_1 = E_2 = I/2\pi a^2 \sigma.$

P.5–5 $I_1 = 0.7 \,(\text{A}), \quad P_{R1} = 0.163 \,(\text{W}); \quad I_2 = 0.140 \,(\text{A}), P_{R2} = 0.392 \,(\text{W});$
$I_3 = 0.093 \,(\text{A}), P_{R3} = 0.261 \,(\text{W}); \quad I_4 = 0.233 \,(\text{A}), P_{R4} = 0.436 \,(\text{W});$
$I_5 = 0.467 \,(\text{A}), P_{R5} = 2.178 \,(\text{W}).$

P.5–7 a) 4.88 (ps).   **b)** $W_i/(W_i)_0 = 10^{-4}$; heat loss.

c) $W_o = 45 \,(\text{kJ}).$

P.5–9 a) $E_2 = \left[\sin^2 \alpha_1 + \left(\dfrac{\sigma_1}{\sigma_2} \cos \alpha_1 \right)^2 \right]^{1/2}, \quad \alpha_2 = \tan^{-1} \left[\dfrac{\sigma_2}{\sigma_1} \tan \alpha_1 \right].$

b) $\rho_s = \left(\dfrac{\sigma_1}{\sigma_2} \epsilon_2 - \epsilon_1 \right) E_1 \sin \alpha_1.$

P.5–11 b) $P = \mathcal{V}^2 S \sigma_1 \sigma_2 / (\sigma_1 d_2 + \sigma_2 d_1).$

P.5–13 a) $J = \dfrac{\sigma_1 \sigma_2 V_0}{r[\sigma_1 \ln (b/c) + \sigma_2 \ln (c/a)]}.$

b) $\rho_{sa} = \dfrac{\epsilon_1 \sigma_2 V_0}{a[\sigma_1 \ln (b/c) + \sigma_2 \ln (c/a)]}, \quad \rho_{sb} = -\dfrac{\epsilon_2 \sigma_1 V_0}{b[\sigma_1 \ln (b/c) + \sigma_2 \ln (c/a)]},$

$\rho_{sc} = \dfrac{(\epsilon_2 \sigma_1 - \epsilon_1 \sigma_2) V_0}{c[\sigma_1 \ln (b/c) + \sigma_2 \ln (c/a)]}.$

P.5–15 $\dfrac{1}{4\pi\sigma} \left(\dfrac{1}{R_1} - \dfrac{1}{R_2} \right).$

P.5–17 $\dfrac{R_2 - R_1}{2\pi\sigma R_1 R_2 (1 - \cos \theta_0)}.$

P.5–19 $\dfrac{1}{4\pi\sigma} \left(\dfrac{1}{b_1} + \dfrac{1}{b_2} - \dfrac{2}{d} \right).$

P.5–21 6.36 (MΩ).

P.5–23 $\mathbf{J} = \mathbf{a}_x J_0 - \dfrac{J_0 b^2}{r^2} (\mathbf{a}_r \cos \phi + \mathbf{a}_\phi \sin \phi).$

Chapter 6

P.6-1 $y^2 + \left(z + \dfrac{u_0}{\omega_0}\right)^2 = \left(\dfrac{u_0}{\omega_0}\right)^2$, $\omega_0 = \dfrac{qB_0}{m}$.

Path of motion of q in magnetic field is a semicircle.

P.6-3 $B_\phi = \dfrac{\mu_0 I r}{2\pi a^2}$, $r \le a$; $\quad B_\phi = \dfrac{\mu_0 I}{2\pi r}$, $a \le r \le b$;

$$B_\phi = \dfrac{\mu_0 I(c^2 - r^2)}{2\pi(c^2 - b^2)r}, \quad b \le r \le c; \quad B_\phi = 0, \quad r \ge c.$$

P.6-5 $\mathbf{a}_z 1.38 I/w$.

P.6-7 $B = \dfrac{\mu_0 NI}{2L}\left[\dfrac{L - z}{\sqrt{(L - z)^2 + b^2}} + \dfrac{z}{\sqrt{z^2 + b^2}}\right]$.

P.6-9 $\mathbf{F}_{12} = \dfrac{\mu_0 q_1 q_2}{4\pi R^2}\mathbf{u}_2 \times (\mathbf{u}_1 \times \mathbf{a}_{12})$.

P.6-11 $\dfrac{\mu_0 I}{2b}\left(\dfrac{1}{\pi} + \dfrac{1}{2}\right)$.

P.6-15 $\mathbf{a}_z \mu_0 J d/2$.

P.6-17 $\mathbf{A}_1 = \mathbf{a}_z\left[-\dfrac{\mu_0 I}{4\pi}\left(\dfrac{r_1}{b}\right)^2 + c\right]$, $r_1 \le b$; $\quad \mathbf{A}_2 = \mathbf{a}_z\left\{-\dfrac{\mu_0 I}{4\pi}\left[\ln\left(\dfrac{r_2}{b}\right)^2 + 1\right] + c\right\}$, $r_2 \ge b$.

P.6-21 a) $\mathbf{a}_z \mu_0 H_0/\mu$. **b)** $\mathbf{a}_z(H_0 - M_i)$.

P.6-27 a) $\mathscr{R}_g = 1.21 \times 10^6 (\text{H}^{-1})$, $\quad \mathscr{R}_c = 6.75 \times 10^4 (\text{H}^{-1})$.

b) $\mathbf{B}_g = \mathbf{B}_c = \mathbf{a}_\phi 5.09 \times 10^{-3} (\text{T})$.

$\mathbf{H}_g = \mathbf{a}_\phi 4.05 \times 10^3 (\text{A/m})$, $\quad \mathbf{H}_c = \mathbf{a}_\phi 1.35 (\text{A/m})$.

c) $I = 25.6 (\text{mA})$.

P.6-33 b) $\mathbf{B} = -\mathbf{a}_x \dfrac{\mu_0 I}{2\pi}\left[\dfrac{y - d}{(y - d)^2 + x^2} + \dfrac{y + d}{(y + d)^2 + x^2}\right]$

$$+ \mathbf{a}_y \dfrac{\mu_0 I}{2\pi} x\left[\dfrac{1}{(y - d)^2 + x^2} + \dfrac{1}{(y + d)^2 + x^2}\right].$$

P.6-35 $L = \mu_0 N^2(r_o - \sqrt{r_o^2 - b^2})$.

P.6-37 $L'_{AA'/BB'} = \dfrac{\mu_0}{2\pi}\ln\left(1 + \dfrac{d^2}{D^2}\right)$.

P.6-39 $L_{12} = \mu_0(d - \sqrt{d^2 - b^2})$.

P.6-41 $I_1/I_2 = -M/L_1$.

P.6-43 $\mathbf{f} = \mathbf{a}_x \dfrac{\mu_0 I^2}{\pi w}\tan^{-1}\left(\dfrac{w}{2D}\right)$.

P.6-45 $\mathbf{F} = \mathbf{a}_x \mu_0 I_1 I_2\left[\dfrac{1}{\sqrt{1 - (b/d)^2}} - 1\right]$, repulsive.

P.6-47 $\mathbf{T} = -\mathbf{a}_x 0.1 (\text{N} \cdot \text{m})$.

P.6-51 Maximum deviation from north-south direction: $55.8°$.

P.6-53 $\mathbf{F} = \mathbf{a}_x \dfrac{\mu_0}{2}(\mu_r - 1)n^2 I^2 S$.

Chapter 7

P.7–3 a) $i_2(t) = -\dfrac{L_{12}}{L}I_1 e^{-(R/L)t}$, $0 < t < T$; $L_{12} = \dfrac{\mu_0 h}{2\pi}\ln\left(1 + \dfrac{w}{d}\right)$.

$i_2(t) = \dfrac{L_{12}}{L}I_1[e^{-(R/L)(t-T)} - e^{-(R/L)T}]$, $t > T$.

P.7–5 a) $0.234\,(\text{A})$ **b)** $48.2°$.

P.7–7 a) $0.0472\mu_0 I\omega b$.   **b)** $0.0469\mu_0 I\omega b$.

P.7–13 a) $V' = V - \dfrac{\partial\psi}{\partial t}$.   **b)** $\nabla^2\psi - \mu\epsilon\dfrac{\partial^2\psi}{\partial t^2} = 0$.

P.7–23 $E_0 = 0.068$, $\theta = -72.8°$.

P.7–25 $\beta = 54.4\,(\text{rad/m})$.

$\mathbf{H}(x, z; t) = -\mathbf{a}_x 2.30 \times 10^{-4}\sin(10\pi x)\cos(6\pi 10^9 t - 54.4z)$

$-\mathbf{a}_z 1.33 \times 10^{-4}\cos(10\pi x)\sin(6\pi 10^9 t - 54.4z)(\text{A/m})$.

P.7–27 $k = \omega\sqrt{\mu_0\epsilon_0}$. $\mathbf{H} = \mathbf{a}_\phi\dfrac{E_0}{R}\sqrt{\dfrac{\epsilon_0}{\mu_0}}\sin\theta\cos\omega(t - \sqrt{\mu_0\epsilon_0}R)$.

Chapter 8

P.8–3 a) $\Delta f = -(2u/c)f$, assuming the vehicle to be moving in the same direction as the direction of the incident wave.

b) $120\,(\text{km/hr})$, or $74.6\,(\text{miles/hr})$.

P.8–5 a) $k_0 = 0.1047\,(\text{rad/m})$, $y = 22.5 \pm n\lambda/2\,(\text{m})$.

b) $\mathbf{E}(y, t) = -\mathbf{a}_x 1.508 \times 10^{-3}\cos\left(10^7\pi t - \dfrac{\pi}{30}y + \dfrac{\pi}{4}\right)(\text{V/m})$.

P.8–7 $\left(\dfrac{E_y}{E_{20}\sin\psi}\right)^2 + \left(\dfrac{E_x}{E_{10}\sin\psi}\right)^2 - 2\dfrac{E_x E_y\cos\psi}{E_{10}E_{20}\sin^2\psi} = 1$, where $E_x = E_{10}\sin(wt - kz)$,

and $E_y = E_{20}\sin(wt - kz + \psi)$.

P.8–11 a) $1.395\,(\text{m})$.

b) $\eta_c = 238(1 + j0.005)\,(\Omega)$, $\lambda = 6.3\,(\text{cm})$, $u_p = 1.8973 \times 10^8\,(\text{m/s})$,

$u_g = 1.8975 \times 10^8\,(\text{m/s})$.

c) $\mathbf{H} = \mathbf{a}_x 0.21\,e^{-0.497x}\sin(6\pi 10^9 t - 31.6\pi x + 1.042)(\text{A/m})$.

P.8–13 a) $0.99 \times 10^5\,(\text{S/m})$.   **b)** $0.175\,(\text{mm})$.

P.8–21 a) Left-hand circularly polarized wave in $-z$ direction.

b) $\dfrac{2E_0}{\eta_0}(\mathbf{a}_x - j\mathbf{a}_y)$.   **c)** $2E_0\sin\beta z(\mathbf{a}_x\sin\omega t - \mathbf{a}_y\cos\omega t)$.

P.8–23 a) $f = 5.73\,(\text{MHz})$, $\lambda = 0.524\,(\text{m})$.

b) $\mathbf{E}_i(y, z; t) = 5(\mathbf{a}_y + \mathbf{a}_z\sqrt{3})\cos(3.6 \times 10^9 t + 6\sqrt{3}y - 6z)\,(\text{V/m})$,

$\mathbf{H}_i(y, z; t) = -\mathbf{a}_x\dfrac{1}{12\pi}\cos(3.6 \times 10^9 t + 6\sqrt{3}y - 6z)\,(\text{A/m})$.

c) $\theta_i = 60°$.

   d) $\mathbf{E}_r(y, z) = 5(-\mathbf{a}_y + \mathbf{a}_z\sqrt{3})e^{j6(\sqrt{3}y+z)}$ (V/m),

      $\mathbf{H}_r(y, z) = -\mathbf{a}_x\dfrac{1}{12\pi}e^{j6(\sqrt{3}y+z)}$ (A/m).

   e) $\mathbf{E}_1(y, z) = (-\mathbf{a}_y j10\sin 6z + \mathbf{a}_z 10\sqrt{3}\cos 6z)e^{j6\sqrt{3}y}$ (V/m),

      $\mathbf{H}_1(y, z) = -\mathbf{a}_x\dfrac{1}{6\pi}(\cos 6z)e^{j6\sqrt{3}y}$ (A/m).

P.8–25 $\mathbf{H}_1(x, z; t) = \mathbf{a}_y\dfrac{2E_{i0}}{\eta_1}\cos(\beta_1 z\cos\theta_i)\sin(\omega t - \beta_1 x\sin\theta_i)$

      $\mathscr{P}_{av} = \mathbf{a}_x\dfrac{2E_{i0}^2}{\eta_1}\sin\theta_i\sin^2(\beta_1 z\cos\theta_i)$.

P.8–27 a) $\mathbf{E}_r(z, t) = \mathbf{a}_x 2.77\cos(1.8\times10^9 t + 6z + 157°)$ (V/m),

      $\mathbf{E}_t(z, t) = \mathbf{a}_x 7.53\,e^{-2.3z}\cos(1.8\times10^9 t - 9.76z - 172°)$ (V/m).

   b) $\mathscr{P}_{av} = \mathbf{a}_z 0.122e^{-4.61z}$ (W/m^2).

P.8–29 a) $E_{r0} = -\dfrac{j(\eta_0^2 - \eta_2^2)\tan\beta_2 d}{\eta_0\eta_2 + j(\eta_0^2 + \eta_2^2)\tan\beta_2 d}E_{i0}$,

      $E_2^+ = \dfrac{\eta_2(\eta_0 + \eta_2)e^{j\beta_2 d}}{\eta_0\eta_2\cos\beta_2 d + j(\eta_0^2 + \eta_2^2)\sin\beta_2 d}E_{i0}$,

      $E_2^- = \dfrac{\eta_2(\eta_0 - \eta_2)e^{-j\beta_2 d}}{\eta_0\eta_2\cos\beta_2 d + j(\eta_0^2 + \eta_2^2)\sin\beta_2 d}E_{i0}$,

      $E_{t0} = \dfrac{2\eta_0\eta_2 e^{j\beta_0 d}}{\eta_0\eta_2\cos\beta_2 d + j(\eta_0^2 + \eta_2^2)\sin\beta_2 d}E_{i0}$.

P.8–31 $\Gamma_0 = \dfrac{(\Gamma_{12} + \Gamma_{23}) + j(\Gamma_{12} - \Gamma_{23})\tan\beta_2 d}{(1 + \Gamma_{12}\Gamma_{23}) + j(1 - \Gamma_{12}\Gamma_{23})\tan\beta_2 d}$.

P.8–33 Assume $|\eta_2| \ll \eta_0$.

   a) $E_2^+ = -j\left(\dfrac{\eta_2}{\eta_0}\right)\dfrac{e^{\alpha_2 d}e^{j\beta_2 d}E_{i0}}{\sin(\beta_2 - j\alpha_2)d}$.   **b)** $E_2^- = -j\left(\dfrac{\eta_2}{\eta_0}\right)\dfrac{e^{-\alpha_2 d}e^{-j\beta_2 d}E_{i0}}{\sin(\beta_2 - j\alpha_2)d}$.

   c) $E_{30} = -j\left(\dfrac{\eta_2}{\eta_0}\right)\dfrac{2e^{j\beta_0 d}E_{i0}}{\sin(\beta_2 - j\alpha_2)d}$.   **d)** $(\mathscr{P}_{av})_3/(\mathscr{P}_{av})_i = 1.839\times10^{-11}$.

P.8–35 a) $\theta_t = 0.03°$ **b)** $\Gamma_\| = 0.0214e^{j\pi/4}$

   c) $(\mathscr{P}_{av})_t/(\mathscr{P}_{av})_i = 1.054\times10^{-3}e^{-0.795z}$.   **d)** 8.69 (m).

P.8–37 a) $\mathbf{E}_t(x, z) = \mathbf{a}_y E_{t0}\,e^{-\alpha_2 z}\,e^{-j\beta_{2x}x}$,

      $\mathbf{H}_t(x, z) = \dfrac{E_{t0}}{\eta_2}\left(\mathbf{a}_x j\alpha_2 + \mathbf{a}_z\sqrt{\dfrac{\epsilon_1}{\epsilon_2}}\sin\theta_i\right)e^{-\alpha_2 z}\,e^{-j\beta_{2x}x}$,

      where $\beta_{2x} = \beta_2\sqrt{\dfrac{\epsilon_1}{\epsilon_2}}\sin\theta_i$, $\alpha_2 = \beta_2\sqrt{\left(\dfrac{\epsilon_1}{\epsilon_2}\right)\sin^2\theta_i - 1}$,

      and $E_{t0} = \dfrac{2\eta_2\cos\theta_i\, E_{i0}}{\eta_2\cos\theta_i - j\eta_1\sqrt{\left(\dfrac{\epsilon_1}{\epsilon_2}\right)\sin^2\theta_i - 1}}$.

P.8–39 a) $6.38°$.   **b)** $e^{j0.66}$.   **c)** $1.89e^{j0.33}$.   **d)** 159 (dB).

P.8–41 a) $\theta_a = \sin^{-1}\left(\dfrac{1}{n_0}\sqrt{n_1^2 - n_2^2}\right)$.   **b)** $80.4°$.

P.8–45 a) $\Gamma_\perp = \dfrac{1.5\cos\theta_i - \sqrt{1 - (1.5\sin\theta_i)^2}}{1.5\cos\theta_i + \sqrt{1 - (1.5\sin\theta_i)^2}}.$

$\Gamma_\parallel = \dfrac{1.5\sqrt{1 - (1.5\sin\theta_i)^2} - \cos\theta_i}{1.5\sqrt{1 - (1.5\sin\theta_i)^2} + \cos\theta_i}.$

P.8–47 a) $\Gamma_\parallel' = \dfrac{\eta_2\cos\theta_t - \eta_1\cos\theta_i}{\eta_2\cos\theta_t + \eta_1\cos\theta_i} = \Gamma_\parallel,$

$\tau_\parallel' = \dfrac{2\eta_2\cos\theta_t}{\eta_2\cos\theta_t + \eta_1\cos\theta_i} = \tau_\parallel\left(\dfrac{\cos\theta_t}{\cos\theta_i}\right).$

Chapter 9

P.9–3 a) $d' = \sqrt{2}\,d,\ u_p' = u_p/\sqrt{2}.$

P.9–7 $\alpha = \sqrt{\dfrac{LC}{2}\left(\dfrac{R}{L} + \dfrac{G}{C}\right)}\left[1 - \dfrac{1}{8\omega^2}\left(\dfrac{R}{L} - \dfrac{G}{C}\right)^2\right],\ \beta = \omega\sqrt{LC}\left[1 + \dfrac{1}{8\omega^2}\left(\dfrac{R}{L} - \dfrac{G}{C}\right)^2\right],$

$R_0 = \sqrt{\dfrac{L}{C}}\left[1 + \dfrac{1}{8\omega^2}\left(\dfrac{R}{L} - \dfrac{G}{C}\right)\left(\dfrac{R}{L} + \dfrac{3G}{C}\right)\right],\ X_0 = -\dfrac{1}{2\omega}\sqrt{\dfrac{L}{C}}\left(\dfrac{R}{L} - \dfrac{G}{C}\right).$

P.9–9 $R = 0.058\,(\Omega/\text{m}),\ L = 0.20\,(\mu\text{H/m}),\ C = 80\,(\text{pF/m}),\ G = 24\,(\mu\text{S/m}).$

P.9–11 Maximum power-transfer efficiency = 50%.

P.9–13 a) $A = D = \dfrac{1}{Z_0}\cosh\gamma\ell,$

$B = Z_0\sinh\gamma\ell,\ C = \dfrac{1}{Z_0}\sinh\gamma\ell.$

P.9–15 a) $V(z, t) = 5.27\,e^{-0.01z}\sin(8000\pi t - 5.55z - 0.322)\,(\text{V}).$
b) $V(50, t) = 3.20\sin(8000\pi t - 0.432\pi)\,(\text{V}).$
c) $0.102\,(\text{W}).$

P.9–17 a) $4Z_0/\alpha\lambda.$ **b)** $Z_0\alpha\lambda/4.$

P.9–19 a) $Z_0 = 289.8 - j77.6\,(\Omega),\ \alpha = 0.139\,(\text{Np/m}),\ \beta = 0.235\,(\text{rad/m}).$
b) $R = 58.6\,(\Omega/\text{m}),\ L = 0.812\,(\mu\text{H/m}),\ G = 0.246\,(\text{mS/m}),$
$C = 12.4\,(\text{pF/m}).$

P.9–21 $\Delta f = \dfrac{\alpha}{\pi\sqrt{LC}} = \dfrac{1}{2\pi}\left(\dfrac{R}{L} + \dfrac{G}{C}\right);\quad Q = \dfrac{\beta}{2\alpha} = \dfrac{1}{[(R/\omega L) + (G/\omega c)]}.$

P.9–27 a) $\Gamma = \tfrac{1}{3}e^{j0.2\pi}.$ **b)** $Z_L = 466 + j206\,(\Omega).$ **c)** $R_m = 150\,(\Omega),\ \ell_m = 0.2\lambda.$

P.9–29 $Z_L = Z_0\left[\dfrac{1 - jS\tan(2\pi z_m'/\lambda)}{S - j\tan(2\pi z_m'/\lambda)}\right].$

P.9–31 a) $P_{\text{inc}} = V_g^2/8R_0.$ **b)** $P_L = \dfrac{V_g^2}{8R_0}(1 - |\Gamma|^2).$ **c)** $\dfrac{P_L}{P_{\text{inc}}} = \dfrac{4S}{(S + 1)^2}.$
d) $P_{\text{inc}} = 25\,(\text{W}),\ \Gamma = 0.243\underline{/-76°},\ S = 1.64,$
$P_L = 23.5\,(\text{W}),\ |V_L| = 54.2\,(\text{V}),\ |I_L| = 0.97\,(\text{A}).$

P.9–33 At $t = 4T$, the voltage and current distributions along the line revert to the conditions at $t = 0$, and the cycle repeats itself.

P.9–35 $\Gamma_g = -1/3$, $\Gamma_L = 1$.

P.9–43 a) $S = 1.77$.   **b)** $\Gamma = 0.28e^{j146°}$.   **c)** $Z_i = 50 + j29\,(\Omega)$.
d) $Y_i = 0.015 - j0.009\,(S)$.

P.9–45 a) $Z_L = 33.75 - j23.75\,(\Omega)$.   **b)** $\Gamma = \frac{1}{3}e^{j252.5°}$.   **c)** $z'_m = 25\,(cm)$.

P.9–47 Line length $= 0.375\,(m)$, Wire radius $= 5.4\,(mm)$.

P.9–49 $d_1/\lambda = 0.074$, $\ell_1/\lambda = 0.347$; $d_2/\lambda = 0.250$, $\ell_2/\lambda = 0.153$.

P.9–51 a) $d_L/\lambda = 0.0113$.   **b)** $\ell_A/\lambda = 0.304$, $\ell_B/\lambda = 0.125$.

Chapter 10

P.10–5 From Eqs. (10–83a, b, & c): $\mathbf{J}_{s\ell} = \mathbf{a}_x B_n$, $\mathbf{J}_{su} = \mathbf{a}_x(-1)^{n+1} B_n$.

P.10–7 $u_{en} = \dfrac{1}{\sqrt{\mu\epsilon}}\sqrt{1 - (f_c/f)^2}$.

P.10–9 a) $\beta = 308\,(rad/m)$, $\alpha_d = 1.28 \times 10^{-8}\,(Np/m)$, $\alpha_c = 1.69 \times 10^{-4}\,(Np/m)$,
$u_p = 2.04 \times 10^8\,(m/s)$, $u_g = 1.96 \times 10^8\,(m/s)$, $\lambda_g = 2.04\,(cm)$.
b) $\beta = 288\,(rad/m)$, $\alpha_d = 1.37 \times 10^{-8}\,(Np/m)$, $\alpha_c = 7.25 \times 10^{-4}\,(Np/m)$,
$u_p = 2.18 \times 10^8\,(m/s)$, $u_g = 1.83 \times 10^8\,(m/s)$, $\lambda_g = 2.18\,(cm)$.

P.10–11 a) $358\,(MW/m)$.   **b)** $207\,(MW/m)$.   **c)** $155\,(MW/m)$.

P.10–13 a) $\mathbf{J}_s(y = 0) = -\mathbf{a}_z \dfrac{j\omega\epsilon}{h^2}\left(\dfrac{\pi}{b}\right) E_0 \sin\left(\dfrac{\pi x}{a}\right) e^{-j\beta_{11}z} = \mathbf{J}_s(y = b)$.

$\mathbf{J}_s(x = 0) = -\mathbf{a}_z \dfrac{j\omega\epsilon}{h^2}\left(\dfrac{\pi}{a}\right) E_0 \sin\left(\dfrac{\pi y}{b}\right) e^{-j\beta_{11}z} = \mathbf{J}_s(x = a)$.

P.10–15 $u_{en} = u\sqrt{1 - (u/2af)^2}$, $u = 1/\sqrt{\mu\epsilon}$.

P.10–17 a) $a > 6\,(cm)$, $b < 4\,(cm)$. Choose $a = 6.5\,(cm)$ and $b = 3.5\,(cm)$.
b) $\beta = 40.1\,(rad/m)$, $u_p = 4.70 \times 10^8\,(m/s)$, $\lambda_g = 15.7\,(cm)$,
$(Z_{TE})_{10} = 590\,(\Omega)$.

P.10–19 a) $f_c = 2.08 \times 10^9\,(Hz)$.   **b)** $\lambda_g = 0.139\,(m)$.
c) $\alpha_c = 2.26 \times 10^{-3}\,(Np/m)$.   **d)** $307\,(m)$.

P.10–21 $1\,(MW)$.

P.10–23 $\alpha_c = \dfrac{2R_s(b/a^2 + a/b^2)}{\eta ab\sqrt{1 - (f_c/f)^2}\,(1/a^2 + 1/b^2)}$.

P.10–29 a) $E_z^0 = C_n J_n(hr)\sin n\phi$.
c) Eigenvalues of TM modes are determined by requiring $J_n(ha) = 0$. The lowest TM mode is TM_{11}.

P.10–31 a) $\alpha = 0.061\,(Np/m)$, $\beta = 4.19\,(rad/m)$.
b) $\alpha = 0.380\,(Np/m)$, $\beta = 10.48\,(rad/m)$.

P.10–33 Even TM modes in the slab: $E_z(y, z; t) = E_e \cos k_y y \cos(\omega t - \beta z)$,

$$E_y(y, z; t) = -\frac{\beta}{k_y} E_e \sin k_y y \sin(\omega t - \beta z), H_x(y, z; t) = \frac{\omega \epsilon_d}{k_y} E_e \sin k_y y \sin(\omega t - \beta z).$$

P.10–37 a) $H_{zi}^0 = C_0 J_0(hr), r \le a$; $\quad H_{zo}^0 = D_0 K_0(\zeta r), r \le a$.

b) $\dfrac{J_0(ha)}{J_0'(ha)} = -\dfrac{\mu_1 \zeta}{\mu_2 h} \dfrac{K_0(\zeta a)}{K_0'(\zeta a)}$.

P.10–39 a) Dominant mode: TE_{101}. $f_{101} = 4.802$ (GHz).

b) $Q = 6869$, $W_e = W_m = 0.07728$ (pJ).

P.10–41 a) $a = d$. $\quad$ b) $1.11 \eta / R_s (1 + a/2b)$.

P10–43 $Q_{110} = \dfrac{\sqrt{\pi f_{110} \mu_0 \sigma}\; abd(a^2 + b^2)}{2d(a^3 + b^3) + ab(a^2 + b^2)}$.

P.10–45 $f = \dfrac{1}{\pi a \sqrt{\dfrac{2h}{d} \mu \epsilon} \ln\left(\dfrac{b}{a}\right)}$.

Chapter 11

P.11–1 $\nabla^2 \mathbf{E} - \mu\epsilon \dfrac{\partial^2 \mathbf{E}}{\partial t^2} = \dfrac{1}{\epsilon}\nabla\rho + \mu\dfrac{\partial \mathbf{J}}{\partial t}$.

P.11–3 a) $\mathbf{A} = \mathbf{a}_\phi \dfrac{\mu_0 m}{4\pi R^2} e^{-j\beta R}(1 + j\beta R)\sin\theta$.

P.11–5 a) $\rho_\ell = -j(I_0/c)\sin\beta z$. $\quad$ b) $\rho_\ell = \begin{cases} -j2I_0/\pi c, & 0 < z \le \lambda/4; \\ +j2I_0/\pi c, & -\lambda/4 \le z < 0. \end{cases}$

P.11–7 a) $\mathbf{E} = \mathbf{a}_\theta \dfrac{j30\beta h}{R} I_0 e^{-j\beta R}\sin\theta$. $\quad$ b) $R_r = 20\pi^2 \left(\dfrac{2h}{\lambda}\right)^2$. $\quad$ c) 1.76 (dB).

P.11–9 a) $R_r = 320\pi^6 (b/\lambda)^4$. $\quad$ b) $\eta_r = \dfrac{R_r}{R_r + (bR_s/a)}$.

P.11–13 $\ell_e(\theta) = \dfrac{2\sin\theta\,[1 - \cos(\beta h \cos\theta)]}{\beta^2 h \cos^2\theta}$; Max. $\ell_e = h = \dfrac{\lambda}{12}$.

P.11–15 a) $E_\theta = \dfrac{j120I\beta h}{R} e^{-j\beta(R - \frac{d}{2}\cos\theta)} F(\theta)$, where $F(\theta) = \sin\theta\cos\left(\dfrac{\beta d}{2}\cos\theta\right)$.

P.11–19 b) $(2\Delta\phi)_{1/2} = 4.23\,(\lambda/d)$ (deg.) $\quad$ c) $(2\Delta\phi)_0 = 46.8\sqrt{\lambda/d}$ (deg.)

P.11–23 $|F(\theta, \phi)| = \dfrac{1}{N_1 N_2}\left|\left[\dfrac{\cos\left(\dfrac{\pi}{2}\cos\theta\right)}{\sin\theta}\right]\dfrac{\sin\left(\dfrac{N_1\psi_x}{2}\right)\sin\left(\dfrac{N_2\psi_y}{2}\right)}{\sin\left(\dfrac{\psi_x}{2}\right)\sin\left(\dfrac{\psi_y}{2}\right)}\right|$,

where $\psi_x = \dfrac{\beta d_1}{2}\sin\theta\cos\phi$ and $\psi_y = \dfrac{\beta d_2}{2}\sin\theta\cos\phi$.

P.11-25 **a)** $W_s = \frac{8\pi}{3}\beta^4 b^6 \left(\frac{\epsilon_r - 1}{\epsilon_r + 2}\right)^2 \left(\frac{E_i^2}{2\eta_0}\right)$.   **b)** $\sigma_s = 1.5\sigma_{bs}$.

P.11-27 **b)** Main-lobe beamwidth $= 4/\sqrt{G_D}$.

P.11-29 **a)** $0.55\,(\text{nW})$.   **b)** $1.25n\,(\text{km}), n = 1, 2, \cdots$.

P.11-31 **a)** $|V_{oc}| = 2\lambda E_0/\pi$ if $p = 1$.   **b)** $V_{oc} = 0$ if $p = 1$.   **c)** $|V_{oc}| = \lambda E_0/\pi$.

P.11-33 $\mathbf{E}_P(\theta, \phi) = \frac{jab}{\lambda R_0} e^{-j\beta R_0} \left[\left(\frac{\sin u}{u}\right)\left(\frac{\sin v}{v}\right)\right] (\mathbf{a}_\theta \cos\phi - \mathbf{a}_\phi \sin\phi)$;

$u = \left(\frac{\pi a}{\lambda}\right)\sin\theta\cos\phi, \quad v = \left(\frac{\pi b}{\lambda}\right)\sin\theta\sin\phi$.

P.11-35 **a)** $F_{xz}(\theta) = \frac{(\pi/2)^2 \cos\psi}{(\pi/2)^2 - \psi^2}, \quad \psi = \frac{\beta a}{2}\sin\theta$.

b) $68\lambda/a$ (degrees).   **c)** $86\lambda/a$ (degrees).   **d)** -23.5 (dB).

Index

Some Useful Vector Identities

$$\mathbf{A} \cdot \mathbf{B} \times \mathbf{C} = \mathbf{B} \cdot \mathbf{C} \times \mathbf{A} = \mathbf{C} \cdot \mathbf{A} \times \mathbf{B}$$

$$\mathbf{A} \times (\mathbf{B} \times \mathbf{C}) = \mathbf{B}(\mathbf{A} \cdot \mathbf{C}) - \mathbf{C}(\mathbf{A} \cdot \mathbf{B})$$

$$\nabla(\psi V) = \psi \nabla V + V \nabla \psi$$

$$\nabla \cdot (\psi \mathbf{A}) = \psi \nabla \cdot \mathbf{A} + \mathbf{A} \cdot \nabla \psi$$

$$\nabla \times (\psi \mathbf{A}) = \psi \nabla \times \mathbf{A} + \nabla \psi \times \mathbf{A}$$

$$\nabla \cdot (\mathbf{A} \times \mathbf{B}) = \mathbf{B} \cdot (\nabla \times \mathbf{A}) - \mathbf{A} \cdot (\nabla \times \mathbf{B})$$

$$\nabla \cdot \nabla V = \nabla^2 V$$

$$\nabla \times \nabla \times \mathbf{A} = \nabla(\nabla \cdot \mathbf{A}) - \nabla^2 \mathbf{A}$$

$$\nabla \times \nabla V = 0$$

$$\nabla \cdot (\nabla \times \mathbf{A}) = 0$$

$$\int_V \nabla \cdot \mathbf{A} \, dv = \oint_S \mathbf{A} \cdot d\mathbf{s} \qquad \text{(Divergence theorem)}$$

$$\int_S \nabla \times \mathbf{A} \cdot d\mathbf{s} = \oint_C \mathbf{A} \cdot d\boldsymbol{\ell} \qquad \text{(Stokes's theorem)}$$

Gradient, Divergence, Curl, and Laplacian Operations

Cartesian Coordinates (x, y, z)

$$\nabla V = \mathbf{a}_x \frac{\partial V}{\partial x} + \mathbf{a}_y \frac{\partial V}{\partial y} + \mathbf{a}_z \frac{\partial V}{\partial z}$$

$$\nabla \cdot \mathbf{A} = \frac{\partial A_x}{\partial x} + \frac{\partial A_y}{\partial y} + \frac{\partial A_z}{\partial z}$$

$$\nabla \times \mathbf{A} = \begin{vmatrix} \mathbf{a}_x & \mathbf{a}_y & \mathbf{a}_z \\ \dfrac{\partial}{\partial x} & \dfrac{\partial}{\partial y} & \dfrac{\partial}{\partial z} \\ A_x & A_y & A_z \end{vmatrix} = \mathbf{a}_x \left(\frac{\partial A_z}{\partial y} - \frac{\partial A_y}{\partial z} \right) + \mathbf{a}_y \left(\frac{\partial A_x}{\partial z} - \frac{\partial A_z}{\partial x} \right) + \mathbf{a}_z \left(\frac{\partial A_y}{\partial x} - \frac{\partial A_x}{\partial y} \right)$$

$$\nabla^2 V = \frac{\partial^2 V}{\partial x^2} + \frac{\partial^2 V}{\partial y^2} + \frac{\partial^2 V}{\partial z^2}$$

Cylindrical Coordinates (r, ϕ, z)

$$\nabla V = \mathbf{a}_r \frac{\partial V}{\partial r} + \mathbf{a}_\phi \frac{\partial V}{r \partial \phi} + \mathbf{a}_z \frac{\partial V}{\partial z}$$

$$\nabla \cdot \mathbf{A} = \frac{1}{r} \frac{\partial}{\partial r}(rA_r) + \frac{\partial A_\phi}{r \partial \phi} + \frac{\partial A_z}{\partial z}$$

$$\nabla \times \mathbf{A} = \frac{1}{r} \begin{vmatrix} \mathbf{a}_r & \mathbf{a}_\phi r & \mathbf{a}_z \\ \dfrac{\partial}{\partial r} & \dfrac{\partial}{\partial \phi} & \dfrac{\partial}{\partial z} \\ A_r & rA_\phi & A_z \end{vmatrix} = \mathbf{a}_r \left(\frac{\partial A_z}{r \partial \phi} - \frac{\partial A_\phi}{\partial z} \right) + \mathbf{a}_\phi \left(\frac{\partial A_r}{\partial z} - \frac{\partial A_z}{\partial r} \right) + \mathbf{a}_z \frac{1}{r} \left[\frac{\partial}{\partial r}(rA_\phi) - \frac{\partial A_r}{\partial \phi} \right]$$

$$\nabla^2 V = \frac{1}{r} \frac{\partial}{\partial r} \left(r \frac{\partial V}{\partial r} \right) + \frac{1}{r^2} \frac{\partial^2 V}{\partial \phi^2} + \frac{\partial^2 V}{\partial z^2}$$

Spherical Coordinates (R, θ, ϕ)

$$\nabla V = \mathbf{a}_R \frac{\partial V}{\partial R} + \mathbf{a}_\theta \frac{\partial V}{R \partial \theta} + \mathbf{a}_\phi \frac{1}{R \sin \theta} \frac{\partial V}{\partial \phi}$$

$$\nabla \cdot \mathbf{A} = \frac{1}{R^2} \frac{\partial}{\partial R}(R^2 A_R) + \frac{1}{R \sin \theta} \frac{\partial}{\partial \theta}(A_\theta \sin \theta) + \frac{1}{R \sin \theta} \frac{\partial A_\phi}{\partial \phi}$$

$$\nabla \times \mathbf{A} = \frac{1}{R^2 \sin \theta} \begin{vmatrix} \mathbf{a}_R & \mathbf{a}_\theta R & \mathbf{a}_\phi R \sin \theta \\ \dfrac{\partial}{\partial R} & \dfrac{\partial}{\partial \theta} & \dfrac{\partial}{\partial \phi} \\ A_R & RA_\theta & (R \sin \theta)A_\phi \end{vmatrix} = \mathbf{a}_R \frac{1}{R \sin \theta} \left[\frac{\partial}{\partial \theta}(A_\phi \sin \theta) - \frac{\partial A_\theta}{\partial \phi} \right]$$
$$+ \mathbf{a}_\theta \frac{1}{R} \left[\frac{1}{\sin \theta} \frac{\partial A_R}{\partial \phi} - \frac{\partial}{\partial R}(RA_\phi) \right]$$
$$+ \mathbf{a}_\phi \frac{1}{R} \left[\frac{\partial}{\partial R}(RA_\theta) - \frac{\partial A_R}{\partial \theta} \right]$$

$$\nabla^2 V = \frac{1}{R^2} \frac{\partial}{\partial R} \left(R^2 \frac{\partial V}{\partial R} \right) + \frac{1}{R^2 \sin \theta} \frac{\partial}{\partial \theta} \left(\sin \theta \frac{\partial V}{\partial \theta} \right) + \frac{1}{R^2 \sin^2 \theta} \frac{\partial^2 V}{\partial \phi^2}$$